ENCYCLOPAEDIA OF VEDIC MATHEMATICS

ENCYCLOPAEDIA OF VEDIC MATHEMATICS

(Vol. 1)

Shyam Priya

ANMOL PUBLICATIONS PVT. LTD.
NEW DELHI - 110 002 (INDIA)

ANMOL PUBLICATIONS PVT. LTD.
H.O.: 4374/4B, Ansari Road, Darya Ganj,
New Delhi-110 002 (India)
Ph.: 23278000, 23261597
B.O.: No. 1015, Ist Main Road, BSK IIIrd Stage
IIIrd Phase, IIIrd Block,
Bangalore - 560 085 (India)
Visit us at: www.anmolpublications.com

Encyclopaedia of Vedic Mathematics

First Published, 2008
ISBN 978-81-261-3504-2 (Set)

PRINTED IN INDIA

Printed at Mehra Offset Press, Delhi.

Contents

Preface

Vedic Mathematics contains many examples of striking methods of calculation and there is a remarkable coherence to the system which makes it very attractive. Also the Vedic system itself suggests a kind of approach that involves going directly to the answer. Vedic mathematics is based on sixteen Sutras and some sub-Sutras which provide links through mathematics: the word 'Sutra' means 'thread'.

Vedic scheme is not simply a collection of rapid methods; it is a system, a unified approach, which can be swiftly learnt. Vedic Mathematics extensively exploits the properties of numbers in every practical applications, particularly in the field of computation. It makes available a whole range of methods ideally suited to these properties. After a detailed introduction to Vedic Mathematics, this book shows applications of Vertically and Crosswise from basic calculations to the evaluation of logarithms, exponentials, trigonometrical functions and the solution of simultaneous, transcendental, polynomial and differential equations. The methods shown are not only exceedingly efficient and new but have a unity and coherence which renders them easy to master and enjoyable to execute.

The methods given in this book are not intended to be a complete or thorough treatment of the topics they deal with. All of the ideas can probably be developed further or applied in other areas, and all can doubtless be improved upon. The mathematician will also observe a certain lack of rigor as an attempt has been made to make the material intelligible to as wide a readership as possible.

—Shyam Priya

1

Indian Mathematics: Redressing the Balance

Mathematics has long been considered an invention of European scholars, as a result of which the contributions of non-European countries have been severely neglected in histories of mathematics. Worse still, many key mathematical developments have been wrongly attributed to scholars of European origin. This has led to so-called Eurocentrism. The neglect of non-European mathematics is no more apparent than when studying the contributions of India. Contrary to Euroscentric belief, scholars from India, over a period of some 4500 years, contributed to some of the greatest mathematical achievements in the history of the subject. From the earliest numerate civilisation of the Indus valley, through the scholars of the 5th to 12th centuries who were conversant in arithmetic, algebra, trigonometry, geometry combinatorics and latterly differential calculus, Indian scholars led the world in the field of mathematics. The peak coming between the 14th and 16th centuries in the far South, where scholars were the first to derive infinite series expansions of trigonometric functions.

In addition to mighty contributions to all the principal areas of mathematics, Indian scholars were responsible for the creation, and refinement of the current decimal place value system of numeration, including the number zero, without which higher mathematics would not be possible. The purpose of my project is to highlight the major mathematical contributions of Indian scholars and further to emphasise where neglect has occurred and hence elucidate why the Eurocentric ideal is an injustice and in some cases complete fabrication.

Introduction

The history of science, and specifically mathematics, is a vast topic and one which can never be completely studied as much of the work of ancient times remains undiscovered or has been lost through time.

Nevertheless there is much that is known and many important discoveries have been made, especially over the last 150 years, which have significantly altered the chronology of the history of mathematics, and the conceptions that had been commonly held prior to that. By the turn of the 21st century it was fair to say that there was definite knowledge of where and when a vast majority of the significant developments of mathematics occurred. However, despite widely available, reliable information, I became aware during the course of more general studies of the history of mathematics, that many discrepancies persist.

I became drawn to the topic of Indian mathematics, as there appeared to be a distinct and inequitable neglect of the contributions of the sub-continent. Thus, during the course of this project I aim to discuss that despite slowly changing attitudes there is still an ideology' which plagues much of the recorded history of the subject.

That is, to some extent very little has changed even in our seemingly enlightened historical and cultural position, and, in specific reference to my study area, many of the developments of Indian mathematics remain almost completely ignored, or worse, attributed to scholars of other nationalities, often European.

It is important for me to clarify at this point that the ideology I refer to is that held by (predominantly) European historians of science and mathematics, that mathematics is a European 'invention'. This ideology leads to an 'intrinsically' Eurocentric bias to the history of the subject. As R Rashed comments the ideology can be summarised as such:

> *...Classical science is European and its origins are directly traceable to Greek philosophy and science. [RR, P 332]*

Thus despite the many discoveries that have been made there has been a great reluctance to acknowledge the contributions of non-European countries.

For several hundred years following the European mathematical renaissance of the late 15th and early 16th century there was a commonly held opinion among commentators and historians that mathematics originated in its entirety from Europe and European scholars.

The basic chronology of the history of mathematics was very simple; it had primarily been the invention of the ancient Greeks, whose work had continued up to the middle of the first millenium A.D. Following which there was a period of almost 1000 years where no work of significance was carried out until the European renaissance, which coincided with the 'reawakening' of learning and culture in Europe following the so called dark ages.

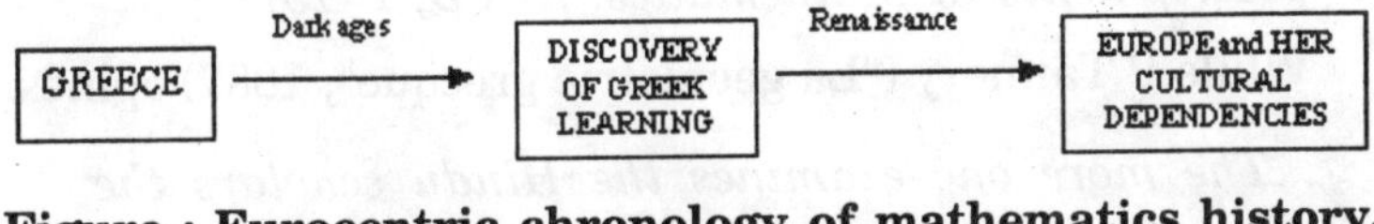

Figure : Eurocentric chronology of mathematics history.

Some historians made some concessions, by acknowledging the work of Egyptian, Babylonian, Indian and Arabic mathematicians (and occasionally the work of the Far East and China). Modified versions of the Eurocentric model commonly took the form seen below.

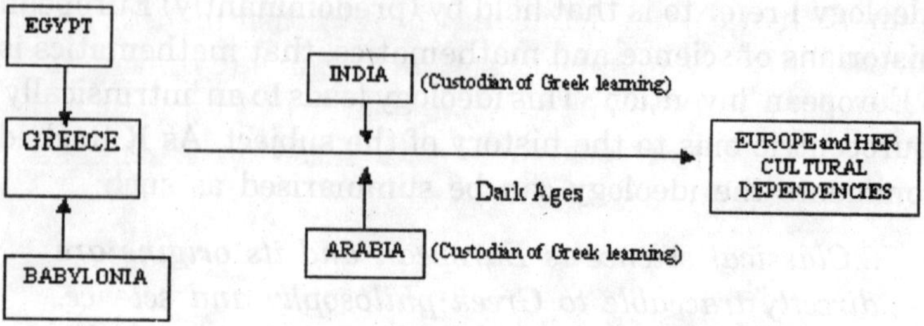

Figure : Modified Eurocentric model.

However, references to the work of these 'others' or 'non-Europeans' were always brief and hazy, and generally concluded that they were merely reconstructions of Greek works and that nothing of significance or importance was contained in them. Indian scholars, on the relatively rare occasions they were discussed, were merely considered to be custodians of Ancient Greek learning. There is a plethora of works and quotes that highlight the attitude that prevailed towards the works of so called non-European mathematics. P Duhem ("Le systeme du monde", 1965) states very simply:

> *...Arabic Science only reproduced the teachings received from Greek science. [RR, P 338]*

Furthermore G Sarton ("Guide to the History of Science", 1927) comments:

> *...One could almost omit Hindu and Chinese developments in Mathematics. [AA'D, P 15]*

While P Tannery ("La geometrie grecque", 1887) opines:

> *...The more one examines the Hindu scholars the more they appear dependent upon the Greeks... (and)...quite inferior to their predecessors on all respects. [RR, P 338]*

It is this underhand and in some cases completely fallacious attitude towards Indian mathematics that I have chosen to focus upon. As said this is a vast topic area and although all non-European 'roots' of mathematics have suffered neglect and miss-representation by many historians I am not going to attempt to focus on all non-European roots of mathematical development. I have chosen to focus on the mathematical developments of the Indian sub-continent, as I consider them not only to be severely neglected in histories of mathematics, but also to have produced some of the most remarkable results of mathematics. Indeed, the research I have conducted has highlighted that many Indian mathematical results, beyond being simply remarkable because of the time in which they were derived, show that several 'key' mathematical topics, and subsequent results, indubitably originate from the Indian subcontinent. Having made these discoveries it seems to me an incredible injustice that the work of Indian scholars is not rewarded with a much more prominent place in the history of mathematics.

Not only have many historians identified Greek influences in all of the work of Indian scholars, (in some cases by suggesting severely inaccurate dates for the works), several others have even attempted to show Arabic influences, which is quite incredible as several Indian works in fact had a significant influence on early Arab works. Although I by no means wish to talk down the incredible developments of Arab scholars, I believe the developments of Indian scholars are on a par with, and occasionally surpass them.

Indian scholars made vast contributions to the field of mathematical astronomy and as a result contributed mightily to the developments of arithmetic, algebra, trigonometry and secondarily geometry (although this topic was well developed by the Greeks) and combinatorics. Perhaps most remarkable were developments in the fields of infinite series expansions of trigonometric expressions and differential calculus.

Surpassing all these achievements however was the development of decimal numeration and the place value system, which without doubt stand together as the most remarkable developments in the history of mathematics, and possibly one of the foremost developments in the history of humankind. The decimal place value system allowed the subject of mathematics to be developed in ways that simply would not have been possible otherwise. It also allowed numbers to be used more extensively and by vastly more people than ever before.

The aim of my work is not just to paint as accurate a picture of the developments of mathematics by the Indian peoples as possible, but also to attempt to give reasons as to why Europe has chosen to neglect the facts of history. Through a detailed discussion of the mathematics of the Indian subcontinent I hope to highlight why this 'neglect' is such an injustice, and briefly discuss the possible consequences of this neglect.

There are several points that I feel it is important to make before progressing further with my discussion. The first is to make it clear that the chronology of the history of mathematics is not entirely linear. One will often find simplified diagrammatic representations where the work of one group of people (or country) is proceeded by the work of another group and so on. In reality things are far more complicated than this. Particularly in the European dark ages (5th-15th centuries) mathematical developments passed between several countries, being constantly refined and improved. G Joseph states:

> ... *A variety of mathematical activity and exchange between a number of cultural areas went on while Europe was in a deep slumber. [GJ, P 9]*

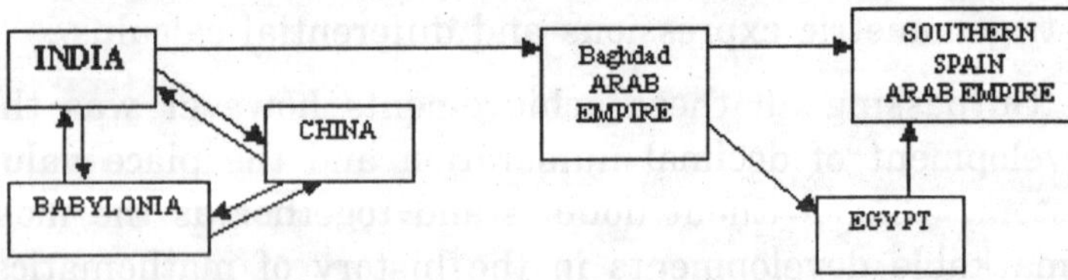

Figure : Non-European mathematics during the dark ages.

Secondly it is worth noting that we view the history of mathematics from:

> *...Our own position of understanding and sophistication. [EFR/JO'C1, P 3]*

While this is unavoidable it is vital to appreciate the uniqueness and ingenuity of the developments of Indian scholars, even if the results are common-place now.

We also view mathematics from an intrinsically European standpoint due largely to the influence of European scholars over the last 500 years and the colonisation of much of the world by European countries. Perhaps this is why many historians find it hard to accept that many results and developments of mathematics are not European in origin. In short, if we are European, somewhat unavoidably we view history from our indigenous standpoint.

There is also a slight methodological problem related to the 'labeling' of the topic of "Indian mathematics" that I will briefly discuss. Using the label "Indian mathematics" is not entirely accurate as much of the earliest "Indian mathematics" was developed in areas, which are now part of Pakistan. The label is used for simplicity and can be justified by stating that developments took place in the Indian subcontinent. Quite often the 'label' Hindu mathematics is used, but it is less accurate as many of the scholars were not Hindus.

Finally before progressing I will specify the time period that my work will cover. The earliest origins of Indian mathematics have been dated to around 3000 BC and this seems a sensible point at which to commence my discussion, while work of a significant nature was still being carried out in the south of India in the 16th century, following which there was an eventual decline. It is hence a vast time scale of almost 5000 years, and indeed it may be greater than that, the estimation of 3000 BC is a slightly crude approximation, and there remains much controversy with regards to the dating of many works prior to 400 AD.

As Gupta states in his paper on the problem of ancient Indian chronology:

> *...In the case of India, the problem of chronology continues to be very serious especially with regard to the prehistoric and ancient periods. [RG1, P 17]*

It is also worth pointing out that this lack of certainty has allowed several unscrupulous scholars to pick dates of choice for certain Indian discoveries so as to justify suggestions for Greek, Arab or other influences. In situations where this has arisen I will attempt to the best of my ability to state fact, although in some cases a well-informed guess will have to suffice.

Early Indian Culture—Indus Civilisation

The first appearance of evidence of the use of mathematics in the Indian subcontinent was in the Indus valley (see Figure) and dates back to at least 3000 BC. Excavations at Mohenjodaro and Harrapa, and the surrounding area of the Indus River, have uncovered much evidence of the use of basic mathematics. The maths used by this early Harrapan civilisation was very much for practical means, and was primarily concerned with weights, measuring scales and a surprisingly advanced 'brick technology', (which utilised ratios). The ratio for brick dimensions 4:2:1 is even today considered optimal for effective bonding.

The discoveries of systems of uniform and decimal weights, over a vast area, are of considerable interest. G Joseph states:

> *...Such standardisation and durability is a strong indication of a numerate culture. [GJ, P 222]*

Also, many of the weights uncovered have been produced in definite geometrical shapes (cuboid, barrel, cone, and cylinder to name a few) which present knowledge of basic geometry, including the circle.

This culture also produced artistic designs of a mathematical nature and there is evidence on carvings that these people could draw concentric and intersecting circles and triangles, leading S Sinha to state:

> *...The civilisation and culture of the inhabitants of the Indus valley...were of a very advanced nature. [SS1, P 71]*

S Srinivasan further comments:

> *...There are many unique features in the construction patterns, which suggest an independent origin of ideas in ancient Indian civilisation. [SSr1, P17]*

Further to the use of circles in 'decorative' design there is indication of the use of bullock carts, the wheels of which may have had a metallic band wrapped round the rim. This clearly points to the possession of knowledge of the ratio of the length of the circumference of the circle and its diameter, and thus values of.

Also of great interest is a remarkably accurate decimal ruler known as the Mohenjodaro ruler. Subdivisions on the ruler have a maximum error of just 0.005 inches and, at a length of 1.32 inches, have been named the Indus inch. Furthermore, a correspondence has been noted between the Indus scale and brick size. Bricks (found in various locations) were found to have dimensions that were integral multiples of the graduations of their respective scales, which suggests advanced mathematical thinking.

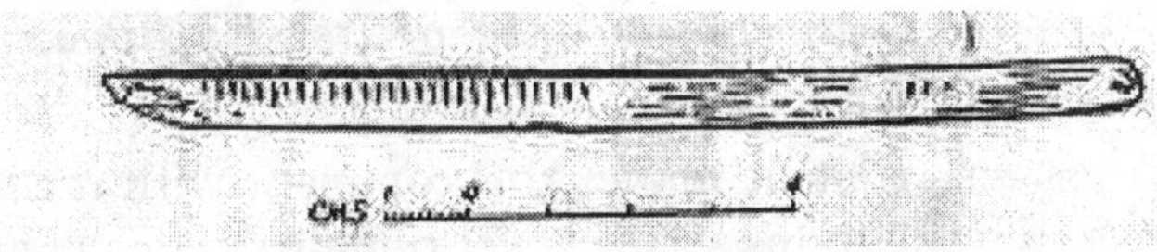

Figure : Ruler found at Lothal. [SSr1, P17]

Above all else there are also brief references to an early decimal system of numeration. The seeds of what were to become the single greatest contribution of the Indian sub-continent to the world (not just of mathematics) had already been sown. My evidence comes from S Sinha who states:

> *...Writers on these civilisations briefly refer to the decimal system of numeration found in these excavations. [SS1, P 71]*

This quote supports the theory that the Brahmi numerals, which were to go on to develop into the numerals we use today, originated in the Indus valley around 2000 BC, however this theory has been rejected by several scholars including Ifrah and Joseph. This quote could be considered a piece of overzealous reporting by the author however, on further investigation I can support the comment with some confidence.

Not only are the markings on all the excavated measuring devices decimal in nature, but there is also research currently being conducted, which is attempting, with success, to show a connection between the Brahmi and Indus scripts. This lends indirect support to suggestions of the existence of early decimal numeral forms. As I will discuss briefly later, the Brahmi numerals undoubtedly developed into the numeral forms we use today.

Although this early mathematics is generally included in histories of mathematics it is often in nothing more than a brief mention, and there is a most curious quote by J Katz who claims:

> *...There is no direct evidence of its (Harappan civilisation) mathematics. [JK, P4]*

It is possible that he makes this comment with regards to the fact that the Indus script as yet remains undeciphered (GJ, P218).

However R Gupta more 'sensibly' states:

> *...In fact the level of mathematical knowledge implied in various geometrical designs, accurate layout of streets and drains and various building constructions etc. was quite high (from a practical point of view). [RG1, P131]*

While Childe claims:

> *...India confronts Egypt and Babylonia by the 3rd millennium with a thoroughly individual and independent civilisation of her own. [EFR/JJO'C2, P 1]*

Some confusion exists as to what caused the decline of this Harrapan culture, there are several theories, the most probable of which in my opinion was the drying up of the Sarasvati River.

This view is supported by S Kak and also S Kalyanaraman who has written an extensive paper on the topic and comments:

> *... The drying-up of the Sarasvati River led to migrations of people eastwards.*

The most commonly held view by historians is that Aryan peoples from the North invaded and destroyed the Harappan culture, this view however is considered increasingly contentious.

In addition to the significance the fledgling decimal system would ultimately have, the most important legacy of this early civilisation is the influence its brick technology may have had on the altar building required by the Vedic religion that followed. A theory of the 'interlinkage' of the Harappan and Vedic cultures has recently arisen from a variety of studies, and it may come to light that there was a greater interaction between the two civilisations than currently thought.

Mathematics in the Service of Religion: Vedas and Vedangas

The Vedic religion was followed by the Indo-Aryan peoples, who originated from the north of the sub-continent. It is through the works of Vedic religion that we gain the first literary evidence of Indian culture and hence mathematics. Written in Vedic Sanskrit the Vedic works, Vedas and Vedangas (and later Sulbasutras) are primarily religious in content, but embody a large amount of astronomical knowledge and hence a significant knowledge of mathematics. The requirement for mathematics was (at least at first) twofold, as R Gupta discusses:

> *...The need to determine the correct times for Vedic ceremonies and the accurate construction of altars led to the development of astronomy and geometry. [RG2, P 131]*

Some chronological confusion exists with regards to the appearance of the Vedic religion. S Kak states in a very recent work that the time period for the Vedic religion stretches back potentially as far as 8000BC and definitely 4000BC. Whereas G Joseph states 1500 BC as the forming of the Hindu civilisation and the recording of Vedas and Vedangas, and later Sulbasutras. However it seems most likely that significant knowledge of astronomy and mathematics first appears in Vedic works around the 2nd millennium BC. The Rg-Veda (fire altar) the earliest extant Vedic work dates from around 1900 BC. R Gupta in his paper on the problem of ancient Indian chronology shows that dates from 26000-200 BC have been suggested for the Vedic 'period'. Having consulted many sources I am confident at placing the period of the Vedas (and Vedangas) at around 1900-1000 BC.

Further mathematical work is found in the Sulbasutras of the later Vedic period, the earliest of which is thought to have been written around 800 BC and the last around

200 BC. I will now move on from this slightly clouded chronological discussion. It is however worth noting that there are serious underlying problems with the chronology of early Indian mathematics which require significant attention.

Although the requirement of mathematics at this time was clearly not for its own sake, but for the purposes of religion and astronomy, it is important not to ignore the secular use of the texts, i.e. by the craftsmen who were building the altars.

Similarly with the earlier Harappan peoples it seems likely that (at least) basic mathematics will have grown to become used by large numbers of the population. Regardless of the fact that at this time mathematics remained for practical uses, some significant work in the fields of geometry and arithmetic were developed during the Vedic period and as L Gurjar states:

> *...The Hindu had made enormous strides in the field of mathematics. [LG, P 2]*

It is also worthwhile briefly noting the astronomy of the Vedic period which, given very basic measuring devices (in many cases just the naked eye), gave surprisingly accurate values for various astronomical quantities. These include the relative size of the planets the distance of the earth from the sun, the length of the day, and the length of the year. For further information, S Kak is an authority on the astronomical content of Vedic works.

Much of the mathematics contained within the Vedas is found in works called Vedangas of which there are six. Of the six Vedangas those of particular significance are the Vedangas Jyotis and Kalpa (the fifth and sixth Vedangas). Jyotis was (at the time) the name for astronomy, while Kalpa contained the rules for the rituals and ceremonies. The Vedangas are best described as an auxiliary to the Vedas.

N Dwary claims, with reference to the Vedanga-Jyotis, that:

> *...Hindus of the period were fully conversant with fundamental operations of arithmetic. [ND1, P 39]*

S Kak suggests a date of around 1350 BC for the Vedanga-Jyotis. I include this as a reminder of the time period being discussed.

Along with the Vedangas there are several further works that contain mathematics, including:

Taittiriya Samhita

Satapatha Brahmana and

Yajur and Atharva-Veda

Rigveda (of which it is thought there are three 'versions') plus additional

Samhitas

Of these the Taittiriya Samhita and Rg-Veda are considered the oldest and contain rules for the construction of great fire altars.

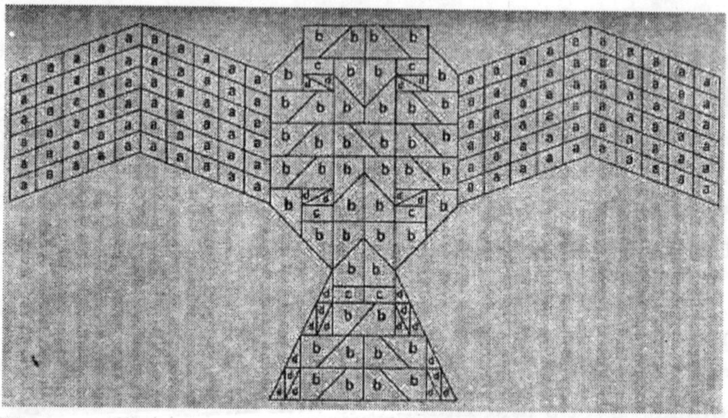

Figure : First layer of a Vedic sacrificial altar (in the shape of a falcon). [GJ, P 227]

As a result of the mathematics required for the construction of these altars, many rules and developments of geometry are found in Vedic works. These include: Use of geometric shapes, including triangles, rectangles, squares, trapezia and circles.

Equivalence through numbers and area.

Equivalence led to the problem of:

Squaring the circle and visa-versa.

Early forms of Pythagoras theorem.

Estimations for π.

S Kak gives three values for from the Satapatha Brahmana. It seems most probable that they arose from transformations of squares into circles and circles to squares. The values are:

$$\pi_1 = \frac{25}{8} (3.125)$$

$$\pi_2 = \frac{900}{289} (3.11418685...)$$

$$\pi_3 = \frac{1156}{361} (3.202216...)$$

Astronomical calculations also leads to a further Vedic approximation:

$$\pi_4 = \frac{339}{108} (3.1389)$$

This is correct (when rounded) to 2 decimal places.

Also found in Vedic works are:

All four arithmetical operators (addition, subtraction, multiplication and division).

A definite system for denoting any number up to 1055 and existence of zero.

Prime numbers.

The Arab scholar Al-Biruni (973-1084 AD) discovered that only the Indians had a number system that was capable of going beyond the thousands in naming the orders in decimal counting.

Evidence of the use of this advanced numerical concept leads S Sinha to comment:

> *...It is fair to agree that a nation with such an advanced and cultured civilisation and which was using the numerical system (decimal place value) knew also how to handle the associated arithmetic. [SS1, P 73]*

It is in Vedic works that we also first find the term "ganita" which literally means "the science of calculation".

It is basically the Indian equivalent of the word mathematics and the term occurs throughout Vedic texts and in all later Indian literature with mathematical content.

Among the other works I have mentioned, mathematical material of considerable interest is found:

Arithmetical sequences, the decreasing sequence 99, 88, ... , 11 is found in the Atharva-Veda.

Pythagoras's theorem, geometric, constructional, algebraic and computational aspects known. A rule found in the Satapatha Brahmana gives a rule, which implies knowledge of the Pythagorean theorem, and similar implications are found in the Taittiriya Samhita.

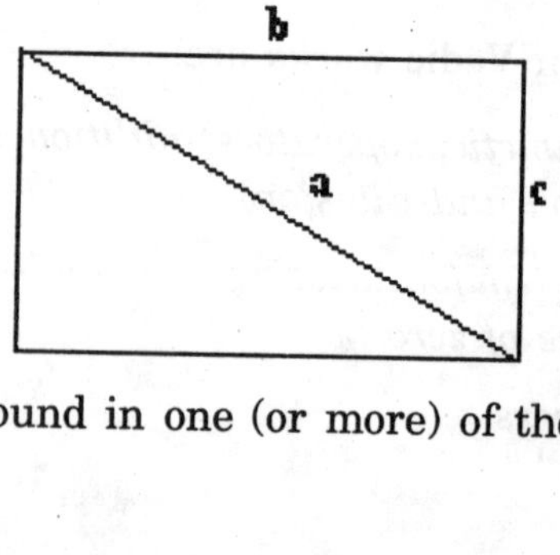

Fractions, found in one (or more) of the Samhitas.

Equations, 972 × 2 = 972 + m for example, found in one of the Samhitas.

The 'rule of three'.

Sulba Sutras

The later Sulba-sutras represent the 'traditional' material along with further related elaboration of Vedic mathematics. The Sulba-sutras have been dated from around 800-200 BC, and further to the expansion of topics in the Vedangas, contain a number of significant developments.

These include first 'use' of irrational numbers, quadratic equations of the form a x2 = c and ax2 + bx = c, unarguable evidence of the use of Pythagoras theorem and Pythagorean triples, predating Pythagoras (c 572 - 497 BC), and evidence of a number of geometrical proofs. This is of great interest as proof is a concept thought to be completely lacking in Indian mathematics.

EXAMPLE 1:

Pythagoras theorem and Pythagorean triples, as found in the Sulba Sutras.

The rope stretched along the length of the diagonal of a rectangle makes an area which the, vertical and horizontal sides make together.

In other words:

$$a^2 = b^2 + c^2$$

Examples of Pythagorean triples given as the sides of right angled triangles:

5, 12, 13

8, 15, 17

12, 16, 20

12, 35, 37

Of the Sulvas so far 'uncovered' the four major and most mathematically significant are those composed by Baudhayana, Manava, Apastamba and Katyayana (perhaps least 'important' of the Sutras, by the time it was composed the Vedic religion was becoming less predominant).

However in a paper written 20 years ago S Sinha claims that there are a further three Sutras, 'composed' by Maitrayana, Varaha and Vadhula (SS1, P 76). I have yet to come across any other references to these three 'extra' sutras.

These men were not mathematicians in the modern sense but they are significant none the less in that they were the first mentioned 'individual' composers. E Robertson and J O'Connor have suggested that they were Vedic priests (and skilled craftsmen).

It is thought that the Sulvas were intended to supplement the Kalpa (the sixth Vedanga), and their primary content remained instructions for the construction of sacrificial altars. The name Sulvasutra means 'rule of chords' which is another name for geometry.

N Dwary states:

...They offer a wealth of geometrical as well as arithmetical results. [ND1, P 40]

R Gupta similarly claims:

...The Sulba-sutras are (quite) rich in mathematical contents. [RG2, P 133]

With reference to the possible appearance of proof is a quote from A Michaels:

...Vedic geometry, though non-axiomatic in character, is provable and indeed proof is implicit in several constructions prescribed in the Sulba-sutras. [RG2, P 133]

This is not particularly compelling evidence but does suggest that the composers of the sulba-sutras may have had a greater depth of knowledge than is generally thought.

Many suggestions for the value of are found within the sutras. They cover a surprisingly wide range of values, from 2.99 to 3.2022.

Pythagoras's theorem and Pythagorean triples arose as the result of geometric rules. It is first found in the Baudhayana sutra - so was hence known from around 800 BC. It is also implied in the later work of Apastamba, and Pythagorean triples are found in his rules for altar construction. Altar construction also led to the discovery of irrational numbers, a remarkable estimation of $\sqrt{2}$ in found in three of the sutras. The method for approximating the value of $\sqrt{2}$ gives the following result:

$$\sqrt{2} = 1 + \frac{1}{3} + \frac{1}{3.4} - \frac{1}{3.434}$$

This is equal to 1.412156..., which is correct to 5 decimal places.

It has been argued by scholars seemingly attempting to deprive Indian mathematics of due credit, that Indians believed that $\sqrt{2} = 1 + \frac{1}{3} + \frac{1}{3.4} - \frac{1}{3.434}$ exactly, which would not indicate knowledge of the concept of irrationality. Elsewhere in Indian works however it is stated that various square root values cannot be exactly determined, which strongly suggests an initial knowledge of irrationality.

Indeed an early method for calculating square roots can be found in some Sutras, the method involves repeated application of the formula: $\sqrt{A} = \sqrt{(a^2 + r)} = a + \frac{r}{2a}$, r being small.

EXAMPLE 2:

Application of formula for calculating square roots.

If A = 10, take a = 9 and r = 1.

Thus $\sqrt{10} = \sqrt{(32+1)} = a + \frac{r}{2a} = 3 + \frac{1}{6} = 3.16667$ in (modern) decimal notation.

$\sqrt{10}$ = 3.162278 to six decimal places when calculated on a calculator. Thus, after only one application of the formula, a moderately accurate value has been calculated.

C Srinivasiengar thus states:

...The credit of using irrational numbers for the first time must go to the Indians. [CS, P 15]

Many of the Vedic contributions to mathematics have been neglected or worse. When it first became apparent that there was geometry contained within works that was not of Greek origin, historians and mathematical commentators went to great lengths to try and claim that this geometry was Greek influenced (to a greater or lesser extent).

It is undeniable that none of the methods of Greek geometry are discernible in Vedic geometry, but this merely serves to support arguments that it is independently developed and not in some way borrowed from Greek sources.

In light of recent evidence and more accurate dating it has been even more strongly claimed by A Seidenberg (in S Kak) that:

...Indian geometry and mathematics pre-dates Babylonian and Greek mathematics. [SK1, P 338]

This may be a somewhat extreme standpoint, and it seems likely that there was traffic of ideas in all directions of the Ancient world, but there is little doubt that the vast majority of Indian work is original to its writers. It may lack the cold logic and truly abstract character of modern mathematics but this observation further helps to identify it as uniquely Indian. Of all the mathematics contained in the Vedangas it is the definite appearance of decimal symbols for numerals and a place value system that should perhaps be considered the most phenomenal.

Before the period of the Sulbasutras was at an end the Brahmi numerals had definitely begun to appear (c. 300BC) and the similarity with modern day numerals is clear to see (see Figures).

More importantly even still was the development of the concept of decimal place value. M Pandit in a recent paper (discussed in RG2) has shown certain rules given by the famous Indian grammarian Panini (c. 500 BC) imply the concept of the mathematical zero. Further to this there is a small amount of evidence of the use of symbols for numbers even earlier in the Harrapan culture. My evidence comes primarily from a paper by S Kak, which analyses some of Panini's work, and there is further support from a paper by S Sinha. B Datta and A Singh also give evidence of an early emergence of numerical forms and the decimal place value system.

Jainism

Following the decline of the Vedic religion around 400BC, the Jaina religion (and Buddhism) became the prominent religion(s) on the Indian subcontinent and gave rise to Jaina mathematics. N Dwary (and others) contend:

> *...According to the religious literature of the Janias, the knowledge of "Sankhyana" (i.e., the science of numbers, which included arithmetic and astronomy) was considered to be one of the principal accomplishments of Jain priests. [ND1, P 39]*

The main Jaina works on mathematics date from around 300BC to 400AD, but the Jaina religion was in its infancy as far back as 500BC. Further, as a point of slight interest, the Jaina religion has still not completely died out, and up till the 19th century some very minor works were produced.

Jaina mathematics played an important role in bridging the gap between 'ancient' Indian mathematics and the so-called 'Classical Period', which was heralded by the work of Aryabhatta I in the 6th century.

Regrettably there are few extant Jaina works, but in the limited material that exists an incredible level of originality is demonstrated. Perhaps the most historically important Jaina contribution to mathematics as a subject is the progression of the subject from purely practical or religious requirements. During the Jaina period mathematics became an abstract discipline to be cultivated "for its own sake". G Joseph states:

> *...The Jaina contribution to this change should be recognised. [GJ, P 250]*

Despite its important historical position, relatively little attention has been paid to Jaina mathematics, and it remains seemingly discarded by many historians. This however is a massive oversight, because within Jaina works there are many remarkable 'new' results, and developments of topics found in Vedic works. It is worth briefly noting here that there is great uncertainty as to whether Vedic works influenced the Jaina mathematicians.

Throughout the history of Indian mathematics it seems very possible each 'leap' was made without knowledge of previous discoveries. This has been suggested to be primarily due to the size of the sub-continent and imperfect channels of communication within it. Another contribution was the 'oral transmission' tradition of Ancient Indian mathematics, which resulted in knowledge being lost over time.

There are several significant Jaina works, including the Surya Prajinapti (4th c. BC) and several Sutras. There is also evidence of individual mathematicians including Bhadrabahu (c. 300 BC, possibly lived in the Mysore State), C Srinivasiengar conjectures he wrote two works (which have yet to be unearthed), and Umaswati, a commentator from 2nd century BC, (possibly lived around 150 BC).

Umaswati is known as a great writer on Jaina metaphysics but also wrote a work Tattwarthadhigama-Sutra Bhashya, which contains mathematics.

Among the important developments of the Jainas are:

Theory of numbers.

There is a great fascination in Jain philosophy with the enumeration of large numbers, selected examples of 'time periods' mentioned include 756 1011 8400000028 days, called shirsa prahelika, and 2588 years.

All numbers were classified into three sets:

Enumerable, Innumerable and Infinite.

Five different types of infinity are recognised in Jaina works:

Infinite in one and two directions, infinite in area, infinite everywhere and infinite perpetually.

This theory is quite incredible and was not realised in Europe until the late 19th century work of George Cantor. Indeed much of the Jaina theory of infinity is extremely advanced for the time in which it was conceived.

Also found in Jaina works:

Knowledge of the fundamental laws of indices.

Arithmetical operations.

Geometry.

Operations with fractions.

Simple equations.

Cubic equations.

Quartic equations (the Jaina contribution to algebra is severely neglected).

Formula for (root 10, comes up almost inadvertently in a problem about infinity).

Operations with logarithms.

Sequences and progressions.

Finally of interest is the appearance of Permutations and Combinations in Jaina works, which resulted in the formation of an early Pascal triangle, called Meru Prastara, many centuries before Pascal himself 'invented' it. This is another case where Indian contributions have been neglected severely.

EXAMPLE 1:

Meru Prastara rule found in Jain works.

Rule is simpler than that of Pascal, and is based on the simple formula:

$$_{n+1}C_r = {}_nC_r + {}_nC_{r-1}$$

EXAMPLE 2:

Formulas for permutations and combinations.

Correct formulas for both permutations and combinations are found in Jaina works:

$$_nC_1 = n,\ _nC_2 = n(n - 1)/1.2,\ _nC_3 = n(n - 1)(n - 2)/1.2.3$$

$$_nP_1 = n,\ _nP_2 = n(n - 1),\ _nP_3 = n(n - 1)(n - 2)$$

The contribution of the Buddhist school should also be briefly discussed. Although in the shadow of Jaina developments, evidence suggests Buddhist scholars were well versed in the use of the decimal place values system and that knowledge of Gainta was considered important.

There was no sudden decline of the Jaina religion as such but from the beginning of the 6th century the work of a mathematician named Aryabhatta surpassed all previous work of the Indian sub-continent and brought about the 'Classical period' of Indian mathematics, (which lasted 600 years). However prior to discussing the work of Aryabhatta there is a major piece of Indian mathematical work yet to be discussed. That is the Bakhshali manuscript.

The Bakhshali Manuscript

The Bakhshali manuscript, which was unearthed in the 19th century, does not appear to belong to any specific period. Although that said, G Joseph classes it as a work of the early 'classical period', while E Robertson and J O'Connor suggest it may be a work of Jaina mathematics, and while this is chronologically plausible there is no proof it was composed by Jain scholars. L Gurjar discusses its date in detail, and concludes it can be dated no more accurately than 'between 2nd century BC and 2nd century AD'. He offers compelling evidence by way of detailed analysis of the contents of the manuscript (originally carried by R Hoernle).

His evidence includes the language in which it was written ('died out' around 300 AD), discussion of currency found in several problems, and the absence of techniques known to have been developed by the 5th century. Further support of these dates is provided by several occurrences of terminology found only in the manuscript, (which form the basis of a paper by M Channabasappa).

The controversy and debate surrounding the date of the Bakhshali manuscript was particularly intense when it was first discovered and highlights the resistance of European historians to accept new discoveries and evidence of the origins of various mathematical results. Clearly establishing a date for the composition of the manuscript is extremely important as it has a vast bearing on the significance of its mathematical content. This is in fact the primary requirement for accurate dating of all mathematical works.

The first translation of the manuscript was carried out by G R Kaye, however he was quite unscrupulous in his work and attempted to date the manuscript as 12th Century in order to justify Arabic and Greek influences on the text. Kaye even went as far as to question the Indian origin of the manuscript.

However the vast majority of his translation and suggestions made about the origins of the manuscript have been debunked by less biased (and more accurate) translations. G Joseph further criticises Kaye and comments:

> *...It is particularly unfortunate that Kaye is still quoted as an authority on Indian mathematics. [GJ, Ps 215-216]*

To slightly confuse the issue, it is now considered (almost without doubt) that the manuscript found at Bakhshali is a copy (of the original work) dating from around the 8th century, and certainly no later than 950 AD. The scholar R Hoernle was the first to reach this conclusion following detailed analysis of the manuscript.

I am content to agree with the (prominent) historians who have placed the date at pre 450 AD and identified the 'current' version as a copy. Avoiding further debate, L Gurjar states that the Bakshali manuscript is the:

> *...Capstone of the advance of mathematics from the Vedic age up to that period. [LG, P49]*

Although, as much work was lost between 'periods', we cannot fully gauge continuity of progress and it is possible the composer(s) of the Bakhshali manuscript were not fully aware of earlier works and had to start from 'scratch'. This would make the work an even more remarkable achievement.

The B. Ms. was written on leaves of birch, in Sarada characters and in Gatha dialect, which is a combination of Sanskrit and Prakrit. This may go some way to explaining the number of inaccurate translations. Many of the historians who have been involved in translating ancient Indian works have done so poorly, due to the obscure script, or alternatively because they have not understood the mathematics fully. More worryingly there could be unscrupulous reasons for poor translating in order to play down the importance of ancient Indian works, because they challenge the Eurocentric ideal.

The B. Ms. highlights developments in Arithmetic and Algebra. The arithmetic contained within the work is of such a high quality that it has been suggested:

> *...In fact [the] Greeks [are] indebted to India for much of the developments in Arithmetic. [LG, P 53]*

This quote 'throws open' the traditional Eurosceptic opinion of the history (and origins) of mathematics. Yet even today histories of mathematics rarely acknowledge this contribution of the Indian sub-continent and the B. Ms. is rarely if ever mentioned.

There are eight principal topics 'discussed' in the B. Ms:

Examples of the rule of three (and profit and loss and interest).

Solution of linear equations with as many as five unknowns.

The solution of the quadratic equation (development of remarkable quality).

Arithmetic (and geometric) progressions.

Compound Series (some evidence that work begun by Jainas continued).

Quadratic indeterminate equations (origin of type $ax/c = y$).

Simultaneous equations.

Fractions and other advances in notation including use of zero and negative sign.

Improved method for calculating square root (and hence approximations for irrational numbers). The improved method (shown below) allowed extremely accurate approximations to be calculated:

$$\sqrt{A} = \sqrt{(a^2+r)} = a + \frac{r}{2a} - \left\{\left(\frac{r}{2a}\right)^2 / 2\left(\frac{a+r}{2a}\right)\right\}$$

EXAMPLE 1:

Application of square root formula.

Again we can calculate $\sqrt{10}$, where a = 3 and r = 1.

$$\sqrt{10} = \sqrt{(32+1)} = 3 + \frac{1}{6} - \{(1/36)/2(3+4/6)\}$$

$$= 3 + 1/6 - \{(1/36)/(19/3)\}$$

$$= 3 + 1/6 - 1/228$$

= 3.16228... in decimal form

$\sqrt{10}$ = 3.16228 when calculated on a calculator and rounded to five decimal places.

EXAMPLE 2:

Quadratic equation as found in B. Ms.

If the equation given is $dn^2 + (2a - d)n - 2s = 0$

Then the solution is found using the equation:

n = (- (2a - d) (2a - d)2 +8ds))/2d

Which is the quadratic equation with a = d, b = 2a - d, and c = 2s.

EXAMPLE 3:

Linear equation with 5 variables.

The following problem is stated in the B. Ms:

"Five merchants together buy a jewel. Its price is equal to half the money possessed by the first together with the money possessed by the others, or one-third the money possessed by the second together with the moneys of the others, or one-fourth the money possessed by the third together with the moneys of the others...etc. Find the price of the jewel and the money possessed by each merchant.

SOLUTION

We have the following systems of equations:

$$x_{1/2} + x_2 + x_3 + x_4 + x_5 = p$$

$$x_1 + x_{2/3} + x_3 + x_4 + x_5 = p$$

$$x_1 + x_2 + x_{3/4} + x_4 + x_5 = p$$

$$x_1 + x_2 + x_3 + x_{4/5} + x_5 = p$$

$$x_1 + x_2 + x_3 + x_4 + x_{5/6} = p$$

Then if $x_{1/2} + x_{2/3} + x_{3/4} + x_{4/5} + x_{5/6} = q$ the equations become $(^{377}/_{60})q = p$.

A number of possible answers can be obtained. This is the origin of the indeterminate equation of the type ax/c = y, the theory of which was greatly developed, and later perfected by Bhaskara II, four hundred years before it was discovered in Europe.

If $q = 60$ then $p = 377$ and $x_1 = 120$, $x_2 = 90$, $x_3 = 80$, $x_4 = 75$ and $x_5 = 72$

With regard to my central discussion of the neglect of Indian mathematics I feel it is important to highlight the presence of moderately advanced algebra in the B. Ms. Historians of mathematics debate whether true algebra 'began' in Greece or Arabia, and little mention is ever made of Indian algebra. In light of my own research I feel that early Arabic algebra (c. 800 AD) in no way surpasses the level of understanding of 6th century Indian scholars.

The Bakhshali manuscript is a unique piece of work and while it not only contains mathematics of a remarkably high standard for the time period, also, in contrast to almost all other Indian works composed before and after, the method of the commentary follows a highly systematic order of:

i. Statement of the rule (sutra)

ii. Statement of the examples (udaharana)

iii. Demonstration of the operation (karana) of the rule.

The work is considerably less concise than other Indian works which were often (if not always) written in a poetic form comprising of short statements of rules, and rarely included examples. This poetic form was favoured, not only as it gave the authors an opportunity to demonstrate their skills but also because of the limited supplies of writing equipment available.

The reasons for the composition of the B. Ms. are unknown but it seems possible that the motivation was to "bring out" the developments of mathematics during the time period. Thus it seems very likely it was composed for solely academic, intellectual and interest ends, in short, mathematics for mathematics sake.

By the end of the 2nd century AD mathematics in India had attained a considerable stature, and had become divorced from purely practical and religious requirements, (although it is worth noting that over the next 1000 years the majority of mathematical developments occurred within works on astronomy). The topics of algebra, arithmetic and geometry had developed significantly and it is widely thought that the decimal place value system of notation had been (generally) perfected by 200 AD, the consequence of which was far reaching.

Decimal Numeration and the Place-value System

I have already mentioned on several occasions the development of a decimal place value system of numeration and there is now very little doubt among historians that this invention originated from the Indian subcontinent. That said, it was considered, until recently, that Arabic scholars were responsible for the system, as C Srinivasiengar writes:

> *...During the earlier decades of this century (20th) attempts were made to credit this invention wholly or in part to the Arabs. [CS, P 2]*

Further attempts have been made to attribute the first use of a place value system to the ancient Babylonian civilisation of Mesopotamia. While it cannot be denied that the Babylonians used a place value system, their's was sexagismal (base 60), and while the concept of place value may have come from Mesopotamia, the Indians were the first to use it with a decimal base (base 10).

All current evidence points towards the Indian system having been influenced by the base 10 Chinese 'counting boards' and the place value system of the Babylonians but combined use of decimal numerals and place value first occurred on the Indian subcontinent. Without doubt the use of a decimal base originates from the most basic human instinct of counting on one's fingers.

The key contribution of the Indians however is not in the development of nine (recognisable) symbols to represent the numbers one to nine, but the invention of the place holder zero.

The great 18th century European mathematician Laplace best described the 'invention' of the decimal place value system as such:

> *...The idea of expressing all quantities by nine figures whereby is imparted to them both an absolute value and one by position is so simple that this very simplicity is the very reason for our not being sufficiently aware how much admiration it deserves. [CS, P 5]*

Beyond not being fully appreciated D Duncan discusses briefly the enduring problem of Eurocentric scholars who long assumed the symbol for zero was a Greek invention, with no proof at all. The claims were based of pure speculations that zero came from the Greek letter omicron (O), the first letter of the Greek word ouden meaning empty. We know this to be untrue, but it serves as a timely reminder of the struggle for recognition of Indian mathematical developments.

There is wide ranging debate as to when the decimal place value system was developed, but there is significant evidence that an early system was in use by the inhabitants of the Indus valley by 3000 BC. Excavations at both Harappa and Mohenjo Daro have supported this theory. At this time however a 'complete' place value system had not yet been developed and along with symbols for the numbers one through nine, there were also symbols for 10, 20, 100 and so on.

The formation of the numeral forms as we know them now has taken several thousand years, and for quite some time in India there were several different forms. These included Kharosthi and Brahmi numerals, the latter were refined into the Gwalior numerals, which are notably similar to those in use today. Study of the Brahmi numerals has also lent weight to claims that decimal numeration was in use by the Indus civilisation as correlations have been noted between the Indus and Brahmi scripts.

It is uncertain how much longer it took for zero to be invented but there is little doubt that such a symbol was in existence by 500 BC, if not in widespread use.

Evidence can be found in the work of the famous Indian grammarian Panini (5th or 6th century BC) and later the work of Pingala a scholar who wrote a work, Chhandas-Sutra (c. 200 BC).

The first documented evidence of the use of zero for mathematical purposes is not until around 2nd century AD (in the Bakhshali manuscript). The first recorded 'non-mathematical' use of zero dates even later, around 680 AD, the number 605 was found on a Khmer inscription in Cambodia.

Despite this it seems certain that a symbol was in use prior to that time. B Datta and A Singh discuss the likelihood that the decimal place value system, including zero had been 'perfected' by 100 BC or earlier. Although there is no concrete evidence to support their claims, they are established on the very solid basis that new number systems take 800 to 1000 years to become 'commonly' used, which the Indian system had done by the 9th century AD.

The inventor of the zero symbol is unknown, but what is known is that it was firstly denoted by a dot, then possibly a circle with a dot in the centre, and later by the oval shape we now use.

Prior to its invention, Indian mathematicians had already taken to leaving an empty column on their counting boards and clearly at some point this empty space was filled. The Indians referred to zero as 'sunya' meaning void. Again, although evidence points towards a Mesopotamian origin for a place holder, their 'zero' (two slanted bars) was not used in conjunction with a decimal base.

Having become firmly established in academic circles in India by the 6th century, the decimal place value system spread across the world. Initially to China and Alexandria, then to the Arab empire where it became the system of choice of the scholars in Baghdad by the 8th century.

Arabic scholars during this time improved the system by introducing decimal fractions. The system also spread into Spain, as has been previously discussed southern Spain was under Arabic rule into the 12th century. It took much longer for the system to be accepted in mainland Europe, but eventually by the 16th century it was widely used. That said, both prejudice and suspicion continued to be widespread, while orthodoxy also played its part in the continued use of Roman numerals. The last significant case of an attempt to abolish the Indian decimal place value system was in Sweden in the early 18th century.

This is clearly a very brief overview of the phenomenal development of the decimal place value system, without which it is accepted 'higher mathematics' would not be possible. It is impossible for me to do justice to its importance in such few words, so I will conclude with a quote from G Halstead who commented:

> *...The importance of the creation of the zero mark can never be exaggerated. This giving to airy nothing, not merely a local habituation and a name, a picture, a symbol but helpful power, is the characteristic of the Hindu race from whence it sprang. No single mathematical creation has been more potent for the general on go of intelligence and power. [CS, P 5]*

The Classical Period

Introduction

Following the Bakhshali manuscript (assuming it was composed no later than 200 AD) Indian mathematics suffered a slump, described by L Gurjar as:

> *...A deep and dire degeneration. [LG, P 78]*

This was in part due to massive communication problems, but also undoubtedly to the huge political upheaval that took place between the 2nd and 4th centuries AD, prior to the capturing of power of most of India by the Imperial Guptas.

During this period however the practice of writing Siddhantas (astronomical works), which had started around 500 BC, continued. Of the Siddhantas the Pitamaha

Siddhanta is the oldest (c. 500 BC) and the Surya Siddhanta (c. 400 AD, author unknown, influenced Aryabhatta) is the best known. The major contribution of these works was the invention of the sine function. R Gupta comments:

> *...Trigonometry based on sine (instead of chord (as in Greek works)) and related functions is another important gift of India to the world of mathematics. [RG2, P 134]*

Following the establishment of the 'Gupta dynasty' a period of political stability arose, which doubtlessly contributed to the ascent of a "galaxy" of mathematician-astronomers, led by Aryabhatta. These men were, first and foremost, astronomers, but the mathematical requirements of astronomy (and no doubt further interest) led to them developing many areas of mathematics, as such, I will refer to them just as mathematicians. The vast majority of the works of the 'classical period' were however, in effect, Siddhantas.

Aryabhatta and His Commentators

Aryabhatta, who is occasionally known as Aryabhatta I, or Aryabhatta the elder to distinguish him from a tenth century astronomer of the same name, stands as a pioneer of the revival of Indian mathematics, and the so called 'classical period', or 'Golden era' of Indian mathematics. Arguably the Classical period continued until the 12th century, although in some respects it was over before Aryabhatta's death following a costly, if ultimately successful, war with invading Huns which resulted in the eroding of the Gupta culture (D Duncan P 171). As mentioned, the classical period arose following a 'dark period' of significant political instability 200-400 AD, which caused the widespread stagnation of mathematical development.

We can accurately claim that Aryabhatta was born in 476 AD, as he writes that he was 23 years old when he wrote his most significant mathematical work the Aryabhatiya (or Arya Bhateeya) in 499 AD. He was a member of the Kusuma Pura School, but is thought to have been a native of Kerala (in the extreme south of India), although unsurprisingly there is some debate.

Further debate surrounds how important the work of Aryabhatta actually was.

In light of discoveries of both Vedic and Jaina mathematics it has been suggested that his work is of less significance mathematically.

However as Gurjar claims it is more than likely that the majority, if not all prior mathematical work may not have been known to him, which makes his contribution remarkable. Either way, Aryabhatta's work was to have massive influence on those who followed, in a similar way that the work of Luca Pacioli influenced the Italian renaissance mathematicians in the late 15th century.

As previously mentioned the only extant work of Aryabhatta is his key work, the Aryabhatiya, a concise astronomical treatise of 118 verses written in a poetic form, of which 33 verses are concerned with mathematical rules.

It is important here to point out that no proofs are contained with his rules, and this is perhaps a primary reason for the neglect by western scholars.

As Indian mathematics is (generally) devoid of proof it is not considered 'true' mathematics in its purest sense. However I believe that adopting this stance is to deny the very origin of remarkable discoveries in mathematics, which may well have been the aim of Eurocentric scholars, as it allowed them to neglect the importance of Indian works in favour of European works.

In the mathematical verses of the Aryabhatiya the following topics are covered:

Arithmetic

Method of inversion.

Various arithmetical operators, including the cube and cube root are though to have originated in Aryabhatta's work.

Aryabhatta can also reliably be attributed with credit for using the relatively 'new' functions of squaring and square rooting.

Algebra

Formulas for finding the sum of several types of series.

Rules for finding the number of terms of an arithmetical progression.

Rule of three - improvement on Bakshali Manuscript.

Rules for solving examples on interest - which led to the quadratic equation, it is clear that Aryabhatta knew the solution of a quadratic equation.

Trigonometry

Tables of sines, not copied from Greek work .

Gupta comments:

> *...The Aryabhatiya is the first historical work of the dated type, which definitely uses some of these (trigonometric) functions and contains a table of sines. [RG3, P 72]*

Spherical trigonometry (some incorrect).

Geometry

Area of a triangle, similar triangles, volume rules.

It has been suggested that Aryabhatta's geometry was borrowed from the Jaina works, but this seems unlikely as it is generally accepted he would not have been familiar with them.

Also of relevance is the use of 'word numerals' and 'alphabet numerals', which are first found in Aryabhatta's work. We can argue that this was not due to the absence of a satisfactory system of numeration but because it was helpful in poetry. C Srinivasiengar quaintly describes it as an:

> *...Exceedingly queer, if original method of enumeration. [CS, P 43]*

However the work of Aryabhatta also affords a proof that:

> *...The decimal system was well in vogue. [CS, P 43]*

Of the mathematics contained within the Aryabhatiya the most remarkable is an approximation for , which is surprisingly accurate.

The value given is: = 3.1416

With little doubt this is the most accurate approximation that had been given up to this point in the history of mathematics.

Aryabhatta found it from the circle with circumference 62832 and diameter 20000. Critics have tried to suggest that this approximation is of Greek origin.

However with confidence it can be argued that the Greeks only used = 10 and = 22/7 and that no other values can be found in Greek texts.

I note with slight concern for the strength of my last comment that the Egyptian scholar Claudius Ptolemy derived the same value 300 years earlier, although there is no suggestion of a link between these two cases.

Further to deriving this highly accurate value for, Aryabhatta also appeared to be aware that it was an 'irrational' number and that his value was an approximation, which shows incredible insight.

Thus even accepting that Ptolemy discovered the 4 decimal place value, there is no evidence that he was aware of the concept of irrationality, which is extremely important.

Inexplicably Aryabhatta preferred to use the approximation = 10 (= 3.1622) in practice!

Aryabhatta's work on astronomy was also pioneering, and was far less tinged with a mythological flavour. Among many theories he was the first to suggest that diurnal motion of the 'heavens' is due to rotation of the earth about its axis, which is incredibly insightful (unsurprisingly he was criticised for this.)

In the field of 'pure' mathematics his most significant contribution was his solution to the indeterminate equation: ax - by = c

R Gupta states:

> *...Aryabhatta (also) made notable contributions to algebra [RG3, P73]*

Although Indian mathematics has become often ignored this was not always the case. The Aryabhatiya was translated into Arabic by Abu'l Hassan al-Ahwazi (before 1000 AD) as Zij al-Arjabhar and it is partly through this translation that Indian computational and mathematical methods were introduced to the Arabs, which will have had a significant effect on the forward progress made by mathematics. The historian A Cajori even goes as far as to suggest that:

> *...Diophantus, the father of Greek algebra, got the first algebraic knowledge from India. [RG4, P 12]*

This theory is supported by evidence that the eminent Greek mathematician Pythagoras visited India, which further 'throws open' the Eurocentric ideal.

EXAMPLE 1:

Solution of 137x + 10 = 60y, as found in the Aryabhatiya.

The general solution is found as follows:

137x + 10 = 60y

```
60) 137 (2  (60 divides into 137 twice with remainder 17, etc.)
    120
    ----
     17) 60 ( 3
         51
        ----
          9) 17 ) 1
              9
            ----
              8) 9 (1
                 8
                ----
                 1
```

The following column of remainders, known as valli (vertical line) form is constructed:

2

3

1

1

The number of quotients, omitting the first one is 3. Hence we choose a multiplier such that on multiplication by the last residue, 1 (in red above), and subtracting 10 from the product the result is divisible by the penultimate remainder, 8 (in blue above). We have 1 × 18 - 10 = 1 × 8. We then form the following table:

	2	2	2	2	297
	3	3	3	130	130
	1	1	37	37	
	1	19	19		
The multiplier	18	18			
Quotient obtained	1				

This can be explained as such: The number 18, and the number above it in the first column, multiplied and added to the number below it, gives the last but one number in the second column.

Thus, $18 \times 1 + 1 = 19$.

The same process is applied to the second column, giving the third column, that is, $19 \times 1 + 18 = 37$.

Similarly $37 \times 3 + 19 = 130$, $130 \times 2 + 37 = 297$.

Then $x = 130$, $y = 297$ are solutions of the given equation.

Noting that $297 \equiv 23 \pmod{137}$ and $130 \equiv 10 \pmod{60}$, we get $x = 10$ and $y = 23$ as simple solutions.

The general solution is $x = 10 + 60m$, $y = 23 + 137m$.

If we stop with the remainder 8 in the process of division above then we can at once get $x = 10$ and $y = 23$. (Working omitted for sake of brevity).

This method was called Kuttaka, which literally means pulveriser, on account of the process of continued division that is carried out to obtain the solution.

The work of Aryabhatta was also extremely influential in India and many commentaries were written on his work (especially his Aryabhatiya). Among the most influential commentators were:

Bhaskara I (c 600-680 AD) also a prominent astronomer, his work in that area gave rise to an extremely accurate approximation for the sine function. His commentary of the Aryabhatiya is of only the mathematics sections, and he develops several of the ideas contained within. Perhaps his most important contribution was that which he made to the topic of algebra.

Lalla (c 720-790 AD) followed Aryabhatta but in fact disagreed with much of his astronomical work. Of note was his use of Aryabhatta's improved approximation of pi to the fourth decimal place. Lalla also composed a commentary on Brahmagupta's Khandakhadyaka.

Govindasvami (c 800-860 AD) his most important work was a commentary on Bhaskara I's astronomical work Mahabhaskariya, he also considered Aryabhatta's sine tables and constructed a table which led to improved values.

Shankara Narayana (c 840-900 AD) wrote a commentary on Bhaskara I's work Laghubhaskariya (which in turn was based on the work of Aryabhatta).

Of note is his work on solving first order indeterminate equations, and also his use of the alternate 'katapayadi' numeration system (as well as Sanskrit place value numerals)

Following Aryabhatta's death around 550 AD the work of Brahmagupta resulted in Indian mathematics attaining an even greater level of perfection.

Between these two 'greats' of the classic period lived Yativrsabha, a little known Jain scholar, his work, primarily Tiloyapannatti, mainly concerned itself with various concepts of Jaina cosmology, and is worthy of minor note as it contained interesting considerations of infinity.

Brahmagupta, and the Influence on Arabia

Brahmagupta was born in 598 AD, possibly in Ujjain (possibly a native of Sind) and was the most influential and celebrated mathematician of the Ujjain school.

It is important here to note that one must not ignore contributions made by Varahamihira, who was an influential figure at the same Ujjain school during the 6th century. He is thought to have lived from 505 AD till 587 AD and made only fairly small contributions to the field of mathematics, he is described by Ifrah as:

...One of the most famous astrologers in Indian history. [EFR/JJO'C18, P 1]

However he increased the stature of the Ujjain school while working there, a legacy that was to last for a long period, and although his contributions to mathematics were small they were of some importance.

They included several trigonometric formulas, improvement of Aryabhatta's sine tables, and derivation of the Pascal triangle by investigating the problem of computing binomial coefficients.

Returning to Brahmagupta, he not only elaborated the mathematical results of Aryabhatta but also made notable contributions to many topics.

L Gurjar describes his results as:

...Unique in the history of world mathematics. [LG, P 91]

His contributions to mathematics are found in two works, the first of which Brahmasphutasiddhanta (BSS) must be considered one of the important mathematical works from this early period, not only of India, but also of the world. Not only was its mathematical content of an exceptional quality, but the work also had a significant influence on the burgeoning scientific awakening in the Arab empire.

I believe that the Indian influence on Arabic work is often ignored or played down and consider this to be unfortunate (at the least). This issue is definitely worthy of discussion as it is noticeable that much is made of the Greek influence on Arabic works but far less of the Indian influence, which in retrospect was quite significant.

His second work was written much later in his life in 665 AD and was titled Khandakhayaka. Although the BSS contains 25 chapters it is generally considered that the first ten chapters make up the first work, and that at a later date Brahmagupta made revisions and additions.

In the BSS among the major developments are those in the areas of:

Arithmetic

Brahmagupta possessed a greater understanding of the number system (and place value system) than anyone to that point. Many rules are given and an advanced technique for multiplication exhibited.

Operations with zero, Brahmagupta was the first to attempt to divide by zero, and while his attempts; showing n /0 = , were not ultimately successful they demonstrate an advanced understanding of an extremely abstract concept.

Operations with negative numbers.

Theory of Arithmetic progressions.

Algorithm for calculating square roots that is equivalent to the Newton-Raphson iterative formula, but clearly pre-dates it by many centuries.

Geometry

Brahmagupta stands in high esteem for his contributions to this topic. A rule that he gives for finding the values of the diagonals of cyclic quadrilaterals is generally known as "Ptolemy's theorem".

Ptolemy 'pre-dated' Brahmagupta by 500 years, so it is wholly reasonable to attribute the 'discovery' of these rules to him. However, Brahmagupta's independent discovery should still be considered a remarkable achievement.

Furthermore, some of his work (regarding right angled triangles, which was later developed by Mahavira, Bhaskara II, et al) is often attributed to Fibonacci (13th c.) and Vieta (16th c.), highlighting the constant European bias.

As L Gurjar quotes:

...(Brahmagupta) derived certain results, which were troubling the brains of Western mathematicians as late as the 17th century. [LG, P 91]

Algebra

Solutions to $N_x^2 + 1 = y^2$, Pell's equation, his most outstanding contribution to mathematics.

He also made many other contributions to solving a variety of algebraic equations, including ax + c = by (which is the focus of a paper by P Majumdar).

Brahmagupta may have been one of the first mathematicians to recognise that the quadratic equation has two solutions.

In his other work, the content is far more 'pure' astronomy, but an interpolation formula used to calculate values of sines bears great similarity to the Newton-Stirling interpolation formula, which is clearly of great historical and mathematical interest.

Without a doubt, Brahmagupta made remarkable contributions to mathematics (and astronomy) and his work continued to be influential for many centuries. In 860 AD an extensive and important commentary on the BSS was written by Prthudakasvami (or Prithudaka Swami). His work was extremely elaborate and unlike many Indian works did not 'suffer' brevity of expression.

The spread of Buddhism (around 500 AD) into China resulted in a period of cultural and scientific exchanges lasting several centuries. Chinese scholars are known to have translated the work of Brahmagupta; this highlights not only the quality of the work but the influence it had on the world outside India. R Gupta mentions four 'Brahminical' translations in a paper. During this time the decimal system and notation was adopted by Chinese scholars and as R Gupta states:

> *...Indian mathematical astronomy exerted a great influence in China during the (glorious) Thang Period (618-907). [RG4, P 11]*

The lasting legacy of the BSS however was its translation by Arab scholars and its contribution to the 'forward progress' of mathematics. These translations, along with translated work of Aryabhatta and (possibly) the Surya Siddhanta were responsible for alerting the Arabs, and the West to Indian mathematics (and astronomy), as G Joseph states:

...This was to have momentous consequences for the development of the two subjects. [GJ, P 267]

Of particular interest is the well told story of the Indian scholar who traveled to Baghdad, at the behest of Caliph al-Mansur (early ruler of the Arab Empire). R Gupta reports the story as such:

...In the year 156 (772/773 AD) there came to Caliph al-Mansur a man (an Ujjain scholar by the name of Kanka) from India, an expert in hisab (computation) bringing with him a work called Sindhind (i.e. Siddhanta) concerning the motions of the planets. [RG, P 12]

A translation of this work, thought to be Brahmagupta's BSS, was subsequently carried out by al-Fazari (and an Indian scholar) and had a far-reaching influence on subsequent Arabic works. The famous Arabic scholar al-Khwarizmi (credited with 'inventing' algebra) is known to have made use of the translation, called Zij al-Sindhind.

Al-Khwarizmi (c. 780-850 AD) is known to have written two subsequent works, one based on Indian astronomy (Zij) and the other on arithmetic (possibly Kitab al-Adad al-Hindi). Later Latin translations of this second work (Algorithmi De Numero Indorum), composed in Spain around the 11th century, are thought to have played a crucial role in introducing the Indian place-value system numerals and the corresponding computational methods into (wider) Europe. Both Indian astronomy and arithmetic had a huge impact in Spain.

This discussion helps to highlight the influence that Indian mathematics had on Arabic mathematics, and ultimately, through Latin translations, on European mathematics, an influence that is considerably neglected. It must be argued that sufficient credit has not been given.

There seems to have been relatively little further 'interaction' between Indian and Arab scholars, and thus Indian works had limited, if any, further influence on mathematical developments in other countries. However Indian mathematics continued to flourish independently throughout the subcontinent for another 400 years, and some of the most outstanding contributions to the history of world mathematics were in fact made during this time period.

Mathematics Over the Next 400 Years (700AD-1100AD)

Mahavira (or Mahaviracharya), a Jain by religion, is the most celebrated Indian mathematician of the 9th century. His major work Ganitasar Sangraha was written around 850 AD and is considered 'brilliant'. It was widely known in the South of India and written in Sanskrit due to his Jaina 'faith'. In the 11th century its influence was still being felt when it was translated into Telegu (a regional language of the south). Mahavira was aware of the works of Jaina mathematicians and also the works of Aryabhatta (and commentators) and Brahmagupta, and refined and improved much of their work. What makes Mahavira unique is that he was not an astronomer, his work was confined solely to mathematics and he stands almost entirely alone in the history of Indian mathematics (at least up to the 14th century) in this respect. He was a member of the mathematical school at Mysore in the south of India and his major contributions to mathematics include:

Arithmetic: GSS was the first text on arithmetic in the present form. He made the classification of arithmetical operators simpler. Detailed operations with fractions (and unit fractions), but no section on decimals (which were not an Indian invention).

Geometric progressions—he gave almost all required formulae.

Permutations and Combinations: Extension and systemisation of Jain works. First to give general formula.

Geometry: Repeated Brahmagupta's construction for cyclic quadrilaterals.

Definitions for most geometric shapes.

Algebra: Work on quadratic, indeterminate and simultaneous equations. Mahavira demonstrated definite understanding of the concept of a quadratic equation having two roots.

Ellipse: Only Indian mathematician to refer to the ellipse, indeed Indian mathematicians did not study conic sections or anything along these lines. Gave incorrect identity for area of ellipse. His formula for the perimeter of an ellipse is worth noting.

Mahavira's work, GSS, could be criticised for being nothing more than an extensive commentary on Jaina works, and the work of Aryabhatta, Brahmagupta (and Bhaskara I). C Srinivasiengar describes his work as containing:

...(No) profoundly fundamental discoveries. [CS, P 70]

To some extent this may be true, but it is also unfair. Mahavira made many subtle contributions and elaborated and revised much of the work of previous mathematicians. Furthermore GSS contained many examples to illustrate his rules, unlike many Indian mathematical works.

The influence of his work (within India) must also not be ignored. The very fact it was still in use more than 250 years after it was written is testimony to its importance as a mathematical work.

EXAMPLE 1:

General formula for combinations, as given by Mahavira.

$_{n}C_{r}$ = {n(n - 1)(n - 2)×××(n - r + 1)}/1.2.3. ... .r

Following Mahavira the most notable mathematician was Prthudakasvami (c. 830-890 AD) a prominent Indian algebraist, who is described by E Robertson and J O'Connor as being:

> *...Best known for his work on solving equations. [EFR/ JJO'C25, P 1]*

He also wrote a commentary on Brahmagupta's Brahma Sputa Siddhanta.

In the early 10th century a mathematician by the name of Sridhara (c. 870-930 AD) may have lived, however there is much debate surrounding his birth and some authors place him as having lived in the 8th century (750 AD). However beyond debate is the fact that he wrote Patiganita Sara a work on arithmetic and mensuration. It contained exactly 300 verses and is hence also known by the name Trisatika. It includes contributions to the following topics:

Rules on extracting square and cube roots, fractions and eight rules for operations involving zero (not division).

Theory of cyclic quadrilaterals with rational sides.

A section concerning rational solutions of various equations of the Pell's type.

Methods for summation of different arithmetic and geometric series. These methods became standard references in later works.

It is thought that Sridhara also composed a text on algebra, which is now lost, and several other works have been attributed to Sridhara, but there is no certainty if they were indeed written by him. The legacy of Sridhara's work was that it had some influence on the work of Bhaskaracharya II, regarded by many as the greatest Indian mathematician.

Prior to Bhaskara II it is worth noting the contributions of Aryabhatta II (c 920-1000 AD) a mathematical-astronomer who notably made important contributions to algebra. In his work Mahasiddhanta he gives twenty verses of detailed rules for solving by = ax + c (and variations of this equation). Also of note, Vijayanandi (c 940-1010 AD) who made several contributions to trigonometry in the course of his astronomical works, and Sripati.

Sripati (c. 1019-1066 AD, the birth year of 1019 is known to be correct) was a follower of the teachings of Lalla and in fact the most important Indian mathematician of the 11th century. He was the author of several astronomical works, including Siddhantasekhara, which contained two chapters devoted to mathematics. His major contributions were in the fields of arithmetic and algebra. His algebra is of particular note; his work includes rules for solving the quadratic equation and simultaneous indeterminate equations.

Further he impressively gave the identity:

$$\sqrt{\left(x+\sqrt{y}\right)}=\sqrt{\left(\left[x+\sqrt{\left(x^2-y\right)}\right]/2\right)}+\sqrt{\left[x-\sqrt{\left(x^2-y\right)}\right]/2}$$

Bhaskaracharya II

Bhaskaracharya, or Bhaskara II, is regarded almost without question as the greatest Hindu mathematician of all time and his contribution to not just Indian, but world mathematics is undeniable. As L Gurjar states:

> *...Because of his work India gave a definite 'quota' to the forward world march of the science. [LG, P 104]*

Born in 1114 AD (in Vijayapura, he belonged to Bijjada Bida) he became head of the Ujjain school of mathematical astronomy (Varahamihira and Brahmagupta had helped to found this school or at least 'build it up').

There is some confusion amongst the texts I have referred to as to the works that he wrote. C Srinivasiengar claims he wrote Siddhanta Siromani in 1150 AD, which contained four sections:

(1) Lilavati (arithmetic)

(2) Bijaganita (algebra)

(3) Goladhyaya (sphere/celestial globe)

(4) Grahaganita (mathematics of the planets)

E Robertson and J O'Connor claim that he wrote 6 works, 1), 2) and SS (which contained two sections) and three further astronomical works, including two commentaries on the SS.

G Joseph claims his mathematically significant works were 1), 2), and SS (which indeed he wrote in 1150 and is a highly influential astronomical work). S Sinha however agrees with C Srinivasiengar that Lilavati was a section (chapter) of the SS, and thus I will agree with the respected Indian historians.

Lilavati (or Leelavati, there is a charming if unlikely story regarding the origin of the name of this work) is divided into 13 chapters (possibly by later scribes) and covers many branches of mathematics, arithmetic, algebra, geometry, and a little trigonometry and mensuration.

More specifically the contents include:

- Definitions.
- Properties of zero (including division).
- Further extensive numerical work, including use of negative numbers and surds.
- Estimation of.

- Arithmetical terms, methods of multiplication, squaring, inverse rule of three, plus rules of 5, 7 and 9.
- Problems involving interest.
- Arithmetical and geometrical progressions.
- Plane geometry.
- Solid geometry.
- Combinations.
- Indeterminate equations (Kuttaka), integer solutions (first and second order) His contributions to this topic are among his most important, the rules he gives are (in effect) the same as those given by the renaissance European mathematicians (17th Century) yet his work was of 12th Century. Method of solving was an improvement of the methods found in the work of Aryabhatta and subsequent mathematicians.
- Shadow of the gnomon.

The Lilivati is written in poetic form with a prose commentary and Bhaskara acknowledges that he has condensed the works of Brahmagupta, Sridhara (and Padmanabha).

However his work is outstanding for its systemisation, improved methods and the new topics that he has introduced. Furthermore the Lilavati contained excellent recreative problems and it is thought that Bhaskara's intention may have been that a student of 'Lilavati' should concern himself with the mechanical application of the method. A student of 'Bijaganita' should however concern himself with the theory underlying the method.

His work Bijaganita is effectively a treatise on algebra and contains the following topics:

- Positive and negative numbers.

- Zero.
- The 'unknown'.
- Surds.
- Kuttaka.
- Simple equations (indeterminate of second, third and fourth degree).
- Simple equations with more than one unknown.
- Indeterminate quadratic equations (of the type $ax^2 + b = y^2$).
- Quadratic equations.
- Quadratic equations with more than one unknown.
- Operations with products of several unknowns.

Bhaskara derived a cyclic, 'Cakraval' method for solving equations of the form $ax^2 + bx + c = y$, which is usually attributed to William Brouncker who 'rediscovered' it around 1657. Bhaskara's method for finding the solutions of the problem $Nx^2 + 1 = y^2$ (so called "Pell's equation") is of considerable interest and importance.

His work the Siddhanta Siromani is an astronomical treatise and contains many theories not found in earlier works. There is not a large mathematical content, but of particular interest are several results in trigonometry and calculus that are found in the work. These include results of differential and integral calculus.

Bhaskara is though to be the first to show that:

$$\delta \sin x = \cos x \ \delta x$$

Evidence suggests Bhaskara was fully acquainted with the principle of differential calculus, and that his researches were in no way inferior to Newton's, asides the fact that it seems he did not understand the utility of his researches, and thus historians of mathematics generally neglect his outstanding achievement, which is extremely regrettable.

Bhaskara also goes deeper into the 'differential calculus' and suggests the differential coefficient vanishes at an extremum value of the function, indicating knowledge of the concept of 'infinitesimals'.

He also gives the (now) well known results for sin(a + b) and sin(a - b). There is also evidence of an early form of Rolle's theorem;

if $f(a) = f(b) = 0$ then $f'(x) = 0$

for some x with $a < x < b$,

in Bhaskara's work.

There have been several unscrupulous attempts to argue that there are traces of Diophantine influence in Bhaskara's work, but this once again seems like an attempt by European scholars to claim European influence on (all) the great works of mathematics. These claims should be ignored. Particularly in the field of algebra, Diophantus only looked at specific cases and did not achieve the general methods of the Indians.

Pell's Equation

Although there were great advancements in many areas of algebra, and the solution of many different forms of equations, from simple forms such as $ax + c = by$, to forms as complex as $ax^2 + bxy + cy^2 = z^2$, I have chosen to look in slightly more detail at solution(s) to the so called Pell's equation, $Nx^2 + 1 = y^2$. The equation was so called due to a mistake on the part of Euler, who attributed the solution of the equation to John Pell, a 17th century English scholar, who actually only referred to the equation in a text he wrote on algebra.

The equation was nearly solved by Brahmagupta (628 AD) and the solution was improved by Bhaskara II (1150 AD), leading some historians, including C Srinivasiengar, to suggest:

...It is therefore fitting that this equation be called the Brahmagupta-Bhaskara equation. [CS, P 110]

The complete theory underlying the solution was expounded by Lagrange in 1767, and rests on the theory of continued fractions. It must be briefly noted how remarkable the achievements of Indian scholars were, given the time period in which equations of the Pell's type were studied. The Indian method involves an element of trail-process but contains no mention of continued fractions.

Further to solving equations of the Pell's type to obtain solutions for the unknowns, Brahmagupta extended his method of solution to find square roots. This contribution is of huge interest as it is essentially the same method rediscovered and used by Newton and Raphson around 1690, which is known as the Newton-Raphson iterative method.

Contained within this brief discussion is a small computer code for the Maple mathematics package, which uses the Brahmagupta type solution of Pell's equation to derive extremely accurate square roots.

The Pell's type of equation was known in India as Varga Prakriti, or "equation of the multiplied square", where prakriti means coefficient and refers to the coefficient N (where N is a positive integer). As previously mentioned, Bhaskara developed a Chakravala or cyclic method of solution.

Regrettably, due to constraints of space, I will have to forego discussion of the general solutions derived by Brahmagupta and Bhaskara, and will include only an example to illustrate Bhaskara's improved 'cyclic' method.

The following example is of great historical interest. It is found in the Bijaganita of Bhaskara and is also of the form of a problem Fermat set as a problem to fellow mathematician Frenicle in 1657. The smallest solution for x and y runs into 4 and 5 digits respectively. The chakravala method is remarkable, as it requires just a few 'easy' steps, while Lagrange's solution required complex use continued fractions.

EXAMPLE 1:

Solution of $67x^2 + 1 = y^2$.

$67x^2 + 1 = y^2$.

Firstly the auxiliary equation $67 \times 12 - 3 = 82$ is taken.

Then using Bhaskara's lemma, where $Na^2 + k = b^2$, where a, b, k are the integers (1, -3 and 8) in the auxiliary equation above, (k being positive or negative) then:

$N((am + b)/k)^2 + ((m^2 - N)/k) = ((bm + Na)/k)^2$

Thus:

$67((1 \times m + 8)/-3)^2 + ((m^2 - 67)/-3) = ((8m + 67\ 1)/-3)^2$ ...(1)

Then, by the method of Kuttaka the solution of $(m + 8)$ -3 = an integer, is $m = -3t + 1$.

Putting $t = -2$, we get $m = 7$, which makes $[m^2 - 67]$ least.

On substituting this value, the equation (1) reduces to:

$67 \times 52 + 6 = 412$

Again, by the lemma:

$67((5n + 41)/6)^2 + ((n^2 - 67)/6) = ((41n + 67\ 5)/6)^2$ ...(2)

Then the solution of $(5n + 41)/6$ = a whole number, is $n = 6t + 5$. $[n^2 - 67]$ will be least for the value $t = 0$, that is, when $n = 5$. The equation (1) then becomes:

$67 \times 112 - 7 = 902$

Now we form:

$67((11p + 90)/-7)^2 + ((p^2 - 67)/-7) = ((90p + 67\ 11/-7)^2$ ...(3)

The solution of $(11p + 90)/-7$ = an integral number, is $p = -7t + 2$. Taking $t = -1$, we have

$p = 9$; and this value makes $[p^2 - 67]$ least. Substituting that into (3) we get:

$67 \times 272 - 2 = 2212$

By the Principle of Composition of Equals, we get from the above equation:

$67(2.27.221)^2 + 4 = (2212 + 67 \times 272)^2$

Or$67(11934)^2 + 4 = (97684)^2$

Dividing out by 4 we have:

$67(5967)^2 + 1 = (48842)^2$

Hence x = 5967, y = 48842 is a solution of the equation $67x^2 + 1 = y^2$. (See C Srinivasiengar Ps 110-133 for further details on Indian solutions to Pell's equation.)

I will conclude my brief discussion of Pell's equation by illustrating Brahmagupta's method for calculating square roots with the following example. The method was used by Brahmagupta to find the square root of the integer N, when $Nx^2 + 1 = y^2$, and solutions for x and y are known.

EXAMPLE 2:

Finding the square root of N, from $Nx^2 + 1 = y2$

If N = 5, then $y^2 = 1 + 5x^2$. It can be observed that $5 = (y^2 - 1)/x^2$ and that $(y^2 - 1)/x^2 >> y^2/x^2$.

This is the key to finding the square root of 5, as 5 = y/x. With ease we can identify y = 9 and x = 4 as solutions to this equation, and we see √5 >> 9/4 = 2.25, (√5 = 2.236067978...).

Clearly the larger y and x are, the better the approximation is. This can shown using the following Maple programme:

```
> n:=5:

> f: = (x,y) ⇒ (2*x*y, y*y + n*x*x):

> m:=0:

> x:=4
```

```
> y=:9
> while m 5 do
> m: = m + 1;
> print (x,y,evalf(y/x,20), evalf(y/x-sqrt(n),50));
> a:=f (x,y);
> x:=a [1];
> y:=a [2];
> end do:
```

The output gives the following:

4, 9 (solutions of x and y)

2.2500000000000000000 (y/x to 20 decimal places)

0.0139320225002103035908263312687237645593816403885 (error between y/x and √5)

By the 5th step the following result is given:

25840354427429161536, 57780789062419261441

(5th pair of solutions for x and y)

2.2360679774997896964 (y/x to 20 d.p.s)

$0.3348791201 \times 10^{-39}$ (error)

This result is extremely accurate, and the method must be considered brilliant given it was derived by Brahmagupta in 628 AD.

The End of the Classic Period

The work of Bhaskara was considered the highest point Indian mathematics attained, and it was long considered that Indian mathematics ceased after that point. Extreme political turmoil through much of the sub-continent shattered the atmosphere of discovery and learning and led to the stagnation of mathematical developments as scholars contented themselves with duplicating earlier works.

Recent discoveries however have found that, despite political turmoil, mathematics continued to a high degree in the south of India up to the 16th century. The South of India avoided the worst of the political upheavals of the subcontinent, and the Kerala School of mathematics flourished for some time, producing some truly remarkable results. These results, the most notable of which are in the field of infinite series expansions of trigonometric functions, are generally inaccurately attributed to great European mathematicians of the 18th century including Newton, Leibniz and Gregory. However, slowly, this rigid position is shifting somewhat.

Before going on to discuss the Kerala contribution to mathematics it is worth noting that by the time of Bhaskara II's death Indian mathematics of the 5th and 6th centuries had exerted a significant influence on mathematics across the world. By the 11th century a number of important Arabic works had been written, based on translations of a number of Indian astronomical works. As the Arab Empire stretched as far as southern Spain, much of this work based on Indian science made its way into southern Europe and was subsequently translated into Latin.

Sadly there is very little recognition of these facts, and even though the Arabic (and hence some Indian) works were prevalent in Spain, they did not transmit any further into Europe, which was still to fully 'awaken and probably 'resisted' the works, and many were subsequently lost. However ultimately a few Latin translations of Indo-Arabic works did flow into wider Europe, causing a step towards the renaissance.

This brief return to this discussion is primarily to serve as a reminder that Indian mathematics has had a far greater influence on the forward progress of mathematics (in conjunction with enlightened Arab scholars, and ultimately a handful of pioneering European scholars) than is generally mentioned. A prime example is the work of Fibonacci, which shows appreciation of Indo-Arabic work as early as the 12th century. In short, around the time of Bhaskara's death (12th c.) Indian mathematics (of the 5th-7th centuries) was still exerting a significant influence throughout the world.

As mentioned, it was long considered that following the 'high point' of the work of Bhaskara that Indian mathematics fell into a steep decline. To some extent this is true, only shortly after Bhaskara's death India was engulfed in war (Mongol invasions) and political turmoil. Consequently the atmosphere of security and tranquillity was lost, which was doubtlessly a major contributory factor to the barrenness of scientific activity and achievement. As C Srinivasiengar quotes:

> *...India suddenly fell into a state of "torpor" and never recovered from this torpor until the advent of mathematicians trained according to Western methods. [CS, P 142]*

There were occasional small developments, and attempts to revive learning, but nothing of the magnitude of the previous millennium.

Worthy of a brief mention are both Kamalakara (c 1616-1700) and Jagannatha Samrat (c. 1690-1750). Both combined traditional ideas of Indian astronomy with Arabic (and some Greek) concepts, Kamalakara gave trigonometric results of interest and Samrat made several Sanskrit translations of Arabic 'versions' of Greek works, including notably Euclid's Elements. However, as C Srinivasiengar comments:

> *...Jagannatha's work was not mere translation. [CS, P 143]*

Indeed his work contained new proofs not given by Euclid.

Under the patronage of monarch Sawai Jayasinha Raja (and his predecessor) Samrat was attempting, along with a group of scholars, to 'reinvigorate' science and learning in

India. It must be admitted that these efforts do not appear to have been wholly successful, although the efforts were in the 'greatest of faith' and should be 'applauded'. To some extent the work of these two scholars only serves to further highlight the lack of originality and indolent attitude that was present by this stage in the north of India, although their efforts should not be completely ignored.

It must be considered most unfortunate that a country, which on reflection was unarguably a world leader in the field of mathematics for several thousands of years, ceased to contribute in any significant way. A theory has been suggested that, had there been definite 'links' between each 'major' period we have discussed, then India would have led the world, unequivocally, in the field of mathematics and may have continued to for much longer. However discoveries that have been made in the last 150 years have significantly altered the chronology of Indian mathematics and the way in which we should view Indian contributions.

Keralese Mathematics I

The south western tip of India escaped the majority of the political upheaval, which engulfed the rest of the country, allowing a generally peaceful existence to continue. Thus the pursuit of scientific development was able to continue 'uninterrupted'. It has only recently come to light that mathematics (and astronomy) continued to flourish in this area for several hundred years. Kerala mathematics was strongly influenced by astronomy, but this led to the derivation of mathematical results of huge importance. As a result of the recency of these discoveries it is quite probable that there are still further discoveries of 'Kerala mathematics' to be made, and a full analysis has yet to be carried out. However several findings have already been made that show several major concepts of renaissance European mathematics were first developed in India.

Indeed G Joseph quotes:

...In Kerala, the period between the 14th and 16th centuries marked a high point in the indigenous development of (astronomy) and mathematics. [GJ, P 287]

The works that have so far been analysed are of such a high level that it is though there may be missing links between the "classical period" and the medieval period of Kerala. There is also interest in the claims that European scholars may have had first hand knowledge of some Kerala mathematics, as the area was a focal point for trading with many parts of the world, including Europe. There is also some evidence of a transfer of technology between Europe and Kerala. I will discuss this issue in a little more depth later.

At this point in time only some interest is being paid to the recent discoveries that have been made, highlighting that in many historical 'circles' Indian developments are still not considered important. A further point of interest is that up to the 10th century there was virtually no mathematical activity of note in the south of India. There are four areas in the south of the country - including Kerala in the south west corner, and with the exception of Mahavira (resident of Karnataka area) there was no mathematical output of significance until the 11th-12th century early Kerala literature. There are however impressions that Aryabhatta I was a Keralite, and indeed, if this were true, then in the words of K Rajagopalan:

...Kerala would find a prominent place in the mathematical map of "our" country. [KR1, P 81]

Bhaskara I is also thought to have possibly been a Keralite.

Of the leading mathematicians of Kerala there is quite possibly more to be discovered but currently there are several whose work is of significant interest.

Mathematicians of Kerala

Narayana Pandit (c. 1340-1400), the earliest of the notable Keralese mathematicians, is known to have definitely written two works, an arithmetical treatise called Ganita Kaumudi and an algebraic treatise called Bijganita Vatamsa. He was strongly influenced by the work of Bhaskara II, which proves work from the classic period was known to Keralese mathematicians and was thus influential in the continued progress of the subject. Due to this influence Narayana is also thought to be the author of an elaborate commentary of Bhaskara II's Lilavati, titled Karmapradipika (or Karma-Paddhati). It has been suggested that this work was written in conjunction with another scholar, Shankara Variyar, while others attribute the work to Madhava (see later).

Although the Karmapradipika contains very little original work, seven different methods for squaring numbers are found within it, a contribution that is wholly original to the author. Narayana's other major works contain a variety of mathematical developments, including a rule to calculate approximate values of square roots, using the second order indeterminate equation $Nx^2 + 1 = y^2$ (Pell's equation). Mathematical operations with zero, several geometrical rules and discussion of magic squares and similar figures are other contributions of note. Evidence also exists that Narayana made minor contributions to the ideas of differential calculus found in Bhaskara II's work.

R Gupta has also brought to light Narayana's contributions to the topic of cyclic quadrilaterals. Subsequent developments of this topic, found in the works of Shankara Variyar and Ganesa interestingly show the influence of work of Brahmagupta.

Parameshvara (c. 1370-1460) is known to have been a pupil of Narayana Pandit, and also Madhava of Sangamagramma, who I will discuss later and is thought to have been a significant influence. He wrote commentaries on the work of Bhaskara I, Aryabhatta I and Bhaskara II, and his contributions to mathematics include an outstanding version of the mean value theorem. Furthermore Parameshvara gave a mean value type formula for inverse interpolation of sine, and is thought to have been the first mathematician to give the radius of circle with inscribed cyclic quadrilateral, an expression that is normally attributed to Lhuilier (1782).

In turn, Nilakantha Somayaji (1444-1544) was a disciple of Parameshvara and was educated by his son Damodra. In his most notable work Tantra Samgraha (which 'spawned' a later anonymous commentary Tantrasangraha-vyakhya and a further commentary by the name Yuktidipaika, written in 1501) he elaborates and extends the contributions of Madhava.

Sadly none of his mathematical works are extant, however it can be determined that he was a mathematician of some note. Nilakantha was also the author of Aryabhatiya-bhasa a commentary of the Aryabhatiya. Of great significance is the presence of mathematical proof (inductive) in Nilakantha's work.

Furthermore, his demonstration of particular cases of the series

$$\tan^{-1} t = t - t^3/3 + t^5/5 - \dots ,$$

when $t = 1$ and $t = 1/\sqrt{3}$, and remarkably good rational approximations of (using another Madhava series) are of great interest. Various results regarding infinite geometrically progressing convergent series are also attributed to Nilakantha.

Citabhanu (1475-1550) has yet to find a place in books on Indian mathematics. His work on the solution of equations is quoted in a work called Kriya-krama-kari, by the scholar Shankara Variar, who is also relatively little known (although R Gupta mentions a further text, written by him).

Jyesthadeva (c. 1500-1575) was a member of the Kerala School, which was founded on the work of Madhava, Nilakantha, Parameshvara and others. His key work was the Yukti-bhasa (written in Malayalam, a regional language of Kerala). Similarly to the work of Nilakantha it is almost unique in the history of Indian mathematics, in that it contains both proofs of theorems and derivations of rules. He also studied various topics found in many previous Indian works, including integer solutions of systems of first degree equations solved using kuttaka.

Madhava of Sangamagramma

Although born in Cochin on the Keralese coast before the previous four scholars I have chosen to save my discussion of Madhava of Sangamagramma (c. 1340 - 1425) till last, as I consider him to be the greatest mathematician-astronomer of medieval India. Sadly all of his mathematical works are currently lost, although it is possible extant work may yet be 'unearthed'. It is vaguely possible that he may have written Karana Paddhati a work written sometime between 1375 and 1475, but this is only speculative. All we know of Madhava comes from works of later scholars, primarily Nilakantha and Jyesthadeva. G Joseph also mentions surviving astronomical texts, but there is no mention of them in any other text I have consulted.

His most significant contribution was in moving on from the finite procedures of ancient mathematics to 'treat their limit passage to infinity', which is considered to be the essence of modern classical analysis. Although there is not complete certainty it is thought Madhava was responsible for the discovery of all of the following results:

(1) $\theta = \tan\theta - (\tan^3\theta)/3 + (\tan^5\theta)/5 - \ldots$, equivalent to Gregory series.

(2) $r\theta = \{r(r\sin\theta)/1(r\cos\theta)\} - \{r(r\sin\theta)^3/3(r\cos\theta)^3\} + \{r(r\sin\theta)^5/5(r\cos\theta)^5\} - \ldots$

(3) $\sin\theta = \theta - \theta^3/3! + \theta^5/5! - ...$, Madhava-Newton power series.

(4) $\cos\theta = 1 - \theta^2/2! + \theta^4/4! - ...$, Madhava-Newton power series.

Remembering that Indian $\sin\theta = r\sin\theta$, and Indian $\cos\theta = r\cos\theta$. Both the above results are occasionally attributed to Maclaurin.

(5) $\pi/4 \cong 1 - 1/3 + 1/5 - ... \pm 1/n \pm(-f_i(n+1))$, $i = 1,2,3$, and where $f_1 = n/2$, $f_2 = (n/2)/(n^2 + 1)$ and $f_3 = ((n/2)^2 + 1)/((n/2)(n^2 + 4 + 1))^2$ (a power series for , attributed to Leibniz)

(6) $\pi/4 = 1 - 1/3 + 1/5 - 1/7 + ... \pm 1/n \pm\{-f(n+1)\}$, Euler's series.

A particular case of the above series when $t = 1/\sqrt{3}$ gives the expression:

(7) $p = \sqrt{12}\ (1 - \{1/(3 \times 3)\} + \{1/(5 \times 3^2)\} - \{1/(7 \times 3^3)\} + ...\}$

In generalisation of the expressions for f2 and f3 as continued fractions, the scholar D Whiteside has shown that the correcting function f(n) which makes 'Euler's' series (of course it is not in fact Euler's series) exact can be represented as an infinite continued fraction. There was no European parallel of this until W Brouncker's celebrated reworking in 1645 of J Wallis's related continued product.

A further expression involving :

(8) $\pi d \cong 2d + 4d/(2^2 - 1) - 4d/(4^2 - 1) + ... \pm 4d/(n^2 + 1)$ etc., this resulted in improved approximations of π, a further term was added to the above expression, allowing Madhava to calculate π to 13 decimal places. The value p = 3.14159265359 is unique to Kerala and is not found in any other mathematical literature. A value correct to 17 decimal places (3.14155265358979324) is found in the work

Sadratnamala. R Gupta attributes calculation of this value to Madhava, (so perhaps he wrote this work, although this is pure conjecture).

Of great interest is the following result:

(9) $\tan^{-1}x = x - x^3/3 + x^5/5 - ...$, Madhava-Gregory series, power series for inverse tangent, still frequently attributed to Gregory and Leibniz.

It is also expressed in the following way:

(10) $r\arctan(y/x) = ry/x - ry^3/3x^3 + ry^5/5x^5 - ...$, where $y/x \leq 1$

The following results are also attributed to Madhava of Sangamagramma:

(11) $\sin(x + h) \cong \sin x + (h/r)\cos x - (h^2/2r^2)\sin x$

(12) $\cos(x + h) \cong \cos x - (h/r)\sin x - (h^2/2r^2)\cos x$

Both the approximations for sine and cosine functions to the second order of small quantities, (see over page) are special cases of Taylor series, (which are attributed to B Taylor).

Finally, of significant interest is a further 'Taylor' series approximation of sine:

(13) $\sin(x + h) \sin x + (h/r)\cos x - (h^2/2r^2)\sin x + (h^3/6r^3)\cos x$.

Third order series approximation of the sine function usually attributed to Gregory.

With regards to this development R Gupta comments:

...It is interesting that a four-term approximation formula for the sine function so close to the Taylor series approximation was known in India more than two centuries before the Taylor series expansion was discovered by Gregory about 1668. [RG5, P 289]

Although these results all appear in later works, including the Tantrasangraha of Nilakantha and the Yukti-bhasa of

Jyesthadeva it is generally accepted that all the above results originated from the work of Madhava. Several of the results are expressly attributed to him, for example Nilakantha quotes an alternate version of the sine series expansion as the work of Madhava. Further to these incredible contributions to mathematics, Madhava also extended some results found in earlier works, including those of Bhaskaracarya.

The work of Madhava is truly remarkable and hopefully in time full credit will be rewarded to his work, as C Rajagopal and M Rangachari note:

> *...Even if he be credited with only the discoveries of the series (sine and cosine expansions, see above, 3) and 4)) at so unexpectedly early a date, assuredly merits a permanent place among the great mathematicians of the world. [CR /MR1, P 101]*

Similarly G Joseph states:

> *... We may consider Madhava to have been the founder of mathematical analysis. Some of his discoveries in this field show him to have possessed extraordinary intuition. [GJ, P 293]*

With regards to Keralese contributions as a whole, M Baron writes (in D Almeida, J John and A Zadorozhnyy):

> *...Some of the results achieved in connection with numerical integration by means of infinite series anticipate developments in Western Europe by several centuries. [DA/JJ/AZ1, P 79]*

There remains a final Kerala work worthy of a brief mention, Sadrhana-Mala an astronomical treatise written by Shankara Varman serves as a summary of most of the results of the Kerala School. What is of most interest is that it was composed in the early 19th century and the author stands as the last notable name in Keralese mathematics.

In recent histories of mathematics there is acknowledgement that some of Madhava's remarkable results were indeed first discovered in India. This is clearly a positive step in redressing the imbalance but it seems unlikely that full 'credit' will be given for some time, as that will possibly require the re-naming of various series, which seems unlikely to happen!

Still in many quarters Keralese contributions go unnoticed, D Almeida, J John and A Zadorozhnyy note that a well known historian of mathematics makes:

...No acknowledgement of the work of the Keralese school. [DA/JJ/AZ1, P 78]

(Despite several Western publications of Keralese work.)

Possible Transmission of Keralese Mathematics to Europe

In addition to my discussion, there is a very recent paper (written by D Almeida, J John and A Zadorozhnyy) of great interest, which goes as far as to suggest Keralese mathematics may have been transmitted to Europe. It is true that Kerala was in continuous contact with China, Arabia, and at the turn of the 16th century, Europe, thus transmission might well have been possible. However the current theory is that Keralese calculus remained localised until its discovery by Charles Whish in the late 19th century. There is no evidence of direct transmission by way of relevant manuscripts but there is evidence of methodological similarities, communication routes and a suitable chronology for transmission.

A key development of pre-calculus Europe, that of generalisation on the basis of induction, has deep methodological similarities with the corresponding Kerala development (200 years before). There is further evidence that John Wallis (1665) gave a recurrence relation and proof of Pythagoras theorem exactly as Bhaskara II did. The only way European scholars at this time could have been aware of the work of Bhaskara would have been through Keralese 'routes'.

The need for greater calendar accuracy and inadequacies in sea navigation techniques are thought to have led Europeans to seek knowledge from their colonies throughout the 16th and 17th centuries. The requirements of calendar reform were imperative with the dating of Easter proving extremely problematic, by the 16th century the European 'Julian' calendar was becoming so inaccurate that without correction Easter would eventually take place in summer! There were significant financial rewards for 'anyone' who could 'assist' in the improvement of navigation techniques. It is thought 'information' was sought from India in particular due to the influence of 11th century Arabic translations of earlier Indian navigational methods.

Events also suggest it is quite possible that Jesuits (Christian missionaries) in Kerala were 'encouraged' to acquire mathematical knowledge while there.

It is feasible that these observations are mere coincidence but if indeed it is true that transmission of ideas and results between Europe and Kerala occurred, then the 'role' of later Indian mathematics is even more important than previously thought.

Conclusions

I wish to conclude initially by simply saying that the work of Indian mathematicians has been severely neglected by western historians, although the situation is improving somewhat. What I primarily wished to tackle was to answer two questions, firstly, why have Indian works been neglected, that is, what appears to have been the motivations and aims of scholars who have contributed to the Eurocentric view of mathematical history. This leads to the secondary question, why should this neglect be considered a great injustice.

I have attempted to answer this by providing a detailed investigation (and analysis) of many of the key contributions of the Indian subcontinent, and where possible, demonstrate how they pre-date European works (whether ancient Greek or later renaissance). I have further developed this 'answer' by providing significant evidence that a number of Indian works conversely influenced later European works, by way of Arabic transmissions. I have also included a discussion of the Indian decimal place value system which is undoubtedly the single greatest Indian contribution to the development of mathematics, and its wider applications in science, economics (and so on).

Discussing my first 'question' is less easy, as within the history of mathematics we find a variety of 'stances'. If the most extreme Eurocentric model is 'followed' then all mathematics is considered European, and even less extreme stances do not give full credit to non-European contributions.

Indeed even in the very latest mathematics histories Indian 'sections' are still generally fairly brief. Why this attitude exists seems to be a cultural issue as much as anything. I feel it important not to be controversial or sweeping, but it is likely European scholars are resistant due to the way in which the inclusion of non-European, including Indian, contributions shakes up views that have been held for hundreds of years, and challenges the very foundations of the Eurocentric ideology. Perhaps what I am trying to say is that prior to discoveries made in technically fairly recent times, and in some cases actually recent times (say in the case of Kerala mathematics) it was generally believed that all science had been developed in Europe. It is almost more in the realms of psychology and culture that we argue about the effect the discoveries of non-European science may have had on the 'psyche' of European scholars.

However I believe this concept of 'late discoveries' is a relatively weak excuse, as there is substantial evidence that many European scholars were aware of some Indian works that had been translated into Latin. All that aside, there was significant resistance to scientific learning in its totality in Europe until at least the 14th/15th c and as a result, even though Spain is in Europe, there was little progression of Arabic mathematics throughout the rest of Europe during the Arab period.

However, following this period it seems likely Latin translations of Indian and Arabic works will have had an influence. It is possible that the scholars using them did not know the origin of these works. There has also been occasional evidence of European scholars taking results from Indian or Arabic works and presenting them as their own. Actions of this nature highlight the unscrupulous character of some European scholars.

Along with cultural reasons there are no doubt religious reasons for the neglect of Indian mathematics, indeed it was the power of the Christian church that contributed to the stagnation of learning, described as the dark ages, in Europe.

Above all, and regardless of the arguments, the simple fact is that many of the key results of mathematics, some of which are at the very 'core' of modern day mathematics, are of Indian origin. The results were almost all independently 'rediscovered' by European scholars during and after the 'renaissance' and while remarkable, history is something that should be complete and to neglect facts is both ignorant and arrogant. Indeed the neglect of Indian mathematical developments by many European scholars highlights what I can best describe as an idea of European "self importance".

In many ways the results of the Indians were even more remarkable because they occurred so much earlier, that is, advanced mathematical ideas were developed by peoples considered less culturally and academically advanced than (late medieval) Europeans. Although this comment is controversial it may have been the motivation of several authors for neglect of Indian works, however, if this is the case, then opinions based on those attitudes should be ignored. Indian culture was of the highest standard, and this is reflected in the works that were produced.

Indian mathematicians made great strides in developing arithmetic (they can generally be credited with perfecting use of the operators), algebra (before Arab scholars), geometry (independent of the Greeks), and infinite series expansions and calculus (attributed to 17th/18th century European scholars). Also Indian works, through a variety of translations, have had significant influence throughout the world, from China, throughout the Arab Empire, and ultimately Europe.

To summarise, the main reasons for the neglect of Indian mathematics seem to be religious, cultural and psychological. Primarily it is because of an ideological choice. R Rashed mentions a concept of modernism vs. tradition. Furthermore Indian mathematics is criticised because it lacks rigour and is only interested in practical aims (which we know to be incorrect). Ultimately it is fundamentally important for historians to be neutral, (that includes Indian historians who may go too far the 'other way') and this has not always been the case, and indeed seems to still persist in some quarters.

In terms of consequences of the Eurocentric stance, it has undoubtedly resulted in a cultural divide and 'angered' non-Europeans scholars. There is an unhealthy air of European superiority, which is potentially quite politically dangerous, and scientifically unproductive.

In order to maximise our knowledge of mathematics we must recognise many more nations as being able to provide valuable input, this statement is also relevant to past works. Eurocentrism has led to an historical 'imbalance', which basically means scholars are not presenting an accurate version of the history of the subject, which I view as unacceptable. Furthermore, it is vital to point out that European colonisation of India most certainly had an extremely negative effect on the progress of indigenous Indian science.

At the very least it must be hoped that the history of Indian mathematics will, in time become as highly regarded, as I believe it should. As D Almeida, J John and A Zadorozhnyy comment:

...Awareness is not widespread. [DA/JJ/AZ1, P 78]

R Rashed meanwhile explains the current problem:

...The same representation is found time and again: classical science, both in modernity and historicity appears in the final count as work of European humanity alone...

He continues:

...It is true that the existence of some scientific activity in other cultures is occasionally acknowledged. Nevertheless, it remains outside history or is only integrated in so far as it contributed to science, which is essentially European. [RR, P 333]

In short, the doctrine of the western essence of classical science does not take objective history into account.

Finally, beyond simply alerting people to the remarkable developments of Indian mathematicians between around 3000 BC and 1600 AD, and challenging the Eurocentric ideology of the history of the subject, it is thought further analysis and research could also have important consequences for future developments of the subject.

It is thought analysis of the difference in the epistemologies of 17th century European and 15th century Keralese calculus could help to provide an answer to the controversial issue of whether mathematics should concern itself with proof or calculation. Furthermore, in terms of the way mathematics is currently 'taught' D Almeida, J John and A Zadorozhnyy elucidate:

> *...The floating point numbers were used by Kerala mathematicians and, using this system of numbers, they were able to investigate and rationalise about the convergence of series. So we (DA/JJ/AZ) believe that a study of Keralese calculus will provide insights into computer-assisted teaching strategies. [DA/JJ/AZ, P 96]*

(N.B. computers use a floating-point number system.)

Clearly there is massive scope for further study in the area of the history of Indian and other non-European mathematics, and it is still a topic on which relatively few works have been written, although slowly significantly more attention is being paid to the contributions of non-European countries.

In specific reference to my own project, I would have liked to have been able to go into more depth in my discussion of Indian algebra, and given many more worked examples, as I consider Indian algebra to be both remarkable and severely neglected. Furthermore there is scope for significant and important study of the transmission of Indian mathematics across the world, especially into Europe, via Arabic and later Keralese routes. It is clear that there are many more discoveries to be made and much more that can be written, as C Srinivasiengar observes:

> *...The last word on the history of ancient civilisation will never be said. [CS, P 1]*

As a final note, many question the worth of historical study, beyond personal interest, but I hope I have shown in the course of my work some of the value and importance of historical study. I will conclude with a quote from the scholar G Miller, who commented:

> *...The history of mathematics is the only one of the sciences to possess a considerable body of perfect and inspiring results which were proved 2000 years ago by the same thought processes as are used today. This history is therefore useful for directing attention to the permanent value of scientific achievements and the great intellectual heritage, which these achievements present, to the world. [AA'D, P 11]*

Contribution to Present Day Education

The system of Vedic Mathematics offers a new approach to mathematics and mathematics education. This system which is beautifully coherent and direct was rediscovered from ancient Sanskrit texts between 1911 and 1918 by Bharati Krsna Tirthaji (1884-1960). He found that all problems in pure and applied mathematics can be solved easily with the aid of sixteen Sutras, or word-formulae: for example By One More than the One Before, or Vertically and Crosswise. This may sound incredible but the methods described in his book "Vedic Mathematics" are so original and yet so simple, offering a very different approach to mathematics that is both powerful and fun.

The word-formulae give a set of natural principles that help to quickly solve all sorts of mathematical problems. The Vedic methods are direct, beautifully interrelated and flexible. The Vedic system is therefore much more unified and flowing than conventional mathematics. Some of the methods are truly amazing in their efficiency and simplicity.

The Vedic system is so easy it is really a system of mental mathematics and this, combined with the coherence and flexibility of the system, encourages the development and use of intuition and creativity.

Mental Mathematics

We all make mental calculations from time to time, though we may not always be aware of it. In deciding at exactly what moment and speed to venture across a busy road, for example, our mind judges continuously the positions and speeds of several vehicles and accurately finds the required gap in which to move forward.

If our mind can make such complex judgements as this it is certainly able to manipulate a few figures. It is the cumbersome calculating devices we have probably been taught- which require pencil and paper or calculator to work out because of their difficulty- and a lack of encouragement for mental calculation which have prevented us from becoming efficient mental calculators.

This need not be so- as Vedic Mathematics shows, mental calculation is easy and to be preferred to pencil and paper or calculator, and has many advantages over these calculating methods.

Most people would probably agree that mathematics holds a special position among subjects of study: that it possess qualities of absolute certainty and precision which cannot be attributed to any other subject. On the other hand however mathematics is seen as difficult and remote by most people- the same people who are also aware of its special absolute qualities. This situation has come about because mathematics education has not been effective enough in bringing out the real nature of mathematics. As young students we glimpse the beauty of mathematics but this is usually a passing phenomenon.

Though mathematics has applications at many levels it is primarily a mental subject. This being so it is likely that lack of mental calculation is partly responsible for the situation described above, and that a system of mental mathematics could provide students with a lasting link with the realm of mathematics and also engender a deeper understanding of the structure and processes of mathematics, as well as helping to develop other important personal qualities.

The following points outline the benefits available from a mental approach to mathematics.

1. Mental calculation sharpens the mind and increases mental agility and intelligence. This will be evident to anyone who has practised or taught mental calculation or who has seen its effects.
2. It enhances the precision of thought. Numbers and other mathematical objects are unbiased, giving only one correct answer to which everyone will agree- there is never a contradiction. This absolute precision is unique to mathematics, so dealing intimately with numbers as we do in mental calculation we cultivate fine and careful thinking.
3. Mental calculation leads naturally to the search for, and discernment of, constancy and law, which are very necessary attributes in a swiftly changing world.
4. Our mind has the ability to retain several ideas at once so that they can be compared, combined and so on. This facility is enhanced by mental calculation as we practise holding the sum in the mind whilst operating with some of the figures.
5. Mental calculation improves the memory. Memory depreciates if it is not exercised. Short term, medium term and long term memory are all stimulated by mental calculation.
6. Because numbers are absolutely dependable and reliable, calculation promotes confidence. In particular, mental calculation creates confidence in oneself and in ones capabilities. To solve a problem, perhaps a difficult one, by mere mental arithmetic without having to rely on some artificial aid is a source of great satisfaction and encouragement.
7. Mental calculation is a delight to the mind: the intrinsic qualities, relationships and beauty of

numbers and the way they create new numbers out of themselves is a source of great enjoyment.

8. Through mental calculation one becomes familiar with numbers and appreciates their various properties. This leads to a real understanding of number.

9. In calculating mentally the subtle properties of numbers and their relationships are appreciated much more readily than if the calculation was written down and thereby fixed. Thus mental calculation leads naturally to innovation and to the invention of new methods, thereby developing the student's natural creativity.

10. Practical uses of mental calculation are many, since we all need to make quick, on the spot, calculations from time to time.

Thus we see that mental calculation has so many advantages and really brings mathematics to life as well as the providing motivation and strengthening and enlivening the mind. This is because numbers are mental concepts, they do not exist on paper. Our unhindered mind operates very fast and has a variety of operational properties. With proper training we can use these properties of the mind to our advantage.

This is not to say that pencil and paper or calculating instruments are to be totally avoided in mathematics: they certainly have their place, but mental calculation should, it is suggested, be the primary method of calculation.

2

India and the World of Mathematics

Secular Education and the Federal Polity

Let me begin by asserting that those who do not understand the past, or refuse to understand it, invariably end up by misunderstanding the present and are unable to move forward into the future. We are faced today with the makers of educational policy in the central government who seem not to understand the Indian past.

There is a constant harking back to a remote past, encapsulated in the phrase 'Vedic'. Irrespective of its historical or civilisational authenticity, this capsule is being forced upon us with the claim that all knowledge is contained in the Vedas and therefore the Vedic capsule amounts to a total education.

There is little recognition of the fact that in the course of Indian history various Indian thinkers discussed the knowledge contained in the Vedic corpus, and some had doubts about various aspects. This process of debate and questioning, the presentation of views and counter-views, both within India and among scholars from other parts of Asia, has been at the root of advances in knowledge in pre-modern times. Much that we pride ourselves on, as Indian contributions to world civilisation, often developed independently of the Vedic corpus and occasionally even in opposition to it. Significant contributions from the past are thus set aside in this obsessive concern with the Vedic capsule.

In saying this, I am not denigrating the study of the Vedic past, but am emphasising that the past has to be assessed in a historical context, and I would further insist that the context has to be, that of critical, rational enquiry. This is now being denied by replacing enquiry with a received version of the past which is then treated as the authentic version.

The claim is made that this is a return to indigenous knowledge, but the new educational curriculum draws its legitimacy from 19th century colonial views of India, and from the priority that European Indologists gave to brahmanical texts and world-view. Indigenous systems drew not only on mainstream texts in the language of learning but also on texts in a variety of regional languages, which could question the former if need be, as also on observed knowledge.

A major pedagogical change in the last few decades has been the professionalising of various subjects, particularly in the social sciences. Each subject is preferably taught in such a way that it also demonstrates its own methodology which draws as much as possible on evidence of proven reliability, on logical analysis and on rational generalisations. This demands an intellectual rigour in setting out the structure of the subject. The training that results from such teaching, as for example in history, enables both the teacher and the student to be aware of the difference between mythology and history.

There is now a retreat from these processes and mythology is taking over from knowledge. Mythology has a role in creative imagination but should not replace knowledge. Instead of further professionalising the subjects taught at school and college, they are being replaced with subjects that have virtually no pedagogical rigour, such as Yoga and Consciousness or cultivating a Spirituality Quotient. These cannot form the core of knowledge and replace subjects with a pedagogical foundation, although Yoga can be an additional activity.

The narrowing of knowledge is being attempted in part by giving a single definition to Indian culture and society, and projecting this through educational channels, and describing it as the sole heritage that is of any consequence to us as a society and a nation. Yet this goes against one of the fundamental concerns of the Indian experience both of the past and of the present.

Among the more significant questions that have continually been at the core of Indian activity, is that of the relations between the needs of the central power in a state and the articulation of variant forms of control manifested by regional and local powers. At the most obvious level in the past, this relationship determined various structures relating to administrative and economic policy. But it is also evident in cultural expression where a distinction was often maintained between the mainstream culture, and the culture and language of the region.

Relations between the two varied from close interlinks on some occasions, to tensions or even confrontations on other occasions. What is relevant to us today is that in the past, cooperation between the centre and the regions needed an immense degree of sensitivity to social and cultural variations and an understanding of why they arose. We are facing a similar problem today. The question is whether we should accept the kind of homogenisation of education and culture that is being imposed on the country or should we attempt to define the modern, educated Indian through an educational policy sensitive to a range of social and economic concerns, and to new systems of knowledge, a sensitivity that will provide us with a worthwhile present and enable us to perceive the inter-connections with the past? Can the interface between the centre and the states in a federal polity, help us in this matter?

Education is not merely about making millions literate, it also involves teaching young people to cope with a changing society, which today means being more aware of the world than ever before, and to creating a worthwhile life for themselves. Therefore, to impose a syllabus oriented to studying an imagined past utopia is to erode the potential of the next generation. Focusing on a utopian past is also a mechanism of diverting attention from having to improve the present in order to provide a better quality of life.

Accountability to the public and transparency in governance is necessary in formulating educational policies. We must know who is drafting educational policy and who have been consulted in doing so, and what has been the participation of professionally qualified persons in the determining of the curriculum in a subject. It requires responsible people and these in turn have to be responsible for what they are doing. Educational policy is both important and sensitive and cannot be left to the whims of a small circle of politicians and bureaucrats.

A sensitive understanding of the interface of centre and region is essential to any educational policy. Two states with high rates of literacy are Kerala and Himachal Pradesh. Each is very different from the other in terms of economic resources and the way they are used; in the hierarchy of castes and the distribution of classes; in religions and religious sects; and in languages. These aspects also undergo change. Can we set aside all this and merely insist on children in both areas studying Sanskrit, Vedic Mathematics, a vague subject called social science, and Yoga and Consciousness? The imposition of the Vedic capsule would be an educational disruption in both regions, educationally negative for many people and resented by others.

But what they do have in common are the aspirations that result from education. Schooling and curriculum would have to relate up to a point to the local conditions and ethos, and these would involve a degree of interest in regional concerns. The question is how best these can be introduced without denying the importance of national concerns - a matter of some sensitivity. Educational policy has to be such that the aspirations at least of regional concerns are recognised as an intrinsic part of those that are of national interest. This would ultimately be more viable than forcing everyone to conform to a top-down policy.

Educational policies in states that do not have a BJP government have a greater responsibility to defend secular education and the continuance of multiple cultures. This is often easier at the state level where multiple cultures are more visible, but would require considerable thinking about education in terms of what is being taught and which groups are appropriating educational facilities. Where parties not belonging to the NDA, tie-up with the Sangh Parivar to harass those supporting secular education, the acts of such parties should also be questioned. Education should not be made the scapegoat for dubious political manoeuvres.

We may well be taking a risk with the future of the next generation by giving them the type of schooling that will not equip them to handle the complexities of our times. These are serious matters that concern the future of an entire generation of young Indians and should be critically discussed and reviewed. But then the Indian middle class is notoriously unconcerned about what it taught to its children through schooling. All that matters is the game of numbers, marks and percentages.

The new policy it is said will reduce social disabilities and the replacing of subjects at school will reduce the burden on the child. Social disabilities can be met to some extent by professionalising what is taught, in other words teaching mainstream subjects as systems of knowledge, without mystifications. The way a subject is taught has a social context and this has a bearing on social disabilities. For example, will Vedic mathematics be taught through memorising shlokas in Sanskrit or essentially as methods of calculation? In the former case, obviously upper caste children will have an advantage; in case of the latter, the quality of what is taught will have to be assessed comparatively with other mathematical methods. If it were to be something more than a slogan, would this kind of mathematics prepare a foundation for the child to handle contemporary technologies requiring mathematics?

Education is related to the social structures that it endorses or wishes to change. The suggested curriculum is essentially intended to construct a middle class ideology producing pliant citizens. The problem will however be intensified when this regimen is imposed on the under-privileged, and when they become aware of this package being a further denial of quality education to them. With this kind of schooling, dalits and scheduled tribes and other marginalised groups such as women will continue to be employed only for lower level jobs. This may be the other agenda of the new curriculum. The package makes no concession in the curriculum to societies that function differently and require adjustments to educational programmes, such as a greater emphasis initially on technical courses.

Those who are privileged and can afford private schools will continue to have a quality education, but the rest, in terms of educational requirements today, will remain virtually uneducated. Such a two-tier system is implicit in the suggestions on educational programmes made by sections of the corporate sector as well, as was discussed on the first day of the conference. This is likely to be a bigger problem in schools run by individual states and at local levels, since the disaffection will be closer to the ground reality in these schools. Yet the curriculum in the states can explore the links between education and the social context more closely, as also the links to aspirations through education. This does not mean a tailor-made education for each caste, religious sect, or tribe, but can consider incorporating variations in the handling of knowledge.

The question of the preferred language for education poses other problems. There is today a relationship between English and the regional languages, between Hindi and English, and among the regional languages. Regional languages have become basic to the administration of the states. But given the aspirations to managerial jobs, corporate sector jobs, technologies of various kinds or employment outside India, English is likely to remain in demand. The greater the investment of the multinationals, the greater will be the requirement for professional English. Economic policies could well dictate educational demands giving rise to yet other problems.

A current misconception lies in the belief that education in the medium of English, or an education in the sciences, automatically results in rational attitudes. But rationality has to be cultivated. Rationality is necessary for the study of technologies, for instance, but those who use it for this purpose do not necessarily apply it to understanding other aspects of knowledge. Those that theoretically should be more enlightened often support outdated views about Indian society without a second thought. Education in the sciences and technologies, is divorced from other areas of thought, and often results in a dichotomy between scientific and technological knowledge on the one hand and knowledge related to the social sciences and the humanities on the other. The necessary integration between the two is frequently absent.

An interesting case in point comes from the Indian diaspora. NRIs often demonstrate this dichotomy, where despite having been trained to enquire critically and rationally into professional areas of work, they make uninformed generalisations about other areas in which they are not trained, such as the Indian past and what goes under the label of Indian culture. This is startlingly evident from their websites. Some of the most untenable theories about the past come from Indian computer scientists, engineers, and such like, and some also claim that because they are scientists, their theories about the past are value-free, a claim that is now much doubted even in scientific circles. Such persons are among the role-models of the Indian middle class. We give legitimacy to their obscurantism because they are technically qualified in the sciences. The reasons for this dichotomy are often explained as due to their lack of adjustment to an alien culture or from the consciousness of being a minority in that culture. If we are now going to have university courses in Vedic rituals, to help create cultural support for such NRIs, then surely we have to concede that the tail is wagging the dog.

If the essentials of critical and rational enquiry in the sciences could be part of a comparative interface with methods in the social sciences, this would result in a far greater awareness of what it means to be educated and how one looks at the world. One is not suggesting that children have to be tutored in the nature of paradigm shifts in all kinds of knowledge, but at least they should know that knowledge changes, that some knowledge becomes out-of-date, and that there are ways of recognising the change. Ultimately the debate between orthodoxies and heterodoxies is essential to advancing knowledge. This debate is a lynchpin in education. The learning process is not a matter of memorising questions and answers but of being encouraged to explore ideas creatively.

To say that educational policy and curriculum should legitimately be replaced with every new government is to make a mockery of the process of education. It is not the monopoly of any government; it is the responsibility of those who are involved in educating the young. What would be useful would be to work out an essential educational structure, incorporating the major debates in various disciplines, and these should not be tampered with whenever there is a change of government. Changing the content of subjects must be recognised as being a professional matter and not a matter of political ideology.

Education widens the mind and encourages an exploration of ideas and actions. But if it is confined to thinking along a single track, then it closes the mind. In the wider context of planning education, if the direction of change points to a federal polity, then the interface between the centre and the states relates not just to education but to many activities, although educational policies often highlight the nature of the connections. If a federal polity is what is required it cannot be brought about accidentally; it has to be forged. Hopefully, this will be a process that strengthens an inclusive nationalism, secular in its essentials and drawing on creative inter-relations between the centre and the states.

From Seattle to Genoa

AS a reporter who has witnessed many anti-globalization protests, from Seattle in 1999 to Washington D.C., to Prague in September 2000, the police brutality seen in Genoa has no equal so far. True, in Seattle hundreds of youngsters were arrested and in Prague the Czech police deployed tanks around the IMF-World Bank meeting (the last time tanks were seen in the streets of Prague was in 1968, during the Soviet invasion).

The police behaviour in Genoa suggests two questions: one is the accountability of security forces in a democratic country. It is useful to remember that reports by human rights groups such as Amnesty International or Human Rights Watch are replete with accounts of violence and harassments by security forces almost everywhere.

The second question is very Italian, but should sound alarm-bells everywhere: a prime minister who has a virtual monopoly over the electronic media in the country has a political ally in the former fascist party, Alleanza Nazionale (National Alliance).

The vice-premier and leader of Alleanza Nazionale was in Genoa to oversee the security operations. This raises questions about democratic control over the security forces. It was a clear announcement of how social movements will be dealt with in this country. The focus on the violent confrontation diverted the public (and the media) from the 'real' news: with 300,000 people, the Genoa mobilization was second only to the Seattle rally against the WTO and its 'millennium round' of negotiations on trade.

Let's look at who was there: Attac, the Association for a global tax on financial transactions (the Tobin tax), came with thousands of activists from France, Italy, Belgium, even Tunisia. There was the French rural trade union Confederation Paysanne, with its leader Jose Bove, who gained international fame in the summer of 1998 when he destroyed a McDonald's restaurant in a small French town - highlighting the commercial war between Europe and the United States, an example of a new WTO-style trade war (the European Union will not import bovine meat produced in the US and fed with a hormone for animal growth that Europeans consider unsafe for human health; the WTO considers this precautionary principle illegal and sanctioned the EU; the US slammed import fees on a number of European imports like Parma Ham and French cheese).

Environmentalists took to the streets in big numbers, from Green peace to the WWF, to the Italian Legambiente (League for Environment). Many came in small, spontaneous groups. The list includes many Catholic NGOs and the many groups of the Lilliput Network, the 'network of the small people' dealing with fair trade or Third World debt. The

League Against Aids (Lila) displayed banners, a reminder of the legal victory of South Africa against 39 pharmaceutical transnationals on drugs and patents.

There were a multitude of 'affinity groups', heirs of the pacifist movements that in the '80s marched or squatted on military sites against the deployment of US nuclear missiles in Italy, the 'Euromissiles' — a reminder of the Cold War. A number of people marched under the banner of the Italian Metal Workers Federation (FLM): it is the first time a big Italian trade union felt compelled to throw its weight behind a global movement.

In Seattle the mobilization against the WTO was largely fed by the American trade unions Afl-Cio, which organized transport to bring American workers to that north-western town. Their main demand was that trade treaties should include clauses on labour protection (and environmental protection, to please the strong environmentalist and consumer movements that also turned out in Seattle: from the Sierra Club to Public Citizen, the consumer watchdog founded in the '70s by Ralph Nader).

What does all this point to? One could point out that NGOs and movements from the South of the world are under represented in these gatherings. True: in Seattle some Asian and Latin American movements sent delegations, but in Prague or in Genoa there were only a few guests to the Public Forums (the Movimiento Sem Terra, the Brazilian movement of the Peasants Without Land; the Ogoni movement of the Niger Delta, an Indonesian environmental NGO, a representative from an indigenous movement in Colombia or from the Bangkok-based think-tank, Focus on the Global South).

But it was the opposite in Porto Alegre, Brazil, in January 2001, when the World Social Forum gathered to discuss experiences and alternatives (the slogan then was 'Another world is possible').

One could also underline that such diversified gatherings bring together different - sometime conflicting - interests and priorities. So for instance when the American Afl-Cio asks for environmental and labour standard clauses in the trade pacts, the Asian NGOs frown, and with reason. A western-style environmentalism took time to notice that poverty and an unequal share of natural resources are the priority (the climate change negotiation is an example). True: with time, links have been forged between, for instance, movements of indigenous people poisoned by Shell in the Niger Delta and environmental groups in London, or social NGOs in the Amazon and environmental lawyers in Washington, so that common actions can be planned.

The main common point in such a diverse patchwork is probably a crisis of legitimacy of a group of 'most industrialized' governments, the lack of accountability of global institutions like the IMF-World Bank and the WTO, their role in pushing unbridled freedom for financial capital in the name of integrating the global economy, and the dismantling of all regulatory mechanisms (both in the North and in the South) in the name of free trade.

The crisis of legitimacy is matched by a deep ecological crisis, and with the crisis of democracy, together with the new 'logic of violence' created by migrations and the re-definition of cultures and people, as Arjun Appadurai analyzes so convincingly in a recent issue of Seminar on Globalization.

Perhaps the economist and director of Focus on the Global South, Walden Bello is right: 'The global economy is so unstable under the menace of so many different crisis, that more and more people from all kinds of backgrounds are directly affected. Of course, each has different reasons to join the protests. ...Now all these different people, with different interests and priorities, are in the process of discovering what they have in common.'

Mathematics appears to have acquired an identity as an independent branch of knowledge early on in human intellectual history. This identity became precise and firmly established thanks to the Greeks in the millennium before Christ. Two characteristics are vital to this identity: abstraction and logical deduction; these are of course present in all scientific enquiries but in mathematics they are defining elements.

Abstraction consists in building mental constructs (often, though not always, born of attempts to understand the concrete). This involves the recognition of common patterns in apparently disparate phenomena and, equally important, the rejection of the irrelevant in an investigation. This is already manifest in the first intimations of mathematical activity: counting. Though almost an involuntary act, what underlies it is profoundly abstract: the human mind recognises that there is an attribute which it can ascribe to a collection of entities - the number of entities in the collection- an attribute which is entirely indifferent to the nature of the individual members of the collection yet can be common to different collections.

What underlies it may be remarkable abstraction, but counting owes its discovery to mundane down to earth compulsions: exchange of goods and barter which required setting values on different commodities. The marketplace was the driving force behind all the arithmetic we learnt at school. And down the ages a great deal of mathematics was born as a response to the needs of diverse human endeavours and has served the cause of these endeavours admirably. This symbiotic relationship is most striking in the case of physics. Attempts to understand motion gave birth to calculus, which has gone on to enhance our understanding not only of motion, but a myriad other natural phenomena in diverse scientific disciplines. The success of mathematics has been so striking that mathematical intervention is what seems to confer the label 'science' on an intellectual discipline.

Mathematics is itself of course a science, yet it stands a little apart from other sciences. The title of Newton's classic, The Mathematical Principles of Natural Philosophy, itself suggests this. Gauss (one of the greatest mathematicians of all time), called mathematics the queen of sciences; indeed, at times, one finds in mathematics the aloofness of the royal personage.

The queen is whimsical - queens are supposed to be - but not quite as arbitrary as the one in Alice's Wonderland. There is a large body of mathematics that has been created as a result of purely aesthetic impulses internal to the discipline, essentially by the fancy of the mathematician. Mathematicians perceive beauty in purely mathematical constructs and the interrelation among them and probe these simply to savour the intellectual delights they offer. Yet time and again it has turned out that such mathematics has proved itself the right vehicle for apprehending nature.

The renowned physicist Wigner has described this pithily as the 'unreasonable effectiveness of mathematics in the natural sciences'. A striking example of this 'unreasonable effectiveness' is the way Descartes' idea of representing spatial entities by numbers - coordinates - has permeated our thinking in diverse contexts. Descartes' motivation was the renovation of geometry, as far away as can be from the company chairman pondering the profit curve.

In every creative endeavour there is a certain tension between imagination and discipline. In the sciences other than mathematics, the discipline is essentially external: theories about natural phenomena have to be in tune with observations. The mathematician's imagination is unfettered by external considerations; the discipline comes from the demands of logical deduction and (the more intangible) aesthetics. In this, mathematics is closer to art than to other sciences.

The glorious days of Greek mathematics ended in the early centuries after Christ and the centre of gravity of mathematical development shifted to the East. From then on right up to the 12th century India had a dominant role. One of the greatest mathematical achievements of all time - the invention of 'zero' and the place value system for representing numbers - emanated from this country. It is at once an absolutely brilliant piece of abstract mathematics and simultaneously an unmatched practical device indispensable in practically every sphere of human activity. Arguably, human progress owes more to this one discovery than to any other (mathematical) innovation.

The practical importance of the place value system stems from the ease with which it enables one to handle large numbers though at the time of its discovery there was rarely a serious need to deal with large numbers. Ancient India seems, however, to have been obsessed with large numbers-names had been given to powers of ten way beyond the billion (the ninth power of ten). This suggests that it was the fascination with numbers for their own sake rather than practical considerations that was the motive force behind the discovery - another example of the 'unreasonable effectiveness' of mathematics.

Algebra - the mathematics of manipulation of symbols representing variable quantities - also originated from India; and that too is a magnificent intellectual leap. Arab scholars who came into contact with India recognised the strength of Indian mathematics, absorbed and built upon it, contributing many great ideas themselves. Eventually they passed it all onto Europe which in a burst of new energy took the lead.

Great as its contributions were, the East nevertheless had not, in its pursuit of mathematics, enforced the rigorous discipline of the Greeks. Europe restored firmly Euclid's postulational paradigm as the framework for mathematics.

Today we recognise a body of knowledge as mathematical only if it can be fitted into that framework. It means that the body of knowledge is derived from a certain number of postulates (called axioms by Euclid) which are accepted without argument through rigorous reasoning, the methods of deduction themselves being governed by set rules which are also to be regarded as postulates.

An amazingly large part of what we regard as mathematical knowledge does meet even this very rigid criterion with only the axioms needed for the elementary arithmetic of natural numbers (and just a little more) as the foundation. However, this last statement has to be tempered by one fact: many mathematical areas which today can be fitted into this remarkably economical postulational scheme could not meet these stringent demands when they were first apprehended. It is only after Dedekind in the 19th century showed how to extend the number system beyond the rational numbers of elementary arithmetic, with only the axioms needed for elementary arithmetic as the starting point, that calculus could be incroporated into the fold of this scheme. Contemporaries of Leibnitz and Newton (inventors of calculus), however, had no hesitation in hailing the birth of some new and wonderful mathematics; and not merely calculus. Alot of other much less sophisticated mathematics could be derived from the axioms of elementary arithmetic by Euclid's deductive method but only after Dedekind had his say. Of course, Newton and his contemporaries as well as other mathematicians did use the deductive principles a la Euclid but based themselves implicitly on a set of much more elaborate unquestioned assumptions than the simple axioms of arithmetic.

The modern history - that of the last 500 years - of mathematics is necessarily Eurocentric: practically all major developments right up to the early 20th century originated in Europe. Descartes and Newton were to be followed by a string of great names: Euler, Gauss, Riemann, and Poincare to name a few. There was a steady and bounteous flow of new ideas through the 18th and 19th centuries, ideas from which emerged whole new areas, and this flow turned into a torrent in the 20th century. Gottingen and Paris were at the heart of this mathematical explosion, an explosion which also fuelled the revolution in physics which was taking place at the same time. The United States of America, not yet a mathematical power, was beginning to display its potential. Czarist Russia was also a participant in the mathematical action in Europe; but with the advent of the Soviet Union, the state came up with a deliberate policy for the promotion of mathematics with resounding effect. By the fifties, Gottingen had declined thanks to the Nazis and was yet to recover from the war while Moscow became a formidable centre rivalling Paris and could boast of a galaxy of extraordinarily creative minds. Sadly, that great school has virtually disintegrated along with the political system: most of the outstanding mathematicians having migrated to the West.

The Americans gave relatively little attention to mathematics in the first half of the 20th century, but the Soviet space programme's first Sputnik jolted them from their benign indifference into eager support for mathematics. Through the sixties and seventies and even into the eighties, support for mathematics was available on a very generous scale in the US and this had a tremendous effect. It produced an array of brilliant mathematicians and much of the most exciting mathematical developments. Europe (Russia included) has now ceded its pre-eminence to the US.

What then of our country? Intellectual activity had certainly taken a back seat for centuries in our country (there was, however, some exceptional mathematical progress in Kerala anticipating later work in Europe). The first stirrings after the long period of dormancy are to be seen in the Bengal renaissance of the 19th century. In the beginning of the reawakening it was the pursuit of humanities that dominated the scene, but in the early 20th century the twin figures of C.V. Raman and Ramanujan blazed new trails in science.

Raman was an outstanding communicator and his leadership provided immense impetus for the development of physics. Mathematics did not have this advantage: Ramanujan's brilliant career was tragically cut short in its prime. Nevertheless, his example inspired many people to pursue mathematics. A career in mathematics was of course unattractive in comparison with many others when viewed in terms of the creature comforts that one could command, but in the first half of the 20th century there was compensation in the kind of respect that learning was accorded.

It must be said that both Raman and Ramanujan received reasonable support from the colonial institutions of that period; in the case of Ramanujan, once the people in positions of power were convinced of his extraordinary talent, they acted with an alacrity that today's bureaucrats would do well to emulate. Of course, Britain was not interested in promoting intellectual activity in this country, but there was some response to sporadic individual achievements. In any event, whatever the rulers thought, Indian society did not have a strong awareness of the importance of science, much less that of mathematics during the colonial days.

With the advent of independence, the national leadership— Jawaharlal Nehru in particular — laid great emphasis on science and propagated the idea of infusing our society with 'scientific temper'. Nehru's vision resulted in the creation of many institutions of scientific research and among them a few which actively promoted mathematics. However, even as there exists a general perception of science as an important human activity, this perception is (understandably) based on the concrete and practical role science plays in industrial development. There is much less understanding of the civilizational role of fundamental science in general, of mathematics in particular. There is little appreciation of the fact that a great deal of today's applicable knowledge was at some period in the past basic science at its frontiers. This applies to mathematics much more than to other sciences.

The glamour attached to physics, thanks to developments in the field of nuclear energy and more recently to biology because of the recent discoveries in genetics, helps attract public attention to these fields. Chemistry too has areas that have captured the public imagination. The Nobel Prize also helped increase public awareness of the importance of these sciences: the prize is a household word and some knowledge of facts about it is practically statutory requirement for school kids taking part in quiz contests. Mathematics does not have the advantage of being able to project such a glamorous image and Alfred Nobel unfortunately did not consider mathematics as worthy of partaking of his huge legacy.

Most people, otherwise well informed, are not aware of the Fields Medal which identifies superior mathematical achievement even as the Nobel does in other fields, perhaps because of the fact that its monetary value is a pittance in comparison with the Nobel. Since its inception about 50 mathematicians (necessarily under the age of 40) have been awarded the medal and if a school kid happens to know the name of one of these recipients, it is a safe bet to assume that his/her family has personal contacts with a mathematician.

Even many of our teachers in schools and colleges are unaware of the existence of this medal for mathematics. Very recently, the Norwegian government has instituted a prize similar to the Nobel for mathematics. The mathematical community hopes that it will bring the same kind of visibility to mathematics as Nobel has given to other sciences.

There is a general feeling that unlike physics or other sciences, mathematics is a somewhat other worldly pursuit. Few realize that esoteric problems of cosmology pursued by astrophysicists or the frontier areas of particle physics are as meaningful to the practical everyday world as Fermat's last theorem. As a mathematician I have come across many people who wonder what there is left in mathematics to discover, an experience that I am sure the physicist does not share.

Despite this general lack of public awareness about mathematics, there is in this country a feeling that we are very good at mathematics and there is a certain pride in its past achievements. We have no doubt major past contributions to mathematics to our credit, as has already been pointed out. The romantic story of Srinivasa Ramanujan no doubt contributes (very justifiably) to this belief. But there are other dubious claims on behalf of Vedic mathematics which are also taken quite seriously and contribute to this confidence in our mathematical proficiency. And there is this feeling that people like Shakuntala Devi who can perform calculating feats represent superior mathematical talent that strengthen this perception.

Since Ramanujan, there have been some substantial contributions to mathematics from Indians (working in India), but by and large the public or even the scientific community outside of mathematics are not aware of them. These have made the international community to sit up and take notice of us. A major mathematical event, the International Congress of Mathematicians (ICM) has been taking place regularly once every four years since 1950 with ever increasing international participation. This is the occasion when the Fields Medals are awarded.

The meet is built around 200 prestigious invited lectures: some 20 'plenary' one-hour lectures address a general mathematical audience and are given by eminent figures who have influenced the evolution of different mathematical areas; and the rest, more specialised talks (of 50-minute duration) by other outstanding contributors to diverse fields within mathematics (the numbers mentioned are from recent congresses).

Some idea of the Indian impact on the world of mathematics may be gleaned from the fact that since 1958, in every Congress except in 1986, there has been at least one invited talk by an Indian mathematician working in India. There have been two plenary talks by Indians but both were non-residents; however, on the whole there have been far fewer talks by non-resident Indians than by those working in India. The Fields Medal has so far eluded us though there are instances of people coming close to winning it. In the developing world, India is certainly in the lead, but China seems poised to overtake us.

That is the 'feel good' side of Indian mathematics, but the peaks do not tell the whole story. The general level of mathematics outside a handful of institutions of higher learning is abysmally low. Much of what is passed off as research in our institutions of higher learning is of shamefully poor quality; nor does one find scholarship of note. For example, if one excludes some half a dozen institutions in the country, it will be impossible to find others even distantly familiar with the mathematical areas in which the fifty odd Field Medallists have worked. Much of the peak achievements have in fact come from two institutions: the Tata Institute of Fundamental Research in Mumbai and the Indian Statistical Institute with branches in Kolkata, Delhi and Bangalore.

The present scenario is indeed depressing, and in many ways things are on a further decline. There has been a steady loss of interest in the pursuit of mathematics (in fact, of all basic science) among the young over the last three decades. In the recent past this problem has become even more acute. This is largely due to socio-economic factors. Career options for mathematics graduates are essentially limited to the academic profession, not attractive in terms of emoluments: a fresh management graduate enters his/her career at a salary at which the academic retires. Unattractive economic status is of course not new to the academic profession, but in an earlier era it commanded considerable respect in our society and that is no longer the case.

In the case of talented youngsters even these factors do not thwart them from pursuing mathematics, but many seek to pursue it in Europe or America. In the advanced countries too there is a decline of interest in mathematics despite career opportunities being available outside the academia. These migrants from India and others like them from the Third World are in fact now the mainstay of American graduate schools, safeguarding the future of mathematics there. One reason of course is that in these countries the academic has a much better socio-economic standing in society than in our country.

The other is the sorry state of most of our institutions of higher learning - an eager youngster cannot get exposure to truly exciting mathematics except in a limited number of places and the best graduate schools we have, while good, are no match for Harvard or Princeton.

Unfortunately, the policies of the state in regard to higher education do not warrant an optimistic prognosis for the future. The state seems to be reducing, even withdrawing completely such support as it provides now for higher education. This is ostensibly because of the contrast with the meagre resources made available for basic education. There can be no two opinions about the need for enhancing several fold the support given to basic education. But this does not mean that the support given by the state to higher education has to be reduced: it is certainly not excessive by itself. However, it is true that most of the support provided for higher education has ended up as a 'subsidy' for business and industry (in the country as well as abroad): the prestigious institutions in the country (run practically entirely with government support) have provided human resources for them.

On the other hand higher studies that do not cater to the needs of commerce get short shrift. State governments have steadily cut back on funds for universities. Many university departments have unfilled positions which cannot be advertised, as the governments are not willing to release the money needed for the purpose. Colleges are run with teachers hired on the basis of contracts on unattractive, humiliating terms. In such a context, it is hardly surprising that a very large number of our college teachers are less than competent. The problem gets more acute with areas like mathematics where the subject has been for several years a last resort option for students and these are the ones with the requisite 'qualification' to teach the subject.

In the context of the enormous socio-economic problems facing us, a low priority for the promotion of mathematics may appear entirely justified. But this is definitely short sighted. Mathematics has an ever-increasing role in every human activity. As was pointed out earlier, it played a vital role in the advancement of science and technology. Industry in the advanced countries has benefited immensely from the intervention of mathematics of increasing levels of sophistication. Biology and medicine which seemed at one time to be impervious to mathematics, now use some advanced mathematics. The social sciences too are increasingly turning to mathematics for help. Wall Street employs mathematics PhDs routinely.

All this suggests that the promotion of mathematics is vital to our future well-being. There is another intangible important benefit that society derives from mathematics. Mathematics inculcates habits of thought that help promote scientific temper. There are many initiatives that can be taken to promote mathematics but they will need political will: we need a major overhaul of the educational system and some social engineering as well. It would appear from our track record that there is mathematical talent and undeniably we have an excellent tradition to revive; what is needed is to create the conditions in which that talent can flower.

Despite all the discouraging factors there is a trickle, small though it is, of headstrong youngsters who pursue mathematics and perform at superior levels working in India. This is the one silver lining in the cloud that keeps optimism alive.

We need to put the astrology debate within the broader framework of the Government's education policy in general, writes Supriya Roy Chowdhury.

He is our quaint R. K. Narayan character, located on the dusty road of Malgudi, sporting a forehead full of sacred ash and vermilion, professional equipment laid out on the pavement - a dozen shells, a square piece of cloth with mystic charts on it, a notebook and a bundle of palymra writing. Narayan forgot to add the inevitable parrot, in a small cage, which invariably accompanies the street side astrologer, ubiquitous anywhere in India. Suddenly this harmless figure becomes a symbol of deep ideological contestation, of serious debates as to whether astrology is science, as the modern Indian academician struggles with the notion of the ash-vermilion-parrot combination making its way into the portals of universities as Reader or Professor. What is a meagre living for an ingenuous, street-smart character, or a harmless pastime for some, suddenly has to become a part of our political discourse. This testifies perhaps that the boundaries of our political culture are limitlessly extendable, that the line between sense and senselessness is indeed thin, and more importantly, that power and intelligence are really at cross-purposes in the present political scenario.

As a recent editorial of Current Science puts it, the battle is not between astrology and science. Rather it is between the UGC - proposing to introduce astrology as a subject of scientific study in universities - and the scientific community. The dynamics of this battle say a great deal not only about a political party using state power to inscribe a particular ideology, but being able to completely ignore democratic procedures and norms in order to push this agenda.

Thus we need to put the astrology debate within the broader framework of the Government's education policy in general. The recently- concluded parliamentary debates on this issue highlighted the essentially undemocratic character in which the National Curriculum Framework for School Education (NCF) was pushed through. It was pointed out, first, that the NCF was enacted without the Minister of Human Resource Development convening a meeting of the Central Advisory Board of Education (CABE). The National Policy of Education, 1986, had laid down the centrality of the CABE in all matters pertaining to changes in education policy. The Minister justified the NCF on grounds that it had been considered by the NCERT and approved by State Education Ministers. In reality it appears that during the NCERT's annual general meeting the NCF was a fait accompli and considered as already passed.

This failure to anchor the process of public policy making in the broad domain of public debate is apparent at every stage of the unfolding drama in education. A large number of academicians has written to the Chairman of the UGC, underlining the non-rational and non-scientific foundations of astrology, and have called for the withdrawal of the scheme to introduce it in universities. The ineffectiveness of these measures of protest stands out. At least two universities have gone ahead with creating astrology departments. More importantly, the UGC chairman has not replied to any of these protest letters.

Ignoring protests in the expectation that they will die down is not an unusual strategy of governance in India. However, the UGC's blanket of silence underlines two things. First, the UGC possibly has very little to offer by way of a substantive defence in the face of the academics' criticism; second, the Government is obviously prepared to shield the UGC chairman's highly inappropriate behaviour in not responding to these criticisms. All of these are unpleasant reminders that the facade of democracy in India conceals a highly undemocratic culture of governance. In a statement released on April 18, more than 100 scientists in top scientific research institutions across the country called for the UGC Chairman's resignation, and for a broad-based movement to press for his dismissal. Expectedly, the Government has done nothing about this.

One may take a moment to look at the Government's broader justificatory framework for introducing courses in Vedic Mathematics and astrology. This framework underlines the need for value-based education, with a stress on our traditional values. The question of values, when anchored in texts such as Vedas and practices such as astrology, obviously refer to values associated with a particular religious tradition, and goes completely against the logic of a multi-religious nation.

But even apart from this fundamental digression from the secular paradigm, it is extremely odd that a Government unashamedly committed to the introduction of a globalised market in the country should at the same time speak of values, traditional or otherwise. The culture that comes with an aggressively consumption oriented, highly elitist, and crudely inegalitarian model of market-based development, in the context of a desperately poor country, is nothing if not valueless. It is this double-speak that highlights the Government's hypocrisy more than anything else.

But then, as we also know, Governments are responsive and accountable to the extent that they are forced to be. It is interesting that not one of the national scientific academies has taken an official position in opposition to the Government on the astrology issue. Fellows of the Indian Academy of Sciences have written to the President of the Academy urging him to convey the critical views of the Fellows to the UGC Chairman. These initiatives have brought forth no action on the part of the President, not even in the form of a reply to these letters. (The only exception to this pattern of official silence has been an open letter to the UGC Chairman by Prof. G. Srinivasan, in his capacity as President of the Astronomical Society of India as well as President of the Division on Space and High Energy Astrophysics of the International Astronomical Union.) This discrepancy between the clamour of protests within the scientific community and the failure of the Academies to express an official critique of the UGC's astrology agenda is obviously due to the reluctance of the scientific establishment to engage in a substantive dispute with the powers that be.

The scientific establishment's close dependence upon the state is not only for research funding, but also for things like high profile positions of power within the Government's scientific agencies and committees, and similar largesse that the state is in a position to offer. This provides a partial window to the ineffectiveness of the academics' critique of the Government's education policy.

To their credit, the struggle goes on, as far as the academic community is concerned, supported by organisations such as SAHMAT who have initiated a national debate on the general issue of communalisation of education, and sought to engage a broad spectrum of critical citizens on this question. In a context where many of the pillars of our democracy are crumbling, the value of such critical dissent cannot be emphasized enough.

One wishes, though, that the academic community's critical discourse was not so imprisoned within the immediacy of the astrology/saffronisation issue. Universities in India have been in a state of decline for many years now. The elite of the academic science community, ensconced in the comfort of top research institutions, and inhabiting, for all practical purposes, an international (read western) professional habitat, has rarely, if ever, reacted to the rapid degeneration of the university system. This curiously ostrich-like approach fails to recognise that a declining university system cuts at the roots of higher education in any context, and in fact potentially destroys the foundations on which these elite scientific research institutions stand. Therefore, if the protection of universities and of science education is the broad concern, one needs to acknowledge that saffronisation maybe the biggest threat, but not the only one. There would be a greater measure of credibility, and perhaps power, in the academics' protest if it was anchored within the broad issue of institutional erosion in education.

Criticism

We, the undersigned, are deeply concerned by the continuing attempts to thrust the so-called `Vedic Mathematics' on the school curriculum by the NCERT. As has been pointed out earlier on several occasions, the so-called `Vedic Mathematics' is neither `Vedic' nor can it be dignified by the name of mathematics. `Vedic Mathematics', as is well-known, originated with a book of the same name by a former Sankracharya of Puri (the late Jagadguru Swami Shri Bharati Krishna Tirthaji Maharaj) published posthumously in 1965. The book assembled a set of tricks in elementary arithmetic and algebra to be applied in performing computations with numbers and polynomials. As is pointed out even in the Foreword to the book by the General Editor, Dr. A.S. Agarwala, the aphorisms in Sanskrit to be found in the book have nothing to do with the Vedas. Nor are these aphorisms to be found in the genuine Vedic literature. The term "Vedic maths" is therefore entirely misleading and factually incorrect.

Further, it is clear from the notation used in the arithmetical tricks in the book that the methods used in this text have nothing to do with the arithmetical techniques of antiquity. Many of the Sanskrit aphorisms in the book are totally cryptic (ancient Indian mathematical writing was anything but cryptic) and often so general as to be devoid of any specific mathematical meaning. There are several authoritative texts on the mathematics of Vedic times that could used in part to teach an authoritative and correct account of ancient Indian mathematics but this book clearly cannot be used for any such purpose. The teaching of mathematics involves both the teaching of the basic concepts of the subject as well as methods of mathematical computation. The so-called "Vedic maths" is entirely inadequate to this task considering that it is largely made up of tricks to do some elementary arithmetic computations. Many of these can be far more easily performed on a simple computer or even an advanced calculator. The book "Vedic maths" essentially deals with arithmetic of the middle and high-school level.

Its claims that "there is no part of mathematics, pure or applied, which is beyond their jurisdiction" is simply ridiculous. In an era when the content of mathematics teaching has to be carefully designed to keep pace with the general explosion of knowledge and the needs of other modern professions that use mathematical techniques, the imposition of "Vedic mathematics" will be nothing short of calamitous. India today has active and excellent schools of research and teaching in mathematics that are at the forefront of modern research in their discipline with some of them recognised as being among the best in the world in their fields of research. It is noteworthy that they have cherished the legacy of distinguished Indian mathematicians like Srinivasa Ramanujam, V. K. Patodi, S. Minakshisundaram, Harish Chandra, K. G. Ramanathan, Hansraj Gupta, Syamdas Mukhopadhyay, Ganesh Prasad, and many others. Including several living Indian mathematicians. But not one of these schools has lent an iota of legitimacy to `Vedic maths'. Nowhere in the world does any school system teach ``Vedic maths" or any form of ancient maths for that matter as an adjunct to modern mathematical teaching. The bulk of such teaching belongs properly to the teaching of history and in particular the teaching of the history of the sciences.

We consider the imposition of `Vedic maths' by a Government agency, as the perpetration of a fraud on our children, condemning particularly those dependent on public education to a sub-standard mathematical education. Even if we assumed that those who sought to impose `Vedic maths' did so in good faith, it would have been appropriate that the NCERT seek the assistance of reknowned Indian mathematicians to evaluate so-called "Vedic mathematics' before making it part of the National Curricular framework for School Education.

Appallingly they have not done so. In this context we demand that the NCERT submit the proposal for the introduction of `Vedic maths' in the school curriculum to recognized bodies of mathematical experts in India, in particular the National Board of Higher Mathematics (under the Dept. of Atomic Energy), and the Mathematics sections of the Indian Academy of Sciences and the Indian National Science Academy, for a thorough and critical examination. In the meanwhile no attempt should be made to thrust the subject into the school curriculum either through the

Centrally-administered school system or by trying to impose it on the school systems of various States.

We are concerned that the essential thrust behind the campaign to introduce the so-called `Vedic mathematics' has more to do with promoting a particular brand of religious majoritarianism and associated obscurantist ideas rather than any serious and meaningful development of mathematics teaching in India.

We note that similar concerns have been expressed about other aspects too of the National Curricular Framework for School Education. We re-iterate our firm conviction that all teaching and pedagogy, not just the teaching of mathematics, must be founded on rational, scientific and secular principles.

Vedic Mathematics—A Briefing

In the previous chapters we have gone through the Vedic Mathematics Sutras and upa-Sutras: their application in solving problems. In this approach we have missed to note some points and merits of one method over the other methods at some instances.

Now we take a few steps in this direction. You may question why this book first gives examples and methods and then once again try to proceed as if an introduction to the Vedic Mathematics has been just started. This is because in this approach the reader first feels that it is easy to solve problems using Vedic Mathematics.

This is clear from the examples given. But the reader may get doubt why we are doing this way or that way some times very close and almost akin to the conventional textual way; and some times very different from these procedures? why new representations and different meanings for the same Sutra (!) in different contexts? But observe that it is not uncommon to Mathematics.

Question some body showing the symbol Π.

Majority may say it is 22 / 7 (is it right?) some may say it is a radian measure. Very few may state it is a function or so.

What does the representation A × B mean?

A boy thinking about numbers may answer that is A multi plied by B and gives the product provided A and B are known. A girl thinking of set notation simply says that it is Cartesian product of the sets A and B. No sort of multiplication at all.

Another may conclude that it is a product of two matrices A and B . No doubt a multiplication but altogether different from above.

Some other may go deep in to elementary number theory and may take ' X ' to be the symbol ' X ' (does not divide) and conclude 'A does not divide B' .

Now the question arises does a student fail to understand and apply contextual meaning and representation of symbols and such forms in mathematical writings? certainly not. In the same way the contextual meanings of the Sutras also can not bring any problem to the practitioners of Vedic Mathematics.

Again a careful observation brings all of us to a conclusion that even though the Sutras are not like mathematical formulae so as to fit in any context under consideration but they are intended to recognize the pattern in the problems and suggest procedures to solve. Now recall the terms, rules and methods once again to fill in some gaps that occur in the previous attempt.

Terms and Operations

a) Ekadhika means 'one more'

e.g: Ekadhika of 0 is 1

Ekadhika of 1 is 2

Ekadhika of 8 is 9

Ekadhika of 23 is 24

Ekadhika of 364 is 365------

b) Ekanyuna means 'one less'

e.g: Ekanyuna of 1 2 3 ..... 8 ..... 14 .....69 ......

is 0 1 2 ..... 7 ......13 .... 68 ......

c) Purak means ' complement'

e.g: purak of 1 2 3 ..... 8., 9 from 10

is 9 8 7 ..... 2 1

d) Rekhank means 'a digit with a bar on its top'. In other words it is a negative number.

e.g: A bar on 7 is $\bar{7}$. It is called rekhank 7 or bar 7. We treat purak as a Rekhank.

e.g.: $\bar{7}$ is 3 and $\bar{3}$ is 7

At some instances we write negative numbers also with a bar on the top of the numbers as

-4 can be shown as $\bar{4}$.

-21 can be shown as $\overline{21}$.

e) Addition and subtraction using Rekhank.

Adding a bar-digit i.e. Rekhank to a digit means the digit is subtracted.

e.g.: $3 + \bar{1} = 2$, $5 + \bar{2} = 3$, $4 + \bar{4} = 0$

Subtracting a bar - digit i.e. Rekhank to a digit means the digit is added.

e.g.: $4 - \bar{1} = 5$, $6 - \bar{2} = 8$, $3 - \bar{3} = 6$

1. Some more examples

e.g.: $3 + 4 = 7$

$(-2) + (-5) = \bar{2} + \bar{5} = \bar{7}$ or -7

f) Multiplication and Division using rekhank.

1. Product of two positive digits or two negative digits (Rekhanks)

   e.g.: $2 \times 4 = 8$; $\bar{4} \times \bar{3} = 12$ i.e. always positive

2. Product of one positive digit and one Rekhank

   e.g.: $3 \times \bar{2} = \bar{6}$ or -6; $\bar{5} \times 3 = \overline{15}$ or -15 i.e. always Rekhank or negative.

3. Division of one positive by another or division of one Rekhank by another Rekhank.

   e.g.: $8 \div 2 = 4$, $\bar{6} \div \bar{3} = 2$ i.e. always positive

4. Division of a positive by a Rekhank or vice versa.

   e.g.: $\overline{10} \div 5 = \bar{2}$, $6 \div \bar{2} = \bar{3}$ i.e. always negative or Rekhank.

g) *Beejank:* The Sum of the digits of a number is called Beejank. If the addition is a two digit number, Then these two digits are also to be added up to get a single digit.

e.g.: Beejank of 27 is $2 + 7 = 9$.

Beejank of 348 is $3 + 4 + 8 = 15$

Further $1 + 5 = 6$. i.e. 6 is Beejank.

Beejank of 1567 $\rightarrow 1 + 5 + 6 + 7 \rightarrow 19$ $1 + 9 \rightarrow 1$

i.e. Beejank of 1567 is 1.

ii) *Easy way of finding Beejank:* Beejank is unaffected if 9 is added to or subtracted from the number. This nature of 9 helps in finding Beejank very quickly, by cancelling 9 or the digits adding to 9 from the number.

e.g. 1: Find the Beejank of 632174.

As above we have to follow

632174 → 6 + 3 + 2 + 1 + 7 + 4 → 23 2 + 3 → 5

But a quick look gives 6 & 3 ; 2 & 7 are to be ignored because 6 + 3 = 9, 2 + 7 = 9. Hence remaining 1 + 4 → 5 is the beejank of 632174.

e.g. 2:

Beejank of 1256847 → 1 + 2 + 5 + 6 + 8 + 4 + 7 → 33 → 3+3 → 6.

But we can cancel 1& 8, 2& 7, 5 & 4 because in each case the sum is 9. Hence remaining 6 is the Beejank.

h) *Check by Beejank Method:*The Vedic sutra - Gunita Samuccayah gives 'the whole product is same'. We apply this sutra in this context as follows. It means that the operations carried out with the numbers have same effect when the same operations are carried out with their Beejanks.

Observe the following examples.

i) 42 + 39

Beejanks of 42 and 39 are respectively 4 + 2 = 6 and 3 + 9 = 12 and 1 + 2 = 3

Now 6 + 3 = 9 is the Beejank of the sum of the two numbers

Further 42 + 39 = 81. Its Beejank is 8+ 1 = 9.

We have checked the correctness.

ii) 64 + 125.

64 → 6 + 4 → 10 → 1 + 0 → 1

125 → 1 + 2 + 5 → 8

Sum of these Beejanks 8 + 1 = 9

Note that

64 + 125 = 189 → 1 + 8 + 9 → 18 1 + 8 → 9

we have checked the correctness.

iii) 134 - 49

134 → 1 + 3 + 4 → 8

49 → 4 + 9 → 13 → 1 + 3 → 4

Difference of Beejanks 8 -4 → 4,

note that 134 - 49 = 85

Beejanks of 85 is 8 + 5 → 85 8 + 5 → 13 → 1 + 3 → 4 verified.

iv) 376 - 284

376 → 7 ∵ (6 + 3 → 9)

284 → 2 + 8 + 4 → 14 → 1 + 4 → 5

Difference of Beejanks = 7 - 5 = 2

376 - 284 = 92

Beejank of 92 → 9 + 2 → 11 1 + 1 → 2

Hence verified.

v) 24 × 16 = 384

Multiplication of Beejanks of

24 and 16 is 6 × 7 = 42 → 4 + 2 → 6

Beejank of 384 → 3 + 8 + 4 → 15 1 + 5 → 6

Hence verified.

vi) 237 × 18 = 4266

Beejank of 237 → 2 + 3 + 7 → 12 → 1 + 2 → 3

Beejank of 18 → 1 + 8 → 9

Product of the Beejanks = 3 × 9 → 27 → 2 + 7 → 9

Beejank of 4266 → 4 + 2 + 6 + 6 → 18 → 1 + 8 → 9

Hence verified.

vii) $242 = 576$

Beejank of $24 \rightarrow 2 + 4 \rightarrow 6$

square of it $62 \rightarrow 36 \rightarrow 9$

Beejank of result $= 576 \rightarrow 5 + 7 + 6 \rightarrow 18 \rightarrow 1 + 8 \rightarrow 9$

Hence verified.

viii) $3562 = 126736$

Beejank of $356 \rightarrow 3 + 5 + 6 \rightarrow 5$

Square of it $= 52 = 25 \rightarrow 2 + 5 \rightarrow 7$

Beejank of result $126736 \rightarrow 1 + 2 + 6 + 7 + 3 + 6 \rightarrow 1 + 6 \rightarrow 7$

$\because (7 + 2 = 9;\ 6 + 3 = 9)$ hence verified.

ix) *Beejank in Division:* Let P, D, Q and R be respectively the dividend, the divisor, the quotient and the remainder.

Further the relationship between them is P = (Q X D) + R

E.g. 1: $187 \div 5$

we know that $187 = (37 \times 5) + 2$ now the Beejank check.

We know that $187 = (37 \times 5) + 2$ now the Beejank check.

$187 \rightarrow 1 + 8 + 7 \rightarrow 7 \because (1 + 8 = 9)$

$(37 \times 5) + 2 \rightarrow$ Beejank $[(3 + 7) \times 5] + 2 \rightarrow 5 + 2 \rightarrow 7$

Hence verified.

E.g. 2: $7986 \div 143$

$7896 = (143 \times 55) + 121$

Beejank of $7986 \rightarrow 7 + 9 + 8 + 6 \rightarrow 21$

($\because$ 9 is omitted) $\rightarrow 2 + 1 \rightarrow 3$

Beejank of $143 \times 55 \rightarrow (1 + 4 + 3)\ (5 + 5)$

$\rightarrow 8 \times 10 \rightarrow 80\ (8 + 0) \rightarrow 8$

Beejank of (143 × 55) + 121 → 8 + (1 + 2 + 1)

→ 8 + 4 → 12 1 + 2 → 3

Hence verified.

Check the following results by Beejank Method

1. 67 + 34 + 82 = 183
2. 4381 - 3216 = 1165
3. 63^2 = 3969
4. $(1234)^2$ = 1522756
5. (86 × 17) + 34 = 1496
6. 2556 ÷ 127 gives Q = 20, R = 16

Vinculum: The numbers which by presentation contains both positive and negative digits are called vinculum numbers.

Conversion of general numbers into vinculum numbers.

We obtain them by converting the digits which are 5 and above 5 or less than 5 without changing the value of that number.

Consider a number say 8.

(Note it is greater than 5).

Use it complement (purak - rekhank) from 10. It is 2 in this case and add 1 to the left (i.e. tens place) of 8.

Thus 8 = 08 = $1\bar{2}$.

The number 1 contains both positive and negative digits

i.e. 1 and $\bar{2}$. Here $\bar{2}$ is in unit place hence it is -2 and value of 1 at tens place is 10.

Thus $1\bar{2}$ = 10 - 2 = 8

Conveniently we can think and write in the following way:

General Number	Conversion	Vinculum Number
6	10 - 4	$1\bar{4}$
97	100 - 3	$10\bar{3}$
289	300 - 11	$10\bar{3}$ etc.

The sutras 'Nikhilam Navatascharamam Dasatah' and 'Ekadhikena purvena' are useful for conversion.

e.g. 1: 289, Edadhika of 2 is 3

Nikhilam from 9 : 8 - 9 = -1 or $\bar{1}$

Charmam from 10 :9 -10 = -1 or $\bar{1}$

i.e. 289 in vinculum form $\overline{311}$

e.g. 2: 47768

'Ekadhika' of 4 is 5

'Nikhilam' from 9 (of 776) $\overline{223}$

'Charmam from 10 (of 8) $\bar{2}$

Vinculum of 47168 is $5\,\overline{2232}$

e.g. 3: 11276

Here digits 11 are smaller. We need not convert. Now apply for 276 the two sutras Ekadhika of 2 is 3

'Nikhilam Navata' for 76 is $\overline{24}$

$$11276 = 113\overline{24}$$

i.e. $113\overline{24} = 11300 - 24 = 11276.$

The conversion can also be done by the sutra sankalana vyavakalanabhyam as follows.

e.g. 4: 315

sankalanam (addition) = 315+315 = 630.

Vyvakalanam (subtraction) = 630 - 315 = $32\bar{5}$

Working steps :

$0 - 5 = \bar{5}$

$3 - 1 = 2$

$6 - 3 = 3$

Let's apply this sutra in the already taken example 47768.

Samkalanam = 47768 + 47768 = 95536

Vyavakalanam = 95536 - 47768.

Steps:
$$\left.\begin{array}{l} 6-8=\bar{2} \\ 3-6=\bar{3} \\ 5-7=\bar{2} \\ 5-7=\bar{2} \\ 9-4=5 \end{array}\right\} = \overline{52232}$$

Consider the convertion by sankalanavyavakalanabhyam and check it by Ekadhika and Nikhilam.

e.g. 5: 12637

1. Sankalana .... gives, 12637 + 12637 = 25274

   25274 - 12637 = (2 - 1)/(5 - 2)/(2 - 6)/(7 - 3)/(4 - 7) = $1 3 \bar{4} 4 \bar{3}$

2. Ekadhika and Nikhilam gives the following.

As in the number 1 2 6 3 7, the smaller and bigger digits (i.e. less than 5 and; 5, greater than 5) are mixed up, we split up in to groups and conversion is made up as given below.

Split 1 2 6 and 3 7

Now the sutra gives 1 2 6 as $13\bar{4}$ and 37 as $4\bar{3}$

Thus 12637 = $1\bar{3}4\bar{4}3$

Now for the number 315 we have already obtained vinculum as 325 by "sankalana ... " Now by 'Ekadhika and Nikhilam ...' we also get the same answer.

315 Since digits of 31 are less than 5,

We apply the sutras on 15 only as

Ekadhika of 1 is 2 and Charman of 5 is $\dot{5}$.

Consider another number which comes under the split process.

e.g. 6: 24173

As both bigger and smaller numbers are mixed up we split the number 24173 as 24 and 173 and write their vinculums by Ekadhika and Nikhilam sutras as

24 = $3\bar{6}$ and 173 = $2\overline{27}$

Thus 24173 = $3\bar{6}2\overline{27}$

Convert the following numbers into viniculum number by i. Ekadhika and Nikhilam sutras, and ii. Sankalana vyavakalana sutra.

Observe whether in any case they give the same answer or not.

1. 64
2. 289
3. 791
4. 2879
5. 19182
6. 823672
7. 123456799
8. 65384738

Conversion of vinculum number into general numbers.

The process of conversion is exactly reverse to the already done. Rekhanks are converted by Nikhilam where as other digits by 'Ekanyunena' sutra. thus:

i) $1\bar{2}$ = (1 - 1) / (10 - 2) Ekanyunena 1 - 1

= 08 = 8 Nikhilam. 2 = 10 - 2

ii) $3\overline{26}$ = (3 - 1) / (9 - 2) / (10 - 6)

= 274

iii) $3\bar{3}4\bar{4}$ = (3 - 1) / (10 - 3) / (4 - 1) / (10 - 4)

= 2736 (note the split)

iv) $20\overline{34}0\overline{12}1$ = 2/(0-1)/(9-3)/(10-4)/(0-1)/(9-1)/(10-2)/1

= $2\bar{1}66\bar{1}881$

= $2\bar{1}$ / 6 / $6\bar{1}$ / 881. once again split

= (2 - 1) / (10 -1) / 6 / (6 -1) / (10 -1) / 881

= 19659881

v) $30\overline{3212}$ = 3 / 0321 / 2

= 3 / (0-1) / (9-3) / (9-2) / (10-1) / 2

3 / $\bar{1}$ / 6792

(3 -1) / (10 -1) / 6792

= 296792.

Single to many conversions.

It is interesting to observe that the conversions can be shown in many ways.

e.g. 1: 86 can be expressed in following many ways

$$86 = 90 - 4 = \overline{94}$$

$$= 100 - 14 = \overline{114}$$

$$= 1000 - 914 = \overline{1914}$$

Thus $86 = \overline{94} = \overline{114} = \overline{1914} = 19914 = \ldots\ldots\ldots\ldots$

e.g. 2 :

$$\overline{07} = -10 + 3 = \bar{1}3$$

$$\overline{36} = -100 + 64 = \bar{1}64$$

$$\overline{978} = -1000 + 22 = \bar{1}022 \text{. etc.,}$$

*** Convert by Vedic process the following numbers into vinculum numbers.**

1) 274
2) 4898
3) 60725
4) 876129.

Convert the following vinculum numbers into general form of numbers (normalized form)

1) 283
2) 3619
3) 27216
4) 364718
5) 60391874

Vedic check for conversion.

The vedic sutra "Gunita Samuctayah" can be applied for verification of the conversion by Beejank method.

Consider a number and find its Beejank. Find the vinculum number by the conversion and find its Beejank. If both are same the conversion is correct.

e.g..

$196 = \bar{2}16$ ·

Now Beejank of 196 $\rightarrow 1 + 6 \rightarrow 7$

Beejank of 216 $\rightarrow 2 + (-1) + 6 \rightarrow 7$.

Thus verified.

But there are instances at which, if beejank of vinculum number is rekhank i.e. negative. Then it is to be converted to +ve number by adding 9 to Rekhank (already we have practised) and hence 9 is taken as zero, or vice versa in finding Beejank.

e.g.:

$\overline{213} = 200 - 13 = 187$

Now Beejank of $21\bar{3} = 2 + (-1) + (-3) = -2$

Beejank of $187 = 1 + 8 + 7 = 16 \rightarrow 1 + 6 = 7$

The variation in answers can be easily understood as

$\bar{2} = \bar{2} + 9 \rightarrow -2 + 9 = 7$.

Hence verified.

Use Vedic check method of the verification of the following result.

1. $24 = \overline{36}$
2. $2736 = 3\bar{3}4\bar{4}$ ·
3. $3\overline{26} = 274$
4. $2\bar{3}2\bar{1}3 = 17187$

Addition and Subtraction Using Vinculum Numbers

e.g. 1: Add 7 and 6 i.e., 7 + 6.

i) Change the numbers as vinculum numbers as per rules already discussed.

{ $7 = 1\bar{3}$ and $6 = 1\bar{4}$ }

ii) Carry out the addition column by column in the normal process, moving from top to bottom or vice versa.

iii) add the digits of the next higher level i.e., 1 + 1 = 2

$$\begin{array}{r} 1\bar{3} \\ 1\bar{4} \\ \hline 2\bar{7} \end{array}$$

iv) the obtained answer is to be normalized as per rules already explained. rules already explained.

i.e., $2\bar{7}$ = (2 - 1) (10- 7) = 13 Thus we get 7 + 6 = 13.

e.g. 2 : Add 973 and 866.

$$973 = 10\bar{3}3 \qquad 10\bar{3}3$$
$$866 = 1\bar{1}\bar{3}4 \qquad \begin{array}{r} 1\bar{1}\bar{3}4 \\ \hline 2\bar{1}\bar{6}\bar{1} \end{array}$$

But $2\bar{1}\bar{6}\bar{1}$ = 2000 - 161 = 1839.

Thus 973 + 866 by vinculum method gives 1839 which is correct.

Observe that in this representation the need to carry over from the previous digit to the next higher level is almost not required.

e.g. 3 : Subtract 1828 from 4247.

i.e.,
$$\begin{array}{r} 4247 \\ -1828 \\ \hline \end{array}$$

Step:

(i) Write -1828 in Bar form i.e., $\overline{1828}$

(ii) Now we can add 4247 and $\overline{1828}$ i.e.,

$$\begin{array}{r} 4247 \\ +\overline{1828} \\ \hline 3\bar{6}2\bar{1} \end{array}$$

Since $7 + \bar{8} = \bar{1}$,

$4 + \bar{2} = 2$,

$2 + \bar{8} = \bar{6}$,

$4 + \bar{1} = 3$

(iii) Changing the answer $3\bar{6}2\bar{1}$ into normal form using Nikhilam, we get

$3\bar{6}2\bar{1}$ = 36 / 21 split

= (3 -1) / (10 - 6) / (2 - 1) / (10 - 1) = 2419

∴ 4247 - 1828 = 2419

Find the following results using Vedic methods and check them

1) 284 + 257
2) 5224 + 6127
3) 582 - 464
4) 3804 - 2612

Addition and Subtraction

Addition

In the convention process we perform the process as follows.

234 + 403 + 564 + 721

write as 234
403
564
721

Step (i): 4 + 3 + 4 + 1 = 12 2 retained and 1 is carried over to left.

Step (ii): 3 + 0 + 6 + 2 = 11 the carried '1' is added

i.e., Now 2 retained as digit in the second place (from right to left) of the answer and 1 is carried over to left.

Step (iii): 2 + 4 + 5 + 7 = 18 carried over '1' is added

i.e., 18 + 1 = 19. As the addition process ends, the same 19 is retained in the left had most part of the answer.

Thus 234
403
564
+721

1922 is the answer

we follow sudhikaran process Recall 'sudha' i.e., dot (.) is taken as an upa-sutra (No: 15)

Consider the same example

234
403
564 ↑
+721

(i) Carry out the addition column by column in the usual fashion, moving from bottom to top.

(a) 1 + 4 = 5, 5 + 3 = 8, 8 + 4 = 12 The final result is more than 9. The tenth place '1' is dropped once number in the unit place i.e., 2 retained. We say at this stage sudha and a dot is above the top 4. Thus column (1) of addition (right to left)

$\dot{4}$

3

4

1

—

2

(b) Before coming to column (2) addition, the number of dots are to be counted, This shall be added to the bottom number of column (2) and we proceed as above. Thus second column becomes

$\dot{3}$ dot = 1, 1 + 2 = 3

0 3 + 6 = 9

6 9 + 0 = 9

2 9 + 3 = 12

—

2

2 retained and '.' is placed on top number 3

(c) proceed as above for column (3)

2 i) dot = 1 ii) 1 + 7 = 8

4 iii) 8 + 5 = 13 iv) Sudha is said.

$\dot{5}$ A dot is placed on 5 and proceed

7 with retained unit place 3.

—

9 v) 3 + 4 = 7, 7 + 2 = 9

Retain 9 in 3rd digit i.e., in 100th place.

(d) Now the number of dots is counted. Here it is 1 only and the number is carried out left side i.e. 1000th place

```
       ..
Thus  234
      403
        .
      564
     +721
    -----
     1922   is the answer.
```

Though it appears to follow the conventional procedure, a careful observation and practice gives its special use.

e.g. (1):

```
   .
  437
  . .
  624
   .
  586
 +162
 ----
 1809
```

Steps 1:

(i) 2 + 6 = 8, 8 + 4 = 12 so a dot on 4 and 2 + 7 = 9 the answer retained under column (i)

(ii) One dot from column (i) treated as 1, is carried over to column (ii), thus 1 + 6 = 7, 7 + 8 = 15 A' dot'; is placed on 8 for the 1 in 15 and the 5 in 15 is added to 2 above.

5 + 2 = 7, 7 + 3 = 10 i.e. 0 is written under column (ii) and a dot for the carried over 1 of 10 is placed on the top of 3.

(iii) The number of dots counted in column (iii) are 2. Hence the number 2 is carried over to column (ii) Now in column (iii) 2 + 1 = 3, 3 + 5 = 8, 8 + 6 = 14 A dot for 1 on the number 6 and 4 is retained to be added 4 above to give 8. Thus 8 is placed under column (iii).

(iv) Finally the number of dots in column (iii) are counted. It is '1' only. So it carried over to 1000th place. As there is no fourth column 1 is the answer for 4th column. Thus the answer is 1809.

EXAMPLE 3:

$$\begin{array}{r} 2\dot{6}4\dot{7} \\ \dot{4}0\dot{8}6 \\ 5\dot{7}1\dot{8} \\ +16\dot{9}2 \\ \hline 15143 \end{array} \uparrow$$

Check the result verify these steps with the procedure mentioned above.

The process of addition can also be done in the downward direction i.e., addition of numbers column wise from top to bottom

EXAMPLE 1:

$$\begin{array}{r} 4326 \\ \dot{5}\dot{6}\dot{9}\dot{4} \\ 3708 \\ 2\dot{7}\dot{8}\dot{4} \\ \hline \end{array} \downarrow$$

Step 1: 6 + 4 = 10, 1 dot ; 0 + 8 = 8; 8 + 4 = 12; 1 dot and 2 answer under first column - total 2 dots.

Step 2: 2+2 (2 dots) = 4; 4 + 9 = 13: 1 dot and 3 + 0 = 3; 3+8 = 11; 1 dot and 1 answer under second column - total 2 dots.

Step 3: 3+2 (2 dots) = 5; 5 + 6 = 11:1 dot and 1 + 7 = 8; 8+7 = 15; 1 dot and 5 under third column as answer - total 2 dots.

Step 4: 4 + 2 (2 dots) = 6; 6 + 5 =11: 1 dot and 1+ 3 = 4; 4 + 2 = 6. - total 1 dot in the fourth 6 column as answer.

Step 5: 1 dot in the fourth column carried over to 5th column (No digits in it) as 1.

Thus answer is from Step 5 to Step 1; 16512

EXAMPLE 2:

$$\begin{array}{r} 3278 \\ 4\dot{6}1\dot{9} \\ \dot{3}08\dot{4} \\ 17\dot{9}1 \\ \hline \end{array} \downarrow$$

Steps

(i) 8 + 9 = 17; 7 + 4 = 11; 1 + 1 = (2) (2dots)

(ii) 7 + 2 = 9; 9 + 1 = 10; 0 + 8 = 8, 8 + 9 = 17, (7) (2dots)

(iii) 2 + 2 = 4; 4 + 6 = 10; 0 + 0 = 0; 0 + 7 = (7) (1 dot)

(iv) 3 + 1 = 4; 4 + 4 = 8; 8 + 3 = 11; 1 + 1 = (2) (1 dot)

(v) 1

Thus answer is 12772.

Add the following numbers use 'Sudhikaran' whereever applicable.

1.	2.	3.
486	5432	968763
395	3691	476509
721	4808	+584376
+609	+6787	

Check up whether 'Sudhkaran' is done correctly. If not write the correct process. In either case find the sums.

1.
$$\begin{array}{r} 4\dot{3} \\ \dot{7}\dot{8} \\ +69 \end{array} \uparrow$$

2.
$$\begin{array}{r} 3\dot{7}\dot{9} \\ 8\dot{5}4 \\ \dot{7}\dot{6}\dot{7} \\ 426 \\ \hline \\ \hline \end{array}$$

3.
$$\begin{array}{r} 7\,8\,\dot{9}\,2\,4 \\ \dot{2}\,\dot{7}\,2\,\dot{7}\,2 \\ +7\,2\,6\,\dot{8}\,4 \end{array} \downarrow$$

Subtraction

The 'Sudha' Sutra is applicable where the larger digit is to be subtracted from the smaller digit. Let us go to the process through the examples.

Procedure:

(i) If the digit to be subtracted is larger, a dot (sudha) is given to its left.

(ii) The purak of this lower digit is added to the upper digit or purak-rekhank of this lower digit is subtracted.

EXAMPLE 1:

34 - 18

$$\begin{array}{r} 34 \\ -\dot{1}8 \\ \hline \end{array}$$

Steps:

(i) Since 8>4, a dot is put on its left i.e. 1

(ii) Purak of 8 i.e. 2 is added to the upper digit i.e. 4

2 + 4 = 6. or Purak-rekhank of 8 i.e. $\bar{2}$ is

Subtracted from i.e. 4 - $\bar{2}$ = 6.

Now at the tens place a dot (means1) makes the '1' in the number into 1+1=2. This has to be subtracted from above digit. i.e. 3 - 2 = 1. Thus

$$\begin{array}{r} 34 \\ -\dot{1}8 \\ \hline 16 \end{array}$$

EXAMPLE 2:

$$\begin{array}{r} 63 \\ -\dot{3}7 \\ \hline \end{array}$$

Steps:

(i) 7>3. Hence a dot on left of 7 i.e., 3

(ii) Purak of 7 i.e. 3 is added to upper digit 3 i.e. 3+3 = 6.

This is unit place of the answer.

Thus answer is 26.

EXAMPLE 3:

$$\begin{array}{r} 3274 \\ -\dot{1}\dot{8}92 \\ \hline \end{array}$$

Steps:

(i) 2 < 4. No sudha . 4-2 = 2 first digit (form right to left)

(ii) 9 > 7 sudha required. Hence a dot on left of 9 i.e. $\dot{8}$

(iii) purak of 9 i.e. 1, added to upper 7 gives 1 + 7 = 8 second digit

(iv) Now means $\dot{8}$ + 1 = 9.

(v) As 9 > 2, once again the same process: dot on left of i.e., $\dot{1}$

(vi) purak of 9 i.e. 1, added to upper 2 gives 1 + 2 = 3, the third digit.

(vii) Now $\dot{1}$ means 1 + 1 = 2

(viii) As 2 < 3, we have 3-2 = 1, the fourth digit

Thus answer is 1382

Vedic Check

E.g. (i) in addition : 437 + 624 + 586 + 162 = 1809.

By beejank method, the Beejanks are

437 ⟶ 4 + 3 + 7 ⟶ 14 1 + 4 ⟶ 5

624 ⟶ 6 + 2 + 4 ⟶ 12 ⟶ 1 + 2 ⟶ 3

586 ⟶ 5 + 8 + 6 ⟶ 19 ⟶ 1 + 9 ⟶

10 ⟶ 1 + 0 ⟶ 1

162 ⟶ 1 + 6 + 2 ⟶ 9

Now

437 + 624 + 586 + 162 ⟶ 5 + 3 + 1 + 9 ⟶ 18

⟶ 1 + 8 ⟶ 9

Beejank of 1809 ⟶ 1 + 8 + 0 + 9 ⟶ 18 1 + 8

⟶ 9 verified.

E.g.(3) in subtraction :

3274 - 1892 = 1382

now beejanks

3274 ⟶ 3 + 2 + 7 + 4 ⟶ 3 + 4 ⟶ 7

1892 $\longrightarrow$ 1 + 8 + 9 + 2 $\longrightarrow$ 2

3292-1892 $\longrightarrow$ 7-2 $\longrightarrow$ 5

1382 $\longrightarrow$ 1 + 3 + 8 + 2 $\longrightarrow$ 5.

Hence verified.

Mixed addition and subtraction using Rekhanks:

EXAMPLE 1 :

423 - 654 + 847 - 126 + 204.

In the conventional method we first add all the +ve terms

423 + 847 + 204 = 1474

Next we add all negative terms

- 654 - 126 = -780

At the end their difference is taken

1474 - 780 = 694

Thus in 3 steps we complete the problem

But in Vedic method using Rekhank we write and directly find the answer.

$$
\begin{array}{c}
4\ 2\ 3 \\
\bar{6}\ \bar{5}\ \bar{4} \\
8\ 4\ 7 \\
\bar{1}\ \bar{2}\ \bar{6} \\
2\ 0\ 4 \\
\hline
7\ \bar{1}\ 4
\end{array}
$$

This gives (7 -1) / (10 - 1) / 4 = 694.

EXAMPLE (2):

6371 - 2647 + 8096 - 7381 + 1234

$= 6371 + \overline{2647} + 8096 + \overline{7381} + 1234$

$= (6 + \bar{2} + 8 + \bar{7} + 1)/(3 + \bar{6} + 0 + \bar{3} + 2)/$

$(\bar{7} + 4 + \bar{9} + 8 + 3)/(\bar{1} + 7 + \bar{6} + 1 + 4)$

$= 6 / \bar{4} / 7 / 3$

= (6 - 1) / (10 - 4) / 73

= 5673

Find the results in the following cases using Vedic methods.

1) 57 -39
2) 1286 -968
3) 384 -127 + 696 -549 +150
4) 7084 + 1232 - 6907 - 3852 + 4286

*** Apply Vedic check for the above four problems and verify the results.**

Multiplication

We have already observed the application of Vedic sutras in multiplication. Let us recall them.

It enables us to have a comparative study of the applicability of these methods, to assess advantage of one method over the other method and so-on.

EXAMPLE 1:

(i) By Ekadhikena purvena, since the number ends up in 5 we write the answer split up into two parts.

The right side part is 52 where as the left side part 19 × (19+1) (Ekhadhikena)

Thus $195^2 = 19 \times 20/5^2 = 380/25 = 38025$

(ii) Find the square of 195.

The Conventional method :

$$195^2 = \begin{array}{r} 195 \\ \times\ 195 \\ \hline 975 \\ 1755 \\ 195 \\ \hline 38025 \\ \hline \end{array}$$

(iii) By Nikhilam Navatascaramam Dasatah, as the number is far from base 100, we combine the sutra with the upa-sutra 'anurupyena' and proceed by taking working base 200.

a) Working Base = 200 = 2 × 100.

Now 195^2 = 195 × 195

195		–5	
195		–5	
(195 – 5)or	/	–5 × –5	
×190	/	= 25	
380	/	025	= 38025

(iv) By the sutras "yavadunam tavadunikritya vargamca yojayet" and "anurupyena"

195^2, base 200 treated as 2 × 100 deficit is 5.

$$195^2 = \frac{(195-5)}{5^2}$$

$$= \frac{190 \times 2}{25} \quad \text{operating on 100}$$

$$= 38025$$

(v) By 'antyayor dasakepi' and 'Ekadhikena' sutras

Since in 195 × 195, 5 + 5 = 10 gives

195^2 = 19 × 20 / 5 x 5 = 380 / 25 = 38025.

EXAMPLE 2 :

98 × 92

(i) 'Nikhilam' sutra

$$\begin{array}{r} 98 \ -2 \\ \times\ 92 \ -8 \\ \hline 90 \ / \ 16 = 9016 \end{array}$$

(ii) urdhava-tiryak sutra

$$\begin{array}{r} 9\,8 \\ \times\ 9\,2 \\ \hline 1\,0\,6 \\ 8\,9\,1 \\ \hline 9\,0\,1\,6 \end{array}$$

(iii) by vinculum method

98 = 100 - 2 = $10\bar{2}$

92 = 100 - 8 = $10\bar{8}$

now

$$\begin{array}{r} 10\bar{2} \\ 10\bar{8} \\ \hline 10\bar{0}06 \\ \bar{1}\ 1 \\ \hline 1\bar{1}\bar{0}16 \end{array} = 9016$$

(iv) 'Antyayordasakepi' and 'Ekadhikena' sutras.

98 × 92 Last digit sum= 8+2 =10 remaining digit (s) = 9 same sutras work.

∴ 98 × 92 = 9 × (9 + 1) / 8 × 2 = 90/16 = 9016.

EXAMPLE 3:

493 × 497.

(1) 'Nikhilam' Method and 'Anurupyena':

(a) Working base is 500, treated as 5 × 100

$$\begin{array}{ll} 493 & -7 \\ 497 & -3 \\ \hline 490 & \\ \times 5 & 21 \quad \text{since base is } 5\times 10 \\ \hline 2450 & 21 = 245021 \end{array}$$

(b) Working base is 500, treated as 1000 / 2

$$\begin{array}{ll} 493 & -7 \\ 497 & -3 \\ \hline \end{array}$$

2) 490 / 021

---

245 / 021 = 245021

(2) Since end digits sum is 3+7 = 10 and remaining part 49 is same in both the numbers, 'antyayordasakepi' is applicable. Further Ekadhikena Sutra is also applicable.

Thus,

$$\begin{aligned} 493 \times 497 &= 49 \times 50 / 3 \times 7 \\ &= 2450 / 21 \\ &= 245021 \end{aligned}$$

(3) With the use of vinculum.

$493 = 500 - 07 = 50\bar{7}$

$497 = 500 - 03 = 50\bar{3}$.

Now 497 x 497 can be taken as $50\bar{7} \times 50\bar{3}$

$$\begin{array}{r} 50\bar{7} \\ \times\ 50\bar{3} \\ \hline 50\bar{0}01 \\ 2\bar{5}2 \\ \hline 25\overline{50}21 = 245021 \end{array}$$

EXAMPLE 4:

99 × 99

(1) Now by urdhva - tiryak sutra.

$$\begin{array}{r} 99 \\ \times\ 99 \\ \hline 8121 \\ 168 \\ \hline 9801 \end{array}$$

(2) By vinculum method

$99 = 100 - 1 = 10\bar{1}$

Now 99 × 99 is

$$\begin{array}{r} 10\bar{1} \\ \times\ 101 \\ \hline 10\bar{2}01 = 9801 \end{array}$$

(3) By Nikhilam method

```
            99   -1
            99   -1
           --------
      98 / 01 = 9801.
```

(4) 'Yadunam' sutra : 99^2 Base = 100

Deficiency is 1 : It indicates $99^2 = (99 - 1) / 1^2$

$= 98 / 01 = 9801.$

In the above examples we have observed how in more than one way problems can be solved and also the variety. You can have your own choice for doing multiplication. Not only that which method suits well for easier and quicker calculations. Thus the element of choice, divergent thinking, insight into properties and patterns in numbers, natural way of developing an idea, resourcefulness play major role in Vedic Mathematics methods.

Division

In the conventional procedure for division, the process is of the following form.

```
           Quotient
          _________
Divisor) Dividend      or   Divisor) Dividend (Quotient
          ----------                  ----------
          ----------                  ----------
          __________                  __________
          Remainder                   Remainder
```

But in the Vedic process, the format is:

```
Divisor) Dividend
          --------
          --------
          ____________________
          Quotient / Remainder
```

The conventional method is always the same irrespective of the divisor. But Vedic methods are different depending on the nature of the divisor.

EXAMPLE 1:

Consider the division 1235 ÷ 89.

(i) Conventional method:

89) 1235 (13

89

345

267 Thus Q = 13 and R = 78.

78

(ii) *Nikhilam method:* This method is useful when the divisor is nearer and less than the base. Since for 89, the base is 100 we can apply the method. Let us recall the nikhilam division already dealt.

Step (i): Write the dividend and divisor as in the conventional method. Obtain the modified divisor (M.D.) applying the Nikhilam formula. Write M.D. just below the actual divisor.

Thus for the divisor 89, the M.D. obtained by using Nikhilam is 11 in the last from 10 and the rest from 9. Now Step 1 gives

89) 1235

$\overline{11}$

Step (ii): Bifurcate the dividend by a slash so that R.H.S of dividend contains the number of digits equal to that of M.D. Here M.D. contains 2 digits hence

89) 12 / 35

$\overline{11}$

Step (iii): Multiply the M.D. with first column digit of the dividend. Here it is 1. i.e. 11 x 1 = 11. Write this product place wise under the 2nd and 3rd columns of the dividend.

89) 12 / 35

—

11 1 1

Step (iv): Add the digits in the 2nd column and multiply the M.D. with that result i.e. 2+1=3 and 11x3=33. Write the digits of this result column wise as shown below, under 3rd and 4th columns. i.e.

89) 12 / 35

—

11 1 1
33

———

13 / 78

Now the division process is complete, giving Q = 13 and R = 78.

EXAMPLE 2:

Find Q and R for 121134 ÷ 8988.

Steps (1+2):

8988) 12 / 1134

—

1012

Step (3):

8988) 12 / 1134

—

1012 1 012

Step(4):

8988) 12 / 1134

—

1012 1 012 [2 + 1 = 3 and 3x1012 = 3036]
3036

Now final Step

8988) 12 / 1134

1012 1 012

3036 (Column wise addition)

13 / 4290

Thus 121134, 8988 gives Q = 13 and R = 4290.

(iii) Paravartya method: Recall that this method is suitable when the divisor is nearer but more than the base.

EXAMPLE 3:

32894 ÷ 1028.

The divisor has 4 digits. So the last 3 digits of the dividend are set apart for the remainder and the procedure follows.

1	0	2	8	32	8	9	4
	0	-2	-8	0	-6	-24	
					0	-4	16
				32	2	-19	-12

(1) 3 × (0, -2, -8) = 0, -6, -24

(2) 2 - 0 = 2 and 2 × (0, -2, -8) = 0, -4, -16

(3) 8 - 6 + 0 = 2

9 - 24 - 4 = -19

4 - 16 = - 12

Now the remainder contains -19, -12 i.e. negative quantities. Observe that 32 is quotient. Take 1 over from the quotient column i.e. 1 × 1028 = 1028 over to the right side and proceed thus: 32 - 1 = 31 becomes the Q and R = 1028 + 200 - 190 - 12 = 1028-2 =1026.

Thus 3289 ÷ 1028 gives Q = 31 and R = 1026.

EXAMPLE 4:

43852 ÷ 54.

Step 1:

Put down the first digit (5) of the divisor (54) in the divisor column as operator and the other digit (4) as flag digit. Separate the dividend into two parts where the right part has one digit. This is because the falg digit is single digit. The representation is as follows.

4 : 4 3 8 5 : 2

5

Step 2:

(i) Divide 43 by the operator 5. Now Q= 8 and R = 3. Write this Q=8 as the 1st Quotient - digit and prefix R=3, before the next digit i.e. 8 of the dividend, as shown below. Now 38 becomes the gross-dividend (G.D.) for the next step.

4 : 4 3 8 5 : 2

5 : 3

: 8

(ii) Subtract the product of falg digit (4) and first quotient digit (8) from the G.D. (38) i.e. 38 - (4 × 8) = 38 - 32 = 6. This is the net - dividend (N.D) for the next step.

Step 3:

Now N.D Operator gives Q and R as follows. 6 ÷ 5, Q = 1, R = 1. So Q = 1, the second quotient-digit and R - 1, the prefix for the next digit (5) of the dividend.

4 : 4 3 8 5 : 2

5 : 3 1

: 8 1

Step 4:

Now G.D = 15; product of flag-digit (4) and 2nd quotient-digit (1) is 4X1=4 Hence N.D=15-4=11 divide N.D by 5 to get 11 ÷ 5, Q = 2, R= 1. The representation is

```
      4 : 4 3 8 5 : 2
  5     :   3 1   : 1
    ─────────────────
        :   8 1 2 :
```

Step 5:

Now the R.H.S part has to be considered. The final remainder is obtained by subtracting the product of falg-digit (4)and third quotient digit (2) form 12 i.e., 12:

Final remainder = 12 - (4 × 2) = 12 - 8 = 4. Thus the division ends into

```
      4 : 4 3 8 5 : 2
        5   : 3 1 :1
    ─────────────────
        : 8 1 2 : 4
```

Thus 43852 ÷ 54 gives Q = 812 and R = 4.

Consider the algebraic proof for the above problem. The divisor 54 can be represented by 5x + 4, where x = 10

The dividend 43852 can be written algebraically as $43x^3 + 8x^2 + 5x + 2$

Since $x^3 = 10^3 = 1000$, $x^2 = 10^2 = 100$.

Now the division is as follows.

$5x + 4)\ 43x^3 + 8x^2 + 5x + 2\ (8x^2 + x + 2$

$43x^3 + 32x^2$

$3x^3 - 24x^2$

$= 6x^2 + 5x \qquad (\because 3x^3 = 3.x.x^2$

$5x^2 + 4x \qquad = 3.10x^2 = 30\ x^2)$

$x^2 + x$

$= 11x + 2 \qquad (\because x^2 = x \,.\, x = 10x)$

$$\begin{array}{r} 10x + 8 \\ \hline x - 6 \end{array}$$

$= 10 - 6 = 4.$

Observe the Following Steps

1. $43x^3 \div 5x$ gives first quotient term 8x2 , remainder $= 3x^3 - 24x^2$ which really mean $30x^2 + 8x^2 - 32x^2 = 6x^2$.

   Thus in step 2 of the problem $43852 \div 54$, we get Q= 8 and N.D = 6.

2. $6x^2 \div 5x$ gives second quotient term x, remainder $= x^2 + x$ which really mean $10x + x = 11x$.

   Thus in step 3 & Step 4, we get Q = 1 and N.D = 11.

3. $11x \div 5x$ gives third quotient term 2, remainder $= x - 6$, which really mean the final remainder $10 - 6 = 4$.

EXAMPLE 5:

Divide $237963 \div 524$

Step 1:

We take the divisor 524 as 5, the operator and 24, the flag-digit and proceed as in the above example. We now separate the dividend into two parts where the RHS part contains two digits for Remainder.

Thus,

24 : 2 3 7 9 : 63

5

Step 2:

(i) $23 \div 5$ gives Q = 4 and R = 3, G.D = 37.

(ii) N.D is obtained as

$$37 - \begin{pmatrix} 2 & 4 \\ 0 & 4 \end{pmatrix} \quad \text{(Flat digits)}$$

$= 37 - (8 + 0) = 29.$

Representation

$$\begin{array}{rcccccl} 24 & : & 2\ 3 & 7 & 9 & : & 63 \\ & & & 5 & & & 3 \\ \hline & & & & & & :4 \end{array}$$

Step 3:

(i) N.D ÷ Operator = 29 ÷ 5 gives Q = 5, R = 4 and G.D = 49.

(ii) N.D is obtained as

$$49 - \begin{pmatrix} 2 & 4 \\ 4 & 5 \end{pmatrix} \text{ (Flat digits)}$$

$$= 49 - (10 + 16)$$

$$= 49 - 26$$

$$= 23.$$

i.e.,

$$\begin{array}{rcccccl} 24 & : & 2\ 3 & 7 & 9 & : & 63 \\ 5 & & : & & 3\ 4 & & : \\ \hline & & : & 4 & 5 & & : \end{array}$$

Step 4:

(i) N.D ÷ Operator = 23 ÷ 5 gives Q = 4, R = 3 and G.D = 363.

Note that we have reached the remainder part thus 363 is total sub-remainder.

$$\begin{array}{rcccccl} 24 & : & 2\ 3 & 7 & 9 & : & 63 \\ 5 & & : & 3 & 4 & : & 3 \\ \hline & & : & 4 & 5 & 4 & : \end{array}$$

Step 5:

We find the final remainder as follows. Subtract the cross-product of the two, falg-digits and two last quotient-digits and then vertical product of last flag-digit with last quotient-digit from the total sub-remainder.

i.e.,

$$363 - \left(\begin{matrix} 2 & 4 \\ 4 & 5 \end{matrix}\right) \Big/ \begin{matrix} 4 \\ \downarrow \\ 4 \end{matrix}$$

$$= \frac{363-(8+20)}{16}$$

$$= \frac{363-28}{16}$$

$$= 363 - 296 = 67.$$

Note that 2, 4 are two falg digits: 5, 4 are two last quotient digits:

$$\begin{pmatrix} 4 \\ 4 \end{pmatrix}$$

represents the last flag - digit and last quotient digit.

Thus the division 237963 ÷ 524 gives Q = 454 and R = 67.

Thus the Vedic process of division which is also called as Straight division is a simple application of urdhva-tiryak together with dhvajanka. This process has many uses along with the one-line presentation of the answer.

Miscellaneous Items

Straight Squaring

We have already noticed methods useful to find out squares of numbers. But the methods are useful under some situations and conditions only. Now we go to a more general formula.

The sutra Dwandwa-yoga (Duplex combination process) is used in two different meanings. They are:

i) by squaring

ii) by cross-multiplying.

We use both the meanings of Dwandwa-yoga in the context of finding squares of numbers as follows:

We denote the Duplex of a number by the symbol D.

We define for a single digit 'a', $D = a^2$. and for a two digit number of the form 'ab', $D = 2(a \times b)$.

If it is a 3 digit number like 'abc', $D = 2(a \times c) + b^2$.

For a 4 digit number 'abcd', $D = 2(a \times d) + 2(b \times c)$ and so on. i.e. if the digit is single central digit, D represents 'square': and for the case of an even number of digits equidistant from the two ends D represent the double of the cross- product.

Consider the examples:

Number	**Duplex D**
3	$3^2 = 9$
6	$6^2 = 36$
23	$2 (2 \times 3) = 12$
64	$2 (6 \times 4) = 48$
128	$2 (1 \times 8) + 22 = 16 + 4 = 20$
305	$2 (3 \times 5) + 02 = 30 + 0 = 30$
4231	$2 (4 \times 1) + 2 (2 \times 3) = 8 + 12 = 20$
7346	$2 (7 \times 6) + 2 (3 \times 4) = 84 + 24 = 108$

Further observe that for a n- digit number, the square of the number contains 2n or 2n-1 digits.

Thus in this process, we take extra dots to the left one less than the number of digits in the given numbers.

EXAMPLES 1:

62^2 Since number of digits = 2, we take one extra dot to the left. Thus

.62 for 2, D . = 22 = 4

644 for 62, D = 2 × 6 × 2 = 24
32 for 62, D = 2(0 × 2) + 62
= 36

3844

∴ $62^2 = 3844$.

EXAMPLES 2:

234^2 Number of digits = 3. extradots = 2 Thus

..234 for 4, D = 4^2 = 16

42546 for 34, D = 2 × 3 × 4 = 24
1221 for 234, D = 2 × 2 × 4 + 32 = 25

54756 for .234, D = 2.0.4 + 2.2.3 = 12

for ..234, D = 2.0.4 + 2.0.3 + 22 = 4

EXAMPLES 3:

1426^2. Number of digits = 4, extra dots = 3

i.e

...1426 6, D = 36

1808246 26, D = 2.2.6 = 24
22523 426, D = 2.4.6 + 22 = 52

2033476 1426, D = 2.1.6 + 2.4.2 = 28

.1426, D = 2.0.6 + 2.1.2 + 42 = 20

..1426, D= 2.0.6 + 2.0.2 + 2.1.4 = 8

...1426, D = 12 = 1

Thus $1426^2 = 2033476$.

With a little bit of practice the results can be obtained mentally as a single line answer.

ALGEBRAIC PROOF:

Consider the first example 62^2

Now $62^2 = (6 \times 10 + 2)^2 = (10a + b)^2$ where $a = 6, b = 2$

$$= 100a^2 + 2.10a.b + b^2$$

$$= a^2 (100) + 2ab (10) + b^2$$

i.e. b^2 in the unit place, 2ab in the 10th place and a^2 in the 100th place i.e. $2^2 = 4$ in units place, $2.6.2 = 24$ in the 10th place (4 in the 10th place and with carried over to 100th place). $6^2=36$ in the 100th place and with carried over 2 the 100th place becomes $36+2=38$.

Thus the answer 3844.

Find the squares of the numbers 54, 123, 2051, 3146.

Applying the Vedic sutra Dwanda yoga.

CUBING

Take a two digit number say 14.

(i) Find the ratio of the two digits i.e. 1:4

(ii) Now write the cube of the first digit of the number i.e. 1^3

(iii) Now write numbers in a row of 4 terms in such a way that the first one is the cube of the first digit and remaining three are obtained in a geometric progression with common ratio as the ratio of the original two digits (i.e. 1:4) i.e. the row is

1 4 16 64.

(iv) Write twice the values of 2nd and 3rd terms under the terms respectively in second row.

i.e.,

1 4 16 64

8 32 (2 × 4 = 8, 2 × 16 = 32)

(v) Add the numbers column wise and follow carry over process.

1	4	16	64	Since 16 + 32 + 6 (carryover) = 54
	8	32	4	written and 5 (carryover) +4+8 = 17
2	7	4	4	7 written and 1 (carryover) + 1 = 2.

This 2744 is nothing but the cube of the number 14

ALGEBRAIC PROOF:

Let a and b be two digits.

Consider the row a^3 a^2b ab^2 b^3

the first is a3 and the numbers are in the ratio a:b

since $a^3:a^2b = a^2b:b^3 = a:b$

Now twice of a^2b, ab^2 are $2a^2b$, $2ab^2$

$$\begin{array}{l} a^3 + a^2b + ab^2 + b^3 \\ \quad\quad\quad 2a^2b + 2ab^2 \\ \hline a^3 + 3a^2b + 3ab^2 + b^3 = (a + b)^3. \end{array}$$

Thus cubes of two digit numbers can be obtained very easily by using the vedic sutra 'anurupyena'. Now cubing can be done by using the vedic sutra 'Yavadunam'.

EXAMPLE 3:

Consider 1063.

(i) The base is 100 and excess is 6. In this context we double the excess and then add.

i.e. 106 + 12 = 118. (∵ 2 × 6 =12)

This becomes the left - hand - most portion of the cube.

i.e. 106^3 = 118 / - - - -

(ii) Multiply the new excess by the initial excess i.e. 18 × 6 = 108 (excess of 118 is 18)

Now this forms the middle portion of the product of course 1 is carried over, 08 in the middle.

i.e. 106^3 = 118 / 08 / - - - - -
1

(iii) The last portion of the product is cube of the initial excess.

i.e. 6^3 = 216.

16 in the last portion and 2 carried over.

i.e. 106^3 = 118 / 081 / 16 = 1191016
1 2

EXAMPLE 4:

Find 1002^3.

(i) Base = 1000. Excess = 2. Left-hand-most portion of the cube becomes 1002 + (2 × 2)=1006.

(ii) New excess x initial excess = 6 × 2 = 12.

Thus 012 forms the middle portion of the cube.

(iii) Cube of initial excess = 23 = 8.

So the last portion is 008.

Thus 10023 = 1006 / 012 / 008 = 1006012008.

EXAMPLE 5:

Find 94^3.

(i) Base = 100, deficit = -6. Left-hand-most portion of the cube becomes 94 + (2x - 6) = 94 - 12 = 82.

(ii) New deficit x initial deficit

= -(100 - 82) × (-6) = -18x - 6 = 108

Thus middle potion of the cube = 08 and 1 is carried over.

(iii) Cube of initial deficit = $(-6)^3 = -216$

$$\text{Now } 943 = 82 \,/\, 08 \,/\, \overline{16} = 83 \,/\, 06 \,/\, \overline{16}$$

$$1 \quad \bar{2}$$

$$= 83 \,/\, 05 \,/\, (100 - 16)$$

$$= 830584.$$

Find the cubes of the following numbers using Vedic sutras.

103, 112, 91, 89, 998, 9992, 1014.

Equation of Straight line passing through two given points:

To find the equation of straight line passing through the points (x1, y1) and (x2, y2) , we generally consider one of the following methods.

1. General equation y = mx + c.

It is passing through (x_1, y_1) then $y_1 = mx_1 + c$.

It is passing through (x_2, y_2) also, then $y_2 = mx_2 + c$.

Solving these two simultaneous equations, we get 'm' and 'c' and so the equation.

2. The formula:

$$y - y_1 \frac{(y_2 - y_1)}{(x_2 - x_1)}(x - x_1) \text{ and substitution.}$$

Some sequence of steps gives the equation. But the paravartya sutra enables us to arrive at the conclusion in a more easy way and convenient to work mentally.

EXAMPLE1 :

Find the equation of the line passing through the points (9,7) and (5,2).

Step 1: Put the difference of the y - coordinates as the x- coefficient and vice - versa.

i.e.x coefficient = 7 - 2 = 5

y coefficient = 9 - 5 = 4.

Thus L.H.S of equation is 5x - 4y.

Step 2: The constant term (R.H.S) is obtained by substituting the coordinates of either of the given points in

L.H.S (obtained through step-1)

i.e. R.H.S of the equation is

5(9) - 4(7) = 45 - 28 = 17

or 5(5) - 4(2) = 25 - 8 = 17.

Thus the equation is 5x - 4y = 17.

EXAMPLE 2:

Find the equation of the line passing through (2, -3) and (4,-7).

Step 1 : x[-3-(-7)] -y[2-4] = 4x + 2y.

Step 2 : 4(2) + 2(-3) = 8 -6 = 2.

Step 3 : Equation is 4x + 2y =2 or 2x +y = 1.

EXAMPLE 3 :

Equation of the line passing through the points (7,9) and (3,-7).

Step 1 : x[9 - (-7)] - y(7 - 3) = 16x - 4y.

Step 2 : 16(7) - 4(9) = 112 - 36 = 76

Step 3 : 16x- 4y = 76 or 4x - y = 19

Find the equation of the line passing through the points using Vedic methods.

1. (1, 2), (4,-3)
2. (5,-2), (5,-4)

3. (-5, -7), (13, 2)
4. (a, o) , (o,b)

Conclusion

After going through the content presented in this book, you may, perhaps, have noted a number of applications of methods of Vedic Mathematics. We are aware that this attempt is only to make you familiar with a few special methods.

The methods discussed, and organization of the content here are intended for any reader with some basic mathematical background. That is why the serious mathematical issues, higher level mathematical problems are not taken up in this volume, even though many aspects like four fundamental operations, squaring, cubing, linear equations, simultaneous equations, factorization, H.C.F, recurring decimals, etc. are dealt with.

Many more concepts and aspects are omitted unavoidably, keeping in view the scope and limitations of the present volume.

Thus the present volume serves as only an 'introduction'. More has to be presented to cover all the issues in Swamiji's 'Vedic Mathematics'. Still more steps are needed to touch the latest developments in Vedic Mathematics.

As a result, serious and sincere work by scholars and research workers continues in this field both in our country and abroad. Sri Sathya Sai Veda Pratisthan intends to bring about more volumes covering the aspects now left over, and also elaborating the content of Vedic Mathematics.

The present volume, even though introductory, has touched almost all the Sutras and sub-Sutras as mentioned in Swamiji's 'Vedic Mathematics'. Further it has given rationale and proof for the methods. As there is a general opinion that the 'so called Vedic Mathematics is only rude, rote, non mathematical and none other than some sort of tricks', the logic, proof and Mathematics behind the 'the so called tricks' has been explained.

An impartial reader can easily experience the beauty, charm and resourcefulness in Vedic Mathematics systems. We feel that the reader can enjoy the diversity and simplicity in Vedic Mathematics while applying the methods against the conventional textbook methods. The reader can also compare and contrast both the methods.

The Vedic Methods enable the practitioner improve mental abilities to solve difficult problems with high speed and accuracy.

3

History of Mathematics in India

In all early civilizations, the first expression of mathematical understanding appears in the form of counting systems. Numbers in very early societies were typically represented by groups of lines, though later different numbers came to be assigned specific numeral names and symbols (as in India) or were designated by alphabetic letters (such as in Rome). Although today, we take our decimal system for granted, not all ancient civilizations based their numbers on a ten-base system. In ancient Babylon, a sexagesimal (base 60) system was in use.

The Decimal System in Harappa

In India a decimal system was already in place during the Harappan period, as indicated by an analysis of Harappan weights and measures. Weights corresponding to ratios of 0.05, 0.1, 0.2, 0.5, 1, 2, 5, 10, 20, 50, 100, 200, and 500 have been identified, as have scales with decimal divisions. A particularly notable characteristic of Harappan weights and measures is their remarkable accuracy. A bronze rod marked in units of 0.367 inches points to the degree of precision demanded in those times. Such scales were particularly important in ensuring proper implementation of town planning rules that required roads of fixed widths to run at right angles to each other, for drains to be constructed of precise measurements, and for homes to be constructed according to specified guidelines. The existence of a gradated system of accurately marked weights points to the development of trade and commerce in Harappan society.

Mathematical Activity in the Vedic Period

In the Vedic period, records of mathematical activity are mostly to be found in Vedic texts associated with ritual activities. However, as in many other early agricultural civilizations, the study of arithmetic and geometry was also impelled by secular considerations. Thus, to some extent early mathematical developments in India mirrored the developments in Egypt, Babylon and China.

The system of land grants and agricultural tax assessments required accurate measurement of cultivated areas. As land was redistributed or consolidated, problems of mensuration came up that required solutions. In order to ensure that all cultivators had equivalent amounts of irrigated and non-irrigated lands and tracts of equivalent fertility — individual farmers in a village often had their holdings broken up in several parcels to ensure fairness. Since plots could not all be of the same shape — local administrators were required to convert rectangular plots or triangular plots to squares of equivalent sizes and so on. Tax assessments were based on fixed proportions of annual or seasonal crop incomes, but could be adjusted upwards or downwards based on a variety of factors. This meant that an understanding of geometry and arithmetic was virtually essential for revenue administrators. Mathematics was thus brought into the service of both the secular and the ritual domains.

Arithmetic operations (Ganit) such as addition, subtraction, multiplication, fractions, squares, cubes and roots are enumerated in the Narad Vishnu Purana attributed to Ved Vyas (pre-1000 BC). Examples of geometric knowledge (rekha-ganit) are to be found in the Sulva-Sutras of Baudhayana (800 BC) and Apasthmaba (600 BC) which describe techniques for the construction of ritual altars in use during the Vedic era. It is likely that these texts tapped geometric knowledge that may have been acquired much earlier, possibly in the Harappan period. Baudhayana's Sutra displays an understanding of basic geometric shapes and techniques of converting one geometric shape (such as a rectangle) to another of equivalent (or multiple, or fractional) area (such as a square). While some of the formulations are approximations, others are accurate and reveal a certain degree of practical ingenuity as well as some theoretical understanding of basic geometric principles. Modern methods of multiplication and addition probably emerged from the techniques described in the Sulva-Sutras.

Pythagoras — the Greek mathematician and philosopher who lived in the 6th CBC, was familiar with the Upanishads and learnt his basic geometry from the Sulva Sutras. An early statement of what is commonly known as the Pythagoras theorem is to be found in Baudhayana's Sutra: The chord which is stretched across the diagonal of a square produces an area of double the size. A similar observation pertaining to oblongs is also noted. His Sutra also contains geometric solutions of a linear equation in a single unknown. Examples of quadratic equations also appear. Apasthamba's sutra (an expansion of Baudhayana's with several original contributions) provides a value for the square root of 2 that is accurate to the fifth decimal place. Apasthamba also looked at the problems of squaring a circle, dividing a segment into seven equal parts, and a solution to the general linear equation. Jain texts from the 6th C BC such as the Surya Pragyapti describe ellipses.

Modern-day commentators are divided on how some of the results were generated. Some believe that these results came about through hit and trial — as rules of thumb, or as generalizations of observed examples. Others believe that once the scientific method came to be formalized in the Nyaya-Sutras — proofs for such results must have been provided, but these have either been lost or destroyed, or else were transmitted orally through the Gurukul system, and only the final results were tabulated in the texts. In any case, the study of Ganit i.e. mathematics was given considerable importance in the Vedic period. The Vedang Jyotish (1000 BC) includes the statement: "Just as the feathers of a peacock and the jewel-stone of a snake are placed at the highest point of the body (at the forehead), similarly, the position of Ganit is the highest amongst all branches of the Vedas and the Shastras."

Many centuries later, Jain mathematician from Mysore, Mahaviracharya further emphasized the importance of mathematics: "Whatever object exists in this moving and non-moving world, cannot be understood without the base of Ganit (i.e. mathematics)".

Panini and Formal Scientific Notation

A particularly important development in the history of Indian science that was to have a profound impact on all mathematical treatises that followed was the pioneering work by Panini (6th C BC) in the field of Sanskrit grammar and linguistics. Besides expounding a comprehensive and scientific theory of phonetics, phonology and morphology, Panini provided formal production rules and definitions describing Sanskrit grammar in his treatise called Asthadhyayi. Basic elements such as vowels and consonants, parts of speech such as nouns and verbs were placed in classes. The construction of compound words and sentences was elaborated through ordered rules operating on underlying structures in a manner similar to formal language theory.

Today, Panini's constructions can also be seen as comparable to modern definitions of a mathematical function. G. G. Joseph, in The crest of the peacock argues that the algebraic nature of Indian mathematics arises as a consequence of the structure of the Sanskrit language. Ingerman in his paper titled Panini-Backus form finds Panini's notation to be equivalent in its power to that of Backus — inventor of the Backus Normal Form used to describe the syntax of modern computer languages. Thus Panini's work provided an example of a scientific notational model that could have propelled later mathematicians to use abstract notations in characterizing algebraic equations and presenting algebraic theorems and results in a scientific format.

Philosophy and Mathematics

Philosophical doctrines also had a profound influence on the development of mathematical concepts and formulations. Like the Upanishadic world view, space and time were considered limitless in Jain cosmology. This led to a deep interest in very large numbers and definitions of infinite numbers. Infinite numbers were created through recursive formulae, as in the Anuyoga Dwara Sutra. Jain mathematicians recognized five different types of infinities: infinite in one direction, in two directions, in area, infinite everywhere and perpetually infinite. Permutations and combinations are listed in the Bhagvati Sutras (3rd C BC) and Sathananga Sutra (2nd C BC).

Jain set theory probably arose in parallel with the Syadvada system of Jain epistemology in which reality was described in terms of pairs of truth conditions and state changes. The Anuyoga Dwara Sutra demonstrates an understanding of the law of indeces and uses it to develop the notion of logarithms. Terms like Ardh Aached , Trik Aached, and Chatur Aached are used to denote log base 2, log base 3 and log base 4 respectively. In Satkhandagama various sets are operated upon by logarithmic functions to base two, by squaring and extracting square roots, and by raising to finite or infinite powers. The operations are repeated to produce new sets. In other works the relation of the number of combinations to the coefficients occurring in the binomial expansion is noted.

Since Jain epistemology allowed for a degree of indeterminacy in describing reality, it probably helped in grappling with indeterminate equations and finding numerical approximations to irrational numbers.

Buddhist literature also demonstrates an awareness of indeterminate and infinite numbers. Buddhist mathematics was classified either as Garna (Simple Mathematics) or Sankhyan (Higher Mathematics). Numbers were deemed to be of three types: Sankheya (countable), Asankheya (uncountable) and Anant (infinite).

Philosophical formulations concerning Shunya — i.e. emptiness or the void may have facilitated in the introduction of the concept of zero. While the zero (bindu) as an empty place holder in the place-value numeral system appears much earlier, algebraic definitions of the zero and it's relationship to mathematical functions appear in the mathematical treatises of Brahmagupta in the 7th C AD. Although scholars are divided about how early the symbol for zero came to be used in numeric notation in India, (Ifrah arguing that the use of zero is already implied in Aryabhatta) tangible evidence for the use of the zero begins to proliferate towards the end of the Gupta period. Between the 7th C and the 11th C, Indian numerals developed into their modern form, and along with the symbols denoting various mathematical functions (such as plus, minus, square root etc.) eventually became the foundation stones of modern mathematical notation.

The Indian Numeral System

Although the Chinese were also using a decimal based counting system, the Chinese lacked a formal notational system that had the abstraction and elegance of the Indian notational system, and it was the Indian notational system that reached the Western world through the Arabs and has now been accepted as universal. Several factors contributed to this development whose significance is perhaps best stated by French mathematician, Laplace: "The ingenious method of expressing every possible number using a set of ten symbols (each symbol having a place value and an absolute value) emerged in India. The idea seems so simple nowadays that its significance and profound importance is no longer appreciated. It's simplicity lies in the way it facilitated calculation and placed arithmetic foremost amongst useful inventions."

Brilliant as it was, this invention was no accident. In the Western world, the cumbersome roman numeral system posed as a major obstacle, and in China the pictorial script posed as a hindrance. But in India, almost everything was in place to favour such a development. There was already a long and established history in the use of decimal numbers, and philosophical and cosmological constructs encouraged a creative and expansive approach to number theory. Panini's studies in linguistic theory and formal language and the powerful role of symbolism and representational abstraction in art and architecture may have also provided an impetus, as might have the rationalist doctrines and the exacting epistemology of the Nyaya Sutras, and the innovative abstractions of the Syadavada and Buddhist schools of learning.

Influence of Trade and Commerce, Importance of Astronomy

The growth of trade and commerce, particularly lending and borrowing demanded an understanding of both simple and compound interest which probably stimulated the interest in arithmetic and geometric series. Brahmagupta's description of negative numbers as debts and positive numbers as fortunes points to a link between trade and mathematical study. Knowledge of astronomy — particularly knowledge of the tides and the stars was of great import to trading communities who crossed oceans or deserts at night. This is borne out by numerous references in the Jataka tales and several other folk-tales. The young person who wished to embark on a commercial venture was inevitably required to first gain some grounding in astronomy. This led to a proliferation of teachers of astronomy, who in turn received training at universities such as at Kusumpura (Bihar) or Ujjain (Central India) or at smaller local colleges or Gurukuls. This also led to the exchange of texts on astronomy and mathematics amongst scholars and the transmission of knowledge from one part of India to another. Virtually every Indian state produced great mathematicians who wrote commentaries on the works of other mathematicians (who may have lived and worked in a different part of India many centuries earlier). Sanskrit served as the common medium of scientific communication.

The science of astronomy was also spurred by the need to have accurate calendars and a better understanding of climate and rainfall patterns for timely sowing and choice of crops. At the same time, religion and astrology also played a role in creating an interest in astronomy and a negative fallout of this irrational influence was the rejection of scientific theories that were far ahead of their time. One of the greatest scientists of the Gupta period — Aryabhatta (born in 476 AD, Kusumpura, Bihar) provided a systematic treatment of the position of the planets in space. He correctly posited the axial rotation of the earth, and inferred correctly that the orbits of the planets were ellipses. He also correctly deduced that the moon and the planets shined by reflected sunlight and provided a valid explanation for the solar and lunar eclipses rejecting the superstitions and mythical belief systems surrounding the phenomenon.

Although Bhaskar I (born Saurashtra, 6th C, and follower of the Asmaka school of science, Nizamabad, Andhra) recognized his genius and the tremendous value of his scientific contributions, some later astronomers continued to believe in a static earth and rejected his rational explanations of the eclipses. But in spite of such setbacks, Aryabhatta had a profound influence on the astronomers and mathematicians who followed him, particularly on those from the Asmaka school.

Mathematics played a vital role in Aryabhatta's revolutionary understanding of the solar system. His calculations on pi, the circumference of the earth (62832 miles) and the length of the solar year (within about 13 minutes of the modern calculation) were remarkably close approximations. In making such calculations, Aryabhatta had to solve several mathematical problems that had not been addressed before, including problems in algebra (beej-ganit) and trigonometry (trikonmiti).

Bhaskar I continued where Aryabhatta left off, and discussed in further detail topics such as the longitudes of the planets; conjunctions of the planets with each other and with bright stars; risings and settings of the planets; and the lunar crescent. Again, these studies required still more advanced mathematics and Bhaskar I expanded on the trigonometric equations provided by Aryabhatta, and like Aryabhatta correctly assessed pi to be an irrational number. Amongst his most important contributions was his formula for calculating the sine function which was 99% accurate. He also did pioneering work on indeterminate equations and considered for the first time quadrilaterals with all the four sides unequal and none of the opposite sides parallel.

Another important astronomer/mathematician was Varahamira (6th C, Ujjain) who compiled previously written texts on astronomy and made important additions to Aryabhatta's trigonometric formulas. His works on permutations and combinations complemented what had been previously achieved by Jain mathematicians and provided a method of calculation of nCr that closely resembles the much more recent Pascal's Triangle. In the 7th century, Brahmagupta did important work in enumerating the basic principles of algebra. In addition to listing the algebraic properties of zero, he also listed the algebraic properties of negative numbers. His work on solutions to quadratic indeterminate equations anticipated the work of Euler and Lagrange.

Emergence of Calculus

In the course of developing a precise mapping of the lunar eclipse, Aryabhatta was obliged to introduce the concept of infinitesimals — i.e. tatkalika gati to designate the infinitesimal, or near instantaneous motion of the moon, and express it in the form of a basic differential equation. Aryabhatta's equations were elaborated on by Manjula (10th C) and Bhaskaracharya (12th C) who derived the differential of the sine function. Later mathematicians used their intuitive understanding of integration in deriving the areas of curved surfaces and the volumes enclosed by them.

Applied Mathematics, Solutions to Practical Problems

Developments also took place in applied mathematics such as in creation of trigonometric tables and measurement units. Yativrsabha's work Tiloyapannatti (6th C) gives various units for measuring distances and time and also describes the system of infinite time measures.

In the 9th C, Mahaviracharya (Mysore) wrote Ganit Saar Sangraha where he described the currently used method of calculating the Least Common Multiple (LCM) of given numbers. He also derived formulae to calculate the area of an ellipse and a quadrilateral inscribed within a circle (something that had also been looked at by Brahmagupta) The solution of indeterminate equations also drew considerable interest in the 9th century, and several mathematicians contributed approximations and solutions to different types of indeterminate equations.

In the late 9th C, Sridhara (probably Bengal) provided mathematical formulae for a variety of practical problems involving ratios, barter, simple interest, mixtures, purchase and sale, rates of travel, wages, and filling of cisterns. Some of these examples involved fairly complicated solutions and his Patiganita is considered an advanced mathematical work. Sections of the book were also devoted to arithmetic and geometric progressions, including progressions with fractional numbers or terms, and formulas for the sum of certain finite series are provided. Mathematical investigation continued into the 10th C. Vijayanandi (of Benares, whose Karanatilaka was translated by Al-Beruni into Arabic) and Sripati of Maharashtra are amongst the prominent mathematicians of the century.

The leading light of 12th C Indian mathematics was Bhaskaracharya who came from a long-line of mathematicians and was head of the astronomical observatory at Ujjain. He left several important mathematical texts including the Lilavati and Bijaganita and the Siddhanta Shiromani, an astronomical text. He was the first to recognize that certain types of quadratic equations could have two solutions. His Chakrawaat method of solving indeterminate solutions preceded European solutions by several centuries, and in his Siddhanta Shiromani he postulated that the earth had a gravitational force, and broached the fields of infinitesimal calculation and integration.

In the second part of this treatise, there are several chapters relating to the study of the sphere and it's properties and applications to geography, planetary mean motion, eccentric epicyclical model of the planets, first visibilities of the planets, the seasons, the lunar crescent etc. He also discussed astronomical instruments and spherical trigonometry. Of particular interest are his trigonometric equations:

sin(a + b) = sin a cos b + cos a sin b;

sin(a - b) = sin a cos b - cos a sin b;

The Spread of Indian Mathematics

The study of mathematics appears to slow down after the onslaught of the Islamic invasions and the conversion of colleges and universities to madrassas. But this was also the time when Indian mathematical texts were increasingly being translated into Arabic and Persian. Although Arab scholars relied on a variety of sources including Babylonian, Syrian, Greek and some Chinese texts, Indian mathematical texts played a particularly important role.

Scholars such as Ibn Tariq and Al-Fazari (8th C, Baghdad), Al-Kindi (9th C, Basra), Al-Khwarizmi (9th C. Khiva), Al-Qayarawani (9th C, Maghreb, author of Kitab fi al-hisab al-hindi), Al-Uqlidisi (10th C, Damascus, author of The book of Chapters in Indian Arithmetic), Ibn-Sina (Avicenna), Ibn al-Samh (Granada, 11th C, Spain), Al-Nasawi (Khurasan, 11th C, Persia), Al-Beruni (11th C, born Khiva, died Afghanistan), Al-Razi (Teheran), and Ibn-Al-Saffar (11th C, Cordoba) were amongst the many who based their own scientific texts on translations of Indian treatises. Records of the Indian origin of many proofs, concepts and formulations were obscured in the later centuries, but the enormous contributions of Indian mathematics was generously acknowledged by several important Arabic and Persian scholars, especially in Spain. Abbasid scholar Al-Gaheth wrote: " India is the source of knowledge, thought and insight". Al-Maoudi (956 AD) who travelled in Western India also wrote about the greatness of Indian science. Said Al-Andalusi, an 11th C Spanish scholar and court historian was amongst the most enthusiastic in his praise of Indian civilization, and specially remarked on Indian achievements in the sciences and in mathematics. Of course, eventually, Indian algebra and trigonometry reached Europe through a cycle of translations, travelling from the Arab world to Spain and Sicily, and eventually penetrating all of Europe. At the same time, Arabic and Persian translations of Greek and Egyptian scientific texts became more readily available in India.

The Kerala School

Although it appears that original work in mathematics ceased in much of Northern India after the Islamic conquests, Benaras survived as a centre for mathematical study, and an important school of mathematics blossomed in Kerala. Madhava (14th C, Kochi) made important mathematical discoveries that would not be identified by European mathematicians till at least two centuries later. His series expansion of the cos and sine functions anticipated Newton by almost three centuries. Historians of mathematics, Rajagopal, Rangachari and Joseph considered his contributions instrumental in taking mathematics to the next stage, that of modern classical analysis.

Nilkantha (15th C, Tirur, Kerala) extended and elaborated upon the results of Madhava while Jyesthadeva (16th C, Kerala) provided detailed proofs of the theorems and derivations of the rules contained in the works of Madhava and Nilkantha. It is also notable that Jyesthadeva's Yuktibhasa which contained commentaries on Nilkantha's Tantrasamgraha included elaborations on planetary theory later adopted by Tycho Brahe, and mathematics that anticipated work by later Europeans. Chitrabhanu (16th C, Kerala) gave integer solutions to twenty-one types of systems of two algebraic equations, using both algebraic and geometric methods in developing his results. Important discoveries by the Kerala mathematicians included the Newton-Gauss interpolation formula, the formula for the sum of an infinite series, and a series notation for pi. Charles Whish (1835, published in the Transactions of the Royal Asiatic Society of Great Britain and Ireland) was one of the first Westerners to recognize that the Kerala school had anticipated by almost 300 years many European developments in the field.

Yet, few modern compendiums on the history of mathematics have paid adequate attention to the often pioneering and revolutionary contributions of Indian mathematicians. But as this essay amply demonstrates, a significant body of mathematical works were produced in the Indian subcontinent.

The science of mathematics played a pivotal role not only in the industrial revolution but in the scientific developments that have occurred since. No other branch of science is complete without mathematics. Not only did India provide the financial capital for the industrial revolution India also provided vital elements of the scientific foundation without which humanity could not have entered this modern age of science and high technology.

Notes: Mathematics and Music: Pingala (3rd C AD), author of Chandasutra explored the relationship between combinatorics and musical theory anticipating Mersenne (1588-1648) author of a classic on musical theory.

Mathematics and Architecture: Interest in arithmetic and geometric series may have also been stimulated by (and influenced) Indian architectural designs - (as in temple shikaras, gopurams and corbelled temple ceilings). Of course, the relationship between geometry and architectural decoration was developed to it's greatest heights by Central Asian, Persian, Turkish, Arab and Indian architects in a variety of monuments commissioned by the Islamic rulers.

Transmission of the Indian Numeral System: Evidence for the transmission of the Indian Numeral System to the West is provided by Joseph (Crest of the Peacock):

Quotes Severus Sebokht (662) in a Syriac text describing the "subtle discoveries" of Indian astronomers as being "more ingenious than those of the Greeks and the Babylonians" and "their valuable methods of computation which surpass description" and then goes on to mention the use of nine numerals.

Quotes from Liber abaci (Book of the Abacus) by Fibonacci (1170-1250): The nine Indian numerals are ...with these nine and with the sign 0 which in Arabic is sifr, any desired number can be written. (Fibonaci learnt about Indian numerals from his Arab teachers in North Africa)

Influence of the Kerala School: Joseph (Crest of the Peacock) suggests that Indian mathematical manuscripts may have been brought to Europe by Jesuit priests such as Matteo Ricci who spent two years in Kochi (Cochin) after being ordained in Goa in 1580. Kochi is only 70km from Thrissur (Trichur) which was then the largest repository of astronomical documents.

Whish and Hyne — two European mathematicians obtained their copies of works by the Kerala mathematicians from Thrissur, and it is not inconceivable that Jesuit monks may have also taken copies to Pisa (where Galileo, Cavalieri and Wallis spent time), or Padau (where James Gregory studied) or Paris (where Mersenne who was in touch with Fermat and Pascal, acted as an agent for the transmission of mathematical ideas).

Indic Mathematics

Math and Ethnocentrism

The study of mathematics in the West has long been characterized by a certain ethnocentric bias, a bias which most often manifests not in explicit racism, but in a tendency toward undermining or eliding the real contributions made by non-Western civilizations. The debt owed by the West to other civilizations, and to India in particular, go back to the earliest epoch of the "Western" scientific tradition, the age of the classical Greeks, and continued up until the dawn of the modern era, the renaissance, when Europe was awakening from its dark ages. This awakening was in part made possible by the rediscovery of mathematics and other sciences and technologies through the medium of the Arabs, who transmitted to Europe both their own lost heritage as well as the advanced mathematical traditions formulated in India.

George Ghevarughese Joseph, in an important article entitled "Foundations of Eurocentrism in Mathematics," argued that "the standard treatment of the history of non-European mathematics is a product of historiographical bias (conscious or otherwise) in the selection and interpretation of facts, which, as a consequence, results in ignoring, devaluing or distorting contributions arising outside European mathematical traditions." (1987:14)

Due to the legacy of colonialism, the exploitation of which was ideologically justified through a doctrine of racial superiority, the contributions of non-European civilizations were often ignored, or, as Joseph argued, even distorted, in that they were often misattributed as European, i.e. Greek, contributions, and when their contributions were so great as to resist such treatment, they were typically devalued, considered inferior or irrelevant to Western mathematical traditions.

This tendency has not only led to the devaluation of non-Western mathematical traditions, but has distorted the history of Western mathematics as well. In so far as the contributions from non-Western civilizations are ignored, there is the problem of accounting for the development of mathematics purely within the Western cultural framework.

This has led, as Sabetai Unguru has argued, toward a tendency to read more advanced mathematical concepts into the relatively simplistic geometrical formulations of Greek mathematicians such as Euclid, despite the fact that the Greeks lacked not only mathematic notation, but even the place-value system of enumeration, without which advanced mathematical calculation is impossible. Such ethnocentric revisionist history resulted in the attribution of more advanced algebraic concepts, which were actually introduced to Europe over a millennium later by the Arabs, to the Greeks.

And while the contributions of the Greeks to mathematics was quite significant, the tendency of some math historians to jump from the Greeks to renaissance Europe results not only in an ethnocentric history, but an inadequate history as well, one which fails to take into account the full history of the development of modern mathematics, which is by no means a purely European development.

4

Indic Mathematics: India and the Scientific Revolution

"The study of mathematics in the West has long been characterized by a certain ethnocentric bias, a bias which most often manifests not in explicit racism, but in a tendency toward undermining or eliding the real contributions made by non-Western civilizations. The debt owed by the West to other civilizations, and to India in particular, go back to the earliest epoch of the "Western" scientific tradition, the age of the classical Greeks, and continued up until the dawn of the modern era, the renaissance, when Europe was awakening from its dark ages."

Dr. Gray goes on to list some of the most important developments in the history of mathematics that took place in India, summarizing the contributions of luminaries such as Aryabhatta, Brahmagupta, Mahavira, Bhaskara and Madhava. He concludes by asserting that "the role played by India in the development (of the scientific revolution in Europe) is no mere footnote, easily and inconsequentially swept under the rug of Eurocentric bias. To do so is to distort history, and to deny India one of its greatest contributions to world civilization." -Indic Mathematics

Math and Ethnocentrism

The study of mathematics in the West has long been characterized by a certain ethnocentric bias, a bias which most often manifests not in explicit racism, but in a tendency toward undermining or eliding the real contributions made by non-Western civilizations. The debt owed by the West to other civilizations, and to India in particular, go back to the earliest epoch of the "Western" scientific tradition, the age of the classical Greeks, and continued up until the dawn of the modern era, the renaissance, when Europe was awakening from its dark ages. This awakening was in part made possible by the rediscovery of mathematics and other sciences and technologies through the medium of the Arabs, who transmitted to Europe both their own lost heritage as well as the advanced mathematical traditions formulated in India.

George Ghevarughese Joseph, in an important article entitled "Foundations of Eurocentrism in Mathematics," argued that "the standard treatment of the history of non-European mathematics is a product of historiographical bias (conscious or otherwise) in the selection and interpretation of facts, which, as a consequence, results in ignoring, devaluing or distorting contributions arising outside European mathematical traditions." (1987:14)

Due to the legacy of colonialism, the exploitation of which was ideologically justified through a doctrine of racial superiority, the contributions of non-European civilizations were often ignored, or, as Joseph argued, even distorted, in that they were often misattributed as European, i.e. Greek, contributions, and when their contributions were so great as to resist such treatment, they were typically devalued, considered inferior or irrelevant to Western mathematical traditions.

This tendency has not only led to the devaluation of non-Western mathematical traditions, but has distorted the history of Western mathematics as well. In so far as the contributions from non-Western civilizations are ignored, there is the problem of accounting for the development of mathematics purely within the Western cultural framework. This has led, as Sabetai Unguru has argued, toward a tendency to read more advanced mathematical concepts into the relatively simplistic geometrical formulations of Greek mathematicians such as Euclid, despite the fact that the Greeks lacked not only mathematic notation, but even the place-value system of enumeration, without which advanced mathematical calculation is impossible. Such ethnocentric revisionist history resulted in the attribution of more advanced algebraic concepts, which were actually introduced to Europe over a millennium later by the Arabs, to the Greeks. And while the contributions of the Greeks to mathematics was quite significant, the tendency of some math historians to jump from the Greeks to renaissance Europe results not only in an ethnocentric history, but an inadequate history as well, one which fails to take into account the full history of the development of modern mathematics, which is by no means a purely European development.

Vedic Altars and the "Pythagorean Theorem"

A perfect example of this sort of misattribution involves the so-called Pythagorean theorem, the well-known theorem which was attributed to Pythagoras who lived around 500 BCE, but which was first proven in Greek sources in Euclid's Geometry, written centuries later. Despite the scarcity of evidence backing this attribution, it is not often questioned, perhaps due to the mantra-like frequency with which it is repeated. However, Seidenberg, in his 1978 article, shows that the thesis that Greece was the origin of geometric algebra was incorrect, "for geometric algebra existed in India before the classical period in Greece." (1978:323)

It is now generally understood that the so-called "Pythagorean theorem" was understood in ancient India, and was in fact proved in Baudhayana's Shulva Sutra, a text dated to circa 600 BCE. (1978:323).

Knowledge of mathematics, and geometry in particular, was necessary for the precise construction of the complex Vedic altars, and mathematics was thus one of the topics covered in the brahmanas. This knowledge was further elaborated in the kalpa sutras, which gave more detailed instructions concerning Vedic ritual. Several of these treat the topic of altar construction. The oldest and most complete of these is the previously mentioned Shulva Sutra of Baudhaayana. As this text was composed about a century before Pythagoras, the theory that the Greeks were the source of Geometric algebra is untenable, while the hypothesis that India was have been a source for Greek geometry, transmitted via the Persians who traded both with the Greeks and the Indians, looks increasingly plausible. On the other hand, it is quite possible that both the Greeks and the Indians developed geometry. Seidenberg has argued, in fact, that both seem to have developed geometry out of the practical problems involving their construction of elaborate sacrificial altars. (See Seidenberg 1962 and 1983)

Zero and the Place Value System

Far more important to the development of modern mathematics than either Greek or Indian geometry was the development of the place value system of enumeration, the base ten system of calculation which uses nine numerals and zero to represent numbers ranging from the most minuscule decimal to the most inconceivably large power of ten. This system of enumeration was not developed by the Greeks, whose largest unit of enumeration was the myriad (10,000) or in China, where 10,000 was also the largest unit of enumeration until recent times. Nor was it developed by the Arabs, despite the fact that this numeral system is commonly called the Arabic numerals in Europe, where this system was first introduced by the Arabs in the thirteenth century.

Rather, this system was invented in India, where it evidently was of quite ancient origin. The Yajurveda Samhitaa, one of the Vedic texts predating Euclid and the Greek mathematicians by at least a millennium, lists names for each of the units of ten up to 10 to the twelfth power (paraardha). (Subbarayappa 1970:49)

Later Buddhist and Jain authors extended this list as high as the fifty-third power, far exceeding their Greek contemporaries, who lacking a system of enumeration were unable to develop abstract mathematical concepts.

The place value system of enumeration is in fact built into the Sanskrit language, where each power of ten is given a distinct name. Not only are the units ten, hundred and thousand (daza, zata, sahasra) named as in English, but also ten thousand, hundred thousand, ten million, hundred million (ayuta, laksa, koti, vyarbuda), and so forth up to the fifty-third power, providing distinct names where English makes use of auxiliary bases such as thousand, million, etc. Ifrah has commented that; by giving each power of ten an individual name, the Sanskrit system gave no special importance to any number. Thus the Sanskrit system is obviously superior to that of the Arabs (for whom the thousand was the limit), or the Greeks and Chinese (whose limit was ten thousand) and even to our own system (where the names thousand, million etc. continue to act as auxiliary bases). Instead of naming the numbers in groups of three, four or eight orders of units, the Indians, from a very early date, expressed them taking the powers of ten and the names of the first nine units individually. In other words, to express a given number, one only had to place the name indicating the order of units between the name of the order of units immediately immediately below it and the one immediately above it. That is exactly what is required in order to gain a precise idea of the place-value system, the rule being presented in a natural way and thus appearing self-explanatory. To put it plainly, the Sanskrit numeral system contained the very key to the discovery of the place-value system. (2000:429)

As Ifrah has shown at length, there is little doubt that our place-value numeral system developed in India (2000:399-409), and this system is embedded in the Sanskrit language, several aspects of which make it a very logical language, well suited to scientific and mathematical reasoning. Nor did this system exhaust Indian ingenuity; as van Nooten has shown, Pingala, who lived circa the first century BCE, developed a system of binary enumeration convertible to decimal numerals, described in his Chandahzaastra. His system is quite similar to that of Leibniz, who lived roughly fourteen hundred years later. (See Van Nooten)

India is also the locus of another closely related an equally important mathematical discovery, the numeral zero. The oldest known text to use zero is a Jain text entitled the Lokavibhaaga, which has been definitely dated to Monday 25 August 458 CE. (Ifrah 2000:417-1 9)

This concept, combined by the place-value system of enumeration, became the basis for a classical era renaissance in Indian mathematics.

The Indian numeral system and its place value, decimal system of enumeration came to the attention of the Arabs in the seventh or eighth century, and served as the basis for the well known advancement in Arab mathematics, represented by figures such as al-Khwarizmi. It reached Europe in the twelfth century when Adelard of Bath translated al-Khwarizmi's works into Latin. (Subbarayappa 1970:49)

But the Europeans were at first resistant to this system, being attached to the far less logical roman numeral system, but their eventual adoption of this system led to the scientific revolution that began to sweep Europe beginning in the thirteenth century.

Luminaries of Classical Indian Mathematics

Aryabhatta

The world did not have to wait for the Europeans to awake from their long intellectual slumber to see the development of advanced mathematical techniques. India achieved its own scientific renaissance of sorts during its classical era, beginning roughly one thousand years before the European Renaissance. Probably the most celebrated Indian mathematicians belonging to this period was Aryabhatta, who was born in 476 CE.

In 499, when he was only 23 years old, Aryabhatta wrote his Aaryabhattiya, a text covering both astronomy and mathematics. With regard to the former, the text is notable for its for its awareness of the relativity of motion. (See Kak p. 16)

This awareness led to the astonishing suggestion that it is the Earth that rotates the Sun. He argued for the diurnal rotation of the earth, as an alternate theory to the rotation of the fixed stars and sun around the earth (Pingree 1981:18).

He made this suggestion approximately one thousand years before Copernicus, evidently independently, reached the same conclusion.

With regard to mathematics, one of Aryabhatta's greatest contributions was the calculation of sine tables, which no doubt was of great use for his astronomical calculations. In developing a way to calculate the sine of curves, rather than the cruder method of calculating chords devised by the Greeks, he thus went beyond geometry and contributed to the development of trigonometry, a development which did not occur in Europe until roughly one thousand years later, when the Europeans translated Indian influenced Arab mathematical texts.

Aryabhatta's mathematics was far ranging, as the topics he covered include geometry, algebra, trigonometry. He also developed methods of solving quadratic and indeterminate equations using fractions. He calculated pi to four decimal places, i.e., 3.1416. (Pingree 1981:57) In addition, Aryabhatta "invented a unique method of recording numbers which required perfect understanding of zero and the place-value system." (Ifrah 2000:419)

Given the astounding range of advanced mathematical concepts and techniques covered in this fifth century text, it should be of no surprise that it became extremely well known in India, judging by the large numbers of commentaries written upon it. It was studied by the Arabs in the eighth century following their conquest of Sind, and translated into Arabic, whence it influenced the development of both Arabic and European mathematical traditions.

Brahmagupta

Born in 598 CE in Rajasthan in Western India, Brahmagupta founded an influential school of mathematics which rivalled Aryabhatta's. His best known work is the Brahmasphuta Siddhanta, written in 628 CE, in which he developed a solution for a certain type of second order indeterminate equation. This text was translated into Arabic in the eighth century, and became very influential in Arab mathematics. (See Kak p. 16)

Mahavira

Mahavira was a Jain mathematician who lived in the ninth century, who wrote on a wide range of mathematical topics. These include the mathematics of zero, squares, cubes, square-roots, cube-roots, and the series extending beyond these. He also wrote on plane and solid geometry, as well as problems relating to the casting of shadows. (Pingree 1981:60)

Bhaskar

Bhaskar was one of the many outstanding mathematicians hailing from South India. Born in 1114 CE in Karnataka, he composed a four-part text entitled the Siddhanta Ziromani. Included in this compilation is the Biijaganita, which became the standard algebra textbook in Sanskrit. It contains descriptions of advanced mathematical techniques involving both positive and negative integers as well as zero, irrational numbers. It treats at length the "pulverizer" (kuttakaara) method of solving indeterminate equations with continued fractions, as well as the so-called "Pell's equation (vargaprakrti) dealing with indeterminate equations of the second degree. He also wrote on the solution to numerous kinds of linear and quadratic equations, including those involving multiple unknowns, and equations involving the product of different unknowns. (Pingree 1981, p. 64)

In short, he wrote a highly sophisticated mathematical text that proceeded by several centuries the development of such techniques in Europe, although it would be better to term this a rediscovery, since much of the Renaissance advances of mathematics in Europe was based upon the discovery of Arab mathematical texts, which were in turn highly influenced by these Indian traditions.

Madhava

The Kerala region of South India was home to a very important school of mathematics. The best known member of this school Madhava (c. 1444-1545), who lived in Sangamagrama in Kerala. Primarily an astronomer, he made history in mathematics with his writings on trigonometry. He calculated the sine, cosine and arctangent of the circle, developing the world's first consistent system of trigonometry. (See Hayashi 1997:784-786) He also correctly calculated the value of p to eleven decimal places. (Pingree 1981:490)

This is by no means a complete list of influential Indian mathematicians or Indian contributions to mathematics, but rather a survey of the highlights of what is, judged by any fair, unbiased standard, an illustrious tradition, important both for its own internal elegance as well as its influence on the history of European mathematical traditions. The classical Indian mathematical renaissance was an important precursor to the European renaissance, and to ignore this fact is to fail to grasp the history of latter, a history which was truly multicultural, deriving its inspiration from a variety of cultural roots.

There are in fact, as Frits Staal has suggested in his important (1995) article, "The Sanskrit of Science", profound similarities between the social contexts of classical India and renaissance Europe. In both cases, important revolutions in scientific thought occurred in complex, hierarchical societies in which certain elite groups were granted freedom from manual labour, and thus the opportunity to dedicate themselves to intellectual pursuits. In the case of classical India, these groups included certain brahmins as well as the Buddhist and Jain monks, while in renaissance Europe they included both the monks as well as their secular derivatives, the university scholars.

Why, one might ask, did Europe take over thousand years to attain the level of abstract mathematics achieved by Indians such as Aryabhatta? The answer appears to be that Europeans were trapped in the relatively simplistic and concrete geometrical mathematics developed by the Greeks. It was not until they had, via the Arabs, received, assimilated and accepted the place-value system of enumeration developed in India that they were able to free their minds from the concrete and develop more abstract systems of thought. This development thus triggered the scientific and information technology revolutions which swept Europe and, later, the world.

The role played by India in the development is no mere footnote, easily and inconsequentially swept under the rug of Eurocentric bias. To do so is to distort history, and to deny India one of it's greatest contributions to world civilization.

Vedic mathematics is a system of mental calculation developed by Shri Bharati Krishna Tirthaji which he claimed he had based on a lost appendix of Atharvaveda, an ancient text of the Indian teachings called Veda.

It has some similarities to the Trachtenberg system in that it speeds up some arithmetic calculations. It claims to have applications to more advanced mathematics, such as calculus and linear algebra. The system was first published in the book Vedic Mathematics ISBN 8120801644 in 1965. The system has since been developed further and there have been several other books released.

Criticism

Critics have questioned whether this subject deserves the name Vedic or indeed mathematics. They point to the lack of evidence of any sutras from the Vedic period consistent with the system, the inconsistency between the topics addressed by the system (such as decimal fractions) and the known mathematics of early India, the substantial extrapolations from a few words of a sutra to complex arithmetic, and the restriction of applications to convenient cases.

They further say that such arithmetic as is speeded up by application of the sutras can be performed on a computer or calculator anyway, making their knowledge rather irrelevant in the modern world.

They are also worried that it deflects attention from genuine achievements of ancient and modern Indian mathematics and mathematicians, and that its promotion by Hindu nationalists may damage mathematics education in India. Actual Indian mathematicians, at leading institutions such as TIFR or IISc, largely share these opinions.

The system is based upon sixteen formulas and their corollaries, some of which are described below.

All from Nine and the Last from Ten

Corollary 1: Whatever the extent of its deficiency, lessen it still further to that very extent; and also set up the square of that deficiency.

For instance, in computing the square of 9 we go through the following steps:

1. The nearest power of 10 to 9 is 10. Therefore, let us take 10 as our base.
2. Since 9 is 1 less than 10, decrease it still further to 8. This is the left side of our answer.
3. On the right hand side put the square of the deficiency, that is 12. Hence the answer is 81.
4. Similarly, $8^2 = 64$, $7^2 = 49$.
5. For numbers above 10, instead of looking at the deficit we look at the surplus. For example:

   and so on.

This is based on the identities $(a + b)(a - b) = a^2 - b^2$ and $(a + b)^2 = a^2 + 2ab + b^2$.

By One More than the One Before

The proposition "by" means the operations this formula concerns are either multiplication or division. [In case of addition/subtraction proposition "to" or "from" is used.] Thus this formula is used for either multiplication or division. It turns out that it is applicable in both operations.

An interesting application of this formula is in computing squares of numbers ending in five. Consider:

$35 \times 35 = (3 \times (3 + 1)), 25 = 12,25$

The latter portion is multiplied by itself (5 by 5) and the previous portion is multiplied by one more than itself (3 by 4) resulting in the answer 1225.

This is a simple application of $(a + b)^2 = a^2 + 2ab + b^2$ when $a = 10c$ and $b = 5$, i.e.

$$(10c + 5)^2 = 100c^2 + 100c + 25 = 100c(c + 1) + 25.$$

It can also be applied in multiplications when the last digit is not 5 but the sum of the last digits is the base (10) and the previous parts are the same. Consider:

$$37 \times 33 = (3 \times 4), 7 \times 3 = 12,21$$

$$29 \times 21 = (2 \times 3), 9 \times 1 = 6,09$$

This uses $(a + b)(a - b) = a^2 - b^2$ twice combined with the previous result to produce:

$$(10c + 5 + d)(10c + 5 - d) = (10c + 5)^2 - d^2$$

$$= 100c(c + 1) + 25 - d^2$$

$$= 100c(c + 1) + (5 + d)(5 - d).$$

We illustrate this formula by its application to conversion of fractions into their equivalent decimal form. Consider fraction 1/19. Using this formula, this can be converted into a decimal form in a single step. This can be done by applying the formula for either a multiplication or division operation, thus yielding two methods.

METHOD 1: USING MULTIPLICATIONS

1/19, since 19 is not divisible by 2 or 5, the fractional result is a purely circulating decimal. (If the denominator contains only factors 2 and 5 is a purely non-circulating decimal, else it is a mixture of the two.)

So we start with the last digit

1

Multiply this by "one more", that is, 2 (this is the "key" digit from Ekadhikena)

21

Multiplying 2 by 2, followed by multiplying 4 by 2

421 → 8421

Now, multiplying 8 by 2, sixteen

68421

1 ← carry

multiplying 6 by 2 is 12 plus 1 carry gives 13

368421

1 ← carry

Continuing

7368421 → 47368421 → 947368421

1

Now we have 9 digits of the answer. There are a total of 18 digits (= denominator ? numerator) in the answer computed by complementing the lower half:

052631578

947368421

Thus the result is .052631578,947368421

METHOD 2: USING DIVISIONS

The earlier process can also be done using division instead of multiplication. We divide 1 by 2, answer is 0 with remainder 1

.0

Next 10 divided by 2 is five

.05

Next 5 divided by 2 is 2 with remainder 1

.052

next 12 (remainder,2) divided by 2 is 6

.0526

and so on.

As another example, consider 1/7, this same as 7/49 which as last digit of the denominator as 9. The previous digit is 4, by one more is 5. So we multiply (or divide) by 5, that is,

...7 ⇒ 57 ⇒ 857 ⇒ 2857 ⇒ 42857 ⇒ 142857 ⇒ .142,857 (stop after 7 ⇒ 1 digits)

3 2 4 1 2

Vertically and Crosswise

This formula applies to all cases of multiplication and is very useful in division of one large number by another large number.

Transpose and Apply

This formula complements "all from nine and the last from ten", which is useful in divisions by large numbers. This formula is useful in cases where the divisor consists of small digits. This formula can be used to derive the Horner's process of Synthetic Division.

When the Samuccaya is the Same, that Samuccaya is Zero

This formula is useful in solution of several special types of equations that can be solved visually. The word samuccaya has various meanings in different applications. For instance, it may mean a term which occurs as a common factor in all the terms concerned. A simple example is equation "12x + 3x = 4x + 5x". Since "x" occurs as a common factor in all the terms, therefore, x = 0 is a solution. Another meaning may be that samuccaya is a product of independent terms. For instance, in (x + 7)(x + 9) = (x + 3)(x + 21), the samuccaya is 7 × 9 = 3 × 21, therefore, x = 0 is a solution. Another meaning is the sum of the denominators of two fractions having the same numerical numerator, for example: 1/(2x - 1) + 1/(3x - 1) = 0 means 5x - 2 = 0.

Yet another meaning is "combination" or total. This is commonly used. For instance, if the sum of the numerators and the sum of denominators are the same then that sum is zero. Therefore,

Therefore, $4x + 16 = 0$ or $x = ?4$.

This meaning ("total") can also be applied in solving quadratic equations. The total meaning can not only imply sum but also subtraction. For instance when given $N_1D_1 = N_2/D_2$, if $N_1 + N_2 = D_1 + D_2$ (as shown earlier) then this sum is zero. Mental cross multiplication reveals that the resulting equation is quadratic (the coefficients of x_2 are different on the two sides). So, if $N_1 - D_1 = N_2 - D_2$ then that samuccaya is also zero. This yield the other root of a quadratic equation.

Yet interpretation of "total" is applied in multi-term RHS and LHS. For instance, consider

Here $D_1 + D_2 = D_3 + D_4 = 2x - 16$.

Thus $x = 8$.

There are several other cases where samuccaya can be applied with great versatility. For instance "apparently cubic" or "biquadratic" equations can be easily solved as shown below:

$(x - 3)^2 + (x - 9)^3 = 2(x - 6)^3$.

Note that $x - 3 + x - 9 = 2(x - 6)$.

Therefore $(x - 6) = 0$ or $x = 6$.

(remark by different author: Note also that:

$(6 - 3)^2 + (6 - 9)^3 = 3^2 - 3^3 = 9 - 27 = - 18$,

whereas $2(6 - 6)^3 = 0$

Thus: $x = 6$ is NOT a solution!)

(Another note. This example does work if one considers all exponents to be cubic, in which case $(6 - 3)^3 + (6 - 9)^3$ will give $27 - 27$.)

Consider

Observe: $N_1 + D_1 = N_2 + D_2 = 2x + 8$.

Therefore, x = -4.

This formula has been extended further.

If One is in Ratio, the Other One is Zero

This formula is often used to solve simultaneous simple equations which may involve big numbers. But these equations in special cases can be visually solved because of a certain ratio between the coefficients. Consider the following example:

$$6x + 7y = 8$$

$$19x + 14y = 16$$

Here the ratio of coefficients of y is same as that of the constant terms. Therefore, the "other" is zero, i.e., x = 0. Hence the solution of the equations is x = 0 and y = 8/7.

(alternatively:

19x + 14y = 16 is equivalent to:

(19/2)x +7y = 8.

Thus it is obvious that x has to be zero, no ratio needed, just div by 2!)

This formula is easily applicable to more general cases with any number of variables. For instance

$$ax + by + cz = a$$

$$bx + cy + az = b$$

$$cx + ay + bz = c$$

which yields x = 1, y = 0, z = 0.

A corollary says By addition and by subtraction. It is applicable in case of simultaneous linear equations where the x- and y-coefficients are interchanged. For instance:

$$45x - 23y = 113$$

$$23x - 45y = 91$$

By addition:

$$68x - 68\,y = 204$$

$$\Rightarrow \quad 68(x - y) = 204$$

$$\Rightarrow \quad x - y = 3.$$

By subtraction:

$$22x + 22y = 22$$

$$\Rightarrow \quad 22(x + y) = 22$$

$$\Rightarrow \quad x + y = 1.$$

5

Science and Technology in Epics

How much hi-tech and science did our forefathers really know? On the one hand, there are emphatic claims that they knew everything that modern science has taught us today and will teach us in the years to come. On the other, there are skeptics who would associate the dawn of science and technology with the West, dating back no more than a few centuries and would thus deny any knowledge to oriental ancients.

The arguments "for" fall into three categories. The first line of argument takes to our puranas and the epics. In the Ramayana, we have the Pushpak Vimana which clearly establishes the existence of helicopters and aeroplanes. What better example of a laser-guided missile that the Shakti with which Karna killed Ghatotkacha in the Mahabharata? And the instant beaming of a living body across vast distances that you see in the futuristic Star Trek was routinely practised by our devas and rishis.

Ancient Work

The next line of argument invokes ancient works like the Brihad Vimanshastra which describes different flying vehicles and methods of constructing them. Then there are interpretations of Vedic Richas to show that they contain details of the nuclear fusion that powers the Sun. There is Vedic mathematics which is claimed to contain advanced mathematical concepts and methods. It has often been said that the philosophers of the Vedic and later times knew about quantum mechanics, gauge theory, unified field theory and so on which the theoretical physicists of today are grappling with.

The third argument brings forth examples of old tests like the Vedanga Jyotisha attributed to Lagadha which has astronomical discussions relating to the solar system, the Shalbha Sutra containing geometric ideas, and the Charaka and Sushruta Samhitas which contain ideas on medicine and surgery. Surely these demonstrate that some science was known and practised in the ancient times.

You don't need a lawyer to tell you that a case is weakened if some of its arguments are demonstrably refutable or are basically ill-formulated. The first two lines of arguments fall into in this category.

Thus to convince a scientist it is not sufficient to refer to descriptions of the epics and puranas. Absorbing and imaginative though they are they are far cry from what constitutes a technical manual.

Imagine that thousands of years later our descendents recover a cassette of Star Trek from an archaeological excavation and manage to view it. Are they justified in concluding on the basis of what they see that we ourselves were in possession of the kind of technology shown there? Therefore, we can credit our ancients with giant leaps of imagination to have produced those works...but no more.

For, if we treat these descriptions as of reality then we run into trouble. In our epics and puranas we see no description of electrical power that is the first significant step of high technology. Why do we find no reference to even the capital cities of Dasharatha or Dhritrashtra enjoying this amenity? Remember that of the four fundamental forces of nature the electrical one has the largest variety of manifestations in nature and should be the first one to be extensively studied, understood and applied.

I had studied the Brihad Vimanshastra and failed to find in it a technical description of aeronautics. For example, a shastra could contain a basic framework of rules governing flying objects which then just translated into technical reality. Basic principles of flying like what is today known as Bernoulli's theorem, the idea of life and streamlines should be described.

Even assuming that the ancients somehow bypassed these basics through some yogic insight no available to modern scientist, the actual description of construction is also vague and certainly cannot be used for making a real aircraft.

Vedic Mathematics

Much is made of Vedic mathematics. Leaving aside the claims and counterclaims about its adjective "Vedic" let us look at the mathematics part. If it is looked upon as a system of carrying out quick numerical manipulation it has its merits. But as any mathematician of distinction will tell you, that does not constitute advanced mathematics.

I love Sanskrit as a language and have enjoyed reading its classics. But its very verstality proves a stumbling block in interpreting ancient descriptions and with Vedic Sanskrit (which I confess, I cannot readily follow) the problem of interpretation is even more tricky.

It is easy for the interpreter to read into a statement the meaning he or she wishes to read. This probably explains the post facto nature of interpretations: only after modern science makes a discovery do claims appear that it is contained in such an such ancient tome. In many cases, the antiquity of the statement may also be questionable in the sense that later writers inserted their own ideas into the old works. Simply because a statement occurs in Sanskrit does not mean it is ancient!

Arguments in the third category, however, are worth taking note of and deserve further scholarly studies. For they are clearly documented, can be interpreted without "personal" ambiguities and do tell us that our ancient civilizations did have thinkers and experimenters of high quality.

Thus Lagadha should be studied with as much care as his (much) later descendents like Aryabhatta (fifth century A.D.) and Bhaskara (12th century A.D.). The works ascribed to Sushruta and Charaka, unlike speculations described above, are experimentally testable. The legacy of yoga as a discipline of mind and body is indisputable.

Old Text

Our old texts are often examined for astronomical events like eclipses, comets and conjunctions. Another useful and hitherto unexplored line of investigation is supernovae. A supernova is an exploding star: an event which can be spectacular if it takes place in our galaxy.

The Chinese and Japanese astronomers have recorded an event that was seen on July 4, 1054... when the exploding star brightened so much as to be visible during daylight. Surely a spectacular even like that would not have gone unnoticed on our subcontinent. With the glorious tradition from Aryabhatta to Bhaskara continuing at that time it is worth looking for records wherever they may be. A finding will be of enormous importance to astronomical archives.

Let me end with an example from social sciences. In his book Labour Management the late R.G. Rajwade has an appendix on Shukra Neeti written mostly in Gupta period.

Shukra had laid down precise rules for the employee and the employer, spelling out details on the type of wages, periods of payment, grades of employment, fair wages, industrial disputes, leave rules, sickness benefits, pension, provident fund and family allowance, bonus etc.

For example, Shukra recommends that an employee who has served for forty years should be paid a pension half his wages throughout his life. For bonus he recommends one eighth of the employee's salary.

I mention this example to illustrate the fact that our ancients could be precise when they chose to be so. Let us stress examples of precision which is the hallmark of science rather than ambiguous personalised interpretations of what they said. Only then will we do justice to the achievements of our ancients.

6

World of Mathematics: The Indian Contribution

The Contribution to Mathematics by India can be divided into ten categories:

1. Zero and the place-value notation for numbers
2. Vedic Mathematics and arithmetical operations
3. Geometry of the Sulba Sutras
4. Jaina contribution to Fundamentals of numbers
5. The anonymous Bakshali manuscript
6. Astronomy
7. Classical contribution to Indeterminate Equations and Algebra
8. Indian Trigonometry
9. Kerala contribution to Infinite Series and Calculus.
10. Modern Contribution: Srinivasa Ramanujan onwards

Section 1: Zero and the Place-Value Notation

The number zero is the subtle gift of the Hindus of antiquity to mankind. The concept itself was one of the most significant inventions in the ascent of Man for the growth of culture and civilization. To it must be credited the enormous usefulness of its counterpart, the place value system of expressing all numbers with just ten symbols. And to these two concepts we owe all the arithmetic and mathematics based upon them, the great ease which it has lent to all computations for two millenia and the binary system which now lies at the foundation of communicating with computers. Already in the first three centuries A.D., the Hindu ancients were using a decimal positional system, that is, a system in which numerals in different positions represent different numbers and in which one of the ten symbols used was a fully functional zero.

They called it 'Sunya'. The word and its meaning 'void' were obviously borrowed from its use in philosophical literature. Though the Babylonians used a special symbol for zero as early as the 3rd century B.C., they used it only as a place holder; they did not have the concept of zero as an actual value. It appears the Maya civilisation of South America had a zero in the first century A.D. but they did not use it in a fixed base system. The Greeks were hampered by their use of letters for the numbers. Before zero was invented, the art of reckoning remained an exclusive and highly skilled profession. It was difficult to distinguish, say, 27, 207, 270, 2007, because the latter three were all written 2 7, with a 'space' in between. The positional system is not possible in the Roman numeral system which had no expression or symbol for zero. A number, say, 101,000, would have to be written only by 101 consecutive M's. The Egyptians had no zero and never reached the idea of expressing all numbers with ten digits. The mathematical climate among the Hindus, however, was congenial for the invention of zero and for its use as the null-value in all facets of calculation, due to four factors:

1. A notation for powers of 10 upto the power 17 was already in use even from vedic times. Single words have been used to denote the powers of the number 10. The numbers one, ten, hundred, thousand, ten thousand, ... are given by the sequence of words in the list:

eka, dasa, Sata, sahasra, ayuta, laksha, prayuta, koTi, arbuda, abja, kharva, nikharva, mahA-padma, Sankha, jaladhi, antya, mahASankha, parArdha. Thus the decimal system was in the culture even in the early part of the first millenium B.C.. The Yajurveda, in its description of rituals and the mantras employed therein, the Mahabharata and the Ramayanaa in their descriptions of statistics and measurements, used all these words, with total abandon.

2. Counting boards with columns representing units and tens were in use from very ancient times. The numberless content of an empty column in course of time was symbolized to be 'nothing'.
3. The thriving activity in astrology, astronomy, navigation and business during the first few centuries A.D. naturally looked forward to a superior numerical system that lent itself to complicated calculations.
4. Distinct symbols for the numbers 1 to 9 already existed and the counting system used the base 10 in all its secular, religious and ritual activities. Compare this with the Babylonian numeration which had only three figures, one for 1, one for 10, and one for 100, so that a number, say, 999, would require 27 symbols, namely, nine of each of the symbols.

Of these, the first and fourth factors are probably unique to Hindu culture and contributed most to the thought process that led to the decimal place value notation as well as zero having a value. When exactly the invention of this most modest of all numerals took place, we do not know. The first time it reached Europe was during the Moorish invasion of Spain around 700 A.D. Later, when massive Latin translations of books from Baghdad took place around the close of the first millenium A.D. the concept was found in an arithmetic book dated 820 A.D., by Muhammad Ibn

Musa al-Khouarizmi who explained the whole system in great detail. The Arabs themselves had no number system of their own. It was the Hindu system he was explaining. That is why the numerals are called the Indo-Arabic numerals even today. It is however a misnomer to call them so; they are already found on the Rock Edits of Ashoka (256 B.C.). In spite of its being so crucial to our living, it took centuries for the western world to appreciate and incorporate this most valuable numeral, zero, in their recording of accounts or in scholarly writings. For by the time 'Zero' reached the West, the Dark Ages of the western world had begun. There are traces, however, of its knowledge in Spain in the tenth century A.D. But the final breakthrough of the introduction to the West was by Leonardo of Pisa, through his popular text Liber Abaci, 1202 A.D. The first European book (in French) that used the zero appeared in 1275.

How the Sunya of the Hindus became the Zero of the modern world is interesting. The 'Sunya' of Sanskrit became the Arabic 'sifr' which means empty space. In medieval Latin it manifested as 'ciphra' , then in middle English as 'siphre', in English as 'cypher' and in American as 'cipher'. In the middle ages, the word 'ciphra' evolved to stand for the whole system. In the wake of this general meaning, the Latin 'zephirum' came to be used to denote the Sunya. And that entered English finally as 'zero'. In medieval Europe some countries banned the positional number system, along with zero, brought by the Arabs whom they considered as heathens. So they took Sunya to be a creation of the devil! As a result 'ciphra' came to mean a secret code. From this came 'deciphering', the resolution of a code!

Section 2. Vedic Mathematics and Arithmetical Operations

Vedic Mathematics provides an original and refreshing approach to subjects which are usually dismissed as mechanical and tedious. Bharati Krishna Tirtha who published his reconstruction of Vedic Mathematics in 1965, maintains that there are 16 aphorisms and 13 secondary aphorisms which forms his base of the so-called Vedic Mathematics. Though the origins of Vedic Mathematics have not yet been historically established, if nothing else, it provides tremendous insights into the place-value system of numbers without which it would not work. It is amazing that Vedic Mathematics does not require of cramming of multiplication tables beyond 5 x 5. One can improvise all the necessary multiplication tables for oneself and with the aid of the relevant Vedic formulae get the required products very easily, speedily, and correctly, almost immediately. The formulae can be used to evaluate determinants, solve simultaneous linear equations, evaluate logarithms and exponentials. Vedic Mathematics recognises that any algebraic polynomial may be expressed in terms of a positional notation without specifying the base. The same algorithmic scheme as applied to arithmetical operations will easily apply to algebraic problems. And this brings it to the Modern Algebra of Polynomials. It is difficult, in a historical introduction like this to get into the details of Vedic Mathematics. Suffice it to say that with today's over-dependence on calculators for even simple arithmetical computations, the Vedic methods have great pedagogical value and, through their revival, the skills of mental arithmetic may not be lost for posterity.

Section 3. Geometry of the Sulba Sutras

Hailing from the times of the Vedas, the ritual literature which gave directions for constructing sacrificial fires at different times of the year dealt with the their measurement and construction in a systematic and logical way, thus giving rise to the Sulba Sutras. The construction of altars (vedi) and the location of sacrificial fires had to conform to clearly laid down instructions about their shapes and areas in order that they may be effective instruments of sacrifice. The

Sulba Sutras provide such instructions for two types of ritual - one for worship at home and one for communal worship. The instructions were mainly for the benefit of craftsmen laying out and building the altars. Bodhayana, Apastamba and Katyayana who have recorded these Sulbasutras were not only priests in the conventional sense but must have been craftsmen themselves. The earliest of them, The Bodhayana Sutras , in three chapters, (800 - 600 B.C.) contains a general statement of the Pythagorean theorem, an approximation procedure for obtaining the square root of two correct to five decimal places and a number of geometric constructions. These latter include an approximate squaring the circle, and construction of rectilinear shapes whose area is equal to the sum or difference of areas of other shapes. The Bodhayana version of the Pythagorean theorem sates as follows:

The rope which is stretched across the diagonal of a square produces an area double the size of the original square.

It is therefore in the fitness of things that the Pythagorean theorem of Mathematics may be renamed as the Bodhayana theorem.! The other sutras are two centuries later but all of them are prior to Panini of the fourth century B.C. The geometry arising from these sutras give several geometric constructions. Some of these are:

1. To merge two equal or unequal squares to obtain a third square.
2. To transform a rectangle into a square of equal area
3. Squaring a circle and circling a square (approximately)

A remarkable achievement was the discovery of a procedure for evaluating square roots to a high degree of approximation. The square root of two is obtained as

1.4142156 ...

the true value being 1.414213... . The fact that such procedures were used successfully by the Sulbasutra geometers to operations with other irrational numbers, is clear proof for negating the western-held opinion that the Sulba sutra geometers borrowed their methods from the Babylonians. The latter's calculation of the square root of two is an isolated instance and further they used the sexagesimal notation for numbers. The achievement of geometrical constructs in Indian mathematics reached its peak later when they arrived at the construction of Sriyantra, which is a complicated diagram, consisting of nine interwoven isosceles triangles, four pointing upwards and four pointing downwards. The triangles are arranged in such a way that they produce 43 subsidiary triangles, at the centre of the smallest of which there is a big dot called the binḍu. The difficult problem is to construct the diagram in such a way that all the intersections are correct and the vertices of the largest triangles fall on the circumference of the enclosing circle. In all cases the base angles of the largest triangles is about 51.5 degrees. This has connections with the two most famous irrational numbers of Mathematics. The quantity , called the golden ratio, has remarkable mathematical properties and is almost a semi-mystical number.

Section 4. Jaina Contribution to Fundamentals of Numbers

By the time of the Jains, the role of rituals in the development of mathematics declined and mathematics began to be pursued also for its own sake. The Jains had a fascination for large numbers. Their definitions of the various types of infinities they comprehended are sophisticated, though lacking in mathematical precision. But it must be said to their credit that they were the first, in the chronology of scientific thinking, to have recognised that all infinities were not the same or equal. In fact this idea was established in the mathematical world only in the latter half of the nineteenth century when Cantor initiated his theory of sets.

The Jains were also aware of the theory of indices, though they did not have the modern notation or any convenient notation for the same. Calling the successive squares and square roots as the first, the second, etc. they make the following statement: The first square root multiplied by the second square root is the cube of the second square root. In modern notation this is nothing but the identity in the theory of indices:

$$a^{1/2} \times a^{1/4} = (a^{1/4})^3$$

They have several such rules for working with powers of a number. They also seem to have had an idea of the logarithm of a number though they don't seem to have put them to practical use in calculation. Another favourite topic with them was the study of permutations and combinations. They had also a great interest in sequences and progressions developed out of their philosophical theory of cosmological structures. A Jain canonical text entitled Triloka prajnApati has a very detailed treatment of arithmetic progressions.

Section 5: The Anonymous Bakshali Manuscript

This manuscript was discovered in 1881 A.D. near a village called Bakshali. It is written in an old form of Sanskrit on seventy leaves of birch bark. It is probably a copy of a manuscript composed in the early centuries A.D.

It is a handbook of rules and illustrative examples together with solutions, all mainly on arithmetic and algebra.

Fractions, Square roots, Profit and Loss, Interest, Rule of Three, Approximation to surds, Simple equations as well as Simultaneous equations, Quadratic equations, Arithmetic and Geometric Progressions -- all these are covered. Very unusually in the entire history of Ancient Indian mathematics, the subject matter is organised in a sequence: first, a rule or a sutra; then a relevant example in word form; the same in notational form; then the solution and finally the demonstration or the proof.

Here for the first time in the history of world mathematics, the Rule of Three is stated in its abstract form. It was from here that the rule was taken to Europe by the Arabs and it was then known as the Golden Rule. It became very popular in Europe after the Renaissance. The rule goes as follows:

If p yields f what will i yield?

Here p stands for pramAna, f for phala and i for icchA.

Here p and i are of the same denomination and

f is of a different denomination.

Write p, f, i in that order. Multiply the middle quantity by the last quantity and divide by the first. The result is fi / p.

The first appearance of indeterminate equations is in the Bakshali mss. This marks the beginning of the continued work on indeterminate equations in India. Such an interest, though originally generated from the demands made by astronomical calculations, was also pursued for its own sake, sometimes even for recreational purposes. The ease with which the place value system is worked in the Bakshali mss. suggests that the system predates the mss. by probably a few centuries.

Section 6. Astronomy

The contribution to Astronomy by ancient Indians is so great that it does not befit it to include it as one of the contributions of Indian Mathematics to the rest of the world. It needs a separate forum all for itself. We shall leave it right there except to add a note on the ancient contribution to the problem of telling time at night by a look at the stars on the meridian. This part is usually not emphasized.

The ancients of India have passed on to us 27 mathematical formulae coded in the Sanskrit language, but not very difficult to remember. In fact, very possibly it has mostly come down to us by oral transmission from generation to generation. For instance, the formula

krittika simhe kaya

says that if you see the asterism krittikA (Pleides, in modern terminology) on the meridian, that is the time the Leo (= simha) constellation (of the zodiac) has risen above the horizon by an amount indicated by the word: kAyA. This latter word interpreted in katapayA sankhyA, which is the notation used by astronomers, astrologers and mathematicians to represent numbers, means in this context that the amount of Leo above the (Eastern) horizon is 27 minutes of time.

From this and the known position of the Sun on the date in question, one mentally calculates the time of night. On November 7 for example, the Sun is in the middle of Scorpio. So if you see krittikA on the meridian it means Leo has risen 27 minutes before and this means the Sun is behind by

93m (remaining portion of simha)

+ 2h (full portion of kanyA)

+ 2h (full portion of tulA)

+ 60m (half portion of vrischika)

that is 6 hours 33 minutes. In other words it is 6h 33m before sunrise. So it is 11-27 P.M.

Suffice it to say these beautiful formulae constitute an intellectual marvel put to the most mundane use. Never perhaps was so much achieved with so little so early in the Ascent of Man.

Section 7. Classical Contribution to Indeterminate Equations and Algebra

The apex of Mathematical achievement of ancient India occurred during the so-called classical period of Indian Mathematics. The great names are: Aryabhatta I (b.476 A.D.); Brahmagupta (b.598 A.D.); Bhaskara I (circa 620 A.D.); Mahavira (circa 850 A.D.); Sridhara (circa 900 A.D.); Bhaskara II (b.1114 A.D.); Nilakantha Somayaji (1445 - 1545 A.D.).

Aryabhatta wrote the famous Aryabhatiyam which is an exhaustive exposition of Astronomy. In addition he gave a unique method of representing large numbers by word forms. He systematized all the knowledge of astronomy and mathematics prior to him. The first one in Indian mathematics to give the formula for the area of a triangle was Aryabhatta. Several results on Triangles and circles and on Progressions, algorithm for finding cube roots, approximation of , all these give him a unique position in the development of mathematics. Aryabhatta ushered in a Renaissance in Indian Mathematics and Astronomy, that resulted in a remarkable flourishing of science and technology in India. Aryabhatta, for the first time, secularised mathematics and astronomy in India and established these as intellectual disciplines in their own right. Excellent commentators followed Aryabhatta and to them are due several modifications and applications, explanation of subtle points, and finally, the proofs of results embodied in the sutras of the Master.

Bhaskara I takes a large share of the credit of explaining the too brief and aphoristic statements of Aryabhatta. On the important topic of indeterminate equations the Kuttaka method was introduced by Aryabhatta and elucidated by Bhaskara I.

Brahmagupta is generally known as the Indian mathematician par excellence. His monumental work

Brahma SiddhAnta has 24 chapters of which the latter 14 contain original results on arithmetic algebra and on astronomical instruments. The 12th chapter is on mensuration. The 18th chapter is on Kuttaka. Among his famous results are those on rational right-angled triangles, and cyclic quadrilaterals. He is the earliest one, in the history of world mathematics, to have discussed cyclic quadrilaterals. There is every reason for us to name cyclic quadrilaterals as Brahmagupta Quadrilaterals. It was partly through a translation of Brahma-siddhAnta that the Arabs became aware of Indian astronomy and mathematics.

Bhaskara II's famous work SiddhAnta Siromani has four parts of which the first two are Mathematics and the latter two are astronomy. The first part, LilAvati is an extremely popular text dealing with arithmetic, algebra, geometry and mensuration. The second part, BIjaganitam is a treatise on Advanced Algebra. It contains problems on determining unknown quantities, evaluating surds and solving simple and quadratic equations.

The sheer ingenuity and versatility of Brahmagupta's approach to indeterminate equations of the second degree of the form

$$N x^2 + 1 = y^2$$

is the climax of Indian work in this area. Bhaskara II's cakravAla method to solve such equations is world-famous. By using this powerful method he solved, as one example, the above equation with N = 61 and gave the least integral solution as

$$x = 226153980 \text{ and } y = 1766319049.$$

The famous French mathematician, Fermat, in 1657 A.D. proposed this equation with N = 61 for solution as a challenge to his contemporaries. None of them succeeded in solving the equation in integers. It was not until 1767 A.D. that the western world through Euler, by Lagrange's method of continued fractions, had a complete solution to such types of equations, wrongly called Pell's equation by Euler. But the very same equation, though coincidentally, was completely solved by Bhaskara II five hundred years earlier.

The problem of determining integer solutions of such equations is called Diophantine Analysis after the Greek Mathematician Diophantus (3rd cen. A.D.). As soon as one finds a non-trivial solution (that is, other than the obvious solution $x = 0$, $y = 1$) an infinite number of new solutions can be found by repeated application of the Principle of Compositions, known as Brahmagupta's Bhavana Principle. It is Bhaskara's cakravAla method that makes the decisive step in determining a non-trivial solution. Under these circumstances it is appropriate to designate these equations as the Brahmagupta-Bhaskara equations.

Before we leave this topic it is important to mention Srinivasa Ramanujan, the 20th century genius, who revelled in such problems - namely, to determine the possible cases in which a number can be broken up into two or more equal sums of like or unlike powers or more generally to solve intermediate problems in rational numbers.

Bhaskara II introduces also the notion of instantaneous motion of planets. He clearly distinguishes between sthUla gati (average velocity) and sUkshma gati (accurate velocity) in terms of differentials. He also gave formulae for the surface area of a sphere and its volume, and volume of the frustum of a pyramid. Suffice it to say that his work on fundamental operations, his rules of three, five, seven, nine and eleven, his work on permutations and combinations and his handling of zero all speak of a maturity, a culmination of five hundred years of mathematical progress.

Section 8. Indian Trigonometry

Though Trigonometry goes back to the Greek period, the character of the subject started to resemble modern form only after the time of Aryabhatta. From here it went to Europe through the Arabs and went into several modifications to reach its present form.

In ancient times Trigonometry was considered a part of astronomy. Three functions were introduced: jya, kojya and ukramajya. The first one is r sin where r is the radius of the circle and is the angle subtended at the centre. The second one is r cos and the third one is r (1 - cos). By taking the radius of the circle to be 1, we get the modern trigonometric functions. Various relationships between the sine of an arc and its integral and fractional multiples were used to construct sine tables for different arcs lying between 0 and 90.

Section 9: Kerala Contribution to Infinite Series and Calculus

Kerala mathematicians produced rules for second order interpolation to calculate intermediate sine values. The Kerala mathematician Madhava may have discovered the sine and cosine series about three hundred years before Newton. In this sense we may consider Madhava to have been the founder of mathematical analysis. Madhava (circa 1340 - 1425 A.D.) was the first to take the decisive step from the finite procedures of ancient Indian mathematics to treat their limit-passage to infinity. His contributions include infinite-series expansions of circular and trigonometric functions and finite-series approximations. His power series for and for sine and cosine functions is referred to reverentially by later writers.

Many later discoveries in European mathematics (for example, the Gregory series for the inverse tangent) were anticipated by Kerala astronomer-mathematicians. Nilakantha was mainly an astronomer but his Aryabhatiya bhashya and tantra-sangraha contain work on infinite-series expansions, problems of algebra and spherical geometry.

Section 10. Modern Contribution: Srinivasa Ramanujan Onwards

The second decade of the 20th century compulsorily turned the attention of the mathematical world to India and the Number Theory genius, Srinivasa Ramanujan. The ideas and innovative genius of Ramanujan have not been surpassed ever before or even 100 years after him. His birth, his super-activity in Madras and Cambridge, his glorious rise to international fame and unfortunate death—all happened almost in a flash. But ever since, India has remained on the mathematical map of the world more and more prominently. The 20th century saw the growth of Mathematics in India in the same style, though not in the same speed, the western world has been taking the race for the quest of mathematical knowledge. May we hope that the new millenium will see not only further speed in the acquisition and spread of mathematical knowledge by Indian-born mathematicians, but also further depth in the intensity and relevance of that knowledge, true to its traditions which are at least three millenia old .

Sacred Geometry

Details on this page is an extraction of some of the more interesting facts that have made an impression on me..

Pi

The circumference of a circle is (pi) times the diameter of the circle, where pi is approximately 3 1/7 (or 22/7 - these sacred numbers turn up frequently in Hindu texts and Rosicrucian literature), or 3.14, or about square root of 10. The value of pi is an irrational number, and pi to 39 decimal places is sufficient accuracy to calculate the circumference of the universe to within the radius of the hydrogen atom! Here it is to 100 decimal places (see here for 5000 decimal places):

3.14159265358979323846264338327950288419716939937510582097494459230781640628620899862803482534211706 79

ga - pa -bha - ya - ma -dha - tha -ta | gna - da - sa-da - cha - ssa - dha - ga

3- 1 - 4- 1 - 5 - 9 - 2 - 6 | 5 - 3 - 5 - 8 - 9 - 7 - 9 - 3

kha - la -ja - va - ta - kha - ta - va | ga - la - ha - la - ra - ssa - dha - ra

2 - 3 - 8 - 4 - 6 - 2 - 6 - 4 | 3 - 3 - 8 - 3 - 2 - 7 - 9 - 2 (? 5 - wrong in the 32nd digit?)

In Vedic Mathematics, Swami Sri Bharati Krsna Tirthaji Majaraja shows the vedic numeric code, and how a hymn in praise of the Lord Shri Krishna, the Lord Shri Shankara, is also the value of /10 to 32 decimal places (and apparently a key to calculate it to any number of decimal place, which I do not know) - the Sanskrit is shown above, which you can decode in to syllables and the decoding key below.

1 ka, ta, pa, ya

2 kha, tha, pha, ra

3 ga, da, ba, la

4 gha, dha, bha, va

5 gna, na, ma, sa

6 sa, ta, sha

7 cha, tha, ssa

8 ja, da, ha

9 jha, dha

0 ksa

The original sanskrit verse can be found in the appendix of Vedic Mathematics by Jagadguru Swami Sri Bharati Krsna Tirthaji Maharaja.

$$= \frac{1}{3}\left[\frac{1}{36} - \frac{1}{14}\right]$$

$$= 0.014550$$

So the displacement of the object from time t = 2 to t = 3 is 0.015 units.

PROBLEM:

Find:

$$\int_0^1 \sqrt{x^2 + 1dx}$$

We put $u = x^2 + 1$.

So du = 2x dx

But the question does not have "x dx" so we cannot solve it using any of the integration methods used above. (It can be done using trigonometric substitution, however).

We need to use numerical approaches. When software like Scientific Notebook, Matlab or Mathcad perform definite integrals, it uses numerical methods.

We use two methods:

- Trapezoidal rule
- Simpson's Rule

We meet these methods in the next 2 sections.

The Trapezoidal Rule

PROBLEM:

Find:

$$\int_0^1 \sqrt{x^2 + 1dx}$$

We put $u = x^2 + 1$ and as $x = 0 \rightarrow 1$, $u = 1 \rightarrow 2$

So du = 2x dx

But the question does not contain an x dx term so we cannot solve it using any of the normal integration methods.

We need to use numerical approaches. When software like Mathcad or graphics calculators perform definite integrals, they use numerical methods.

We can use one of two methods:

- Trapezoidal rule
- Simpson's Rule

The Trapezoidal Rule

We saw the basic idea in our first attempt at solving the area under the arches problem above.

Instead of using rectangles, we see that trapezoids (trapeziums) give a better approximation to the area.

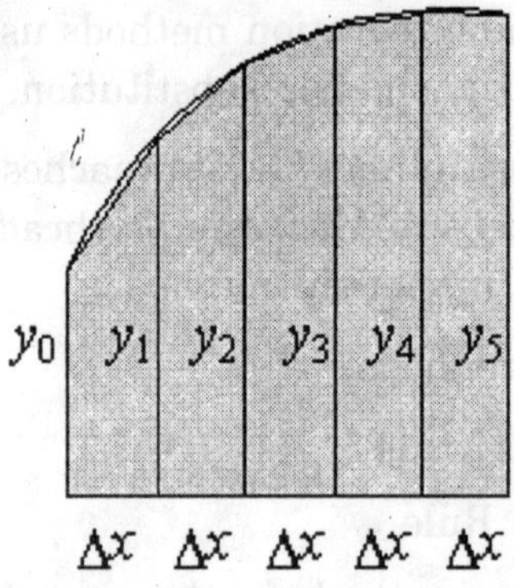

Now, the area of a trapezoid (trapezium) is given by:

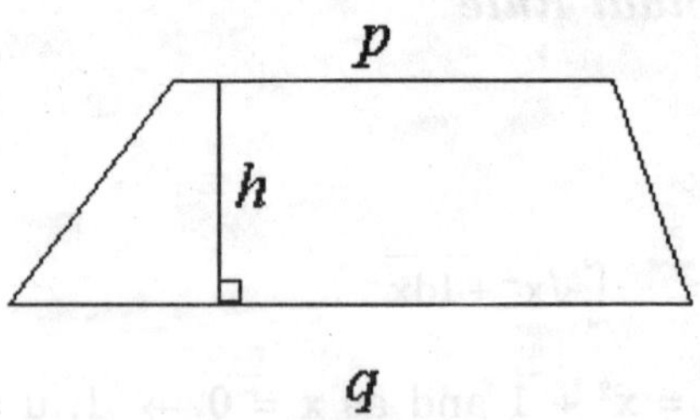

$$\text{Area} = \frac{h}{2}(p+q)$$

So the approximate area under the curve is found by adding the area of the trapezoids. (Our trapezoids are rotated 90° so that their new base is actually the height. So h = ?x.)

$$\text{Area} \approx \frac{1}{2}(y_0 + y_1)\Delta x + \frac{1}{2}(y_1 + y_2)\Delta x + \frac{1}{2}(y_2 + y_3)\Delta x...$$

We can simplify this to give us the Trapezoidal Rule, for n trapezoids:

$$\text{Area} \approx \Delta x\left(\frac{y_0}{2} +_1 + y_2 + y_3 + ... + \frac{y_n}{2}\right)$$

To find Δx for the area from x = a to x = b, we use:

$$\Delta x = \frac{b - a}{n}$$

and we also need

$$y_0 = f(a)$$

$$y_1 = f(a + \Delta x)$$

$$y_2 = f(a + 2\Delta x)$$

$$...$$

$$y_n = f(b)$$

Note:

- We get a better approximation if we take more trapezoids [up to a limit!].
- The more trapezoids we take, we have: $\Delta x \to 0$.
- We can write (if the curve is above the x-axis only between x = a and x = b):

$$\text{Area} = \int_a^b f(x)dx \approx \Delta x\left(\frac{y_0}{2} + y_1 + ... + \frac{y_n}{2}\right)$$

EXERCISE:

Using n = 5, approximate:

$$\int_0^1 \sqrt{x^2+1}\, dx$$

Here, a = 0 and b = 1.

$$\Delta x = \frac{b-a}{n} = \frac{1-0}{5} = 0.2$$

$$y_0 = f(a) = f(0) = \sqrt{0^2+1} = 1$$

$$y_1 = f(a + \Delta x) = f(0.2) = \sqrt{0.2^2+1} = 1.0198039$$

$$y_2 = f(a + 2\Delta x) = f(0.4) = \sqrt{0.4^2+1} = 1.0770330$$

$$y_3 = f(a + 3\Delta x) = f(0.6) = \sqrt{0.6^2+1} = 1.1661904$$

$$y_4 = f(a + 4\Delta x) = f(0.8) = \sqrt{0.8^2+1} = 1.2806248$$

$$y_5 = f(b) = f(1) = \sqrt{1^2+1} = 1.4142136$$

So the area ≈

$$0.2\left(\frac{1}{2}\times 1 + 1.0198039 + 1.0770330 + 1.1661904 + 1.2806248\ldots + \frac{1}{2}\times 1.4142136\right)$$

$$= 1.150$$

So $\int_0^1 \sqrt{x^2+1}\, dx \approx 1.150$

Simpson's Rule

The Trapezoidal Rule is an improvement over using rectangles because we have much less "missing" from our calculations. We used straight lines to model the curve in trapezoidal Rule.

We seek an even better approximation. In Simpson's Rule, we use parabolas to approximate each part of the curve. This proves to be very efficient.

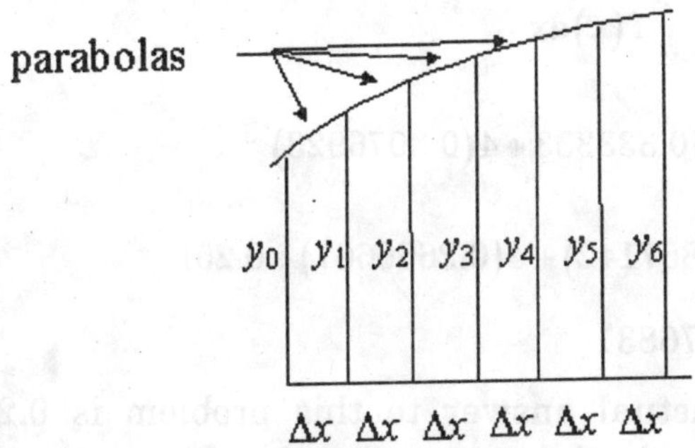

We can show (by integrating the area under each parabola and adding these areas) that the approximate area is given by Simpson's Rule:

$$\text{Area} = \int_a^b f(x)dx$$

$$\approx \frac{\Delta 3}{3}(y_0 + 4y_1 + 2y_2 + 4y_3 + 2y_4 + \ldots + 4y_{n-1} + y_n)$$

Note: **In Simpson's Rule, n must be EVEN.**

In the LiveMath document, note how the approximation is very good, even with a small number of divisions.

EXERCISE:

Approximate $\int_2^3 \frac{dx}{x+1}$ **using Simpson's Rule with n = 4.**

ANSWER

$\Delta x = (3 - 2)/4 = 0.25$

$y_0 = f(a) = f(2) = 1/(2 + 1) = 0.3333333$

$y_1 = f(a + \Delta x) = f(2.25) = 1/(2.25+1) = 0.3076923$

$y_2 = f(a + 2\Delta x) = f(2.5) = 1/(2.5+1) = 0.2857142$

$y_3 = f(a + 3\Delta x) = f(2.75) = 1/(2.75+1) = 0.2666667$

$y_4 = f(b) = f(3) = 1/(3+1) = 0.25$

$$\text{Area} = \int_a^b f(x)dx$$

$$\approx \frac{0.25}{3}(0.333333 + 4(0.3076923)$$

$$+2(0.2857142) + 4(0.2666667) + 0.25)$$

$$= 0.2876831$$

[The actual answer to this problem is 0.287682 so our Simpson's Rule approximation has an error of only 3.6 × 10-4%.]

Mental Math Methods from Asia

First of all, let us figure out what exactly is mental math. Today if you search the phrase "mental math" you will probably end up with millions of options. Not exactly that makes your life easy; instead it builds up and strengthens your curiosity. Put in simplest terms, mental math can be defined as calculations performed in your head - mentally - without help of any external device be it as simple as pen and paper or any modern day device such as calculator, computer or any other electronic gadget.

We humans perform mental mathematical calculations everyday, consciously and unconsciously. When you are driving you figure out when to apply brakes to bring the vehicle to stop before hitting something. You figure out time difference between east coast and west coast. But where we falter is at the simplest and most mundane of calculations. Go to a restaurant and figure out 18% gratuity.

Abacus Mental Mathematics

What is abacus mental mathematics? Origin of Abacus is highly disputed today, some say it originated in

Mesopotamia and some claim to be in China. Over centuries, abacus has evolved in to various different forms and sizes. The most commonly used is the Japanese Soroban Abacus.

The Soroban Abacus consists of one upper row and four lower rows and columns vary from thirteen, fifteen, seventeen or twenty one. It is claimed and proven by many researchers in Asia that Abacus stimulates whole brain development. When children use both hands to move the abacus beads to perform arithmetic calculations, there is quick communication between the hands and the brain that stimulates both the right and left hemispheres of the brain. This promotes rapid, balanced whole brain development.

If a child starts learning the abacus before being taught traditional arithmetic, there is minimal conflict and the child will easily work within both systems. If a child starts the program later, having already received traditional foundations, there may be a slightly extended learning period for the child to accept and integrate the abacus method.

Vedic Mental Mathematics

What is Vedic mental mathematics? Origin of Vedic Mathematics is in Atharva Veda (Holy Scripture from Hinduism). Vedic mathematics is a system based on sixteen sutras (aphorisms) which are actually word-formulae describing natural ways of solving a whole range of mathematical problems. These formulae describe the way the mind naturally works and are therefore a great help in directing the student to the appropriate method of solution.

It is claimed and proven by many researchers in Asia that practice and use of Vedic mathematics helps the person in many different aspects of decision making. From intelligent guessing to thinking outside the box ability. Vedic mathematics has its applications to much advanced mathematics, such as calculus and linear algebra. The sixteen sutras are: By one more than the one before, All from 9 and the last from 10, Vertically and crosswise, Transpose and apply, If the Samuccaya is the same it is zero, If one is in ratio the other is zero, By addition and by subtraction, By the completion or non-completion, Differential calculus, By the deficiency, Specific and general, The remainders by the last digit, The ultimate and twice the penultimate, By one less than the one before, The product of the sum, and All the multipliers

Today, both these methods have made a come back in Asia. Abacus Mental Math method is extremely popular in nations of China, Taiwan, Japan, Malaysia, Korea and India whereas Vedic Mental Math method is extremely popular only in India.

Vedas and Mathematics

Early age mathematicians had made Mathematics a Religion and created a simple set of rules to handle complex problems. This ancient science was lost for a long time only to have been discovered recently.

In this blog I intend to share with you the wisdom behind the sutras written long back by some Brilliant Rishis (learned men). I will try to give as many examples as possible to help you quickly understand the concept and also practice the same.

Please give me sometime to develop the complete material. However you can keep visiting the blog for interesting notes and updates.

To start off, I will try to show how easy it is to solve complex mathematical calculations with the use of a few sutras (rules)...

Example 1: How much time would you take to multiply 67 * 69? Did you need to use a pen and paper for this?

The traditional way to solve this would be:

```
     61
   * 69
 ------
    549  ...1 (Multiply 9 to 61)
   366X  ...2 (Multiply 6 to 61, with space on the right)
 ------
   4209       (Sum up the lines 1 and 2)
```

I could have never done this calculation without pen and paper.

But try this Sutra: BY ONE MORE THAN THE ONE BEFORE

The application of the same:

61 * 69 = (6 * 7) + (9 * 1) = 4209

Try this now: 47 * 43 :

Check 7+3 = 10 and the tens digit is same for both numbers being multiplied: So:

(4 * 5) + (7 * 3) = 2021

Could you reach to the answer before you could type the digits into your computer/calculator?

Interested

Well the question I generally get after this introduction is how many such Sutras (rules) does one have to remember to solve all types of mathematical problems?

The answer is merely 16 Sutras and 13 Sub-Sutras!! Thes sutras help solve numerous mathematical problems including multiplications, square roots, fractions, HCF, LCM, etc. Compare that to how many tables you had to remember and how many complex procedures you have remembered to solve mathematical problems??

Vedic Mathematics applies to arithmetic, algebra, geometry - plane and solid, trigonometry - plane and spherical, conics- geometrical and analytical, astronomy, calculus - differential and integral etc., etc.

Vedic Mathematics uses elegant, coherent and unified structures that are direct and easy unlike the methods taught in school. I was afraid of problems due to the rigour involved in doing calculations. Vedic Mathematics opens a fresh way to look at old problems or is it an old way to look at fresh problems!!

Ancient India's Contribution to Mathematics

> *"India was the motherland of our race and Sanskrit the mother of Europe's languages. India was the mother of our philosophy, of much of our mathematics, of the ideals embodied in Christianity... of self-government and democracy. In many ways, Mother India is the mother of us all." - Will Durant - American Historian 1885-1981*

Mathematics represents a high level of abstraction attained by the human mind. In India, mathematics has its roots in Vedic literature which is nearly 4000 years old. Between 1000 BC and 1000 AD, various treatises on mathematics were authored by Indian mathematicians in which were set forth for the first time, the concept of zero, the techniques of algebra and algorithm, square root and cube root.

A Search for Our Present in History

As in the applied sciences like production technology, architecture and shipbilding, Indians in ancient times also made advances in abstract sciences like Mathematics and Astronomy. It has now been generally accepted that the technique of algebra and the concept of zero originated in India.

But it would be surprising for us to know that even the rudiments of Geometry, called Rekha-Ganita in ancient India, were formulated and applied in the drafting of Mandalas for architectural purposes. They were also displayed in the geometric patterns used in many temple motifs.

Many motifs in Hindu temples and Palaces display a mix of floral and Geometric patterns.

Even the technique of calculation, called algorithm, which is today widely used in designing soft ware programs (instructions) for computers was also derived from Indian mathematics. In this chapter we shall examine the advances made by Indian mathematicians in ancient times.

Algebra—The Other Mathematics?

In India around the 5th century AD, a system of mathematics that made astronomical calculations easy was developed. In those times its application was limited to astronomy as its pioneers were Astronomers. As tronomical calculations are complex and involve many variables that go into the derivation of unknown quantities. Algebra is a short-hand method of calculation and by this feature it scores over conventional arithmetic.

In ancient India conventional mathematics termed Ganitam was known before the development of algebra. This is borne out by the name - Bijaganitam, which was given to the algebraic form of computation. Bijaganitam means 'the other mathematics' (Bija means 'another' or 'second' and Ganitam means mathematics). The fact that this name was chosen for this system of computation implies that it was recognised as a parallel system of computation, different from the conventional one which was used since the past and was till then the only one. Some have interpreted the term Bija to mean seed, symbolizing origin or beginning. And the inference that Bijaganitam was the original form of computation is derived. Credence is lent to this view by the existence of mathematics in the Vedic literature which was also shorthand method of computation. But whatever the origin of algebra, it is certain that this technique of computation Originated in India and was current around 1500 years back. Aryabhatta an Indian mathematican who lived in the 5th century AD has referred to Bijaganitam in his treatise on Mathematics, Aryabhattiya. An Indian mathematician - astronomer, Bhaskaracharya has also authored a treatise on this subject, the treatise which is dated around the 12th century AD is entitled 'Siddhanta-Shiromani' of which one section is entitled Bijaganitam.

Thus the technique of algebraic computation was known and was developed in India in earlier times. From the 13th century onwards, India was subject to invasions from the Arabs and other Islamised communities like the Turks and Afghans. Alongwith these invader: came chroniclers and critics like Al-Beruni who studied Indian society and polity.

The Indian system of mathematics could no have escaped their attention. It was also the age of the Islamic Renaissance and the Arabs generally improved upon the arts and sciences that they imbibed from the land they overran during their great Jehad. The system of mathematics they observed in India was adapted by them and given the name 'Al-Jabr' meaning 'the reunion of broken parts'. 'Al' means 'The' & 'Jabr' mean 'reunion'. This name given by the Arabs indicates that they took it from an external source and amalgamated it with their concepts about mathematics.

Between the 10th to 13th centuries, the Christian kingdoms of Europe made numerous attempts to reconquer the birthplace of Jesus Christ from its Mohammedan-Arab rulers. These attempts called the Crusades failed in their military objective, but the contacts they created between oriental and occidental nations resulted in a massive exchange of ideas. The technique of algebr could have passed on to the west at this time.

During the Renaissance in Europe, followed by the industrial revolution, the knowledge received from the east was further developed. Algebra as we know it today has lost any characteristics that betray it eastern origin save the fact that the tern 'algebra' is a corruption of the term 'Al jabr' which the Arabs gave to Bijaganitam Incidentally the term Bijaganit is still use in India to refer to this subject.

In the year 1816, an Englishman by the name James Taylor translated Bhaskara's Leelavati into English. A second English translation appeared in the following year (1817) by the English astronomer Henry Thomas Colebruke. Thus the works of this Indian mathematician astronomer were made known to the western world nearly 700 years after he had penned them, although his ideas had already reached the west through the Arabs many centuries earlier.

In the words of the Australian Indologist A.L. Basham (A.L. Basham; The Wonder That was India.) "... the world owes most to India in the realm of mathematics, which was developed in the Gupta period to a stage more advanced than that reached by any other nation of antiquity. The success of Indian mathematics was mainly due to the fact that Indians had a clear conception of the abstract number as distinct from the numerical quantity of objects or spatial extension."

Thus Indians could take their mathematical concepts to an abstract plane and with the aid of a simple numerical notation devise a rudimentary algebra as against the Greeks or the ancient Egyptians who due to their concern with the immediate measurement of physical objects remained confined to Mensuration and Geometry.

Geometry and Algorithm

But even in the area of Geometry, Indian mathematicians had their contribution. There was an area of mathematical applications called Rekha Ganita (Line Computation). The Sulva Sutras, which literally mean 'Rule of the Chord' give geometrical methods of constructing altars and temples. The temples layouts were called Mandalas. Some of important works in this field are by Apastamba, Baudhayana, Hiranyakesin, Manava, Varaha and Vadhula.

The Buddhist Pagodas borrowed their plan of construction from the geometric grid of the Mandala used for constructing temples in India (A majestic Pagoda at Bangkok)

The Arab scholar Mohammed Ibn Jubair al Battani studied Indian use of ratios from Retha Ganita and introduced them among the Arab scholars like Al Khwarazmi, Washiya and Abe Mashar who incorporated the newly acquired knowledge of algebra and other branches of Indian mathema into the Arab ideas about the subject.

The chief exponent of this Indo-Arab amalgam in mathematics was Al Khwarazmi who evolved a technique of calculation from Indian sources. This technique which was named by westerners after Al Khwarazmi as "Algorismi" gave us the modern term Algorithm, which is used in computer software.

Algorithm which is a process of calculation based on decimal notation numbers. This method was deduced by Khwarazmi from the Indian techniques geometric computation which he had stied. Al Khwarazmi's work was translated into Latin under the title "De Numero Indico" which means 'of Indian Numerals' thus betraying its Indian origin. This translation which belong to the 12th century A.D credited to one Adelard who lived in a town called Bath in Britian.

Thus Al Khwarazmi and Adelard could looked upon as pioneers who transmit Indian numerals to the west. Incidents according to the Oxford Dictionary, word algorithm which we use in the English language is a corruption of the name Khwarazmi which literally means '(a person) from Khawarizm', which was the name of the town where Al Khwarazmi lived. To day unfortunately', the original Indian texts that Al Khwarazmi studied arelost to us, only the translations are avail able. The Arabs borrowed so much from India the field of mathematics that even the subject of mathematics in Arabic came to known as Hindsa which means 'from India and a mathematician or engineer in Arabic is called Muhandis which means 'an expert in Mathematics'. The word Muhandis possibly derived from the Arabic term mathematics viz. Hindsa.

The Concept of Zero

The concept of zero also originated inancient India. This concept may seem to be a very ordinary one and a claim to its discovery may be viewed as queer. But if one gives a hard thought to this concept it would be seen that zero is not just a numeral. Apart from being a numeral, it is also a concept, and a fundamental one at that. It is fundamental because, terms to identify visible or perceptible objects do not require much ingenuity.

But a concept and symbol that connotes nullity represents a qualitative advancement of the human capacity of abstraction. In absence of a concept of zero there could have been only positive numerals in computation, the inclusion of zero in mathematics opened up a new dimension of negative numerals and gave a cut off point and a standard in the measurability of qualities whose extremes are as yet unknown to human beings, such as temperature.

In ancient India this numeral was used in computation, it was indicated by a dot and was termed Pujyam. Even today we use this term for zero along with the more current term Shunyam meaning a blank. But queerly the term Pujyam also means holy. Param-Pujya is a prefix used in written communication with elders. In this case it means respected or esteemed. The reason why the term Pujya - meaning blank - came to be sanctified can only be guessed.

Indian philosophy has glorified concepts like the material world being an illusion Maya), the act of renouncing the material world (Tyaga) and the goal of merging into the void of eternity (Nirvana). Herein could lie the reason how the mathematical concept of zero got a philosophical connotation of reverence.

In a queer way the concept of 'Zero' or Shunya is derived from the concept of a void. The concept of void existed in Hindu Philosophy hence the derivation of a symbol for it. The concept of Shunyata, influenced Southeast Asian culture through the Buddhist concept of Nirvana 'attaining salvation by merging into the void of eternity' (Ornate Entrance of a Buddhist temple in Laos).

It is possible that like the technique of algebra; the concept of zero also reached the west through the Arabs. In ancient India the terms used to describe zero included Pujyam, Shunyam, Bindu the concept of a void or blank was termed as Shukla and Shubra. The Arabs refer to the zero as Siphra or Sifr from which we have the English terms Cipher or Cypher. In English the term Cipher connotes zero or any Arabic numeral. Thus it is evident that the term Cipher is derived from the Arabic Sifr which in turn is quite close to the Sanskrit term Shubra.

The ancient India astronomer Brahmagupta is credited with having put forth the concept of zero for the first time: Brahmagupta is said to have been born the year 598 AD at Bhillamala (today's Bhinmal) in Gujarat, Western India.] much is known about Brahmagupta's early life. We are told that his name as a mathematician was well established when K Vyaghramukha of the Chapa dyansty m him the court astronomer. Of his two treatises, Brahma-sputa siddhanta and Karanakhandakhadyaka, first is more famous. It was a corrected version of the old Astronomical text, Brahma siddhanta. It was in his Brahma-sphu siddhanta, for the first time ever had be formulated the rules of the operation zero, foreshadowing the decimal system numeration. With the integration of zero into the numerals

Magic Squares

Translating Numbers Into Art 6 1 8 7 5 3 2 9 4 Imagine in front of you some playing cards and the ancient puzzle was to create a 3x3 square such that all columns, rows and diagonals that have the same sum. As shown above, the Magic Sum, aka the Magic Square Constant = 15. There is a treasure house of knowledge encoded within this harmonic matrix that the ancients knew and hid. Magic Squares for thousands of years have fascinated many cultures and many great minds like Benjamin Franklin and Albrecht Durer. Why was it at the centre of the Tibetan Calendar/Cosmolgy (Time Code / Applications in Time Travel)? Why did Ambaji, of the Hindu pantheon, the ancient goddesses of Creation (the universe flows through her breast), have as her tantric symbol, this Magic Square of 3x3, which has 8 possible permutations? and which Buddha, in his enlightenment, described in detail the 8 sided nature of sub-atomic molecules: the "Acta-Kalapos". Durga and Kali and Shakti etc. have variations of the same.

Why is it that when you translate NUMBERS INTO ART, by drawing a long connecting line, from 1 to 2 to 3 etc. to the last number, that elegant and phi-ratioed patterns appear? that when tiled or tessellated, produce the atomic structure of Diamond (which is verified by the clairvoyant drawings of the Occult Chemists: Charles Leadbeater and Annie Besant. Why is it that the pattern for the Magic Square of 11x11 is almost identical to the Ley Line Grid developed by Bruce Cathy, a New Zealand pilot, which was originally the jig-saw-puzzle-map of UFO sightings in his area? Why did Professor Searl, of the UK, have his work on levity discs and over-unity generators banned?

The metals of his generators for the space-craft he built were intimately related, by proportion, to the atomic table of elements and to the opposing pairs of the above Magic Square of 3x3, eg: notice that all pairs around the central cell have a sum of 10, thus 2 parts gold was alloyed with

8 parts silver, or 3 parts titanium to 7 parts nickel. Why did King Solomon the Rabbi, design his Seal based on the 3 rows of the above Magic Square being translated into a Natural Square of 3 (that has its numbers in Counting Order)? In this day lesson, you will superimpose a Magic Square of 7x7, by rotation of 90 degrees upon itself, to reveal ancient symbols at its centre: the Hindu Peace Symbol of the Swaztika nesting within the Christian Cross!

The ancient Essenes adored the richness of symbolism within these Mystic Squares and gave accounts or Parables of Jesus (Isa or Iesu) that related to our sense of community symbolized by the magic square, that each person, so positioned, has a divine role to play, that if one number was taken away or moved, it would no longer be magical, to realise 'Order amidst the Chaos', and the ancient axiom: 'As Above.. so below'.

Some questions to ponder upon:

* How could the Aryan civilisation, near the Hunza Valley, 3,000 years ago, mentally compute mathematical operations that today only a calculator could achieve?
* And if the maths could not be done mentally, how could they do everything we know today, IN ONE-LINE? eg: converting the fraction 1/19 into a decimal is an 18 digit answer that requires 35 lines of exhausting long division, how could they do it in a one simple line solution?
* Why is NASA secretly using this MENTAL ONE-LINE SYSTEM OF ARITHMETIC in the field of Advanced Artificial Intelligence? and not incorporating it into our current mathematics curriculum?
* Why are many of Bill Gate's Mathematicians Indians?
* Is the decimal form of Pi a sonic code for dedications to Krishna?

* What does the Bhagavad Gita have to say about this lost science of Clairvoyant Calculation that the ancient Temple Builders employed?
* How can you become a Human Bio-Calculator?
* Is the current Western Maths Curriculum keeping our Children in a state of limited intelligence? The answer to all these questions are found by understanding the ancient system of Vedic Mathematics as rediscovered by Bharati Krsna Tirthaji (1884- 1960), of Puri Govardhan Math, dux of India in Mathematics and a Sanskrit Philosopher who wrote and spoke perfect English, he reached the position of a Shankacharya (the spiritual head of India 100 years ago, like a Dalai Lama or Vatican Pope), renunciated several high offical positions in society to retreat into a forest for 7 years as a yogi, between 1911 and 1918 in which time he collected all the lost mathematical knowledge (Ganita Sutras), improved and reconstructed them to 16 basic Sutras or word formulae that solve every known mathematical problem.

These 16 Volumes relating to the 16 Sutras was our legacy from the ancient past. This highly intelligent and sophisticated system of Mental, One-Line Arithmetic was taught by The Master to his main student, who today is the leading exponent in the world: Maharishi Mahesh Yogi, the long white-haired and bearded yogi that many of the 60s generation know of as the Beatle´s guru. He is considered as the greatest human bio-calculator on the planet today, holding the vision of raising the collective consciousness of all nations by re-introducing the lost Vedic Mathematics Sutras early in a child s spiritual and educational development. Imagine a maths class in India today , like in the times of old when knowlege was passed on as an oral tradition, and the teacher quickly calls or drills out certain numbers to multiply or divide and the children instantly reply with the correct intuitive answers!

Today, in schools where Maharishi s Transcendental Mediatation technique are used students perform remarkably well academically, viewing Mathematics as a Universal Language and therefore often winning the highest competitve prizes in mathematics. Maharishi organised the scattered Vedic Literature into a complete science of consciousness for both its theoretical and practical value, describing the 16 Sutras as fundamental impulses of intelligence that structure all the Laws of Nature and those that govern the Universe. Maharishi, the Founder of TM, once said: "Vedic Mathematics is that one field of knowledge which fulfils the purpose of education by developing the total creative genius of the individual, giving him/her the ability to be always spontaneously right and automatically precise so that his/her action, supported by Natural Law, is always effortlessly fulfilling."

Unfortunately, in 1958 after Bharati Krsna Tirthaji returned from the USA tour, sponsored by Paramahansa Yogananda (who introduced Yoga to the West and created the Self-Realisation Foundation) all of Bharati Krsna's 16 Volumes for the 16 Sutras were confirmed lost. Disciples observed that people in the streets of India were imitating Tirthaji's lecture's and dazzling audiences with mathematically mis-used pschic powers, and begged the Master to write a book just before his death in 1960. This single volume, called VEDIC MATHEMATICS, has become the bible on this subject and spawned a thousand web-sites worldwide. Maharishi schools disseminated the teachings to England where certain schools have adopted the Vedic curriculum but the irony is that today India is struggling to release the British curriculum and NASA have adopted it fully in the realms of advanced robotics.

Calculations that can be solved as quick as lightning are a great tool to adopt, but you would nt want to teach it worldwide in the fear that you may churn out a generation of child geniuses that may threaten the intellectual status quo. The gift that the Hindus gave to world, thousands of years ago, and that which is currently responsible for a global silicon chip technology, was none other than the invention of ZERO and the use of the decimal point. We call our common numbers "Arabic Numerals" but really they extend back to the Hindu concept of creation and void known as "Bindu" or The Zero Point. In Maths, it makes the magic work eg: if you wish to multiply 98 x 98 we need to understand Zero and Unity.

All Vedic Maths is based on the understanding of Unity Consciousness which means they utilise processes or Number Bases that correspond to: 0, 10, 100, 1000, 10000 etc. all of which add to 1. Let us look at some practical examples, in light of the fact that the Vedas, literally "the illimitable storehouse of All Knowledge" came under 4 headings or categories like Ayurveda (Healing), and Vedas for Music and Astronomy etc. and one relevant sub-Veda that related to Temple Building and Engineering and therefore Mathematics called SthapathyVeda.

Thus a Vedic Mathematician was also an astronomer, healer and poet. It was a total system. As a temple builder, there was no pen and paper, you simply calculated in your head. You are out in the field and you need to tile a floor that is, say 98 units square. How do you do it with mental ease? The Squaring of Numbers near a Base To solve 98 Squared (98 × 98) we must first determine what Base we are in. It is close to 100, therefore we say Base 100.

We must now choose one of the 16 Sutras to effectively solve the problem. It is called: "By The Deficiency": "By whatever the deficiency, lessen it further by that much and set up the square thereof" says one of the 16 Sutras. Sounds cryptic and meaningless yet it quickly solves the problem.

We get our answer by merely knowing how much is 100 less 98. Knowing that the deficiency is 2, we merely lessen

98 by 2 and then we tag on the squaring of that 2. As a one-line answer, the setting out would appear as thus:

98 Squared = 98 - 2/2×2. Simplifying it: = 96 / _ 4 We almost have our answer.

What we need to know is that since our Base is 100, it has 2 zeroes, therefore this fact governs the need for 2 spaces or 2 digits after the "forward slash" symbol (/). By inserting or inventing The Zero as a "Place Marker", the answer is achieved: 98 Squared = 96 / 04 = 9,604. (Perhaps the greatest invention of all time, the Hindu creation of the Zero was to change the world unimaginably as time passed on leading to the ability to send Vimana U.F.O craft and rockets into space, to supercomputers and thus the ancient and continuing battle between Michael and Lucifer: Archangel Michael's Internal Light Vehicle Merkabah Time/Space Travel and Natural Spiritual Powers Versus Lucifer's daring invention of the External Merkabah Technological Metal Spacecraft).

Observe similar examples:

97 Squared = 97 - 3 / 3x3 = 94 / 09

96 Squared = 96 - 4 / 4x4 = 92 / 16.

When the number being squared is above the base, of 100 here, we add the Excess and Square the Excess:

104 Squared = 104 + 4 / 4x4 = 108 / 16 = 10, 816

104 x 105 = 104 + 5 / 4x5 = 109 / 20 = 10, 920

What if we enlarged our numbers to 998 Squared?

It is close to 1,000 so we say Base 1,000 and know to have 3 zeroes on the right hand side of the (/).

998 Squared = 998 - 2 / 2x2 = 996 / _ _ 4 = 996 / 004.
= 996,004

Understanding this, you can be calculating digits in the millions: 9998 Squared = 9998 - 2 / 2x2 = 9996 / _ _ _ 4

(Since we are in Base 10,000 the 4 Zeroes determine the need for 4 digits after the (/). = 9996 / 0004 = 99,960,004.

There is a world-wide debate currently raging about the efficacy of Vedic Mathematics versus the crumbling foundations of Western Mathematics. Generally speaking, the theorems we all learned at school are not wrong but clumsy.

Some of the Western geometrical formulae are certainly wrong or inadequate: for example, the formulae for sphere packing in the higher dimensions increase up to the 6th Dimension then suddenly decrease for higher dimensions, which is simply absurd.

Unfortunately, some die-hard senior mathematicians in an attempt to protect the crumbling foundations that they now stand on feel threatened by the lightning quick mental calcualations of the Vedic seers, and go to great lengths to deride Vedic maths as a "bag of tricks". Here is another example illustrating its utter simplicity: "the Path of Least Resistance".

Sutra: Vilokanam, Means by Mere Intuition

Perhaps the greatest of all Sutras is the one known as Vilokanam which literally means: "By Mere Inspection" or "By Mere Observation" or simply "By Mere Intuition".

By relying on modern electronic calculators for the next 20 years or so, without cultivating mental calculations which build the "Mental Muscle", will sadly produce a generation of young adults whose brain will have undoubtedly atrophied! Vedic Mathematics promotes clarity and sharpness of mind as well as improving memory skills and self-confidence, by awakening the Pineal Gland (or Third Eye) to clairvoyantly "see" or intuitively know the predicted answer.

Mystics say that the Pineal gland, located in the exact centre of the brain, was once the size of large cherry, but now it is the size of a pea. Two reasons are that we drink denatured defunct water and we no longer build the Mental Muscle, over-relying on calculators etc. to solve simple problems. The brain is a muscle, being 90% water, it needs to be hydrated with pure alive waters and regularly exercised. By application of the key Sutra called Vilokanam which means: "By Mere Inspection", a child can know for example that the Cube Root of 35,937 without a calculator, to be equal to 33. By Mere Observation! No working out, no pen or paper.

Mathematics without Tears. "On seeing this kind of work actually being performed by the little children, the doctors, professors and other "big guns" of mathematics are wonder struck and exclaim:- "Is this mathematics or magic? And we invariably answer and say: "It is both. It is magic until you understand it; and it is mathematics thereafter"...

" *Bharati Krsna Tithaji (1884-1960)*

Discoverer of Vedic Mathematics.

"WATCHING A CHILD GO THROUGH OUR EDUCATION SYSTEM IS LIKE WATCHING REVERSE METAMORPHOSIS: IN FLIES A BEAUTIFUL BUTTERFLY, OUT CRAWLS A CATERPILLAR."

If you disliked Mathematics in your school years, do not be disalarmed. It is actually a blessing as your Higher Self or Intelligence refused to embrace a system that was anything less that pure Star Language. If you observe Fig 3, relating to the Sutra "Vertically and Crosswise" for multiplication of 2 Digits by 2 Digits or 3 Digits by Digits, etc. the pictures actually look like star glyphs or codes. The Brain thinks in symbols and pictures. Maths must be taught laden with Shapes, Colour, Aromatherapy and Music for effective Whole-Brain Learning.

Vedic Numerical Codified Knowledge

How the ancient seers sang the long decimal form of Pi. In ancient India, they would sing songs to memorise long decimals. The Brahmins or scholarly caste, secretive of their knowledge, sonically encrypted mathematical formulae into their devotional praises or hymns to Lord Shri Krishna and also recorded historical data in codified lyrics. Similar to numerology, by ascribing the values of numbers to consonant letters as in A = 1, B = 2, C = 3, D = 4, but in Sanskrit, the Vedic Numerical Code was so sophisticated that it possessed 3 layers and therefore triple meanings.

It turns out that decimal form of the transcendental number: Pi = 3.14159265358979323846264338327 92 etc. was hidden or codified in these following syllables, by chanting: "Gopi Bhagyamaduv rata Shringishodadi Sandiga Kala Jeevitarava Tava Galaddhalara Sangara" (Observing the top line, "go" = 3, "pi" = 1, "bha = 4, "ya" = 1, "ma" = 5, "dhu" = 9, "ra" =2, "ta" = 6 etc. giving the first 8 figures of Pi, the ratio of the circumference of a circle to its diameter). Not only did it give Pi to 32 decimal places, but there was a secret Master Key within the patterning of the 32 that could unlock the next 32 decimals of Pi, and so on, a ticket to infinity. Not only did it praise Krishna, it operated on another level as a dedication to Shankara: (800A.D, the celebrated Acharya or Teacher, and founder of 4 monastic orders.

He could read at the age of 2 and mastered the Vedas at the age of 8. He wrote scholarly commentaries on the Bhagavad-Gita and the Upanishads which led to the decline of Buddhism in India. He is considered as a partial incarnation of Shiva the Avatara). To conclude this article, the time has come that all past secret knowledge can no longer be kept secret. The Brahmins, by keeping this knowledge complex and inaccessible to the masses, and therefore not shareable, did not live in Truth. The Truth is that we have entered an Age where if something is not Simple and Shareable, then it has no value. It is therefore ripe time to reintroduce the ancient seers infallible Mental and One-Line System of Vedic Mathematics.

Topics Include:

* Magic Squares and their Atomic Artforms.
* The Vedic Square and it's Digital Sums.
* The Divine PHI Proportion or The Living Mathematics of Nature,
* The 5 Platonic Solids and The 13 Archimedean Solids
* Pythagorean Mystery School.
* Vedic Mathematics. He works with Gifted and Talented children and lectures at Maths Camps, Universities and Maths Conferences. He is currently writing a series of 4 books for the Australian Curriculum on Vedic Mathematics. As an Authorised School Performer of a Travelling Show called "Mathemagics", Jain has taught thousands of children and adults in Australia, the wonder of the pure mathematical principles relating to Atomic Art and Discovery. In 1993 this travelling Sacred Geometry Show was banned in Queensland schools, loved by mathematicians but shunned by the top officials in the Department of Education, who are Fundamental Christian-based and antagonistic towards the New Maths.

Bibliography

A B Chace, L S Bull, H P Manning and R C Archibald: *The Rhind Mathematical Papyrus,* Oberlin, Ohio, 1927.

A.L. Basham: *The Wonder That Was India*, Rupa & Co., Calcutta, 1967.

Aaboe, Asger: *Episodes from the Early History of Mathematics*, New York, 1964.

Amma, Saraswarti: *Geometry in Ancient and Medieval India*, Motilal Banarsidas, 1979.

Armstrong, Robert G.: *Yoruba Numerals*, London, Oxford University Press, 1962.

Arnett, William S.: *The Predynastic Origin of Egyptian Hieroglyphs*, Washington, University Press of America, 1982.

Ascher, Marcia: *Ethnomathematics*, Pacific Grove, CA: Brooks/Cole, 1991.

Bag, A. K.: *Mathematics in Ancient and Medieval India*, Varanasi: Chaukhambha Orientalia, 1979.

Barnard, F. P.: *The Casting-Counter and the Counting Board. A Chapter in the History of Numismatics and Early Arithmetic*, Oxford, Oxford University Press, 1916.

Barrow, John D.: *Pi in the Sky*, Oxford, Clarendon Press, 1992.

Basham, A. L.: *The Wonder That Was India*, Calcutta, Rupa & Co., 1967.

Beckmann, Petr: *A History of Pi*, New York, St. Martin's Press, 1974.

Bell, Eric T.: *The Magic of Numbers*, New York, McGraw-Hill, 1946.

Berlin, Brent: Tzeltal Numeral Classifiers, The Hague, Mouton, 1968.

Bharati Krishna Tirthaji Majaraja.: *Vedic Mathematics*. Delhi, Motilal Banarsidass, 1988.

Bidyaranya, Swami: *History of Hindu Mathematics*,Bombay, Asia Pub. House, 1962.

Bischoff, Bernard: *Latin Palaeography*, Cambridge, Cambridge University Press, 1990.

Bonfante, Larissa: *Etruscan*. Berkeley, University of California Press, 1990.

Bowditch, Charles P.: *The Numeration, Calendar Systems and Astronomical knowledge of the Mayas*, Cambridge, Cambridge University Press, 1910.

Brainerd, Charles J.: *The Origins of the Number Concept*, New York, Praeger, 1979.

Brunschwig, Jacques and Geoffrey E.R. Lloyd:. *Greek Thought: A Guide to Classical Knowledge*, Cambridge, Harvard University Press, 2000.

Bunt, L. N. H., Jones, P. S., & Bedient, J. D.: *The Historical Roots of Elementary Mathematics*, New York, Dover, 1988.

Chattopadhyaya, D. P.: *Mathematics, Astronomy, and Biology in Indian Tradition: Some Conceptual Preliminaries*, New Delhi, Project of History of Indian Science, Philosophy, and Culture, 1995.

Clawson, Calvin: *The Mathematical Traveller*, New York, Plenum Press, 1994.

Closs, Michael P.: *Native American Mathematics*, Austin, University of Texas Press, 1986.

Conant, Levi L.: *The Number Concept*, New York, MacMillan, 1896.

Crossley, John N: *The Emergence of Number*, Singapore, World Scientific, 1987.

Crump, Thomas: *The Anthropology of Numbers*, Cambridge, Cambridge University Press, 1990.

Dantzig, Tobias: *Number, the Language of Science*, London, George Allen and Unwin, 1940.

De Villiers, Melius: *The Numeral-Words: Their Origin, Meaning, History and Lesson*, London, Witherby, 1923.

Dehaene, Stanislas: *The Number Sense*, New York, Oxford University Press, 1997.

Dilke, Oswald Ashton Wentworth: *Mathematics and Measurement*, Berkeley, CA: U. of Cal. Press, 1987.

Eves, Howard: *An Introduction to the History of Mathematics*, New York, Rinehart and Company Inc., 1953.

Flegg, Graham: *Numbers: Their History and Meaning*, London, André Deutsch, 1983.

Fowler, David: *The Mathematics of Plato's Academy*, Oxford, Clarendon, 1999.

Fuson, K.C.: *Children's Counting and Concepts of Number*, New York, Springer-Verlag, 1988.

Ganeri, Anita: *The Story of Numbers and Counting*, New York, Oxford University Press, 1996.

Gokhale, Shebhana Laxman: *Indian Numerals*, Poona, Deccan College, 1966.

Griffith, Francis Llewellyn: *Hieratic Papyri from Kahun and Gurob*, London, Bernard Quaritch, 1898.

Grover, Sudarshan: *History of Development of Mathematics in India*, Delhi, Atma Ram & Sons, 1994.

Haldane, John Burdon Sanderson: *Science and Indian Culture*, Calcutta, New Age Publishers, 1965.

Halliwell, James Orchard: *Rara Mathematica*, London, J.W. Parker, 1839.

Herbert Meschkowski: *Ways of Thought of Great Mathematicians*, Holden-Day Inc., San Francisco, 1964.

Hogben, Lancelot: *Mathematics for the Million*, New York, W. W. Norton, 1937.

Howard Eves: *An Introduction to the History of Mathematics*, Rinehart and Company Inc., New York, 1953.

Ifrah, Georges: *From One to Zero: A Universal History of Numbers*, New York, Viking Penguin, 1985.

Jagadguru Swami Shri Bharati Krishna Tirthaji Maharaja: *Vedic Mathematics*, Motilal Banarsidass, Delhi, 1988.

Jaggi, O. P.: *Science and Technology in Medieval India*, New Delhi, Atma Ram & Sons, 1977.

Kline, Morris: *Mathematical Thought from Ancient to Modern Times*, New York, Oxford U. Press, 1990.

Kuppuram, G & K. Kumudamani: *History of Science and Technology in India*, Delhi, Sundeep Prakashan, 1990.

Lancy, David F.: *Cross-Cultural Studies in Cognition and Mathematics*, New York, Academic, 1983.

Marcus, Joyce: *Mesoamerican Writing Systems*, Princeton, Princeton University Press, 1992.

Marshack, Alexander: *The Roots of Civilization*, New York, McGraw-Hill, 1972.

Martzloff, Jean-Claude: *A History of Chinese Mathematics*, New York, Springer, 1997.

McLeish, John: *The Story of Numbers*, New York, Fawcett Columbine, 1991.

Menninger, Karl: *Number-Words and Number-Symbols*. London, MIT, 1969.

Murthy, T. S. Bhanu: *A Modern Introduction to Ancient Indian Mathematics*, Wiley Eastern Ltd., New Delhi, 1992.

Neugebauer, Otto and A. Sachs: *Mathematical Cuneiform Texts*, New Haven, Yale University Press, 1945.

Neugebauer, Otto: *The Exact Sciences in Antiquity*, Princeton, Princeton University Press, 1952.

North, John: *The Norton History of Astronomy and Cosmology*, New York, W.W.Norton, 1995.

Pandit, M. D.: *Mathematics as known to the Vedic Samhitas*, Delhi, Sri Satguru Publications, 1993.

Pappas, Theoni: *The Magic of Mathematics*, San Carlos, Wide World Publishing / Tetra, 1994.

Parpola, Asko: *Deciphering the Indus Script*, Cambridge, Cambridge University Press, 1994.

Peet, T. Eric: *The Rhind Mathematical Papyrus*, London, British Museum, 1923.

Piaget, Jean: *The Child's Conception of Number*, New York, W.W. Norton, 1965.

Pullan, J.M: *The History of the Abacus*, New York, Praeger, 1968.

Rao, S. Balaachandra: *Indian Mathematics and Astronomy*, Bangalore, Jnana Deep Publications, 1994.

Restivo, Sal: *The Social Relations of Physics, Mysticism and Mathematics*, Dordrecht, Kluwer, 1983.

Robins, G. and C. Shute: *The Rhind Mathematical Papyrus*, London, British Museum, 1987.

Russell, Bertrand: *The Principles of Mathematics*, Cambridge, Cambridge University Press, 1903.

Salomon, Richard: *Indian Epigraphy*, New York, Oxford University Press, 1998.

Saraswati Amma: *Geometry in Ancient and Medieval India*, Motilal Banarsidas, 1979.

Sarton, George: *The Study of the History of Mathematics*, Cambridge, Cambridge University Press, 1936.

Satterthwaite, Linton, Jr.: *Concepts and Structures of Maya Calendrical Arithmetics*, Philadelphia, University of Pennsylvania, 1947.

Schaaf, W.L.: *Mathematics, Our Great Heritage*, New York, Harper and Row, 1948.

Schimmel, Annemarie: *The Mystery of Numbers*, New York, Oxford University Press, 1993.

Scriba, C.J.: *The Concept of Number*, Zurich, Mannheim, 1968.

Seidenberg, A.: *The Zero in the Mayan Numerical Notation*, Austin, University of Texas Press, 1986.

Smeltzer, Donald: *Man and Number*, New York, Emerson, 1959.

Smith, D. E.: *History of Mathematics*, New York, Dover Publications Inc., 1951.

Srinivasiengar, C. N.: *The History of Ancient Indian Mathematics*. Calcutta, World Press, 1967.

Struik, Dirk J.: *A Concise History of Mathematics*, New York, Dover, 1987.

Thompson, Sir J. Eric S.: *Maya Hieroglyphic Writing: An Introduction*, Norman, University of Oklahoma Press, 1971.

Todd, R.R., P.J. Barber, and D. Jones: *The Internal Representation of Number: Analogue or Digital?*, Oxford, Clarendon, 1987.

Urton, Gary: *The Social Life of Numbers*, Austin, University of Texas Press, 1997.

Volodarsky, A.I.: *Mathematics in Ancient India*, Tokyo, Science Council of Japan, 1974.

Woolner, Alfred C.: *Introduction to Prakrit*, Calcutta, Baptist Mission Press, 1917.

Index

A

B

N

O

P

R

S

T

U

V

□□□

ENCYCLOPAEDIA OF VEDIC MATHEMATICS

ENCYCLOPAEDIA OF
VEDIC MATHEMATICS
(Vol. 2)

Shyam Priya

ANMOL PUBLICATIONS PVT. LTD.
NEW DELHI - 110 002 (INDIA)

ANMOL PUBLICATIONS PVT. LTD.
H.O.: 4374/4B, Ansari Road, Darya Ganj,
New Delhi-110 002 (India)
Ph.: 23278000, 23261597
B.O.: No. 1015, Ist Main Road, BSK IIIrd Stage
IIIrd Phase, IIIrd Block,
Bangalore - 560 085 (India)
Visit us at: www.anmolpublications.com

Encyclopaedia of Vedic Mathematics

First Published, 2008
ISBN 978-81-261-3504-2 (Set)

PRINTED IN INDIA

Printed at Mehra Offset Press, Delhi.

Contents

Preface

Vedic Mathematics contains many examples of striking methods of calculation and there is a remarkable coherence to the system which makes it very attractive. Also the Vedic system itself suggests a kind of approach that involves going directly to the answer. Vedic mathematics is based on sixteen Sutras and some sub-Sutras which provide links through mathematics: the word 'Sutra' means 'thread'.

Vedic scheme is not simply a collection of rapid methods; it is a system, a unified approach, which can be swiftly learnt. Vedic Mathematics extensively exploits the properties of numbers in every practical applications, particularly in the field of computation. It makes available a whole range of methods ideally suited to these properties. After a detailed introduction to Vedic Mathematics, this book shows applications of Vertically and Crosswise from basic calculations to the evaluation of logarithms, exponentials, trigonometrical functions and the solution of simultaneous, transcendental, polynomial and differential equations. The methods shown are not only exceedingly efficient and new but have a unity and coherence which renders them easy to master and enjoyable to execute.

The methods given in this book are not intended to be a complete or thorough treatment of the topics they deal with. All of the ideas can probably be developed further or applied in other areas, and all can doubtless be improved upon. The mathematician will also observe a certain lack of rigor as an attempt has been made to make the material intelligible to as wide a readership as possible.

—Shyam Priya

1

Great Hindu Mathematicians

Early Mathematicians of India

The earliest expression of mathematical understanding is inextricably linked with the origin of Hinduism. Mathematics formed an important part of the Vedas - the original Hindu scriptures - associated with ritual activities and the study of 'ganit' or mathematics was given special importance in the Vedic period.

Mathematics in Early Hindu Texts

Arithmetic operations including squares, cubes and roots are mentioned in the 'Narad Vishnu Purana' attributed to Ved Vyas (c 1000 BC), and the 1000 BC Vedic text 'Vedang Jyotish' says: "Just as the feathers of a peacock and the jewel-stone of a snake are placed at the highest point of the body, so is the position of mathematics the highest amongst all branches of the Vedas and the Shastras."

Seminal Figures in Mathematics

Hindu mathematicians, from Vedic times to the modern age have been in the forefront of making seminal contributions in the field mathematics.

Early Hindu Mathematicians

- Brahmagupta (c. 598-c. 670)
- Govindaswami (c. 800-850)

- Mahavira (Mahaviracharya) (850)
- Pruthudakaswami (850)
- Sridhara (900)
- Manjula (930)
- Aryabhatta II (950)
- Prashastidhara (958)
- Halayudha (975)
- Jayadeva (1000)
- Sripathi (1039)
- Hemachandra Suri (b. 1089)
- Bhaskara (1114-c. 1185)
- Cangadeva (1205)
- Madhava of Sangamagramma (c. 1340-1425)
- Narayama Pandit (1350)
- Parameshvara (1360-1455)
- Nilakantha Somayaji (1455-1555)
- Shankara Variar (c. 1500-1560)
- Narayana (c. 1500-1575)
- Jyesthadeva (550)
- Acyuta Pisarati (c. 1550-1621)
- Putumana Somayaji (c. 1660-1740)
- Jaganath Pandit (1700)

Ancient Indian Mathematics

800 BC	Baudhayana
750 BC	Manava
600 BC	Apastamba
520 BC	Panini

200 BC	Katyayana
120 AD	Yavanesvara
476	Aryabhatta I
500	Yativrsabha
505	Varahamihira
598	Brahmagupta
600	Bhaskara I
720	Lalla
800	Govindasvami
800	Mahavira
830	Prthudakasvami
840	Shankara
870	Sridhara
920	Aryabhatta II
940	Vijayanandi
1060	Brahmadeva
1114	Bhaskara II
1340	Mahendra Suri
1340	Narayana
1350	Madhava
1370	Parameshvara
1444	Nilakantha
1500	Jyesthadeva
1616	Kamalakara
1019	Sripati
1690	Jagannatha

An Overview of Indian Mathematics

It is without doubt that mathematics today owes a huge debt to the outstanding contributions made by Indian mathematicians over many hundreds of years. What is quite surprising is that there has been a reluctance to recognise this and one has to conclude that many famous historians of mathematics found what they expected to find, or perhaps even what they hoped to find, rather than to realise what was so clear in front of them.

We shall examine the contributions of Indian mathematics in this article, but before looking at this contribution in more detail we should say clearly that the "huge debt" is the beautiful number system invented by the Indians on which much of mathematical development has rested. Laplace put this with great clarity:-

The ingenious method of expressing every possible number using a set of ten symbols (each symbol having a place value and an absolute value) emerged in India. The idea seems so simple nowadays that its significance and profound importance is no longer appreciated. Its simplicity lies in the way it facilitated calculation and placed arithmetic foremost amongst useful inventions the importance of this invention is more readily appreciated when one considers that it was beyond the two greatest men of Antiquity, Archimedes and Apollonius.

We shall look briefly at the Indian development of the place-value decimal system of numbers later in this article and in somewhat more detail in the separate article Indian numerals. First, however, we go back to the first evidence of mathematics developing in India.

Histories of Indian mathematics used to begin by describing the geometry contained in the Sulbasutras but research into the history of Indian mathematics has shown that the essentials of this geometry were older being contained in the altar constructions described in the Vedic mythology text the Shatapatha Brahmana and the Taittiriya Samhita. Also it has been shown that the study of mathematical astronomy in India goes back to at least the third millennium BC and mathematics and geometry must have existed to support this study in these ancient times.

The first mathematics which we shall describe in this article developed in the Indus valley. The earliest known urban Indian culture was first identified in 1921 at Harappa in the Punjab and then, one year later, at Mohenjo-daro, near the Indus River in the Sindh. Both these sites are now in Pakistan but this is still covered by our term "Indian mathematics" which, in this article, refers to mathematics developed in the Indian subcontinent. The Indus civilisation (or Harappan civilisation as it is sometimes known) was based in these two cities and also in over a hundred small towns and villages. It was a civilisation which began around 2500 BC and survived until 1700 BC or later. The people were literate and used a written script containing around 500 characters which some have claimed to have deciphered but, being far from clear that this is the case, much research remains to be done before a full appreciation of the mathematical achievements of this ancient civilisation can be fully assessed.

We often think of Egyptians and Babylonians as being the height of civilisation and of mathematical skills around the period of the Indus civilisation, yet V G Childe in New Light on the Most Ancient East (1952) wrote:-

> *India confronts Egypt and Babylonia by the 3rd millennium with a thoroughly individual and independent civilisation of her own, technically the peer of the rest. And plainly it is deeply rooted in Indian soil. The Indus civilisation represents a very perfect adjustment of human life to a specific*

environment. And it has endured; it is already specifically Indian and forms the basis of modern Indian culture.

We do know that the Harappans had adopted a uniform system of weights and measures. An analysis of the weights discovered suggests that they belong to two series both being decimal in nature with each decimal number multiplied and divided by two, giving for the main series ratios of 0.05, 0.1, 0.2, 0.5, 1, 2, 5, 10, 20, 50, 100, 200, and 500. Several scales for the measurement of length were also discovered during excavations. One was a decimal scale based on a unit of measurement of 1.32 inches (3.35 centimetres) which has been called the "Indus inch". Of course ten units is then 13.2 inches which is quite believable as the measure of a "foot". A similar measure based on the length of a foot is present in other parts of Asia and beyond. Another scale was discovered when a bronze rod was found which was marked in lengths of 0.367 inches. It is certainly surprising the accuracy with which these scales are marked. Now 100 units of this measure is 36.7 inches which is the measure of a stride. Measurements of the ruins of the buildings which have been excavated show that these units of length were accurately used by the Harappans in construction.

It is unclear exactly what caused the decline in the Harappan civilisation. Historians have suggested four possible causes: a change in climatic patterns and a consequent agricultural crisis; a climatic disaster such flooding or severe drought; disease spread by epidemic; or the invasion of Indo-Aryans peoples from the north. The favourite theory used to be the last of the four, but recent opinions favour one of the first three. What is certainly true is that eventually the Indo-Aryans peoples from the north did spread over the region. This brings us to the earliest literary record of Indian culture, the Vedas which were composed in Vedic Sanskrit, between 1500 BC and 800 BC. At first these texts, consisting of hymns, spells, and ritual observations, were transmitted orally. Later the texts became written works for use of those practicing the Vedic religion.

The next mathematics of importance on the Indian subcontinent was associated with these religious texts. It consisted of the Sulbasutras which were appendices to the Vedas giving rules for constructing altars. They contained quite an amount of geometrical knowledge, but the mathematics was being developed, not for its own sake, but purely for practical religious purposes. The mathematics contained in the these texts is studied in some detail in the separate article on the Sulbasutras.

The main Sulbasutras were composed by Baudhayana (about 800 BC), Manava (about 750 BC), Apastamba (about 600 BC), and Katyayana (about 200 BC). These men were both priests and scholars but they were not mathematicians in the modern sense. Although we have no information on these men other than the texts they wrote, we have included them in our biographies of mathematicians. There is another scholar, who again was not a mathematician in the usual sense, who lived around this period. That was Panini who achieved remarkable results in his studies of Sanskrit grammar. Now one might reasonably ask what Sanskrit grammar has to do with mathematics. It certainly has something to do with modern theoretical computer science, for a mathematician or computer scientist working with formal language theory will recognise just how modern some of Panini's ideas are.

Before the end of the period of the Sulbasutras, around the middle of the third century BC, the Brahmi numerals had begun to appear.

Here is one style of the Brahmi numerals..

These are the earliest numerals which, after a multitude of changes, eventually developed into the numerals 1, 2, 3, 4, 5, 6, 7, 8, 9 used today. The development of numerals and place-valued number systems are studied in the article Indian numerals.

The Vedic religion with its sacrificial rites began to wane and other religions began to replace it. One of these was Jainism, a religion and philosophy which was founded in India around the 6th century BC. Although the period after the decline of the Vedic religion up to the time of Aryabhatta I around 500 AD used to be considered as a dark period in Indian mathematics, recently it has been recognised as a time when many mathematical ideas were considered. In fact Aryabhatta is now thought of as summarising the mathematical developments of the Jaina as well as beginning the next phase.

The main topics of Jaina mathematics in around 150 BC were: the theory of numbers, arithmetical operations, geometry, operations with fractions, simple equations, cubic equations, quartic equations, and permutations and combinations. More surprisingly the Jaina developed a theory of the infinite containing different levels of infinity, a primitive understanding of indices, and some notion of logarithms to base 2. One of the difficult problems facing historians of mathematics is deciding on the date of the Bakhshali manuscript. If this is a work which is indeed from 400 AD, or at any rate a copy of a work which was originally written at this time, then our understanding of the achievements of Jaina mathematics will be greatly enhanced. While there is so much uncertainty over the date, a topic discussed fully in our article on the Bakhshali manuscript, then we should avoid rewriting the history of the Jaina period in the light of the mathematics contained in this remarkable document.

You can see a separate article about Jaina mathematics.

If the Vedic religion gave rise to a study of mathematics for constructing sacrificial altars, then it was Jaina cosmology which led to ideas of the infinite in Jaina mathematics.

Later mathematical advances were often driven by the study of astronomy. Well perhaps it would be more accurate to say that astrology formed the driving force since it was that "science" which required accurate information about the planets and other heavenly bodies and so encouraged the development of mathematics. Religion too played a major role in astronomical investigations in India for accurate calendars had to be prepared to allow religious observances to occur at the correct times. Mathematics then was still an applied science in India for many centuries with mathematicians developing methods to solve practical problems.

Yavanesvara, in the second century AD, played an important role in popularising astrology when he translated a Greek astrology text dating from 120 BC. If he had made a literal translation it is doubtful whether it would have been of interest to more than a few academically minded people. He popularised the text, however, by resetting the whole work into Indian culture using Hindu images with the Indian caste system integrated into his text.

By about 500 AD the classical era of Indian mathematics began with the work of Aryabhatta. His work was both a summary of Jaina mathematics and the beginning of new era for astronomy and mathematics. His ideas of astronomy were truly remarkable. He replaced the two demons Rahu, the Dhruva Rahu which causes the phases of the Moon and the Parva Rahu which causes an eclipse by covering the Moon or Sun or their light, with a modern theory of eclipses. He introduced trigonometry in order to make his astronomical calculations, based on the Greek epicycle theory, and he solved with integer solutions indeterminate equations which arose in astronomical theories.

Aryabhatta headed a research centre for mathematics and astronomy at Kusumapura in the northeast of the Indian subcontinent. There a school studying his ideas grew up there but more than that, Aryabhatta set the agenda for mathematical and astronomical research in India for many centuries to come. Another mathematical and astronomical centre was at Ujjain, also in the north of the Indian subcontinent, which grew up around the same time as Kusumapura. The most important of the mathematicians at this second centre was Varahamihira who also made important contributions to astronomy and trigonometry.

The main ideas of Jaina mathematics, particularly those relating to its cosmology with its passion for large finite numbers and infinite numbers, continued to flourish with scholars such as Yativrsabha. He was a contemporary of Varahamihira and of the slightly older Aryabhatta. We should also note that the two schools at Kusumapura and Ujjain were involved in the continuing developments of the numerals and of place-valued number systems. The next figure of major importance at the Ujjain school was Brahmagupta near the beginning of the seventh century AD and he would make one of the most major contributions to the development of the numbers systems with his remarkable contributions on negative numbers and zero. It is a sobering thought that eight hundred years later European mathematics would be struggling to cope without the use of negative numbers and of zero.

These were certainly not Brahmagupta's only contributions to mathematics. Far from it for he made other major contributions in to the understanding of integer solutions to indeterminate equations and to interpolation formulas invented to aid the computation of sine tables.

The way that the contributions of these mathematicians were prompted by a study of methods in spherical astronomy is described in :-

The Hindu astronomers did not possess a general method for solving problems in spherical astronomy, unlike the Greeks who systematically followed the method of Ptolemy, based on the well-known theorem of Menelaus. But, by means of suitable constructions within the armillary sphere, they were able to reduce many of their problems to comparison of similar right-angled plane triangles. In addition to this device, they sometimes also used the theory of quadratic equations, or applied the method of successive approximations. ... Of the methods taught by Aryabhatta and demonstrated by his scholiast Bhaskara I, some are based on comparison of similar right-angled plane triangles, and others are derived from inference. Brahmagupta is probably the earliest astronomer to have employed the theory of quadratic equations and the method of successive approximations to solving problems in spherical astronomy.

Before continuing to describe the developments through the classical period we should explain the mechanisms which allowed mathematics to flourish in India during these centuries. The educational system in India at this time did not allow talented people with ability to receive training in mathematics or astronomy. Rather the whole educational system was family based. There were a number of families who carried the traditions of astrology, astronomy and mathematics forward by educating each new generation of the family in the skills which had been developed. We should also note that astronomy and mathematics developed on their own, separate for the development of other areas of knowledge.

Now a "mathematical family" would have a library which contained the writing of the previous generations. These writings would most likely be commentaries on earlier works such as the Aryabhatiya of Aryabhatta. Many of the commentaries would be commentaries on commentaries on commentaries etc. Mathematicians often wrote commentaries on their own work. They would not be aiming to provide texts to be used in educating people outside the family, nor would they be looking for innovative ideas in astronomy. Again religion was the key, for astronomy was considered to be of divine origin and each family would remain faithful to the revelations of the subject as presented by their gods. To seek fundamental changes would be unthinkable for in asking others to accept such changes would be essentially asking them to change religious belief. Nor do these men appear to have made astronomical observations in any systematic way. Some of the texts do claim that the computed data presented in them is in better agreement with observation than that of their predecessors but, despite this, there does not seem to have been a major observational programme set up. Parameshvara in the late fourteenth century appears to be one of the first Indian mathematicians to make systematic observations over many years.

Mathematics however was in a different position. It was only a tool used for making astronomical calculations. If one could produce innovative mathematical ideas then one could exhibit the truths of astronomy more easily. The mathematics therefore had to lead to the same answers as had been reached before but it was certainly good if it could achieve these more easily or with greater clarity. This meant that despite mathematics only being used as a computational tool for astronomy, the brilliant Indian scholars were encouraged by their culture to put their genius into advances in this topic.

A contemporary of Brahmagupta who headed the research centre at Ujjain was Bhaskara I who led the Asmaka school. This school would have the study of the works of Aryabhatta as their main concern and certainly Bhaskara was commentator on the mathematics of Aryabhatta. More than 100 years after Bhaskara lived the astronomer Lalla, another commentator on Aryabhatta.

The ninth century saw mathematical progress with scholars such as Govindasvami, Mahavira, Prthudakasvami,

Shankara, and Sridhara. Some of these such as Govindasvami and Shankara were commentators on the text of Bhaskara I while Mahavira was famed for his updating of Brahmagupta's book. This period saw developments in sine tables, solving equations, algebraic notation, quadratics, indeterminate equations, and improvements to the number systems. The agenda was still basically that set by Aryabhatta and the topics being developed those in his work.

The main mathematicians of the tenth century in India were Aryabhatta II and Vijayanandi, both adding to the understanding of sine tables and trigonometry to support their astronomical calculations. In the eleventh century Sripati and Brahmadeva were major figures but perhaps the most outstanding of all was Bhaskara II in the twelfth century. He worked on algebra, number systems, and astronomy. He wrote beautiful texts illustrated with mathematical problems, some of which we present in his biography, and he provided the best summary of the mathematics and astronomy of the classical period.

Bhaskara II may be considered the high point of Indian mathematics but at one time this was all that was known:-

> *For a long time Western scholars thought that Indians had not done any original work till the time of Bhaskara II. This is far from the truth. Nor has the growth of Indian mathematics stopped with Bhaskara II. Quite a few results of Indian mathematicians have been rediscovered by Europeans. For instance, the development of number theory, the theory of indeterminates infinite series expressions for sine, cosine and tangent, computational mathematics, etc.*

Following Bhaskara II there was over 200 years before any other major contributions to mathematics were made on the Indian subcontinent. In fact for a long time it was thought that Bhaskara II represented the end of mathematical developments in the Indian subcontinent until modern times. However in the second half of the fourteenth century Mahendra Suri wrote the first Indian treatise on the astrolabe and Narayana wrote an important commentary on Bhaskara II, making important contributions to algebra and magic squares.

The most remarkable contribution from this period, however, was by Madhava who invented Taylor series and rigorous mathematical analysis in some inspired contributions. Madhava was from Kerala and his work there inspired a school of followers such as Nilakantha and Jyesthadeva.

Some of the remarkable discoveries of the Kerala mathematicians are described. These include: a formula for the ecliptic; the Newton-Gauss interpolation formula; the formula for the sum of an infinite series; Lhuilier's formula for the circumradius of a cyclic quadrilateral. Of particular interest is the approximation to the value of π which was the first to be made using a series. Madhava's result which gave a series for π, translated into the language of modern mathematics, reads

$$\pi R = 4R - 4R/3 + 4R/5 - ...$$

This formula, as well as several others referred to above, were rediscovered by European mathematicians several centuries later. Madhava also gave other formulae for π, one of which leads to the approximation 3.14159265359.

The first person in modern times to realise that the mathematicians of Kerala had anticipated some of the results of the Europeans on the calculus by nearly 300 years was Charles Whish in 1835. Whish's publication in the Transactions of the Royal Asiatic Society of Great Britain and Ireland was essentially unnoticed by historians of mathematics. Only 100 years later in the 1940s did historians of mathematics look in detail at the works of Kerala's mathematicians and find that the remarkable claims made by Whish were essentially true. Indeed the Kerala mathematicians had, as Whish wrote:-

... laid the foundation for a complete system of fluxions...

and these works:-

... abound with fluxional forms and series to be found in no work of foreign countries.

There were other major advances in Kerala at around this time. Citrabhanu was a sixteenth century mathematicians from Kerala who gave integer solutions to twenty-one types of systems of two algebraic equations. These types are all the possible pairs of equations of the following seven forms:

$x + y = a,$

$x - y = b,$

$xy = c,$

$x^2 + y^2 = d,$

$x^2 - y^2 = e,$

$x^3 + y^3 = f,$ and

$x^3 - y^3 = g.$

For each case, Citrabhanu gave an explanation and justification of his rule as well as an example. Some of his explanations are algebraic, while others are geometric.

Now we have presented the latter part of the history of Indian mathematics in an unlikely way. That there would be essentially no progress between the contributions of Bhaskara II and the innovations of Madhava, who was far more innovative than any other Indian mathematician producing a totally new perspective on mathematics, seems unlikely. Much more likely is that we are unaware of the contributions made over this 200 year period which must have provided the foundations on which Madhava built his theories.

Our understanding of the contributions of Indian mathematicians has changed markedly over the last few decades. Much more work needs to be done to further our understanding of the contributions of mathematicians whose work has sadly been lost, or perhaps even worse, been ignored. Indeed work is now being undertaken and we should soon have a better understanding of this important part of the history of mathematics.

Baudhayana

Born: about 800 BC in India

Died: about 800 BC in India

To write a biography of Baudhayana is essentially impossible since nothing is known of him except that he was the author of one of the earliest Sulbasutras. We do not know his dates accurately enough to even guess at a life span for him, which is why we have given the same approximate birth year as death year.

He was neither a mathematician in the sense that we would understand it today, nor a scribe who simply copied manuscripts like Ahmes. He would certainly have been a man of very considerable learning but probably not interested in mathematics for its own sake, merely interested in using it for religious purposes. Undoubtedly he wrote the Sulbasutra to provide rules for religious rites and it would appear an almost certainty that Baudhayana himself would be a Vedic priest.

The mathematics given in the Sulbasutras is there to enable the accurate construction of altars needed for sacrifices. It is clear from the writing that Baudhayana, as well as being a priest, must have been a skilled craftsman. He must have been himself skilled in the practical use of the mathematics he described as a craftsman who himself constructed sacrificial altars of the highest quality.

The Sulbasutras are discussed in detail in the article Indian Sulbasutras. Below we give one or two details of Baudhayana's Sulbasutra, which contained three chapters, which is the oldest which we possess and, it would be fair to say, one of the two most important.

The Sulbasutra of Baudhayana contains geometric solutions (but not algebraic ones) of a linear equation in a single unknown. Quadratic equations of the forms $ax^2 = c$ and $ax^2 + bx = c$ appear.

Several values of p occur in Baudhayana's Sulbasutra since when giving different constructions Baudhayana uses different approximations for constructing circular shapes. Constructions are given which are equivalent to taking p equal to $\frac{676}{225}$ (where $\frac{676}{225} = 3.004$), $\frac{900}{289}$ (where $\frac{900}{289} = 3.114$) and to $\frac{1156}{361}$ (where $\frac{1156}{361} = 3.202$). None of these is particularly accurate but, in the context of constructing altars they would not lead to noticeable errors.

An interesting, and quite accurate, approximate value for O2 is given in Chapter 1 verse 61 of Baudhayana's Sulbasutra. The Sanskrit text gives in words what we would write in symbols as

$$\sqrt{2} = 1 + \frac{1}{3} + \frac{1}{(3+4)} - \frac{1}{(3\times4\times34)} = 577/408$$

which is, to nine places, 1.414215686. This gives O2 correct to five decimal places. This is surprising since, as we mentioned above, great mathematical accuracy did not seem necessary for the building work described. If the approximation was given as:

$$\sqrt{2} = 1 + \frac{1}{3} + \frac{1}{(3\times4)}$$

then the error is of the order of 0.002 which is still more accurate than any of the values of π. Why then did Baudhayana feel that he had to go for a better approximation?

The Indian Sulbasutras

The Vedic people entered India about 1500 BC from the region that today is Iran. The word Vedic describes the religion of these people and the name comes from their collections of sacred texts known as the Vedas. The texts date from about the 15th to the 5th century BC and were used for sacrificial rites which were the main feature of the religion. There was a ritual which took place at an altar where food, also sometimes animals, were sacrificed. The Vedas contain recitations and chants to be used at these ceremonies. Later prose was added called Brahmanas which explained how the texts were to be used in the ceremonies. They also tell of the origin and the importance of the sacrificial rites themselves.

The Sulbasutras are appendices to the Vedas which give rules for constructing altars. If the ritual sacrifice was to be successful then the altar had to conform to very precise measurements. The people made sacrifices to their gods so that the gods might be pleased and give the people plenty food, good fortune, good health, long life, and lots of other material benefits. For the gods to be pleased everything had to be carried out with a very precise formula, so mathematical accuracy was seen to be of the utmost importance. We should also note that there were two types of sacrificial rites, one being a large public gathering while the other was a small family affair. Different types of altars were necessary for the two different types of ceremony.

All that is known of Vedic mathematics is contained in the Sulbasutras. This in itself gives us a problem, for we do not know if these people undertook mathematical investigations for their own sake, as for example the ancient Greeks did, or whether they only studied mathematics to solve problems necessary for their religious rites. Some historians have argued that mathematics, in particular geometry, must have also existed to support astronomical work being undertaken around the same period.

Certainly the Sulbasutras do not contain any proofs of the rules which they describe. Some of the rules, such as the method of constructing a square of area equal to a given rectangle, are exact. Others, such as constructing a square of area equal to that of a given circle, are approximations. We shall look at both of these examples below but the point we wish to make here is that the Sulbasutras make no distinction between the two. Did the writers of the Sulbasutras know which methods were exact and which were approximations?

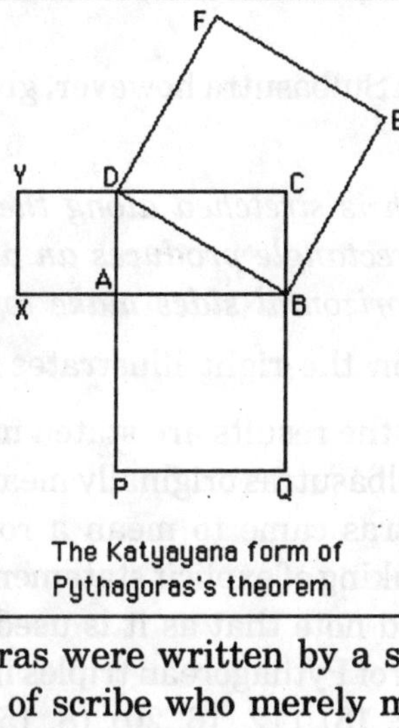

The Katyayana form of Pythagoras's theorem

The Sulbasutras were written by a scribe, although he was not the type of scribe who merely makes a copy of an existing document but one who put in considerable content and all the mathematical results may have been due to these scribes. We know nothing of the men who wrote the

Sulbasutras other than their names and a rough indication of the period in which they lived. Like many ancient mathematicians our only knowledge of them is their writings. The most important of these documents are the Baudhayana Sulbasutra written about 800 BC and the Apastamba Sulbasutra written about 600 BC. Historians of mathematics have also studied and written about other Sulbasutras of lesser importance such as the Manava Sulbasutra written about 750 BC and the Katyayana Sulbasutra written about 200 BC.

Let us now examine some of the mathematics contained within the Sulbasutras. The first result which was clearly known to the authors is Pythagoras's theorem. The Baudhayana Sulbasutra gives only a special case of the theorem explicitly:-

The rope which is stretched across the diagonal of a square produces an area double the size of the original square.

The Katyayana Sulbasutra however, gives a more general version:-

The rope which is stretched along the length of the diagonal of a rectangle produces an area which the vertical and horizontal sides make together.

The diagram on the right illustrates this result.

Note here that the results are stated in terms of "ropes". In fact, although sulbasutras originally meant rules governing religious rites, sutras came to mean a rope for measuring an altar. While thinking of explicit statements of Pythagoras's theorem, we should note that as it is used frequently there are many examples of Pythagorean triples in the Sulbasutras. For example (5, 12, 13), (12, 16, 20), (8, 15, 17), (15, 20, 25), (12, 35, 37), (15, 36, 39), $\left(\frac{5}{2}, 6, \frac{13}{2}\right)$, and $\left(\frac{15}{2}, 10, \frac{25}{2}\right)$ all occur.

Now the Sulbasutras are really construction manuals for geometric shapes such as squares, circles, rectangles, etc. and we illustrate this with some examples.

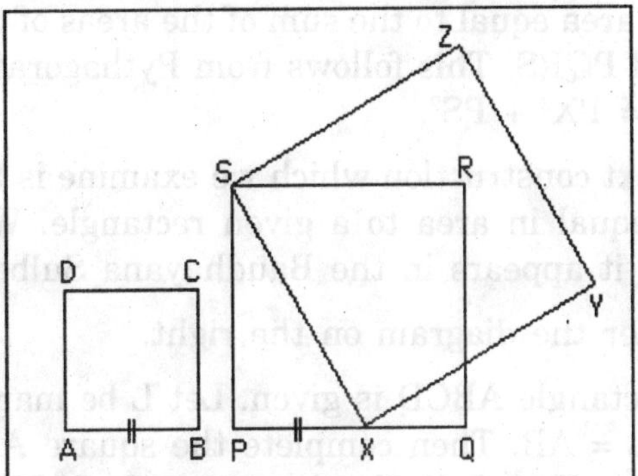

To find a square equal to the sum of two given squares - given in all Sulbasutras

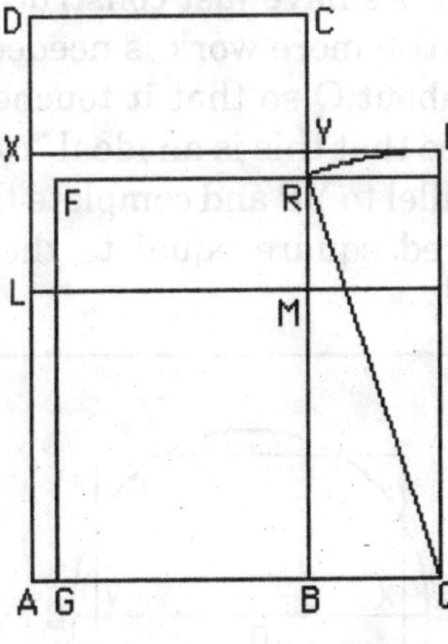

Finding a square equal to a given rectangle

The first construction we examine occurs in most of the different Sulbasutras. It is a construction, based on Pythagoras's theorem, for making a square equal in area to two given unequal squares.

Consider the diagram on the right.

ABCD and PQRS are the two given squares. Mark a point X on PQ so that PX is equal to AB. Then the square on SX has area equal to the sum of the areas of the squares ABCD and PQRS. This follows from Pythagoras's theorem since $SX^2 = PX^2 + PS^2$.

The next construction which we examine is that to find a square equal in area to a given rectangle. We give the version as it appears in the Baudhayana Sulbasutra.

Consider the diagram on the right.

The rectangle ABCD is given. Let L be marked on AD so that AL = AB. Then complete the square ABML. Now bisect LD at X and divide the rectangle LMCD into two equal rectangles with the line XY. Now move the rectangle XYCD to the position MBQN. Complete the square AQPX.

Now the square we have just constructed is not the one we require and a little more work is needed to complete the work. Rotate PQ about Q so that it touches BY at R. Then QP = QR and we see that this is an ideal "rope" construction. Now draw RE parallel to YP and complete the square QEFG. This is the required square equal to the given rectangle ABCD.

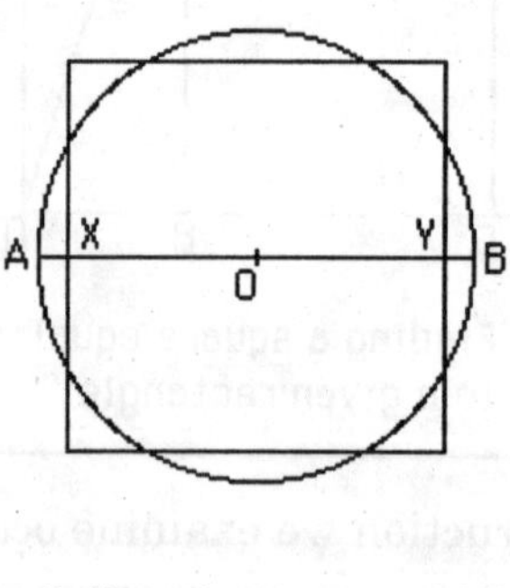

The Sulbasutras 13/15 method of "circling the square"

The Baudhayana Sulbasutra offers no proof of this result (or any other for that matter) but we can see that it is true by using Pythagoras's theorem.

$$\begin{aligned} EQ^2 &= QR^2 - RE^2 \\ &= QP^2 - YP^2 \\ &= ABYX + BQNM \\ &= ABYX + XYCD \\ &= ABCD. \end{aligned}$$

All the Sulbasutras contain a method to square the circle. It is an approximate method based on constructing a square of side $^{13}/_{15}$ times the diameter of the given circle as in the diagram on the right. This corresponds to taking

$\pi = \left(\frac{13}{15}\right)^2 = \frac{676}{225} = 3.00444$ so it is not a very good approximation and certainly not as good as was known earlier to the Babylonians.

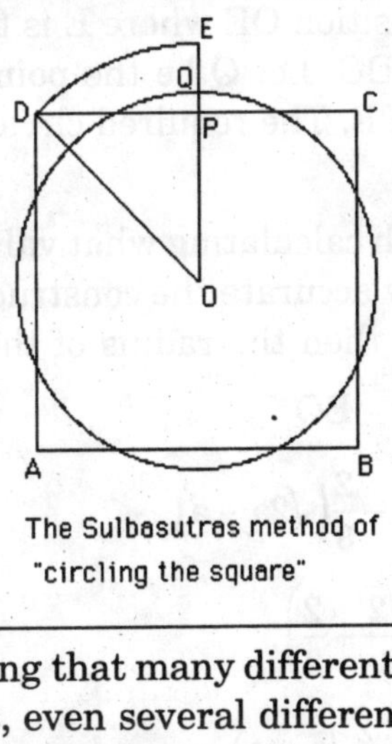

The Sulbasutras method of "circling the square"

It is worth noting that many different values of p appear in the Sulbasutras, even several different ones in the same text. This is not surprising for whenever an approximate construction is given some value of p is implied. The authors thought in terms of approximate constructions, not in terms of exact constructions with p but only having an approximate value for it. For example in the Baudhayana Sulbasutra, as well as the value of $\frac{676}{225}$, there appears $\frac{900}{289}$ and $\frac{1156}{361}$. In different Sulbasutras the values 2.99, 3.00, 3.004, 3.029, 3.047, 3.088, 3.1141, 3.16049 and 3.2022 can all be found; see. In the value $\pi = \frac{25}{8} = 3.125$ is found in the Manava Sulbasutras.

In addition to examining the problem of squaring the circle as given by Apastamba, the authors examine the problem of dividing a segment into seven equal parts which occurs in the same Sulbasutra.

Consider the diagram on the right.

The Sulbasutras also examine the converse problem of finding a circle equal in area to a given square. The following construction appears. Given a square ABCD find the centre O. Rotate OD to position OE where E is the midpoint of the side of the square DC. Let Q be the point on PE such that PQ is one third of PE. The required circle has centre O and radius OQ.

Again it is worth calculating what value of p this implies to get a feel for how accurate the construction is. Now if the square has side 2a then the radius of the circle is r where

$$r = OE - EQ$$

$$= \sqrt{2a} - \frac{2}{3}\left(\sqrt{2a} - a\right)$$

$$= a\left(\frac{\sqrt{2}}{3} + \frac{2}{3}\right).$$

Then $4a^2 = \pi a^2 \left(\frac{\sqrt{2}}{3} + \frac{2}{3}\right)^2$

which gives $\pi = \frac{36}{\left(\sqrt{2}+2\right)^2} = 3.088.$

As a final look at the mathematics of the Sulbasutras we examine what may be the most remarkable. Both the Apastamba Sulbasutra and the Katyayana Sulbasutra give the following approximation to $\sqrt{2}$:

Increase a unit length by its third and this third by its own fourth less the thirty-fourth part of that fourth.

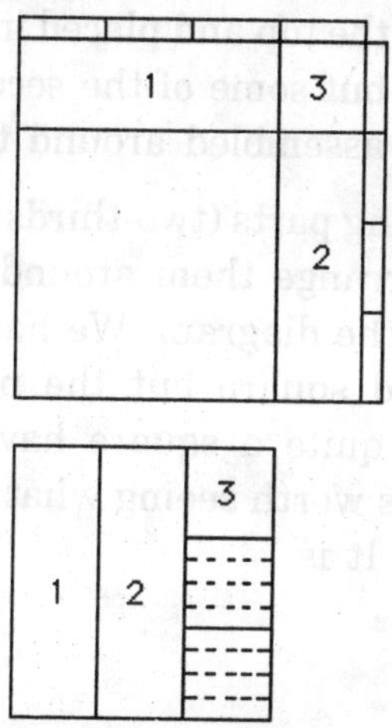

Datta's suggestion of how Apastamba found his approximation of √2

Now this gives

$$\sqrt{2} = 1 + \frac{1}{3} + \frac{1}{(3\times4)} - \frac{1}{(3\times4\times34)} = \frac{577}{408}$$

which is, to nine places, 1.414215686. Compare the correct value $\sqrt{2}$ = 1.414213562 to see that the Apastamba Sulbasutra has the answer correct to five decimal places. Of course no indication is given as to how the authors of the

Sulbasutras achieved this remarkable result. Datta, in 1932, made a beautiful suggestion as to how this approximation may have been reached.

The most likely reason for the construction was to build an altar twice the size of one already built. Datta's suggestion involves taking two squares and cutting up the second square and assembling it around the first square to give a square twice the size, thus having side $\sqrt{2}$. The second square is cut into three equal strips, and strips 1 and 2 placed around the first square as indicated in the diagram. The third strip has a square cut off the top and placed in position 3. We now have a new square but some of the second square remains and still has to be assembled around the first.

Cut the remaining parts (two-thirds of a strip) into eight equal strips and arrange them around the square we are constructing as in the diagram. We have now used all the parts of the second square but the new figure we have constructed is not quite a square having a small square corner missing. It is worth seeing what the side of this "not quite a square" is. It is

$$1+\frac{1}{3}+\frac{1}{(3\times 4)}$$

which, of course, is the first three terms of the approximation. Now Datta argues in that to improve the "not quite a square" the Sulbasutra authors could have calculated how broad a strip one needs to cut off the left hand side and bottom to fill in the missing part which has area $\left(\frac{1}{12}\right)^2$. If x is the width one cuts off then

$$2\times x\times\left(1+\frac{1}{3}+\frac{1}{12}\right)=\left(\frac{1}{12}\right)^2.$$

This has the solution $x = \frac{1}{(3 \times 4 \times 34)}$ which is approximately 0.002450980392. We now have a square the length of whose sides is

$$1+\frac{1}{3}+\frac{1}{(3\times4)}-\frac{1}{(3\times4\times34)}$$

which is exactly the approximation given by the Apastamba Sulbasutra.

Of course we have still made an approximation since the two strips of breadth x which we cut off overlapped by a square of side x in the bottom left hand corner. If we had taken this into account we would have obtained the equation

$$2\times x\times\left(1+\frac{1}{3}+\frac{1}{12}\right)-x^2=\left(\frac{1}{12}\right)^2$$

for x which leads to $x = \frac{17}{12} - \sqrt{2}$ which is approximately equal to 0.002453105. Of course we cannot take this route since we have arrived back at a value for x which involves $\sqrt{2}$ which is the quantity we are trying to approximate!

Gupta gives a simpler way of obtaining the approximation for $\sqrt{2}$ than that given by Datta. He uses linear interpolation to obtain the first two terms, he then corrects the two terms so obtaining the third term, then correcting the three terms obtaining the fourth term. Although the method given by Gupta is simpler (and an interesting contribution) there is certainly something appealing in Datta's argument and somehow a feeling that this is in the spirit of the Sulbasutras.

Of course the method used by these mathematicians is very important to understanding the depth of mathematics being produced in India in the middle of the first millennium BC. If we follow the suggestion of some historians that the writers of the Sulbasutras were merely copying an approximation already known to the Babylonians then we might come to the conclusion that Indian mathematics of this period was far less advanced than if we follow Datta's suggestion.

Manava

Born: about 750 BC in India

Died: about 750 BC in India

Manava was the author of one of the Sulbasutras. The Manava Sulbasutra is not the oldest (the one by Baudhayana is older) nor is it one of the most important, there being at least three Sulbasutras which are considered more important. We do not know Manava's dates accurately enough to even guess at a life span for him, which is why we have given the same approximate birth year as death year. Historians disagree on 750 BC, and some would put this Sulbasutra later by one hundred or more years.

Manava would have not have been a mathematician in the sense that we would understand it today. Nor was he a scribe who simply copied manuscripts like Ahmes. He would certainly have been a man of very considerable learning but probably not interested in mathematics for its own sake, merely interested in using it for religious purposes. Undoubtedly he wrote the Sulbasutra to provide rules for religious rites and it would appear an almost certainty that Manava himself would be a Vedic priest.

The mathematics given in the Sulbasutras is there to enable accurate construction of altars needed for sacrifices. It is clear from the writing that Manava, as well as being a priest, must have been a skilled craftsman.

Manava's Sulbasutra, like all the Sulbasutras, contained approximate constructions of circles from rectangles, and squares from circles, which can be thought of as giving approximate values of p. There appear therefore different values of p throughout the Sulbasutra, essentially every construction involving circles leads to a different such approximation. The paper is concerned with an interpretation of verses 11.14 and 11.15 of Manava's work which give

$$\pi = \frac{25}{8} = 3.125.$$

Apastamba

Born: about 600 BC in India

Died: about 600 BC in India

To write a biography of Apastamba is essentially impossible since nothing is known of him except that he was the author of a Sulbasutra which is certainly later than the Sulbasutra of Baudhayana. It would also be fair to say that Apastamba's Sulbasutra is the most interesting from a mathematical point of view. We do not know Apastamba's dates accurately enough to even guess at a life span for him, which is why we have given the same approximate birth year as death year.

Apastamba was neither a mathematician in the sense that we would understand it today, nor a scribe who simply copied manuscripts like Ahmes. He would certainly have been a man of very considerable learning but probably not interested in mathematics for its own sake, merely interested in using it for religious purposes. Undoubtedly he wrote the Sulbasutra to provide rules for religious rites and to improve and expand on the rules which had been given by his predecessors. Apastamba would have been a Vedic priest instructing the people in the ways of conducting the religious rites he describes.

The mathematics given in the Sulbasutras is there to enable the accurate construction of altars needed for sacrifices. It is clear from the writing that Apastamba, as well as being a priest and a teacher of religious practices, would have been a skilled craftsman. He must have been himself skilled in the practical use of the mathematics he described as a craftsman who himself constructed sacrificial altars of the highest quality.

The Sulbasutras are discussed in detail in the article Indian Sulbasutras. Below we give one or two details of Apastamba's Sulbasutra. This work is an expanded version of that of Baudhayana. Apastamba's work consisted of six chapters while the earlier work by Baudhayana contained only three.

The general linear equation was solved in the Apastamba's Sulbasutra. He also gives a remarkably accurate value for $\sqrt{2}$ namely

$$1+\frac{1}{3}+\frac{1}{(3\times 4)}-\frac{1}{(3\times 4\times 34)}.$$

which gives an answer correct to five decimal places. A possible way that Apastamba might have reached this remarkable result is described in the article Indian Sulbasutras.

As well as the problem of squaring the circle, Apastamba considers the problem of dividing a segment into 7 equal parts. The article looks in detail at a reconstruction of Apastamba's version of these two problems.

Panini

Born: about 520 BC in Shalatula (near Attock), now Pakistan

Died: about 460 BC in India

Panini was born in Shalatula, a town near to Attock on the Indus river in present day Pakistan. The dates given for

Panini are pure guesses. Experts give dates in the 4th, 5th, 6th and 7th century BC and there is also no agreement among historians about the extent of the work which he undertook. What is in little doubt is that, given the period in which he worked, he is one of the most innovative people in the whole development of knowledge. We will say a little more below about how historians have gone about trying to pinpoint the date when Panini lived.

Panini was a Sanskrit grammarian who gave a comprehensive and scientific theory of phonetics, phonology, and morphology. Sanskrit was the classical literary language of the Indian Hindus and Panini is considered the founder of the language and literature. It is interesting to note that the word "Sanskrit" means "complete" or "perfect" and it was thought of as the divine language, or language of the gods.

A treatise called Astadhyayi (or Astaka) is Pańini's major work. It consists of eight chapters, each subdivided into quarter chapters. In this work Panini distinguishes between the language of sacred texts and the usual language of communication. Panini gives formal production rules and definitions to describe Sanskrit grammar. Starting with about 1700 basic elements like nouns, verbs, vowels, consonants he put them into classes. The construction of sentences, compound nouns etc. is explained as ordered rules operating on underlying structures in a manner similar to modern theory. In many ways Panini's constructions are similar to the way that a mathematical function is defined today. Joseph writes:-

> *[Sanskrit's] potential for scientific use was greatly enhanced as a result of the thorough systemisation of its grammar by Panini. ... On the basis of just under 4000 sutras [rules expressed as aphorisms], he built virtually the whole structure of the Sanskrit language, whose general 'shape' hardly changed for the next two thousand years. ... An indirect consequence*

of Panini's efforts to increase the linguistic facility of Sanskrit soon became apparent in the character of scientific and mathematical literature. This may be brought out by comparing the grammar of Sanskrit with the geometry of Euclid - a particularly apposite comparison since, whereas mathematics grew out of philosophy in ancient Greece, it was ... partly an outcome of linguistic developments in India.

Joseph goes on to make a convincing argument for the algebraic nature of Indian mathematics arising as a consequence of the structure of the Sanskrit language. In particular he suggests that algebraic reasoning, the Indian way of representing numbers by words, and ultimately the development of modern number systems in India, are linked through the structure of language.

Panini should be thought of as the forerunner of the modern formal language theory used to specify computer languages. The Backus Normal Form was discovered independently by John Backus in 1959, but Panini's notation is equivalent in its power to that of Backus and has many similar properties. It is remarkable to think that concepts which are fundamental to today's theoretical computer science should have their origin with an Indian genius around 2500 years ago.

At the beginning of this article we mentioned that certain concepts had been attributed to Panini by certain historians which others dispute. One such theory was put forward by B. Indraji in 1876. He claimed that the Brahmi numerals developed out of using letters or syllables as numerals. Then he put the finishing touches to the theory by suggesting that Panini in the eighth century BC (earlier than most historians place Panini) was the first to come up with the idea of using letters of the alphabet to represent numbers.

There are a number of pieces of evidence to support Indraji's theory that the Brahmi numerals developed from letters or syllables. However it is not totally convincing since, to quote one example, the symbols for 1, 2 and 3 clearly do not come from letters but from one, two and three lines respectively. Even if one accepts the link between the numerals and the letters, making Panini the originator of this idea would seem to have no more behind it than knowing that Panini was one of the most innovative geniuses that world has known so it is not unreasonable to believe that he might have made this step too.

There are other works which are closely associated with the Astadhyayi which some historians attribute to Panini, others attribute to authors before Panini, others attribute to authors after Panini. This is an area where there are many theories but few, if any, hard facts.

We also promised to return to a discussion of Panini's dates. There has been no lack of work on this topic so the fact that there are theories which span several hundreds of years is not the result of lack of effort, rather an indication of the difficulty of the topic. The usual way to date such texts would be to examine which authors are referred to and which authors refer to the work. One can use this technique and see who Panini mentions.

There are ten scholars mentioned by Panini and we must assume from the context that these ten have all contributed to the study of Sanskrit grammar. This in itself, of course, indicates that Panini was not a solitary genius but, like Newton, had "stood on the shoulders of giants". Panini must have lived later than these ten but this is absolutely no help in providing dates since we have absolutely no knowledge of when any of these ten lived.

What other internal evidence is there to use? Well of course Panini uses many phrases to illustrate his grammar any these have been examined meticulously to see if anything is contained there to indicate a date. To give an example of what we mean: if we were to pick up a text which contained as an example "I take the train to work every day" we would know that it had to have been written after railways became common. Let us illustrate with two actual examples from the Astadhyayi which have been the subject of much study. The first is an attempt to see whether there is evidence of Greek influence. Would it be possible to find evidence which would mean that the text had to have been written after the conquests of Alexander the Great? There is a little evidence of Greek influence, but there was Greek influence on this north east part of the Indian subcontinent before the time of Alexander. Nothing conclusive has been identified.

Another angle is to examine a reference Panini makes to nuns. Some argue that these must be Buddhist nuns and therefore the work must have been written after Buddha. A nice argument but there is a counter argument which says that there were Jaina nuns before the time of Buddha and Panini's reference could equally well be to them. Again the evidence is inconclusive.

There are references by others to Panini. However it would appear that the Panini to whom most refer is a poet and although some argue that these are the same person, most historians agree that the linguist and the poet are two different people. Again this is inconclusive evidence. Let us end with an evaluation of Panini's contribution by Cardona:-

> *Panini's grammar has been evaluated from various points of view. After all these different evaluations, I think that the grammar merits asserting ... that it is one of the greatest monuments of human intelligence.*

Katyayana

Born: about 200 BC in India

Died: about 200 BC in India

We cannot attempt to write a biography of Katyayana since essentially nothing is known of him except that he was the author of a Sulbasutra which is much later than the Sulbasutras of Baudhayana and Apastamba. It would also be fair to say that Katyayana's Sulbasutra is the least interesting from a mathematical point of view of the three best known Sulbasutras. It adds very little to that of Apastamba written several hundreds of years earlier. We do not know Katyayana's dates accurately enough to even guess at a life span for him, which is why we have given the same approximate birth year as death year.

Katyayana was neither a mathematician in the sense that we would understand it today, nor a scribe who simply copied manuscripts like Ahmes. He would certainly have been a man of very considerable learning but probably not interested in mathematics for its own sake, merely interested in using it for religious purposes. Undoubtedly he wrote the Sulbasutra to provide rules for religious rites and to improve and expand on the rules which had been given by his predecessors. Katyayana would have been a priest instructing the people in the ways of conducting the religious rites he describes.

Katyayana lived in a period when the religious rites that the Sulbasutras were written to support were becoming less influential. People were turning to other religions and perhaps this lack of vigour in the religion at this time partly explains why several hundreds of years after Apastamba Katyayana adds little of importance to the Sulbasutra which he wrote.

Yavanesvara

Born: about 120 in Western India

Died: about 180 in India

Indian astrology was originally known as Jyotisha, which means "science of the stars". Until around the first century AD no real distinction was made between astrology and astronomy and in fact most astronomical theories were propounded to support the theory that the positions of the heavenly bodies directly influenced human events.

The Indian methods of computing horoscopes all date back to the translation of a Greek astrology text into Sanskrit prose by Yavanesvara in 149 AD. Yavanesvara (or Yavanaraja) literally means "Lord of the Greeks" and it was a name given to many officials in western India during the period 130 AD - 390 AD. During this period the Ksatrapas ruled Gujarat (or Madhya Pradesh) and these "Lord of the Greeks" officials acted for the Greek merchants living in the area.

The particular "Lord of the Greeks" official Yavanesvara who we are interested in here worked under Rudradaman. Rudradaman became ruler of the Ksatrapas in around 130 AD and it was during the period of his rule that Yavanesvara worked as an official and made his translation. We know of Rudradaman because information is recorded in a lengthy Sanskrit inscription at Junagadh written around 150 AD.

The Greek astrology text in question was written in Alexandria some time round about 120 BC. Yavanesvara did far more than just translate the Greek text for such a translation would have had little relevance to the Indians. He therefore not only translated the language but he translated the context too. Instead of the Greek gods who appear in the original, Yavanesvara used Hindu images. Again he worked the Indian caste system into the work and made the work one which would fit well with the Indian thought.

The work was written with the aim of letting Indians became astrologers so it had to present astronomy in a form in which it could be used for astrology. In order to do this Yavanesvara put into his work an explanation of the Greek version of the Babylonian theory of the motions of the planets. All this he wrote in Sanskrit prose but sadly the original has not survived. We do have, however, a version written in Sanskrit verse 120 years after Yavanesvara's work appeared.

Yavanesvara had an important influence on the whole of astrology in India for centuries after he made his popular translation. Although the influence was more than on astrology, as the science of astronomy split from astrology, the influence of Yavanesvara's work reached into astronomy too.

Aryabhatta the Elder

Born: 476 in Kusumapura (now Patna), India

Died: 550 in India

Aryabhatta is also known as Aryabhatta I to distinguish him from the later mathematician of the same name who lived about 400 years later. Al-Biruni has not helped in understanding Aryabhatta's life, for he seemed to believe that there were two different mathematicians called Aryabhatta living at the same time. He therefore created a confusion of two different Aryabhatas which was not clarified until 1926 when B Datta showed that al-Biruni's two Aryabhatas were one and the same person.

We know the year of Aryabhatta's birth since he tells us that he was twenty-three years of age when he wrote Aryabhatiya which he finished in 499. We have given Kusumapura, thought to be close to Pataliputra (which was refounded as Patna in Bihar in 1541), as the place of Aryabhatta's birth but this is far from certain, as is even the location of Kusumapura itself. As Parameswaran writes:-

> *... no final verdict can be given regarding the locations of Asmakajanapada and Kusumapura.*

We do know that Aryabhatta wrote Aryabhatiya in Kusumapura at the time when Pataliputra was the capital of the Gupta empire and a major centre of learning, but there have been numerous other places proposed by historians as his birthplace. Some conjecture that he was born in south India, perhaps Kerala, Tamil Nadu or Andhra Pradesh, while others conjecture that he was born in the north-east of India, perhaps in Bengal. It is claimed that Aryabhatta was born in the Asmaka region of the Vakataka dynasty in South India although the author accepted that he lived most of his life in Kusumapura in the Gupta empire of the north. However, giving Asmaka as Aryabhaṭta's birthplace rests on a comment made by Nilakantha Somayaji in the late 15th century. It is now thought by most historians that Nilakantha confused Aryabhatta with Bhaskara I who was a later commentator on the Aryabhatiya.

We should note that Kusumapura became one of the two major mathematical centres of India, the other being Ujjain. Both are in the north but Kusumapura (assuming it to be close to Pataliputra) is on the Ganges and is the more northerly. Pataliputra, being the capital of the Gupta empire at the time of Aryabhatta, was the centre of a communications network which allowed learning from other parts of the world to reach it easily, and also allowed the mathematical and astronomical advances made by Aryabhatta and his school to reach across India and also eventually into the Islamic world.

As to the texts written by Aryabhatta only one has survived. However Jha claims that:-

> *... Aryabhatta was an author of at least three astronomical texts and wrote some free stanzas as well.*

The surviving text is Aryabhatta's masterpiece the Aryabhatiya which is a small astronomical treatise written in 118 verses giving a summary of Hindu mathematics up to that time. Its mathematical section contains 33 verses giving 66 mathematical rules without proof. The Aryabhatiya contains an introduction of 10 verses, followed by a section on mathematics with, as we just mentioned, 33 verses, then a section of 25 verses on the reckoning of time and planetary models, with the final section of 50 verses being on the sphere and eclipses.

There is a difficulty with this layout which is discussed in detail by van der Waerden. Van der Waerden suggests that in fact the 10 verse Introduction was written later than the other three sections. One reason for believing that the two parts were not intended as a whole is that the first section has a different meter to the remaining three sections. However, the problems do not stop there. We said that the first section had ten verses and indeed Aryabhatta titles the section Set of ten giti stanzas. But it in fact contains eleven giti stanzas and two arya stanzas. Van der Waerden suggests that three verses have been added and he identifies a small number of verses in the remaining sections which he argues have also been added by a member of Aryabhatta's school at Kusumapura.

The mathematical part of the Aryabhatiya covers arithmetic, algebra, plane trigonometry and spherical trigonometry. It also contains continued fractions, quadratic equations, sums of power series and a table of sines. Let us examine some of these in a little more detail.

First we look at the system for representing numbers which Aryabhatta invented and used in the Aryabhatiya. It consists of giving numerical values to the 33 consonants of the Indian alphabet to represent 1, 2, 3, ... , 25, 30, 40, 50, 60, 70, 80, 90, 100. The higher numbers are denoted by these consonants followed by a vowel to obtain 100, 10000, ... In fact the system allows numbers up to 1018 to be represented with an alphabetical notation. Ifrah argues that Aryabhatta was also familiar with numeral symbols and the place-value system. He writes:-

> *... it is extremely likely that Aryabhatta knew the sign for zero and the numerals of the place value system. This supposition is based on the following two facts:*

first, the invention of his alphabetical counting system would have been impossible without zero or the place-value system; secondly, he carries out calculations on square and cubic roots which are impossible if the numbers in question are not written according to the place-value system and zero.

Next we look briefly at some algebra contained in the Aryabhatiya. This work is the first we are aware of which examines integer solutions to equations of the form by =ax+c and by = ax - c, where a, b, c are integers. The problem arose from studying the problem in astronomy of determining the periods of the planets. Aryabhatta uses the kuttaka method to solve problems of this type. The word kuttaka means "to pulverise" and the method consisted of breaking the problem down into new problems where the coefficients became smaller and smaller with each step. The method here is essentially the use of the Euclidean algorithm to find the highest common factor of a and b but is also related to continued fractions.

Aryabhatta gave an accurate approximation for p. He wrote in the Aryabhatiya the following:

Add four to one hundred, multiply by eight and then add sixty-two thousand the result is approximately the circumference of a circle of diameter twenty thousand. By this rule the relation of the circumference to diameter is given.

This gives $\pi = \frac{62832}{20000} = 3.1416$ which is a surprisingly accurate value. In fact $\pi = 3.14159265$ correct to 8 places. If obtaining a value this accurate is surprising, it is perhaps even more surprising that Aryabhatta does not use his accurate value for p but prefers to use O10 = 3.1622 in practice. Aryabhatta does not explain how he found this accurate value but, for example, Ahmad considers this value as an approximation to half the perimeter of a regular polygon of 256 sides inscribed in the unit circle. However, Bruins shows that this result cannot be obtained from the doubling of the number of sides. Another interesting paper discussing this accurate value of π by Aryabhatta where Jha writes:

> *Aryabhatta I's value of π is a very close approximation to the modern value and the most accurate among those of the ancients. There are reasons to believe that Aryabhatta devised a particular method for finding this value. It is shown with sufficient grounds that Aryabhatta himself used it, and several later Indian mathematicians and even the Arabs adopted it. The conjecture that Aryabhatta's value of π is of Greek origin is critically examined and is found to be without foundation. Aryabhatta discovered this value independently and also realised that π is an irrational number. He had the Indian background, no doubt, but excelled all his predecessors in evaluating π. Thus the credit of discovering this exact value of π may be ascribed to the celebrated mathematician, Aryabhatta I.*

We now look at the trigonometry contained in Aryabhatta's treatise. He gave a table of sines calculating the approximate values at intervals of 90°/24 = 3°45'. In order to do this he used a formula for sin(n+1)x - sin nx in terms of sin nx and sin (n-1)x. He also introduced the versine (versin = 1 - cosine) into trigonometry.

Other rules given by Aryabhatta include that for summing the first n integers, the squares of these integers and also their cubes. Aryabhatta gives formulae for the areas of a triangle and of a circle which are correct, but the formulae for the volumes of a sphere and of a pyramid are claimed to be wrong by most historians. For example Ganitanand describes as "mathematical lapses" the fact that Aryabhatta gives the incorrect formula V = Ah/2 for the volume of a pyramid with height h and triangular base of area A. He also appears to give an incorrect expression for the volume of a sphere. However, as is often the case, nothing is as straightforward as it appears and Elfering argues that this is not an error but rather the result of an incorrect translation.

This relates to verses 6, 7, and 10 of the second section of the Aryabhatiya and Elfering produces a translation which yields the correct answer for both the volume of a pyramid and for a sphere. However, in his translation Elfering translates two technical terms in a different way to the meaning which they usually have. Without some supporting evidence that these technical terms have been used with these different meanings in other places it would still appear that Aryabhatta did indeed give the incorrect formulae for these volumes.

We have looked at the mathematics contained in the Aryabhatiya but this is an astronomy text so we should say a little regarding the astronomy which it contains. Aryabhatta gives a systematic treatment of the position of the planets in space. He gave the circumference of the earth as 4 967 yojanas and its diameter as $1\ 581\frac{1}{24}$ yojanas.

Since 1 yojana = 5 miles this gives the circumference as 24 835 miles, which is an excellent approximation to the currently accepted value of 24 902 miles. He believed that the apparent rotation of the heavens was due to the axial rotation of the Earth. This is a quite remarkable view of the nature of the solar system which later commentators could not bring themselves to follow and most changed the text to save Aryabhatta from what they thought were stupid errors!

Aryabhatta gives the radius of the planetary orbits in terms of the radius of the Earth/Sun orbit as essentially their periods of rotation around the Sun. He believes that the Moon and planets shine by reflected sunlight, incredibly he believes that the orbits of the planets are ellipses. He correctly explains the causes of eclipses of the Sun and the Moon. The Indian belief up to that time was that eclipses were caused by a demon called Rahu. His value for the length of the year at 365 days 6 hours 12 minutes 30 seconds is an overestimate since the true value is less than 365 days 6 hours.

Bhaskara I who wrote a commentary on the Aryabhatiya about 100 years later wrote of Aryabhatta:-

> *Aryabhatta is the master who, after reaching the furthest shores and plumbing the inmost depths of the sea of ultimate knowledge of mathematics, kinematics and spherics, handed over the three sciences to the learned world.*

Yativrsabha

Born: about 500 in India

Died: about 570 in India

Yativrsabha (or Jadivasaha) was a Jaina mathematician who studied under Arya Manksu and Nagahastin. We know nothing of Yativrsabha's dates except for a reference which he makes to the end of the Gupta dynasty which he says was after 231 years of ruling. This ended in 551 so we must assume that 551 AD is a date which occurred during Yativrsabha's lifetime. This fits with the only other information regarding his dates which are that his work is referenced by Jinabhadra Ksamasramana in 609 and that Yativrsabha himself refers to a work written by Sarvanandin in 458.

Yativrsabha's work Tiloyapannatti gives various units for measuring distances and time and also describes the system of infinite time measures. It is a work which describes Jaina cosmology and gives a description of the universe which is of historical importance in understanding Jaina science and mathematics. The Jaina belief was in an infinite world, both infinite in space and in time. This led the Jainas to devise ways of measuring larger and larger distances and longer and longer intervals of time. It led them to consider different measures of infinity, and in this respect the Jaina mathematicians would appear to be the only ones before the time when Cantor developed the theory of infinite cardinals to envisage different magnitudes of infinity.

Varahamihira

Born: 505 in Kapitthaka, India

Died: 587 in India

Our knowledge of Varahamihira is very limited indeed. According to one of his works, he was educated in Kapitthaka. However, far from settling the question this only gives rise to discussions of possible interpretations of where this place was. Dhavale discusses this problem. We do not know whether he was born in Kapitthaka, wherever that may be, although we have given this as the most likely guess.

We do know, however, that he worked at Ujjain which had been an important centre for mathematics since around 400 AD. The school of mathematics at Ujjain was increased in importance due to Varahamihira working there and it continued for a long period to be one of the two leading mathematical centres in India, in particular having Brahmagupta as its next major figure.

The most famous work by Varahamihira is the Pancasiddhantika (The Five Astronomical Canons) dated 575 AD. This work is important in itself and also in giving us information about older Indian texts which are now lost. The work is a treatise on mathematical astronomy and it summarises five earlier astronomical treatises, namely the Surya, Romaka, Paulisa, Vasistha and Paitamaha siddhantas. Shukla states:-

The Pancasiddhantika of Varahamihira is one of the most important sources for the history of Hindu astronomy before the time of Aryabhatta II.

One treatise which Varahamihira summarises was the Romaka-Siddhanta which itself was based on the epicycle theory of the motions of the Sun and the Moon given by the Greeks in the 1st century AD. The Romaka-Siddhanta was based on the tropical year of Hipparchus and on the Metonic cycle of 19 years. Other works which Varahamihira summarises are also based on the Greek epicycle theory of the motions of the heavenly bodies. He revised the calendar by updating these earlier works to take into account precession since they were written. The Pancasiddhantika also contains many examples of the use of a place-value number system.

There is, however, quite a debate about interpreting data from Varahamihira's astronomical texts and from other similar works. Some believe that the astronomical theories are Babylonian in origin, while others argue that the Indians refined the Babylonian models by making observations of their own. Much needs to be done in this area to clarify some of these interesting theories.

Ifrah notes that Varahamihira was one of the most famous astrologers in Indian history. His work Brihatsamhita (The Great Compilation) discusses topics such as:-

... descriptions of heavenly bodies, their movements and conjunctions, meteorological phenomena, indications of the omens these movements, conjunctions and phenomena represent, what action to take and operations to accomplish, sign to look for in humans, animals, precious stones, etc.

Varahamihira made some important mathematical discoveries. Among these are certain trigonometric formulae which translated into our present day notation correspond to:

$\sin x = \cos(\pi/2 - x)$,

$\sin^2 x + \cos^2 x = 1$, and

$(1 - \cos 2x)/2 = \sin^2 x$.

Another important contribution to trigonometry was his sine tables where he improved those of Aryabhatta I giving more accurate values.

It should be emphasised that accuracy was very important for these Indian mathematicians since they were computing sine tables for applications to astronomy and astrology. This motivated much of the improved accuracy they achieved by developing new interpolation methods.

The Jaina school of mathematics investigated rules for computing the number of ways in which r objects can be selected from n objects over the course of many hundreds of years. They gave rules to compute the binomial coefficients nCr which amount to

$${}_nC_r = n(n-1)(n-2)...(n-r+1)/r!$$

However, Varahamihira attacked the problem of computing ${}_nC_r$ in a rather different way. He wrote the numbers n in a column with n = 1 at the bottom. He then put the numbers r in rows with r = 1 at the left-hand side. Starting at the bottom left side of the array which corresponds to the values n = 1, r = 1, the values of ${}_nC_r$ are found by summing two entries, namely the one directly below the (n, r) position and the one immediately to the left of it. Of course this table is none other than Pascal's triangle for finding the binomial coefficients despite being viewed from a different angle from the way we build it up today. Full details of this work by Varahamihira is given.

Hayashi, examines Varahamihira's work on magic squares. In particular he examines a pandiagonal magic square of order four which occurs in Varahamihira's work.

Brahmagupta

Born: 598 in (possibly) Ujjain, India

Died: 670 in India

Brahmagupta, whose father was Jisnugupta, wrote important works on mathematics and astronomy. In particular he wrote Brahmasphutasiddhanta (The Opening of the Universe), in 628. The work was written in 25 chapters and Brahmagupta tells us in the text that he wrote it at Bhillamala which today is the city of Bhinmal. This was the capital of the lands ruled by the Gurjara dynasty.

Brahmagupta became the head of the astronomical observatory at Ujjain which was the foremost mathematical centre of ancient India at this time. Outstanding mathematicians such as Varahamihira had worked there and built up a strong school of mathematical astronomy.

In addition to the Brahmasphutasiddhanta Brahmagupta wrote a second work on mathematics and astronomy which is the Khandakhadyaka written in 665 when he was 67 years old. We look below at some of the remarkable ideas which Brahmagupta's two treatises contain. First let us give an overview of their contents.

The Brahmasphutasiddhanta contains twenty-five chapters but the first ten of these chapters seem to form what many historians believe was a first version of Brahmagupta's work and some manuscripts exist which contain only these chapters. These ten chapters are arranged in topics which are typical of Indian mathematical astronomy texts of the period.

The topics covered are: mean longitudes of the planets; true longitudes of the planets; the three problems of diurnal rotation; lunar eclipses; solar eclipses; risings and settings; the moon's crescent; the moon's shadow; conjunctions of the planets with each other; and conjunctions of the planets with the fixed stars.

The remaining fifteen chapters seem to form a second work which is major addendum to the original treatise. The chapters are: examination of previous treatises on astronomy; on mathematics; additions to chapter 1; additions to chapter 2; additions to chapter 3; additions to chapter 4 and 5; additions to chapter 7; on algebra; on the gnomon; on meters; on the sphere; on instruments; summary of contents; versified tables.

Brahmagupta's understanding of the number systems went far beyond that of others of the period. In the Brahmasphutasiddhanta he defined zero as the result of subtracting a number from itself. He gave some properties as follows:-

When zero is added to a number or subtracted from a number, the number remains unchanged; and a number multiplied by zero becomes zero.

He also gives arithmetical rules in terms of fortunes (positive numbers) and debts (negative numbers):-

A debt minus zero is a debt.

A fortune minus zero is a fortune.

Zero minus zero is a zero.

A debt subtracted from zero is a fortune.

A fortune subtracted from zero is a debt.

The product of zero multiplied by a debt or fortune is zero.

The product of zero multipliedby zero is zero.

The product or quotient of two fortunes is one fortune.

The product or quotient of two debts is one fortune.

The product or quotient of a debt and a fortune is a debt.

The product or quotient of a fortune and a debt is a debt.

Brahmagupta then tried to extend arithmetic to include division by zero:-

Positive or negative numbers when divided by zero is a fraction the zero as denominator.

Zero divided by negative or positive numbers is either zero or is expressed as a fraction with zero as numerator and the finite quantity as denominator.

Zero divided by zero is zero.

Really Brahmagupta is saying very little when he suggests that n divided by zero is n/0. He is certainly wrong when he then claims that zero divided by zero is zero. However it is a brilliant attempt to extend arithmetic to negative numbers and zero.

We can also describe his methods of multiplication which use the place-value system to its full advantage in almost the same way as it is used today. We give three examples of the methods he presents in the Brahmasphuta siddhanta and in doing so we follow Ifrah. The first method we describe is called "gomutrika" by Brahmagupta. Ifrah translates "gomutrika" to "like the trajectory of a cow's urine". Consider the product of 235 multiplied by 264. We begin by setting out the sum as follows:

2 235
6 235
4 235

Now multiply the 235 of the top row by the 2 in the top position of the left hand column. Begin by 2 5 = 10, putting 0 below the 5 of the top row, carrying 1 in the usual way to get

2 235
6 235
4 235

470

Now multiply the 235 of the second row by the 6 in the left hand column writing the number in the line below the 470 but moved one place to the right

2 235

6 235

4 235

470

1410

Now multiply the 235 of the third row by the 4 in the left hand column writing the number in the line below the 1410 but moved one place to the right

2 235

6 235

4 235

470

1410

940

Now add the three numbers below the line

2 235

6 235

4 235

470

1410

940

62040

The variants are first writing the second number on the right but with the order of the digits reversed as follows

235 4
235 6
235 2

940
1410
470

62040

The third variant just writes each number once but otherwise follows the second method

235
----
940 4
1410 6
470 2

62040

Another arithmetical result presented by Brahmagupta is his algorithm for computing square roots. This algorithm is discussed where it is shown to be equivalent to the Newton-Raphson iterative formula.

Brahmagupta developed some algebraic notation and presents methods to solve quardatic equations. He presents methods to solve indeterminate equations of the form $ax + c = by$. Majumdar writes:-

Brahmagupta perhaps used the method of continued fractions to find the integral solution of an indeterminate equation of the type $ax + c = by$.

Majumdar gives the original Sanskrit verses from Brahmagupta's Brahmasphuta siddhanta and their English translation with modern interpretation.

Brahmagupta also solves quadratic indeterminate equations of the type $ax^2 + c = y^2$ and $ax^2 - c = y^2$. For example he solves $8x^2 + 1 = y^2$ obtaining the solutions (x,y) = (1,3), (6,17), (35,99), (204,577), (1189,3363), ... For the equation $11x^2 + 1 = y^2$ Brahmagupta obtained the solutions (x,y) = (3,10), (161/5,534/5), ... He also solves $61x^2 + 1 = y^2$ which is particularly elegant having x = 226153980, y = 1766319049 as its smallest solution.

A example of the type of problems Brahmagupta poses and solves in the Brahmasphutasiddhanta is the following:-

> *Five hundred drammas were loaned at an unknown rate of interest, The interest on the money for four months was loaned to another at the same rate of interest and amounted in ten mounths to 78 drammas. Give the rate of interest.*

Rules for summing series are also given. Brahmagupta gives the sum of the squares of the first n natural numbers as n(n+1)(2n+1)/6 and the sum of the cubes of the first n natural numbers as $(n(n+1)/2)^2$. No proofs are given so we do not know how Brahmagupta discovered these formulae.

In the Brahmasphutasiddhanta Brahmagupta gave remarkable formulae for the area of a cyclic quadrilateral and for the lengths of the diagonals in terms of the sides. The only debatable point here is that Brahmagupta does not state that the formulae are only true for cyclic quadrilaterals so some historians claim it to be an error while others claim that he clearly meant the rules to apply only to cyclic quadrilaterals.

Much material in the Brahmasphutasiddhanta deals with solar and lunar eclipses, planetary conjunctions and positions of the planets. Brahmagupta believed in a static Earth and he gave the length of the year as 365 days 6 hours 5 minutes 19 seconds in the first work, changing the value to 365 days 6 hours 12 minutes 36 seconds in the second book the Khandakhadyaka. This second values is not, of course, an improvement on the first since the true length of the years if less than 365 days 6 hours. One has to wonder whether Brahmagupta's second value for the length of the year is taken from Aryabhatta I since the two agree to within 6 seconds, yet are about 24 minutes out.

The Khandakhadyaka is in eight chapters again covering topics such as: the longitudes of the planets; the three problems of diurnal rotation; lunar eclipses; solar eclipses; risings and settings; the moon's crescent; and conjunctions of the planets. It contains an appendix which is some versions has only one chapter, in other versions has three.

Of particular interest to mathematics in this second work by Brahmagupta is the interpolation formula he uses to compute values of sines. This is studied in detail where it is shown to be a particular case up to second order of the more general Newton-Stirling interpolation formula.

Bhaskara I

Born: about 600 in (possibly) Saurastra (modern Gujarat state), India

Died: about 680 in (possibly) Asmaka, India

We have very little information about Bhaskara I's life except what can be deduced from his writings. Shukla deduces from the fact that Bhaskara I often refers to the Asmakatantra instead of the Aryabhatiya that he must have been working in a school of mathematicians in Asmaka which was probably in the Nizamabad District of Andhra Pradesh. If this is correct, and it does seem quite likely, then the school in Asmaka would have been a collection of scholars who were followers of Aryabhatta I and of course this fits in well with the fact that Bhaskara I himself was certainly a follower of Aryabhatta I.

There are other references to places in India in Bhaskara's writings. For example he mentions Valabhi (today Vala), the capital of the Maitraka dynasty in the 7th century, and Sivarajapura, which were both in Saurastra which today is the Gujarat state of India on the west coast of the continent. Also mentioned are Bharuch (or Broach) in southern Gujarat and Thanesar in the eastern Punjab which was ruled by Harsa for 41 years from 606. Harsa was the preeminent ruler in north India through the first half of Bhaskara I's life. A reasonable guess would be that Bhaskara was born in Saurastra and later moved to Asmaka.

Bhaskara I was an author of two treatises and commentaries to the work of Aryabhatta I. His works are the Mahabhaskariya, the Laghubhaskariya and the Aryabhatiyabhasya. The Mahabhaskariya is an eight chapter work on Indian mathematical astronomy and includes topics which were fairly standard for such works at this time. It discusses topics such as: the longitudes of the planets; conjunctions of the planets with each other and with bright stars; eclipses of the sun and the moon; risings and settings; and the lunar crescent.

Bhaskara I included in his treatise the Mahabhaskariya three verses which give an approximation to the trigonometric sine function by means of a rational fraction. These occur in Chapter 7 of the work. The formula which Bhaskara gives is amazingly accurate and use of the formula leads to a maximum error of less than one percent. The formula is

$$\sigma\iota\nu\ \xi = 16\xi\ (\pi - x)/[5\pi^2 - 4x\ (\pi - x)]$$

and Bhaskara attributes the work as that of Aryabhatta I. We have computed the values given by the formula and compared it with the correct value for sin x for x from 0 to p/2 in steps of $\pi/20$.

$x = 0$ formula = 0.00000 sin x = 0.00000 error = 0.00000

$x = \pi/20$ formula = 0.15800 sin x = 0.15643 error = 0.00157

$x = \pi/10$ formula = 0.31034 sin x = 0.30903 error = 0.00131

$x = 3\pi/20$ formula = 0.45434 sin x = 0.45399 error = 0.00035

$x = \pi/5$ formula = 0.58716 sin x = 0.58778 error = -0.00062

$x = \pi/4$ formula = 0.70588 sin x = 0.70710 error = -0.00122 x = π/10 formula = 0.80769 sin x = 0.80903 error = -0.00134

x = 7π/20 formula = 0.88998 sin x = 0.89103 error = -0.00105

x = 2π/5 formula = 0.95050 sin x = 0.95105 error = -0.00055

x = 9π/20 formula = 0.98753 sin x = 0.98769 error = -0.00016

x = π/2 formula = 1.00000 sin x = 1.00000 error = 0.00000

In 629 Bhaskara I wrote a commentary, the Aryabhatiyabhasya, on the Aryabhatiya by Aryabhatta I. The Aryabhatiya contains 33 verses dealing with mathematics, the remainder of the work being concerned with mathematical astronomy. The commentary by Bhaskara I is only on the 33 verses of mathematics. He considers problems of indeterminate equations of the first degree and trigonometric formulae. In the course of discussions of the Aryabhatiya, Bhaskara I expressed his idea on how one particular rectangle can be treated as a cyclic quadrilateral. He was the first to open discussion on quadrilaterals with all the four sides unequal and none of the opposite sides parallel.

One of the approximations used for p for many centuries was O10. Bhaskara I criticised this approximation. He regretted that an exact measure of the circumference of a circle in terms of diameter was not available and he clearly believed that p was not rational.

Shukla discusses some features of Bhaskara's mathematics such as: numbers and symbolism, the classification of mathematics, the names and solution methods of equations of the first degree, quadratic equations, cubic equations and equations with more than one unknown, symbolic algebra, unusual and special terms in Bhaskara's work, weights and measures, the Euclidean algorithm method of solving linear indeterminate equations, examples given by Bhaskara I illustrating Aryabhatta I's rules, certain tables for solving an equation occurring in astronomy, and reference made by Bhaskara I to the works of earlier Indian mathematicians.

Lalla

Born: about 720 in India

Died: about 790 in India

Lalla's father was Trivikrama Bhatta and Trivikrama's father, Lalla's paternal grandfather, was named Samba. Lalla was an Indian astronomer and mathematician who followed the tradition of Aryabhatta I. Lalla's most famous work was entitled Shishyadhividdhidatantra. This major treatise was in two volumes. The first volume, On the computation of the positions of the planets, was in thirteen chapters and covered topics such as: mean longitudes of the planets; true longitudes of the planets; the three problems of diurnal rotation; lunar eclipses; solar eclipses; syzygies; risings and settings; the shadow of the moon; the moon's crescent; conjunctions of the planets with each other; conjunctions of the planets with the fixed stars; the patas of the moon and sun, and a final chapter in the first volume which forms a conclusion.

The second volume was On the sphere. In this volume Lalla examined topics such as: graphical representation; the celestial sphere; the principle of mean motion; the terrestrial sphere; motions and stations of the planets; geography; erroneous knowledge; instruments; and finally selected problems.

In Shishyadhividdhidatantra Lalla uses Sanskrit numerical symbols. Ifrah writes:-

> *... over the centuries, Sanskrit has lent itself admirably to the rules of prosody and versification. This explains why Indian astronomers [like Lalla] favoured the use of Sanskrit numerical symbols, based on a complex symbolism which was extraordinarily fertile and sophisticated, possessing as it did an almost limitless choice of synonyms.*

Despite writing the most famous treatise giving the views of Aryabhatta I, Lalla did not accept his theory given in the Aryabhatiya that the earth rotated. Lalla argues in his commentary, like many other Indian astronomers before him such as Varahamihira and Brahmagupta, that if the earth rotated then the speed would have to be such that one would have to ask how do the bees or birds flying in the sky come back to their nests? In fact Lalla misinterpreted some of Aryabhatta I's statements about the rotating earth. One has to assume that the idea appeared so impossible to him that he just could not appreciate Aryabhatta I's arguments. As Chatterjee writes, Lalla in his commentary:-

> *... did not interpret the relevant verses in the way meant by Aryabhatta I.*

Astrology at this time was based on astronomical tables and often the horoscopes allow one to identify the tables used. Some Arabic horoscopes were based on astronomical tables calculated in India. The most frequently used tables were by Aryabhatta I. Lalla improved on these tables and he produced a set of corrections for the Moon's longitude. One aspect of Aryabhatta I's work which Lalla did follow was his value of π. Lalla uses $\pi = \frac{62832}{20000}$, i.e. $\pi = 3.1416$ which is a value correct to the fourth decimal place.

Lalla also wrote a commentary on Khandakhadyaka, a work of Brahmagupta. Lalla's commentary has not survived but there is another work on astrology by Lalla which has survived, namely the Jyotisaratnakosa. This was a very popular work which was the main one on the subject in India for around 300 years.

Govindasvami

Born: about 800 in India

Died: about 860 in India

Govindasvami (or Govindasvamin) was an Indian mathematical astronomer whose most famous treatise was a commentary on the Mahabhaskariya of Bhaskara I.

Bhaskara I wrote the Mahabhaskariya in about 600 A.D. It is an eight chapter work on Indian mathematical astronomy and includes topics which were fairly standard for such works at this time. It discussed topics such as the longitudes of the planets, conjunctions of the planets with each other and with bright stars, eclipses of the sun and the moon, risings and settings, and the lunar crescent.

Govindasvami wrote the Bhasya in about 830 which was a commentary on the Mahabhaskariya. In Govindasvami's commentary there appear many examples of using a place-value Sanskrit system of numerals. One of the most interesting aspects of the commentary, however, is Govindasvami's construction of a sine table.

Indian mathematicians and astronomers constructed sine table with great precision. They were used to calculate the positions of the planets as accurately as possible so had to be computed with high degrees of accuracy. Govindasvami considered the sexagesimal fractional parts of the twenty-four tabular sine differences from the Aryabhatiya. These lead to more correct sine values at intervals of 90/24 = 3 45'. In the commentary Govindasvami found certain other empirical rules relating to computations of sine differences in the argumental range of 60 to 90 degrees.

Mahavira

Born: about 800 in possibly Mysore, India

Died: about 870 in India

Mahavira (or Mahaviracharya meaning Mahavira the Teacher) was of the Jaina religion and was familiar with Jaina mathematics. He worked in Mysore in southern Indian where he was a member of a school of mathematics. If he was not born in Mysore then it is very likely that he was born close to this town in the same region of India. We have essentially no other biographical details although we can gain just a little of his personality from the acknowledgement he gives in the introduction to his only known work, see below. However Jain mentions six other works which he credits to Mahavira and he emphasises the need for further research into identifying the complete list of his works.

The only known book by Mahavira is Ganita Sara Samgraha, dated 850 AD, which was designed as an updating of Brahmagupta's book. Filliozat writes:-

> *This book deals with the teaching of Brahmagupta but contains both simplifications and additional information. ... Although like all Indian versified texts, it is extremely condensed, this work, from a pedagogical point of view, has a significant advantage over earlier texts.*

It consisted of nine chapters and included all mathematical knowledge of mid-ninth century India. It provides us with the bulk of knowledge which we have of Jaina mathematics and it can be seen as in some sense providing an account of the work of those who developed this mathematics. There were many Indian mathematicians before the time of Mahavira but, perhaps surprisingly, their work on mathematics is always contained in texts which discuss other topics such as astronomy. The Ganita Sara Samgraha by Mahavira is the earliest Indian text which we possess which is devoted entirely to mathematics.

In the introduction to the work Mahavira paid tribute to the mathematicians whose work formed the basis of his book. These mathematicians included Aryabhatta I, Bhaskara I, and Brahmagupta. Mahavira writes:-

> *With the help of the accomplished holy sages, who are worthy to be worshipped by the lords of the world ... I glean from the great ocean of the knowledge of numbers a little of its essence, in the manner in which gems are picked from the sea, gold from the stony rock*

and the pearl from the oyster shell; and I give out according to the power of my intelligence, the Sara Samgraha, a small work on arithmetic, which is however not small in importance.

The nine chapters of the Ganita Sara Samgraha are:

1. Terminology
2. Arithmetical operations
3. Operations involving fractions
4. Miscellaneous operations
5. Operations involving the rule of three
6. Mixed operations
7. Operations relating to the calculations of areas
8. Operations relating to excavations
9. Operations relating to shadows

Throughout the work a place-value system with nine numerals is used or sometimes Sanskrit numeral symbols are used. Of interest in Chapter 1 regarding the development of a place-value number system is Mahavira's description of the number 12345654321 which he obtains after a calculation. He describes the number as:-

... beginning with one which then grows until it reaches six, then decreases in reverse order.

Notice that this wording makes sense to us using a place-value system but would not make sense in other systems. It is a clear indication that Mahavira is at home with the place-value number system.

Among topics Mahavira discussed in his treatise was operations with fractions including methods to decompose integers and fractions into unit fractions. For example:

2/17 = 1/12 + 1/51 + 1/68.

He examined methods of squaring numbers which, although a special case of multiplying two numbers, can be computed using special methods. He also discussed integer solutions of first degree indeterminate equation by a method called kuttaka. The kuttaka (or the "pulveriser") method is based on the use of the Euclidean algorithm but the method of solution also resembles the continued fraction process of Euler given in 1764. The work kuttaka, which occurs in many of the treatises of Indian mathematicians of the classical period, has taken on the more general meaning of "algebra".

An example of a problem given in the Ganita Sara Samgraha which leads to indeterminate linear equations is the following:

> *Three merchants find a purse lying in the road. One merchant says "If I keep the purse, I shall have twice as much money as the two of you together". "Give me the purse and I shall have three times as much" said the second merchant. The third merchant said "I shall be much better off than either of you if I keep the purse, I shall have five times as much as the two of you together". How much money is in the purse? How much money does each merchant have?*

If the first merchant has x, the second y, the third z and p is the amount in the purse then

$$p + x = 2(y + z),\ p + y = 3(x + z),\ p + z = 5(x + y).$$

There is no unique solution but the smallest solution in positive integers is $p = 15, x = 1, y = 3, z = 5$. Any solution in positive integers is a multiple of this solution as Mahavira claims.

Mahavira gave special rules for the use of permutations and combinations which was a topic of special interest in Jaina mathematics. He also described a process for calculating the volume of a sphere and one for calculating the cube root of a number. He looked at some geometrical results including right-angled triangles with rational sides.

Mahavira also attempts to solve certain mathematical problems which had not been studied by other Indian mathematicians. For example, he gave an approximate formula for the area and the perimeter of an ellipse. Hayashi writes:-

The formulas for a conch-like figure have so far been found only in the works of Mahavira and Narayana.

It is reasonable to ask what a "conch-like figure" is. It is two unequal semicircles (with diameters AB and BC) stuck together along their diameters. Although it might be reasonable to suppose that the perimeter might be obtained by considering the semicircles, Hayashi claims that the formulae obtained:-

... were most probably obtained not from the two semicircles AB and BC.

Prthudakasvami

Born: about 830 in India

Died: about 890 in India

Prthudakasvami is best known for his work on solving equations.

The solution of a first-degree indeterminate equation by a method called kuttaka (or "pulveriser") was given by Aryabhatta I. This method of finding integer solutions resembles the continued fraction process and can also be seen as a use of the Euclidean algorithm.

Brahmagupta seems to have used a method involving continued fractions to find integer solutions of an indeterminate equation of the type ax + c = by. Prthudakasvami's commentary on Brahmagupta's work is helpful in showing how "algebra", that is the method of calculating with the unknown, was developing in India. Prthudakasvami discussed the kuttaka method which he renamed as "bijagnita" which means the method of calculating with unknown elements.

To see just how this new idea of algebra was developing in India, we look at the notation which was being used by Prthudakasvami in his commentary on Brahmagupta's Brahma Sputa Siddhanta. In this commentary Prthudakasvami writes the equation $10x + 8 = x^2 + 1$ as:

yava 0 ya 10 ru 8

yava 1 ya 0 ru 1

Here yava is an abbreviation for yavat avad varga which means the "square of the unknown quantity", ya is an abbreviation for yavat havat which means the "unknown quantity", and ru is an abbreviation for rupa which means "constant term". Hence the top row reads

$0x^2 + 10x + 8$

while the second row reads

$x^2 + 0x + 1$

The whole equation is therefore

$0x^2 + 10x + 8 = x^2 + 0x + 1$

or

$10x + 8 = x^2 + 1.$

Shankara Narayana

Born: about 840 in India

Died: about 900 in India

Shankara Narayana (or Shankaranarayana) was an Indian astronomer and mathematician. He was a disciple of the astronomer and mathematician Govindasvami. His most famous work was the Laghubhaskariyavivarana which was a commentary on the Laghubhaskariya of Bhaskara I which in turn is based on the work of Aryabhatta I.

The Laghubhaskariyavivarana was written by Shankara Narayana in 869 AD for the author writes in the text that it is written in the Shaka year 791 which translates to a date AD by adding 78. It is a text which covers the standard mathematical methods of Aryabhatta I such as the solution of the indeterminate equation $by = ax \quad c$ (a, b, c integers) in integers which is then applied to astronomical problems. The standard Indian method involves using the Euclidean algorithm. It is called kuttakara ("pulveriser") but the term eventually came to have a more general meaning like "algebra". The paper examines this method. The reader who is wondering what the determination of "mati" means in the title of the paper then it refers to the optional number in a guessed solution and it is a feature which differs from the original method as presented by Bhaskara I.

Perhaps the most unusual feature of the Laghubhaskariyavivarana is the use of katapayadi numeration as well as the place-value Sanskrit numerals which Shankara Narayana frequently uses. Shankara Narayana is the first author known to use katapayadi numeration with this name but he did not invent it for it appears to be identical to a system invented earlier which was called varnasamjna. The numeration system varnasamjna was almost certainly invented by the astronomer Haridatta, and it was explained by him in a text which many historians believe was written in 684 but this would contradict what Shankara Narayana himself writes. This point is discussed below. First we should explain ideas behind Shankara Narayana's katapayadi numeration.

The system is based on writing numbers using the letters of the Indian alphabet. Let us quote from:-

> *... the numerical attribution of syllables corresponds to the following rule, according to the regular order of succession of the letters of the Indian alphabet: the first nine letters represent the numbers 1 to 9 while*

the tenth corresponds to zero; the following nine letters also receive the values 1 to 9 whilst the following letter has the value zero; the next five represent the first five units; and the last eight represent the numbers 1 to 8.

Under this system 1 to 5 are represented by four different letters. For example 1 is represented by the letters ka, ta, pa, ya which give the system its name (ka, ta, pa, ya becomes katapaya). Then 6, 7, 8 are represented by three letters and finally nine and zero are represented by two letters.

The system was a spoken one in the sense that consonants and vowels which are not vocalised have no numerical value. The system is a place-value system with zero but one may reasonably ask why such an apparently complicated numeral system might ever come to be invented. Well the answer must be that it lead to easily remembered mnemonics. In fact many different "words" could represent the same number and this was highly useful for works written in verse as the Indian texts tended to be.

Let us return to the interesting point about the date of Haridatta. Very unusually for an Indian text, Shankara Narayana expresses his thanks to those who have gone before him and developed the ideas about which he is writing. This in itself is not so unusual but the surprise here is that Shankara Narayana claims to give the list in chronological order. His list is

Aryabhatta I

Varahamihira

Bhaskara I

Govindasvami

Haridatta

[Note that we have written Bhaskara I where Shankara Narayana simply wrote Bhaskara. The more famous

Bhaskara II lived nearly 300 years after Shankara Narayana.]

The chronological order in the list agrees with the dates we have for the first four of these mathematicians. However, putting Haridatta after Govindasvami would seem an unlikely mistake for Shankara Narayana to make if Haridatta really did write his text in 684 since Shankara Narayana was himself a disciple of Govindasvami. If the dating given by Shankara Narayana is correct then katapayadi numeration had been invented only a few years before he wrote his text.

Sridhara

Born: about 870 in possibly Bengal, India

Died: about 930 in India

Sridhara is now believed to have lived in the ninth and tenth centuries. However, there has been much dispute over his date and in different works the dates of the life of Sridhara have been placed from the seventh century to the eleventh century. The best present estimate is that he wrote around 900 AD, a date which is deduced from seeing which other pieces of mathematics he was familiar with and also seeing which later mathematicians were familiar with his work. We do know that Sridhara was a Hindu but little else is known. Two theories exist concerning his birthplace which are far apart. Some historians give Bengal as the place of his birth while other historians believe that Sridhara was born in southern India.

Sridhara is known as the author of two mathematical treatises, namely the Trisatika (sometimes called the Patiganitasara) and the Patiganita. However at least three other works have been attributed to him, namely the Bijaganita, Navasati, and Brhatpati. Information about these books was given the works of Bhaskara II (writing around 1150), Makkibhatta (writing in 1377), and Raghavabhatta (writing in 1493). We give details below of Sridhara's rule for solving quadratic equations as given by Bhaskara II. There is another mathematical treatise Ganitapancavimsi which some historians believe was written by Sridhara. Hayashi, however, argues that Sridhara is unlikely to have been the author of this work in its present form.

The Patiganita is written in verse form. The book begins by giving tables of monetary and metrological units. Following this algorithms are given for carrying out the elementary arithmetical operations, squaring, cubing, and square and cube root extraction, carried out with natural numbers. Through the whole book Sridhara gives methods to solve problems in terse rules in verse form which was the typical style of Indian texts at this time. All the algorithms to carry out arithmetical operations are presented in this way and no proofs are given. Indeed there is no suggestion that Sridhara realised that proofs are in any way necessary. Often after stating a rule Sridhara gives one or more numerical examples, but he does not give solutions to these example nor does he even give answers in this work.

After giving the rules for computing with natural numbers, Sridhara gives rules for operating with rational fractions. He gives a wide variety of applications including problems involving ratios, barter, simple interest, mixtures, purchase and sale, rates of travel, wages, and filling of cisterns. Some of the examples are decidedly non-trivial and one has to consider this as a really advanced work. Other topics covered by the author include the rule for calculating the number of combinations of n things taken m at a time. There are sections of the book devoted to arithmetic and geometric progressions, including progressions with a fractional numbers of terms, and formulae for the sum of certain finite series are given.

The book ends by giving rules, some of which are only approximate, for the areas of a some plane polygons. In fact the text breaks off at this point but it certainly was not the end of the book which is missing in the only copy of the work which has survived.

We do know something of the missing part, however, for the Patiganitasara is a summary of the Patiganita including the missing portion.

Shukla examines Sridhara's method for finding rational solutions of $Nx^2 \pm 1 = y^2$, $1 - Nx^2 = y^2$, $Nx^2 \pm C = y^2$, and $C - Nx^2 = y^2$ which Sridhara gives in the Patiganita.

Shukla states that the rules given there are different from those given by other Hindu mathematicians.

Sridhara was one of the first mathematicians to give a rule to solve a quadratic equation. Unfortunately, as we indicated above, the original is lost and we have to rely on a quotation of Sridhara's rule from Bhaskara II:-

> *Multiply both sides of the equation by a known quantity equal to four times the coefficient of the square of the unknown; add to both sides a known quantity equal to the square of the coefficient of the unknown; then take the square root.*

To see what this means take

$ax^2 + bx = c.$

Multiply both sides by 4a to get

$4a^2x^2 + 4abx = 4ac$

then add b2 to both sides to get

$4a^2x^2 + 4abx + b^2 = 4ac + b^2$

and, taking the square root

$2ax + b = v(4ac + b^2).$

There is no suggestion that Sridhara took two values when he took the square root.

Aryabhatta II

Born: about 920 in India

Died: about 1000 in India

Essentially nothing is known of the life of Aryabhatta II. Historians have argued about his date and have come up with many different theories. Pingree gives the date for his main publications as being between 950 and 1100. This is deduced from the usual arguments such as which authors Aryabhatta II refers to and which refer to him. G R Kaye argued in 1910 that Aryabhatta II lived before al-Biruni but Datta in 1926 showed that these dates were too early.

The article argues for a date of about 950 for Aryabhatta II's main work, the Mahasiddhanta, but R Billiard has proposed a date for Aryabhatta II in the sixteenth century. Most modern historians, however, consider the most likely dates for his main work as around 950 and we have given very approximate dates for his birth and death based on this hypothesis.

The most famous work by Aryabhatta II is the Mahasiddhanta which consists of eighteen chapters. The treatise is written in Sanskrit verse and the first twelve chapters form a treatise on mathematical astronomy covering the usual topics that Indian mathematicians worked on during this period. The topics included in these twelve chapters are: the longitudes of the planets, eclipses of the sun and moon, the projection of eclipses, the lunar crescent, the rising and setting of the planets, conjunctions of the planets with each other and with the stars.

The remaining six chapters of the Mahasiddhanta form a separate part entitled On the sphere. It discusses topics such as geometry, geography and algebra with applications to the longitudes of the planets.

In Mahasiddhanta Aryabhatta II gives in about twenty verses detailed rules to solve the indeterminate equation:

by = ax + c. The rules apply in a number of different cases such as when c is positive, when c is negative, when the number of the quotients of the mutual divisions is even, when this number of quotients is odd, etc.

Aryabhatta II also gave a method to calculate the cube root of a number, but his method was not new, being based on that given many years earlier by Aryabhatta I.

Aryabhatta II constructed a sine table correct up to five decimal places when measured in decimal parts of the radius. Indian mathematicians were very interested in giving accurate sine tables since they were used to calculate the planetary positions as accurately as possible.

Vijayanandi

Born: about 940 in Benares (now Varanasi), India

Died: about 1010 in India

Vijayanandi (or Vijayanandin) was the son of Jayananda. He was born into the Brahman caste which meant he was from the highest ranking caste of Hindu priests. He was an Indian mathematician and astronomer whose most famous work was the Karanatilaka.

We should note that there was another astronomer named Vijayanandi who was mentioned by Varahamihira in one of his works. Since Varahamihira wrote around 550 and the Karanatilaka was written around 966, there must be two astronomers both named "Vijayanandi".

The Karanatilaka has not survived in its original form but we know of the text through an Arabic translation by al-Biruni. It is a work in fourteen chapters covering the standard topics of Indian astronomy.

It deals with the topics of: units of time measurement; mean and true longitudes of the sun and moon; the length of daylight; mean longitudes of the five planets; true longitudes of the five planets; the three problems of diurnal rotation; lunar eclipses, solar eclipses; the projection of eclipses; first visibility of the planets; conjunctions of the planets with each other and with fixed stars; the moon's crescent; and the patas of the moon and sun.

The Indians had a cosmology which was based on long periods of time with astronomical events occurring a certain whole number of times within the cycles. This system led to much work on integer solutions of equations and their application to astronomy.

In particular there was, according to Aryabhatta I, a basic period of 4320000 years called a mahayuga and it was assumed that the sun, the moon, their apsis and node, and the planets reached perfect conjunctions after this period. Hence it was assumed that the periods were rational multiples of each other.

All the planets and the node and apsis of the moon and sun had to have an integer number of revolutions in the mahayuga. Many Indian astronomers proposed different values for these integral numbers of revolutions.

For the number of revolutions of the apsis and node of the moon per mahayuga, Aryabhatta I proposed 488219 and 232226, respectively.

However Vijayanandi changed these numbers to 488211 and 232234. The reasons for giving the new numbers is unclear. Like other Indian astronomers, Vijayanandi made contributions to trigonometry and it appears that his calculation of the periods was computed by using tables of sines and versed sines.

It is significant that accuracy was need in trigonometric tables to give accurate astronomical theories and this motivated many of the Indian mathematicians to produce more accurate methods of approximating entries in tables.

Sripati

Born: 1019 in (probably) Rohinikhanda, Maharashtra, India

Died: 1066 in India

Sripati's father was Nagadeva (sometimes written as Namadeva) and Nagadeva's father, Sripati's paternal grandfather, was Kesava. Sripati was a follower of the teaching of Lalla writing on astrology, astronomy and mathematics. His mathematical work was undertaken with applications to astronomy in mind, for example a study of spheres. His work on astronomy was undertaken to provide a basis for his astrology. Sripati was the most prominent Indian mathematicians of the 11th Century.

Among Sripati's works are: Dhikotidakarana written in 1039, a work of twenty verses on solar and lunar eclipses; Dhruvamanasa written in 1056, a work of 105 verses on calculating planetary longitudes, eclipses and planetary transits; Siddhantasekhara a major work on astronomy in 19 chapters; and Ganitatilaka an incomplete arithmetical treatise in 125 verses based on a work by Sridhara.

The titles of Chapters 13, 14, and 15 of the Siddhantasekhara are Arithmetic, Algebra and On the Sphere. Chapter 13 consists of 55 verses on arithmetic, mensuration, and shadow reckoning. It is probable that the lost portion of the arithmetic treatise Ganitatilaka consisted essentially of verses 19-55 of this chapter. The 37 verses of Chapter 14 on algebra state various rules of algebra without proof. These are given in verbal form without algebraic symbols. In verses 3, 4 and 5 of this chapter Sripati gave the rules of signs for addition, subtraction, multiplication, division, square, square root, cube and cube root of positive and negative quantities. His work on equations in this chapter contains the rule for solving a quadratic equation and, more impressively, he gives the identity:

$$O(x + Oy) = O[(x + O(x^2 - y)]/2 + O[(x - O(x^2 - y)]/2)$$

Other mathematics included in Sripati's work includes, in particular, rules for the solution of simultaneous indeterminate equations of the first degree that are similar to those given by Brahmagupta.

Sripati obtained more fame in astrology than in other areas and it is fair to say that he considered this to be his most important contributions. He wrote the Jyotisaratnamala which was an astrology text in twenty chapters based on the Jyotisaratnakosa of Lalla.

Sripati wrote a commentary on this work in Marathi and it is one of the oldest works to have survived that is written in that language. Marathi is the oldest of the regional languages in Indo-Aryan, dating from about 1000.

Another work on astrology written by Sripati is the Jatakapaddhati or Sripatipaddhati which is in eight chapters :-

> *... one of the fundamental textbooks for later Indian genethlialogy, contributing an impressive elaboration to the computation of the strengths of the planets and astrological places. It was enormously popular, as the large number of manuscripts, commentaries, and imitations attests.*

Genethlialogy was the science of casting nativities and it was the earliest branch of astrology which claimed to be able to predict the course of a person's life based on the positions of the planets and of the signs of the zodiac at the moment the person was born or conceived.

There is one other work on astrology the Daivajnavallabha which some historians claim was written by Sripati while other claim that it is the work of Varahamihira. As yet nobody has come up with a definite case to show which of these two is the author, or even whether the author is another astrologer.

Brahmadeva

Born: about 1060 in (possibly) Mathura, India

Died: about 1130 in India

Brahmadeva was the son of Candrabudha. The family came from the Mathura district of Uttar Pradesh in northern India. He was born into the Brahman caste which meant he was from the highest ranking caste of Hindu priests.

The Brahman caste had a strong tradition of education so Brahmadeva would have received one of the best educations of men of his time.

We have only one work by Brahmadeva and this is Karanaprakasa which is a commentary on the Aryabhatiya by Aryabhatta I.

Brahmadeva's work is in nine chapters and it follows the contents of the original Aryabhatiya.

Topics covered include the longitudes of the planets, problems relating to the daily rotation of the heavens, eclipses of the sun and the moon, risings and settings, the lunar crescent, and conjunctions of the planets.

The work contains some contributions to trigonometry, motivated by its application to mathematical astronomy. It is this aspect of the work which is mentioned by Gupta.

Different commentaries on the Aryabhatiya achieved popularity in different parts of India. Brahmadeva's commentary seems to have been particularly popular in Madras, Mysore and Maharastra.

The more important commentaries on the Aryabhatiya became the basis for further commentaries and indeed this is what happened to the Karanaprakasa.

Commentaries on Brahmadeva's work continued to appear up to the seventeenth century.

Bhaskara

Born: 1114 in Vijayapura, India

Died: 1185 in Ujjain, India

Bhaskara is also known as Bhaskara II or as Bhaskaracharya, this latter name meaning "Bhaskara the Teacher". Since he is known in India as Bhaskaracharya we will refer to him throughout this article by that name. Bhaskaracharya's father was a Brahman named Mahesvara. Mahesvara himself was famed as an astrologer. This happened frequently in Indian society with generations of a family being excellent mathematicians and often acting as teachers to other family members.

Bhaskaracharya became head of the astronomical observatory at Ujjain, the leading mathematical centre in India at that time. Outstanding mathematicians such as Varahamihira and Brahmagupta had worked there and built up a strong school of mathematical astronomy.

In many ways Bhaskaracharya represents the peak of mathematical knowledge in the 12th century. He reached an understanding of the number systems and solving equations which was not to be achieved in Europe for several centuries.

Six works by Bhaskaracharya are known but a seventh work, which is claimed to be by him, is thought by many historians to be a late forgery. The six works are: Lilavati (The Beautiful) which is on mathematics; Bijaganita (Seed Counting or Root Extraction) which is on algebra; the Siddhantasiromani which is in two parts, the first on mathematical astronomy with the second part on the sphere; the Vasanabhasya of Mitaksara which is Bhaskaracharya's own commentary on the Siddhantasiromani; the Karanakutuhala (Calculation of Astronomical Wonders) or Brahmatulya which is a simplified version of the Siddhantasiromani; and the Vivarana which is a commentary on the Shishyadhividdhidatantra of Lalla. It is the first three of these works which are the most interesting, certainly from the point of view of mathematics, and we will concentrate on the contents of these.

Given that he was building on the knowledge and understanding of Brahmagupta it is not surprising that Bhaskaracharya understood about zero and negative numbers. However his understanding went further even than that of Brahmagupta. To give some examples before we examine his work in a little more detail we note that he knew that $x^2 = 9$ had two solutions.

He also gave the formula:

$$\sqrt{a \pm \sqrt{b}} = \sqrt{\frac{a + \sqrt{a^2 - b}}{2}} \pm \sqrt{\frac{a - \sqrt{a^2 - b}}{2}}$$

Bhaskaracharya studied Pell's equation $px^2 + 1 = y^2$ for $p = 8, 11, 32, 61$ and 67. When $p = 61$ he found the solutions $x = 226153980$, $y = 1776319049$. When $p = 67$ he found the solutions $x = 5967$, $y = 48842$. He studied many Diophantine problems.

Let us first examine the Lilavati. First it is worth repeating the story told by Fyzi who translated this work into Persian in 1587. We give the story as given by Joseph:-

Lilavati was the name of Bhaskaracharya's daughter. From casting her horoscope, he discovered that the auspicious time for her wedding would be a particular hour on a certain day. He placed a cup with a small hole at the bottom of the vessel filled with water, arranged so that the cup would sink at the beginning of the propitious hour. When everything was ready and the cup was placed in the vessel, Lilavati suddenly out of curiosity bent over the vessel and a pearl from her dress fell into the cup and blocked the hole in it. The lucky hour passed without the cup sinking.

Bhaskaracharya believed that the way to console his dejected daughter, who now would never get married, was to write her a manual of mathematics!

This is a charming story but it is hard to see that there is any evidence for it being true. It is not even certain that Lilavati was Bhaskaracharya's daughter. There is also a theory that Lilavati was Bhaskaracharya's wife. The topics covered in the thirteen chapters of the book are: definitions; arithmetical terms; interest; arithmetical and geometrical progressions; plane geometry; solid geometry; the shadow of the gnomon; the kuttaka; combinations.

In dealing with numbers Bhaskaracharya, like Brahmagupta before him, handled efficiently arithmetic involving negative numbers. He is sound in addition, subtraction and multiplication involving zero but realised that there were problems with Brahmagupta's ideas of dividing by zero. Madhukar Mallayya argues that the zero used by Bhaskaracharya in his rule (a.0)/0 = a, given in Lilavati, is equivalent to the modern concept of a non-zero "infinitesimal". Although this claim is not without foundation, perhaps it is seeing ideas beyond what Bhaskaracharya intended.

Bhaskaracharya gave two methods of multiplication in his Lilavati. We follow Ifrah who explains these two methods due to Bhaskaracharya. To multiply 325 by 243 Bhaskaracharya writes the numbers thus:

243 243 243

3 2 5

Now working with the rightmost of the three sums he computed 5 times 3 then 5 times 2 missing out the 5 times 4 which he did last and wrote beneath the others one place to the left. Note that this avoids making the "carry" in ones head.

243 243 243
3 2 5

1015
20

Now add the 1015 and 20 so positioned and write the answer under the second line below the sum next to the left.

243 243 243
3 2 5

1015
20

1215

Work out the middle sum as the right-hand one, again avoiding the "carry", and add them writing the answer below the 1215 but displaced one place to the left.

243 243 243
3 2 5
----
4 6 1015
8 20

1215
486

Finally work out the left most sum in the same way and again place the resulting addition one place to the left under the 486.

243 243 243
3 2 5
----
6 9 4 6 1015
12 8 20

1215
486
729

Finally add the three numbers below the second line to obtain the answer 78975.

243 243 243
3 2 5
----
6 9 4 6 1015
12 8 20

1215
486
729

78975

Despite avoiding the "carry" in the first stages, of course one is still faced with the "carry" in this final addition.

The second of Bhaskaracharya's methods proceeds as follows:

325
243

Multiply the bottom number by the top number starting with the left-most digit and proceeding towards the right. Displace each row one place to start one place further right than the previous line.

First step

325
243

729

Second step

325
243

729
486

Third step, then add

325
243

729
486
1215

78975

Bhaskaracharya, like many of the Indian mathematicians, considered squaring of numbers as special cases of multiplication which deserved special methods. He gave four such methods of squaring in Lilavati.

Here is an example of explanation of inverse proportion taken from Chapter 3 of the Lilavati. Bhaskaracharya writes:-

In the inverse method, the operation is reversed. That is the fruit to be multiplied by the augment and divided by the demand. When fruit increases or decreases, as the demand is augmented or diminished, the direct rule is used. Else the inverse.

Rule of Three Inverse: If the fruit diminish as the requisition increases, or augment as that decreases, they, who are skilled in accounts, consider the rule of three to be inverted. When there is a diminution of fruit, if there be increase of requisition, and increase of fruit if there be diminution of requisition, then the inverse rule of three is employed.

As well as the rule of three, Bhaskaracharya discusses examples to illustrate rules of compound proportions, such as the rule of five (Pancarasika), the rule of seven (Saptarasika), the rule of nine (Navarasika), etc. Bhaskaracharya's examples of using these rules are discussed.

An example from Chapter 5 on arithmetical and geometrical progressions is the following:-

Example: On an expedition to seize his enemy's elephants, a king marched two yojanas the first day. Say, intelligent calculator, with what increasing rate of daily march did he proceed, since he reached his foe's city, a distance of eighty yojanas, in a week?

Bhaskaracharya shows that each day he must travel $\frac{22}{7}$ yojanas further than the previous day to reach his foe's city in 7 days.

An example from Chapter 12 on the kuttaka method of solving indeterminate equations is the following:-

Example: Say quickly, mathematician, what is that multiplier, by which two hundred and twenty-one being multiplied, and sixty-five added to the product, the sum divided by a hundred and ninety-five becomes exhausted.

Bhaskaracharya is finding integer solution to $195x = 221y + 65$. He obtains the solutions $(x,y) = (6,5)$ or $(23,20)$ or $(40, 35)$ and so on.

In the final chapter on combinations Bhaskaracharya considers the following problem. Let an n-digit number be represented in the usual decimal form as

(*) $d_1 d_2 ... d_n$

where each digit satisfies $1 \le d_j \le 9, j = 1, 2, ..., n$. Then Bhaskaracharya's problem is to find the total number of numbers of the form (*) that satisfy

$d_1 + d_2 + ... + d_n = S.$

In his conclusion to Lilavati Bhaskaracharya writes:-

Joy and happiness is indeed ever increasing in this world for those who have Lilavati clasped to their throats, decorated as the members are with neat reduction of fractions, multiplication and involution, pure and perfect as are the solutions, and tasteful as is the speech which is exemplified.

The Bijaganita is a work in twelve chapters. The topics are: positive and negative numbers; zero; the unknown; surds; the kuttaka; indeterminate quadratic equations; simple equations; quadratic equations; equations with more than one unknown; quadratic equations with more than one unknown; operations with products of several unknowns; and the author and his work.

Having explained how to do arithmetic with negative numbers, Bhaskaracharya gives problems to test the abilities of the reader on calculating with negative and affirmative quantities:-

Example: Tell quickly the result of the numbers three and four, negative or affirmative, taken together; that is, affirmative and negative, or both negative or both affirmative, as separate instances; if thou know the addition of affirmative and negative quantities.

Negative numbers are denoted by placing a dot above them: The characters, denoting the quantities known and unknown, should be first written to indicate them generally; and those, which become negative should be then marked with a dot over them.

Example: Subtracting two from three, affirmative from affirmative, and negative from negative, or the contrary, tell me quickly the result ...

In Bijaganita Bhaskaracharya attempted to improve on Brahmagupta's attempt to divide by zero (and his own description in Lilavati) when he wrote:-

A quantity divided by zero becomes a fraction the denominator of which is zero. This fraction is termed an infinite quantity. In this quantity consisting of that which has zero for its divisor, there is no alteration, though many may be inserted or extracted; as no change takes place in the infinite and immutable God when worlds are created or destroyed, though numerous orders of beings are absorbed or put forth.

So Bhaskaracharya tried to solve the problem by writing n/0 = 8. At first sight we might be tempted to believe that Bhaskaracharya has it correct, but of course he does not. If this were true then 0 times 8 must be equal to every number n, so all numbers are equal. The Indian mathematicians could not bring themselves to the point of admitting that one could not divide by zero.

Equations leading to more than one solution are given by Bhaskaracharya:-

Example: Inside a forest, a number of apes equal to the square of one-eighth of the total apes in the pack are playing noisy games. The remaining twelve apes, who are of a more serious disposition, are on a nearby hill and irritated by the shrieks coming from the forest. What is the total number of apes in the pack?

The problem leads to a quadratic equation and Bhaskaracharya says that the two solutions, namely 16 and 48, are equally admissible.

The kuttaka method to solve indeterminate equations is applied to equations with three unknowns.

The problem is to find integer solutions to an equation of the form $ax + by + cz = d$. An example he gives is:-

Example: The horses belonging to four men are 5, 3, 6 and 8. The camels belonging to the same men are 2, 7, 4 and 1. The mules belonging to them are 8, 2, 1 and 3 and the oxen are 7, 1, 2 and 1, all four men have equal fortunes. Tell me quickly the price of each horse, camel, mule and ox.

Of course such problems do not have a unique solution as Bhaskaracharya is fully aware. He finds one solution, which is the minimum, namely horses 85, camels 76, mules 31 and oxen 4.

Bhaskaracharya's conclusion to the Bijaganita is fascinating for the insight it gives us into the mind of this great mathematician:-

A morsel of tuition conveys knowledge to a comprehensive mind; and having reached it, expands of its own impulse, as oil poured upon water, as a secret entrusted to the vile, as alms bestowed upon the worthy, however little, so does knowledge infused into a wise mind spread by intrinsic force.

It is apparent to men of clear understanding, that the rule of three terms constitutes arithmetic and sagacity constitutes algebra. Accordingly I have said ... The rule of three terms is arithmetic; spotless understanding is algebra. What is there unknown to the intelligent? Therefore for the dull alone it is set forth.

The Siddhantasiromani is a mathematical astronomy text similar in layout to many other Indian astronomy texts of this and earlier periods. The twelve chapters of the first part cover topics such as: mean longitudes of the planets; true longitudes of the planets; the three problems of diurnal rotation; syzygies; lunar eclipses; solar eclipses; latitudes of the planets; risings and settings; the moon's crescent; conjunctions of the planets with each other; conjunctions of the planets with the fixed stars; and the patas of the sun and moon.

The second part contains thirteen chapters on the sphere. It covers topics such as: praise of study of the sphere; nature of the sphere; cosmography and geography; planetary mean motion; eccentric epicyclic model of the planets; the armillary sphere; spherical trigonometry; ellipse calculations; first visibilities of the planets; calculating the lunar crescent; astronomical instruments; the seasons; and problems of astronomical calculations.

There are interesting results on trigonometry in this work. In particular Bhaskaracharya seems more interested in trigonometry for its own sake than his predecessors who saw it only as a tool for calculation. Among the many interesting results given by Bhaskaracharya are:

sin(a + b) = sin a cos b + cos a sin b

and

sin(a - b) = sin a cos b - cos a sin b.

Bhaskaracharya rightly achieved an outstanding reputation for his remarkable contribution. In 1207 an educational institution was set up to study Bhaskaracharya's works. A medieval inscription in an Indian temple reads:-

Triumphant is the illustrious Bhaskaracharya whose feats are revered by both the wise and the learned. A poet endowed with fame and religious merit, he is like the crest on a peacock.

It is from this quotation that the title of Joseph's book comes.

Mahendra Suri

Born: about 1340 in Western India

Died: about 1410 in India

Mahendra Suri was a Jain. Jainism began around the sixth century BC and the religion had a strong influence on mathematics particularly in the last couple of centuries BC. By the time of Mahendra Suri, however, Jainism had lost support as a national religion and was much less vigorous.

It had been influenced by Islam and in particular Islamic astronomy came to form a part of the background. However, Pingree writes that this filtering of Islamic astronomy into Indian culture was:-

> *... not allowed to affect in any way the structure of the traditional science.*

Mahendra Suri was a pupil of Madana Suri. He is famed as the first person to write a Sanskrit treatise on the astrolabe. Ohashi writes of the early history of the astrolabe in the Delhi Sultanate in India:-

The astrolabe was introduced into India at the time of Firuz Shah Tughluq (reign AD 1351 - 88), and Mahendra Suri wrote the first Sanskrit treatise on the astrolabe entitled Yantraraja (AD 1370).

The Delhi Sultanate was established around 1200 and from that time on Muslim culture flourished in India. The ideas of Islamic astronomy began to appear in works in the Sanskrit language and it is the Islamic ideas on the astrolabe which Mahendra Suri wrote on in his famous text. It is clear from the various references in the text and also from the particular values that Mahendra Suri uses for the angle of the ecliptic etc. that his work is based on Islamic rather than traditional Indian astronomy works.

Narayana Pandit

Born: about 1340 in India

Died: about 1400 in India

Narayana was the son of Narsimha (sometimes written Narasimha). We know that he wrote his most famous work Ganita Kaumudi on arithmetic in 1356 but little else is known of him. His mathematical writings show that he was strongly influenced by Bhaskara II and he wrote a commentary on the Lilavati of Bhaskara II called Karmapradipika. Some historians dispute that Narayana is the author of this commentary which they attribute to Madhava.

In the Ganita Kaumudi Narayana considers the mathematical operation on numbers. Like many other Indian writers of arithmetics before him he considered an algorithm for multiplying numbers and he then looked at the special case of squaring numbers. One of the unuṣual features of Narayana's work Karmapradipika is that he gave seven methods of squaring numbers which are not found in the work of other Indian mathematicians.

He discussed another standard topic for Indian mathematicians namely that of finding triangles whose sides had integral values. In particular he gave a rule of finding integral triangles whose sides differ by one unit of length and which contain a pair of right-angled triangles having integral sides with a common integral height. In terms of geometry Narayana gave a rule for a segment of a circle. Narayana :-

> *... derived his rule for a segment of a circle from Mahavira's rule for an 'elongated circle' or an ellipse-like figure.*

Narayana also gave a rule to calculate approximate values of a square root. He did this by using an indeterminate equation of the second order, $Nx^2 + 1 = y^2$, where N is the number whose square root is to be calculated. If x and y are a pair of roots of this equation with $x < y$ then vN is approximately equal to y/x. To illustrate this method Narayana takes N = 10. He then finds the solutions x = 6, y = 19 which give the approximation 19/6 = 3.1666666666666666667, which is correct to 2 decimal places. Narayana then gives the solutions x = 228, y = 721 which give the approximation 721/228 = 3.1622807017543859649, correct to four places. Finally Narayana gives the pair of solutions x = 8658, y = 227379 which give the approximation 227379/8658 = 3.1622776622776622777, correct to eight decimal places. Note for comparison that v10 is, correct to 20 places, 3.16227766016837933320.

The thirteenth chapter of Ganita Kaumudi was called Net of Numbers and was devoted to number sequences. For example, he discussed some problems concerning arithmetic progressions.

The fourteenth chapter (which is the last one) of Naryana's Ganita Kaumudi contains a detailed discussion of magic squares and similar figures. Narayana gave the rules for the formation of doubly even, even and odd perfect magic squares along with magic triangles, rectangles and circles. He used formulae and rules for the relations between magic squares and arithmetic series. He gave methods for finding "the horizontal difference" and the first term of a magic square whose square's constant and the number of terms are given and he also gave rules for finding "the vertical difference" in the case where this information is given.

Madhava of Sangamagramma

Born: 1350 in Sangamagramma (near Cochin), Kerala, India

Died: 1425 in India

Madhava of Sangamagramma was born near Cochin on the coast in the Kerala state in southwestern India. It is only due to research into Keralese mathematics over the last twenty-five years that the remarkable contributions of Madhava have come to light. Rajagopal and Rangachari put his achievement into context when they write:-

[Madhava] took the decisive step onwards from the finite procedures of ancient mathematics to treat their limit-passage to infinity, which is the kernel of modern classical analysis. All the mathematical writings of Madhava have been lost, although some of his texts on astronomy have survived. However his brilliant work in mathematics has been largely discovered by the reports of other Keralese mathematicians such as Nilakantha who lived about 100 years later.

Madhava discovered the series equivalent to the Maclaurin expansions of sin x, cos x, and arctan x around 1400, which is over two hundred years before they were rediscovered in Europe. Details appear in a number of works written by his followers such as Mahajyanayana prakara which means Method of computing the great sines. In fact this work had been claimed by some historians such as Sarma to be by Madhava himself but this seems highly unlikely and it is now accepted by most historians to be a 16th century work by a follower of Madhava.

Jyesthadeva wrote Yukti-Bhasa in Malayalam, the regional language of Kerala, around 1550. In [9] Gupta gives a translation of the text and this is also given and a number of other sources. Jyesthadeva describes Madhava's series as follows:-

> *The first term is the product of the given sine and radius of the desired arc divided by the cosine of the arc. The succeeding terms are obtained by a process of iteration when the first term is repeatedly multiplied by the square of the sine and divided by the square of the cosine. All the terms are then divided by the odd numbers 1, 3, 5, ... The arc is obtained by adding and subtracting respectively the terms of odd rank and those of even rank. It is laid down that the sine of the arc or that of its complement whichever is the smaller should be taken here as the given sine. Otherwise the terms obtained by this above iteration will not tend to the vanishing magnitude.*

This is a remarkable passage describing Madhava's series, but remember that even this passage by Jyesthadeva was written more than 100 years before James Gregory rediscovered this series expansion. Perhaps we should write down in modern symbols exactly what the series is that Madhava has found. The first thing to note is that the Indian meaning for sine of q would be written in our notation as r sin q and the Indian cosine of would be r cos q in our notation, where r is the radius. Thus the series is

$$r\,q = r(r \sin q)/1(r \cos q) - r(r \sin q)^3/3r(r \cos q)^3 + r(r \sin q)^5/5r(r \cos q)^5 - r(r \sin q)^7/7r(r \cos q)^7 + \ldots$$

putting tan = sin/cos and cancelling r gives

$$q = \tan q - (\tan^3 q)/3 + (\tan^5 q)/5 - \ldots$$

which is equivalent to Gregory's series

$$\tan^{-1} q = q - q^3/3 + q^5/5 - \ldots$$

Now Madhava put $q = \pi/4$ into his series to obtain

$$\pi/4 = 1 - 1/3 + 1/5 - \ldots$$

and he also put $q = \pi/6$ into his series to obtain

$$\pi = O12(1 - 1/(3 \times 3) + 1/(5 \times 32) - 1/(7 \times 33) + \ldots$$

We know that Madhava obtained an approximation for p correct to 11 decimal places when he gave

$$\pi = 3.14159265359$$

which can be obtained from the last of Madhava's series above by taking 21 terms. Gupta gives a translation of the Sanskrit text giving Madhava's approximation of π correct to 11 places.

Perhaps even more impressive is the fact that Madhava gave a remainder term for his series which improved the approximation. He improved the approximation of the series for π/4 by adding a correction term R_n to obtain

$$p/4 = 1 - 1/3 + 1/5 - \ldots 1/(2n-1)\ \ Rn$$

Madhava gave three forms of Rn which improved the approximation, namely:

$R_n = 1/(4n)$ or

$R_n = n/(4n^2 + 1)$ or

$R_n = (n^2 + 1)/(4n^3 + 5n)$.

There has been a lot of work done in trying to reconstruct how Madhava might have found his correction terms. The most convincing is that they come as the first three convergents of a continued fraction which can itself be derived from the standard Indian approximation to π namely $\frac{62832}{20000}$.

Madhava also gave a table of almost accurate values of half-sine chords for twenty-four arcs drawn at equal intervals in a quarter of a given circle. It is thought that the way that he found these highly accurate tables was to use the equivalent of the series expansions:

$\sin q = q - q^3/3! + q^5/5! - \ldots$

$\cos q = 1 - q^2/2! + q^4/4! - \ldots$

Jyesthadeva in Yukti-Bhasa gave an explanation of how Madhava found his series expansions around 1400 which are equivalent to these modern versions rediscovered by Newton around 1676. Historians have claimed that the method used by Madhava amounts to term by term integration.

Rajagopal's claim that Madhava took the decisive step towards modern classical analysis seems very fair given his remarkable achievements. In the same vein Joseph writes:-

> *We may consider Madhava to have been the founder of mathematical analysis. Some of his discoveries in this field show him to have possessed extraordinary intuition, making him almost the equal of the more recent intuitive genius Srinivasa Ramanujan, who spent his childhood and youth at Kumbakonam, not far from Madhava's birthplace.*

Parameshvara

Born: about 1370 in Alattur, Kerala, India

Died: about 1460 in India

Parameshvara was an Indian astronomer and mathematician who wrote many commentaries on earlier works as well as making many observations. Although his father has not been identified, we know that Parameshvara was born into a Namputiri Brahmana family who were astrologers and astronomers. The family home was Vatasseri (sometimes called Vatasreni) in the village of Alattur. This village was in Kerala and Parameshvara himself gives its coordinates with respect to Ujjain. This puts it at latitude 10 51' north. It is on the north bank of the river Nila at its mouth.

From Parameshvara's writing we know that Rudra was his teacher, and Nilakantha, who knew Parameshvara personally, tells us that Parameshvara's teachers included Madhava and Narayana. We can be fairly confident that the dates we have given for Parameshvara are roughly correct since he made eclipe observations over a period of 55 years. We will say a little more about these observations below. He played an important part in the remarkable developments in mathematics which took place in Kerala in the late 14th and early part of the 15th century.

The commentaries by Parameshvara on a number of works have been published. For example the Karmadipika is a commentary on the Mahabhaskariyam, an astronomical and mathematical work by Bhaskara I. Parameshvara's commentary on the Laghubhaskariyam of Bhaskara I is given. Munjala wrote the astronomical work Laghumanasam in the year 932 and Parameshvara wrote a commentary. It is a work containing typical topics for Indian mathematical astronomy works of this period: the mean motions of the heavenly bodies; the true motions of the heavenly bodies; miscellaneous mathematical rules; the systems of coordinates, direction, place and time; eclipses of the sun and the moon; and the operation for apparent longitude.

Aryabhatta gave a rule for determining the height of a pole from the lengths of its shadows in the Aryabhatiya. Parameshvara gave several illustrative examples of the method in his commentary on the Aryabhatiya.

Like many mathematicians from Kerala, Madhava clearly had a very strong influence on Parameshvara. One can see throughout his work that it is teachings by Madhava which direct much of Parameshvara's mathematical ideas. One of Parameshvara's most remarkable mathematical discoveries, no doubt influenced by Madhava, was a version of the mean value theorem. He states the theorem in his commentary Lilavati Bhasya on Bhaskara II's Lilavati. There are other examples of versions of the mean value theorem in Parameshvara's work which we now consider.

The Siddhantadipika by Parameshvara is a commentary on the commentary of Govindasvami on Bhaskara I's Mahabhaskariya. Parameshvara gives some of his eclipse observations in this work including one made at Navaksetra in 1422 and two made at Gokarna in 1425 and 1430. This work also contains a mean value type formula for inverse interpolation of the sine. It presents a one-point iterative technique for calculating the sine of a given angle. In the Siddhantadipika Parameshvara also gives a more efficient approximation that works using a two-point iterative algorithm which turns out to be essentially the same as the modern secant method.

The expression for the radius of the circle in which a cyclic quadrilateral is inscribed, given in terms of the sides of the quadrilateral, is usually attributed to Lhuilier in 1782. However Parameshvara described the rule 350 years earlier. If the sides of the cyclic quadrilateral are a, b, c and d then the radius r of the circumscribed circle was given by Parameshvara as:

$r^2 = x/y$ where

$x = (ab + cd)(ac + bd)(ad + bc)$

and $y = (a + b + c - d)(b + c + d - a)(c + d + a - b)(d + a + b - c)$.

The original text by Parameshvara describing the rule is given.

Parameshvara made a series of eclipse observations between 1393 and 1432 which we have referred to above. The last observation which we know he made was in 1445 but Nilakantha quotes a verse by Parameshvara in which he claims to have made observations spanning 55 years. The known observatons by Parameshvara do not quite square with this statement, there being a discrepancy of three years. Although we do not know when Parameshvara died we do know, again from Nilakantha, that the two knew each other personally. Since we have a definite date for Nilakantha's birth of 1444 it is hard to believe that Parameshvara died before 1460.

Using his observations, Parameshvara made revisions of the planetary parameters and, like many other Indian astronomers, he constantly attempted to compare the theoretically computed positions of the planets with those which he actually observed. He was keen to improve the theoretical model to bring it into as close an agreement with observations as possible.

Nilakantha Somayaji

Born: 14 June 1444 in Trkkantiyur (near Tirur), Kerala, India

Died: 1544 in India

Nilakantha was born into a Namputiri Brahmin family which came from South Malabar in Kerala. The Nambudiri is the main caste of Kerala. It is an orthodox caste whose members consider themselves descendants of the ancient Vedic religion.

He was born in a house called Kelallur which it is claimed coincides with the present Etamana in the village of Trkkantiyur near Tirur in south India. His father was Jatavedas and the family belonged to the Gargya gotra, which was a Indian caste that prohibits marriage to anyone outside the caste. The family followed the Ashvalayana sutra which was a manual of sacrificial ceremonies in the Rigveda, a collection of Vedic hymns. He worshipped the personified deity Soma who was the "master of plants" and the healer of disease. This explains the name Somayaji which means he was from a family qualified to conduct the Soma ritual.

Nilakantha studied astronomy and Vedanta, one of the six orthodox systems of Indian Hindu philosophy, under the teacher Ravi. He was also taught by Damodra who was the son of Parameshvara. Parameshvara was a famous Indian astronomer and Damodra followed his father's teachings. This led Nilakantha also to become a follower of Parameshvara. A number of texts on mathematical astronomy written by Nilakantha have survived. In all he wrote about ten treatises on astronomy.

The Tantrasamgraha is his major astronomy treatise written in 1501. It consists of 432 Sanskrit verses divided into 8 chapters, and it covers various aspects of Indian astronomy. It is based on the epicyclic and eccentric models of planetary motion. The first two chapters deal with the motions and longitudes of the planets. The third chapter Treatise on shadow deals with various problems related with the sun's position on the celestial sphere, including the relationships of its expressions in the three systems of coordinates, namely ecliptic, equatorial and horizontal coordinates.

The fourth and fifth chapters are Treatise on the lunar eclipse and On the solar eclipse and these two chapters treat various aspects of the eclipses of the sun and the moon. The sixth chapter is On vyatipata and deals with the complete deviation of the longitudes of the sun and the moon. The seventh chapter On visibility computation discusses the rising and setting of the moon and planets. The final chapter On elevation of the lunar cusps examines the size of the part of the moon which is illuminated by the sun and gives a graphical representation of it.

The Tantrasamgraha is very important in terms of the mathematics Nilakantha uses. In particular he uses results discovered by Madhava and it is an important source of the remarkable mathematical results which he discovered. However, Nilakantha does not just use Madhava's results, he extends them and improves them.

An anonymous commentary entitled Tantrasangraha-vakhya appeared and, somewhat later in about 1550, Jyesthadeva published a commentary entitled Yuktibhasa that contained proofs of the earlier results by Madhava and Nilakantha. This is quite unusual for an Indian text in giving mathematical proofs.

The series $\pi/4 = 1 - 1/3 + 1/5 - 1/7 + ...$ is a special case of the series representation for arctan, namely

$$\tan^{-1}x = x - x^3/3 + x^5/5 - x^7/7 + ...$$

It is well known that one simply puts x = 1 to obtain the series for p/4. The author of reports on the appearance of these series in the work of Leibniz and James Gregory from the 1670s. The contributions of the two European mathematicians to this series are well known but the results on this series in the work of Madhava nearly three hundred years earlier as presented by Nilakantha in the Tantrasamgraha is also discussed.

Nilakantha derived the series expansion

$$\tan^{-1}x = x - x^3/3 + x^5/5 - x^7/7 + ...$$

by obtaining an approximate expression for an arc of the circumference of a circle and then considering the limit. An interesting feature of his work was his introduction of several additional series for $\pi/4$ that converged more rapidly than

$\pi/4 = 1 - 1/3 + 1/5 - 1/7 + \dots$.

The author of provides a reconstruction of how he may have arrived at these results based on the assumption that he possessed a certain continued fraction representation for the tail series

$1/(n+2) - 1/(n+4) + 1/(n+6) - 1/(n+8) + \dots$.

The Tantrasamgraha is not the only work of Nilakantha of which we have the text. He also wrote Golasara which is written in fifty-six Sanskrit verses and shows how mathematical computations are used to calculate astronomical data. The Siddhanta Darpana is written in thirty-two Sanskrit verses and describes a planetary model. The Candracchayaganita is written in thirty-one Sanskrit verses and explains the computational methods used to calculate the moon's zenith distance.

The head of the Nambudiri caste in Nilakantha's time was Netranarayana and he became Nilakantha's patron for another of his major works, namely the Aryabhatiyabhasya which is a commentary on the Aryabhatiya of Aryabhatta I. In this work Nilakantha refers to two eclipses which he observed, the first on 6 March 1467 and the second on 28 July 1501 at Anantaksetra. Nilakantha also refers in the Aryabhatiyabhasya to other works which he wrote such as the Grahanirnaya on eclipses which have not survived.

Jyesthadeva

Born: about 1500 in Kerala, India

Died: about 1575 in Kerala, India

Jyesthadeva lived on the southwest coast of India in the district of Kerala. He belonged to the Kerala school of mathematics built on the work of Madhava, Nilakantha Somayaji, Parameshvara and others.

Jyesthadeva wrote a famous text Yuktibhasa which he wrote in Malayalam, the regional language of Kerala. The work is a survey of Kerala mathematics and, very unusually for an Indian mathematical text, it contains proofs of the theorems and gives derivations of the rules it contains. It is one of the main astronomical and mathematical texts produced by the Kerala school. The work was based mainly on the Tantrasamgraha of Nilakantha.

The Yuktibhasa is a major treatise, half on astronomy and half on mathematics, written in 1501. The Tantrasamgraha on which it is based consists of 432 Sanskrit verses divided into 8 chapters, and it covers various aspects of Indian astronomy. It is based on the epicyclic and eccentric models of planetary motion.

The first two chapters deal with the motions and longitudes of the planets. The third chapter Treatise on shadow deals with various problems related with the sun's position on the celestial sphere, including the relationships of its expressions in the three systems of coordinates, namely ecliptic, equatorial and horizontal coordinates.

The fourth and fifth chapters are Treatise on the lunar eclipse and On the solar eclipse and these two chapters treat various aspects of the eclipses of the sun and the moon. The sixth chapter is On vyatipata and deals with the complete deviation of the longitudes of the sun and the moon. The seventh chapter On visibility computation discusses the rising and setting of the moon and planets. The final chapter On elevation of the lunar cusps examines the size of the part of the moon which is illuminated by the sun and gives a graphical representation of it.

The Yuktibhasa is very important in terms of the mathematics Jyesthadeva presents. In particular he presents results discovered by Madhava and the treatise is an important source of the remarkable mathematical theorems which Madhava discovered. Written in about 1550, Jyesthadeva's commentary contained proofs of the earlier results by Madhava and Nilakantha which these earlier authors did not give. Gupta gives a translation of the text and this is also given and a number of other sources. Jyesthadeva describes Madhava's series as follows:-

The first term is the product of the given sine and radius of the desired arc divided by the cosine of the arc. The succeeding terms are obtained by a process of iteration when the first term is repeatedly multiplied by the square of the sine and divided by the square of the cosine. All the terms are then divided by the odd numbers 1, 3, 5, ... The arc is obtained by adding and subtracting respectively the terms of odd rank and those of even rank. It is laid down that the sine of the arc or that of its complement whichever is the smaller should be taken here as the given sine. Otherwise the terms obtained by this above iteration will not tend to the vanishing magnitude.

This is a remarkable passage describing Madhava's series, but remember that even this passage by Jyesthadeva was written more than 100 years before James Gregory rediscovered this series expansion. To see how this description of the series fits with Gregory's series for arctan(x) see the biography of Madhava. Other mathematical results presented by Jyesthadeva include topics studied by earlier Indian mathematicians such as integer solutions of systems of first degree equation solved by the kuttaka method, and rules of finding the sines and the cosines of the sum and difference of two angles.

Not only does the mathematics anticipate work by European mathematicians a century later, but the planetary theory presented by Jyesthadeva is similar to that adopted by Tycho Brahe.

Kamalakara

Born: about 1616 in Benares (now Varanasi), India

Died: about 1700 in India

Kamalakara was an Indian astronomer and mathematician who came from a family of famous astronomers. Kamalakara's father was Narsimha who was born in 1586. Two of Kamalakara's three brothers were also famous astronomer/mathematicians, these being Divakara, who was the eldest of the brothers born in 1606, and Ranganatha who was younger than Kamalakara.

As was common throughout the classical period of Indian mathematics, members of the family acted as teachers to other family members. In particular Kamalakara was taught by his elder brother Divakara while Divakara himself had been taught by their uncle Shiva. Pingree writes :-

[Kamalakara] combined traditional Indian astronomy with Aristotelian physics and Ptolemaic astronomy as presented by Islamic scientists (especially Ulugh Beg). Following his family's tradition he wrote a commentary, Manorama, on Ganesa's Grahalaghava and, like his father, Narsimha, another commentary on the Suryasiddhanta, called the Vasanabhasya ...

Kamalakara's most famous work, the Siddhanta-tattva-viveka, was commented on by Kamalakara himself. The work was completed in 1658. It is a work of fifteen chapters covering standard topics for Indian astronomy texts at this time. It deals with the topics of: units of time measurement; mean motions of the planets; true longitudes of the planets; the three problems of diurnal rotation; diameters and distances of the planets; the earth's shadow; the moon's crescent; risings and settings; syzygies; lunar eclipses, solar eclipses; planetary transits across the sun's disk; the patas of the moon and sun; the "great problems"; and a final chapter which forms a conclusion.

The third chapter of the Siddhanta-tattva-viveka contains some of the most interesting mathematical results. In that chapter Kamalakara used the addition and subtraction theorems for the sine and the cosine to give trigonometric formulae for the sines and cosines of double, triple, quadruple and quintuple angles. In particular he gives formulae for sin(A/2) and sin(A/4) in terms of sin(A) and iterative formulae for sin(A/3) and sin(A/5).

The Siddhanta-tattva-viveka is a Sanskrit text and in it Kamalakara makes frequent use of the place-value number system with Sanskrit numerals. This and many other aspects of the work are discussed.

Jagannatha Samrat

Born: about 1690 in Amber (now Jaipur), India

Died: about 1750 in India

Jagannatha had Jai Singh Sawai as his patron. Jai Singh Sawai was the ruler of Amber, now Jaipur, in eastern Rajasthan. He began his rule in 1699 and by clever use of tax rights on land that was rented by the state to an individual person he became the most important ruler in the region. His financial success let him finance the scholarly works of people such as Jagannatha. It is worth noting that Jai Singh's importance was recognised by Amber which was then called Jaipur in his honour.

Jai Singh ruled Amber throughout the period in which Jagannatha was producing his scientific work. He realised that the health of the country required Indian culture and science to be revitalised and returned to its position of leading importance which it had possessed. So Jai Singh employed Jagannatha to make Sanskrit translations of the important Greek scientific works which at that time were only available in Arabic translations. Jagannatha translated Euclid's Elements from the Arabic translation by Nasir al-Din al-Tusi made nearly 500 years earlier. His Sanskrit version was called Rekhaganita and it was completed by 1727. We know this date since a copy was made by a scribe and he dated the start of his work as 1727.

Ptolemy's Almagest had been one of the works which Arabic scientists had studied intently and, in 1247, al-Tusi wrote Tahrir al-Majisti (Commentary on the Almagest) in which he introduced various trigonometrical techniques to calculate tables of sines. Jagannatha translated al-Tusi's Arabic version but he did more than this for he included in the same work, which he called Siddhantasamrat, his own comments on related work of other Arabic mathematical astronomers such as Ulugh Beg and al-Kashi.

It is clear from Jagannatha's work that he is working as one of a group of mathematicians and astronomers gathered by Jai Singh in his scheme to bring the best in scientific ideas from outside India to reinvigorate the scientific scene in India.

Gupta looks at the history of the result

$$\sin(\pi/10) = (O5 - 1)/4$$

in Indian mathematics. The result appears for the first time in the work of Bhaskara II, but there were a number of interesting proofs of the result by later Indian mathematicians. One of the proofs presented by Gupta was by Jagannatha who gave a proof which was essentially geometric in nature but, interestingly, contained an analytic procedure in terms of trigonometric and algebraic steps.

Indian Numerals

It is worth beginning this article with the same quote from Laplace which we give in the article Overview of Indian mathematics. Laplace wrote:-

The ingenious method of expressing every possible number using a set of ten symbols (each symbol having a place value and an absolute value) emerged in

India. The idea seems so simple nowadays that its significance and profound importance is no longer appreciated. Its simplicity lies in the way it facilitated calculation and placed arithmetic foremost amongst useful inventions, the importance of this invention is more readily appreciated when one considers that it was beyond the two greatest men of Antiquity, Archimedes and Apollonius.

The purpose of this article is to attempt the difficult task of trying to describe how the Indians developed this ingenious system. We will examine two different aspects of the Indian number systems in this article. First we will examine the way that the numerals 0, 1, 2, 3, 4, 5, 6, 7, 8, 9 evolved into the form which we recognise today. Of course it is important to realise that there is still no standard way of writing these numerals. The different fonts on this computer can produce many forms of these numerals which, although recognisable, differ markedly from each other. Many hand-written versions are even hard to recognise.

The second aspect of the Indian number system which we want to investigate here is the place value system which, as Laplace comments in the quote which we gave at the beginning of this article, seems "so simple that its significance and profound importance is no longer appreciated." We should also note the fact, which is important to both aspects, that the Indian number systems are almost exclusively base 10, as opposed to the Babylonian base 60 systems.

Beginning with the numerals themselves, we certainly know that today's symbols took on forms close to that which they presently have in Europe in the 15th century. It was the advent of printing which motivated the standardisation of the symbols. However we must not forget that many countries use symbols today which are quite different from 0, 1, 2, 3, 4, 5, 6, 7, 8, 9 and unless one learns these symbols they are totally unrecognisable as for example the Greek alphabet is to someone unfamiliar with it.

One of the important sources of information which we have about Indian numerals comes from al-Biruni. During the 1020s al-Biruni made several visits to India. Before he went there al-Biruni already knew of Indian astronomy and mathematics from Arabic translations of some Sanskrit texts. In India he made a detailed study of Hindu philosophy and he also studied several branches of Indian science and mathematics.

Al-Biruni wrote 27 works on India and on different areas of the Indian sciences. In particular his account of Indian astronomy and mathematics is a valuable contribution to the study of the history of Indian science. Referring to the Indian numerals in a famous book written about 1030 he wrote:-

> *Whilst we use letters for calculation according to their numerical value, the Indians do not use letters at all for arithmetic. And just as the shape of the letters that they use for writing is different in different regions of their country, so the numerical symbols vary.*

It is reasonable to ask where the various symbols for numerals which al-Biruni saw originated. Historians trace them all back to the Brahmi numerals which came into being around the middle of the third century BC. Now these Brahmi numerals were not just symbols for the numbers between 1 and 9. The situation is much more complicated for it was not a place-value system so there were symbols for many more numbers. Also there were no special symbols for 2 and 3, both numbers being constructed from the symbol for 1.

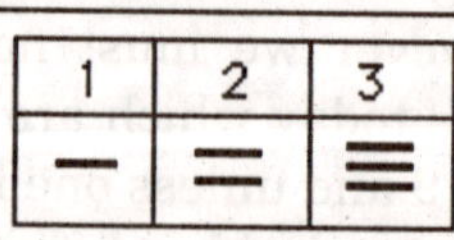

Brahmi one, two, three

Here is the Brahmi one, two, three.

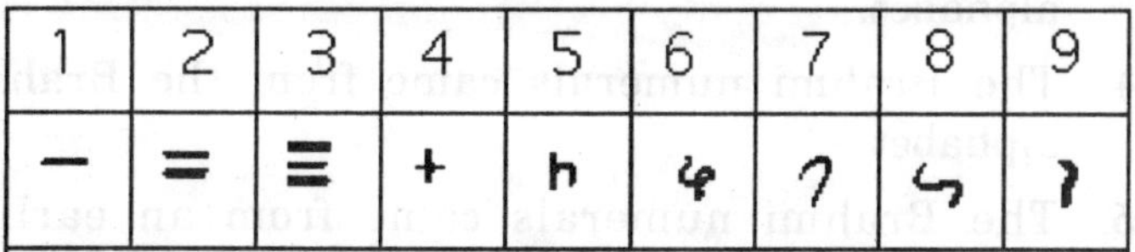

Brahmi numerals around 1st century A.D.

There were separate Brahmi symbols for 4, 5, 6, 7, 8, 9 but there were also symbols for 10, 100, 1000, ... as well as 20, 30, 40, ... , 90 and 200, 300, 400, ..., 900.

The Brahmi numerals have been found in inscriptions in caves and on coins in regions near Poona, Bombay, and Uttar Pradesh. Dating these numerals tells us that they were in use over quite a long time span up to the 4th century AD. Of course different inscriptions differ somewhat in the style of the symbols.

Here is one style of the Brahmi numerals.

We should now look both forward and backward from the appearance of the Brahmi numerals. Moving forward leads to many different forms of numerals but we shall choose to examine only the path which has led to our present day symbols. First, however, we look at a number of different theories concerning the origin of the Brahmi numerals.

There is no problem in understanding the symbols for 1, 2, and 3. However the symbols for 4, ... , 9 appear to us to have no obvious link to the numbers they represent. There have been quite a number of theories put forward by historians over many years as to the origin of these numerals. Ifrah lists a number of the hypotheses which have been put forward.

1. The Brahmi numerals came from the Indus valley culture of around 2000 BC.
2. The Brahmi numerals came from Aramaean numerals.

3. The Brahmi numerals came from the Karoshthi alphabet.
4. The Brahmi numerals came from the Brahmi alphabet.
5. The Brahmi numerals came from an earlier alphabetic numeral system, possibly due to Panini.
6. The Brahmi numerals came from Egypt.

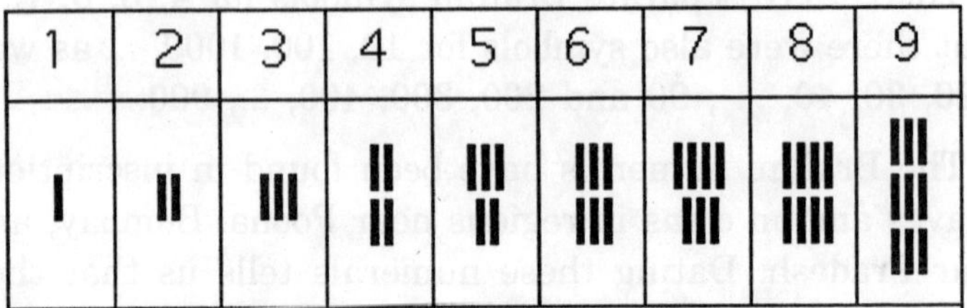

Ifrah's guess for pre-Brahmi numbers

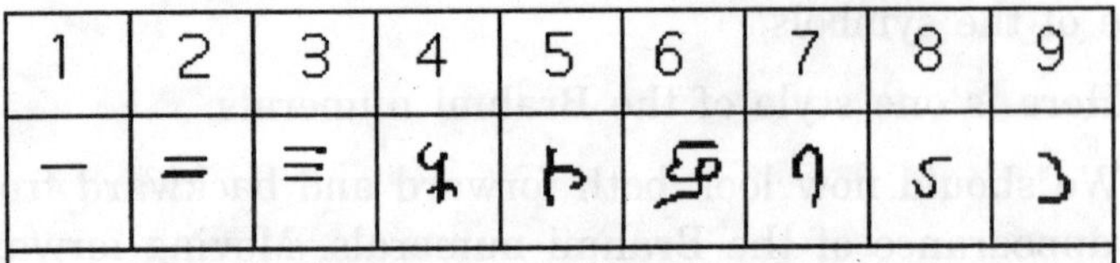

Gupta numerals around 4th century A.D.

Basically these hypotheses are of two types. One is that the numerals came from an alphabet in a similar way to the Greek numerals which were the initial letters of the names of the numbers. The second type of hypothesis is that they derive from an earlier number system of the same broad type as Roman numerals. For example the Aramaean numerals of hypothesis 2 are based on I (one) and X (four):

I, II, III, X, IX, IIX, IIIX, XX.

Ifrah examines each of the six hypotheses in turn and rejects them, although one would have to say that in some cases it is more due to lack of positive evidence rather than to negative evidence.

Ifrah proposes a theory of his own, namely that:-

> *... the first nine Brahmi numerals constituted the vestiges of an old indigenous numerical notation, where the nine numerals were represented by the corresponding number of vertical lines ... To enable the numerals to be written rapidly, in order to save time, these groups of lines evolved in much the same manner as those of old Egyptian Pharonic numerals. Taking into account the kind of material that was written on in India over the centuries (tree bark or palm leaves) and the limitations of the tools used for writing (calamus or brush), the shape of the numerals became more and more complicated with the numerous ligatures, until the numerals no longer bore any resemblance to the original prototypes.*

It is a nice theory, and indeed could be true, but there seems to be absolutely no positive evidence in its favour. The idea is that they evolved from:

> *One might hope for evidence such as discovering numerals somewhere on this evolutionary path. However, it would appear that we will never find convincing proof for the origin of the Brahmi numerals.*

If we examine the route which led from the Brahmi numerals to our present symbols (and ignore the many other systems which evolved from the Brahmi numerals) then we next come to the Gupta symbols. The Gupta period is that during which the Gupta dynasty ruled over the Magadha state in northeastern India, and this was from the early 4th century AD to the late 6th century AD. The Gupta numerals developed from the Brahmi numerals and were spread over large areas by the Gupta empire as they conquered territory.

1	2	3	4	5	6	7	8	9	0
१	२	३	४	५	६	७	८	९	०

Nagari numerals around 11th century A.D.

The Gupta numerals evolved into the Nagari numerals, sometimes called the Devanagari numerals. This form evolved from the Gupta numerals beginning around the 7th century AD and continued to develop from the 11th century onward. The name literally means the "writing of the gods" and it was the considered the most beautiful of all the forms which evolved. For example al-Biruni writes:-

What we [the Arabs] use for numerals is a selection of the best and most regular figures in India.

These "most regular figures" which al-Biruni refers to are the Nagari numerals which had, by his time, been transmitted into the Arab world. The way in which the Indian numerals were spread to the rest of the world between the 7th to the 16th centuries in examined in detail. In this paper, however, Gupta claims that Indian numerals had reached Southern Europe by the end of the 5th century but his argument is based on the Geometry of Boethius which is now known to be a forgery dating from the first half of the 11th century. It would appear extremely unlikely that the Indian numerals reach Europe as early as Gupta suggests.

We now turn to the second aspect of the Indian number system which we want to examine in this article, namely the fact that it was a place-value system with the numerals standing for different values depending on their position relative to the other numerals. Although our place-value system is a direct descendant of the Indian system, we should note straight away that the Indians were not the first to develop such a system. The Babylonians had a place-value system as early as the 19th century BC but the Babylonian systems were to base 60. The Indians were the first to develop a base 10 positional system and, considering the date of the Babylonian system, it came very late indeed.

The oldest dated Indian document which contains a number written in the place-value form used today is a legal document dated 346 in the Chhedi calendar which translates to a date in our calendar of 594 AD. This document is a donation charter of Dadda III of Sankheda in the Bharukachcha region. The only problem with it is that some historians claim that the date has been added as a later forgery. Although it was not unusual for such charters to be modified at a later date so that the property to which they referred could be claimed by someone who was not the rightful owner, there seems no conceivable reason to forge the date on this document. Therefore, despite the doubts, we can be fairly sure that this document provides evidence that a place-value system was in use in India by the end of the 6th century.

Many other charters have been found which are dated and use of the place-value system for either the date or some other numbers within the text. These include:

1. a donation charter of Dhiniki dated 794 in the Vikrama calendar which translates to a date in our calendar of 737 AD.
2. an inscription of Devendravarman dated 675 in the Shaka calendar which translates to a date in our calendar of 753 AD.
3. a donation charter of Danidurga dated 675 in the Shaka calendar which translates to a date in our calendar of 737 AD.
4. a donation charter of Shankaragana dated 715 in the Shaka calendar which translates to a date in our calendar of 793 AD.
5. a donation charter of Nagbhata dated 872 in the Vikrama calendar which translates to a date in our calendar of 815 AD.
6. an inscription of Bauka dated 894 in the Vikrama calendar which translates to a date in our calendar of 837 AD.

All of these are claimed to be forgeries by some historians but some, or all, may well be genuine.

The first inscription which is dated and is not disputed is the inscription at Gwalior dated 933 in the Vikrama calendar which translates to a date in our calendar of 876 AD. Further details of this inscription is given in the article on zero.

There is indirect evidence that the Indians developed a positional number system as early as the first century AD. The evidence is found from inscriptions which, although not in India, have been found in countries which were assimilating Indian culture. Another source is the Bakhshali manuscript which contains numbers written in place-value notation. The problem here is the dating of this manuscript, a topic which is examined in detail in our article on the Bakhshali manuscript.

We are left, of course, with asking the question of why the Indians developed such an ingenious number system when the ancient Greeks, for example, did not. A number of theories have been put forward concerning this question. Some historians believe that the Babylonian base 60 place-value system was transmitted to the Indians via the Greeks. We have commented in the article on zero about Greek astronomers using the Babylonian base 60 place-value system with a symbol o similar to our zero. The theory here is that these ideas were transmitted to the Indians who then combined this with their own base 10 number systems which had existed in India for a very long time.

A second hypothesis is that the idea for place-value in Indian number systems came from the Chinese. In particular the Chinese had pseudo-positional number rods which, it is claimed by some, became the basis of the Indian positional system. This view is put forward by, for example, Lay Yong Lam. Lam argues that the Chinese system already contained what he calls the:-

> *... three essential features of our numeral notation system: (i) nine signs and the concept of zero, (ii) a place value system and (iii) a decimal base.*

A third hypothesis is put forward by Joseph. His idea is that the place-value in Indian number systems is something which was developed entirely by the Indians. He has an interesting theory as to why the Indians might be pushed into such an idea. The reason, Joseph believes, is due to the Indian fascination with large numbers. Freudenthal is another historian of mathematics who supports the theory that the idea came entirely from within India.

To see clearly this early Indian fascination with large numbers, we can take a look at the Lalitavistara which is an account of the life of Gautama Buddha. It is hard to date this work since it underwent continuous development over a long period but dating it to around the first or second century AD is reasonable. In Lalitavistara Gautama, when he is a young man, is examined on mathematics. He is asked to name all the numerical ranks beyond a koti which is 107. He lists the powers of 10 up to 1053. Taking this as a first level he then carries on to a second level and gets eventually to 10421. Gautama's examiner says:-

You, not I, are the master mathematician.

It is stories such as this, and many similar ones, which convince Joseph that the fascination of the Indians with large numbers must have driven them to invent a system in which such numbers are easily expressed, namely a place-valued notation. He writes:-

> *The early use of such large numbers eventually led to the adoption of a series of names for successive powers of ten. The importance of these number names cannot be exaggerated. The word-numeral system, later replaced by an alphabetic notation, was the logical outcome of proceeding by multiples of ten. ... The decimal place-value system developed when a decimal scale came to be associated with the value of the places of the numbers arranged left to right or right to left and this was precisely what happened in India ...*

However, the same story in Lalitavistara convinces Kaplan that the Indians' ideas of numbers came from the Greeks, for to him the story is an Indian version of Archimedes' Sand-reckoner. All that we know is that the place-value system of the Indians, however it arose, was transmitted to the Arabs and later into Europe to have, in the words of Laplace, profound importance on the development of mathematics.

Jaina Mathematics

It is a little hard to define Jaina mathematics. Jainism is a religion and philosophy which was founded in India around the 6th century BC. To a certain extent it began to replace the Vedic religions which, with their sacrificial procedures, had given rise to the mathematics of building altars. The mathematics of the Vedic religions is described in the article Indian Sulbasutras.

Now we could use the term Jaina mathematics to describe mathematics done by those following Jainism and indeed this would then refer to a part of mathematics done on the Indian subcontinent from the founding of Jainism up to modern times. Indeed this is fair and some of the articles in the references refer to fairly modern mathematics. For example, Jha looks at the contributions of Jainas from the 5th century BC up to the 18th century AD.

This article will concentrate on the period after the founding of Jainism up to around the time of Aryabhatta in around 500 AD. The reason for taking this time interval is that until recently this was thought to be a time when there was little mathematical activity in India. Aryabhatta's work was seen as the beginning of a new classical period for Indian mathematics and indeed this is fair. Yet Aryabhatta did not work in mathematical isolation and as well as being seen as the person who brought in a new era of mathematical investigation in India, more recent research has shown that there is a case for seeing him also as representing the end-product of a mathematical period of which relatively little is known. This is the period we shall refer to as the period of Jaina mathematics.

There were mathematical texts from this period yet they have received little attention from historians until recent times. Texts, such as the Surya Prajnapti which is thought to be around the 4th century BC and the Jambudvipa Prajnapti from around the same period, have recently received attention through the study of later commentaries. The Bhagabati Sutra dates from around 300 BC and contains interesting information on combinations. From about the second century BC is the Sthananga Sutra which is particularly interesting in that it lists the topics which made up the mathematics studied at the time. In fact this list of topics sets the scene for the areas of study for a long time to come in the Indian subcontinent. The topics are listed as:-

> *... the theory of numbers, arithmetical operations, geometry, operations with fractions, simple equations, cubic equations, quartic equations, and permutations and combinations.*

The ideas of the mathematical infinite in Jaina mathematics is very interesting indeed and they evolve largely due to the Jaina's cosmological ideas. In Jaina cosmology time is thought of as eternal and without form. The world is infinite, it was never created and has always existed. Space pervades everything and is without form. All the objects of the universe exist in space which is divided into the space of the universe and the space of the non-universe. There is a central region of the universe in which all living beings, including men, animals, gods and devils, live. Above this central region is the upper world which is itself divided into two parts. Below the central region is the lower world which is divided into seven tiers. This led to the work described on a mathematical topic in the Jaina work,

Tiloyapannatti by Yativrsabha. A circle is divided by parallel lines into regions of prescribed widths. The lengths of the boundary chords and the areas of the regions are given, based on stated rules.

This cosmology has strongly influenced Jaina mathematics in many ways and has been a motivating factor in the development of mathematical ideas of the infinite which were not considered again until the time of Cantor. The Jaina cosmology contained a time period of 2^{588} years. Note that 2^{588} is a very large number!

2^{588} = 1013 065324 433836 171511 818326 096474 890383 898005 918563 696288 002277 756507 034036 354527 929615 978746 851512 277392 062160 962106 733983 191180 520452 956027 069051 297354 415786 421338 721071 661056.

So what are the Jaina ideas of the infinite. There was a fascination with large numbers in Indian thought over a long period and this again almost required them to consider infinitely large measures. The first point worth making is that they had different infinite measures which they did not define in a rigorous mathematical fashion, but nevertheless are quite sophisticated. The paper describes the way that the first unenumerable number was constructed using effectively a recursive construction.

The Jaina construction begins with a cylindrical container of very large radius r^q (taken to be the radius of the earth) and having a fixed height h. The number $n^q = f(r^q)$ is the number of very tiny white mustard seeds that can be placed in this container. Next, $r_1 = g(r^q)$ is defined by a complicated recursive subprocedure, and then as before a new larger number $n_1 = f(r_1)$ is defined. The text the Anuyoga Dwara Sutra then states:-

Still the highest enumerable number has not been attained.

The whole procedure is repeated, yielding a truly huge number which is called jaghanya- parita- asamkhyata meaning "unenumerable of low enhanced order". Continuing the process yields the smallest unenumerable number.

Jaina mathematics recognised five different types of infinity:-

> *... infinite in one direction, infinite in two directions, infinite in area, infinite everywhere and perpetually infinite.*

The Anuyoga Dwara Sutra contains other remarkable numerical speculations by the Jainas. For example several times in the work the number of human beings that ever existed is given as 2^{96}.

By the second century AD the Jaina had produced a theory of sets. In Satkhandagama various sets are operated upon by logarithmic functions to base two, by squaring and extracting square roots, and by raising to finite or infinite powers. The operations are repeated to produce new sets.

Permutations and combinations are used in the Sthananga Sutra. In the Bhagabati Sutra rules are given for the number of permutations of 1 selected from n, 2 from n, and 3 from n. Similarly rules are given for the number of combinations of 1 from n, 2 from n, and 3 from n. Numbers are calculated in the cases where n = 2, 3 and 4. The author then says that one can compute the numbers in the same way for larger n. He writes:-

> *In this way, 5, 6, 7, ..., 10, etc. or an enumerable, unenumerable or infinite number of may be specified. Taking one at a time, two at a time, ... ten at a time, as the number of combinations are formed they must all be worked out.*

Interestingly here too there is the suggestion that the arithmetic can be extended to various infinite numbers. In other works the relation of the number of combinations to the coefficients occurring in the binomial expansion was noted. In a commentary on this third century work in the tenth century, Pascal's triangle appears in order to give the coefficients of the binomial expansion.

Another concept which the Jainas seem to have gone at least some way towards understanding was that of the logarithm. They had begun to understand the laws of indices. For example the Anuyoga Dwara Sutra states:-

The first square root multiplied by the second square root is the cube of the second square root.

The second square root was the fourth root of a number. This therefore is the formula

$(Oa).(OOa) = (OOa)^3$.

Again the Anuyoga Dwara Sutra states:-

... the second square root multiplied by the third square root is the cube of the third square root.

The third square root was the eighth root of a number. This therefore is the formula

$(OOa).(OOOa) = (OOOa)^3$.

Some historians studying these works believe that they see evidence for the Jainas having developed logarithms to base 2.

The value of p in Jaina mathematics has been a topic of a number of research papers. As with much research into Indian mathematics there is interest in whether the Indians took their ideas from the Greeks. The approximation $\pi = O10$ seems one which was frequently used by the Jainas.

Finally let us comment on the Jaina's astronomy. This was not very advanced. It was not until the works of Aryabhatta that the Greek ideas of epicycles entered Indian astronomy. Before the Jaina period the ideas of eclipses were based on a demon called Rahu which devoured or captured the Moon or the Sun causing their eclipse. The Jaina school assumed the existence of two demons Rahu, the

Dhruva Rahu which causes the phases of the Moon and the Parva Rahu which has irregular celestial motion in all directions and causes an eclipse by covering the Moon or Sun or their light. The author of points out that, according to the Jaina school, the greatest possible number of eclipses in a year is four.

Despite this some of the astronomical measurements were fairly good. The data in the Surya Prajnapti implies a synodic lunar month equal to 29 plus $\frac{16}{31}$ days; the correct value being nearly 29.5305888. There has been considerable interest in examining the data presented in these Jaina texts to see if the data originated from other sources. For example in the Surya Prajnapti data exists which implies a ratio of 3:2 for the maximum to the minimum length of daylight. Now this is not true for India but is true for Babylonia which makes some historians believe that the data in the Surya Prajnapti is not of Indian origin but is Babylonian. However, Sharma and Lishk present an alternative hypothesis which would allow the data to be of Indian origin. One has to say that their suggestion that 3:2 might be the ratio of the amounts of water to be poured into the water-clock on the longest and shortest days seems less than totally convincing.

The Bakhshali Manuscript

The Bakhshali manuscript is an early mathematical manuscript which was discovered over 100 years ago. We shall discuss in a moment the problem of dating this manuscript, a topic which has aroused much controversy, but for the moment we will examine how it was discovered. The paper describes this discovery along with the early history of the manuscript. Gupta writes:-

The Bakhshali Manuscript is the name given to the mathematical work written on birch bark and found in the summer of 1881 near the village Bakhshali (or Bakhshalai) of the Yusufzai subdivision of the Peshawar district (now in Pakistan). The village is in Mardan tahsil and is situated 50 miles from the city of Peshawar.

An Inspector of Police named Mian An-Wan-Udin (whose tenant actually discovered the manuscript while digging a stone enclosure in a ruined place) took the work to the Assistant Commissioner at Mardan who intended to forward the manuscript to Lahore Museum. However, it was subsequently sent to the Lieutenant Governor of Punjab who, on the advice of General A Cunningham, directed it to be passed on to Dr. Rudolf Hoernle of the Calcutta Madrasa for study and publication. Dr. Hoernle presented a description of the BM before the Asiatic Society of Bengal in 1882, and this was published in the Indian Antiquary in 1883. He gave a fuller account at the Seventh Oriental Conference held at Vienna in 1886 and this was published in its Proceedings. A revised version of this paper appeared in the Indian Antiquary of 1888. In 1902, he presented the Bakhshali Manuscript to the Bodleian Library, Oxford, where it is still (Shelf mark: MS. Sansk. d. 14).

A large part of the manuscript had been destroyed and only about 70 leaves of birch-bark, of which a few were only scraps, survived to the time of its discovery.

To show the arguments regarding its age we note that F R Hoernle, referred to in the quotation above, placed the Bakhshali manuscript between the third and fourth centuries AD. Many other historians of mathematics such as Moritz Cantor, F Cajori, B Datta, S N Sen, A K Bag, and R C Gupta agreed with this dating. In 1927-1933 the Bakhshali manuscript was edited by G R Kaye and published with a comprehensive introduction, an English translation, and a transliteration together with facsimiles of the text. Kaye claimed that the manuscript dated from the twelfth century AD and he even doubted that it was of Indian origin.

Channabasappa gives the range 200 - 400 AD as the most probable date. The same author identifies five specific mathematical terms which do not occur in the works of Aryabhatta and he argues that this strongly supports a date for the Bakhshali manuscript earlier than the 5th century. Joseph suggests that the evidence all points to the:-

> *... manuscript [being] probably a later copy of a document composed at some time in the first few centuries of the Christian era.*

L V Gurjar claims that the manuscript is no later than 300 AD. On the other hand T Hayashi claims that the date of the original is probably from the seventh century, but he also claims that the manuscript itself is a later copy which was made between the eighth and the twelfth centuries AD.

I [EFR] feel that if one weighs all the evidence of these experts the most likely conclusion is that the manuscript is a later copy of a work first composed around 400 AD. Why do I believe that the actual manuscript was written later? Well our current understanding of Indian numerals and writing would date the numerals used in the manuscript as not having appeared before the ninth or tenth century. To accept that this style of numeral existed in 400 AD. would force us to change greatly our whole concept of the time-scale for the development of Indian numerals. Sometimes, of course, we are forced into major rethinks but, without supporting evidence, everything points to the manuscript being a tenth century copy of an original from around 400 AD. Despite the claims of Kaye, it is essentially certain that the manuscript is Indian.

The attraction of the date of 400 AD for the Bakhshali manuscript is that this puts it just before the "classical period" of Indian mathematics which began with the work of Aryabhatta I around 500. It would then fill in knowledge we have of Indian mathematics for, prior to the discovery of this manuscript, we had little knowledge of Indian mathematics between the dates of about 200 BC, and 500 AD. This date would make it a document near the end of the period of Jaina mathematics and it can be seen as, in some sense, marking the achievements of the Jains.

What does the manuscript contain? Joseph writes:-

The Bakhshali manuscript is a handbook of rules and illustrative examples together with their solutions. It is devoted mainly to arithmetic and algebra, with just a few problems on geometry and mensuration. Only parts of it have been restored, so we cannot be certain about the balance between different topics.

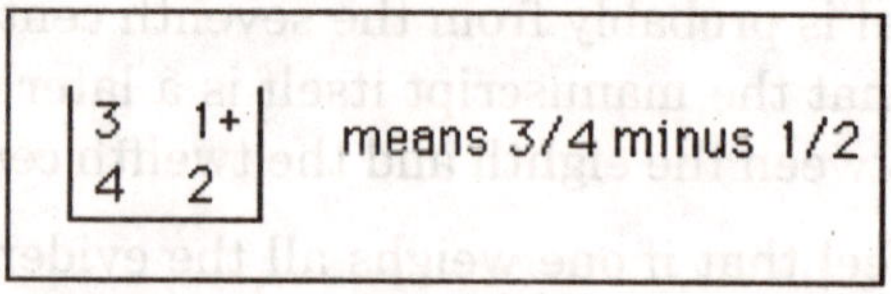

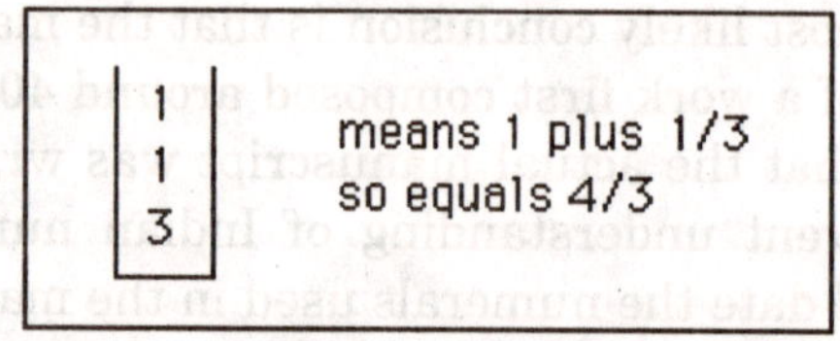

Now the way that the manuscript is laid out is quite unusual for an Indian document (which of course leads people like Kaye to prefer the hypothesis that it is not Indian at all - an idea in which we cannot see any merit). The Bakhshali manuscript gives the statement of a rule. There then follows an example given first in words, then using mathematical notation. The solution to the example is then given and finally a proof is set out.

The notation used is not unlike that used by Aryabhatta but it does have features not found in any other document. Fractions are not dissimilar in notation to that used today, written with one number below the other. No line appears between the numbers as we would write today, however. Another unusual feature is the sign + placed after a number to indicate a negative. It is very strange for us today to see our addition symbol being used for subtraction. As an example, here is how $\frac{3}{4} - \frac{1}{2}$ would be written.

$\frac{3}{4}$ min us $\frac{1}{2}$

1
1
3+

means 1 minus 1/3
so equals 2/3

5	2	yu	pha 7
1	1		

This means 5/1 plus 2/1
equals 7

Compound fractions were written in three lines. Hence 1 plus 1/3 would be written thus

1 plus 1/3

and 1 minus 1/3 = 2/3 in the following way

1 minus 1/3 = 2/3

Sums of fractions such as 5/1 plus 2/1 are written using the symbol yu (for yuta)

5/1 plus 2/1

•	1	1	1		
1	1	1	1	bha 32	phala 108
	3+	3+	3+		

Find number 32 divided by
$(2/3)^3$. It equals 108

```
| 1         |
| 1  bha 8  |   pha 12
| 3+        |

This means 8 divided by
2/3 equals 12
```

Division is denoted by bha, an abbreviation for bhaga meaning "part". For example

8 divided by 2/3

Equations are given with a large dot representing the unknown. A confusing aspect of Indian mathematics is that this notation was also often used to denote zero, and sometimes this same notation for both zero and the unknown are used in the same document. Here is an example of an equation as it appears in the Bakhshali manuscript.

EQUATION

The method of equalisation is found in many types of problems which occur in the manuscript. Problems of this type which are found in the manuscript are examined and some of these lead to indeterminate equations. Included are problems concerning equalising wealth, the positions of two travellers, wages, and purchases by a number of merchants. These problems can all be reduced to solving a linear equation with one unknown or to a system of n linear equations in n unknowns. To illustrate we give the following indeterminate problem which, of course, does not have a unique solution:-

> *One person possesses seven asava horses, another nine haya horses, and another ten camels. Each gives two animals, one to each of the others. They are then equally well off. Find the price of each animal and the total value of the animals possesses by each person.*

The solution, translated into modern notation, proceeds as follows. We seek integer solutions x_1, x_2, x_3 and k (where

x_1 is the price of an asava, x_2 is the price of a haya, and x_3 is the price of a horse) satisfying

$$5x_1 + x_2 + x_3 = x_1 + 7x_2 + x_3 = x_1 + x_2 + 8x_3 = k.$$

Then $4x_1 = 6x_2 = 7x_3 = k - (x_1 + x_2 + x_3)$.

For integer solutions $k - (x_1 + x_2 + x_3)$ must be a multiple of the lcm of 4, 6 and 7. This is the indeterminate nature of the problem and taking different multiples of the lcm will lead to different solutions. The Bakhshali manuscript takes $k - (x_1 + x_2 + x_3) = 168$ (this is 4 6 7) giving $x_1 = 42$, $x_2 = 28$, $x_3 = 24$. Then $k = 262$ is the total value of the animals possesses by each person. This is not the minimum integer solution which would be $k = 131$.

If we use modern methods we would solve the system of three equation for x_1, x_2, x_3 in terms of k to obtain

$x_1 = 21k/131,$

$x_2 = 14k/131,$

$x_3 = 12k/131$

so we obtain integer solutions by taking $k = 131$ which is the smallest solution. This solution is not given in the Bakhshali manuscript but the author of the manuscript would have obtained this had he taken $k - (x_1 + x_2 + x_3) = \text{lcm}(4, 6, 7) = 84$.

Here is another equalisation problem taken from the manuscript which has a unique solution:-

Two page-boys are attendants of a king. For their services one gets $\frac{13}{6}$ dinaras a day and the other $\frac{3}{2}$. The first owes the second 10 dinaras. calculate and tell me when they have equal amounts.

Now I would solve this by saying that the first gets $\frac{13}{6} - \frac{3}{2} = \frac{2}{3}$ dinaras more than the second each day. He needs

20 dinaras more than the second to be able to give back his 10 dinaras debt and have them with equal amounts. So 30 days are required when each has 13 $\frac{30}{6}$ - 10 = 55 dinaras. This is not the method of the Bakhṣhali maṇuscript which uses the "rule of three".

The rule of three is the familiar way of solving problems of the type: if a man earns 50 dinaras in 8 days how much will he earn in 12 days. The Bakhshali manuscript describes the rule where the three numbers are written

8 50 12

The 8 is the "pramana", the 50 is the "phala" and the 12 is the "iccha". The rule, according to the Bakhshali manuscript gives the answer as

phala × iccha/pramana

or in the case of the example $50 \times \frac{12}{8} = 75$ dinaras.

Applying this to the page-boy problem we obtain equal amounts for the page-boys after n days where

$$13 \times \frac{n}{6} = 3 \times \frac{n}{2} + 20$$

so n = 30 and each has $13 \times \frac{30}{6} - 10 = 55$ dinaras.

Another interesting piece of mathematics in the manuscript concerns calculating square roots. The following formula is used

$$OQ = O(A^2 + b) = A + b/2A - (b/2A)2/(2(A + b/2A))$$

This is stated in the manuscript as follows:-

In the case of a non-square number, subtract the nearest square number, divide the remainder by twice this nearest square; half the square of this is divided by the sum of the approximate root and the fraction this is subtracted and will give the corrected root.

Taking Q = 41, then A = 6, b = 5 and we obtain 6.403138528 as the approximation to O41 = 6.403124237. Hence we see that the Bakhshali formula gives the result correct to four decimal places.

The Bakhshali manuscript also uses the formula to compute O105 giving 10.24695122 as the approximation to O105 = 10.24695077. This time the Bakhshali formula gives the result correct to five decimal places.

The following examples also occur in the Bakhshali manuscript where the author applies the formula to obtain approximate square roots:

O487

Bakhshali formula gives 22.068076490965

Correct answer is 22.068076490713

Here 9 decimal places are correct

O889

Bakhshali formula gives 29.816105242176

Correct answer is 29.8161030317511

Here 5 decimal places are correct

[Note. If we took 889 = 30^2 - 11 instead of 29^2 + 48 we would get

Bakhshali formula gives 29.816103037078

Correct answer is 29.8161030317511

Here 8 decimal places are correct]

O339009

Bakhshali formula gives 582.2447938796899

Correct answer is 582.2447938796876

Here 11 decimal places are correct

It is interesting to note that Channabasappa derives from the Bakhshali square root formula an iterative scheme for approximating square roots. He finds that it is 38% faster than Newton's method in giving v41 to ten places of decimals.

A History of Zero

One of the commonest questions which the readers of this archive ask is: Who discovered zero? Why then have we not written an article on zero as one of the first in the archive? The reason is basically because of the difficulty of answering the question in a satisfactory form. If someone had come up with the concept of zero which everyone then saw as a brilliant innovation to enter mathematics from that time on, the question would have a satisfactory answer even if we did not know which genius invented it. The historical record, however, shows quite a different path towards the concept. Zero makes shadowy appearances only to vanish again almost as if mathematicians were searching for it yet did not recognise its fundamental significance even when they saw it.

The first thing to say about zero is that there are two uses of zero which are both extremely important but are somewhat different. One use is as an empty place indicator in our place-value number system. Hence in a number like 2106 the zero is used so that the positions of the 2 and 1 are correct. Clearly 216 means something quite different. The second use of zero is as a number itself in the form we use it as 0. There are also different aspects of zero within these two uses, namely the concept, the notation, and the name. (Our name "zero" derives ultimately from the Arabic sifr which also gives us the word "cipher".)

Neither of the above uses has an easily described history. It just did not happen that someone invented the ideas, and then everyone started to use them. Also it is fair to say that the number zero is far from an intuitive concept. Mathematical problems started as 'real' problems rather than abstract problems. Numbers in early historical times were thought of much more concretely than the abstract concepts which are our numbers today. There are giant mental leaps from 5 horses to 5 "things" and then to the abstract idea of "five". If ancient peoples solved a problem about how many horses a farmer needed then the problem was not going to have 0 or -23 as an answer.

One might think that once a place-value number system came into existence then the 0 as an empty place indicator is a necessary idea, yet the Babylonians had a place-value number system without this feature for over 1000 years. Moreover there is absolutely no evidence that the Babylonians felt that there was any problem with the ambiguity which existed. Remarkably, original texts survive from the era of Babylonian mathematics. The Babylonians wrote on tablets of unbaked clay, using cuneiform writing. The symbols were pressed into soft clay tablets with the slanted edge of a stylus and so had a wedge-shaped appearance (and hence the name cuneiform). Many tablets from around 1700 BC survive and we can read the original texts. Of course their notation for numbers was quite different from ours (and not based on 10 but on 60) but to translate into our notation they would not distinguish between 2106 and 216 (the context would have to show which was intended). It was not until around 400 BC that the Babylonians put two wedge symbols into the place where we would put zero to indicate which was meant, 216 or 21 " 6.

The two wedges were not the only notation used, however, and on a tablet found at Kish, an ancient Mesopotamian city located east of Babylon in what is today south-central Iraq, a different notation is used. This tablet, thought to date from around 700 BC, uses three hooks to denote an empty place in the positional notation. Other tablets dated from around the same time use a single hook for an empty place. There is one common feature to this use of different marks to denote an empty position. This is the fact that it never occurred at the end of the digits but always between two digits. So although we find 21 " 6 we never find 216 ". One has to assume that the older feeling that the context was sufficient to indicate which was intended still applied in these cases.

If this reference to context appears silly then it is worth noting that we still use context to interpret numbers today. If I take a bus to a nearby town and ask what the fare is then I know that the answer "It's three fifty" means three pounds fifty pence. Yet if the same answer is given to the question about the cost of a flight from Edinburgh to New York then I know that three hundred and fifty pounds is what is intended.

We can see from this that the early use of zero to denote an empty place is not really the use of zero as a number at all, merely the use of some type of punctuation mark so that the numbers had the correct interpretation.

Now the ancient Greeks began their contributions to mathematics around the time that zero as an empty place indicator was coming into use in Babylonian mathematics. The Greeks however did not adopt a positional number system. It is worth thinking just how significant this fact is. How could the brilliant mathematical advances of the Greeks not see them adopt a number system with all the advantages that the Babylonian place-value system possessed? The real answer to this question is more subtle than the simple answer that we are about to give, but basically the Greek mathematical achievements were based on geometry. Although Euclid's Elements contains a book on number theory, it is based on geometry. In other words Greek mathematicians did not need to name their numbers since they worked with numbers as lengths of lines. Numbers which required to be named for records were used by merchants, not mathematicians, and hence no clever notation was needed.

Now there were exceptions to what we have just stated. The exceptions were the mathematicians who were involved in recording astronomical data. Here we find the first use of the symbol which we recognise today as the notation for zero, for Greek astronomers began to use the symbol O. There are many theories why this particular notation was used. Some historians favour the explanation that it is omicron, the first letter of the Greek word for nothing namely "ouden". Neugebauer, however, dismisses this explanation since the Greeks already used omicron as a number - it represented 70 (the Greek number system was based on their alphabet). Other explanations offered include the fact that it stands for "obol", a coin of almost no value, and that it arises when counters were used for counting on a sand board. The suggestion here is that when a counter was removed to leave an empty column it left a depression in the sand which looked like O.

Ptolemy in the Almagest written around 130 AD uses the Babylonian sexagesimal system together with the empty place holder O. By this time Ptolemy is using the symbol both between digits and at the end of a number and one might be tempted to believe that at least zero as an empty place holder had firmly arrived. This, however, is far from what happened. Only a few exceptional astronomers used the notation and it would fall out of use several more times before finally establishing itself. The idea of the zero place (certainly not thought of as a number by Ptolemy who still considered it as a sort of punctuation mark) makes its next appearance in Indian mathematics.

The scene now moves to India where it is fair to say the numerals and number system was born which have evolved into the highly sophisticated ones we use today. Of course that is not to say that the Indian system did not owe something to earlier systems and many historians of mathematics believe that the Indian use of zero evolved from its use by Greek astronomers. As well as some historians who seem to want to play down the contribution of the Indians in a most unreasonable way, there are also those who make claims about the Indian invention of zero which seem to go far too far. For example Mukherjee claims:-

> *... the mathematical conception of zero ... was also present in the spiritual form from 17 000 years back in India.*

What is certain is that by around 650 AD the use of zero as a number came into Indian mathematics. The Indians also used a place-value system and zero was used to denote an empty place. In fact there is evidence of an empty place holder in positional numbers from as early as 200 AD in India but some historians dismiss these as later forgeries. Let us examine this latter use first since it continues the development described above.

In around 500 AD Aryabhatta devised a number system which has no zero yet was a positional system. He used the word "kha" for position and it would be used later as the name for zero. There is evidence that a dot had been used in earlier Indian manuscripts to denote an empty place in positional notation. It is interesting that the same documents sometimes also used a dot to denote an unknown where we might use x. Later Indian mathematicians had names for zero in positional numbers yet had no symbol for it. The first record of the Indian use of zero which is dated and agreed by all to be genuine was written in 876.

We have an inscription on a stone tablet which contains a date which translates to 876. The inscription concerns the town of Gwalior, 400 km south of Delhi, where they planted a garden 187 by 270 hastas which would produce enough flowers to allow 50 garlands per day to be given to the local temple. Both of the numbers 270 and 50 are denoted almost as they appear today although the 0 is smaller and slightly raised.

We now come to considering the first appearance of zero as a number. Let us first note that it is not in any sense a natural candidate for a number. From early times numbers are words which refer to collections of objects. Certainly the idea of number became more and more abstract and this abstraction then makes possible the consideration of zero and negative numbers which do not arise as properties of collections of objects. Of course the problem which arises when one tries to consider zero and negatives as numbers is how they interact in regard to the operations of arithmetic, addition, subtraction, multiplication and division. In three important books the Indian mathematicians Brahmagupta, Mahavira and Bhaskara tried to answer these questions.

Brahmagupta attempted to give the rules for arithmetic involving zero and negative numbers in the seventh century. He explained that given a number then if you subtract it from itself you obtain zero. He gave the following rules for addition which involve zero:-

The sum of zero and a negative number is negative, the sum of a positive number and zero is positive, the sum of zero and zero is zero.

Subtraction is a little harder:-

A negative number subtracted from zero is positive, a positive number subtracted from zero is negative, zero subtracted from a negative number is negative, zero subtracted from a positive number is positive, zero subtracted from zero is zero.

Brahmagupta then says that any number when multiplied by zero is zero but struggles when it comes to division:-

A positive or negative number when divided by zero is a fraction with the zero as denominator. Zero divided by a negative or positive number is either zero or is expressed as a fraction with zero as numerator and the finite quantity as denominator. Zero divided by zero is zero.

Really Brahmagupta is saying very little when he suggests that n divided by zero is n/0. Clearly he is struggling here. He is certainly wrong when he then claims that zero divided by zero is zero. However it is a brilliant attempt from the first person that we know who tried to extend arithmetic to negative numbers and zero.

In 830, around 200 years after Brahmagupta wrote his masterpiece, Mahavira wrote Ganita Sara Samgraha which was designed as an updating of Brahmagupta's book. He correctly states that:-

> *... a number multiplied by zero is zero, and a number remains the same when zero is subtracted from it.*

However his attempts to improve on Brahmagupta's statements on dividing by zero seem to lead him into error. He writes:-

> A number remains unchanged when divided by zero.

Since this is clearly incorrect my use of the words "seem to lead him into error" might be seen as confusing. The reason for this phrase is that some commentators on Mahavira have tried to find excuses for his incorrect statement. Bhaskara wrote over 500 years after Brahmagupta. Despite the passage of time he is still struggling to explain division by zero. He writes:-

> *A quantity divided by zero becomes a fraction the denominator of which is zero. This fraction is termed an infinite quantity. In this quantity consisting of that which has zero for its divisor, there is no alteration, though many may be inserted or extracted; as no change takes place in the infinite and immutable God when worlds are created or destroyed, though numerous orders of beings are absorbed or put forth.*

So Bhaskara tried to solve the problem by writing n/0 = 8. At first sight we might be tempted to believe that Bhaskara has it correct, but of course he does not. If this were true then 0 times 8 must be equal to every number n, so all numbers are equal. The Indian mathematicians could not bring themselves to the point of admitting that one could not divide by zero. Bhaskara did correctly state other properties of zero, however, such as $0^2 = 0$, and O0 = 0.

Perhaps we should note at this point that there was another civilisation which developed a place-value number system with a zero. This was the Maya people who lived in central America, occupying the area which today is southern Mexico, Guatemala, and northern Belize.. This was an old civilisation but flourished particularly between 250 and 900. We know that by 665 they used a place-value number system to base 20 with a symbol for zero. However their use of zero goes back further than this and was in use before they introduced the place-valued number system. This is a remarkable achievement but sadly did not influence other peoples.

You can see a separate article about Mayan mathematics.

The brilliant work of the Indian mathematicians was transmitted to the Islamic and Arabic mathematicians further west. It came at an early stage for al-Khwarizmi wrote Al'Khwarizmi on the Hindu Art of Reckoning which describes the Indian place-value system of numerals based on 1, 2, 3, 4, 5, 6, 7, 8, 9, and 0. This work was the first in what is now Iraq to use zero as a place holder in positional base notation. Ibn Ezra, in the 12th century, wrote three treatises on numbers which helped to bring the Indian symbols and ideas of decimal fractions to the attention of some of the learned people in Europe. The Book of the Number describes the decimal system for integers with place values from left to right. In this work ibn Ezra uses zero which he calls galgal (meaning wheel or circle). Slightly later in the 12th century al-Samawal was writing:-

If we subtract a positive number from zero the same negative number remains. ... if we subtract a negative number from zero the same positive number remains.

The Indian ideas spread east to China as well as west to the Islamic countries. In 1247 the Chinese mathematician Ch'in Chiu-Shao wrote Mathematical treatise in nine sections which uses the symbol O for zero. A little later, in 1303, Zhu Shijie wrote Jade mirror of the four elements which again uses the symbol O for zero. Fibonacci was one of the main people to bring these new ideas about the number system to Europe. As the authors write:-

An important link between the Hindu-Arabic number system and the European mathematics is the Italian mathematician Fibonacci.

In Liber Abaci he described the nine Indian symbols together with the sign 0 for Europeans in around 1200 but it was not widely used for a long time after that. It is significant that Fibonacci is not bold enough to treat 0 in the same way as the other numbers 1, 2, 3, 4, 5, 6, 7, 8, 9 since he speaks of the "sign" zero while the other symbols he speaks of as numbers. Although clearly bringing the Indian numerals to Europe was of major importance we can see that in his treatment of zero he did not reach the sophistication of the Indians Brahmagupta, Mahavira and Bhaskara nor of the Arabic and Islamic mathematicians such as al-Samawal.

One might have thought that the progress of the number systems in general, and zero in particular, would have been steady from this time on. However, this was far from the case. Cardan solved cubic and quartic equations without using zero. He would have found his work in the 1500's so much easier if he had a zero but it was not part of his mathematics. By the 1600's zero began to come into widespread use but still only after encountering a lot of resistance.

Of course there are still signs of the problems caused by zero. Recently many people throughout the world celebrated the new millennium on 1 January 2000. Of course they celebrated the passing of only 1999 years since when the calendar was set up no year zero was specified. Although one might forgive the original error, it is a little surprising that most people seemed unable to understand why the third millennium and the 21st century begin on 1 January 2001. Zero is still causing problems!

A History of Pi

A little known verse of the Bible reads:

And he made a molten sea, ten cubits from the one brim to the other: it was round all about, and his height was five cubits: and a line of thirty cubits did compass it about. (I Kings 7, 23)

The same verse can be found in II Chronicles 4, 2. It occurs in a list of specifications for the great temple of Solomon, built around 950 BC and its interest here is that it gives $\pi = 3$. Not a very accurate value of course and not even very accurate in its day, for the Egyptian and Mesopotamian values of $\frac{25}{8} = 3.125$ and O10 = 3.162 have been traced to much earlier dates: though in defence of Solomon's craftsmen it should be noted that the item being described seems to have been a very large brass casting, where a high degree of geometrical precision is neither possible nor necessary. There are some interpretations of this which lead to a much better value.

The fact that the ratio of the circumference to the diameter of a circle is constant has been known for so long that it is quite untraceable. The earliest values of p including the 'Biblical' value of 3, were almost certainly found by measurement. In the Egyptian Rhind Papyrus, which is dated about 1650 BC, there is good evidence for $4 \times (8/9)^2 = 3.16$ as a value for π. The first theoretical calculation seems to have been carried out by Archimedes of Syracuse (287-212 BC). He obtained the approximation

$$\frac{223}{71} < \pi < \frac{22}{7}.$$

Before giving an indication of his proof, notice that very considerable sophistication involved in the use of inequalities here. Archimedes knew, what so many people to this day do not, that p does not equal $\frac{22}{7}$, and made no claim to have discovered the exact value. If we take his best estimate as the average of his two bounds we obtain 3.1418, an error of about 0.0002.

Here is Archimedes' argument.

Consider a circle of radius 1, in which we inscribe a regular polygon of $3 \times 2n^{-1}$ sides, with semiperimeter bn, and superscribe a regular polygon of $3 \times 2n^{-1}$ sides, with semiperimeter an.

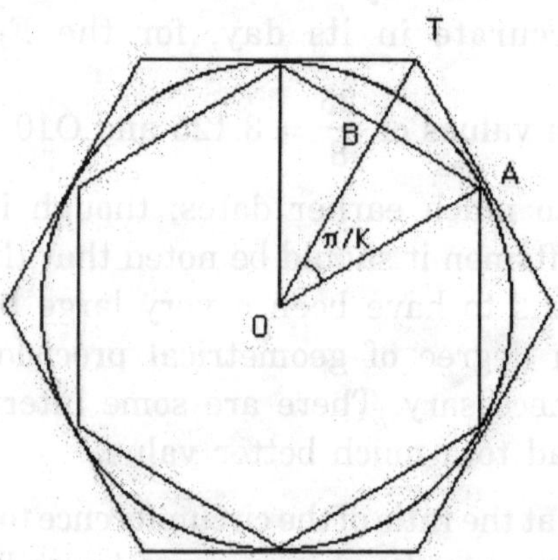

OA = 1
AB = sin(π/K)
AT = tan(π/K)
where $K = 3 \times 2^{n-1}$

The diagram for the case n = 2 is on the right.

The effect of this procedure is to define an increasing sequence b_1 , b_2 , b_3 , ...

and a decreasing sequence a_1 , a_2 , a_3 , ...

such that both sequences have limit p.

Using trigonometrical notation, we see that the two semiperimeters are given by

a_n = K tan(π/K), bn = K sin(π/K),

where K = 3 × 2n^{-1}. Equally, we have

a_{n+1} = 2K tan(π/2K), b_{n+1} = 2K sin(π/2K),

and it is not a difficult exercise in trigonometry to show that.

$$\left(\frac{1}{a_n}+\frac{1}{b_n}\right)=\frac{2}{a_{n+1}} \quad ...(1)$$

$$a_{n+1}b_n=\left(b_{n+1}\right)^2 \quad ...(2)$$

Archimedes, starting from a_1 = 3 tan(π/3) = 3O3 and b_1 = 3 sin(π/3) = 3O3/2, calculated a_2 using (1), then b_2 using (2), then a_3 using (1), then b_3 using (2), and so on until he had calculated a_6 and b_6. His conclusion was that

$b_6 < \pi < a_6$.

It is important to realise that the use of trigonometry here is unhistorical: Archimedes did not have the advantage of an algebraic and trigonometrical notation and had to derive (1) and (2) by purely geometrical means. Moreover he did not even have the advantage of our decimal notation for numbers, so that the calculation of a6 and b6 from (1) and (2) was by no means a trivial task. So it was a pretty stupendous feat both of imagination and of calculation and the wonder is not that he stopped with polygons of 96 sides, but that he went so far.

For ofcourse there is no reason in principle why one should not go on. Various people did, including:

Ptolemy	(c. 150 AD)	3.1416
Zu Chongzhi	(430-501 AD)	$^{355}/_{113}$
al-Khwarizmi	(c. 800)	3.1416
al-Kashi	(c. 1430)	14 places
Viète	(1540-1603)	9 places
Roomen	(1561-1615)	17 places
Van Ceulen	(c. 1600)	35 places

Except for Zu Chongzhi, about whom next to nothing is known and who is very unlikely to have known about Archimedes' work, there was no theoretical progress involved in these improvements, only greater stamina in calculation. Notice how the lead, in this as in all scientific matters, passed from Europe to the East for the millennium 400 to 1400 AD.

Al-Khwarizmi lived in Baghdad, and incidentally gave his name to 'algorithm', while the words al jabr in the title of one of his books gave us the word 'algebra'. Al-Kashi lived still further east, in Samarkand, while Zu Chongzhi, one need hardly add, lived in China.

The European Renaissance brought about in due course a whole new mathematical world. Among the first effects of this reawakening was the emergence of mathematical formulae for π. One of the earliest was that of Wallis (1616-1703)

$2/\pi = (1.3.3.5.5.7. \ldots)/(2.2.4.4.6.6. \ldots)$.

and one of the best-known is

$$\frac{\pi}{4} = 1 - \frac{1}{3} + \frac{1}{5} - \frac{1}{7} + \ldots$$

This formula is sometimes attributed to Leibniz (1646-1716) but is seems to have been first discovered by James Gregory (1638- 1675).

These are both dramatic and astonishing formulae, for the expressions on the right are completely arithmetical in character, while p arises in the first instance from geometry. They show the surprising results that infinite processes can achieve and point the way to the wonderful richness of modern mathematics.

From the point of view of the calculation of p, however, neither is of any use at all. In Gregory's series, for example, to get 4 decimal places correct we require the error to be less than $0.00005 = \frac{1}{20000}$, and so we need about 10000 terms of the series. However, Gregory also showed the more general result

$$\tan^{-1} x = x - \frac{x^3}{3} + \frac{x^5}{5} - \ldots (-1 \le x \le 1) \quad . . . (3)$$

from which the first series results if we put x = 1. So using the fact that

$$\tan^{-1}\left(\frac{1}{\sqrt{3}}\right) = \frac{\pi}{6} \text{ we get}$$

$$\frac{\pi}{6} = \left(\frac{1}{\sqrt{3}}\right)\left(1 - \frac{1}{(3.3)} + \frac{1}{(5.3.3)} - \frac{1}{(7.3.3.3)}\right) + \ldots$$

which converges much more quickly. The 10th term is $\frac{1}{\left(19 \times 3^9 \sqrt{3}\right)}$, which is less than 0.00005, and so we have at least 4 places correct after just 9 terms.

An even better idea is to take the formula

$$\frac{\pi}{4} = \tan^{-1}\left(\frac{1}{2}\right) + \tan^{-1}\left(\frac{1}{3}\right) \quad . . . (4)$$

and then calculate the two series obtained by putting first $\frac{1}{2}$ and the $\frac{1}{3}$ into (3).

Clearly we shall get very rapid convergence indeed if we can find a formula something like

$$\frac{\pi}{4} = \tan^{-1}\left(\frac{1}{a}\right) + \tan^{-1}\left(\frac{1}{b}\right)$$

with a and b large. In 1706 Machin found such a formula:

$$\frac{\pi}{4} = 4\tan^{-1}\left(\frac{1}{5}\right) - \tan^{-1}\left(\frac{1}{239}\right) \quad \ldots (5)$$

Actually this is not at all hard to prove, if you know how to prove (4) then there is no real extra difficulty about (5), except that the arithmetic is worse. Thinking it up in the first place is, of course, quite another matter.

With a formula like this available the only difficulty in computing p is the sheer boredom of continuing the calculation. Needless to say, a few people were silly enough to devote vast amounts of time and effort to this tedious and wholly useless pursuit. One of them, an Englishman named Shanks, used Machin's formula to calculate p to 707 places, publishing the results of many years of labour in 1873. Shanks has achieved immortality for a very curious reason which we shall explain in a moment.

Here is a summary of how the improvement went:

1699: Sharp used Gregory's result to get 71 correct digits

1701: Machin used an improvement to get 100 digits and the following used his methods:

1719: de Lagny found 112 correct digits

1789: Vega got 126 places and in 1794 got 136

1841: Rutherford calculated 152 digits and in 1853 got 440

1873: Shanks calculated 707 places of which 527 were correct.

A more detailed Chronology is available.

Shanks knew that p was irrational since this had been proved in 1761 by Lambert. Shortly after Shanks' calculation it was shown by Lindemann that p is transcendental, that is, p is not the solution of any polynomial equation with integer coefficients. In fact this result of Lindemann showed that 'squaring the circle' is impossible. The transcendentality of π implies that there is no ruler and compass construction to construct a square equal in area to a given circle.

Very soon after Shanks' calculation a curious statistical freak was noticed by De Morgan, who found that in the last of 707 digits there was a suspicious shortage of 7's. He mentions this in his Budget of Paradoxes of 1872 and a curiosity it remained until 1945 when Ferguson discovered that Shanks had made an error in the 528th place, after which all his digits were wrong. In 1949 a computer was used to calculate p to 2000 places. In this and all subsequent computer expansions the number of 7's does not differ significantly from its expectation, and indeed the sequence of digits has so far passed all statistical tests for randomness.

You can see 2000 places of π.

We should say a little of how the notation p arose. Oughtred in 1647 used the symbol d/p for the ratio of the diameter of a circle to its circumference. David Gregory (1697) used p/r for the ratio of the circumference of a circle to its radius. The first to use p with its present meaning was an Welsh mathematician William Jones in 1706 when he states "3.14159 andc. = π". Euler adopted the symbol in 1737 and it quickly became a standard notation.

We conclude with one further statistical curiosity about the calculation of p, namely Buffon's needle experiment. If we have a uniform grid of parallel lines, unit distance apart and if we drop a needle of length k < 1 on the grid, the probability that the needle falls across a line is 2k/p. Various people have tried to calculate p by throwing needles. The most remarkable result was that of Lazzerini (1901), who made 34080 tosses and got

$$\pi = \frac{355}{113} = 3.1415929$$

which, incidentally, is the value found by Zu Chongzhi. This outcome is suspiciously good, and the game is given away by the strange number 34080 of tosses. Kendall and Moran comment that a good value can be obtained by stopping the experiment at an optimal moment. If you set in advance how many throws there are to be then this is a very inaccurate way of computing π. Kendall and Moran comment that you would do better to cut out a large circle of wood and use a tape measure to find its circumference and diameter.

Still on the theme of phoney experiments, Gridgeman, in a paper which pours scorn on Lazzerini and others, created some amusement by using a needle of carefully chosen length k = 0.7857, throwing it twice, and hitting a line once. His estimate for π was thus given by

$$\frac{2 \times 0.7857}{\pi} = \frac{1}{2}$$

from which he got the highly creditable value of $\pi = 3.1428$. He was not being serious!

It is almost unbelievable that a definition of π was used, at least as an excuse, for a racial attack on the eminent mathematician Edmund Landau in 1934. Landau had defined p in this textbook published in GOttingen in that year by the, now fairly usual, method of saying that $\pi/2$ is the value of x between 1 and 2 for which cos x vanishes. This unleashed an academic dispute which was to end in Landau's dismissal from his chair at GOttingen. Bieberbach, an eminent number theorist who disgraced himself by his racist views, explains the reasons for Landau's dismissal:-

Thus the valiant rejection by the GOttingen student body which a great mathematician, Edmund Landau, has experienced is due in the final analysis to the fact that the un-German style of this man in his research and teaching is unbearable to German feelings. A people who have perceived how members of another race are working to impose ideas foreign to its own must refuse teachers of an alien culture.

G H Hardy replied immediately to Bieberbach in a published note about the consequences of this un-German definition of π :

There are many of us, many Englishmen and many Germans, who said things during the War which we scarcely meant and are sorry to remember now. Anxiety for one's own position, dread of falling behind the rising torrent of folly, determination at all cost not to be outdone, may be natural if not particularly heroic excuses. Professor Bieberbach's reputation excludes such explanations of his utterances, and I find myself driven to the more uncharitable conclusion that he really believes them true.

Not only in Germany did p present problems. In the USA the value of p gave rise to heated political debate. In the State of Indiana in 1897 the House of Representatives unanimously passed a Bill introducing a new mathematical truth.

Be it enacted by the General Assembly of the State of Indiana: It has been found that a circular area is to the square on a line equal to the quadrant of the circumference, as the area of an equilateral rectangle is to the square of one side. (Section I, House Bill No. 246, 1897)

The Senate of Indiana showed a little more sense and postponed indefinitely the adoption of the Act!

Open Questions about the Number π

1. Does each of the digits 0, 1, 2, 3, 4, 5, 6, 7, 8, 9 each occur infinitely often in π?
2. Brouwer's question: In the decimal expansion of π, is there a place where a thousand consecutive digits are all zero?
3. Is p simply normal to base 10? That is does every digit appear equally often in its decimal expansion in an asymptotic sense?
4. Is p normal to base 10? That is does every block of digits of a given length appear equally often in its decimal expansion in an asymptotic sense?
5. Is p normal ? That is does every block of digits of a given length appear equally often in the expansion in every base in an asymptotic sense? The concept was introduced by Borel in 1909.
6. Another normal question! We know that π is not rational so there is no point from which the digits will repeat. However, if π is normal then the first million digits 314159265358979... will occur from some point. Even if π is not normal this might hold! Does it? If so from what point? Note: Up to 200 million the longest to appear is 31415926 and this appears twice.

As a postscript, here is a mnemonic for the decimal expansion of π. Each successive digit is the number of letters in the corresponding word.

How I want a drink, alcoholic of course, after the heavy lectures involving quantum mechanics. All of thy geometry, Herr Planck, is fairly hard...:

3.14159265358979323846264...

You can see more about the history of p in the History topic: Squaring the circle and you can see a Chronology of how calculations of p have developed over the years.

Squaring the Circle

There are three classical problems in Greek mathematics which were extremely influential in the development of geometry. These problems were those of squaring the circle, doubling the cube and trisecting an angle. Although these are closely linked, we choose to examine them in separate articles. The present article studies what has become the most famous for these problems, namely the problem of squaring the circle or the quadrature of the circle as it is sometimes called.

One of the fascinations of this problem is that it has been of interest throughout the whole of the history of mathematics. From the oldest mathematical documents known up to the mathematics of today the problem and related problems concerning p have interested both professional mathematicians and amateur mathematicians.

You can see more about the history of π in the History topics: Pi through the ages and A chronology of Pi.

One of the oldest surviving mathematical writings is the Rhind papyrus, named after the Scottish Egyptologist A Henry Rhind who purchased it in Luxor in 1858. It is a scroll about 6 metres long and 1/3 of a metre wide and was written around 1650 BC by the scribe Ahmes who copied a document which is 200 years older. This gives date for the original papyrus of about 1850 BC but some experts believe that the Rhind papyrus is based on a work going back to 3400 BC.

You can see more about the Rhind papyrus in the History topic article Egyptian papyri.

In the Rhind papyrus Ahmes gives a rule to construct a square of area nearly equal to that of a circle. The rule is to cut $\frac{1}{9}$ off the circle's diameter and to construct a square on the remainder. Although this is not really a geometrical construction as such it does show that the problem of constructing a square of area equal to that of a circle goes back to the beginnings of mathematics. This is quite a good approximation, corresponding to a value of 3.1605, rather than 3.14159, for π.

The problem of squaring the circle in the form which we think of it today originated in Greek mathematics and it is not always properly understood. The problem was, given a circle, to construct geometrically a square equal in area to the given circle. The methods one was allowed to use to do this construction were not entirely clear, for really the range of methods used in geometry by the Greeks was enlarged through attempts to solve this and other classical problems. Pappus, writing in his work Mathematical collection at the end of the period of Greek development of geometry, distinguishes three types of methods used by the ancient Greeks:-

> *There are, we say, three types of problem in geometry, the so-called 'plane', 'solid', and 'linear' problems. Those that can be solved with straight line and circle are properly called 'plane' problems, for the lines by which such problems are solved have their origin in a plane. Those problems that are solved by the use of one or more sections of the cone are called 'solid' problems. For it is necessary in the construction to use surfaces of solid figures, that is to say, cones. There remain the third type, the so-called 'linear' problem. For the construction in these cases curves other than those already mentioned are required, curves having a more varied and forced origin and arising from more irregular surfaces and from complex motions.*

Now we usually think of the problem of squaring the circle to be a problem which has to be solved using a ruler and compass. This is really asking whether squaring the circle is a 'plane' problem in the terminology of Pappus given above (we shall often refer to a 'plane solution' rather than use the more cumbersome 'solutions using ruler and compass"). The ancient Greeks, however, did not restrict themselves to attempting to find a plane solution (which we now know to be impossible), but rather developed a great variety of methods using various curves invented specially for the purpose, or devised constructions based on some mechanical method.

The first mathematician who is on record as having attempted to square the circle is Anaxagoras. Plutarch, in his work On Exile which was written in the first century AD, says :-

> *There is no place that can take away the happiness of a man, nor yet his virtue or wisdom. Anaxagoras, indeed, wrote on the squaring of the circle while in prison.*

Now the problem must have become quite popular shortly after this, not just among a small number of mathematicians, but quite widely, since there is a reference to it in a play Birds written by Aristopenes in about 414 BC. Two characters are speaking, Meton is the astronomer (see D Barrett (trs.), Aristophanes, Birds (London, 1978) for a shorter quote):-

Meton: I propose to survey the air for you: it will have to be marked out in acres.

Peisthetaerus: Good lord, who do you think you are?

Meton: Who am I? Why Meton. THE Meton. Famous throughout the Hellenic world - you must have heard of my hydraulic clock at Colonus?

Peisthetaerus (eyeing Meton's instruments): And what are these for?

Meton: Ah! These are my special rods for measuring the air. You see, the air is shaped - how shall I put it? - like a sort of extinguisher: so all I have to do is to attach this flexible rod at the upper extremity, take the compasses, insert the point here, and - you see what I mean?

Peisthetaerus: No.

Meton: Well I now apply the straight rod - so - thus squaring the circle: and there you are. In the centre you have your market place: straight streets leading into it, from here, from here, from here. Very much the same principle, really, as the rays of a star: the star itself is circular, but sends out straight rays in every direction.

Peisthetaerus: Brilliant - the man's a Thales.

Now from this time the expression 'circle-squarers' came into usage and it was applied to someone who attempts the impossible. Indeed the Greeks invented a special word which meant 'to busy oneself with the quadrature'. For references to squaring the circle to enter a popular play and to enter the Greek vocabulary in this way, there must have been much activity between the work of Anaxagoras and the writing of the play. Indeed we know of the work of a number of mathematicians on this problem during this period: Oenopides, Antiphon, Bryson, Hippocrates, and Hippias.

Oenopides is thought by Heath to be the person who required a plane solution to geometry problems. Proclus attributes two theorems to Oenopides , namely to draw a perpendicular to a line from a given point not on the line, and to construct from a given point on a given line, a line at a given angle to the given line. Heath believes that the significance of these elementary results was that Oenopides set out for the first time the explicit 'plane' or 'ruler and compass' type of construction. Heath writes :-

... [Oenopides] may have been the first to lay down the restriction of the means permissible in

constructions t the ruler and compasses which became a canon of Greek geometry for all plane constructions...

There is no record of any attempt by Oenopides to square the circle by plane methods. In fact it is a rather remarkable fact that the Greeks did not produce fallacious 'proofs' that the circle could be squared by plane methods. The few claims for such false proofs rather seem to result from less able mathematicians failing to understand exactly what some of the more brilliant contributions to the problem were intended to show.

Sadly later mathematicians did not follow the good example shown by the ancient Greeks and indeed many claimed incorrectly to have discovered a 'ruler and compass' proof. Amateur mathematicians, greatly attracted to the classical problems, have produced (and still continue to produce) thousands of false proofs.

Antiphon and Bryson both produced arguments relating to squaring the circle which were to prove important in the future development of mathematics. Antiphon inscribed a square in a circle, then a regular polygon with eight sides, then one with sixteen sides and he continued the process continually doubling the number of sides. It appears that Bryson improved the argument of Antiphon by not only inscribing polygons in a circle but also circumscribed polygons. Themistius states :-

... that Bryson declared the circle to be greater than all inscribed, and less than all circumscribed polygons.

Hippocrates was the first to actually use a plane construction to find a square with area equal to a figure with circular sides. He squared certain lunes, and also the sum of a lune and a circle.

Now although he squared certain lunes, he had not shown that every lune can be squared. In particular the lune that he squared in his plane construction of a square of area equal to that of a certain lune and a circle was one he could not square by plane methods. Of course this lune cannot be squared by plane methods otherwise Hippocrates would have squared the circle. Although some, such as Aristotle, seemed to fail to understand the logic of Hippocrates argument, there seems little doubt that Hippocrates was perfectly aware that his methods had failed to square the circle. Examples of Hippocrates' methods of squaring lunes are given in his biography in this archive.

Hippias and Dinostratus are associated with the method of squaring the circle using a quadratrix. The curve it thought to be the invention of Hippias while its application to squaring the circle appears to be due to Dinostratus. The construction of this curve with a diagram is given in the biography of Hippias in this archive. Now this curve certainly solves the problem of squaring the circle but, as given by Hippias, the curve is constructed by mechanical means given by a uniform motion of a line in a time equal to the rotating radius of a circle. The construction was rightly criticised as requiring a knowledge of the ratio of a line and an arc of a circle, so one assumed as known the property required to square the circle in the first place. It is clear that Dinostratus never claimed that the quadratrix gave a plane method to square the circle. Nicomedes many years later also used the quadratrix to square the circle.

Aristotle did not seem to appreciate the contributions of those who had attempted to square the circle. He wrote in his work Physics :-

> *The exponent of any science is not called upon to solve every kind of difficulty that may be raised, but only such as arise through false deductions from the principles of the science: with others than these he need not concern himself. For example, it is for the geometer to expose the quadrature by means of segments, but it is not the business of the geometer to refute the arguments of Antiphon.*

In this quote "quadrature by means of segments" refers to Hippocrates quadrature of lunes which Aristotle mistakenly thinks was intended as a proof that the circle can be squared by plane methods. Antiphon's methods come in for even more criticism from Aristotle, but all credit to Antiphon whose methods contained important ideas which would lead eventually to integration. Aristotle also wrote in similar terms in Sophistical refutations again probably having had handed down to him an incorrect interpretation of what Antiphon and Bryson had attempted to show:-

The method by which Bryson tried to square the circle, were it ever so much squared thereby, is yet made sophistical by the fact that it has no relation to the matter in hand. ... The squaring of the circle by means of lunes is not eristic, but the quadrature of Bryson is eristic. The reasoning used by the former cannot be applied to any subject other than geometry alone, whereas Bryson's argument is directed to the mass of people who do not know what is possible and what is impossible in each department, for it will fit any. And the same is true of Antiphon's quadrature.

Next we should consider the contributions of Archimedes to the problem of squaring the circle. Now Archimedes is famed for his introduction of the spiral curve, but why did he introduced this curve? The authors suggest three reasons:-

Is it for purely geometric reasons because he studied this curve as a means of calculating p, and squaring the circle? Is it because of his astronomical interests, trying to calculate geometrically the spiral movements of the planets? Or is it finally through the interest of a mechanical mind in a curve which results from the combination of two regular uniform movements, one in a straight line the other in a circle? these three reasons are evident at one and the same time...

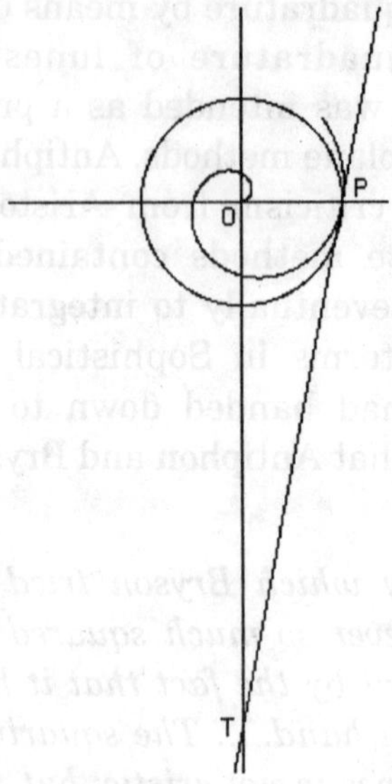

Archimedes gives the following definition of the spiral in his work On spirals:-

> *If a straight line drawn in a plane revolves uniformly any number of times about a fixed extremity until it returns to its original position, and if, at the same time as the line revolves, a point moves uniformly along the straight line beginning at the fixed extremity, the point will describe a spiral in the plane.*

To square the circle Archimedes gives the following construction. Let P be the point on the spiral when it has completed one turn. Let the tangent at P cut the line perpendicular to OP at T. Then Archimedes proves in Proposition 19 of On spirals that OT is the length of the circumference of the circle with radius OP.

Now it may not be clear that this is solved the problem of squaring the circle but Archimedes had already proved as the first proposition of Measurement of the circle that the area of a circle is equal to a right-angled triangle having the two shorter sides equal to the radius of the circle and the circumference of the circle. So the area of the circle with radius OP is equal to the area of the triangle OPT.

Both Apollonius and Carpus used curves to square the circle but it is not clear exactly what these curves were. The one used by Apollonius is called by Iamblichus 'sister of the cochloid' and this has led to various guesses as to what the curve might have been. Again the curve used by Carpus of Antioch is called the 'curve of double motion' which Paul Tannery argued was the cycloid.

Now we leave the ancient Greek period and look at later developments but the first comment we should make is that the Greeks were certainly not the only ones to be interested in squaring the circle at this time. Mathematicians in India were interested in the problem while in China mathematicians such as Liu Hsiao of the Han Dynasty showed himself to be one of the prominent of those attempting to square the circle in around 25 AD.

Some time later the Arab mathematicians were, like the Greeks, fascinated by the problem. The work of al-Haytham on squaring the circle is discussed. Now al-Haytham aimed to convince people that squaring the circle was possible by a plane construction but since his promised treatise on the topic never appeared he must at least have realised that he could not solve the problem.

Not long after the work of al-Haytham, Franco of Liège in 1050 wrote a treatise De quadratura circuli on squaring the circle. The text is reproduced and in it Franco examines three earlier methods based on the assumption that p is 25/8 , 49/16 or 4. Franco states (reasonably enough) that these are false, then gives his own construction which is based on the assumption that p is 22/7 .

Although this treatise is of great historical interest, it does show how European mathematics at the time was far behind the ancient Greeks in depth of understanding.

Moving forward to about 1450, Cusa attempted to prove that the circle could be squared by a plane construction. Although his method of averaging certain inscribed and circumscribed polygons is quite fallacious, it is one of the first serious attempts in 'modern' Europe to solve the problem. Again it is worth commenting that the ancient Greeks basically knew that the circle could not be squared by plane methods, although they stood no chance of proving it. Regiomontanus, who brought a new impetus to European mathematics, was quick to point out the error in Cusa's arguments.

The mechanical methods of the Greeks certainly appealed to Leonardo who thought about mathematics in a very mechanical way. He devised several new mechanical methods to square the circle. Many mathematicians in the sixteenth century studied the problem, including Oronce Fine and Giambattista della Porta. The 'proof' by Fine was shown to be incorrect by Pedro Nunes soon after he produced it. The beginnings of the differential and integral calculus led to an increased interest in squaring the circle, but the new era of mathematics still produced fallacious 'proofs' of plane methods to square the circle. One such false proof, given by Saint-Vincent in a book published in 1647, was based on an early type of integration. The problem was still providing much impetus for mathematical development.

James Gregory developed a deep understanding of infinite sequences and convergence. He applied these ideas to the sequences of areas of the inscribed and circumscribed polygons of a circle and tried to use the method to prove that there was no plane construction for squaring the circle. His proof essentially attempted to prove that p was transcendental, that is not the root of a rational polynomial equation. Although he was correct in what he tried to prove, his proof was certainly not correct. However, others such as Huygens, believed that p was algebraic, that is that it is the root of a rational polynomial equation.

There was still an interest in obtaining methods to square the circle which were not plane methods. For example Johann Bernoulli gave a method of squaring the circle through the formation of evolvents and this method is described .

The historian of mathematics, Montucla, made squaring the circle the topic of his first historical work published in 1754. This was written at a time long before the problem was finally resolved, so is necessarily very outdated. The work is, however, a classic and still well worth reading.

A major step forward in proving that the circle could not be squared using ruler and compasses occurred in 1761 when Lambert proved that p was irrational. This was not enough to prove the impossibility of squaring the circle with ruler and compass since certain algebraic numbers can be constructed with ruler and compass. It only led to a greater flood of amateur solutions to the problem of squaring the circle and in 1775 the Paris Académie des Sciences passed a resolution which meant that no further attempted solutions submitted to them would be examined. A few years later the Royal Society in London also banned consideration of any further 'proofs' of squaring the circle as large numbers of amateur mathematicians tried to achieve fame by presenting the Society with a solution. This decision of the Royal Society was described by De Morgan about 100 years later as the official blow to circle-squarers.

The popularity of the problem continued and there are many amusing stories told by De Morgan on this topic in his book Budget of Paradoxes which was edited and published by his wife in 1872, the year after his death. De Morgan suggests that St. Vitus be made the patron saint of circle-squarers. This is a reference to St. Vitus' dance, a wild leaping dance in which people screamed and shouted and which led to a kind of mass hysteria. De Morgan also suggested the term 'morbus cyclometricus' as being the 'circle squaring disease'. Clearly De Morgan found himself having to try to persuade these circle-squarers that their methods were incorrect, yet many stubbornly held to their views despite the best efforts of the professional mathematicians. For example a certain Mr James Smith wrote several books attempting to prove that $\pi = \frac{25}{8}$.. Of course Mr Smith was able to deduce from this that the circle could be squared but neither Hamilton, De Morgan nor others could convince him of his errors.

The final solution to the problem of whether the circle could be squared using ruler and compass methods came in 1880 when Lindemann proved that p was transcendental, that is it is not the root of any polynomial equation with rational coefficients. The transcendentality of p finally proves that there is no ruler and compass construction to square the circle.

One might imagine that this would be the end of interest in the problem of squaring the circle, but this was certainly not the case. It neither prevented the stream of publications claiming that p had some simple rational value, nor did it prevent the stream of publications of quite correct constructions to approximately square the circle with ruler and compass. As an example of the former type of claim, the New York Tribune published a letter in 1892 in which the author claimed to have rediscovered a secret going back to Nicomedes which proved that $\pi = 3.2$. Perhaps more surprising is the fact that there were many who were totally convinced by this letter and firmly believed thereafter that $\pi = 3.2$.

Among the correct approximate constructions to square the circle was one by Hobson in 1913. This was a fairly accurate construction which was based on constructing the approximate value of 3.14164079... for p instead of 3.14159265.... . More remarkable, however, was the ruler and compass constructions published by Ramanujan. In the

Journal of the Indian Mathematical Society in 1913 in a paper named Squaring the circle Ramanujan gave a construction which was equivalent to giving the approximate value of 355/113 for p, which differs from correct value only in the seventh decimal place. He ended the paper with the following:-

Note. If the area of the circle be 140,000 square miles, then [the side of the square] is greater than the true length by about an inch.

Among other constructions given by Ramanujan in 1914 (Approximate geometrical constructions for p, Quarterly Journal of Mathematics XLV (1914), 350-374) was a ruler and compass construction which was equivalent to taking the strange yet remarkable approximate value for π to be $(9^2 + 19^2/22)^{1/4}$. Now this is 3.1415926525826461253.... which differs from π only in the ninth decimal place (π = 3.1415926535897932385...). For a circle of diameter 8000 miles, the error in the length of the side of the square constructed was only a fraction of an inch.

2

FAQ in Ancient Indian Mathematics

Strengths of the Ancient Indian Mathematicians

The Indian Mathematicians of the ancient era were primarily number theorists. Their interest in this field arose from a need to do astronomical calculations, which in turn were needed to advise the farmers on the proper timing to plant crops and for performing rituals . In particular they excelled in Diophantine Equations (probably erroneously credited to Diophantus , because the Vedics were adapt at it long before Diophantus), algebraic equations in which only solutions in integers are permitted. Examples of Diophantine equations that the ancients dealt with are

- $ax + by = 1$: this is a linear Diophantine.
- $x^n + y^n = z^n$: For $n = 2$ there are infinitely many solutions (x,y,z), the Pythagorean triples. For larger values of n, Fermat's last theorem states that no positive integer solutions x, y, z satisfying the above equation exist.
- $x^2 - n y^2 = 1$: (Pell's equation) which is named, mistakenly, after the English mathematician John Pell. It was studied by Brahmagupta and much later by Fermat.

Another field or approach which the ancients favoured was algebraic geometry, where geometry was studied predominantly using algebraic equations. Again the need was for designing ritual altars and their proper orientation with respect to the heavenly bodies and for following the rules of Vastu shastra (architecture) .

The use of concise mathematical symbolism (<, >, = ,*, (), inf., sup., the integral sign, the derivative sign, the sigma sign etc. is a relatively recent revolution in mathematics, When calculus was invented by Newton and Leibniz independently, they used different notations. Newton used dots on top of the quantity and called them fluxions, while Leibniz used the 'd' notation. Unfortunately when all the major advancements in science and mathematics were happening in Europe in the 17th and 18th century, India was stagnant, caught in interminable wars and conquests.

Ironically it was the Indian place value notation that triggered the advance in Europe. Prior to this it was common to express all problems in wordy sentences and verse.

As to the value of PI (cannot be expressed as a fraction) it is impossible to calculate it without the use of some variant of the 'method of exhaustion'. In this particular instance it was a matter of using increasingly larger number of isosceles triangles(forming an n sided polygon) within the circle. This is the germ of the method of series expansions.

The ancient Indians were well aware of this technique and used it for several trigonometric calculations. The felicity with which Indians did series expansions extended to Srinivasa Ramanujam in recent times. He was the incomparable master and there may be none like him on the face of this earth again.

Again there is no claim that the Indians did everything. For example there is no evidence that the Indians were familiar with the representation of complex numbers and complex variables or advanced topics such as the calculus of variations. The ancient vedic Indics were interested in very practical aspects of mathematics, namely the positions of the stars, developing a Panchanga (calendar), ordinary mathematics for everyday use, measurements, such as that of land and weights etc.. There is no evidence that they were familiar with the science of mechanics for instance, which developed in Europe in the 18th century after Newton.

Of course, the most important single concept that they developed was the decimal place value system. While other civilizations like the Maya and the Oaxaca of ancient Mexico developed place value systems independently, there is no evidence that they mastered the technique of using numbers with such facility as the Vedics did. The Vedics also developed facility with areas such as Trigonometry and there is mention that they manipulated arrays of numbers in the same manner as we use matrices today.

Even with all these skills, it must be recognized that this forms only a small part of what constitutes the corpus of modern mathematics today. It must also be recognized that the Greeks and the Babylonians also had developed mathematical prowess, if one accepts that the axiomatic approach to geometry as primarily a Greek development.

Despite all these caveats, it must be admitted that for the era in which they lived the Vedic contributions to the sum of human knowledge especially in mathematics was considerable and should therefore be a matter of great inspiration for those of us who consider themselves part of the Indic civilization.

Further there is much study yet to be done and many more manuscripts which remain uninvestigated so the quest for what was the state of knowledge of the ancients in fields such as mathematics has just begun.

Use the Term Vedic Mathematics

Strictly speaking the later contributors like Varahamihira did not live in the Vedic era, but the methods used show a continuity in development till the Modern age and hence the use of the term Vedic , signifying techniques which came into use in the distant past seems justified.

Vedic Mathematics Progressed much Farther in the Intervening Millennia?

I will quote the answer I gave in a discussion forum (Bharat-Rakshak)a few years ago.

First, do we understand the corpus of VM and Vedic Science (i certainly don't know enough about VM to make an authoritative statement). What part of what we know to be contributions, were purely Vedic and what part were contributions by mathematicians like Aryabhatta and Bhaskaracharya who came much later ?

Second what part of it is already accepted in modern mathematics (and I am not talking about arithmetical tricks to do multiplication and division). I remember in my Hall and Stevens text on geometry which i used in India, the proof of Pythagoras theorem took up a whole page. I would have loved to have the 4 alternate proofs offered by the VMB. I am interested for example in the development of Astronomy.

Ptolemy used a concept called ecliptics (if i recall the use of epicycles - pl. don't harangue me on this, as this is of the top of my head) to get around the fact that the earth went around the sun and not the other way around. Did he borrow that from Vedic or was that a different stream of thought (Ptolemy of course predates al-Khwarismi and his 'Sindhind zij'). Given that Ptolemy post dates Vedic why did al-Khwarismi choose to rely more on Vedic (a more ancient technology) when he presumably had access to both ?

Third the point made by James is valid. The point is not to judge Vedic Science by the standards of 21st century and then trash it saying it is outmoded. Of course it is outmoded, in certain respects, you should expect that after 6 millennia. If it was not, that is tantamount to admitting that we have made no progress in the intervening millennia. The point is, does it give an alternate model at looking at nature (e.g. Ayurveda) that is equally valid, useful and perhaps more elegant.

Fourth, these efforts at understanding our past, should not be restricted to Vedic systems but to other systems in India's past such as Yunani (etymology Ionian = Greek) which is attributed to Islamic savants. I am interested for example in tracing through the development of algebra (coined by al Khwarismi as al-gibr wal Maquaballah) from its ancient Vedic origins because of the efforts of Islamic savants in the middle ages, at a time when Europe was in the dark ages and was struggling with Roman numerals.

My own view is that Vedic science and math is a forgotten science and for the most part does not contradict what has been discovered subsequently. First let us understand what it says before getting an anxiety neurosis that it is going to replace Western science in Indian schools.

Results to be Found in the Ancient Texts

The earliest works on Mathematics by the Vedic savants are recorded in the Sulvasutras, the sacred books of altar construction in the Vedas, in particular the Apastambha Sulvasutra, the Baudhayana Sulvasutra and the Katyayana Sulvasutra (named after the mathematicians who developed the mathematics and wrote the sutras. The Sulvasutras are appended to a particular Veda (see FAQ on Hinduism for the typical contents of a Veda) can be translated as rope rules or "manuals of measurement", the modern term for which would be metrology. But in reality the scope of the investigations in the Sulvasutras is far broader and comprises among other fields, number theory, trigonometric, algebra, algebraic geometry, series expansions, the concept of a rational number, etc., etc.. The dating of these Sulvasutras, while occurring after the main corpus of the Veda was compiled, is of great antiquity, greater than that of Babylonian mathematics. It is only now that we are beginning to understand the extent of the antiquity of these ancient mathematicians. See for instance The origin of Mathematics by Lakshmikantham and Leela, in the list of references later in this page.

A word needs to be said about the use of Sutras (Aphorisms) as a means of communication and recording of results. The dictionary says that a sutra is Any of various aphoristic doctrinal summaries produced for memorization generally during the millennia before the common era and later incorporated into Hindu literature. We must recall that writing materials during that period of history, were not plentiful and had to be laboriously produced probably by the author of the sutras himself . Whatever needed to be communicated had to be in as brief a form as possible. Hence the need for economy in language and the use of sutras. In fact there is evidence from which one can infer that the Vedics were the first to use symbols and mathematical equations and hence their prowess with Algebra .

Brevity has its downside however and the charge has been made by Europeans that the Vedics rarely provided proof. The reason was that the proof was generally terse and incomprehensible to the majority of the people. The notion that those who mastered these topics used brevity as a means to restricting this knowledge, overlooks the fact few in the populace would have the capability to pursue this rigorously without the intellectual discipline that comes from years of study. Then as now Mathematics had the reputation of being a difficult subject to master.

Main Contributors to Vedic Mathematics

The main contributors during the Vedic period were, as we mentioned earlier Baudhayana, Apastambha, and Katyayana. The following list of Mathematicians who were born in the geographical region corresponding to present day India was compiled by researchers at the University of St. Andrews, Scotland, who have done an outstanding job in compiling a lot of information (this list includes some modern day mathematicians born in the Indian sub continent also).

Contributions of the Vedic Savants not as Widely known as those of Greek, Arab Mathematicians

In the early years of colonial rule by the British (the attitude persists among Western Indologists even today, although less so among mathematicians in the West) there was great reluctance to believe the sacred texts, after they were first recognized by Sir William Jones in the 1770's. A typical reaction was that of W. W Rouse Ball in History of Mathematics . I posted the following in the Bharat Rakshak forum in 2000.

"Typical of the racism exhibited by the Brits and other Europeans is W.W. Rouse Ball in 'A short account of the History of mathematics' Dover Publications, 1960, (originally appeared in 1908), page 146 'The Arabs had considerable commerce with India, and a knowledge of one or both of the two great Hindu works on algebra had been obtained in the Caliphate of Al-Mansur (754-775 AD) though it was not until fifty or seventy years later that they attracted much attention. The algebra and arithmetic of the Arabs were largely founded on these treatises, and I therefore devote this section to the consideration of Hindu mathematics. The Hindus like the Chinese have pretended that they are the most ancient people on the face of the earth, and that to them all sciences owe their creation. But it is probable that these pretensions have no foundation; and in fact no science or useful art (except a rather fantastic architecture and sculpture) can be definitely traced back to the inhabitants of the Indian peninsula prior to the Aryan invasion. This seems to have taken place at some time in the fifth century or in the sixth century when a tribe of Aryans entered India by the north west part of their country. Their descendants, wherever they have kept their blood pure, may still be recognized by their superiority over the races they originally conquered; but as is the case with the modern Europeans, they found the climate trying and gradually degenerated. Note the blatant racism in the second paragraph and the venom that this author exhibits. [This message has been edited by Kaushal (edited 15-06-2000).]

There are a lot of facile and unsubstantiated assumptions that are made here, later to be proved false, but we will deal with them on the chapter where we discuss AIT. Thus there was great reluctance to admit that the dark skinned natives of the Indian subcontinent could be capable of intellectual effort. Even after the advent of the legendary Srinivasa Ramanujam, the great number theorist in the early years of the 20th century, from what is now Chennai, Tamil Nadu , such attitudes among British and European scholars were hard to dispel. We will have a lot more to say about Srinivasa Ramanujam later in these pages.

With the coming of the internet, and the great proficiency of the Indics in matters related to Information Technology, this state of affairs has begun to change. Both the Indics and Western savants have begun to realize the profound importance of these early developments in mathematics to the advancement of human civilization. See for instance a recent column on Place Value systems.

Can we see Examples of what you Mean by Vedic Mathematics

First we will make a distinction between the corpus of Vedic Mathematics (VM) from the 16 extant sutras that are still available to us courtesy of the book on Vedic mathematics (VMB) by Jagadguru Bhaarathi Krishnatirtha, the late Shankaracharaya of Puri, here we will restrict ourselves to VMB, which is largely a text at High school level, although I will wager very few school children even in India are familiar with these techniques. So what does VMB comprise of?

The Natural Calculator

We have probably been taught methods, which require pencil and paper or calculator, to work out because of the difficulty and a lack of encouragement for mental calculation, which have prevented us form becoming efficient mental calculators. Though mathematics has applications at many levels, it is primarily a mental subject. A system of mental mathematics could provide students with a lasting link with the realm of mathematics and also engender a deeper understanding of the structure and processes of mathematics. The beautiful system of Vedic Mathematics is far more unified and direct than conventional mathematics. It is so easy that it is really a system of mental mathematics, and this combined with the coherence and flexibility of the system, encourages the development and use of intuition and creativity. These features make Vedic Mathematics very attractive. Vedic Mathematics deals with all the fields in Mathematics. The simplicity of Vedic Mathematics means that calculations can be carried out mentally and this is very much encouraged in the Vedic system. Being naturally creative students is to devise their own methods of solution. The Vedic system seeks to cultivate intuition having a conscious proof or explanation of a method beforehand is not considered essential in the Vedic methodology. Students are shown general methods and also special methods, which apply in special cases. This means they do not rigidly have to follow a certain procedure but have a choice and are encouraged to invent their own methods. Every problem is unique and invites its own style of solution. The Vedic system does not impose any unnecessary constraints. For e.g. we can work from left to right or vice-versa with both positive and negative numbers; more than two numbers can be operated simultaneously.

The Vedic system uses a collection of sixteen Sutras (or formulae), given in word form, and some sub-Sutras. They express fundamental principles, which run like threads ('Sutra' means 'thread') through the hole of mathematics, unifying diverse topics. They are extremely useful in education for this reason. We use our mind in certain specific ways - we might extend an idea or reverse it or compare or combine it with another.

Each of these types of mental activity is described by one of the Vedic Sutras. They describe the ways in which the mind can work and so they tell the student, how to go about solving a problem. It is not necessary for the teacher to be familiar with these. It is best not to stress the Sutras. They become familiar as you go through and seem quite natural. The best part of Vedic Mathematics is revealed in it's applications. Let us consider one of the applications using "Nikhilam Navatascaramam Dasatah". 'Nikhilam Navatascaramam Dasatah', literally translated in English, means 'All from nine and last from ten'. This means that we are supposed to take into consideration, the base to be that power of ten, which is closest to the number.

Everyone knows the universal fact that for multiplication and division, by the current method, a pupil of mathematics is forced to learn the multiplication tables of 20*20 and so on. But, according to Vedic System, the multiplication required to be learnt is maximum 5*5 and all that he has to know more than this is simple addition and subtraction (of single-digit numbers). So these formulae enable him to get at the required products, very easily and speedily, practically and immediately.

The important applications of this Sutra are: Multiplication and Division. Since examples would make things clearer, let us consider multiplication first.

Say we have to multiply 9 by 7.

Here, we consider 10 (that power of ten, which is closest to the number to be multiplied - 9 - in this case). The modus-operandi chart is as shown, where the method is clearly understood.

The base will be that power of 10 which is closest to 9. Therefore the base is 10.	10
Put the numbers 9 and 7 above and below on the left-hand side.	9 7
Connecting with a minus sign on the right-hand side is the 10's complement of the corresponding left-hand side number.	9 - (10 - 9) = 9 - 1 7 - (10 - 7) = 7 - 3
The left digit is obtained by adding the above 2 left-hand side numbers and subtracting 10 from this sum.	A = (9 +7)-10 = 6

Vertical multiplication of the two deficit right-side numbers gives the right-side result.

3

History of Vedic Mathematics

Calculating Prodigies

People who can perform astonishingly long or complex calculations in their head are often called calculating prodigies or lightning calculators. There are many examples from the past, from Jedediah Buxton who worked as a labourer and remained illiterate all his life to Karl Friedrich Gauss a mathematician who is ranked with Newton and Archimedes. Since Vedic Mathematics is the most efficient mathematical system it would be interesting to ask whether these calculating prodigies may have used it.

It is peculiar to arithmetic that once the basic facts are understood any intelligent person can construct the whole science: number science develops in a totally natural way. This is what seems to have happened in most cases of calculating prodigies. From a basic understanding the young calculator realises that the power of their mind to calculate seems to be almost unlimited. Once they understand a problem they can solve it. Ampere was an amazing mental calculator at the age of three and Gauss at the same age corrected his father's mistake in a payroll reckoning. These people are self-taught, they construct their own methods of calculation.

In some cases however the methods they devise are extraordinarily complicated, especially in view of the ultra-efficient methods of Vedic Mathematics. Jedediah Buxton, for example, could multiply huge numbers together in his head but seems to use the most awkward methods, multiplying parts of the number and adding them. This may not have been difficult for him as he had an extraordinary memory. He could remember many long numbers and recite them from the left, from the right or from any other point. He would also remember all the steps in a calculation for a long time afterwards.

All mental calculators have a good memory and they make use of previous results they have found out. Some also work two or more figures at a time (a feature of the Vedic system, as is calculating from right to left or left to right). Calculating prodigies do not seem to necessarily always seek the most efficient method - only one that works and that they are happy with. But calculations can often become unconscious and the calculator cannot then describe their method. And furthermore what may have been an effective method can get altered and simplified with practice so that many calculating prodigies may be using Vedic methods without knowing it.

When Bharati Krsna Tirthaji gave talks in India after discovering the Vedic system some people took advantage by learning the Vedic methods and giving demonstrations, sometimes pretending to be divinely inspired or to have the help of some spirit. This demonstrates the effectiveness of Vedic Mathematics as an efficient mental system. The Introduction to book "The Natural Calculator" describes the many advantages of mental mathematics and the Vedic system and supports the view that everyone is potentially a lightning calculator.

What is Vedic Mathematics?

Vedic Mathematics is the name given to the ancient system of Mathematics which was rediscovered from the

Vedas between 1911 and 1918 by Sri Bharati Krishna Tirthaji (1884-1960). According to his research all of mathematics is based on sixteen Sutras or word-formulae. For example, 'Vertically and Crosswise` is one of these Sutras. These formulae describe the way the mind naturally works and are therefore a great help in directing the student to the appropriate method of solution.

Perhaps the most striking feature of the Vedic system is its coherence. Instead of a hotch-potch of unrelated techniques the whole system is beautifully interrelated and unified: the general multiplication method, for example, is easily reversed to allow one-line divisions and the simple squaring method can be reversed to give one-line square roots. And these are all easily understood. This unifying quality is very satisfying, it makes mathematics easy and enjoyable and encourages innovation.

In the Vedic system 'difficult' problems or huge sums can often be solved immediately by the Vedic method. These striking and beautiful methods are just a part of a complete system of mathematics which is far more systematic than the modern 'system'. Vedic Mathematics manifests the coherent and unified structure of mathematics and the methods are complementary, direct and easy.

The simplicity of Vedic Mathematics means that calculations can be carried out mentally (though the methods can also be written down). There are many advantages in using a flexible, mental system. Pupils can invent their own methods, they are not limited to the one 'correct' method. This leads to more creative, interested and intelligent pupils.

Interest in the Vedic system is growing in education where mathematics teachers are looking for something better and finding the Vedic system is the answer. Research is being carried out in many areas including the effects of learning Vedic Maths on children; developing new, powerful but easy applications of the Vedic Sutras in geometry, calculus, computing etc. But the real beauty and effectiveness of Vedic Mathematics cannot be fully appreciated without actually practising the system. One can then see that it is perhaps the most refined and efficient mathematical system possible.

Origin

The origin of "Magical Methods of Fast Calculations" is from Vedic sutras. There are sixteen Sutras and thirteen Sub-Sutras scattered in the writings of "Atharva Vedas" the fourth Veda. "Atharva Veda" deals with the branches like Engineering, Mathematics, sculpture, Medicine, and all other sciences with which we are today aware of.

The first book on these wonderful methods was written by Late Swami Bharati Krishna Tirtha ji Maharaj, Shankaracharya of Goverdhan Peeth. He was the person who collected lost formulae from the writings of "Atharwa Vedas" and wrote them in the form of Sixteen Sutras and thirteen sub-sutras.

These Sutras and Sub-Sutras can be applied to Arithmetical computations, theory of numbers, compound multiplication, algebraic operations, factorisations, simple quadratic and higher order equations, simultaneous quadratic equations, partial fractions, calculus, squaring, cubing, square root, cube root, coordinate geometry etc.

Magical Methods of Fast Calculations has been written in very simple and lucid language, which everybody can understand.

"India was the motherland of our race and Sanskrit the mother of Europe's languages. India was the mother of our philosophy, of much of our mathematics, of the ideals embodied in Christianity...of self-government and democracy. In many ways, Mother India is the mother of us all." — Will Durant, American Historian 1885-1981

What is an ancient system of mathematics that is being taught in some of the most prestigious institutions in England and Europe and not in India? It is Vedic mathematics—a long forgotten technique for mathematical calculations! It is amazing how with the help of 16 sutras and 16 upa-sutras you will be able to solve/calculate complex mathematical problems mentally! The basic roots of Vedic mathematics lie in Vedas jus as basic roots of Hinduism. Vedic Maths form part of Jyothisha which is one of the six Vedangas. To many Indians Vedic and Sanskrit slokas/manthras are relevant only for religious purposes/occasions. But Vedas (written around 1500-900 BCE) in fact are a treasure house of knowledge and human experience—both secular and spiritual. Here you will get an idea about the power of Vedic Mathematics.

Vedic Number Representation

Vedic knowledge is in the form of slokas or poems in Sanskrit verse. A number was encoded using consonant groups of the Sanskrit alphabet, and vowels were provided as additional latitude to the author in poetic composition. The coding key is given as Kaadi nav, taadi nav, paadi panchak, yaadashtak ta ksha shunyam Translated as below·

VARNMALA

ka kha ga gha gna

cha chha ja jha inja

ta tha da dha na

pa pha ba bha ma

ya ra la va sha

sha sa ha chjha tra gna

- letter "ka" and the following eight letters
- letter "ta" and the following eight letters
- letter "pa" and the following four letters

- letter "ya" and the following seven letters, and
- letter "ksha". for zero.

In other words,

- ka, ta, pa, ya = 1
- kha, tha, pha, ra = 2
- ga, da, ba, la = 3
- gha, dha, bha, va = 4
- gna, na, ma, scha = 5
- cha, ta, sha = 6
- chha, tha, sa = 7
- ja, da, ha = 8
- jha, dha = 9
- ksha = 0

Thus pa pa is 11, ma ra is 52. Words kapa, tapa , papa, and yapa all mean the same that is 11.

Tirthaji and the Rediscovery of Vedic Mathematics

The ancient system of Vedic Mathematics was rediscovered from the Sanskrit texts known as the Vedas, between 1911 and 1918 by Sri Bharati Krsna Tirthaji (1884-1960). At the beginning of the twentieth century, when there was a great interest in the Sanskrit texts in Europe, Bharati Krsna tells us some scholars ridiculed certain texts which were headed 'Ganita Sutras'— which means mathematics. They could find no mathematics in the translation and dismissed the texts as rubbish. Bharati Krsna, who was himself a scholar of Sanskrit, Mathematics, History and Philosophy, studied these texts and after lengthy and careful investigation was able to reconstruct the mathematics of the Vedas. According to his research all of mathematics is based on sixteen Sutras, or word-formulae.

Bharati Krsna wrote sixteen volumes expounding the Vedic system but these were unaccountably lost and when the loss was confirmed in his final years he wrote a single book: Vedic Mathematics, currently available. It was published in 1965, five years after his death.

Development of Further Material

A copy of the book was brought to London a few years later and some English mathematicians (Kenneth Williams, Andrew Nicholas, Jeremy Pickles) took an interest in it. They extended the introductory material given in Bharati Krsna's book and gave many courses and talks in London. A book (now out of print), Introductory Lectures on Vedic Mathematics, was published in 1981. Between 1981 and 1987 Andrew Nicholas made four trips to India initially to find out what further was known about it. Following these journeys a renewed interest was taken by scholars and teachers in India. It seems that once they saw that some people in the West took Vedic Mathematics seriously they realised they had something special. St. James' School, then in Queensgate, London, and other schools began to teach the Vedic system, with notable success. Today Vedic Mathematics is taught widely in schools in India and a great deal of research is being done. Three further books appeared in 1984, the year of the centenary of the birth of Sri Bharati Krsna Tirthaji. These were published by The Vedic Mathematics Research Group.

Maharishi Schools

When Maharishi Mahesh Yogi began to explain the significance and marvellous qualities of Vedic Mathematics in 1988, Maharishi Schools around the world began to teach it. At the school in Skelmersdale, Lancashire a full course was written and trialled for 11 to 14 year old pupils, called The Cosmic Computer. (Maharishi had said that the Sutras of Vedic Mathematics are the software for the cosmic computer- the cosmic computer runs the entire universe on every level and in every detail). This course was published in March 1998.

Neither Vedic Nor Mathematics

As has been pointed out earlier on several occasions, the so-called 'Vedic Mathematics' is neither 'Vedic' nor can it be dignified by the name of mathematics. 'Vedic Mathematics', as is well-known, originated with a book of the same name by a former Shankracharya of Puri (the late Jagadguru Swami Shri Bharati Krishna Tirthaji Maharaj) published posthumously in 1965. The book assembled a set of tricks in elementary arithmetic and algebra to be applied in performing computations with numbers and polynomials. As is pointed out even in the foreword to the book by the General Editor, Dr. A. S. Agarwala, the aphorisms in Sanskrit to be found in the book have nothing to do with the Vedas. Nor are these aphorisms to be found in the genuine Vedic literature.

The term "Vedic mathematics" is therefore entirely misleading and factually incorrect. Further, it is clear from the notation used in the arithmetical tricks in the book that the methods used in this text have nothing to do with the arithmetical techniques of antiquity. Many of the Sanskrit aphorisms in the book are totally cryptic (ancient Indian mathematical writing was anything but cryptic) and often so generalize to be devoid of any specific mathematical meaning. There are several authoritative texts on the mathematics of Vedic times that could used in part to teach an authoritative and correct account of ancient Indian mathematics but this book clearly cannot be used for any such purpose. The teaching of mathematics involves both the teaching of the basic concepts of the subject as well as methods of mathematical computation. The so-called "Vedic mathematics" is entirely inadequate to this task considering that it is largely made up of tricks to do some elementary arithmetic computations. Many of these can be far more easily performed on a simple computer or even an advanced calculator.

The book "Vedic mathematics" essentially deals with arithmetic of the middle and high-school level. Its claims that "there is no part of mathematics, pure or applied, which is beyond their jurisdiction" is simply ridiculous. In an era when the content of mathematics teaching has to be carefully designed to keep pace with the general explosion of knowledge and the needs of other modern professions that use mathematical techniques, the imposition of ``Vedic mathematics" will be nothing short of calamitous.

India today has active and excellent schools of research and teaching in mathematics that are at the forefront of modern research in their discipline with some of them recognised as being among the best in the world in their fields of research. It is noteworthy that they have cherished the legacy of distinguished Indian mathematicians like Srinivasa Ramanujam, V. K. Patodi, S. Minakshisundaram, Harish Chandra, K. G. Ramanathan, Hansraj Gupta, Syamdas Mukhopadhyay, Ganesh Prasad, and many others including several living Indian mathematicians. But not one of these schools has lent an iota of legitimacy to `Vedic mathematics'. Nowhere in the world does any school system teach "Vedic mathematics" or any form of ancient mathematics for that matter as an adjunct to modern mathematical teaching. The bulk of such teaching belongs properly to the teaching of history and in particular the teaching of the history of the sciences.

We consider the imposition of `Vedic mathematics' by a Government agency, as the perpetration of a fraud on our children, condemning particularly those dependent on public education to a sub-standard mathematical education. Even if we assumed that those who sought to impose `Vedic mathematics' did so in good faith, it would have been appropriate that the NCERT seek the assistance of renowned Indian mathematicians to evaluate so-called "Vedic mathematics" before making it part of the National Curricular framework for School Education. Appallingly they have not done so. In this context we demand that the NCERT submit the proposal for the introduction of `Vedic mathematics in the school curriculum to recognized bodies of mathematical experts in India, in particular the National Board of Higher Mathematics (under the Dept. of Atomic Energy), and the Mathematics sections of the Indian Academy of Sciences and the Indian National Science Academy, for a thorough and critical examination. In the meanwhile no attempt should be made to thrust the subject into the school curriculum either through the centrally administered school system or by trying to impose it on the school systems of various States.

We are concerned that the essential thrust behind the campaign to introduce the so-called 'Vedic mathematics' has more to do with promoting a particular brand of religious majoritarianism and associated obscurantist ideas rather than any serious and meaningful development of mathematics teaching in India. We note that similar concerns have been expressed about other aspects too of the National Curricular Framework for School Education. We re-iterate our firm conviction that all teaching and pedagogy, not just the teaching of mathematics, must be founded on rational, scientific and secular principles.

Myths and Reality : On 'Vedic Mathematics'

School of Mathematics

We in India have good reasons to be proud of a rich heritage in science, philosophy and culture in general, coming to us down the ages. In mathematics, which is my own area of specialisation, the ancient Indians not only took great strides long before the Greek advent, which is a standard reference point in the Western historical perspective, but also enriched it for a long period making in particular some very fundamental contributions such as the place-value system for writing numbers as we have today, introduction of zero and so on. Further, the sustained development of mathematics in India in the post-Greek period was indirectly instrumental in the revival in Europe after "its dark ages".

Notwithstanding the enviable background, lack of adequate attention to academic pursuits over a prolonged period, occasioned by several factors, together with about two centuries of Macaulayan educational system, has unfortunately resulted, on the one hand, in a lack of awareness of our historical role in actual terms and, on the other, an empty sense of pride which is more of an emotional reaction to the colonial domination rather than an intellectual challenge. Together they provide a convenient ground for extremist and misguided elements in society to "reconstruct history" from nonexistent or concocted source material to whip up popular euphoria.

That this anti-intellectual endeavour is counter-productive in the long run and, more important, harmful to our image as a mature society, is either not recognised or ignored in favour of short-term considerations. Along with the obvious need to accelerate the process of creating an awareness of our past achievements, on the strength of authentic information, a more urgent need has also arisen to confront and expose such baseless constructs before it is too late. This is not merely a question of setting the record straight. The motivated versions have a way of corrupting the intellectual processes in society and weakening their very foundations in the long run, which needs to be prevented at all costs.

The so-called " Vedic mathematics" is a case in point. A book by that name written by Jagadguru Swami Shri Bharati Krishna Tirthaji Maharaja (Tirathji, 1965) is at the centre of this pursuit, which has now acquired wide following; Tirthaji was the Shankaracharya of Govardhan Math, Puri, from 1925 until he passed away in 1960. The book was published posthumously, but he had been carrying out a campaign on the theme for a long time, apparently for several decades, by means of lectures, blackboard demonstrations, classes and so on. It has been known from the beginning that there is no evidence of the contents of the book being of Vedic origin; the Foreword to the book by the General Editor, Dr. A. S. Agrawala, and an account of the genesis of the work written by Manjula Trivedi, a disciple of the swamiji, make this clear even before one gets to the text of the book. No one has come up with any positive evidence subsequently either.

There has, however, been a persistent propaganda that the material is from the Vedas. In the face of a false sense of national pride associated with it and the neglect, on the part of the knowledgeable, in countering the propaganda, even educated and well meaning people have tended to accept it uncritically. The vested interests have also involved politicians in the propaganda process to gain state support. Several leaders have lent support to the "Vedic mathematics" over the years, evidently in the belief of its being from ancient scriptures. In the current environment, when a label as ancient seems to carry considerable premium irrespective of its authenticity or merit, the purveyors would have it going easy.

Large sums have been spent both by the Government and several private agencies to support this "Vedic mathematics", while authentic Vedic studies continue to be neglected. People, especially children, are encouraged to learn and spread the contents of the book, largely on the baseless premise of their being from the Vedas. With missionary zeal several "devotees" of this cause have striven to take the "message" around the world; not surprisingly, they have even met with some success in the West, not unlike some of the gurus and yogis peddling their own versions of "Indian philosophy". Several people are also engaged in "research" in the new "Vedic mathematics."

To top it all, when in the early nineties the Uttar Pradesh Government introduced "Vedic mathematics" in school text books, the contents of the swamiji's book were treated as if they were genuinely from the Vedas; this also naturally seems to have led them to include a list of the swamiji's sutras on one of the opening pages (presumably for the students to learn them by heart and recite!) and to accord the swamiji a place of honour in the " brief history of Indian mathematics" described in the beginning of the textbook, together with a chart, which curiously has Srinivasa Ramanujan's as the only other name from the twentieth century!

For all their concern to inculcate a sense of national pride in children, those responsible for this have not cared for the simple fact that modern India has also produced several notable mathematicians and built a worthwhile edifice in mathematics (as also in many other areas). Harish Chandra's work is held in great esteem all over the world and several leading seats of learning of our times pride themselves in having members pursuing his ideas; (see, for instance, Langlands, 1993).

Even among those based in India, several like Syamdas Mukhopadhyay, Ganesh Prasad, B.N.Prasad, K. Anand Rau, T. Vijayaraghavan, S. S. Pillai, S. Minakshisundaram, Hansraj Gupta, K. G. Ramanathan, B. S. Madhava Rao, V. V. Narlikar, P. L. Bhatnagar and so on and also many living Indian mathematicians have carved a niche for themselves on the international mathematical scene (see Narasimhan, 1991). Ignoring all this while introducing the swamiji's name in the "brief history" would inevitably create a warped perspective in children's minds, favouring gimmickry rather than professional work.

What does the swamiji's "Vedic mathematics" seek to do and what does it achieve? In his preface of the book, grandly titled" A Descriptive Prefatory Note on the astounding Wonders of Ancient Indian Vedic Mathematics," the swamiji tells us that he strove from his childhood to study the Vedas critically " to prove to ourselves (and to others) the correctness (or otherwise)" of the "derivational meaning" of "Veda" that the " Vedas should contain within themselves all the knowledge needed by the mankind relating not only to spiritual matters but also those usually described as purely 'secular', 'temporal' or 'worldly'; in other words, simply because of the meaning of the word 'Veda', everything that is worth knowing is expected to be contained in the vedas and the swamiji seeks to prove it to be the case!

It may be worthwhile to point out here that there would be room for starting such an enterprise with the word 'science'! He also describes how the " contemptuous or at best patronising " attitude of Orientalists, Indologists and so on strengthened his determination to unravel the too-long-hidden mysteries of philosophy and science contained in ancient India's Vedic lore, with the consequence that, "after eight years of concentrated contemplation in forest solitude, we were at long last able to recover the long lost keys which alone could unlock the portals thereof."

The mindset revealed in this can hardly be said to be suitable in scientific and objective inquiry or pursuit of knowledge, but perhaps one should not grudge it in someone from a totally different milieu, if the outcome is positive. One would have thought that with all the commitment and grit the author would have come up with at least a few new things which can be attributed to the Vedas, with solid evidence.

This would have made a worthwhile contribution to our understanding of our heritage. Instead, all said and done there is only the author's certificate that "we were agreeably astonished and intensely gratified to find that exceedingly though mathematical problems can be easily and readily solved with the help of these ultra-easy Vedic sutras (or mathematical aphorisms) contained in the Parishishta (the appendix portion) of the Atharva Veda in a few simple steps and by methods which can be conscientiously described as mere 'mental arithmetic' (paragraph 9 in the preface). That passing reference to the Atharva Veda is all that is ever said by way of source material for the contents. The sutras, incidentally, which appeared later scattered in the book, are short phrases of just about two to four words in Sanskrit, such as Ekadhikena Purvena or Anurupye Shunyam Anyat. (There are 16 of them and in addition there are 13 of what are called sub-sutras, similar in nature to the sutras).

The first key question, which would occur to anyone, is where are these sutras to be found in the Atharva Veda. One does not mean this as a rhetorical question. Considering that at the outset the author seemed set to send all doubting Thomases packing, the least one would expect is that he would point out where the sutras are, say in which part, stanza, page and so on, especially since it is not a small article that is being referred to. Not only has the author not cared to do so, but when Prof. K. S. Shukla, a renowned scholar of ancient Indian mathematics, met him in 1950, when the swamiji visited Lucknow to give a blackboard demonstration of his "Vedic mathematics", and requested him to point out the sutras in question in the Parishishta of the Atharva Veda, of which he even carried a copy (the standard version edited by G. M. Bolling and J. Von Negelein), the swamiji is said to have told him that the 16 sutra demonstrated by him were not in those Parishishtas and that "they occurred in his own Parishishta and not any other" (Shukla, 1980, or Shukla, 1991). What justification the swamiji thought he had for introducing an appendix in the Atharva Veda, the contents of which are nevertheless to be viewed as from the Veda, is anybody's guess. In any case, even such a Parishishta, written by the swamiji, does not exist in the form of a Sanskrit text.

Let us suppose for a moment that the author indeed found the sutras in some manuscript of the Atharva Veda, which he came across. Would he not then have preserved the manuscript? Would he not have shown at least to some people where the sutras are in the manuscript? Would he not have revealed to some cherished students how to look for sutras with such profound mathematical implications as he attributes to the sutras in question, in that or other manuscripts that may be found? While there is a specific mention in the write-up of Manjula Trivedi, in the beginning of the book, about some 16-volume manuscript written by the swamiji having been lost in 1956, there is no mention whatever (let alone any lamentation that would be due in such an event) either in her write-up nor in the swamiji's preface about any original manuscript having been lost. No one certainly has come forward with any information received from the swamiji with regard to the other questions above. It is to be noted that want of time could not be a factor in any of this, since the swamiji kindly informs us in the preface that "

Ever since (i.e. since several decades ago), we have been carrying on an incessant and strenuous campaign for the India-wide diffusion of all this scientific knowledge".

The only natural explanation is that there was no such manuscript. It has in fact been mentioned by Agrawala in his general editor's foreword to the book, and also by Manjula Trivedi in the short account of the genesis of the work, included in the book together with a biographical sketch of the swamiji, that the sutras do not appear in hitherto known Parishishtas. The general editor also notes that the style of language of the sutras "point to their discovery by Shri Swamiji himself" (emphasis added); the language style being contemporary can be confirmed independently from other

Sanskrit scholars as well. The question why then the contents should be considered" Vedic" apparently did not bother the general editor, as he agreed with the author that "by definition" the Vedas should contain all knowledge (never mind whether found in the 20th century, or perhaps even later)! Manjula Trivedi, the disciple has of course no problem with the sutras not being found in the Vedas as she in fact says that they were actually reconstructed by her beloved " Gurudeva," on the basis of intuitive revelation from material scattered here and there in the Atharva Veda, after " assiduous research and 'Tapas' for about eight years in the forests surrounding Shringeri." Isn't that adequate to consider them to be "Vedic"? Well, one can hardly argue with the devout! There is a little problem as to why the Gurudeva himself did not say so (that the sutras were reconstructed) rather than referring to them as sutras contained in the Parishishta of the Atharva Veda, but we will have to let it pass. Anyway the fact remains that she was aware that they could not actually be located in what we lesser mortals consider to be the Atharva Veda.

The question of the source of the sutras is merely the first that would come to mind, and already on that there is such a muddle. Actually, even if the sutras were to be found, say in the Atharva Veda or some other ancient text, that still leaves open another fundamental question as to whether they mean or yield, in some cognisable way, what the author claims; in other words, we would still need to know whether such a source really contains the mathematics the swamiji deals with or merely the phrases, may be in some quite different context. It is interesting to consider the swamiji's sutras in this light. One of them, for instance, is Ekadhikena Purvena which literally just means" by one more than the previous one." In chapter I, the swamiji tells us that it is a sutra for finding the digits in the decimal expansion of numbers such as 1/19, and 1/29, where the denominator is a number with 9 in the unit's place; he goes on to give a page-long description of the procedure to be followed, whose only connection with the sutra is that it involves, in particular, repeatedly multiplying by one more than the previous one, namely 2, 3 and so on, respectively, the "previous one" being the number before the unit's place; the full procedure involves a lot more by way of arranging the digits which can in no way be read off from the phrase.

In Chapter II, we are told that the same sutra also means that to find the square of a number like 25 and 35, (with five in unit's place) multiply the number of tens by one more than itself and write 25 ahead of that; like 625, 1,225 and so on. The phrase Ekanyunena Purvena which means " by one less than the previous one" is however given to mean something which has neither to do with decimal expansions nor with squaring of numbers but concerns multiplying together two numbers, one of which has 9 in all places (like 99,999, so on.)! Allowing oneself such unlimited freedom of interpretation, one can also interpret the same three-word phrase to mean also many other things not only in mathematics but also in many other subjects such as physics, chemistry, biology, economics, sociology and politics. Consider, for instance, the following " meaning": the family size may be allowed to grow, at most, by one more than the previous one. In this we have the family-planning message of the 1960s; the "previous one" being the couple, the prescription is that they should have no more than three children.

Thus the lal trikon (red triangle) formula may be seen to be " from the Atharva Veda," thanks to the swamiji's novel technique (with just a bit of credit to yours faithfully). If you think the three children norm now outdated, there is no need to despair. One can get the two-children or even the one-child formula also from the same sutra; count only the man as the "previous one" (the woman is an outsider joining in marriage, isn't she) and in the growth of the family either count only the children or include also the wife, depending on what suits the desired formula! Another sutra is Yavadunam, which means "as much less;" a lifetime may not suffice to write down all the things such a phrase could "mean," in the spirit as above. There is even a sub-sutra, Vilokanam (observation) and that is supposed to mean various mathematical steps involving observation! In the same vein one can actually suggest a single sutra adequate not only for all of mathematics but many subjects:

It may be argued that there are, after all, ciphers which convey more information than meets the eye. But the meaning in those cases is either arrived at from the knowledge of the deciphering code or deduced in one or other way using various kinds of contexual information. Neither applies in the present case. The sutras in the swamiji's book are in reality mere names for various steps to be followed in various contexts; the steps themselves had to be known independently. In other words, the mathematical step is not arrived at by understanding or interpreting what are given as sutras; rather, sutras somewhat suggestive of the meaning of the steps are attached to them like names. It is like associating the 'sutra' VIBGYOR to the sequence of colours in rainbow (which make up the white light). Usage of words in Sanskrit, a language which the popular mind unquestioningly associates with the distant past(!), lend the contents a bit of antique finish!

An analysis of the mathematical contents of Tirthaji's book also shows that they cannot be from the Vedas. Though unfortunately there is considerable ignorance about the subject, mathematics from the Vedas is far from being an unexplored area. Painstaking efforts have been made for well over a century to study the original ancient texts from the point of view of understanding the extent of mathematical knowledge in ancient times. For instance, from the study of Vedic Samhitas and Brahamanas it has been noted that they had the system of counting progressing in multiples of 10 as we have today and that they considered remarkably large numbers, even up to 14 digits, unlike other civilizations of those times. From the Vedanga period there is in fact available a significant body of mathematical literature in the form of Shulvasutras, from the period between 800 BC and 500 BC, or perhaps even earlier, some of which contain expositions of various mathematical principles involved in construction of sacrificial 'vedi's needed in performing 'yajna's (see, for instance, Sen and Bag 1983).

Baudhyana Shulvasutra, the earliest of the extant Shulvasutras, already contains, for instance, what is currently known as Pythagoras' Theorem (Sen and Bag, 1983, page 78, 1.12). It is the earliest known explicit statement of the theorem in the general form (anywhere in the world) and precedes Pythagoras by at least a few hundred years. The texts also show a remarkable familiarity with many other facts from the so-called Euclidean Geometry and it is clear that considerable use was made of these, long before the Greeks formulated them. The work of George Thibaut in the last century and that of A.Burk around the turn of the century brought to the attention of the world the significance of the mathematics of the Shulvasutras. It has been followed up in this century by both foreign and Indian historians of mathematics. It is this kind of authentic work, and not some mumbo-jumbo that would highlight our rich heritage. I would strongly recommend to the reader to peruse the monograph, The Sulbasutras by S. N. Sen and A. K. Bag (Sen and Bag, 1983), containing the original sutras, their translation and a detailed commentary, which includes a survey of a number of earlier works on the subject. There are also several books on ancient Indian mathematics from the Vedic period.

The contents of the swamiji's book have practically nothing in common with what is known of the mathematics from the Vedic period or even with the subsequent rich tradition of mathematics in India until the advent of the modern era; incidentally, the descriptions of mathematical principles or procedures in ancient mathematical texts are quite explicit and not in terms of cryptic sutras. The very first chapter of the book (as also chapters XXVI to XXVIII) involves the notion of decimal fractions in an essential way. If the contents are to be Vedic, there would have had to be a good deal of familiarity with decimal fractions, even involving several digits, at that time. It turns out that while the Shulvasutras make extensive use of fractions in the usual form, nowhere is there any indication of fractions in decimal form. It is inconceivable that such an important notion would be left out, had it been known, from what are really like users manuals of those times, produced at different times over a prolonged period.

Not only the Shulvasutras and the earlier Vedic works, but even the works of mathematicians such as Aryabhatta, Brahmagupta and Bhaskara, are not found to contain any decimal fractions. Is it possible that none of them had access to some Vedic source that the swamiji could lay his hands on (and still not describe it specifically)? How far do we have to stretch our credulity? The fact is that the use of decimal fractions started only in the 16th century, propagated to a large extent by Francois Viete; the use of the decimal point (separating the integer and the fractional parts) itself, as a notation for the decimal representation, began only towards the end of the century and acquired popularity in the 17th century following their use in John Napier's logarithm tables (see, for instance, Boyer, 1968, page 334).

Similarly, in chapter XXII the swamiji claims to give " sutras relevant to successive differentiation, covering the theorems of Leibnitz, Maclaurin, Taylor, etc. and a lot of other material which is yet to be studied and decided on by the great mathematicians of the present-day Western world;" it should perhaps be mentioned before we proceed that the chapter does not really deal with anything of the sort that would even remotely justify such a grandiloquent announcement, but rather deals with differentiation as an operation on polynomials, which is a very special case reducing it all to elementary algebra devoid of the very soul of calculus, as taught even at the college level.

Given the context, we shall leave Leibnitz and company alone, but consider the notions of derivative and successive differentiation. Did the notions exist in the Vedic times? While certain elements preliminary to calculus have been found in the works of Bhaskara II from the 12th century and later Indian mathematicians in the pre-calculus era in international mathematics, such crystallised notions as the derivative or the integral were not known. Though a case may be made that the developments here would have led to the discovery of calculus in India, no historians of Indian mathematics would dream of proposing that they actually had such a notion as the derivative, let alone successive differentiation; the question here is not about performing the operation on polynomials, but of the concept. A similar comment applies with regard to integration, in chapter XXIV. It should also be borne in mind that if calculus were to be known in India in the early times, it would have been acquired by foreigners as well, long before it actually came to be discovered, as there was enough interaction between India and the outside world.

If this is not enough, in Chapter XXXIX we learn that analytic conics has an " important and predominating place for itself in the Vedic system of mathematics," and in Chapter XL we find a whole list of subjects such as dynamics, statics, hydrostatics, pneumatics and applied mathematics listed alongside such elementary things as subtractions, ratios, proportions and such money matters as interest and annuities (!), discounts (!) to which we are assured, without going into details, that the Vedic sutras can be applied. Need we comment any further on this?

The remaining chapters are mostly elementary in content, on account of which one does not see such marked incongruities in their respect. It has, however, been pointed out by Shukla that many of the topics considered in the book are alien to the pursuits of ancient Indian mathematicians, not only form the Vedic period but until much later (Shukla, 1989 or Shukla, 1991).

These include many such topics as factorisation of algebraic expressions, HCF (highest common factor) of algebraic expressions and various types of simultaneous equations. The contents of the book are akin to much later mathematics, mostly of the kind that appeared in school books of our times or those of the swamiji's youth, and it is unthinkable, in the absence of any pressing evidence, that they go back to the Vedic lore. The book really consists of a compilation of tricks in elementary arithmetic and algebra, to be applied in computations with numbers and polynomials. By a "trick" I do not mean a sleight of hand or something like that; in a general sense a trick is a method or procedure which involves observing and exploring some special features of a situation, which generally tend to be overlooked; for example, the trick described for finding the square of numbers like 15 and 25 with 5 in the unit's place makes crucial use of the fact of 5 being half of 10, the latter being the base in which the numbers are written. Some of the tricks given in the book are quite interesting and admittedly yield quicker solutions than by standard methods (though the comparison made in the book are facetious and misleading). They are of the kind that an intelligent hobbyist experimenting with numbers might be expected to come up with. The tricks are, however, based on well-understood mathematical principles and there is no mystery about them.

Of course to produce such a body of tricks, even using the well-known is still a non-trivial task and there is a serious question of how this came to be accomplished. It is sometimes suggested that Tirthaji himself might have invented the tricks. The fact that he had a M.A.degree in mathematics is notable in this context. It is also possible that he might have learnt some of the tricks from some elders during an early period in his life and developed on them during those "eight years of concentrated contemplation in forest solitude:" this would mean that they do involve a certain element of tradition, though not to the absurd extent that is claimed. These can, however, be viewed only as possibilities and it would not be easy to settle these details. But it is quite clear that the choice is only between alternatives involving only the recent times.

It may be recalled here that there have also been other instances of exposition and propagation of such faster methods of computation applicable in various special situations (without claims of their coming from ancient sources). Trachtenberg's Speed System (see Arther and McShane, 1965) and Lester Meyers' book, High-Speed Mathematics (Meyers, 1947) are some well-known examples of this. Trachtenberg had even set up an Institute in Germany to provide training in high-speed mathematics. While the swamiji's methods are independent of these, for the most part they are similar in spirit. One may wonder why such methods are not commonly adopted for practical purposes. One main point is that they turn out to be quicker only for certain special classes of examples. For a general example the amount of effort involved (for instance, the count of the individual operations needed to be performed with digits, in arriving at the final answer) is about the same as required by the standard methods; in the swamiji's book, this is often concealed by not writing some of the steps involved, viewing it as "mental arithmetic." Using such methods of fast arithmetic involves the ability or practice to recognise various patterns which would simplify the calculations. Without that, one would actually spend more time, in first trying to recognise patterns and then working by rote anyway, since in most cases it is not easy to find useful patterns.

People who in the course of their work have to do computations as they arise, rather than choose the figures suitably as in the demonstrations, would hardly find it convenient to carry them out by employing umpteen different ways depending on the particular case, as the methods of fast arithmetic involve. It is more convenient to follow the standard method, in which one has only to follow a set procedure to find the answer, even though in some cases this might take more time. Besides, equipment such as calculators and computers have made it unnecessary to tax one's mind with arithmetical computations. Incidentally, the suggestion that this "Vedic mathematics" of the Shankaracharya could lead to improvement in computers is totally fallacious, since the underlying mathematical principles involved in it were by no means unfamiliar in pro- fessional circles. One of the factors causing people not to pay due attention to the obvious questions about "Vedic mathematics" seems to be that they are overwhelmed by a sense of wonderment by the tricks. The swamiji tells us in the preface how "the educationists, the cream of the English educated section of the people including highest officials (e.g. the high court judges, the ministers etc.) and the general public as such were all highly impressed; nay thrilled, wonder-struck and flabbergasted!" at his demonstrations of the "Vedic mathematics." Sometimes one comes across reports about similar thrilling demonstrations by some of the present-day expositors of the subject. Though inevitably they have to be taken with a pinch of salt, I do not entirely doubt the truth of such reports. Since most people have had a difficult time with their arithmetic at school and even those who might have been fairly good would have lost touch, the very fact of someone doing some computations rather fast can make an impressive sight. This effect may be enhanced with well-chosen examples, where some quicker methods are applicable.

Even in the case of general examples where the method employed is not really more efficient than the standard one, the computations might appear to be fast, since the demonstrator would have a lot more practice than the people in the audience. An objective assessment of the methods from the point of view of overall use can only be made by comparing how many individual calculations are involved in working out various general examples, on an average, and in this respect the methods of fast arithmetic do not show any marked advantage which would offset the inconvenience indicated earlier. In any case, it would be irrational to let the element of sur- prise interfere in judging the issue of origin of " Vedic mathematics" or create a dreamy and false picture of its providing solutions to all kinds of problems.

It should also be borne in mind that the book really deals only with some middle and high school level mathematics; this is true despite what appear to be chapters dealing with some notions in calculus and coordinate geometry and the mention of a few, little more advanced topics, in the book. The swamiji's claim that "there is no part of mathematics, pure or applied, which is beyond their jurisdiction" is ludicrous. Mathematics actually means a lot more than arithmetic of numbers and algebra of polynomials; in fact multiplying big numbers together, which a lot of people take for mathematics, is hardly something a mathematician of today needs to engage himself in. The mathematics of today concerns a great variety of objects beyond the high school level, involving various kinds of abstract objects generalising numbers, shapes, geometries, measures and so on and several combinations of such structures, various kinds of operations, often involving infinitely many entities; this is not the case only about the frontiers of mathematics but a whole lot of it, including many topics applied in physics, engineering, medicine, finance and various other subjects. Despite all its pretentious verbiage page after page, the swamiji's book offers nothing worthwhile in advanced mathematics whether concretely or by way of insight.

Modern mathematics with its multitude of disciplines (group theory, topology, algebraic geometry, harmonic analysis, ergodic theory, combinatorial mathematics-to name just a few) would be a long way from the level of the swamiji's book. There are occasionally reports of some "re- searchers" applying the swamiji's "Vedic mathematics" to advanced problems such as Kepler's problem, but such work involves nothing more than tinkering superficially with the topic, in the manner of the swamiji's treatment of calculus, and offers nothing of interest to professionals in the area. Even at the school level "Vedic mathematics" deals only with a small part and, more importantly, there too it concerns itself with only one particular aspect, that of faster computation. One of the main aims of mathematics education even at the elementary level consists of developing familiarity with a variety of concepts and their significance. Not only does the approach of " Vedic mathematics" not contribute anything towards this crucial objective, but in fact might work to its detriment, because of the undue emphasis laid on faster computation. The swamiji's assertion "8 months (or 12 months) at an average rate of 2 or 3 hours per day should suffice for completing the whole course of mathematical studies on these Vedic lines instead of 15 or 20 years required according to the existing systems of the Indian and also foreign universities," is patently absurd and hopefully nobody takes it seriously, even among the activists in the area. It would work as a cruel joke if some people choose to make such a substitution in respect of their children.

It is often claimed that " Vedic mathematics" is well-appreciated in other countries, and even taught in some schools in UK etc.. In the normal course one would not have the means to examine such claims, especially since few details are generally supplied while making the claims. Thanks to certain special circumstances I came to know a few things about the St. James Independent School, London which I had seen quoted in this context. The School is run by the 'School of Economic Science' which is, according to a letter to me from Mr. James Glover, the Head of Mathematics at the School, "engaged in the practical study of Advaita philosophy". The people who run it have had substantial involvement with religious groups in India over a long period. Thus in essence their adopting " Vedic mathematics" is much like a school in India run by a religious group adopting it; that school being in London is beside the point. (It may be noted here that while privately run schools in India have limited freedom in choosing their curricula, it is not the case in England). It would be interesting to look into the background and motivation of other institutions about which similar claims are made. At any rate, adoption by institutions abroad is another propaganda feature, like being from ancient source, and should not sway us.

It is not the contention here that the con- tents of the book are not of any value. Indeed, some of the observations could be used in teaching in schools. They are entertaining and could to some extent enable children to enjoy mathematics. It would, however, be more appropriate to use them as aids in teaching the related concepts, rather than like a series of tricks of magic. Ultimately, it is the understanding that is more important than the transient excitement, By and large, however, such pedagogical application has limited scope and needs to be made with adequate caution, without being carried away by motivated propaganda.

It is shocking to see the extent to which vested interests and persons driven by misguided notions are able to exploit the urge for cultural self-assertion felt by the Indian psyche. One would hardly have imagined that a book which is transparently not from any ancient source or of any great mathematical significance would one day be passed off as a store- house of some ancient mathematical treasure. It is high time saner elements joined hands to educate people on the truth of this so-called Vedic mathematics and prevent the use of public money and energy on its propagation, beyond the limited extent that may be deserved, lest the intellectual and educational life in the country should get vitiated further and result in wrong attitudes to both history and mathematics, especially in the coming generation.

4

The Sutra Era of Vedic Mathematics

The first glimpses of Mathematical activity in the Indian peninsula appear in the Sulva Sutras. The commonly accepted dates for such a dating are not without controversy. These conventional dates place the sutras, about 800 years before the common era. Here is one alternative chronology compiled by Lakshmikantham and Leela;

Proposed Skeleton of Indian Chronology 3374 BC

Event	Individual	Date
Birth		3374BCE
Birth	Veda Vyaasa	
Birth	Apasthambha	3200 BCE
Birth	Baudhayana	2765 BCE
Coronation of	Aryabhatta	1888-1807 BCE
Coronation of	Gautama Buddha	1554 BCE
Reign of	Chandragupta Maurya	1472 BCE
Coronation of	Ashoka Maurya	1294 - 1234 BCE
Pancha Siddhantas,	Kanishka	327 BCE
authored by	Chandragupta of Imperial	123 BCE
Reign of	Gupta dynasty	102 BCE to 78 CE
Birth of	Varahamihira	30 BCE
Era of	Vikramaaditya	78 CE
Writings	Brahmagupta	486 CE
	Salivahana	486 CE
	Bhaskara II	
	Siddhanta Siromani	

The politically correct date of the birth of Aryabhatta is 476 CE. Why the difference of 23 centuries. Apparently this resulted from a mistranslation of a Sanskrit word, where in fact one entire word has been replaced with another.

To quote Lakshmikantham and Leela, Page 27, Chapter 4, Chronological Aspects

Was Sir William indulging in subterfuge or was he woefully incompetent that he did not recognize the difference between shasti and shabdi

In the modern period of Bharat (India), Aryabhatta is the first famous mathematician and astronomer. In his book Aryabhatteeyam, Aryabhatta clearly provides his birth data. In the 10th stanza, he says that when 60 × 6 = 360 years elapsed in this Kali Yuga, he was 23 years old. The stanza of the sloka starts with "Shastyabdanam Shadbhiryada vyateetastra yascha yuga padah." "Shastyabdanam Shadbhi" means 60 × 6 = 360. While printing the manuscript, the word "Shadbhi" was altered to "Shasti", which implies 60 × 60 = 3600 years after Kali Era. As a result of this intentional arbitrary change, Aryabhatta's birth time was fixed as 476 CE Since in every genuine manuscript, we find the word "Shadbhi" and not the altered "Shasti", it is clear that Aryabhatta was 23 years old in 360 Kali Era or 2742 BCE. This implies that Aryabhatta was born in 337 Kali Era or 2765 BCE. and therefore could not have lived around 500 AD, as manufactured by the Indologists to fit their invented framework. Bhaskara I is the earliest known commentator of Aryabhatta's works. His exact time is not known except that he was in between Aryabhatta (2765 BCE) and Varahamihira (123 BCE).

Is this another case of Napoleon''s Dictum 'attribute not to malice that which can be explained away by sheer incompetence. You be the judge.

The authors feel that such an error could not have crept in (remember it was a oral tradition) until the wording was changed after the British arrived in India and may have been done at the behest of Sir William.

The Indian Sulvasutras (also Pronounced Sulbasutras)

Cut the remaining parts (two-thirds of a strip) into eight equal strips and arrange them around the square we are constructing as in the diagram. We have now used all the parts of the second square but the new figure we have constructed is not quite a square having a small square corner missing. It is worth seeing what the side of this "not quite a square" is. It is

$$1 + 1/3 + 1/(3 \times 4)$$

which, of course, is the first three terms of the approximation. Now Datta argues in that to improve the "not quite a square" the Sulbasutra authors could have calculated how broad a strip one needs to cut off the left hand side and bottom to fill in the missing part which has area $\left(\frac{1}{12}\right)^2$. If x is the width one cuts off then

$$2 \times x \times \left(1 + \frac{1}{3} + \frac{1}{12}\right) = \left(\frac{1}{12}\right)^2.$$

This has the solution x = 1/(3 × 4 × 34) which is approximately 0.002450980392. We now have a square the length of whose sides is

$$1 + \frac{1}{3} + \frac{1}{(3 \times 4)} - \frac{1}{(3 \times 4 \times 34)}.$$

which is exactly the approximation given by the Apastamba Sulbasutra.

Of course we have still made an approximation since the two strips of breadth x which we cut off overlapped by a square of side x in the bottom left hand corner. If we had taken this into account we would have obtained the equation

$$2\times x\times\left(1+\frac{1}{3}+\frac{1}{12}\right)-x^2=\left(\frac{1}{12}\right)^2$$

for x which leads to $x = \frac{17}{2} - v^2$ which is approximately equal to 0.002453105. Of course we cannot take this route since we have arrived back at a value for x which involves v2 which is the quantity we are trying to approximate!

In [4] Gupta gives a simpler way of obtaining the approximation for v2 than that given by Datta in [1]. He uses linear interpolation to obtain the first two terms, he then corrects the two terms so obtaining the third term, then correcting the three terms obtaining the fourth term. Although the method given by Gupta is simpler (and an interesting contribution) there is certainly something appealing in Datta's argument and somehow a feeling that this is in the spirit of the Sulbasutras.

Of course the method used by these mathematicians is very important to understanding the depth of mathematics being produced in India in the middle of the first millennium BC. If we follow the suggestion of some historians that the writers of the Sulbasutras were merely copying an approximation already known to the Babylonians then we might come to the conclusion that Indian mathematics of this period was far less advanced than if we follow Datta's suggestion.

Muhammad Ben Musa al-Khuwarizmi (circa 783-850)

Introduction

We are often told by western historians and scientists that Indians of the pre-Christian era were poor historians and even poorer at record keeping and hence that we know very little of the identities of the mathematicians and their contributions to the subject of mathematics. Our contention is precisely the opposite. Not only were the Indics superb record keepers , but they reported on the discoveries of their predecessors as well as contemporaries without the slightest sign of condescension or attempts to purloin the credit for themselves, a trait that appears to be singularly rare among those studying the field of Indology which seeks to obfuscate anything emanating out of India. In many cases the Vedic mathematicians were also the pre-eminent astronomers of their day. The problem lay not so much in the much bandied poor record keeping of the Indics, but in the fact that these records were either ignored or in certain egregious instances were in fact altered.

In Part I we furnished sufficient quotes to establish the proposition that throughout the ancient as well as the medieval eras Mediterranean, Arab and European savants went to great length to acknowledge the contribution of the Indics in various fields such as number theory, geometry ,astronomy, and medicine. It was only in the colonial era beginning with the discovery of Sir William Jones of the antiquity of Sanskrit, which had far reaching implications on the roots of their own civilization, that racial prejudice towards the Indics took on a dominant role and began to affect the quality as well as the accuracy of the scholarship in Europe. The roots of this prejudice and its progression over the 2 centuries of British rule in India are chronicled by Thomas Trautmann Everything, including the truth became subordinate to the paramount goal of maintaining a dominant role in India and Asia. We do not know for certain that there was deliberate falsification of records to support their versions of Indic history, but what we do know is that in certain key instances, the text has been altered to suit the pre-conceived notions of the Europeans on Indic history and that the general approach was to reject key pieces of data and dub them as being unreliable when it did not fit in with their overall paradigm of ancient Indic chronology. Before we get into specific acts of misdating, let us list chronologically the cast of characters who studied India under the general rubric of Indology.

List of Indologists Who Worked in the Area of Indology

Among these there are quite a few who neither harboured preconceived notions nor would they indulge in the dishonest act of altering documents. Typical of these were names like Playfair, Jacobi, Schopenhauer, Alain Danielou, Heinrich Zimmer and Joseph Campbell. But even with the best of intentions it is difficult to translate accurately from a language and culture which is alien to ones own. When that language is over 4000 years old, the difficulties are multiplied in manifold ways. When Europeans studied Sanskrit and the Vedas the paradigm they followed was that of studying insects in a jar. The science of studying insects is known as entomology. The insects for obvious reasons have little say in the matter. Such was the case also when they studied the civilization of the Indics.

The opinion of the Indics hardly mattered and they were rarely consulted and in many instances such as that of Max Mueller and Franz Bopp (max Mueller's professor) they had never set foot in India or conversed with a pundit. In just as many instances such as that of max Mueller they could not converse or chant a single sloka in Sanskrit much less understand one when it was chanted in front of them. But that did not stop them from claiming to be Sanskritists of the first rank. Neither did their dilettante status in Sanskrit stop Bopp and Sir William from deciding that there must have been an ancestral language (which they called Proto Indo European (PIE for short) spoken anywhere but in India. Now that I ponder on the reluctance of Max Mueller to visit India, the suspicion is overwhelming that the real reason he never wished to set foot in India was that he would thereby be spared the embarrassment of facing a real pundit in Sanskrit and have to then admit how shallow his knowledge of Sanskrit was.

Motivation for Present dating of Ancients

The original preoccupation of the European (to some extent true to this day) was to find the roots of his/her own language, which till the advent of Sanskrit was assumed to have been derived from Hebrew. The notion of a Hebraic origin was hardly very popular in Europe steeped as it was in anti Semitism. Therefore, when Sanskrit was first discovered, the notion that there was a race of noble Aryans who were their putative ancestors was then greeted with a great deal of enthusiasm once they had disposed of the prior suspicion that they were descended from the teeming millions of India.

The corollary to this proposition was that the denizens of the Indian subcontinent could not possibly have an antiquity greater than that of Greece or that of Pericles and the Shakespearean vision of the golden age of the Hellenic civilization. It was the conflict with his strict belief in the Creation theory postulated in the bible that led Sir William to lop of 1200 years from the Puranic history of India and to further postulate that the contemporary of Megasthenes , the Greek historian who visited India around 300 BCE was Chandragupta Maurya and not the Chandragupta of the Imperial Gupta dynasty.

We have already alluded to the postulate adhered to by almost all Indologists in the western world, that the Saraswati Sindhu civilization had little to do with the Vedic civilization in Part I of this series and we have described in great detail the efforts by the Colonial Power to undermine the Civilizational unity of India in the South Asia File. We wish to emphasize once again that the net result of all these efforts was to change the complexion of the debate. What was once a search for the roots of their own languages has now been transformed into, in their words, an obsession on the part of the Hindu right wing to prove that the Aryans were indigenous to India. This is undoubtedly a very astute strategy on their part since it takes the limelight of their own obsession to find a Urheimat for their group of languages and the fact that from the inception the postulates of dating Indic history have been political rather than academic in nature. The entire dating of the revisionist Indic history by the British and the Europeans has been a political enterprise right from the start. So, now when the argument is made that political considerations are driving the Hindu right wing in their opposition to theories such as the AIT, regardless of the truth of such an allegation, it ignores the glaring fact that it has always been so.

It is sad that a section of the Indic populace has internalized this revisionist view of Indic History propounded primarily by Europeans and it is important to remember that AIT is crucial to validate their racial view of civilizations and people and their sense of self esteem. They had to reconcile what was obviously a vast dependence on the contributions of the Semitic speaking people primarily along the Mediterranean Sea. In AIT they saw their opportunity to portray themselves as the progenitors of a vast Eurasian civilization without aligning themselves too closely to the brown skinned people of India. The conclusion is inescapable that while validating the AIT is not crucial to the pride of the Indic other than that it robbed him of his own authentic history, the debunking of AIT would have a devastating effect on the European weltanschauung of the roots of his own civilization. Indics in general for understandable reasons tend to be Indocentric and look to their own psyche to comprehend the nature of this paradigm. We, the Indics, would be much better served if we sought to understand the motivations and psyche of the European, at least in this instance or in other words to understand why there is such a constant emphasis on Eurocentrism in the European psyche.

What do we expect to achieve by removing such misconceptions created essentially by a Eurocentric world. Every people need to have pride in their own traditions in order for there to be harmony, mutual respect and dignity among the peoples and civilizations of this world . Robbing them of their essential contributions is tantamount to breeding the seeds of racial superiority. What we are not seeking is retrospective privileges or an apology for such past misrepresentations, merely an acknowledgement that legitimate discoveries be attributed accurately to their rightful civilizations and that no single civilization or group of civilizations has a monopoly on the capability to advance the sum of human knowledge.

We will now discuss the anomalies in the assumptions made by European writers of Indic history and the evidence supporting our contention and see how devastating they were to a proper understanding and appreciation of Indic history and to a proper understanding of the Indic contribution to the sciences of antiquity.

Misdating of Aryabhatta the Elder

Aryabhatta is without doubt the Astronomer/Mathematician non-pareil of the Post Vedic/Post Epic era in the historical narrative, especially so since his magnum opus The Aryabhattium, which packs a lot of information in the terse aphoristic style characteristic, of that era, has survived intact from the mists of a distant past when he first developed his thesis in 4 Chapters covering the subject of Mathematics and Astronomy. His work and the prior work in the Vedic area form an important sheet anchor for the entire chronology that follows, important by virtue of the fact that it attests to the state of the language prior to his contribution, and refers to the beginning of Kaliyuga as he reveals his own age. But first we list the main mathematicians during the period in question

Some of the Vedic personalities that we will meet here are:

Yajnavalkya who wrote the Shatapatha Brahmana (as well as the Brihadaranyaka Upanishad) in which he describes the motion of the sun and the moon and advances a 95 year cycle to synchronize the motions of the sun and the moon

Lagadha who authored the Jyotisha Vedanga

Baudhayana the author of the Sulvasutra named after him

Apastambha the author of the Sulvasutra named after him

Katyayana the author of the Sulvasutra named after him

Panini the Grammarian for the Indo Europeans

Pingala Binary System of number representation:

Aryabhatta the astronomer laureate of ancient India

Varahamihira who synthesized the knowledge

The author of the Jaina treatises the Suryaprajnapati, Chandraprajnapati and the seventh section of Jambudvipaprajnapati

We will tackle first the misdating of Aryabhatta for the reasons stated above . Details of the life and contributions of this veritable genius are presented in the linked site. What concerns us most here is the episode relating to the misdating.

"Aryabhatta is the first famous mathematician and astronomer of Ancient India. In his book Aryabhatteeyam, Aryabhatta clearly provides his birth data. In the 10th stanza, he says that when 60 × 6 = 360 years elapsed in this Kali Yuga, he was 23 years old. The stanza of the sloka starts with "Shastyabdanam Shadbhiryada vyateetastra yascha yuga padah."

"Shastyabdanam Shadbhi" means 60 × 6 = 360. While printing the manuscript, the word "Shadbhi" was altered to "Shasti", which implies 60 × 60 = 3600 years after Kali Era. As a result of this intentional arbitrary change, Aryabhatta's birth time was fixed as 476 AD. Since in every genuine manuscript, we find the word "Shadbhi" and not the altered "Shasti", it is clear that Aryabhatta was 23 years old in 360 Kali Era or 2742 BC. This implies that Aryabhatta was born in 337 Kali Era or 2765 BC, and therefore could not have lived around 500 AD, as manufactured by the Indologists to fit their invented framework.

Bhaskara I is the earliest known commentator of Aryabhatta's works. His exact time is not known except that he was in between Aryabhatta (2765 BC) and Varahamihira (123 BC)."

The implications are profound, if indeed this is the case. The zero is by then in widespread use and if he uses Classical Sanskrit then he antedates Panini.

"How the beginning of Kaliyuga is Linked with the Dates of Indian astronomers? The ancient Indian astronomers perhaps purposely linked the determination of their dates of birth, composition of their works; calculation of number of years elapsed, etc., based on two eras Kali and Saka. Therefore, without the significance of these two eras, the dates cannot be determined specifically.

> *Shastabdhanam shastardha vyatitastrashyam yugapadha* |
>
> *Trayadhika vimsatirabdhastdheha mama janmanoatita* | |
>
> *"When sixty times sixty years and three quarter yugas (of the current yuga) had elapsed, twenty three years had then passed since by birth" (K. S. Shukla).*
>
> *"Now when sixty times sixty years and three quarter Yugas also have passed, twenty increased by three years have elapsed since my birth" (P. C. Sengupta).*

"I was born at the end of Kali 3600; I write this work when I am 23 years old i.e., at the end of Kali 3623" "(T. S. Kuppanna Sastry11).

Here, though only Yuga is mentioned, Kaliyuga is implied and its starting of 3102 BCE is taken for reckoning purpose. Thus, the date of Aryabhatta is determined as follows:

The year of birth = 3600 - 3102 = 488/499 - 23 = 476 CE. This has been accepted by most of the scholars and generally considered as accepted date. Had the commencement year 3102 BCE is a myth or not astronomical one, the year of Aryabhatta cannot be historical date or could be determined like this using 3102 BCE.

Bhaskara I in his commentary to Aryabhatteeyam mentions as follows (Ch.I.verse.9):

Kalpadherabdhnirodhadhayam abdharashiritiritaha:
khagnyadhriramarkarasavasurandhrenadhavaha: te
cangkkairapi 1986123730 |

"Since the beginning of the current Kalpa, the number of years elapsed is this: zero, three, seven, three, twelve, six, eight, nine, one (proceeding from right to left) years. The same (years) in figures are 1986123730".

"The time elapsed, in terms of years, since the commencement of the current kalpa is zero, three, seven, three, twelve, six, eight, nine, one (years written in figures) are 1986123730".

Aryabhatta gives the number of years elapsed since the beginning of the current yuga

= 6 Manus + 27 ¾ yugas

= 6 × 72 yugas + ¾ yugas

= 6 × 72 + 27 ¾) × 43,20,000 years

= (1866240000 + 119880000) years

= 198612000 years.

From this, we can calculate the number of years elapsed since Bhaskara wrote his commentary

= 1986123730 -198612000 = 3730 years

= 3730 - 3102

= 628/629 CE.

Bhaskaracharya and others too imply Kaliyua/era in their works as revealed through commentators, as their dates are determined with the calculations reckoning the date of starting of Kaliyuga/era as 3102 BCE.

Mahabharata and Kaliyuga: The starting of Kaliyuga has been associated with the following events12:

The end of Mahabharat war.

The death of Sri Krishna.

The deluge, which made Dwaraka, submerged.

Coronation of Yudhistira.

The renouncement of Yuddhistira.

In Indian astronomical works including Tantras and Karanas, the word yuga has been taken as Kaliyuga for calculating, illustrative and explanatory purposes. In a Tantra, the epoch is the beginning of Kaliyuga or 3102 BCE. In a Karana, any convenient epoch is selected by the astronomer.

"The Saka year (when the civil days are required) added to 3179 gives the solar years elapsed since the beginning of the Kaliyuga" (Sisyadhivrddhida -Tantra13 - hereinafter mentioned as ST. - of Lalla.I.12).

Here, that the Saka year began 3179 after the beginning of the Kaliyuga is specifically mentioned. Moreover, in the calculation of days elapsed, solar years elapsed, Suddhi ertc., Kaliyuga is repeatedly mentioned and used for illustrations.

"...the solar months elapsed since the beginning of the Kaliyuga multiplied by 22,26,389 and divided by 21,60,0000 give the corresponding lunar months" (ST.I.15).

The commentators Bhaskara I, Somesvara, Suryadeva Yajvan and others have pointed out the relation between Mahabharata and Kaliyuga.

Mahabharata and Kali Era: The usage of Kali era by the astronomers with the Mahabharata, that too, with Mahabharat war in particular, has been consistent. Many astronomers mention Kali era and Saka era together.

"Since the birth of Brahma up to the beginning of the Saka era, 8 ½ years (of Brahma), ½ month (of Brahma), 6 Manus of the (current) day (of Brahma), 27¾ yugas, and 3179 years of the (current) Kali era had gone by" (Vatesvara Siddhanta 14.I.10, K. S. Shukla).

Here, the number of years 3179 specifically mentioned is to obtain any year in terms of Saka, but it has been derived from the Kali era i.e, 3102/3101 + 78 = 3180/3179. As Vatesvara (c.880-960 CE) uses the notation, it is evident that even during the 9th century it had been very popular among the astronomers and established one. He also records his year of birth in that fashion as explained below.

Kali Era and Saka Era: After Aryabhatta, astronomers use the computation of years in Saka and as well as Kali Eras. The number of years to be reckoned in Saka with respect to Kali is given as 3179 and this is obtained by adding 78 to 3101/3102, thus, 3101 + 78 = 3179.

Thus, the Tantra directs: Navadhrirupagniyuttam mahibhujam shakendratnam gatavarshadaraham (I.4) meaning, "Add 3179 to the Saka years elapsed, the Kali years elapsed are obtained". Thus, it is evident that such method of reckoning of years in Saka Era related to Kali Era and vice versa had been in vogue before 6th century.

Vateswara says: "When 802 years had elapsed since the commencement of the Saka era, my birth took place; and when 24 years had passed since my birth, this Siddhanta was written by me by the grace of the heavenly bodies" (I.21).

Thus, the year of birth = 802 + 78 = 880 CE and that of his work = 880 + 24 = 904 CE.

The astronomers use certain Sakas as illustrative examples in their works. For example Mallikarjuna Suri and Candesvara, an astronomer of Mitthila use 1100 and 1107 Sakas for illustrating rules.

Mahabharata and 3102 BCE: The above discussion about Kaliyuga and Kali era amply proves its astronomical importance in time reckoning. It also points to the well established date of such reckoning starting with 3102 BCE and its connection with Mahabharata. The Saka Era has also been associated with it as starting 3179 after the starting of Kali Era. The date 3102/3101 BCE is very often used by the epigraphists, numismatists, archaeologists, historians, astronomers and others, but, assert that Kali Era/Yuga is a myth! Therefore, the application of 3102 BCE to determine and calculate other dates should be explained properly, because, the modern scholars use the same date, but even condemn and criticize it unwittingly at many places.

Bhaskara mentions the names of Latadeva, Nisanku and Panduranga Svami as disciples of Aryabhatta.

So the question is which version of the sloka in Aryabhattium is the correct one

Is it this one?

Shastyabdanam Shadbhịryada vyateetastra yascha yuga padah."

"Shastyabdanam Shadbhi" means 60 × 6 = 360. Which places his birth at 2765 BCE (360 -23 - 3102)

Or this one

Shastabdhanam shastardha vyatitastrashyam yugapadha |

Trayadhika vimsatirabdhastdheha mama janmanoatita | |

Shastabdhanam shastardha means 60 × 60 = 3600

Which places his birth at (3600 - 23 - 3102) = 475 CE

The resulting shift in the date of Aryabhatta of 3240 years, makes him roughly 3 times higher in chronology , and has profound consequences for the Indic contributions relative to those of Babylonian mathematicians

The Vedanga Period (Vyakarana, Jyotisha, Chandas, Kalpa Sutra)

The Vedanga (IAST vedanga, "member of the Veda") are six auxiliary disciplines for the understanding and tradition of the Vedas.

1. Shiksha (siksa): phonetics and phonology (sandhi)
2. Chandas (chandas): meter Pingala
3. Vyakarana (vyakarana): grammar Panini
4. Nirukta (nirukta): etymology Yaska
5. Jyotisha (jyotisa): astrology Lagadha
6. Kalpa (kalpa): ritual Apastambha, Baudhayana, Katyayana , Manava

The Vedangas are first mentioned in the Mundaka Upanishad as topics to be observed by students of the Vedas. Later, they developed into independent disciplines, each with its own corpus of Sutras.

During this period we will consider the timeline of the following contributors:

Panini - Vyakarana (Grammar, Place value system) was one of the many all time great savants that India has produced in such abundance. The scope and scale of his vast contributions to language, grammar, computing science, and place value system is mind numbing at a time when scripts were in an embryonic stage of development.

It is no hyperbole to say that there was no study of grammar as a codified set of rules until the west discovered Pannini's Ashtadhyayi

Pingala

Yaska

Apastambha

Baudhayana

Katyayana

Ashvalayana

Timeline According to Max Mueller

Chandas Rg Veda	1200 to 1000 BCE
Mantras later Vedas	1000 to 800 BCE
Brahmanas	800 to 600 BCE
Sutras	600 to 200 BCE

Timeline According to Keith

Taittiriya Samhita	500 BCE
Baudhayana	400 BCE
Ashvalayana	350 BCE
Sankhayana	350 BCE
Yaska	300 BCE
Apastambha	300 BCE
Pratisakhya	300 BCE
Panini	250 BCE
Katyayana	800to 600 BCE

Assumptions

The Sulvasutras precede the developments in Babylon and Egypt and must therefore date at least to 2000 BCE.

The Rigveda must have been fully composed prior to 2000 BCE because of the drying up of the Saraswati River prior to 1900 BCE.

The Vedangas indicate a knowledge of the use of zero and the place value system.

Seidenberg on the significance of the Sulvasutras.

Seidenberg, A. On the volume of a sphere. Arch. Hist. Exact Sci. 39 (1988), no. 2, 97--119. (Reviewer: K.-B. Gundlach.) SC: 01A20 (01A15 01A17 01A25 01A32), MR: 89j:01012.

Abraham Seidenberg argues that there is a common source for Pythagorean and Chinese (or Chinese-like) mathematics. He suggests that Old-Babylonian mathematics is a derivative of a more ancient mathematics having a much clearer geometric component (p. 104), and is "in some respects ... is derivative of a Chinese-like mathematics" (p. 109). Van der Waerden holds a similar view on this, and tells us that the mathematics of the Chiu Chang Suan Shu represents the common source more faithfully than the Babylonian does. Seidenberg believes that the common source is most similar to the Sulvasutras. He discusses how questions of the sphere and the circle were treated by the Greeks, Chinese, Egyptians, and to a lesser extent Indians. He discusses the some similarities and differences in the work on the sphere in Greece (Archimedes, with a very brief account of the application of his Method), and in Chinese (first in the Chiu Chang Suan Shu, improved by Liu Hui or perhaps Tsu Ch'ung-Chih, and then further improved by the Tsu Ch'ung-Chih's son Tsu Keng-Chih). He believes that the problem of the volume of a sphere goes back to the common source, to the first part of the second millennium BC, or earlier. An interesting and related topic is the topic of the equality of the proportionality constants pi that occur in the formulas for the area and circumference of a circle. Seidenberg examines the Moscow Papyrus, Chinese sources, and an Old-Babylonian text and finds that this fact seemed to be recognized in all three groups. He argues that the Egyptian, Babylonian, and Chinese approaches to the volume of a truncated pyramid may have derived from the same common source. He believe that the common source also used infinitesimal, Cavalieri-type, arguments as well. It is interesting as well that Heron, who as Seidenberg notes is sometimes considered to be continuing the Babylonian tradition, gives the formula 1/2(s+p)p+1/14(1/2s)2 for the area of a segment of a circle with chord s and height (sagita, arrow) p (with an Archimedean value of 22/7 for pi), and "that the 'ancients' took [the area as] 1/2(s+p)p and even conjectured that they did so because they took pi = 3." The paper is also interesting in that he discusses the development of some of his ideas from his early papers in the 60s until much later, (the paper was received soon before his death). Closely related topics: The Sphere, The Circle, The Pythagoreans, China, The Chiu Chang Suan Shu (Nine Chapters on the Mathematical Art), Sumerians and Babylonians, The Sulvasutras, Archimedes, Archimedes' Method, The Moscow Mathematical Papyrus, and Heron.

Mathews, Jerold. A Neolithic oral tradition for the van der Waerden/Seidenberg origin of mathematics. Arch. Hist. Exact Sci. 34 (1985), no. 3, 193--220. (Reviewer: M. Folkerts.) SC: 01A10 (01A25), MR: 88b:01005.

Abraham Seidenberg advanced a theory that mathematics arose from a common origin, and that some the mathematics was preserved by an oral tradition, and very likely a religious tradition, perhaps one like the one seen in the Indian Sulvasutras. Van der Waerden's book Geometry and Algebra in Ancient Civilizations takes a similar views, and in fact van der Waerden credits Seidenberg for making him look at the history of mathematics a new way. As Mathews notes, the Chinese Chiu Chang Suan Shu is very important in van der Waerden's work. Mathews relies heavily on this work as well to "give a small, coherent, and basic core of geometry concerning rectangles and their parts, ..., which may serve as what van der Waerden has called an 'oral tradition current in the Neolithic age.'" He states the he hoped "to give this hypothesized ancient core some credence through its relation to the Chiu Chang and its explanatory power. After giving a thorough discussion of this geometry, he then carefully analyzes the ninth chapter of the Chiu Chang in terms of this core. He is able to find a strong match, though his conclusions on one of the problems (Problem 20) are not consistent with those of some other researchers, who find in problem 20 instead suggestions of something like Horner's method. A very interesting article. Hopefully future papers will discuss how well the author's geometry agrees with the ancient geometry of other cultures. As he notes, "Until I can thoroughly test his conjecture on, say, the Babylonian corpus, I can argue for the merits of my conjecture only on such grounds as the simplicity of explanation it allows, or its congruence with received results or figures." Closely related topics: The Neolithic Era, Religion, Geometry, The Chiu Chang Suan Shu (Nine Chapters on the Mathematical Art), and Abraham Seidenberg.

5

Vedic Maths, a Forgotten Science

It is being taught in some of the most prestigious institutions in England and Europe. NASA scientists applied its principles in the area of artificial intelligence. And yet, in the country of its birth it languishes as a forgotten science.

Vedic mathematics, which simplifies arithmetic and algebraic operations, has increasingly found acceptance the world over. Experts suggest that it could be a handy tool for those who need to solve mathematical problems faster by the day.

What is Vedic Mathematics?

It is an ancient technique, which simplifies multiplication, divisibility, complex numbers, squaring, cubing, square and cube roots. Even recurring decimals and auxiliary fractions can be handled by Vedic mathematics.

Who Brought Vedic Maths to Limelight?

The subject was revived largely due to the efforts of Jagadguru Swami Bharathikrishna Tirthaji of Govardhan Peeth, Puri Jaganath (1884-1960). Having researched the subject for years, even his efforts would have gone in vain but for the enterprise of some disciples who took down notes during his last days. That resulted in the book, Vedic

Mathematics, in the 1960s. These are now available in a book called "VEDIC MATHEMATICS".

What is the Basis of Vedic Mathematics?

The basis of Vedic mathematics, are the 16 sutras, which attribute a set of qualities to a number or a group of numbers. The ancient Hindu scientists (Rishis) of Bharat in 16 Sutras (Phrases) and 120 words laid down simple steps for solving all mathematical problems in easy to follow 2 or 3 steps.

Vedic Mental or one or two line methods can be used effectively for solving divisions, reciprocals, factorisation, HCF, squares and square roots, cubes and cube roots, algebraic equations, multiple simultaneous equations, quadratic equations, cubic equations, bi-quadratic equations, higher degree equations, differential calculus, Partial fractions, Integrations, Pythogorus theoram, Apollonius Theoram, Analytical Conics and so on.

What is the Speciality of Vedic Mathematics?

Vedic scholars did not use figures for big numbers in their numerical notation. Instead, they preferred to use the Sanskrit alphabets, with each alphabet constituting a number. Several mantras, in fact, denote numbers; that includes the famed Gayatri mantra, which adds to 108 when decoded.

Is it Useful Today?

Given the initial training in modern maths in today's schools, students will be able to comprehend the logic of Vedic mathematics after they have reached the 8th standard. It will be of interest to every one but more so to younger students keen to make their mark in competitive entrance exams.

India's past could well help them make it in today's world.

It is amazing how with the help of 16 Sutras and 16 sub-sutras, the Vedic seers were able to mentally calculate complex mathematical problems.

1. Why did the Vedic Seers need Vedic Mathematics or Mental Mathematics?

   The Vedic Seers were highly ritualistic in practice. All the four Vedas namely Rig Veda, Yajur Veda, Sama Veda and Atharva Veda, consist of Samhitas, Brahmanas, Aranyakas and Upanishads. Of these four, the first three namely Samhitas, Brahmanas and Aranyakas contain several thousand Mantras or Hymns, Ritual practices and their interpretations. The Vedic Seers were very particular about the time of doing the rituals. Hence they needed a very scientific and accurate calendar and time measurements. Hence we have the adage IST which once meant Indian Standard Time. But in recent times due to our laziness and carelessness, it has come to mean Indian Stretchable Time.

   The Vedic culture was that of Yajna and the Vedic Purohits or priests were also very particular about the shape and size of the Yajna Kund. For this they needed a very highly developed Geometry and Trigonometry.

2. Where is Vedic Mathematics found?

   Vedic Mathematics forms part of Jyotish Shastra which is one of the six parts of Vedangas. The Jyotish Shastra or Astronomy is made up of three parts called Skandas. A Skanda means the big branch of a tree shooting out of the trunk

Some of the Sutras (Phrases) used in Vedic Mathematics:

1. Ekadhikena Purvena (One More than the Previous) is useful in solving Special Multiplications like 25 × 25, 95 × 95, 105 × 105 etc.

Special Divisons like 1 divided by 19, 29, 39, ... 199 etc.

2. Eka Nunena Purvena (One less than the Previous) is useful in solving Special Multiplications like 777 × 999, 123456789 × 999 999 999.
3. Urdhva Tiryak bhyam (Vertically and Crosswise) is useful in General Multiplication of any number by any number.
4. Paravartya Yojayet (P-64) (Transpose and Apply) is useful in solving Algebraic factors and divisions of some numbers etc.
5. Anurupyena (P-87) (Proportionately).
6. Adhyam-Adhyena, Antyam-antyena (P-87) (first by the first and the last by the last) are useful in solving Quadratics.
7. Lopana-stapana-bhyam (P-90) (by (alternate) Elimination and Retention) is useful in factorizing long and harder quadratics.
8. Gunita-Samuchhayah Samuchhaya-gunitah which means "The product of the sum of the coefficients in the factors is equal to the sum of the coefficients in the product" is a Sub-sutra of immense utility for the purpose of verifying the correctness of our answers in multiplications, divisions and factorisations.
9. Sunyam Samya samuccaye P-107 (when Samuccaya is the same, that Samuccaya is zero) Samuccaya is a technical term which has several meanings. This is useful in solving many complex factors and equations.

1st Example—1 Divided by 19, 29, 39, ... 129, etc.

1. To Divide 1 by numbers ending in 9 like 1 divided by 19, 29, 39, ... 119, etc.

Some of these numbers like 19, 29, 59 are prime numbers and so cannot be factorised and division becomes all the more difficult and runs into many pages in the present conventional method and the chances of making mistakes are many.

The Vedic Solution is obtained by applying the Sutra (theorem) Ekadhikena Purvena which when translated means "One more than the Previous".

Take for example 1 divided by 19. In the divisor 19, the previous is 1 and the factor is obtained by adding 1 to it which is 2. Similarly when we have to divide by 29, 39, ... 119 the factors shall be 3,4,... 12 respectively. (Add 1 to the previous term in the divisor). After this divide 1 by the factor in a typical Vedic way and the answer is obtained in 1 step. Thus:

1 divided by 19 = 0.0 5 2 6 3 1 5 7 8 9 4 7 3 6 8 4 2 1

1 divided by 29 = 0.0 3 4 4 8 2 7 5 8 6 2 0 6 8 9 6 5 5 1 7 2 4 1 3 7 9 3 1

2nd Example—Square of Numbers ending in 5

Squares of 25, 35, 45, 85, 95, 105, 195, etc. can be worked out mentally.

Again the Sutra used here is Ekadhikena Purvena which means, "One more than the previous."

The last term is always 5 and the Previous terms are 2, 3, 4, 8, 9, 10, 19 etc. and we have to add 1 to them. Square of the last term 5 is always 25.

Thus the Square of 25 is 2 × 3/25 = 625

the Square of 35 is 3 × 4/25 = 1225

the Square of 45 is 4 × 5/25 = 2025

the Square of 85 is 8 × 9/25 = 7225

the Square of 95, 105, 195 can be obtained in the same way.

Use the above formula to find the products of 23 multiplied by 27; 44 multiplied by 46; 192 multiplied by 198 and so on.

3rd Example—Multiplier-digits Consist Entirely of Nines

The Sutra (theorem) used here is Ekanyunena Purvena, sound as if it were the converse of Ekadhik Sutra i.e "one less"

777 multiplied by 999 = 776,223

(776 is one less than multiplicand 777 and 223 is the compliment of 776 from 9)

120 35 79 multiplied by 999 99 99 = 120 35 78, 879 64 21

1234 5678 09 multiplied 9999 9999 99
= 1234 5678 08 8765 4321 91

Such multiplications come up in advanced astronomy.

4th Example—General Multiplication of any Number by any Number

The Sutra used is Urdhva-Tiryagbhyam which means, "Vertically and crosswise" (P-39)

To multiply 12 by 13 mentally multiply

1 by 1, 3 by 1 and 2 by 1 and finally 2 by 3 and write the answer as 1 56

To multiply 37 by 42 mentally multiply

3 by 4, 3 by 2 and 7 by 4 and finally 7 by 2 and the answer is 12/34/14 = 1554

To multiply 1021 by 2103 mentally multiply as follows

1 by2 1by1+0by2 1by0+2by2+1by0 1by3 + 0by0 + 2by1 + 1by2 0by3 + 2by0 + 1by1 2by3 + 1by0 1by3 = 2147163

Multiplying 8 7 2 6 5 by 3 2 1 1 7
gives 2 8 0 2 6 9 0 0 0 5

5th Example—Algebraic Divisions

Divide ($12x^2$ -8x -32) by (X -2) using Urdhva-Tiryak, (Vertically and Crosswise)

Just by observation we can say the quotient must be (12X+k)

We also know that -8x = kx - 24x Hence k = 16 cross-check -2k = -32 third term

Divide (X^3 + $7X^2$ +6X +5) by (X-2)

X^3 divided by X gives X2 which is therefore the 1st term of the equation.

X^2 multiplied by -2 gives -2x2, But we have $7x^2$ in the Dividend. This means we have to get $9X^2$ more. This must result from the multiplication of X by 9X. Hence the 2nd term of the divisor must be 9X. Hence the quotient Q= X^2+ 9x + k...

For the third term we already have 6X = kX - 18X.

Hence k = 24

To find the Remainder R: 2k+5 = 53

6th Example—Division

Sutra: Paravartya Yojayet "Transpose and Apply" (P-63)

Divide ($12x^2$ - 8x - 32) by (x - 2), factor is +2

(X - 2)/2 $12x^2$ - 8x -32

+24x + 32

12x + 16 R = 0

Divide $7x^2$ + 5x + 3 by x - 1 Q = 7x + 12 R = 15

Divide $7x^2$ + 5x + 3 by x + 1 Q = 7x -2 R = 5

Divide x^3 + $7x^2$ + 6x + 5 by x - 2 Q = x^2 + 9x + 24 R = 53

Divide x^4 - x^3 + x^2 + 3x +5 by x^2 - x - 1 Factors are x + 1

Q = x^2 + 0x + 2 R = 5x + 7

Divide $6x^4$ + $13x^3$ + $39x^2$ + 37x + 45 by X^2 - 2x - 9

Factors 2x + 9 Q = $6x^2$ + 25x + 143 R = 548x + 1332

Divide $x^4 + x^2 + 1$ by $x^2 - x + 1$ (add $0x^3$ & $0x$)

$X^4 + 0x^3 + x^2 + 0x + 1$

Factors are: x - 1

Q = $x^2 + x + 1$ R = 0

Divide 1 2 3 4 by 1 1 2

Factors -1 -2 Q = 11 R = 2

Divide 1 3 4 5 6 by 1 1 2 3 Q = 12 R = - 20

The Reminder cannot be negative.

Hence Q = 11 R = 1 1 0 3

Divide 1 3 9 0 5 by 1 1 3 Q = 1 24 R = -107

Q = 123 R = 6

7th Example—Factorisations of Quadratics

Sutras: Anurupyena (P-84) "Proportionately."

Adhyam-Adhyena, Antyam-antyena (P-84) "first by the first and the last by the last"

8th Example—Verifying Correctness of Answers

A Sub-sutra of immense utility for the purpose of verifying the correctness of our answers in multiplications, divisions and factorisations:

Gunita-Samuchhayah Samuchhaya-gunitah (P-86) means

"The product of the sum of the coefficients in the factors is equal to the sum of the coefficients in the product"

S_c of the product = Product of the S_c in the factors

For example (x+7) (x+9) = (x^2 + 16x + 63)

(1+7) (1+9) = (1 + 16 + 63) = 80

or(x+1) (x+2) (x+3) = ($x^3 + 6x^2$ + 11x + 6)

(1+1) (1+2) (1+3) = (1 + 6 + 11 + 6) = 24

9th Example—Factorisations of Harder Quadratics

Lopana-stapana-bhyam (P-87) "by (alternate) Elimination and Retention"

It is very difficult to factorise the long quadratic

$(2x^2 + 6y^2 + 3z^2 + 7xy + 11yz + 7zx)$

But "Lopana-Sthapana" removes the difficulty. Eliminate z by putting $z = 0$.

Hence the given expression

$E = 2x^2 + 6y^2 + 7xy = (x+2y)(2x+3y)$

Similarly, if $y=0$, then $E = 2x^2 + 3z^2 + 7zx = (x+3z)(2x+z)$

Hence $E = (x + 2y+3z)(2x+3y+z)$

Factorise $2x^2 + 2y^2 + 5xy + 2x - 5y - 12 = (x+3)(2x-4)$ and $(2y+3)(y-4)$

Hence, $E = (x+2y+3)(2x+y-4)$

*This "Lopana-sthapana" method (of alternate elimination and retention) will be found highly useful in HCF, in Solid Geometry and in Co-ordinate Geometry of the straight line, the Hyperbola, the conjugate Hyperbola, the Asymptotes etc.

10th Example—Factorisations of Harder Quadratics-Special Cases

Sunyam Samya samuccaye P-107 (when Samuccaya is the same, that Samuccaya is zero) Samuccaya is a technical term which has several meanings.

First Meaning: It is a term which occurs as a common factor in all the terms concerned

Thus $12x + 3x = 4x + 5x$ x is common.

Hence $x = 0$

Or $9(x+1) = 7(x+1)$ here $(x+1)$ is common.

Hence $x + 1 = 0$

Second Meaning: Here Samuccaya means "the product of the independent terms"

Thus, (x +7) (x +9) = (x +3) (x + 21)

Here 7 x9 = 3 x 21. Therefore x = 0

Third Meaning: Samuccaya thirdly means the sum of the Denominators of two fractions having the same numerical numerator

Thus, 1/(2x -1) + 1/(3x -1) = 0.

Hence 5x - 2 =0 or x = 2/5

Fourth Meaning: Here Samuccaya means combination (or TOTAL).

In this context it is used in different contexts. These are

If the sum of the Numerators and the sum of the Denominators be the same, then that sum = 0

(2x +9)/ (2x +7) = (2x +7)/ (2x +9)

N1 + N2 = D1 + D2 = 2x + 9 + 2x + 7 = 0

Hence 4x + 16 = 0.

Hence x = -4

Note: If there is a numerical factor in the algebraic sum, that factor should be removed.

(3x +4)/ (6x +7) = (x +1)/ (2x +3)

Here N1 +N2 = 4x +5; D1 +D2 = 8x + 10;

4x +5 =0 x= -5/4

Fifth Meaning: Here Samuccaya means TOTAL i.e Addition & subtraction

Thus, (3x +4)/ (6x +7) = (5x +6)/ (2x +3)

Here N1+N2 = D1 + D2 = 8x + 10 = 0.

Hence x = - 5/4

D1 - D2 = N2 - N1 = 2x + 2 = 0 x = -1

Sixth Meaning: Here Samuccaya means TOTAL; used in Harder equations

Thus, 1/ (x-7) + 1/(x-9) = 1/(x-6) + 1/(x-10)

Vedic Sutra says, (other elements being equal), the sum-total of the denominators on LHS and the total on the RHS are the same, then the total is zero.

Here, D1 + D2 = D3 + D4 = 2x-16 = 0.

Hence x = 8

Examples 1/(x+7) + 1/(x+9) = 1/(x+6) + 1/(x+10) x = - 8

1/(x-7) + 1(x+9) = 1/(x+11) + 1/(x-9) x = - 1

1/(x-8) + 1/(x-9) = 1/(x-5) + 1/(x-12) x = 8-1/2

1/(x-b) - 1/(x-b-d) = 1/(x-c+d) - 1/(x-c) x = 1/2(b+c)

Special Types of seeming Cubics

(x- 3)3 + (x -9)3 = 2(x -6)3

current method is very lengthy, but Vedic method says,

(x-3) + (x-9) = 2x - 12 Hence x = 6

$(x-149)^3 + (x-51)^3 = 2(x-100)^3$

Hence 2x-200 =0 & x = 100

$(x+a+b-c)^3 + (x+b+c-a)^3 = 2(x+b)^3$ x = -b

Sixteen Simple Mathematical Sutras (Phrases) From the Vedas

Sixteen Sutras and Their Corollaries

Sutras or Formulae	*Sub-Sutras or Corollaries*
Ekadhikena Purvena (One more than the previous) P-2 Division eg. 1/19	Anurupyena (Proportionately) P-20, 87 Multiplications
Nikhilam Navatascaram Dashtah (All from 9 last from 10) P-13 Multiplication eg 9927 X 9999	Sishyate Seshasanjnah

Urdhva-tiryaghyam P- (Vertically and crosswise)	Adhyam-adhyena antyam-antyena P-87 (First by the first and last by the last)
Paravartya Yojayet P- (Transpose and apply)	Kevalaih Saptakam Gunyat
Sunyam Samya samuccaye P-107 (when Samuccaya is the same, that Samuccaya is zero)	Veshtanam
(Anurupye) Sunyamanyat	Yavadunam Tavadunam
Sankalana-vyavakalana	Yavadunam Tavadunikrtya Vargamca Yojayet
Purna purnabhyam	Antyayor Dasakepi (also for two numbers whose last digits together total 10)
Calana-Kalanabhyam	Antyayoreva
Yavadunam	Samuccayagunitah
Vyashti samashti	Lopana Sthapanabhyam "by (alternate) Elimination and Retention" very useful in HCF, Solid Geometry, Coordinate Geometry,
Seshanyankena caramena	Vikolanam
Sopantya dvayamantyam	Gunita samuccayah Samucchaya gunitah P-89 *13
Ekanyunena purvena (one less than the previous) P-35 one of the multiplier-digits is all 9	
Gunita Samuccayah P-	
16. Gunaka Samuccayah	

Notes and Quiz

To many Indians Vedic and Sanskrit Mantras are relevant only for religious occasions like weddings, Grahapraveshas and Upanayanams. Some who are aware of Upanishads and Bhagwad Gita, are fascinated by the depth of their philosophical teachings. But Vedas in fact are a store-house of knowledge - both secular and spiritual eg. Ayur Veda is a treatise on medicine and health. Dhanur Veda on war techniques and martial arts, Stapathya Veda on buildings and constructions, Gandharva Veda on Music and art, and Artha Shastra on political systems and economics.

- According to scholars like Max Muller, Vedas are the oldest texts of mankind. They are the treasure house of both secular and spiritual knowledge.
- Schaupenhopher, the German scholar said,

  "The Upanishads are the solace of my life and they shall be the solace of my death too".

  * Upanishads are the Philosophical portions of the Vedas.
- Once a great Vedic scholar Mandana Mishra remarked, "If the Vedas are true nothing shall happen to me" and he jumped from a tall building. To every one's surprise he landed safely but one of his legs was bruised little bit. He questioned the Ved Bhagwan why he was hurt and he got the answer why he was punished.

Can you guess the answer?

- Once a German Indologist remarked "India's contribution in Science in zero"

Do you think he was: (a) right (b) wrong (c) ill-informed (d) biased against India

- If you were to keep one rice grain on the 1st square of the chess board, 2 grains on the 2nd square, 4 grains on the 3rd, 8 on the 4th, 16 on the 5th, 32 on the 6th and so on, ... how many kilograms of rice would you need to fill all the squares in a chess board?

Vedic Sciences

As it is the Rigveda, which pointed to the Cosmic Law first, I feel aggrieved to note that all the ancient wisdom that the vedic seers discovered has been attributed to other nations. The discovery of the Natural Law is credited to the Stoics of Greece who existed circa 300 B.C. whereas the Rigveda was composed at least in the 3500 B.C., if not earlier. This is the price of political weakness, and the Hindus have still not stopped paying for it. Until they learn to be strong, the erosion of their cultural greatness will not only continue, but is likely to accelerate its pace.

'Earliest writing' found

The first known examples of writing may have been unearthed at an archaeological dig in Pakistan.

So-called 'plant-like' and 'trident-shaped' markings have been found on fragments of pottery dating back 5500 years.

They were found at a site called Harappa in the region where the great Harappan or Indus civilisation flourished four and a half thousand years ago.

Harappa was originally a small settlement in 3500 BC but by 2600 BC it had developed into a major urban centre.

The earliest known writing was etched onto jars before and after firing. Experts believe they may have indicated the contents of the jar or be signs associated with a deity.

According to Dr. Richard Meadow of Harvard University, the director of the Harappa Archaeological Research Project, these primitive inscriptions found on pottery may pre-date all other known writing.

Last year it was suggested that the oldest writing might have come from Egypt.

Clay tablets containing primitive words were uncovered in southern Egypt at the tomb of a king named Scorpion.

They were carbon-dated to 3300-3200 BC. This is about the same time, or slightly earlier, to the primitive writing developed by the Sumerians of he Mesopotamian civilisation around 3100 BC.

"It's a big question as to if we can call what we have found true writing," he told BBC News Online, "but we have found symbols that have similarities to what became Indus script.

"One of our research aims is to find more examples of these ancient symbols and follow them as they changed and became a writing system," he added.

One major problem in determining what the symbols mean is that no one understands the Indus language. It was unique and is now dead.

Dr. Meadow points out that nothing similar to the 'Rosetta stone' exists for the Harappan text. The Rosetta stone contained Egyptian hieroglyphics and a Greek translation and thus helped early language researchers decipher the meaning.

The Harappan language died out and did not form the basis of other languages.

"So probably we will never know what the symbols mean," Dr. Meadow told BBC News Online from Harappa.

What historians know of the Harappan civilisation makes them unique. Their society did not like great differences between social classes or the display of wealth by rulers. They did not leave behind large monuments or rich graves.

They appear to be a peaceful people who displayed their art in smaller works of stone.

Their society seems to have petered out. Around 1900 BC Harappa and other urban centres started to decline as people left them to move east to what is now India and the Ganges.

This discovery will add to the debate about the origins of the written word.

It probably suggests that writing developed independently in at least three places - Egypt, Mesopotamia and Harappa between 3500 BC and 3100 BC.

About Multiple Methods of Multiplication

Why look at other ways to multiply when our usual algorithm works just fine? First, this provides alternatives for kids who struggle with the usual algorithm. Second, thinking about how these other algorithms work makes us think more deeply about numbers and operations. Third, it's kind of fun to figure out how people from other times and places using other numeration systems computed. The Egyptian method for multiplication, in particular, makes us appreciate the ease of calculation in the number system which comes to us from the Hindu Vedic tradition.

Ancient Egyptian Multiplication

The Egyptian numeration system was radically different from the one we use today. Both that system and our current system represent numerals with powers of 10. Beyond that there is not much similarity. Our current system has a different symbol to represent each number from 0 through 9, and the placement of the symbol in the numeral determines whether that symbol represents the number of ones, or tens, or hundreds, etc. The Egyptian system had a different symbol for each power of ten, and the number of copies of the symbol indicated how many of that power of ten was in the numeral. This is an awkard numeration system in which to create an easy algorithm for multiplication. Multiplication is just repeated addition of a number. The multiplication algorithm which the Egyptians came up was remarkably efficient use of repeated addition to get around the limitations of a number system without place value. It is probably the first application of binary numbers.

Vedic Mathematics

The history of mathematics in India is ancient. The Hindu Vedic tradition is an oral tradition of knowledge passed down in short verses, dating to before the invention of paper. The Vedas encompass a broad spectrum of knowledge, including the sutras (verses) pertaining to mathematics. In the early 20th century Swami Shri Bharati Krishna Tirthaji Maharaja claimed to have rediscovered a collection of 16 ancient mathematical sutras from the Vedas and published it in a book called Vedic Mathematics. Historians do not agree on whether or not these were truly part of the Vedic tradition. If these sutras date back to the Vedic era they were certainly part of an oral rather than a written tradition.

However, they are a novel and useful approach to computation: they are flexible in application and easy to remember. They can often be applied in algebraic contexts as well as in simple arithmetic.. "Vertically and Crosswise" bridges the gap between arithmetic and algebra: the algorithm is very similar to the standard algorithm used in the US and also is similar to the "FOIL" (first, outer, inner, last) rule used for multiplying binomials in algebra.

The oldest surviving math texts from India include a text treating math from a geometric perspective, the Sulbasutra written by Baudhayana around the 9th century BC, and the Aryabhatiya by Aryabhatta, written around the 5th century AD. If the Vedic Sutras collected by Bharati Krishna have a historical source, they are probably older than either of these texts. They are in any case an interesting alternative approach.

Our number system is derived from the Hindu numeration system which evolved from Vedic mathematics. The concept of zero and the Hindu-Arabic numerals found their way to Europe in the 12th century with Fibonacci, 5 centuries after the work by Hindu mathematician

Brahmagupta. It is the place value system with zero which makes it possible for us to use our familiar multiplication algorithm and spares us from coping with the Egyptian algorithm or the abacus.

Why have a picture of a lamp on this display? The lamp is a symbol for the Hindu festival of Diwali, an Indian festival of lights. One of many interpretations of Diwali is a celebration of the light of knowledge scattering the darkness of ignorance.

Napier's Bones

Napier's main contribution to mathematics, logarithms, arose from his desire to take the tedium out of computation. The difficult computations of his time were multiplication, division, and square and cube roots. Accurate computation was important for applications in navigation--the stars were the global positioning system of Renaissance Europe, and locating a position on the globe required careful computations involving trigonometry. The bones, of course, were an aid to quick and accurate multiplication. Using the bones for multiplication is similar to using the lattice, or Gelosia, method of multiplication.

Materials Preparation for this Module

The materials needed for this are a set of Napier's Bones and handouts for Egyptian multiplication. The Printable material includes a set of paper Napier's Bones which are intended to be cut out and glued onto 3/4 inch craft sticks (extra large popsicle sticks). I like to have three copies of each "bone".

Running the Activity

The multiplication problems are not provided on this website. The person in charge of the display can make up multiplication problems and encourage the visitors to the display to try one of the other methods of working the problem. He or she will of course want to practice doing a few problems using each of these methods before kids show up. If there are difficulties producing problems, the problems given in Notes to the Volunteer can be used.

Mathevedics Frequently Asked Questions

Q: What is Mathevedics?

Answer: Mathevedics is a unique mathematics teaching method. This method has elements of Vedic mathematics and mathematical methods specially developed by well qualified and experienced individuals. The method developed this way provides a unique tool to our students to "fly through" mathematics an alternate way which is extremely fast and a lot of fun! Our method is patent pending and it is illegal to copy either our method or material even in parts. We constantly work on developing and renewing our material. Our purpose is to provide a tool to students to look at Mathematics in a more fundamental and intelligent way whether it is arithmetic, algebra, geometry or calculus.

Q: What is Vedic Mathematics?

Answer: Vedic Mathematics is a form of mathematics developed in ancient times by mathematicians and thinkers from India. These methods are very succinct and if used with necessary explanations provide solutions to complicated mathematical problems in very simplified forms. It has been in use in India on a small scale all along. People trying to get an edge over others in speed of solving problems have been successfully using it in India .

Q: Is Mathevedics Vedic Mathematics?

Answer: Mathevedics uses elements of Vedic Mathematics. It is not completely Vedic mathematics. Our course material has been developed with a lot of effort by very qualified individuals using many original ideas and new methods in addition to Vedic Mathematics.

Q: Does it apply to school mathematics?

Answer: Yes, it applies to School mathematics. We have ensured to bridge the gap between our alternate methods and traditional school methods in such a way that transition appears to be very seamless and natural to our students and it enhances their ability to appreciate mathematics.

Q: Will it confuse children as it applies to the same curriculum as School?

Answer: From our experience, it does not confuse children. We demonstrate a seamless transition between the school method and the alternate fast method to students. This re-enforces what is learnt in the school in addition to be able to look at the school method in a more basic way. As a result the student develops mathematical intelligence.

Q: Does it also cover algebra, geometry, trigonometry and calculus?

Answer: Yes, it very much covers algebra, geometry, trigonometry and calculus. Some people not totally familiar with Vedic techniques may think that Vedic techniques applied only to arithmetic. It is not true. Vedic Methods are originally in the form of verses which provide much generalized solutions to problems. Arithmetic is only an easy implementation and verification of Vedic methods. Mathevedics methods use elements of Vedic methods in many cases and are not limited by any limitation of Vedic Mathematics.

Q: How is Mathevedics different from Vedic Math?

Answer: Vedic Mathematics provides generalized solutions to many mathematical problems. Without proper interpretation of the generalized solutions as stated in Vedic Mathematics, it may appear to be very limited in scope and in most cases useless as many people have been thinking.

Mathevedics methods use elements of Vedic Mathematics which other wise seem not very useful to many people. In addition to Vedic elements our method uses modern mathematics and mostly specially developed mathematical methods. Mathevedics combines these in such a way that the gap between these are bridged. Most of our method is developed from fundamentals by our team of experts. It provides a broader outlook of mathematics and is intended to ingrain the fundamentals of mathematics into the students' minds and provide the student a source of mathematical intelligence.

Q: How different is this method from Kumon, Sylvan Learning, Score, etc.?

Answer: We teach only Mathematics. We do not focus on learning by repetition. We focus on improving intelligence by teaching fundamentals and alternate methods. Our purpose is not limited to improving performance in the school or tests (which of course is one of our objectives), but we focus on providing a broader outlook resulting in improved mathematical intelligence. We provide one on one attention to our students. No other institute or learning centre does that.

Q: Some schools allow students to use calculator in the class? What is the need of Mathevedics?

Answer: Yes, it is true that some schools allow calculator in the class or even tests. One of the objectives of Mathevedics is providing a method to do complicated arithmetic calculations fast and in many cases in minds. Therefore we do provide a tool to reduce dependence on calculators.

However our main objective is to make the students mathematically intelligent by providing alternate methods and giving a broader perspective of standard problems. In addition to making the student intelligent, this boosts their confidence. The methods apply to Algebra, geometry and calculus as well.

Therefore in order to go that extra mile in Mathematics the students will still need us even though they have calculators.

Calculators are not permitted in SAT etc. SAT in many cases is a career determining test and it is important to perform well there.

Q: Some one is selling Vedic Mathematics material for less than $200 in India without restrictions. Why would anyone try to get a Mathevedics Franchise if he can do the same thing for less money?

Answer: We are aware of Vedic Mathematics teaching resources in India and also at other places in the world. We only use elements of Vedic Mathematics. Most of our material is developed from fundamentals by a team of experts independent of Vedic Mathematics to provide to the student a source of Mathematical intelligence. We are confident that no one else can provide to the student what we provide and the objective of no Vedic Mathematics teaching centre or resource is to improve mathematical intelligence. Our method is patent pending and our material is copyrighted. Mathevedics will take legal action against anyone trying to steal the concept or the teaching material even in part.

Bibliography

A.L. Basham: *The Wonder That Was India*, Rupa & Co., Calcutta, 1967.

Aaboe, Asger: *Episodes from the Early History of Mathematics*, New York, 1964.

Alfred, C.: *Introduction to Prakrit*, Calcutta, Baptist Mission Press, 1917.

Amma, S: *Geometry in Ancient and Medieval India*, Motilal Banarsidas, 1979.

Archibald, R.C.: *The Rhind Mathematical Papyrus,* Oberlin, Ohio, 1927.

Armstrong, G.: *Yoruba Numerals*, London, Oxford University Press, 1962.

Arnett, S.: *The Predynastic Origin of Egyptian Hieroglyphs*, Washington, University Press of America, 1982.

Ascher, Marcia: *Ethnomathematics*, Pacific Grove, CA: Brooks/Cole, 1991.

Bag, A. K.: *Mathematics in Ancient and Medieval India*, Varanasi: Chaukhambha Orientalia, 1979.

Barnard, F. P.: *The Casting-Counter and the Counting Board. A Chapter in the History of Numismatics and Early Arithmetic*, Oxford, Oxford University Press, 1916.

Barrow, D.: *Pi in the Sky*, Oxford, Clarendon Press, 1992.

Basham, A. L.: *The Wonder That Was India*, Calcutta, Rupa & Co., 1967.

Beckmann, Petr: *A History of Pi*, New York, St. Martin's Press, 1974.

Bedient, J. D.: *The Historical Roots of Elementary Mathematics*, New York, Dover, 1988.

Bell, Eric T.: *The Magic of Numbers*, New York, McGraw-Hill, 1946.

Berlin, Brent: *Tzeltal Numeral Classifiers*, The Hague, Mouton, 1968.

Bharati Krishna Tirthaji Majaraja.: *Vedic Mathematics*. Delhi, Motilal Banarsidass, 1988.

Bidyaranya, Swami: *History of Hindu Mathematics*,Bombay, Asia Pub. House, 1962.

Bischoff, Bernard: *Latin Palaeography*, Cambridge, Cambridge University Press, 1990.

Bonfante, Larissa: *Etruscan*. Berkeley, University of California Press, 1990.

Brainerd, Charles J.: *The Origins of the Number Concept*, New York, Praeger, 1979.

Brunschwig, Jacques : *Greek Thought: A Guide to Classical Knowledge*, Cambridge, Harvard University Press, 2000.

Bunt, L. N. H.: *The Historical Roots of Elementary Mathematics*, New York, Dover, 1988.

Charles, P.: *The Numeration, Calendar Systems and Astronomical knowledge of the Mayas*, Cambridge, Cambridge University Press, 1910.

Charles, P.: *The Numeration, Calendar Systems and Astronomical knowledge of the Mayas*, Cambridge, Cambridge University Press, 1910.

Chattopadhyaya, D. P.: *Mathematics, Astronomy, and Biology in Indian Tradition: Some Conceptual Preliminaries*, New Delhi, Project of History of Indian Science, Philosophy, and Culture, 1995.

Clawson, Calvin: *The Mathematical Traveller*, New York, Plenum Press, 1994.

Closs, P.: *Native American Mathematics*, Austin, University of Texas Press, 1986.

Conant, L.: *The Number Concept*, New York, MacMillan, 1896.

Crossley, N: *The Emergence of Number*, Singapore, World Scientific, 1987.

Crump, Thomas: *The Anthropology of Numbers*, Cambridge, Cambridge University Press, 1990.

Dantzig, Tobias: *Number, the Language of Science*, London, George Allen and Unwin, 1940.

David, F.: *Cross-Cultural Studies in Cognition and Mathematics*, New York, Academic, 1983.

De Villiers, Melius: *The Numeral-Words: Their Origin, Meaning, History and Lesson*, London, Witherby, 1923.

Dehaene, Stanislas: *The Number Sense*, New York, Oxford University Press, 1997.

Dilke, Wentworth: *Mathematics and Measurement*, Berkeley, CA: U. of Cal. Press, 1987.

Dirk, J.: *A Concise History of Mathematics*, New York, Dover, 1987.

Eric, S.: *Maya Hieroglyphic Writing: An Introduction*, Norman, University of Oklahoma Press, 1971.

Eves, Howard: *An Introduction to the History of Mathematics*, New York, Rinehart and Company Inc., 1953.

Flegg, Graham: *Numbers: Their History and Meaning*, London, André Deutsch, 1983.

Fowler, David: *The Mathematics of Plato's Academy*, Oxford, Clarendon, 1999.

Francis, Llewellyn: *Hieratic Papyri from Kahun and Gurob*, London, Bernard Quaritch, 1898.

Fuson, K.C.: *Children's Counting and Concepts of Number*, New York, Springer-Verlag, 1988.

Ganeri, Anita: *The Story of Numbers and Counting*, New York, Oxford University Press, 1996.

Geoffrey E.R. Lloyd:. *Greek Thought: A Guide to Classical Knowledge*, Cambridge, Harvard University Press, 2000.

Gokhale, Laxman: *Indian Numerals*, Poona, Deccan College, 1966.

Griffith, Llewellyn: *Hieratic Papyri from Kahun and Gurob*, London, Bernard Quaritch, 1898.

Grover, Sudarshan: *History of Development of Mathematics in India*, Delhi, Atma Ram & Sons, 1994.

Haldane, Sanderson: *Science and Indian Culture*, Calcutta, New Age Publishers, 1965.

Halliwell, Orchard: *Rara Mathematica*, London, J.W. Parker, 1839.

Herbert Meschkowski: *Ways of Thought of Great Mathematicians*, Holden-Day Inc., San Francisco, 1964.

Hogben, Lancelot: *Mathematics for the Million*, New York, W. W. Norton, 1937.

Howard Eves: *An Introduction to the History of Mathematics*, Rinehart and Company Inc., New York, 1953.

Ifrah, Georges: *From One to Zero: A Universal History of Numbers*, New York, Viking Penguin, 1985.

Jagadguru Swami Shri Bharati Krishna Tirthaji Maharaja: *Vedic Mathematics*, Motilal Banarsidass, Delhi, 1988.

Jaggi, O. P.: *Science and Technology in Medieval India*, New Delhi, Atma Ram & Sons, 1977.

James, Orchard: *Rara Mathematica*, London, J.W. Parker, 1839.

John, Burdon Sanderson: *Science and Indian Culture*, Calcutta, New Age Publishers, 1965.

John, D.: *Pi in the Sky*, Oxford, Clarendon Press, 1992.

John, N: *The Emergence of Number*, Singapore, World Scientific, 1987.

Jones, D.: *The Internal Representation of Number: Analogue or Digital?*, Oxford, Clarendon, 1987.

Kline, Morris: *Mathematical Thought from Ancient to Modern Times*, New York, Oxford U. Press, 1990.

Kuppuram, G & K. Kumudamani: *History of Science and Technology in India*, Delhi, Sundeep Prakashan, 1990.

Lancy, F.: *Cross-Cultural Studies in Cognition and Mathematics*, New York, Academic, 1983.

Levi, L.: *The Number Concept*, New York, MacMillan, 1896.

Manning, H.P. and R C Archibald: *The Rhind Mathematical Papyrus,* Oberlin, Ohio, 1927.

Marcus, Joyce: *Mesoamerican Writing Systems*, Princeton, Princeton University Press, 1992.

Marshack, Alexander: *The Roots of Civilization*, New York, McGraw-Hill, 1972.

Martzloff, Jean-Claude: *A History of Chinese Mathematics*, New York, Springer, 1997.

McLeish, John: *The Story of Numbers*, New York, Fawcett Columbine, 1991.

Menninger, Karl: *Number-Words and Number-Symbols*. London, MIT, 1969.

Michael, P.: *Native American Mathematics*, Austin, University of Texas Press, 1986.

Murthy, T. S. Bhanu: *A Modern Introduction to Ancient Indian Mathematics*, Wiley Eastern Ltd., New Delhi, 1992.

Neugebauer, Otto : *Mathematical Cuneiform Texts*, New Haven, Yale University Press, 1945.

Neugebauer, Otto: *The Exact Sciences in Antiquity*, Princeton, Princeton University Press, 1952.

North, John: *The Norton History of Astronomy and Cosmology*, New York, W.W.Norton, 1995.

Oswald Ashton Wentworth: *Mathematics and Measurement*, Berkeley, CA: U. of Cal. Press, 1987.

Pandit, M. D.: *Mathematics as known to the Vedic Samhitas*, Delhi, Sri Satguru Publications, 1993.

Pappas, Theoni: *The Magic of Mathematics*, San Carlos, Wide World Publishing / Tetra, 1994.

Parpola, Asko: *Deciphering the Indus Script*, Cambridge, Cambridge University Press, 1994.

Peet, T. Eric: *The Rhind Mathematical Papyrus*, London, British Museum, 1923.

Piaget, Jean: *The Child's Conception of Number*, New York, W.W. Norton, 1965.

Pullan, J.M: *The History of the Abacus*, New York, Praeger, 1968.

Rao, S. Balaachandra: *Indian Mathematics and Astronomy*, Bangalore, Jnana Deep Publications, 1994.

Restivo, Sal: *The Social Relations of Physics, Mysticism and Mathematics*, Dordrecht, Kluwer, 1983.

Robert, G.: *Yoruba Numerals*, London, Oxford University Press, 1962.

Robins, G. and C. Shute: *The Rhind Mathematical Papyrus*, London, British Museum, 1987.

Russell, Bertrand: *The Principles of Mathematics*, Cambridge, Cambridge University Press, 1903.

Sachs, A.: *Mathematical Cuneiform Texts*, New Haven, Yale University Press, 1945.

Salomon, Richard: *Indian Epigraphy*, New York, Oxford University Press, 1998.

Saraswarti, A.: *Geometry in Ancient and Medieval India*, Motilal Banarsidas, 1979.

Saraswati Amma: *Geometry in Ancient and Medieval India*, Motilal Banarsidas, 1979.

Sarton, George: *The Study of the History of Mathematics*, Cambridge, Cambridge University Press, 1936.

Satterthwaite, Jr.: *Concepts and Structures of Maya Calendrical Arithmetics*, Philadelphia, University of Pennsylvania, 1947.

Satterthwaite, Jr.: *Concepts and Structures of Maya Calendrical Arithmetics*, Philadelphia, University of Pennsylvania, 1947.

Schaaf, W.L.: *Mathematics, Our Great Heritage*, New York, Harper and Row, 1948.

Schimmel, Annemarie: *The Mystery of Numbers*, New York, Oxford University Press, 1993.

Scriba, C.J.: *The Concept of Number*, Zurich, Mannheim, 1968.

Seidenberg, A.: *The Zero in the Mayan Numerical Notation*, Austin, University of Texas Press, 1986.

Shebhana, Laxman: *Indian Numerals*, Poona, Deccan College, 1966.

Smeltzer, Donald: *Man and Number*, New York, Emerson, 1959.

Smith, D. E.: *History of Mathematics*, New York, Dover Publications Inc., 1951.

Srinivasiengar, C. N.: *The History of Ancient Indian Mathematics*. Calcutta, World Press, 1967.

Struik, J.: *A Concise History of Mathematics*, New York, Dover, 1987.

Thompson, Sir J. : *Maya Hieroglyphic Writing: An Introduction*, Norman, University of Oklahoma Press, 1971.

Todd, R.R.: *The Internal Representation of Number: Analogue or Digital?*, Oxford, Clarendon, 1987.

Urton, Gary: *The Social Life of Numbers*, Austin, University of Texas Press, 1997.

Volodarsky, A.I.: *Mathematics in Ancient India*, Tokyo, Science Council of Japan, 1974.

William, S.: *The Predynastic Origin of Egyptian Hieroglyphs*, Washington, University Press of America, 1982.

Woolner, C.: *Introduction to Prakrit*, Calcutta, Baptist Mission Press, 1917.

Index

A

B

C

D

J

K

L

M

T

U

V

❑❑❑

ENCYCLOPAEDIA OF VEDIC MATHEMATICS

ENCYCLOPAEDIA OF VEDIC MATHEMATICS

(Vol. 3)

Shyam Priya

ANMOL PUBLICATIONS PVT. LTD.
NEW DELHI - 110 002 (INDIA)

ANMOL PUBLICATIONS PVT. LTD.

H.O.: 4374/4B, Ansari Road, Darya Ganj,
New Delhi-110 002 (India)
Ph.: 23278000, 23261597

B.O.: No. 1015, Ist Main Road, BSK IIIrd Stage
IIIrd Phase, IIIrd Block,
Bangalore - 560 085 (India)

Visit us at: www.anmolpublications.com

Encyclopaedia of Vedic Mathematics

First Published, 2008

ISBN 978-81-261-3504-2 (Set)

PRINTED IN INDIA

Printed at Mehra Offset Press, Delhi.

Contents

Preface

Vedic Mathematics contains many examples of striking methods of calculation and there is a remarkable coherence to the system which makes it very attractive. Also the Vedic system itself suggests a kind of approach that involves going directly to the answer. Vedic mathematics is based on sixteen Sutras and some sub-Sutras which provide links through mathematics: the word 'Sutra' means 'thread'.

Vedic scheme is not simply a collection of rapid methods; it is a system, a unified approach, which can be swiftly learnt. Vedic Mathematics extensively exploits the properties of numbers in every practical applications, particularly in the field of computation. It makes available a whole range of methods ideally suited to these properties. After a detailed introduction to Vedic Mathematics, this book shows applications of Vertically and Crosswise from basic calculations to the evaluation of logarithms, exponentials, trigonometrical functions and the solution of simultaneous, transcendental, polynomial and differential equations. The methods shown are not only exceedingly efficient and new but have a unity and coherence which renders them easy to master and enjoyable to execute.

The methods given in this book are not intended to be a complete or thorough treatment of the topics they deal with. All of the ideas can probably be developed further or applied in other areas, and all can doubtless be improved upon. The mathematician will also observe a certain lack of rigor as an attempt has been made to make the material intelligible to as wide a readership as possible.

—***Shyam Priya***

1

Vedic Mathematics Re-discovered

Vedic Mathematics is the name given to the ancient system of mathematics which was rediscovered between 1911 and 1918 by Sri Bharati Krishna Tirthali (1884-1960).

According to his research, all of mathematics is based on 16 sutras or word formulas. For example 'Vertically and Crosswise' is one of these sutras. These formulas are compatible with the natural workings of the mind and are therefore a great help in directing the student to the appropriate method of solution.

Perhaps the most striking feature of the Vedic system is it's coherence. Instead of a hotchpotch of unrelated techniques. the whole system is beautifully interrelated and unified. The general multiplication method for example, is easily reversed to allow one- line divisions, and the simple squaring method can be reversed to give one-line square roots... and these are all easily understood. This unifying quality is very satisfying... it makes mathematics easy and enjoyable while encouraging innovation.

Difficult problems or huge sums can often be mentally solved immediately by the Vedic method. These striking and beautiful methods are just a part of a complete system of mathematics which is far more systematic than the modern "system'. Vedic mathematics manifests the coherent and unified structure of mathematics, the methods are complementary, direct and easy.

The simplicity of Vedic mathematics means that calculations can be carried out mentally (although the methods can also be written down). There are many advantages in using a flexible mental system. Pupils can invent their own methods; they are not limited to the one "correct' method. This leads to more creative, interested and intelligent students. Interest in the Vedic system is growing, mathematics teachers everywhere are looking for something better and are finding that the Vedic system really works.

Research is being carried out in many areas, including the effects on children who learn Vedic maths and the development of new, powerful but easy applications of the Vedic sutras in geometry, calculus, computing etc. But the real beauty and effectiveness of Vedic mathematics cannot be fully appreciated without actually practising the system. One can then see that it is perhaps the most refined and efficient mathematical system possible.

What is Vedic Mathematics?

The term Vedic Mathematics refers to the unique technique of Calculations based on a set of 16 sutras or aphorisms or formulae and their upa-sutras or corollaries derived from these sutras. Any mathematical problems (algebra, arithmetic, geometry or trigonometry) can be solved mentally with these sutras. Vedic Mathematics is more coherent than modern mathematics.

The sixteen sutras and their corollaries are as follows:

1. Ekadhikina Purvena -By one more than the previous one (Cor: Anurupyena)
2. Nikhilam Navatashcaramam Dashatah -All from 9 and the last from 10 (Cor: Sisyate Sesasamjnah)

3. Urdhva-Tiryagbyham-Vertically and crosswise (Cor: Adyamadyenantyamantyena)
4. Paraavartya Yojayet-Transpose and adjust (Cor: Kevalaih Saptakam Gunyat)
5. Shunyam Saamyasamuccaye-When the sum is the same, that sum is zero. (Cor: Vestanam)
6. (Anurupye) Shunyamanyat-If one is in ratio, the other is zero (Cor: Yavadunam Tavadunam)
7. Sankalana-vyavakalanabhyam-By addition and by subtraction (Cor:Yavadunam Tavadunikritya Varga Yojayet)
8. Puranapuranabyham-By the completion or non-completion (Cor: Antyayordashake)
9. Chalana-Kalanabyham-Differences and Similarities (Cor: Antyayoreva)
10. Yaavadunam-Whatever the extent of its deficiency (Cor: Samuccayagunitah)
11. Vyashtisamanstih-Part and Whole (Cor: Lopanasthapanabhyam)
12. Shesanyankena Charamena-The remainders by the last digit (Cor: Vilokanam)
13. Sopaantyadvayamantyam-The ultimate and twice the penultimate (Cor: Gunitasamuccayah Samuccayagunitah)
14. Ekanyunena Purvena-By one less than the previous one (Cor: Dhvajanka)
15. Gunitasamuchyah-The product of the sum is equal to the sum of the product (Cor: Dwandwa Yoga)
16. Gunakasamuchyah-The factors of the sum is equal to the sum of the factors

EXAMPLE 1

Ekadhikena Purvena or "By one more than the previous one".

1. Square of numbers ending in 5

65 × 65 = (6 × (6+1)) 25 = (6x7) 25 = 4225

45 × 45 = (4 × (4+1)) 25 = (4x5) 25 = 2025

105 × 105 = (10 × (10+1) 25 = (10 × 11) 25 = 11025

2. When sum of the last digits is the base(10) and previous parts are the same

44 × 46 = (4 × (4+1)) (4 × 6) = (4 × 5) (4 × 6) = 2024

37 × 33 = (3 × (3+1)) (7 × 3) = (3 × 4) (7 × 3) = 1221

11 × 19 = (1 × (1+1)) (1 × 9) = (1 × 2) (1 × 9) = 209

3. 1 divided by 19, 29, 39,............

Consider 1/19 since 19 is not divisible by 2 or 5 it is a purely a recurring decimal take last digit 1

multiply this with 1+1 (one more) i.e. 2 (this is the key digit) ⇒ 21

multiply 2 by 2 ⇒ 421 multiplying 4 by 2 ⇒ 8421
multiply 8 by 2 ⇒ 68421 carry 1

multiply 6 by 2 = 12 + carry 1= 13 ⇒ 368421 carry 1 continuing (till 18 digits = denominator-numerator) the result is 0.052631578947368421

4. 1/19 using divisions

divide 1 by 2, answer is 0 with remainder 1 ⇒ .0

next 10 divided by 2 is 5 ⇒ .05

next 5 divided by 2 is 2 remainder 1 ⇒ 0.052

next 12 (remainder 2) divided by 2 is 6 ⇒ 0.0526

next 6 divided by 2 is 3 ⇒ 0.05263

next 3 divided by 2 is 1 remainder 1 ⇒ 0.052631

next 11 divided by 2 is 5 remainder 1 ⇒ 0.0526315

and so on...

5. 1/7 = 7/49 previous digit is 4 so multiply by 4+1 i.e. by 5

7 → 57 → 857 → 42857 → 0.142857 (stop after 7-1= 6 digits)

EXAMPLE 2

Nikhilam Navatashcaramam Dashatah or "All from nine and the last from ten"

1. 1000-457

(9-4)(9-5)(10-7) = 543

2. 1000-54 = 1000-054 = (9-0)(9-5)(10-4) = 946

Corollary: Yavdunam Jaavdunikritya Varga Yojayet or "Whatever the extent of its deficiency, lessen it still further to that very extend; and also setup the square of the deficiency."

3. Computing square of 8

power of 10 nearest to 8 is 10, so take 10 as base

8 is 2 less than 10 so 8-2 or 6 is the left side of the answer

right hand side of the answer is square of 2(deficiency) i.e., 4

so the answer is 64

similarly square of 7 is (7-3) (3x3) = 49

** if the number is greater than 10 take the surplus

square of 12 is (12+2) (2x2) = 144

square of 13 is (13+3) (3x3) = 169

EXAMPLE 3

Urdhva-tiryagbhyam or "vertically and crosswise"

1. 31 × 12

Multiply vertically on the left: 3 × 1 = 3 (first figure)

Multiply crosswise and add: (3 × 2) + (1 × 1) = 7 (middle figure)

Multiply vertically on the right 2 × 1 = 2 (last figure)

So the answer is 372

2. 275 × 513

Multiply vertically 5 × 3 = 15 write 5 and carry 1 □⇒ 5

Multiply and add crosswise last two digits

(3x7)+(1x5) = 26 + carry 1 = 27

write 7 and carry 2 ⇒ 75

Multiply and add vertically and cross wise all digits

(2x3) + (5x5) + (7x1)=38+ carry 2 = 40 write 0 and carry 4 ⇒ 075

Multiply and add crosswise first two digits

(1 x 2) + (5 x 7) = 37+ carry 4 = 41

write 1 and carry 4 ⇒ 1075

Multiply vertically 2 x 5 = 10 + carry 4 ⇒ 14

write 141075

So 275x513 = 141075

3. Divide ($12X^2$ -8X-32) by (X-2) using Urdhva-Tiryak

The quotient can be written as (12X+k)

We also know that -8x = kx - 24x Hence k = 16 cross-check -2k = -32 third term

Some Rules of Thumb

1. Multiplying a number by 11

To multiply any two digit number by 11 we just put the total of the two figures between the 2 digits.

36 × 11 = 3 (3+6) 6 = 396

74 × 11 = 7 (7+4) 4 = 7+carry 1 (1) 4 = 814

234 × 11 = 2 (2+3) (3+4) 4 = 2574

2. Diving by 9.

23/9 = 2 remainder 5

The first digit of 23 i.e., 2 is the answer and remainder is sum of 2 and 3!

134/9 = 14 remainder 8

Answer is (1st digit of 134)(1+3) remainder (1+3+4)

EXAMPLE 4

Paraavartya Yojayet or "Transpose and apply"

Divide ($12x^2$ - 8x - 32) by (x - 2), factor is +2

(X - 2)/2 $12x^2$ - 8x -32

+24x + 32

12x + 16 R = 0

Divide $7x^2$ + 5x + 3 by × - 1 Q = 7x + 12 R = 15

Divide $7x^2$ + 5x + 3 by x + 1 Q = 7x -2 R = 5

Divide x^3 + $7x^2$ + 6x + 5 by x - 2 Q = x^2 + 9x + 24 R = 53

Divide x^4 - x^3 + x^2 + 3x +5 by x^2 - x - 1

Factors are x + 1

Q = x^2 + 0x + 2 R = 5x + 7

Divide $6x^4$ + 13 x^3 + 39 x^2 + 37x + 45 by X2 - 2x - 9

Factors 2x + 9 Q = $6x^2$ + 25x + 143 R = 548x + 1332

Divide x^4 + x^2 + 1 by x^2 - x +1 (add 0 x 3 & 0x)

X^4 + $0x^3$ + x^2 + 0x + 1

Factors are: x - 1

Q = x^2 + x + 1 R = 0

Divide 1 2 3 4 by 1 1 2 Factors -1 -2 Q = 11 R = 2

Divide 1 3 4 5 6 by 1 1 2 3 Q = 12 R = - 20

The Reminder cannot be negative.

Hence Q = 11 R = 1 1 0 3

Divide 1 3 9 0 5 by 1 1 3 Q = 1 24 R = -107

Q = 123 R = 6

The following examples will help you to understand how to apply the above sutra to get partial equations in a simple and short manner.

1. $\dfrac{x^2+5x+3}{(x+2)(x-7)(x-4)}$ Eq. (1)

This can be written in the form

$$\frac{A}{(x+2)}+\frac{B}{(x-7)}+\frac{C}{(x-4)} \quad \text{Eq. (2)}$$

Here A , B , C are constants to be determined.

Now equate each term of the denominator to zero to get three different values of x. Now leaving out the term containing the particular value of x, substitute for x in all the other terms of the given equation to obtain respective constant for the term in the second equation.

i.e. Equating (x + 2) to zero we obtain x + 2 = 0

$\Rightarrow x = -2$

Now substituting for x in the first equation by (-2) we obtain the constant A.

$$\text{Therefore A} = \frac{(-2)^2+5(-2)+3}{(-2-7)(-2-4)} = \frac{-11}{54}$$

$$\text{Similarly B} = \frac{(7)^2+5(7)+3}{(7+2)(7-4)} = \frac{52}{27}$$

Similarly $C = \frac{(4)^2 + 5(4) + 3}{(4-7)(4+2)} = \frac{31}{-18}$

The above steps can be put in the form of a few simple formulae as follows :

If the equation is of the form : $\frac{Lx^2 + Mx + N}{(x-a)(x-b)(x-c)}$

Then the partial fractions can be written as

$$\frac{A}{(x-a)} + \frac{B}{(x-b)} + \frac{C}{(x-c)}$$

where the constants A, B, C are given as

$$A = \frac{La^2 + Ma + N}{(a-b)(a-c)}$$

$$B = \frac{Lb^2 + Mb + N}{(b-a)(b-c)}$$

$$C = \frac{Lc^2 + Mc + N}{(c-a)(c-b)}$$

2. $\frac{8x^2 + 9x + 11}{(x-12)(x+13)(x+14)}$

This can be written in the form $\frac{A}{(x-12)} + \frac{B}{(x+13)} + \frac{C}{(x+14)}$

Using the above explained formulas we get

A = 1271/650

B = -1246/25

C = 1453/26

EXAMPLE 5

Shunyam Saamyasamuccaye or "When the samuccaya is the same, that samuccaya is zero".

This sutra is useful in solution of several special types of equations that can be solved visually. The word samuccaya has various meanings in different applicatins.

1. It is a term which occurs as a common factor in all the terms concerned

Thus 12x + 3x = 4x + 5x x is common, hence x = 0

Or 9 (x+1) = 7 (x+1) here (x+1) is common; hence x +1= 0

2. Here Samuccaya means "the product of the independent terms"

Thus, (x + 7) (x + 9) = (x + 3) (x + 21)

Here 7 × 9 = 3 × 21. Therefore x = 0

3. Samuccaya thirdly means the sum of the Denominators of two fractions having the same numerical numerator

Thus, 1/(2x -1) + 1/(3x -1) = 0

Hence 5x - 2 = 0 or x = 2/5

4. Here Samuccaya means combination (or TOTAL).

If the sum of the Numerators and the sum of the Denominators be the same, then that sum = 0

(2x +9)/(2x +7) = (2x +7)/(2x +9)

N1 + N2 = D1 + D2 = 2x + 9 + 2x + 7 = 0

Hence 4x + 16 = 0 hence x = -4

Note: If there is a numerical factor in the algebraic sum, that factor should be removed.

(3x +4)/(6x +7) = (x +1)/(2x +3)

Here N1 +N2 = 4x +5; D1 +D2

= 8x + 10; 4x +5 =0 x= -5/4

5. Here Samuccaya means TOTAL i.e. Addition & subtraction

Thus, (3x +4)/(6x +7) = (5x +6)/(2x +3)

Here N1+N2 = D1 + D2 = 8x + 10 =0 hence x = - 5/4

D1 - D2 = N2 - N1 = 2x + 2 = 0 x = -1

6. Here Samuccaya means TOTAL; used in Harder equations

Thus, 1/(x-7) + 1/(x-9) = 1/(x-6) + 1/(x-10)

Vedic Sutra says, (other elements being equal), the sum-total of the denominators on LHS and the total on the RHS are the same, then the total is zero.

Here, D1 + D2 = D3 + D4 = 2x-16 =0 hence x = 8

Examples 1/(x+7) + 1/(x+9) = 1/(x+6) + 1/(x+10) x = - 8

1/(x-7) + 1(x+9) = 1/(x+11) + 1/(x-9) x = - 1

1/(x-8) + 1/(x-9) = 1/(x-5) + 1/(x-12) x = 8-1/2

1/(x-b) - 1/(x-b-d) = 1/(x-c+d) - 1/(x-c) x = 1/2(b+c)

Special Types of seeming Cubics

(x- 3)3 + (x -9)3 = 2(x -6)3

current method is very lengthy, but Vedic method says,

(x-3) + (x-9) = 2x - 12

Hence x = 6

(x-149)3 + (x-51)3 = 2(x-100)3

Hence 2x-200 = 0 & x = 100

$(x+a+b-c)^3 + (x+b+c-a)^3 = 2(x+b)^3$ x = -b

EXAMPLE 6

(Anurupye) Shunyamanyat or "If one is in ratio, the other one is zero".

This sutra is often used to solve simultaneous simple equations which may involve big numbers. But these equations in special cases can be visually solved because of a certain ratio between the coefficients. Consider the following example:

$6x + 7y = 8$

$19x + 14y = 16$

Here the ratio of coefficients of y is same as that of the constant terms.

Therefore, the "other" is zero, i.e., x = 0. Hence the solution of the equations is x = 0 and y = 8/7.

This sutra is easily applicable to more general cases with any number of variables. For instance

$ax + by + cz = a$

$bx + cy + az = b$

$cx + ay + bz = c$

which yields x = 1, y = 0, z = 0.

A corollary (upsutra) of this sutra says Sankalana-Vyavakalanaabhyam or By addition and by subtraction. It is applicable in case of simultaneous linear equations where the x- and y-coefficients are interchanged. For instance:

$45x - 23y = 113$

$23x - 45y = 91$

By addition: $68x - 68y = 204$

$\Rightarrow 68(x-y) = 204$

$\Rightarrow x - y = 3$

By subtraction: $22x + 22y = 22$

$\Rightarrow 22(x+y) = 22$

$\Rightarrow x + y = 1$

EXAMPLE 7

Yaavadunam-"Whatever the extent of its deficiency"

1. Compute 13^3

Step 1 : Consider nearest base (here 10).

Step 2 : As 13 has a excess of '3' (13 - 10 = 3), we double the excess and add the original number (13) to it, and put it on the LHS. Therefore we get 13 + 6 = 19

Step 3 : Now find the new excess. In this case it is 19-10 = 9. Now multiply this with the original excess to get the middle part of the answer. Therefore we get 9 * 3 = 27

Step 4 : Now cube the original excess and put it as the last part.

Carry over any big numbers and total to get the answer.

$$\begin{array}{rrr} 19 & 7 & 7 \\ 2 & 2 & \\ \hline 21 & 9 & 7 \end{array}$$

Therefore 13^3 = 2197

2. 47^3

As in 'Nikhilam' and Squaring, we use 'Aanurupyena' here.

1) Let the main base be 10 and the working base be 50 therefore the ratio

   x = (Main Base)/(Working Base) = 10/50 = 1/5

2) Excess is -3 (47 - 50 = -3). Double the excess and add the original number (here 47) to it.

   We get 47 - 6 = 41.

   The Base correction for this part is achieved by dividing by x^2.

   therefore we get 41/(1/25) = 41 * 25 = 1025

3) Excess in the new uncorrected number (41 - 50 = -9) is multiplied by the original excess (-3) to obtain the second part.

Therefore we get -9 * -3 = 27

The Base correction for this part is achieved by dividing by x.

therefore we get 27 * 5 = 135

4) The third part is obtained by cubing the excess.

$(-3)^3 = -27$

5) Carry over the extra numbers and total to obtain the final answer

1025	0	0
13	5	0
	-2	7
1038	2	3

Therefore the final answer is 103823

Sutra : Yaavadunam Taavaduunikruthya vargam cha yogayet

Meaning : "Whatever the extent of its deficiency, lessen it further to that very extent; and also set up the square of that deficiency".

This sutra is a corollary of the Nikhilam sutra.

1. Consider a simple example 92

Step 1 : Consider the nearest base (here 10).

Step 2 : As 9 has a deficiency of 1 (10 - 9 = 1), we should decrease it further by 1, and set down our LHS of the Answer as '8'.

Step 3 : On the RHS put the square of the deficiency (here 1).

we get 9^2 = 81.

2. Consider 102

1) Base is 100

2) Deficiency is '-2' (100 - 102 = -2)

   Therefore we subtract '-2' from 102

   102 - (-2) = 104

   This is our RHS

3) Our LHS now becomes (-2)2 which is 4

Since the base is 100 we write it as '04', so that we get 1022 = 10404

If we have multiples or sub multiples of a base, we employ the same technique as in 'Aanurupyena'. (See Nikhilam Multiplication)

3. Consider 28^2

1) Let 20 be the Working Base and 10 as the Main Base.

   Therefore x = (Main Base)/(Working Base) = 10/20 = 1/2

2) Here the deficiency = 20 - 28 = -8

   Therefore RHS = 28 - (-8) = 36

   Divide by x i.e. by (1/2).

   We get 36/(1/2) = 72. This is the required RHS.

3) LHS = $(-8)^2 = 64$

Since Main Base is 10, we put only '4' on the LHS and carry over '6' to the RHS

Therefore we get

282 = 72+6 | 4 == 784

Compute: 8 × 7

8 is 2 below 10 and 7 is 3 below 10.

You subtract crosswise 8-3 or 7 - 2 to get 5, the first figure of the answer.

And you multiply vertically: 2 x 3 to get 6, the last figure of the answer

The answer is 56.

Multiply 88 by 98

Both 88 and 98 are close to 100.

88 is 12 below 100 and 98 is 2 below 100

As before the 86 comes from subtracting crosswise: 88 - 2 = 86 (or 98 - 12 = 86: you can subtract either way, you will always get the same answer).

And the 24 in the answer is just 12 x 2: you multiply vertically.

So 88 × 98 = 8624

Multiply 103 × 104 = 10712

The answer is in two parts: 107 and 12, 107 is just 103 + 4 (or 104 + 3), and 12 is just 3 × 4.

EXAMPLE 8

Vyashtisamanstih- "Part and Whole"

Corollary : Lopanasthapanabhyam

It is very difficult to factorise the long quadratic $(2x^2 + 6y^2 + 3z^2 + 7xy + 11yz + 7zx)$

But "Lopana-Sthapana" removes the difficulty. Eliminate z by putting z = 0.

Hence the given expression $E = 2x^2 + 6y^2 + 7xy = (x+2y)(2x+3y)$

Similarly, if y=0, then $E = 2x^2 + 3z^2 + 7zx = (x+3z)(2x+z)$

Hence $E = (x+2y+3z)(2x+3y+z)$

Factorise $2x^2 + 2y^2 + 5xy + 2x - 5y - 12 = (x+3)(2x-4)$ and $(2y+3)(y-4)$

Hence, $E = (x+2y+3)(2x+y-4)$

EXAMPLE 9

Shesanyankena Charamena- "The remainders by the last digit"

Consider: 1/7. When divising 1(0) by 7 the remainder is 3. Therefore, dividing 3 by 7 will subsequently lead to remainder 9 (= 3x3). But since 9 is more than 7 the remainder would be 2, so the remainder sequence is:

3, 2

Now 2 divided by 7 will have remainder of 6 (3x2), that is

3, 2, 6

Continuing

3, 2, 6, 4, 5, 1

We stop when the remainder sequence starts to repeat. Now, multiply these remainders by the last digit (7) of the denominator and keep only the first digit (LSD). So we have:

7×3 = 21 ⇒ put down 1

.1

3, 2, 6, 4, 5, 1

7×2 = 14 ⇒ put down 4

.1 4

3, 2, 6, 4, 5, 1

7×6 = 42 ⇒ put down 2

.1 4 2

3, 2, 6, 4, 5, 1

Continuing

.1 4 2 8 5 7

3, 2, 6, 4, 5, 1

So the answer is 1/7 = .142857142857..........

EXAMPLE 10

Sopaantyadvayamantyam- "The ultimate and twice the penultimate"

Corollary : Gunitasamuccayah Samuccayagunitah—"The product of the sum of the coefficients in the factors is equal to the sum of the coefficients in the product".

Sc of the product = Product of the Sc in the factors

For example (x+7) (x+9) = (x2 + 16x + 63)

(1+7) (1+9) = (1 + 16 + 63) = 80

or (x+1) (x+2) (x+3) = (x3 + 6X2 + 11x + 6)

(1+1) (1+2) (1+3) = (1 + 6 + 11 + 6) = 24

EXAMPLE 11

Ekanyunena Purvena-"By one less than the previous one".

777 multiplied by 999 = 776,223

(776 is one less than multiplicand 777 and 223 is the compliment of 776 from 9)

120 35 79 multiplied by 999 99 99 = 120 35 78, 879 64 21

1234 5678 09 multiplied 9999 9999 99 = 1234 5678 08 8765 4321 91

The Ekanyunena sutra can be used to derive the following results:

- Kevalaih Saptakam Gunyaat, or in the case of seven the multiplicand should be 143
- Kalau Kshudasasaih, or in the case of 13 the multiplicand should be 077
- Kamse Kshaamadaaha-khalairmalaih, or in the case of 17 the multiplicand should be 05882353 (by the way, the literal meaning of this result is—"In king Kamsa's reign famine, and unhygenic conditions prevailed"—not immediately obvious what it had to do with Mathematics. These multiple meanings of these sutras were one of the reasons why some of the early translations of Vedas missed discourses on vedaangas.)

These are used to correctly identify first half of a recurring decimal number, and then applying Ekanyuna to arrive at the complete answer mechanically.

Consider for example the following visual computations:

1/7 = 143×999/999999 = 142857/999999 = 0.142857

1/13 = 077×999/999999 = 076923/999999 = 0.076923

1/17 = 05882353×99999999/9999999999999999 = 0.05882352 94117647

Note that

7×142857 = 999999

13×076923 = 999999

17×05882352 94117647 = 9999999999999999

which says that if the last digit of the denominator is 7 or 3 then the last digit of the equivalent decimal fraction is 7 or 3 respectively.

Research and Reconstruction

At the beginning of the 20th century, when there was a great interest in the Sanskrit texts in Europe, Bharati Krsna tells us some scholars ridiculed certain texts which were headed Ganita Sutraswhich means mathematics. They could find no mathematics in the translation and dismissed the texts as rubbish.

However, Bharati Krsna, who was himself a scholar of Sanskrit, mathematics, history and philosophy, studied these texts and after lengthy and careful investigation was able to reconstruct the mathematics of the Vedas. According to his research, all of mathematics is based on 16 sutras, or word formulas.

Bharati Krsna wrote 16 volumes expounding the Vedic system, but these were unaccountably lost... when the loss was confirmed in his final years, he wrote a single book...

Vedic Mathematics. It was published in 1965 five years after his death, and is still available.

"On seeingthis kind of work actually being performed by little children... the doctors, professors and other 'bigguns' of mathematics are wonder struck and exclaim: 'Is this mathematics or magic?' And we invariably answer and say: 'It is both. It is magic until you understand it, and it is mathematics thereafter" Bharati Krishna Tithaji (1884-1960)

Rekindled Interest in Vedic Maths

A copy of the book was brought to London a few years later and some English mathematicians (Kenneth Williams, Andrew Nicholas, Jeremy Pickles) tookan interest in it. They extended the introductory material given in Bharati Krsna's book and gave many courses and talks in London. A book (now out of print) Introductory Lectures on Vedic Mathematics, was published in 1981.

Between 1981 and 1987, Andrew Nicholas made four trips to India, initially to find out what else was known about Vedic mathematics. As a result of these journeys, scholars and teachers in India showed renewed interest. It seems that once they saw that some people in the West took Vedic maths seriously, they realised they had something special.

St. James's School, then in Queensgate London also other schools, began to teach the Vedic system with notable success. Today Vedic mathematics is taught widely in schools in India and a great deal of research is being done. Three further books appeared in 1984, the year of the centenary of the birth of Sri Bharati Krsna Tirthaji. These were published by The Vedic Mathematics Research Group.

The Cosmic Computer

When Maharishi Mahesh Yogi began to explain the significance and marvellous qualities of Vedic mathematics in 1988, Maharishi schools around the world began to teach it. At the school in Skelmersdale, Lancashire, UK, a full course, called "The Cosmic Computer', was trialled for 11-14year-old pupils. Maharishi said that the sutras of Vedic mathematics are the software for the cosmic computer which runs the entire Universe on every level and in every detail.

The Vedic Mathematics Sutras

This list of sutras is taken from the book Vedic Mathematics, which includes a full list of the 16 main sutras and 14 sub-sutras in Sanskrit, but in some cases a translation of the Sanskrit is not given in the text and comes from elsewhere. Sutra:

Translation:

By One More than the One Before

All from 9 and the Last from 10

Vertically and Cross-wise

Transpose and Apply

If the Samuccaya is the Same,it is Zero

If One is in Ratio, the Other is Zero

By Addition and by Subtraction

By the Completion or Non-completion

Differential Calculus

By the Deficiency

Specific and General

The Remainders by the last Digit

The Ultimate and Twice the Penultimate

By One less than the One Before

The Product of the Sum

All the Multipliers

-Watching a child go through our education system is like watching reverse metamorphosis: in flies a beautiful butterfly; out crawlsacater pillar".

(If You Want To Be Rich And Happy, Don't Go To School, by Robert Kiyosaki, 1992)

The gift that the Hindus gave to world, thousands of years ago, and which is currently employed in our global silicon chip technology, was none other than the invention of zero and the use of the decimal point. We call our common numbers "Arabic numerals' but, really they extend back to the Hindu concept of creation and void known as Bindu or the "Zero Point'. All Vedic mathematics is based upon the understanding of Unity consciousness, which means the utilisation of number bases that correspond to:

0, 10, 100, 1,000, 10,000 etc...all of which add to 1.

There are 16 sutras, or simple Sanskrit word formulas which solve all known mathematical problems in the branches of arithmetic, algebra, geometry and calculus. They are easy to understand, easy to apply and easy to remember.

This Vedic one-line mental arithmetic is very helpful in stimulating modern mathematicians because of its simplicity and speed. Once initiated into the Vedic rules, students of all ages tend to appreciate and enjoy the enhancement of mathematics. Vedic mathematics is a previously long-hidden treasure trove of intelligent mathematical knowledge and should be within easy reach of everyone who wishes to obtain and benefit by it.

Vedic mathematics is a total system. The Vedic mathematician was also an astronomer, an engineer, a musician a healer and a poet. The temple builder had no pen and paper; he simply calculated in his head. So, you are out in the field and you need to tile a floor that is say 98 units square. How do you do it with such mental ease? Let's look at some practical examples.

The Squaring of Numbers Near a Base

To solve 98 squared (98 x 98, or 98), we must first determine what base we are in. It is close to 100, therefore we say Base 100. We must now choose one of the 16 major sutras to solve the problem.

The one to use here is called "By the Deficiency-by whatever the deficiency, lessen it further by that much and set up the square thereof'. Sounds cryptic and meaningless, yet it quickly solves the problem. We get our answer by merely knowing how much is 100 less 98. Knowing that the deficiency is 2, we merely lessen 98 by 2 and then we tag on the squaring of that 2. As a one-line answer, the setting out would appear thus:

98 squared = 98 - 2 / 2 x 2

Simplifying it: = 96 / 4

We almost have our answer. What we need to know is that since our base is 100, it has two zeroes; therefore this fact governs the need for two spaces for two zeroes or digits after the "forward slash symbol (/). By inserting or inventing the zero as a "place marker', the answer is achieved:

98 squared = 96 / _04 = 9604

Observe similar examples:

97 squared = 97-3 / 3x3 = 94 / 09 = 9409

96 squared = 96-4 / 4x4 = 92 / 16 = 9216

When the number being squared is above the base _ of 100, here _ we add the excess and square the excess:

104 × 104 = 104 + 4 / 4 x 4 = 108 / 16 = 10,816

104 × 105 = 104 + 5 / 4 x 5 = 109 / 20 = 10,920

What if we enlarged our numbers to 998 squared? It is close to 1,000, so we say Base 1,000 and know to have three spaces (for zeroes or digits) on the right hand side of the (/).

998 squared = 998 - 2 / 2 x 2 = 996 / _ _ 4

= 996 / 004 = 996,004

Understanding this, you can be calculating digits in the millions:

9998 squared = 9998 - 2 / 2 × 2 = 9996 / _ _ _ 4

Since we are in Base 10,000, the four zeroes determine the need for four spaces (zeroes or digits) after the (/).

= 9996 / 0004 = 99,960,004

There is a worldwide debate currently raging about the efficiency of Vedic mathematics versus the crumbling foundations of Western mathematics.

Generally speaking, the theorems we all learned at school are not wrong, but clumsy. Some of the Western geometrical formulas are certainly inadequate. For example, the formulas for sphere packing in the higher dimensions increase up to the sixth dimension, then suddenly decrease for higher dimensions, which is simply absurd. Unfortunately, some die-hard senior mathematicians, in an attempt to protect the crumbling foundations that they now stand on, feel threatened by the lightning-quick mental calculations of Vedic mathematicians and go to great lengths to deride Vedic maths as a "bag of tricks'.

The Squaring of Numbers Ending in Five

Here is another example illustrating the Vedic mathematical system's utter simplicity in demonstrating "the path of least resistance'. If we want to square the number 25, i.e., 25 × 25, we would conventionally take three lines of working out. Vedic mathematics merely looks at the question, applies one of the 16 sutras, and solves it mentally in one line. In this case, the sutra at work is "By One More than the One Before', that is... the previous digit.

We observe that 25 is a two-digit number and 5 is the last digit. However we are mainly interested in "the previous digit' which is 2. Mentally we say -What is one more than two?

It is three". The word "By' in the sutra really means "to multiply'. The setting out for the first half of the answer is thus:

25 squared = 2 by 3 /.... = 2 × 3 / ....

To this we tag on the last digit "5' squared:

= 2 × 3 / 5 × 5 = 6 / 25 = 625

Similarly, all other numbers that end in 5, when squared, can be calculated instantly:

15 squared = 1 × 2 / 5 × 5 = 2 / 25 = 225

35 squared = 3 × 4 / 5 × 5 = 12 / 25 = 1,225

45 squared = 4 × 5 / 5 × 5 = 20 / 25 = 2,025

95 squared = 9 × 10/ 5 × 5 = 90 / 25 = 9,025

Sutra: 'Vertically and Crosswise'

Here is another simple sutra, the one that Bharati Krsna Tirthaji refers to as most widely used, called "Vertically and Crosswise'. It solves all multiplication by application of a pattern, which is registered by the right brain as feminine-natured mathematics... in contrast to the logical, male, left-brain style of mathematics generally taught at school.

0 0 0 0 0 0

0 0 0 0 0 0

This sutra shows we will have a three-digit answer, represented by the three short horizontal lines. Here is how we traditionally write the setting out for "26 x 31'

26

× 31

Notice there are four digits involved. Let each digit be represented by a small circle or dot, as in the format shown in the diagram above. This will help you understand "cross-addition' which is shown as the middle part [(2 x 1)+(6 x 3)] and uses both multiplication and addition in the form of the letter "X', corresponding with the crossover of the optical nerve in the brain. Below, the small letter "x' stands for multiplication.

$$= 2 \times 3 \quad (2 \times 1) + (6 \times 3) \quad 6 \times 1$$

$$= 6 \quad 20 \quad 6$$

("2' is carried over)

$$= 8\ 0\ 6 = 806$$

Sutra: Digital Sums for Multiplication by Eleven

When computer users need to move large volumes of electronic data efficiently, the solution employed is invariably compression. "Digital Compression' (or "If the Samuccaya is the Same,it is Zero') is a powerful sutra that solves multiplication by 11 very quickly. If we want to multiply 25 by 11, we merely add the two digits of the 25 and say "2 +5', which equals 7, and insert that digit between the other two digits.

Thus the answer is 275.

Another way of showing this is to separate the two digits and insert their digital sum:

$$25 \times 11 = 2 \quad (2 + 5) \quad 5$$

$$= 2 \quad 7 \quad 5$$

$$275$$

In this example, the "1' of the "12' gets carried over to the left:

$$39 \times 11 = 3 \quad (3 + 9) \quad 9$$

$$= 3 \quad 12 \quad 9$$

$$429$$

Vedic Numerical Codified Knowledge

How the ancient seers sangthe longdecimal form of Pi In ancient India, they would sing songs to memorise long decimals.

The Brahmans or scholarly cast, secretive of their knowledge, sonically encrypted mathematical formulas into their devotional praises or hymns to Lord Sri Krishna and also recorded historical data in codified lyrics.

The system has similarities with numerology where values of numbers are ascribed to consonants (as in A = 1, B = 2, C = 3, D = 4)...but the Vedic numerical code was so sophisticated in Sanskrit because it possessed three layers and therefore has triple meanings.

It turns out that the decimal form of the transcendental number:

Pi = 3.14159265358979323846264338327 92....

was hidden or codified in the syllables in the following chant:

Gopi Bhagyamaduvrata - Shringishodadi Sandiga -
Kala Jeevitarava Tava - Galaddhalara Sangara

The top line, go = 3, pi= 1, bha = 4, ' ya = 1, ma = 5, duv = 9, ra = 2, ta = 6 etc., which gives the first eight figures of pi the circumference of a circle to its diameter.

Not only did the code give pi to 32 decimal places, but there was a secret Master Key within the patterning of the 32 that could unlock the next 32 decimals of pi, and so on... a ticket to infinity!

The code not only praised Krishna, it operated on another level as a dedication to Shankara. (Around AD 800. Shankara was a celebrated acharya, or teacher...who founded four monastic orders. He could read at the age of two and had mastered the Vedas by the time he was eight.

He wrote scholarly commentaries on The Bhagawad-Gita and Upanishads, which led to the decline of Buddhism in India. He is considered a partial incarnation of the avatar Shiva.

An Ancient System for the Modern Age

In conclusion, I believe the time has come for all secret knowledge of the past to be kept secret no longer.

It is time for the ancient seers' infallible mental and one-line system of Vedic Mathematics to be reintroduced.

2

The Mysteries of Vedic Mathematics

How to multiply large numbers...in your head! Find out how to accomplish rapid calculations - without a calculator. Do your children really understand the fragmented and confusing mathematical data taught to them? Would you be interested in another method that boosts confidence, increases memory skills and is presented in a fun and exciting manner?

After this course of study, you will never multiply, divide or add in the same manner that you are currently using to solve arithmetic problems. You will walk out of Jain's class confidently multiplying double digit numbers without a calculator. You do not need to know any mathematics to do this course, in fact, which has been proven many times, children who have had learning problems with the way we teach them numbers, can perform these exercises such faster than the adults in the class!

AN EXTENSIVE EXPLANATION:

Do your children really understand the fragmented and confusing mathematical data taught to them? Would you be interested in another method that boosted their confidence, increased their memory skills and is presented in a fun and exciting manner?

After this course of study, you will never multiply, divide or add in the same manner again. And no more times tables!

Jain expands and demonstrates the real ability of your mind by showing you how to complete complex mathematics... in your head!

"WATCHING A CHILD GO THROUGH OUR EDUCATION SYSTEM IS LIKE WATCHING REVERSE METAMORPHOSIS:

IN FLIES A BEAUTIFUL BUTTERFLY, OUT CRAWLS A CATERPILLAR."

"If You Want To Be Rich And Happy Don't Go To School"

If you disliked Mathematics in your school years, do not be alarmed. It is actually a blessing as your Higher Self or Intelligence refused to embrace a system that was anything less that pure Star Language. The Brain thinks in symbols and pictures. Math must be taught laden with Shapes, Color, Aromatherapy and Music for effective Whole-Brain Learning. Most children are visual, and the ability to tap into Pattern Recognition' is the key to rapid mental calculation.

In this one day workshop you will know how to accomplish rapid calculations, mentally. No more calculators which make the brain lazy. The brain is a mental muscle and needs to be exercised. Jain predicts that if we continue to teach children with an overemphasis of using calculators in the classroom, over a 20 year period, the Global Brain will deteriorate.

An understanding of the 16 basic Sutras or "Word Formulae" solves all known mathematical problems by using an advanced and highly intelligent system of MENTAL, ONE LINE ARITHMETIC used by the Hindus 2,000 years.

Jain reveals ancient yet simple Math formulas which allow ordinary people to perform complex equations. At the same time he shows you how the world as we know it is more than we think... his teachings can for some people, bring about major shifts in the way we think about the world!

Learn also how the Hindu Invention of Zero (and the decimal system) led to modern technological advancements.

This system was reconstructed by Bharati Krsna Tirthaji (1884-1960, aka Shankarcharya of Puri) who was invited by Paramahansa Yogananda to take this knowledge, for the first time, to the western world, in 1958. The Beatles' guru: Maharishi Mahesh Yogi, who introduced Transcendental Meditation to the west, was also a student of this master and currently teaches this system in India in the spiritual Maharishi schools, whose teenage students are highly distinguished mathematical geniuses.

Some questions to ponder upon:

How could the Aryan civilization, near the Hunza Valley, 3,000 years ago, mentally compute mathematical operations that today only a calculator could achieve?

And if the maths could not be done mentally, how could they do everything we know today, IN ONE-LINE? e.g.: converting the fraction 1/19 into a decimal is an 18 digit answer that requires 35 lines of exhausting long division, how could they do it in a one simple line solution?

Why is NASA secretly using this MENTAL ONE-LINE SYSTEM OF ARITHMETIC in the field of Advanced Artificial Intelligence? and not incorporating it into our current mathematics curriculum?

Why are many of Bill Gate's Mathematicians Indians?

Is the decimal form of Pi a sonic code for dedications to Krishna?

What does the Bhagawad Gita have to say about this lost science of Clairvoyant Calculation that the ancient Temple Builders employed?

How can you become a Human Bio-Calculator?

Is the current Western Math Curriculum keeping our Children in a state of limited intelligence?

The answer to all these questions are found by understanding the ancient system of Vedic Mathematics.

If you feel like some enlightenment and want to experience Math as Art, and wish to develop your mental muscle and expand your consciousness - join us on this day for what could be one of the most exciting discoveries of your life.

The gift that the Hindus gave to world, thousands of years ago, and that which is currently responsible for a global silicon chip technology, was none other than the invention of ZERO and the use of the decimal point. We call our common numbers "Arabic Numerals" but really they extend back to the Hindu concept of creation and void known as "Bindu" or The Zero Point. In Math, it makes the magic work e.g.: if you wish to multiply 98 x 98 we need to understand Zero and Unity. All Vedic Math is based on the understanding of Unity Consciousness which means they utilize processes or Number Bases that correspond to: 0, 10, 100, 1000, 10000 etc. all of which add to 1.

There are 16 Jaw-Dropping Sutras or simple Sanskrit Word Formulae which solve all known mathematical problems in the branches of Arithmetic, Algebra, Geometry and Calculus are easy to understand, easy to apply and easy to remember.

This Vedic One-Line Mental Arithmetic is very helpful in stimulating Modern Mathematicians to adopt its simplicity and quickness. Once initiated to these Vedic rules, students of all ages will appreciate and enjoy the enhanced subject.

The aim of Vedic Mathematics is to bring this long-hidden treasure trove of intelligent mathematical knowledge within easy reach of everyone who wishes to obtain it and benefit by it. This material shakes the whole foundations of the current Western mathematics curriculum, will rewrite our current world view on Mathematics and will require a new definition of the Golden Mean: that its decimal does carry on infinitely but within its infinity there is recursive or self-similar beauty. Scientists like Marco Rodin wind toroidal coils in sync with this recursion or rhythm or pulse or pattern of 24 digits to get more output than their input! The sum of these 24 digits is 108 (108 degrees being the external angles of the pentagram and the number of beads in a rosary) which is sonically disguised as the GAYATRI MANTRA, the most famous Eastern prayer for Enlightenment.

By relying on modern electronic calculators for the next 20 years or so, without cultivating mental calculations which build the "Mental Muscle", will sadly produce a generation of young adults whose brain will have undoubtedly atrophied!

Vedic Mathematics promotes clarity and sharpness of mind as well as improving memory skills and self-confidence, by awakening the Pineal Gland (or Third Eye) to clairvoyantly "see" or intuitively know the predicted answer. Mystics say that the Pineal gland, located in the exact centre of the brain, was once the size of large cherry, but now it is the size of a pea. Two reasons are that we drink denatured defunct water and we no longer build the Mental Muscle, over-relying on calculators etc. to solve simple problems. The brain is a muscle, being 90% water, it needs to be hydrated with pure alive waters and regularly exercised.

> *"This Cosmic Manager, the Cosmic Mathematician is eternally free from mistakes. The Cosmic Mathematician is present within everyone, it is the light of life of everyone; it is present within everyone in the structure of Rigveda and the Vedic Literature; it is the light of pure intelligence." (Maharishi Mahesh Yogi)*

Mental, One-Line Calculation

The Indians have an ancient system of Clairvoyant Mathematics claiming that they can literally "see" the "shape" of the Answer to any Problem, like 106 x 108. The highest Sutra or Formula of all the 16 possible Sutras, is called "Vilokanam" which means, "By mere Intuition" or "By Mere Observation". Just by looking at the Question: What is 106 × 108" they can see the answer within one second. In some cases, the answer is not resolved mentally, so it must be written down, but only by using one line of working out. So these Yogis, 2,000 years ago, had intellectual contests or spiritual games of outwitting one another with mathematical "Shortcuts". It was the spiritual head of India: Bharati Krsna Tirthaji, (1884-1960) who reconstructed all the knowledge from the time of Buddha circa BC500 and evolved it into Mental, One-Line Arithmetic that we are just catching onto today.

So to solve 106 × 108, think of the Base 100 and consider only how much these 2 numbers are above the Base of 100. As soon as I mentally flag those two excesses of 6 and 8, I have my answer instantly, by adding the 6 to the other number 108, making it 108 + 6, and by multiplying those two deficiencies which is 6 x 8, and viola, there is the magical, instant, mental answer:

106 × 108 = 108 + 6 / 6 × 8 = 114 / 48 =11,448.

Vedic Philosophy

According to the spiritual head of India, Bharati Krsna Tirthaji (1884-1960) the "Vedas" are "the illimitable storehouse of all Knowledge". This meant that you had to become accomplished in many fields of Knowledge. There was no such thing as just a Vedic Mathematician, you also had to study the Vedas of the Stars and Astronomy, you had to know the AyurVeda (Ayur means "Life") which is knowledge of Medicine and Healing; and the Vedas of Music, and Engineering and the Art of Cooking etc. Knowledge of the Vedas evolved you into a Total Being.

Some scholars believe that when Atlantis fell, when the Oceans rose only the mountain-top dwellers survived and those that arrived to the Himalayan (and Andes) areas were known as the Aryan Race and with them they brought the gifts of the Vedas.

Vedic Mathematics

This is a "Mental, One-Line System of Arithmetic" based on 16 Universal Sutras or Formulas or Shapes that solve all known mathematical problems. Developed over 3,000 years ago, originating with the Zero concept and the Decimal System, it has become a highly intelligent and sophisticated system of mathematical Short Cuts. E.g.: To find the reciprocal of 19, in decimal form, the high-school student is required to labour with 35 lines of working out going down the page, and is bound to make a mistake, as you are required to know your 19 Times table; whereas in India today, a teenage student can accomplish this 18 digit solution in one swift, horizontal and elegant line of arithmetic, in one line, whereas, today, we need a computer to divide 1 by 19. Vedic Mathematics therefore encodes the Law of Efficiency, the Path of Least Resistance. Traditional Mathematics is not wrong, it is only clumsy. The Indians claimed that they invented the Zero and gave us the decimal, not the Arabs whose books falsely claim they invented it; but research now indicates that the Babylonians, 12,000 years ago used The Zero to build their stepped pyramids known as 'ziggurats'.

> *"Vedic Mathematics is that one field of knowledge which fulfills the purpose of education by developing the total creative genius of the individual, giving him/her the ability to be always spontaneously right and automatically precise so that his/her action, supported by Natural Law, is always effortlessly fulfilling".*

Art and the Magic Squares

There is a treasure house of knowledge encoded within these harmonic matrixes that the ancients knew and hid. Magic Squares for thousands of years have fascinated many cultures and many great minds like Benjamin Franklin and Albrecht Durer. The ancient mystery schools adored their richness of symbolism. The 3x3 square is known as the Lo Shu in Tibetan and Taoist Calendars and Cosmology. The ancient Indian seers referred to as a Space Matrix or Code that governed the sacred architecture of their Temple's. In effect, the Lo-Shu is the simplest blueprint that unifies all complex Life components. Not only is it linked to the workings of the D.N.A molecule but it is part of our Ascension Process.

- Learn how to transform numbers into Art.
- Understanding Higher Knowledge through the Language of Shape.
- Learn the patterned language of our Universe!
- Learn Ancient symbols of Order amidst Chaos.

6	1	8
7	5	3
2	9	4

Magic Squares are ancient boxes of numbers whose sums of rows, columns and diagonals all add up to the same sum. This matrix represents Perfect Equality in all directions or dimensions. Elegant or aesthetically pleasing patterns can be created when a long, continuous line is joined in arithmetical order from 1 to 2 to 3 to 4 etc. to the last number. This is how Mathematics is transformed into Art. This line connecting such consecutive or sequential numbers is the backdoor into the world of Atomic Art or Nuclear

Geometry, arriving at scientifically verified atomic structures. Order is created out of Chaos.

The Latin axiom: e pluribus, unum, applies in Feng Shui and Magic Squares, meaning: Out of the Many, the One. Each individual part or cell of any system, whether it be crystal or human must be honored in respect to the Greater Whole, in all Time, in all Space unconditionally.

The ancient Magic Square of 3 × 3 (also known as the "Lo-Shu" in ancient China, C17th BC) is also the centre of the Tibetan Calendar/Cosmology. It is a timeless mathematical harmonic whose sums of the columns, rows and diagonals all add up to and vibrate to 15. Symbolically the Magic Square creates Order amidst Chaos, and Equality in All Dimensions.

One definition of Feng Shui details the aesthetic placement of objects; this is similar to the divine order of the placement of numbers in a magic square. Notice that all the pairs around the centre of the magic square of 3x3 have a sum of double the central cell. That is, 1 and 9 =10, 2 and 8 =10, 3 and 7 =10, 4 and 6 =10.

Ancient Card Puzzle: There are 9 playing cards in front of you.

1, 2, 3, 4, 5, 6, 7, 8, 9

How can you align them so that the 3 vertical columns, the 3 horizontal rows and the 2 diagonals all add up to the same number?

The solution below is one of 8 possible arrangements:

6 1 8

7 5 3

2 9 4

The oldest known record in print is from China, 5,000 years, known as The Lo-Shu and is the emblem or badge for the Feng Shui community who seek harmony in the placement of things.

Symbolic of searching for Order amidst the Chaos, draw a long and continuous line from 1, to 2 to 3 , etc., to the last number 9, in sequential or consecutive or natural counting order. A specific Yantra or Pattern is created. This mere act of Transforming Numbers into Art has subtly shifted you from the logical Left Brain realm of Numbers to the creative, picturesque Right Brain realm. This simple activity has created fusion of the left and right brain, now called Whole-Brain Learning.

It transpires, upon further superimposition of the same pattern upon itself, at various angles like 90 and 45 degrees etc. that the patterns begin to resemble specific Atomic Structures of Rutile, and Platinum Crystal, and Beryllium etc. . For this reason, a new realm of knowledge has emerged known as Atomic Art. What this does is to put side by side the Scientist's Tunneling Microscope diagrams alongside the Mystic's Yantras or Magic Square diagrams and conclude without doubt that they are identical. Importantly, this bridges the Physics with the Metaphysics, and East with West. This places Mathematics as a strong candidate for the worlds' evolution towards a Universal Language, all through the hidden secrets of the humble Magic Square.

Magic Squares have appeared in old Essene Bibles, where Jesus apparently referred to the Divine Kingdom in terms of the Magic Square of 7 × 7, where The Above is the same as The Below, where The Left is the same as The Right, where everyone is equal, for if you were to take out only one number, the magic square would become disharmonic. In this sense, it is a symbol of community, of society, where we need the baker, the plumber, the nurse and encouraged to develop a sense of Gratitude for the underlying structural Unity of all Life. The Magic Square therefore defines perfect Equality in all aspects, in all Dimensions, "e pluribus unum" meaning, "Out of the Many, the One.

There is a treasure house of knowledge encoded within this harmonic matrix that the ancients knew and hid. Magic Squares for thousands of years have fascinated many cultures and many great minds like Benjamin Franklin and Albrecht Durer.

Many sacred symbols emerge when the original Magic Square Pattern or Yantram (singular of Yantra which is the plural) is rotated upon itself. Like snowflakes, millions of Magic Squares can be created. There are 8 permutations of the primal 3x3 Magic Square known also as The Lo-Shu from China 4,000 years ago. There are 880 permutations of the 4x4 Magic Square, but as we get to the 5x5s and 6x6s there are millions! Mathematics is a Universal Language. That means that it will always exist, without decay or corruption.

As Pythagorus said, in the 6th century BC, "NUMBER IS THE BOND OF ALL THINGS".

— Why was it at the centre of the Tibetan Calendar/ Cosmology (Time Code / Applications in Time Travel)?

— Why did Ambaji, of the Hindu pantheon, the ancient goddesses of Creation (the universe flows through her breast), have as her tantric symbol, this Magic Square of 3x3, which has 8 possible permutations?

— And which Buddha, in his enlightenment, described in detail the 8 sided nature of subatomic molecules: the "Acta-Kalapos". Durga and Kali and Shakti etc. have variations of the same.

— Why is it that when you translate NUMBERS INTO ART, by drawing a long connecting line, from 1 to 2 to 3 etc. to the last number, that elegant and phi-ratioed patterns appear? That when tiled or tessellated, produce the atomic structure of Diamond (which is verified by the clairvoyant drawings of the Occult Chemists: Charles Leadbeater and Annie Besant.

— Why is it that the pattern for the Magic Square of 11x11 is almost identical to the Ley Line Grid developed by Bruce Cathy, a New Zealand pilot, which was originally the jig-saw-puzzle-map of UFO sightings in his area?

— Why did Professor Searl, of the UK, have his work on levity discs and over-unity generators banned? The metals of his generators for the spacecraft he built were intimately related, by proportion, to the atomic table of elements and to the opposing pairs of the above Magic Square of 3×3, e.g.,: notice that all pairs around the central cell have a sum of 10, thus 2 parts gold was alloyed with 8 parts silver, or 3 parts titanium to 7 parts nickel.

— Why did King Solomon the Rabbi, design his Seal based on the 3 rows of the above Magic Square being translated into a Natural Square of 3 (that has its numbers in Counting Order)?

In this day lesson, you will superimpose a Magic Square of 7×7, by rotation of 90 degrees upon itself, to reveal ancient symbols at its centre: the Hindu Peace Symbol of the Swaztika nesting within the Christian Cross! The ancient Essenes adored the richness of symbolism within these Mystic Squares and gave accounts or Parables of Jesus (Isa or Iesu) that related to our sense of community symbolized by the magic square, that each person, so positioned, has a divine role to play, that if one number was taken away or moved, it would no longer be magical, to realize 'Order amidst the Chaos', and the ancient axiom:

"As Above.. so below"

Sacred Mandalas, Magic Squares and Atomic Art

"Mandal" is the "world", as you know from other languages like in French: le Monde. Most mandalas have a geometric centre, the Source of Creation, and from this radiates all Life. Eastern teachers state that the Mandala is really a 3-Dimensional object descriptive of the World, and what westerners call 'mandala' is really a Yantra. A Yantra is the 2-Dimensional view, or the shadow of the Mandala. A Yantra is the Artwork, the Power-Lines, the Black and the White of the 3D Mandala. Yantra's psychoactive lines conducts psychic essences infused by Yogi or designer or artist in a sacred space, so the goal of the simple Yantra is to make the complex and Invisible Mandala, visible. Thus the Yantra of the Triangle is really the Mandala of the Tetrahedron, and the Circle is the Sphere. By understanding this concept, this connection of 2D to 3D, the neophyte can tap into the 4D, and the 5-Dimensional realities, to ultimately conclude that there is no separation, no dimensions, only Oneness as symbolize by the Mandala.

A Yantra is charged by one's thoughts and intentions. That is why yogis and yoginis construct them on the full-moon, whilst fasting or celibate, and create the actual diagram in datura juice, at specific times of the day and year determined by astrological highlights. Its sacred in the sense that all your energies are focused on just that activity of Yantra Creation, as all Manifestation arises through pure thought.

The Divine Proportion

Discover the GOLDEN MEAN SPIRAL, the nature of ALL of creation, the fabric of LIFE itself! Nested within your Heart is the sacred geometry golden mean ratio, it is the centre of a greater Galactic Grid connecting us to all Memory of the Universe and all Atomic Levels. Embrace part of the ascension process already morphologically and geometrically encoded into our bio-magnetic and bio-electrical fields.

— Learn how to recognize this proportion in your body, in nature, in Art and Architecture and in Universal forces.

— Discover why this knowledge is important to survival after death and plays a central role in immortality.

The Divine Proportion:

~ *your map* ~ *your DNA* ~ *your universe*

Learn why the shape of the Pine Cone is optically similar to the Human Heart, why we are awestruck when we view sacred architecture like the Parthenon, why we are attracted to the proportions of the famous Mona Lisa. Essentially, the human body is in resonance with the Living Mathematics of Nature.

A fascinating day that will help you to remember this Lost Knowledge of the PHI CODE that is already morphologically and geometrically encoded in your bio-magnetic and bio-electrical fields.

Ultimately, this Pythagorean Knowledge is not an intellectual process, but rather that these essential geometries are invisibly constructed and nested within the Heart as the Centre of a greater Galactic Grid connecting us to all Memory of all Universes and Atomic Levels.

This is the study of the Universe as found in all parts of the Human Body. This is the mathematics of infinity; Phi expressed as 1:1.618033 etc. The decimal part goes on forever without any detectable repetitious pattern. The Golden Mean, sometimes referred to as the The Sacred Cut or The Phi Ratio has connections with Galactic and Atomic mathematics, as revealed in ancient temples and pyramids; in classical paintings like the Mona Lisa; in the cross-sectional view of the D.N.A. molecule; in the nesting of the 5 Platonic Solids; its occurrence encrypted in the Fibonacci Sequence: 1, 1, 2, 3, 5, 8, 13, 21, 34, 55, 89, 144 etc.; in the counter-rotating fields or spirals of the sunflower and pine cones.

This Transcendental Number has fascinated many cultures and individuals for millennia. This Phi Ratio is part of our Ascension Process and is an important clue to accessing or Multidimensionality. The Golden ratio (the divine proportion) is morphologically and geometrically encoded in your bio-magnetic and bio-electrical fields. Today, in this course, you will study the importance of the Pentacle and the Golden Phi Spiral and learn how to draw these 2 psychoactive diagrams with compasses and rulers only. You will also learn about the 'Vesica Picses' as the mother of all symbols.

The DNA molecule is based precisely on the living mathematics of the nautilus shell, flowers and all parts of the human body.

This is the same ratio where the elbow is to the whole of the arm, or where the knee bends. It is the ideal aesthetic proportion of all parts of the human anatomy, of flower's petals and leaves, of the distance of the planets from the sun. The only mathematical sequence that encodes this is called the Fibonacci Sequence: 1, 1, 2, 3, 5, 8, 13, 21, 34, 55 89, 144 a trinity relationship of 3 numbers, where the previous number like 2, the past, is added to the present, number 3, to give the future, number 5.

A lesson in x-raying the Fibonacci Sequence : 0, 1, 1, 2, 3, 5, 8, 13, 21, 34, 55, 89, 144 etc. to discover where the infinite phi expression has its underlying rhythm, a distinct periodicity of 24 recurring digits that cannot be seen by the Western Mathematical Eye but only by the Truth of Vortex Mathematics expressed as Vedic Mathematics! (Really this Knowledge predates the Vedas and Atlantis and is better referred to as Galactic Mathematics).

The Golden Mean Ratio is an infinite number that connects atoms to the stars. We are the Phi Ratio, whose 4th dimensional emblem is the imploded sphere known as the Tube Torus Doughnut that has no inside or outside, creates vorticity and allows waves to travel infinitely without destruction as they embed in this vital Phi relationship, the Golden Mean Spiral.

The key word for the Divine Proportion or Phi Spiral is "Shareable" in that the key for waves to survive is to be aligned to the Phi Ratio of 1:1.618, this shareability of wavelengths allows a passport into the multidimensional worlds. Ultimately, if something is not shareable, it has no value, it will die, it has no spin, no life-force. The Divine Proportion is the Living Mathematics of Nature that is our template for survival.

Learn why the history of the Pentacle and the '3-Dimensional Phi Spiral' have been suppressed over the last 2000 years. Have you wondered why the following 3 subjects: Magic Squares, Vedic Mathematics and the Phi Ratio (1:1.618) have not been taught in your math classes? They are keys to your ascension process.

Sacred Geometry

The 5 Platonic Solids is the Key to Atomic Structure and Alchemy. This is the Language of Shape. A language that developed from our Ancestors original measurements of the Earth's circumference. It involved a study of the Stars and therefore connected us to the big Picture or Macrocosm which our Ancestors believed was a mirror of the Small, Atomic, D.N.A. Picture or Microcosm.

Multidimensional Geometry

This material shakes the whole foundation of current Western Quantum Physics.

The Five Platonic Solids, the only 5 shapes that fit inside a sphere with equal angles, faces and edges, are the key to Atomic Structures, as expounded by Charles Leadbeater and Buckminster Fuller.

Explore the hidden mysteries within the 5 Platonic Solids which are the key to nuclear geometry. What is the importance of the Star Tetrahedron (the 3-Dimensional version of the Star of David) having 24 faces and 24 edges?

How does this link to the 24 edges of the Cuboctahedron, the compressible and alchemical shape that was revered by Buckminster Fuller (who designed the first geodesic domes).

A day for exploring the hidden mysteries within the 5 Platonic Solids. Learn how the ancient masters, like Pythagoras, 2,500 years ago were aware that all atomic structure was essential based upon the nesting of these 5 universal shapes. e.g.: the atomic structure of Oxgygen is the Icosahedron (20 faces) with a Tetrahedron (4 faces) nested within.

Explore also the 13 Archimedean Solids which are a subset of the Platonic Solids. You have been kicking around one of these Shapes, the Truncated Icosahedron which you know as a Soccer Ball: the Fusion of Pent and Hex as found in the D.N.A. Molecule!

This was also known by the Theosophists 100 years ago who promulgated the teachings of Charles Leadbeater, a siddha yogi who could travel into the Macrocosm eg: to the moons of Jupiter and describe it in full detail, or travel into the Microcosm e.g.: to describe atomic structure based purely upon the 5 Platonic Solids. Modern Technology today has verified his findings of 100 years ago to be true! How did he do this? Inner Merkabah (the inner spiritual vehicle of Light) versus Outer Merkabah (Technological Space Craft made of heavy metals).

You will understand that the Dodecahedron (having 12 pentagonal faces) is the 3-Dimensional view of the Pentacle. That the Star Tetrahedron is the 3-Dimensional view of the Star of David. This is the cubic shape of our first 8 original cells of life.

As we begin to view our own D.N.A. and the shapes of our first mitotic divisions since fertilization we realize that all these sacred shapes are actually part of our Genetic Memory. e.g.: our first 4 cells pack into a Tetrahedron of 4 spheres.

Why did the symbol of the Pentacle, the Swastika, the Number 13 and the Ram's Horn (which is the 3-Dimensional view of the Nautilus Shell Phi Spiral) all become "demonized". Was it to create fear around these shapes so that you could not access the Memory of who you really are?

Every child at school could be learning the origin of where "A,B,C and 1,2,3" come from, as the marriage of the male Tetrahedral Pyramid and the 3-Dimensional feminine Golden Mean Spiral of the Nautilus Shell or Ram's Horn. When we shine a laser upon this union of primal male and female shapes, the shadow produces the Language of Light, that is, every main language has its origin in Sacred Geometry.

These shapes are psychoactive and raise consciousness by merely gazing upon them. The Hands-on Activity for this Module is the drawing of the Pentacle within the Circle, the Mystical Squaring of the Circle, and the true angle of the Cheops Pyramid, all in one important diagram.

Ancient Puzzle: How many spheres can pack around a central sphere, all the same size, touching or kissing one another. If you were to join the centre of those surrounding spheres, what shape would you get?

Is "SHAPE" the key to all Future Technologies? Does SHAPE STORE MEMORY? "Mem" = "Water" and "Ori" = "Light" in ancient Hebrew. i.e: Memory is Liquid Light! and our Brain is 90% Water.

Bridging Physics to Metaphysics by understanding the work of Charles Leadbeater and Annie Besant who, 100 years ago, drew all known atomic structures of the elements of the periodic table. In fact some recently discovered isotopes have been drawn 100 years ago! Having associated with yogis, this Pair of Clairvoyant Chemists were also able to successfully describe the moons of Jupiter, confirmed now by NASA! This material actually validates Jain's life's work: that when the numbers of certain Magic Squares, like the

Magic Square of 3x3, which is the centre of the Tibetan Calendar, when tiled or tessellated into a larger matrix actually shows the shadow form of the 3-Dimensional view of the Atomic Structure of Diamond-Lattice!

This material shakes the whole foundations of the current Western mathematics curriculum, will rewrite our current world view on Mathematics and will require a new definition of the Golden Mean: that its decimal does carry on infinitely but within its infinity there is recursive or self-similar beauty. Scientists like Marco Rodin wind toroidal coils in sync with this recursion or rhythm or pulse or pattern of 24 digits to get more output than their input! The sum of these 24 digits is 108 (108 degrees being the external angles of the pentagram and the number of beads in a rosary) which is sonically disguised as the GAYATRI MANTRA, the most famous Eastern prayer for Enlightenment.

Sacred Geometry: This is the Language of Shape. A language that developed from our Ancestor's original measurements of the Earth's circumference. It involved a study of the Stars and therefore connected us to the big Picture or Macrocosm which our Ancestors believed was a mirror of the Small, Atomic, D.N.A Picture or Microcosm. It involved a Universal Language of symbols that anyone on any planet or any time, could read.

A Language of Light operating via the Right Hemisphere of the Brain that understands Pictures, Colors, Shape, Holographs. "Shape" will be future technology, the ability to visualize in your Pineal Gland a certain shape that will create that image visualized to manifest. Sacred Geometry, relating to this concept of "Shape" poses the next advancement in our quest to ask: "How can we use this Knowledge of 'Sacred Geometry' and 'Shape' to solve all our mathematical multiplications, mentally, in our head, without using the calculator? This now leads to the next door, called: Vedic Mathematics. (Definition By Jain, 2005)

5 Platonic Solids

An Ancient Puzzle: What shapes can fit inside of a sphere, where every line or edge, every face or polygon and every corner or vertex are the same? There are only 5 such shapes, the most common one being the Hexahedron or Cube. They relate to the 5 elements: Fire is the Tetrahedron (having 4 equiangular triangles), Earth is the Cube (have 6 squares), Air is the Octahedron (having 8 equiangular triangles), Water is the Icosahedron (having 20 equiangular triangles), and Ether is the Dodecahedron having 12 (pentagonal faces). These 5 shapes have been found carved in stone from 10,000 years ago and can be seen in a museum in Oxford.

Charles Leadbeater, the occult chemist and Theosophist, described all atomic structures as the nesting, like the Chinese dolls, of these Platonic Solid symmetries, e.g.: Oxygen has a small tetrahedron inside of a larger Icosahedron. Many scientists see these shapes in their experiments and in the natural world: e.g.: when a wet straw is inserted into a large spherical soap bubble, and someone sucks into the straw, forcing the bubble to collapse or compress or minimize its surface area, it does this optimally to form the dodecahedron. (Definition By Jain, 2005)

Pythagorean Mystery School

Pythagoras was contemporary to Buddha and Lao Tzu who lived accomplished lives 2,500 years ago. Born in Sidon, Lebanon, Pythagoras studied 25 years alone with the Egyptian priests, then set up school in Greece but due to political unrest he was forced to move to Crotona in northern Italy where he set up his Mystery School that accommodated both men and women, promoted fasting, and the community sharing of goods, and other strict morals.

Oaths were taken in high secrecy not to reveal the true knowledge of the 3-Dimensional form of the Pentacle, known as the Dodecahedron ("do" in Greek means 2, and "dec" means 10, thus "dodec" means 12, "hedron" means face). The Dodecahedron was the 5th Element symbolizing the Ethers or Spirit, and is linked to the DNA molecule. He is known as Father of Music, having developed the structure of musical intervals that we now use today.

Pythagoras' 'Monochord' was a unit string that was plucked at its midpoint to give the 'octave' and other proportions like thirds and quarters and fifths, these ratios became known as "Logos" or divine 'proportions' and captured the planetary "Music of the Spheres".

Due to further political problems, rumors have it that the whole Pythagorean community were burnt alive. It is for this reason that many modern lecturers today, who have some memory of this time, still register the fear associated to teaching "sacred Geometry", as the witch hunt or destruction of Ancient Knowledge, is still carried on in this modern age, where all sacred symbols like pentacles and swastikas and magic squares are demonized.

Divisibility Rules

1	Two (2)	If a number has the last digit on the right hand side as 0, 2, 4, 6 or 8, the whole number is divisible by 2.
2	Three (3)	If the sum of the digits of a number is divisible by 3, the whole number is divisible by 3.
3	Four (4)	If the last two digits on the right hand side of a number are divisible by 4, the whole number is divisible by 4.
4	Five (5)	If the last digit on the right hand side of a number is 5 or 0 then the number is divisible by 5.
5	Six (6)	If a number is even and also divisible by 3 then the number is divisible by 6.

6	Eight (8)	If the last three digits on the right hand side of a number are divisible by 8 then the whole number is divisible by 8.
7	Nine (9)	If the sum of the digits of a number is divisible by 9 then the whole number is divisible by 9.
8	Ten (10)	If a number ends with 0 the number is divisible by 10.
9	Eleven (11)	If the sums of the alternate digits of a number are equal, the number is divisible by 11.
10	Twelve (12)	If a number is divisible by 4 as well as 3 the number is divisible by 12.
11	Fifteen (15)	If a number is divisible by 3 as well as 5 the number is divisible by 15.
12	Eighteen (18)	An even number divisible by 9 is also divisible by 18.
13	Twenty-five (25)	If a number ends with 25, 50, 75 or 00 the number is divisible by 25.
14	Thirty-three (33)	If a number is divisible by 3 as well as 11, it is divisible by 33.

Today you can define mental math in various different ways. Some would say, memorizing times table and remembering the solutions can form the part of mental mathematics. Some would say ability to perform simple calculations in your head can be mental mathematics. The web dictionary defines mental mathematics as "Computing an exact answer without using pencil and paper or other physical aids." Today there are five methods available to learn and practice mental mathematics.

Let's begin with the first one called 'Learning by Heart' or better known as the rote memorizing method where your teachers ask you to mug up boring multiplication tables. It not only kills the interest of the child in mathematics but also makes sure that he develops hatred towards the subject for the rest of the years he studies it. This system gives its ardent devotee some degree of success initially as he is able to answer easy problems but then when the supposedly bigger application problems come the steam is almost over. The second one gives you a good degree of success and I would highly recommend it to the younger lot out there.

It hails from China and is popular by the name of The Abacus (also known as the Soroban in Japan). An abacus is a calculating tool, often constructed as a wooden frame with beads sliding on wires. With the use of this tool one can perform calculations relating to addition, subtraction, multiplication and division with ease. Gradually one practices with the tool in one's hand and later on when experienced he learns to do it without the tool. This tool is then fitted into the mind mentally and he can then add, subtract multiply and divide in seconds. This tool also enhances a child's concentration levels.

The main drawback of this system is that it focuses only on the 4 mathematical operations. Concepts beyond these operations such as Algebra, Square Roots, Cubes, Squares, Calculus, and Geometry etc. cannot be solved using it at all. Also one needs a longer time to be able to fully get a grasp of the system hence you see courses in the abacus stretching to over 2 years which leads the child to boredom and then quitting from the course. Another Chinese system mainly collected from the book The Nine Chapters on the Mathematical Art lays out an approach to mathematics that centres on finding the most general methods of solving problems.

Entries in the book usually take the form of a statement of a problem, followed by the statement of the solution, and an explanation of the procedure that led to the solution. The methods explained in this system can hardly be termed mental and they lack speed to top it all. The Chinese were definitely the most advanced of the civilization thanks to the

Yangtze and Yellow Rivers but if I were to choose out of the two methods given by this culture It would be the abacus. If wars have a 99.99% downside, sometimes they can have an upside too for they give birth to stories of hope and creativity.

The next mental math system was developed during the Second World War in the Nazi Concentration Camp by a Ukrainian Mathematician Jakow Trachtenberg to keep his mind occupied. What resulted is now known as the Trachtenberg Speed System of Mathematics and consists of Rapid Mental Methods of doing Mathematics.

The system consists of a number of readily memorized patterns that allow one to perform arithmetic computations very quickly. It has wider applications than the Abacus and apart from the four basic operation methods it covers Squares and Square Roots.

The method focuses mostly on Multiplication and it even gives patterns for multiplication by particular number say 5,6,7 and even 11 and 12. It then gives a general method for rapid multiplication and a special two finger method. After practicing the method myself I realized that the multiplication was a very applicable mental method but the other methods covered to solve division and square roots were not very friendly and were impossible to be done mentally. I was in search of a much better wholesome method where I could easily perform other operations also.

Another drawback of this system was that it too like the abacus failed to have a wider scope i.e to encompass other fields like Algebra, Calculus, Trignometry, Cube Roots etc. A Recommendation by a friend of mine from America introduced me to what is known as the Kumon Math Method. It was founded by a Japanese educator Toru Kumon in 1950s and as of 2007 over 4 million children were studying under the Kumon Method in over 43 different countries. Students do not work together as a class but progress through the curriculum at their own pace, moving on to the next level when they have achieved mastery of the previous level. This sometimes involves repeating the same set of worksheets until the student achieves a satisfactory score within a specified time limit.

In North American Kumon Centres, the mathematics program starts with very basic skills, such as pattern recognition and counting, and progresses to increasingly challenging subjects, such as calculus, probability and statistics. The Kumon Method does not cover geometry as a separate topic but provides sufficient geometry practice to meet the prerequisites for trigonometry, which is covered within the Kumon math program. I was much impressed with the glamour around Kumon but a glimpse of its curriculum deeply disappointed me.

It is not mental at all. It does not offer any special methods to do mathematics and one does not improve one's speed by doing Kumon Math. There is a set curriculum of worksheets which one does till one achieves mastery in the subject. So say for example a sheet on Divison- one would continue to do division by the conventional method till he gets a satisfactory score and then he moves on to a higher level. This certainly doesn't make division any faster and the process is certainly not mental. A deep thought on the reason of its tremendous popularity in America led me to conclude was the lack of a franchisee business model of the abacus and the Trachtenberg speed system in the 1950s.

The franchisee model was essential in taking the course from country to country. This is where Toru Kumon thrived. Dissapointed with other cultures in the world, my search made me look in my own Indian culture. What I found astonished and amazed me so much that I fell in love with the system and started coaching neighbourhood students in it. This is easily the World's Fastest Mental Mathematics System called High Speed Vedic Mathematics. It has its roots in Ancient Indian Scriptures called the Vedas meaning 'the fountain head of knowledge'.

With it not only you can add, subtract, multiply or divide which is the limiting factor of the abacus but you can also solve complex mathematics such as algebra, geometry, Calculus, and Trigonometry. Some of the most advanced, complex and arduous problems can be solved using the Vedic Maths method with extreme ease. And all this with just 16 word formulas written in Sanskrit.

High Speed Vedic Mathematics was founded by Swami Sri Bharati Krishna Tirthaji Maharaja who was the Shankaracharya (Monk of the Highest Order) of Govardhan Matha in Puri between 1911 and 1918. They are called "Vedic" as because the sutras are contained in the Atharva Veda - a branch of mathematics and engineering in the Ancient Indian Scriptures. High Speed Vedic Mathematics is far more systematic, simplified and unified than the conventional system.

It is a mental tool for calculation that encourages the development and use of intuition and innovation, while giving the student a lot of flexibility, fun and satisfaction . For your child, it means giving them a competitive edge, a way to optimize their performance and gives them an edge in mathematics and logic that will help them to shine in the classroom and beyond.

Therefore it's direct and easy to implement in schools - a reason behind its enormous popularity among academicians and students. It complements the Mathematics curriculum conventionally taught in schools by acting as a powerful checking tool and goes to save precious time in examinations. The Trachtenberg Method is often compared to Vedic Mathematics. Infact even some of the multiplication methods are strikingly similar. The Trachtenberg system comes the closest to the Vedic System in comparison and ease of the methods. But the ease and mental solvability of the other method especially division, square roots, cube roots, Algebraic Equations, Trigonometry, Calculus etc. clearly gives the Vedic System an edge. Even NASA is said to be using some of this methods applications in the field of artificial intelligence.

There are just 16 Vedic Math sutras or word formulas which one needs to practice in order to be efficient in Vedic Math system. Sutras or Word Math Formulas such as the Vertically and Crosswise, All from Nine and Last from ten helps to solve complex problems with ease and also a single formula can be applied in two or more fields at the same time. The Vertically and Crosswise formula is one such gem by which one can multiply, find squares, solve simultaneous equations and find the determinant of a matrix all at the same time. If either of these methods is learned at an early age, a student aged 14 can perform lightening fast calculations easily during his examinations and ace through them. Vedic Mathematics is fast gaining popularity in this millennium. It is being considered as the only mental math system suited for a child as it helps to develop his numerical as well as mental abilities. The methods are new and practical and teach only Mental Rapid Mathematics. The system does not focus on learning by repetition as in the Kumon Method. The system focuses on improving intelligence by teaching fundamentals and alternate methods. The purpose is not limited to improving performance in the school or tests, but on providing a broader outlook resulting in improved mathematical intelligence and mental sharpness.

3

Glimpses of Vedic Mathematics

Vedic mathematics was lost to the modern world over several thousand years. Fortunately, the renowned scholar Jagadguru Swami Bharti Krishna Tirthahji Maharaj, Shankracharya of Sharda Peeth and Goverdham Math discovered 16 Vedic Sutras and 13 Upsutras in the Parishistha of Atharvved. After continuous Sadhana and hardwork, he was able to decode them and get wide ranging mathematical principles and applications from them. He found that these Sturas cover all aspects of mathematical science in depth. He wrote sixteen volumes on Vedic mathematics covering all aspects of these sutras in depth. Unfortunately, the manuscript was lost which fact was finally confirmed in 1956.

During his last days when his health was failing and his eyesight was weak, he rewrote from his memory an introductory account of the subject. He attained Mahasmadhi in 1960. Later in 1965 a book entitled "Vedic Mathematics" was published. This book is edited by Dr. V.S.Agarwala with the help of Smt. Manula Trivedi on the basis of manuscript and notes left by Swamijee. This book is therefore, the first book to give a glimpse of Vedic mathematics and to reestablished the subject. The book attracted the attention of several Indians and foreigners. Dr. Narinder Puri of

Roorkee University took up the cause of Vedic Mathematics in right earnest. His Holiness Maharishi Mahesh Yogi spread the message of Vedic mathematics throughout the world by sending Dr. Puri to several countries and making it possible for many schools all over the world start teaching Vedic mathematics. A former minister of Human Resource Development, in the Government of India, Sh. P.V.Narasima Rao made a reference in Parliament about the work on Vedic mathematics which created a wider interest in the nation. Rashtritya Ved Vidya Pratishthan started looking into the matter and appointed a committee on Vedic Mathematics. During the same period a controversy arose about the source of these 16 Sutras and the level of mathematics contained in them. One view was that these Sutras do not have a Vedic source and that they deal only with simple arithmetical operations.

In the meantime, Dr. Sant Kumar Kapoor, inspired by integrated impact of the Ganita Sutras, started working on basic aspects of the mathematical structure of Vedic literature. He got several clues from Ganita Sutras, Maheshwara Sutra and many other Vedic Mantras. He was inspired by the divine blessing from Yograj Sri Sripad Babaji of Vrindavan and His Holiness Maharishi Mahesh Yogiji at Maharishi Nagar. He has been awarded Ph.D. degree by Kurukshetra University, in December 1990 on his thesis titled "Mathematical Basis of Vedic literature". His work shows that the potential of Ganita Sutras is much higher than what appears in the book "Vedic Mathematics" and the mathematical structure of Ganita Sutras is similar to that of Samved indicating that Ganita Sturas belong to Vedic family. His work shows that the Devnagri alphabet and the Vedic scriptures are mathematically organized. Perhaps a more important aspect of his work in the existence of Real Geometric Spaces of higher dimensions in the form of a continuum. A comprehensive description of his research is given in the next section. Dr. Sant Kumar Kapoor is presently working as a Visiting Professor in the Indian Institute of Maharishi Vedic Science & Technology which is a part of Maharishi Ved Vigyan Vishwa Vidyapeeth, Maharishi Nagar. Three research students are working in our institution on the structure and properties of fourth, fifth and sixth dimensional spaces.

Why Vedic Mathematics?

The questions which should be addressed are as to why the modern mathematics is held up, why its logic recoils upon itself and why there are mathematical problems, logical knots and mental blocks at all in the modern mathematical approach?

Well known problems of modern mathematics may be cited as:

1. Everywhere continuous but nowhere differentiable functions
2. Hypercubes 1 to 7 increase but hypercube 8 onwards decrease
3. Space Filling Curves
4. Riemann Hypothesis
5. Goldbach's conjecture
6. Fermat's Last Theorem

Isn't it that these problems are there because of the axioms accepted by the modern mathematics?

And then follows a question as to whether Vedic mathematics is in a position to help the modern mathematics to come out of its mental block and to untie its logical knots and to solve the problems?

The Vedic geometric concepts worked out in the books of Dr. Kapoor promise us geometric comprehensions of our existence phenomenon transcending our existing three space format. The real four and higher spaces formats of Vedic comprehensions are new wonderful worlds of very rich mathematics which may ensure us powerful technologies and much potentialised disciplines of knowledge. The basic comprehension pointed out is the way the cosmic surface constitutes and binds the solid granules as synthetic solids manifesting in the cosmos.

Dr. Kapoor is attempting to reconstruct the discipline of geometry as a discipline based on Vedic concepts. He has designated this discipline as Vedic Geometry. His results has added a new dimension to the dialogue initiated with the interpretation of the Ganita Sutras and their potentialities brought to focus by Swami Bharti Krishna Tirthaji Maharaj.

Dr. Kapoor's conclusion is that this all is there only because of the acceptance of the geometric entity (monad) admitting no parts, and "1" has no predecessor. To overcome this, as per him, the modern mathematics needs Vedic mathematics' help to shift from monad without parts to a monad admitting parts. The eliptic equations format $y^2 = x^3$ is bound to give a conceptual slip and this, as per him, can be well glimpsed by chasing the format of this equation on simplex format to see how it is deceptive to appear to be so while as whole numbers artifices parallel to the dimensional frames is well evident inequality. As such, there is a need for the modern mathematics to re-address to itself about the need for re-settlement of the basics to come out of the mental blocks and logical knots to untie the knots and to transcend the blocks and to be face to face with the wonderful worlds of reality awaiting ahead with all potentialities of their structural richness. The parallelism between artifices of whole numbers 1 to 26 and 26 sporadic groups is there because of the cosmic surface within the solids.

The recent academic research attempts and teaching experiments with the help of Vedic mathematical operations demonstrate their potentialities to provide the desired help.

The research results are bringing us nearer the traditional acceptance as that Vedas are written on the rays of the Sun. Vedic mathematics, science & technology is the mathematics, science & technology of the way the nature maintains grand unification of the existence phenomenon on the Earth through the rays of the Sun. It is in this grand design of the nature the individual Vedic mantras are impulses of consciousness. This design maintains the continuity of the life within human frame and beyond through the natural intelligence embedded in the human mind and in the rays of Sun. This continuity and parallelism when chased promises new wonderful experiential domains about new realities and the wonderful domains to unfold for us new disciplines of mathematics, science & technology.

Vedic sounds are multidimensional domain frequencies from within the particular dimensional frame as the structure of that domain. When the sounds are pronounced, the frozen frequencies get initiated and the self-organizing power of the Vedic sounds set the frequency's potentialisation process into action. It is this process whose utilization is the aim of different Vedic scriptures. Rig Ved Samhita is the first Vedic scripture. It is the first book of the mankind. The mathematics precedes the composition of Rig Ved Samhita. Vedic Mathematics helped to transform the universal set of knowledge as a speaking language and in the process it itself as well transformed as such and assimilated its identity into the Vedas. Within Vedas, all discipline of knowledge transform their identity and get assimilated into the single discipline of organization of knowledge on geometric formats. Vedic geometry and mathematics as such help us to work out these formats.

Vedic Mathematical Research

Studies reveal that knowledge and organisation of knowledge are two distinct disciplines. The thesis of Dr. S.K.Kapoor is substantively covering the organisational aspect of Vedic knowledge. The main results of the thesis and his subsequent research can be summed as under: 1. Real 4 and higher dimensional spaces do exist. Real 4 and higher dimensional spaces were not only known to the Vedic seers but those were precisely availed by them for organisation of vedic knowledge. Vedic alphabet format is a precise setup of 4-space. The alphabet letters of Devnagri script, their forms formulations, combinations and rules etc. are precisely mathematical. Sakala Rigvedic Samhita is the oldeset scripture of mankind. The organisational format of this Samhita is precisely that of real 6-space. Vedic knowledge accepts a measuring rod constituted by regular bodies of the first six real dimensional spaces which admit formulations as:

Interval	Square	Cube	Hypercube-4	Hypercube-5	Hypercube-6
		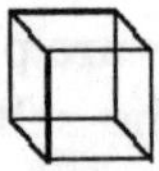	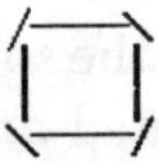		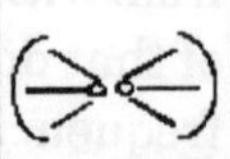
1-space body	2-space body	3-space body	4-space body	5-space body	6-space body

The organisational format of Srimad Bhagwad Gita emerges to be of prime importance as it is parallel to the organising power of the knowledge content of the scripture. This format tallies with the format of human body on the one hand and the sun on the other hand as of real 6-space. The Trinity of Gods namely Brahma, Mahesh, Vishnu are the overlords of real 4, 5 & 6 spaces respectively. The geometrical continuum expressed as manifestation layers of 4 folds of consecutive dimensional spaces contents. These folds of the nth manifestation layer can be represented as under:

First Fold	Second Fold	Third Fold	Fourth Fold
Dimension	Frame	Domain	Origin
(n-2) space content	(n-1) space content	n-space content	(n+1) space content

The transcendence from one manifestation layer to another manifestation layer giving rise to the following (five steps) chain reaction or five steps which are possible within the setup of Panch Mahabhut.

Manifestation Layer	Dimension	Frame	Domain	Origin
nth	Space fold (n-2)	Space fold (n-1)	Space fold n	Space fold (n+1)
(n+1)th	Space fold (n-1)	Space fold n	Space fold (n+1)	Space fold (n+2)

Transition from one space to another space is to be had in terms of unlocking of the seals of the origin points of all the four folds of the manifestation. The modern mathematical models of transition from straight line to plane deserve serious reexamination. In particular the axioms of space filling curves and the axioms of 'one' without a predecessor deserve close scrutiny as their rationale emerges to be without basis.

The role of real numbers additive group (R,+) and real numbers field (R,+,×) with reference to straight line deserve to be differentiated.

The plane deserves to be studied as four geometrically distinct quarters. One faced plane and two faced plane are two distinct geometrical setups and they deserve to be taken up as such.

The concepts of origin and dimension are two concepts with respect to which the modern geometrical models are not up to date. These two concepts deserves to be studied in detail as transcendence to the higher dimensional spaces is possible only in terms of their understanding.

Human body is a compactified phenomenon of multi-layer physiological existence. The start with state of existence is that of waking state which is parallel to the expression of 1-space as dimension into 3-space domain. Sequentially, the existence phenomenon unfolds until seventh state of consciousness which would be corresponding to the 7-space as dimension into 9 space domain. The origin point of the 6-space, being the 7-space setup, the human body, geometrically, turns out to be hypercube-6 and this would explain how the primordial sound, the planetary effects, the

Yajna oblations etc. operate and precisely influence the individual existence patterns.

Srimad Bhagwad Gita is one such scripture whose organisational format precisely workout for us the structural set up and frames of the 6-space. The study zone of Srimad Bhagad Gita can be worked as under:

Srimad Bhagwad Gita
Study Zone

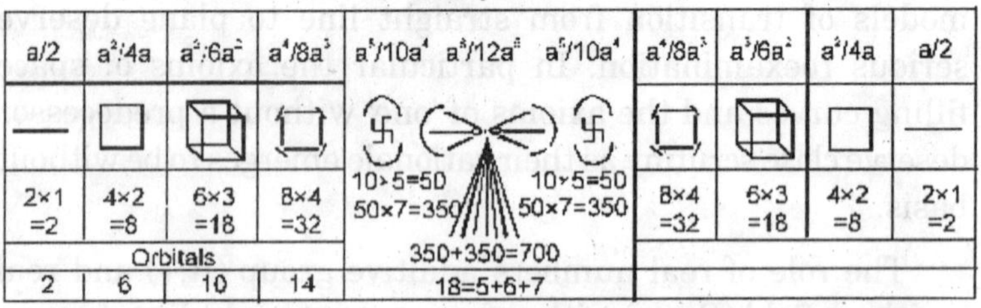

The organisational setups of Ganita Sutras, Maheshwara Sutras, Saraswati Mantras, Gyatri Mantra and Om formulation deserve interdisciplinary explorations.

Sankhay Nistntha and Yoga Nistha are complementary and supplementary of each other and as such their complementary nature and supplementary nature deserve to be distinguished well. Non-differentiation of the same is bound to deprive us of most of the results in specific forms.

Ganita Sutras

Ganita Sutras is a complete system in itself. These are Vedic aphorisms. These are 29 in number; 16 of them are designated as Sutras and other 13 as Upsutras. Credit goes to Swami Bharti Krishna Tirthaji Maharaj, a Shankracharya of Kanshipeeth to focus the attention of the present generation about potentialities of Sutras.

The information contained in the book titled "Vedic Mathematics" compiled by Prof. V. S. Agarwala, published (1965) from the manuscript papers of Swamiji make us known as how Swamiji had to devote many years to decode the working rules of these Sutras.

From the demonstrations of the working rules of these Sutras in this book it has convinced many about the potentialities of the Sutras. Since then many scholars have further demonstrated about their potentialities to make the mathematics much easier and more effectively within the comprehensions of much larger population and also about their utility and academic values in many ways. Dr. S. K. Kapoor has approached the organization format of this system on the whole and as individual Sutras. His approach and the results give us further insight of this wonderful system.

Here Sutras 1, 2 & 3 and Upsutras 1 & 2 are being introduced from the book "Goldbach Theorem" to give insight about their formats and to give idea about the basics of the different processes and the way this system is approaching different branches of mathematics in a unified way as a single discipline. The parallel text of the Sutras adopted in different scriptures is being searched. Dr. Kapoor points out Shatpath Brahman. Hereunder, the illustrative reference with working rule is indicated in reference to Ganita Sutra 1. The other Sutras may be taken up in due course of time.

GANITA SUTRA 1

Format:

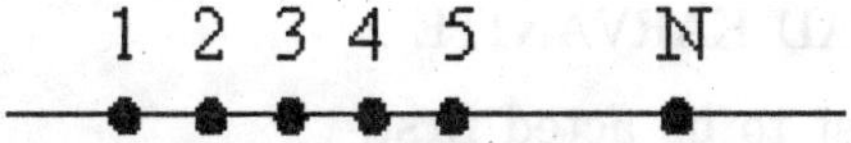

Basics/Technical Words

Text avails three technical words which take us to the basics of the working domain of the Sutra. These are: (1) Eka (one), (2) Edhikena (more than), and (3) Purvena (previous). These three technical words take us to the three basic concepts: (1) the concept of "one", (2) the concept of "one more than", and (3) the concept of "previous one". The first concept is the wholesome concept. Here it manifests in many ways and can be availed as a unit, unit entity, unit measure, the counting number 1, the close interval of unit length, etc., etc. The second concept "more than" is the concept of increase, the concept of comparison of one entity being bigger than/ more than other entity. The third concept is the concept of "order". It is the concept of a queue, a pair of entities of a queue of which one is previous as comparison to the second and the second automatically being subsequent to the first. Therefore, the Sutra 1 has many applied values which would work out for us (1) counting, (2) arrangement of counting pebbles, (3) arrangement of beads along a thread, (4) the queue of points along with line, etc. If we try to chase the repeated applications of "one more than the previous one", we can have a countably infinite steps. Not only that, we can by a suitable choice of "one" and the application of the working rule of Sutra 1, have a chase for uncountable infinity as the line is by having sequential placements of points as counts with index values allotted as per the working rule of Sutra 1 from the set of natural numbers (whole numbers/ counting numbers).

Parallel text and application of Sutra 1 in Shatpath Brahman:

1.1.2.5 (Kanda 1, Parpathaka 1, Brahmana 2, Kandika 5) Page 10

YADEV AGRE TAD KARVANITE

That which is first to be acted first.

2.5.2.22 (Kanda 2, Parpathaka 5, Brahmana 2, Kandika 22) Page 300

EKAIN EKEIN YA ASYE PRAJA .... EKE NATI RIKTANI

For everybody there is one utensil. As many are the members of the family,

so many are the utensils and there is one extra.

2.6.2.4 (Kanda 2, Parpathaka 6, Brahmana 2, Kandika 4) Page 328)

EKAIN EKEIN YA ASYE PRAJA .... EKE NATI RIKTANI

For everybody there is one utensil. As many are the members of the family,

so many are the utensils and there is one extra.

Let there be a family F of members m1, m2, ...mf. Let this family has a kitchen K with utensils

u1, u2, ...uk.

For one-to-one association of family members of F with the utensils of K, the first member m1 is to be associated first, the first utensil of K i.e. u1.

To each member of F is associated from K.

The requirement is that after such association we should have one extra utensil in K. This makes the cardinality of K as f+1.

The set F (of cardinality f) and the set K (of cardinality f+1) makes a pair of sets (F, K) and they supply us the pair of numbers (f, f+1). This is precisely the basics and formation of Ganita Sutra 1.

GANITA UPSUTRA 1

Format:

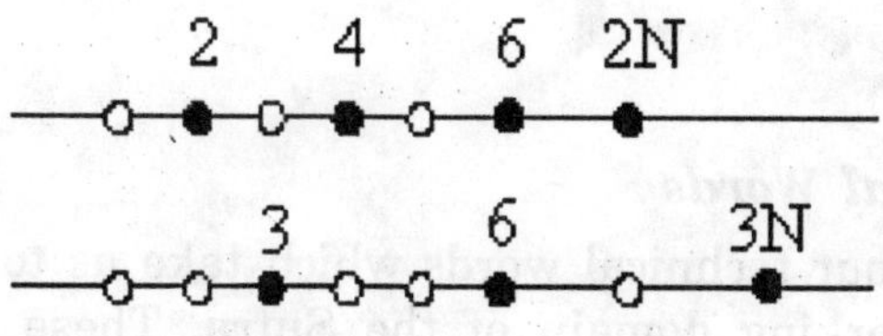

Basics/Technical Words

The text of Upsutra 1 avails a pair of technical words: (1) Anu (to follow), and (2) Rupyena (the form of). When 'one' is viewed as synthetic i.e. composed of parts, then each of its parts would be symmetrical to the whole and it would be proportionate of the whole. As such, (1) part as symmetrical to the whole, and (2) part as proportionate of the whole would be two distinct aspects of the same organisation. Infinity is another word in which melt away all finiteness. In infinite domains or self-contained domains, the uniformity and patterns are other keywords in terms of which the concept of oneness can be approached as a symmetry and proportions. The applied values of Upsutra 1 would help us reach at unit interval as replica of infinite line, the universality of Yatha pinde tatha brahmande (as in the body so in the universe) and Yatha brahmande tatha binde (as in the universe so in the body) as well as the universality of the Upanishadic comprehensions as that even when full is taken out of full, there still remains full. With this, when one would work Sutra 1 with double of a unit instead of a unit or with any multiple of a unit instead of a unit, one would see that the rule works. With this, the Sutra 1 coupled with Upsutra 1 makes the potentialities of Sutra 1 manyfold.

GANITA SUTRA 2

Format:

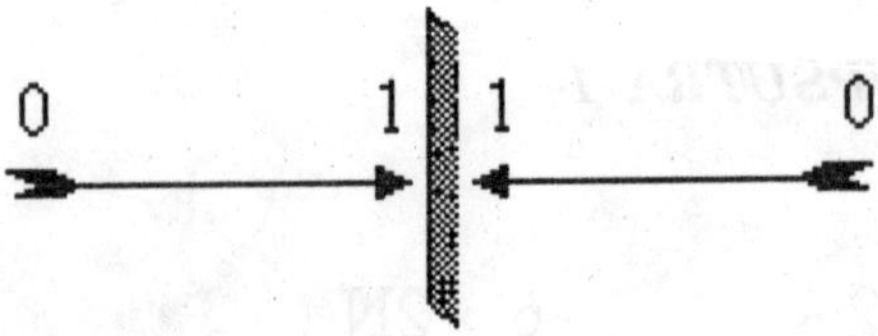

Basics/Technical Words

Text avails four technical words which take us to the basics of the working domain of the Sutra. These are: (1) Nikhilam (all), (2) Navastasca (from nine), (3) Caramam (last), and (4) Dastah (ten).

Nine numerals and ten place value system are the two basic steps in terms of which the infinitely long line stands tamed. The counting pebbles technique requires infinite pebbles for coverage of infinitely long line. However, symmetry / proportions / replica rules suggest that proper system can be constructed to express "infinity in terms of finite". Vedic seers comprehended ten place value system. From the method of expression tenth counting pebble as 10 as in two digit form, we can notice that here lies a shift from single digit expression for first 9 counts to the next multiple digit expression initiation for expression of tenth count onwards. We also know that first 9 counts as well can be expressed in double digit forms as: 01, 02, 03, 04, 05, 06, 07, 08 & 09. Therefore, the first thing about this transition is the introduction of the tenth symbol for the zero count as "0". Further, it also would be interesting to note that the double digit form for 1 as 01 provides us an expression for the tenth count as 10 simply by having swapping of places for first and the second digit. Further, the above interlink of 01 and 10 takes us to their link and inerlocking as reflection pair of numbers as much as that the formats beneath 01 and 10 get interconnected like image of an arrow through looking mirror gets interlocked with the arrow itself. The placements of arrow, mirror, and the image of the arrow work out a dimonad format with mirror playing the role of joint of the diamond. This as such is a big leap forward from the organisation format of Sutra 1. This had become possible with the transition stage of Upsutra 1. Therefore, Sutra 1, Upsutra 1 and Sutra 2 together make much interesting dimensions to the mathematics.

The above organisation of Sutra 1, Upsutra 1 and Sutra 2 takes us to a stage where the middle of interval, which otherwise had the privilege of uncountably infinite placements, gets tamed as a unique mirror placement to have half unit in front of the mirror as first half and its image through the mirror as the second half. It is this way unique fixation of the middle makes the mathematics workable in terms of "half unit", like the organisation in terms of "half dimension". With this, it becomes possible to go to a number which exists prior to "one" i.e. half. It is a big achievement over the mathematics of "one" without predecessor.

However, simultaneously, it makes the working very delicate as we shall be permitting the second half to remain dormant while structures because of the second half also shall be silently marking their presence. We cannot afford to ignore the existence of the structures because of the second dormant half.

If one is to see the type of difficulties in which the mathematics when worked without taking care for the second dormant half, one is simply to pose to oneself as to why from hypercircle-8 onwards the values start decreasing. It simply happens as the linear order worked in terms of half-dimension helps us sequentially increase only uptil seventh step parallel to the possible seven geometries of 3-space of linear order.

The organisation of pairing of two halves with the help of a mirror at the middle is unique organisation of many features, of which the most important is the feature of common remainder for both the halves while both halves are divided by a common divisor.

Though this property may appear to be obvious but in fact when the same is viewed in the context of uncountably infinite points being handled in terms of counting numbers, one would realise, how important this property of constant remainders emerges to be. This is precisely Upsutra 2.

GANITA UPSUTRA 2

Format:

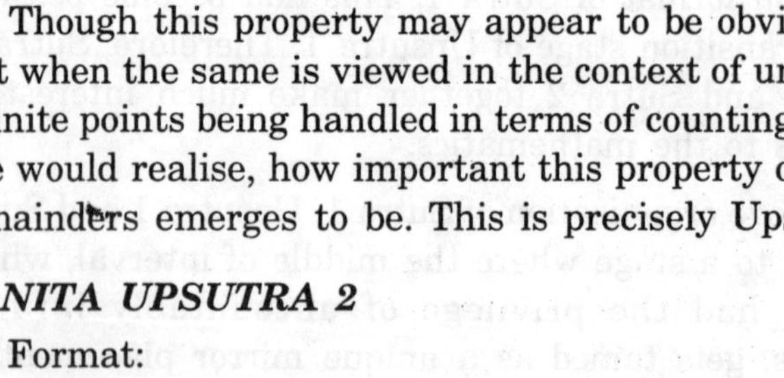

Basics/Technical words:

The basic / technical word availed by the text of Upsutra 2 is "Sesa" (remainder). The text as such is the definition of remainder: remainder is that which remains (as uncovered part of the divisor, on division).

The basic operation is the operation of "division". This Upsutra along with Upsutra 1 makes the mathematics of Upsutras a very interesting branch in itself.

The "one" which permitted itself as a pair of halves, sequentially by the rule of symmetry of Upsutra 1 and the rule of equality of remainders of division of equals takes the process of partitioning of "one" into as many equal parts as one would like to have.

This makes the external expansion of counts from one to two, two to three and so on parallel to the internal partitioning from wholesome one to a pair of parts, then to three parts, four parts and so on. With this, the arrays of mathematical tools become sufficiently large to work out very rich structures only in terms of "parts", of which the working with "half" is most rewarding.

GANITA SUTRA 3

Format:

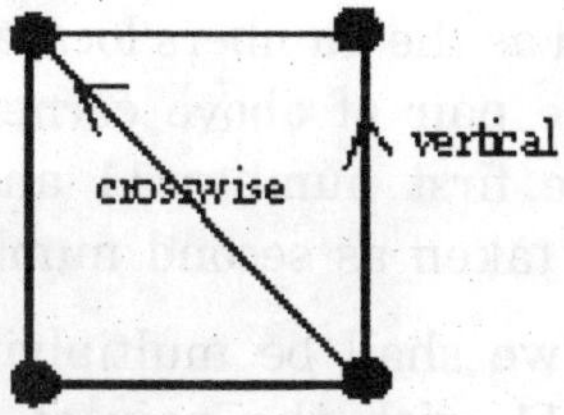

Basics/Technical words

The basic / technical words availed by the text are: (1) Urdhva (vertical) and (2) Tiryag (crosswise).

We may illustrate one of the working rules of this Sutra, which would help us to reach at its organisational format as well. Let us multiply 11 by 11. We shall be getting the product equal to 121. Let us have a close look at this number. The central digit of this number is 2. It admits expression as 1+1. Let us replace the central digit 2 with its above expression 1+1. The three parts of the number 121 would permit depiction as: 1, 1+1, 1. Now, let us chase the multiplication the Sutra way:

The first Sutra gives us the working rule to proceed step by step as "one more than the previous one". When we are working only with "half", therefore, in this case of 121 we shall be requiring the last two digits 12 or the first two digits 21. Therefore, when we shall be having 1 as start with position, as a next step under the rule of Sutra 1, we shall be reaching at 12. Then the reflection image of 12 would give us 21 and we shall be obtaining 121. If we would have been multiplying 111×111, we would be obtaining in three steps 123 and its image 321 which when combined would give us 12321 as the product of 111×111. Likewise if we would have been multiplying 1111×1111, we would be obtaining in four steps 1234 and its image 4321 which when combined would give us the required product as 1234321. The process can be extended.

The product 11×11 being the product of pair of numbers of double digits, therefore, these four digits, here each being 1, can be expressed as the numbers located at the four corners of a square. The pair of above corners may be taken as representing the first number 11 and the pair of lower corners may be taken as second number 11.

Now, when we shall be multiplying the first digit of lower number 11 with the pair of digits of the above number 11 one by one, we shall be approaching as is depicted above in terms of the arrows.

The depiction is of two parts. The first part is Urdhva (vertical) and the second part is Tiryag (crosswise). This as such, would help us comprehend as that geometric format of square is being divided into two triangles and only 1 out of two triangles is being utilised by the organisation format of Sutra 3.

Here it would be relevant to note that the triangle is the first close organisation of lines. In other words we require a minimum of three lines to have a close organisation to enclose planes (area) / surface within its lines.

The triangle as such is the printout of 3-space on 2-space. The sum of all the three internal angles of a triangle isp. The sum of all the three external angles of a triangle is 2p. We can express p as (3-2)p. The sum of all the four internal angles of a quadrilateral is 2p. We can express it as (4-2)p. In general, the printout of n-space on 2-space is n-polygon and the sum of the internal angles n-polygon is (n-2)p. This in general can be seen as that the printouts of n-space in the role of dimension on 2-space is n-polygon.

Therefore, the organisation format of Sutra 3 is that of a printout of 3-space in the role of dimension on 2-space format.

Like that, one can proceed to reach at the organisation formats of the entire range of the text of 16 Ganita Sutras and 13 Upsutras but the context and the space at hand would not permit to proceed ahead with it.

However, the organisation formats achieved uptil this stage well indicate about the mathematics of linear order of Sutra 1, spatial order of Sutra 2 and of solid order of Sutra 3.

Multi-Dimensions of Time and Space in Manasara

Vedic Wisdom

Vedic wisdom is lively in the consciousness of living saints. It is also well preserved in the Vedic literature. The inner evidence of the available Vedic knowledge of the available Vedic literature makes it out that originally the whole range of the pure knowledge i.e. Vedic knowledge was vibrating from a single wholesome Ved. Subsequently this knowledge was organized by Maharishi Ved Vyas as four Veds namely. Rigved, Yajurved, Samved and Atharvved.

The knowledge of Rigved admitted 21 branches while the knowledge of other three Veds namely, Yajurved, Samved and Atharvved respectively admitted 101, 1000 & 9 branches. The organizational format of each Vedic branch consisted of four folds designated as Samhita, Brahmana, Aryanak and Upanishad. As such 21+101+1000+9=1131 Vedic branches had 1131 Samhitas, 1131 Brahmanas, 1131 Arynaks and 1131 Upanishads. These 1131x4 = 4524 scriptures together came to be known as Vedic wisdom. In addition corresponding to each Ved, as applied value of the pure knowledge of the Ved is Upved. The four Upveds are Ayurved, Dhanurved, Gandharvved and Sthapathyaved. Our present day mathematics, science and technology come within the range of Sthapatyaupved. Manasara is one such scripture of Sthapatyaupved.

Manasara : Scripture of Sthapatyaupved

Sri Prasana Kumar Acharya had done a wonderful job of reconstructing the text of Manasara and preparing its translation in English and by drawing the plates.

The scripture begins with the prayer to Lord Brahma, the Creator, the supreme and ends with the chiseling of third eye of the idol of Lord Shiv. The broad organization of the scripture is of the range of 70 chapters with first eight chapters constituting a primary group-I. The next ten chapters i.e. chapter 9 to 18 constitute a primary group-II. Then comes the central part of the Mansara. Chapter 19 to 30 cover single story building to 12 storied buildings. These 12 chapters constitute a central group-I. Next 20 chapters i.e. chapters 31 to chapter 50 cover central group-II and with it the subject of architecture as such is completed. Then follows the subject of sculptures. The science of sculptures covered in chapters 51 to 70 can be organizationally divided in two sculptural groups. Chapter 51 to 65 constitute sculptural group-I and chapters 66 to 70 constitute sculptural group-II.

Space Time Frame

The above topical division of the text has an organizational message of great importance as in terms of it we may reach at the geometric format of the organization of the knowledge of the scripture.

As the scripture begins with the prayer to Lord Brahma, the overlord of real 4-space and ends with the chiseling of third eye of Lord Shiv, the overlord of real 5-space, therefore, geometric format at the dimensional level is bound to be spatial with the flux of time being solid.

In short, the spacetime frame at the dimensional level is going to be E2 (space) × E3 (time). In terms of this dimensional order we shall be manifesting working geometric domain within the spacetime frame (E2)4 (space) x E3 (solid time). Here (today) we shall be concentrating upon the concepts and comprehensions of dimensions of space and time in Manasara. In a way, we shall be taking up the topic of space, time and spacetime in the light of Vedic wisdom. In the context, it may be relevant to note that the modern thought, mathematics, science and technology is speculating the general spacetime frame as E3 (space) × E1 (linear time).

The modern thought, mathematics, science and technology centre around linear dimensional reality but the Vedic systems avail multidimensional reality. The studies of the organizational formats of various Vedic scriptures reveal that higher dimensional geometric formats are being availed to organize the pure knowledge. Illustratively, we may take the case of the oldest book of mankind namely, Sakla Rigved Samhita.

Rig Ved Samhita

Fortunately Rigved Samhita is intact with us from first syllable to the last syllable and as the tradition goes, the whole range of Vedic knowledge is lively in this scripture of 432000 syllables, out of which 397265 syllables are manifest text while remaining 34735 syllables go deep as organisational format of the text and as such remain un-manifest.

For the present, we may accept it an axiom that knowledge and organisation of knowledge are two distinct aspects of knowledge. Being scriptural text, we get the organised knowledge and as such both organisation format and the text are to be accepted as the knowledge content of the scripture. It is like a truck with goods yielding weight of the truck as well as of the goods loaded in the truck.

Organisation of Rigveda Samhita

Total knowledge contents	432000 Syllables
Manifest text	397265 Syllables
Mandals	10
Ashtaks	08
Chapters	64
Anuvaks	85
Suktas	1028
Vargas	2024
Richas	10552

Mathematical Basis

Mathematical basis of the organisational format of the Rigveda Samhita reveals that the Vedic knowledge is organised on geometric format of real 6-space. It admits 4-space in the role of dimension while modern thought, mathematics, science and technology centre around 3-space reality and as such Veds are invincible fort for the modern mind. As such, we have to learn and understand the Vedic wisdom. For this we have to re-examine the rationale and basis of our axioms and postulates for accepting the reality as linear dimensional one. It is only by approaching the Vedic knowledge, the Vedic way, that we may have real bliss of Vedic wisdom.

Higher Dimensional Reality

Mathematical Basis of Vedic Literature

Studies of mathematical basis of available Vedic literature reveal that 4 and higher dimensional reality was not only known to the Vedic seers rather the great use thereof was made by them for organization of pure knowledge.

Let us concentrate upon the nature of space around us. We can see that straight line is a track of a moving point while plane is a track of a moving (straight) line. Likewise, moving plane shall be creating solid space. More precisely, if we reinvestigate this phenomenon with the help of a point, interval, square and cube, we shall be noticing that moving point accepts line as its track, moving interval accepts square as its track and moving square accepts cube as its track. This would pose a question: What is the nature of the track of moving cube/solid /3-space body/ 3-space setup or in the general, 3-space it self?

Hypercubes 4, 5, & 6

To answer the same let us mathematise this situation as:

Taking point as a representative regular body of 0 space, interval, square and cube respectively as representative regular bodies of 1, 2 & 3 space, we may conclude that moving 0-space (body) accepts 1-space format, moving 1-space (body) accepts 2-space format; moving 2-space body accepts 3-space format, and as a logical consequence the moving 3-space body shall be requiring 4-space format. And in general, moving n-space body shall be requiring (n+1) space format. For convenient handling.

We may define and designate four and higher dimensional bodies in continuation of interval, square and cube as hypercubes. To be precise hypercube 4 shall be a representative regular body of 5-space and so on. Further to facilitate comprehension of main properties of hypercube and for symbolic representation of the set of properties synthesizing hypercube it would be desirable to have suitable symbols for them, particularly for hypercube 4, 5 and 6 for reaching at the concepts and comprehension of multi-dimensional spacetime frame being availed for organization of the knowledge of Manasara.

Let these 3 special symbols of hypercube 4, 5 & 6 be as:

Interval	Square	Cube	Hypercube-4	Hypercube-5	Hypercube-6
			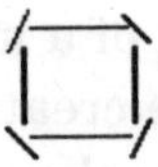		
1-space body	2-space body	3-space body	4-space body	5-space body	6-space body

To have these symbols in continuity of the geometric setup of interval, square and cube, we have to first comprehend these geometric setups. As such, let us have a close look at the geometric setup of interval, square and cube. If we have close look at the geometric setup of interval, square and cube, we may comprehend that interval has length (A^1) and 2 boundary points ($2A^0$) Square has area (A^2) and 4 boundary line ($4A^1$) and cube has volume A^3 and 6 boundary surfaces ($6A^2$).

$A^1{:}2A^0$, $A^2{:}4A^1$ and $A^3.6A^2$ suggest a common formulation $A^n.2nA^{n-1}$, n=1, 2, 3.

This formulation would hold for all values of n. In particular, for n=4 5 & 6 we shall be getting $A^4{:}8A^3$, $A^5{:}10A^4$ and $A^6{:}12A^5$. The geometric message is that boundary of 4-

space body is constituted by 8 cubes while the boundary of 5-space body is constituted by 10 hypercubes-4 and so on.

The boundary components of interval, square, cube and hypercubes 4, 5 & 6 are in the ratio 2:4:6:8:12.This would help us comprehend, appreciate and have the symbols of hypercube 4, 5 & 6 in continuity of and in that sequential order, hypercube 4, 5 & 6 with boundaries as:

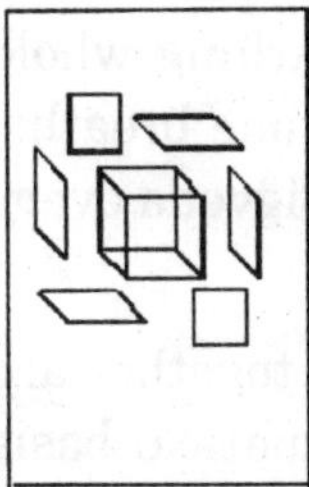	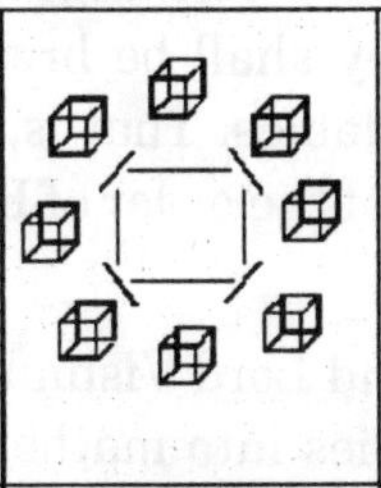	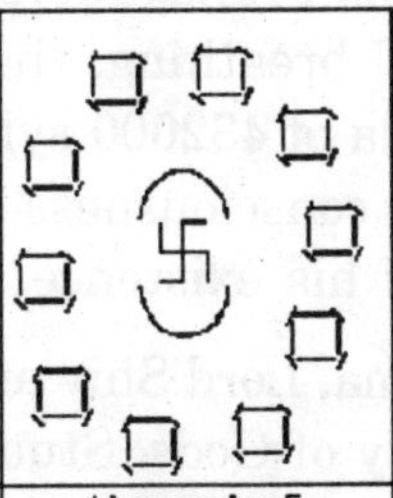	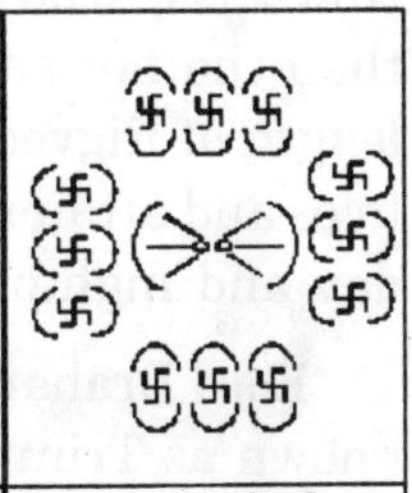
Cube	Hypercube-4	Hypercube-5	Hypercube-6

Shandchakra Format of Human Body

Further it would help us comprehend and appreciate the Upanishad knowledge regarding the Shadchakra format of human body as that:

अथ बहिर्लक्षणाम्। नासिकाकग्रे चतुर्भि षड्भिरष्टभि. दशभि. द्वादशभि. कमात्।

Here are being taken up the external structural characteristics (of shad chakras format of Human body) These with respect to the chakras (second to six) which are ahead of the tip of the nose (first chakra with two characteristics) are four, six, eight ten and twelve respectively, and in that order.

(Advet-tarko Upanishad)

The message of great importance is that Vedic Comprehension of human body is of Shadchakra format which runs parallel to the geometric formats of representative bodies of first six spaces and hence this makes Purusha, the being of 6-space.

This will help us comprehend and appreciate the Vedic tradition which accepts Vedic Richas / Mantras as impulses of one's own consciousness. Here there is another aspect of our existence phenomenon. One breaths 21600 times a day. Equally one would breath 21600 times during night.

This makes day and night breathings as of 43200 units. Now those who do Pranayam (deep berating) 10 times deep than the normal breathing, they shall be breathing whole length of Rigveda of 432000 syllables. That is, one breaths, lives and attains consciousness of the order of Rigveda every day and nigh of his existence.

Lord Brahama, Lord Shiv and Lord Vishnu together are known as Trinity of Gods. Studies into mathematical basis of the knowledge of Trinity of Gods preserved in the available Vedic Literature show a complete parallelism between four heads of Lord Brahma and four dimensions of 4-space; five heads of Lord Shiv and five dimensions of 5-space; and six heads of incarnations of Lord Vishnu and six dimensions of 6-space. The detailed studies of idols of Trinity of Gods further take us to their geometric formats respectively being of hypercube 4, 5 & 6.

Dimensions and Spaces

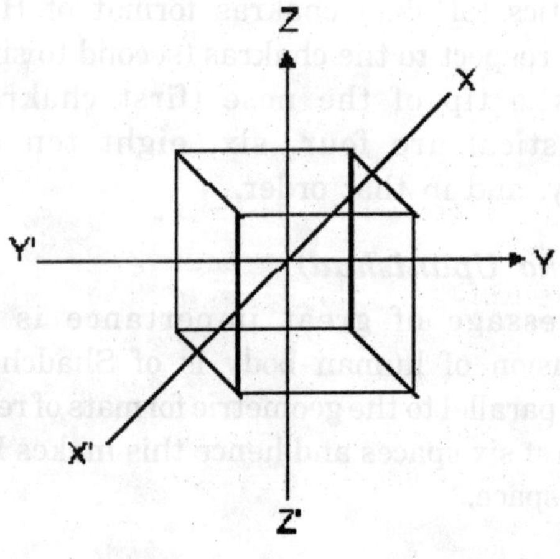

Since our school mathematics, we draw three axes as depicted below as three dimensions of three dimensional frame of three dimensional space.

The message is that we are accepting line as a dimension of 3-space and as such three lines together constitute a three dimensional frame. This in fact is a particular case of general rule that (n-2) space plays the role of dimension of n-space. In modern mathematics as well mental constructs of higher dimensional spaces have been speculated but their dimension always remains a line i.e. 1-space. Obviously this concept is bound to recoil as it would make dimension and space indistinguishable at least in 1-space where 1-space is to be accepted as its own dimension. On the contrary, the rule of difference of space and its dimension as being of two units is well preserved in Ramayana as well as in Mahrabharatam. The narrations are that when the Arm of Lord Rama chased Kag-Bhushandi a difference remained only of two Anguls till Brahman-lok. Likewise when Mother Yashoda tried to tie child Krishna, She found that every string was short by two Anguls.

The great message is that the 'measures' for the measuring rod are bound to be of two units less. The Manasara, the scripture of Sathapatyaupved of Atharvved, accordingly accepts Lord Vishnu (six heads lord) as lord of measuring rod and Lord Brahma (four heads lord) as lord of measure. In geometric language it means that 4-space plays the role of dimension of 6-space.

Mythologically manifestations as idols, accepts head of the idol as format of dimension and eyes of the head together to synthesize a dimensional frame for the dimension.

Format of Idols

We may tabulate the following parallelism between the manifested idols of Trinity of Gods and the geometric set ups of hypercubes 4, 5 & 6 as geometric formats.

S.N.	Idol	Representative Geometric body
1	Number of heads	Number of dimensions of head
2	Head	Dimension of the space
3	Number of eyes in the head	Dimensional frame for the dimensions
4	Lotus (of 8 petals)	Hypercube-4
5	Lotus	Four space
6	Lotus seat	Hypercube-4 as domain
7	Lotus feet	Hypercube-4 as boundary
8	Lotus eye	Hypercube-4 as dimension

Upnishadic Enlightnment

Upanishads and Purans particularly Sri Sri Brahmpuran, Shivpuran and Vishnupuran enlighten us about Trinity of Gods as:

Lord Brahma : Lord Brahma is a four heads lord with two eyes in His each head. He sits majestically on a lotus seat of eight petals. He meditates on His lord (Lord Shiv) in heart and transcends to Shivlok where He multiplies Himself as ten Brahmas of four heads each.

The geometric message of the enlightenment can be decoded in terms of the following complete parallelism between the idol of Lord Brahma and geometric setup of Hypercube-4.

S.N.	Idol of Lord Brahma	Geometrical set up of hypercube-5
1	Four heads	Four dimensions
2	Two eyes	Spatial dimensions 2-space as dimension of 4-space
3	Eight petals	Eight solid boundary components (A4:8A3)
4	Heart	Origin / Centre
5	Lord of Lord Brahma	5-space as origin of 4-space
6	Ten Brahmas	10 hypercubes-4 as boundary of hypercube-5 (A5:10A4)

(Dr. Dharmesh Shukla has worked out the details in his doctoral thesis titled "Some properties of real 4-space in the light of Vedic knowledge accepted (1994) by H.N. Bahuguna University, Uttar Pradesh).

Lord Shiv: Lord Shiv is a five heads lord with three eyes in His each head. He has ten beautiful arms. Sri Sri Shiv Mahanpuran enlightens us that Hanumanji is incarnation of Lord Shiv. Ramayana enlightens us that Lord Rama, the incarnation of Lord Vishnu, resides in the heart of Hanumanji.

The geometric message of this enlightenment can be decoded in terms of the following complete parallelism between the idol of Lord Shiv and geometric setup of Hypercubes-5.

S.N.	Idol of Lord Shiv	Geometrical set up of hypercube-5
1	Five heads	Five dimensions
2	Three eyes	Solid dimensions / 3-space as dimension of 5-space
3	Ten arms	Ten hypersolid boundary components (A5:10A4)
4	Heart	Origin / Centre
5	Lord of Lord Shiv	6-space as origin of 5-space
6	Dwadas Adityas	12 hypercubes-5 as (12 Suns) boundary of hypercube-6 (A6:12A5)

The Thesis of Sh. Bidyadhar Panda titled "Some Properties of real five space in the light of Vedic knowledge works out the details of geometric seup of hypercube 5 and real 5 space of Vedic Comprehension and establishes Sri Sri Shiv Mahapuran as text book of real 5 space. These two these deserve to be published for research scholars.

Manifestation Layers

In the light of the above, if we may have a fresh look at the geometric set up of a cube. We shall be noticing that the 3-space content/lump manifesting as "volume" of the cube is wrapped within 2-space surface (2-space content/sheets). Further, we know that this is the organization which admits linear dimensional order i.e. 1-space in the role of dimension. Now, if we concentrate upon the centre of the cube, we shall be noticing that 8 sub-cube are synthesized as a cube. We may view centre of the cube as being wrapped within 8 cubes. This is parallel to the cut of 3-space by dimensional frame into 8 octants. As such, the formulation A4:8A3 of hypercube-4 being availed by Lord Brahma to manifest as an idol would help us comprehend that at centre of the cube like at the origin of the 3 dimensional frame is sealed 4-space.

Manifestation Layers			
First Fold	Second Fold	Third Fold	Fourth Fold
Dimension Part	Frame Part	Domain Part	Origin Part
		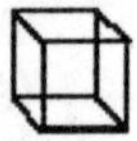	4-space

The origin/centre, the domain/volume, the boundary/surface and linear dimension of cube are four distinct folds of four distinct spaces which together synthesis and manifest geometric setup of a cube. Therefore, the great geometric message is that our physical workd/3-space/cube/macro world is the first manifestation layer with 1-space playing the role of dimension, 2 -pace playing the role of boundary, 3-space playing the role of domain and 4-space playing the role of origin.

Likewise, the world of Lord Brahma i.e. the real 4-space, is the micro world manifestation layer of second order as 2-space plays the role of dimension, 3-space plays the role of boundary, 4-space plays the role of domain and 5-space plays the role of origin. In general, the manifestation layer of order n shall be accepting n-space in the role of dimension and (n+3) space in the role of origin.

The Lord Brahma, the Creator of Supreme, creates reality as manifestation layer of 4 consecutive spaces synthesizing together. This creation accepts origin in a sealed state. It is a dormant state. The transcendence from one manifestation layer to another manifestation layer requires unlocking of the seal of origin fold. This as such becomes possible only on 5-space format. Therefore, Lord Shiv, the transformer the Supreme, is the Lord of transcendence. Accordingly, in Shivlok exists vibrant synthetic setup of 5 folds where transcendence from manifestation layer to another manifestation layer becomes possible.

Vibrant Synthetic Reality

Vedic Model of vibrant synthetic reality of 5-space format is a multi-dimensional format of space and time. It takes account of transcendence through manifestation layers as flux of time. The transition from spatial dimensional order to the solid dimensional order is worked out as a flux of time of 4-space reality which takes us to 5-space reality. In concrete terms, the dimensional order of manifested world being spatial, it attains transition in terms of solid dimensional order of time, being of solid dimensional order. The time flux crystallizes as a tri-monad i.e. the dimensional constituent i.e. monads as of three parts. This tri-monad format of time when supplied to the manifested bodies of the manifestation layers, it works out (2n+1) versions of the manifested bodies. In geometric language, it simply means that the interval, square, cube and hypercube, as representative regular bodies of the respective dimensional space, in the roles of domain fold of respective manifestation layers, shall be accepting (2n+1) versions or in other words, (2n+1) geometries.

Illustratively for n=1, we shall be having 3 geometries of 1-space or in other words, we shall be having 3 versions of an interval designated as close interval, half-open interval and open interval. These versions of intervals are having common domain (length) and they differ only in terms of the boundary components (points). Close interval accepts 2 boundary points, while half-open interval accepts one boundary component and open interval is devoid of both the boundary points. For n=2, we shall be having 5 geometries of 2-space or in other words, we shall be having 5 version of a square in terms, of 4, 3, 2, 1 & 0 boundary lines and for n=3, we shall be having 7 geometries / versions of 3 space/ cube in terms of 6, 5, 4, 3, 2, 1, 0 surface plates and so on.

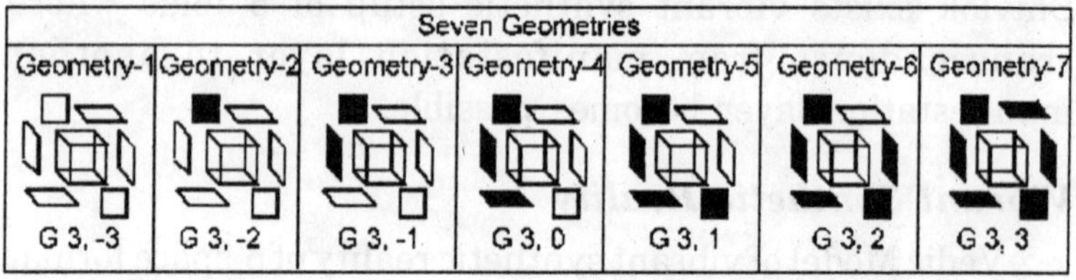

In 3-space, we shall be having 1-space as dimension. The flux of time would be of spatial order. 2-space admits 5 geometries. The 3 dimensions plus 5 versions would give us 3 + 5=8 order. In terms of it, we shall be attaining 8 = 2 × 4, i.e., 4 spatial dimensions which shall be constituting dimensional order of 4-space. This is how the spatial time flux shall be working out transition from linear dimensional order / space i.e. 3-space to spatial dimensional order / space, i.e., 4-space. Likewise, in 4-space, we shall be having 2-space as dimension and 3-space as flux of time. 3-space admits 7 versions. The dimensional order of 3-space admits 7 versions. The dimensional order of 4-space would yield 2 × 4 order. This 2 × 4 order plus 7 versions shall be yielding 8 + 7 = 15 order. In terms of it we shall be attaining 15 = 3 × 5, i.e., 5 solid dimensions which shall be constituting dimensional order of 5-space. This is how the solid time flux shall be working out transition from spatial dimensional order space, i.e., 4-space to solid dimensional order space, i.e., 5-space.

Therefore for proper appreciation of vibrant synthetic reality of Vedic comprehension, we have to work out the

Vedic Knowledge, the Vedic way. The geometric continuum is at play around us. The organizational format of Srimad Bhagwad Gita, the eternal song of Lord Krishna, makes complete use of real 6-space. This may be depicted as study-zone of Srimad Bhagwad Gita as under:

Srimad Bhagwad Gita
Study Zone

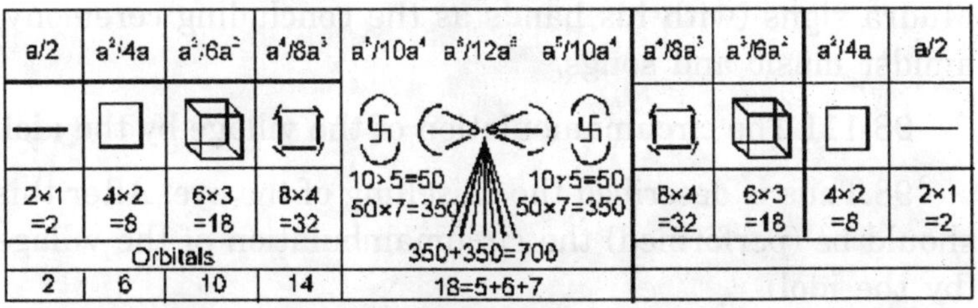

The organization of Manasara runs parallet to the study zone of Srimad Bhagwad Gita. This also runs parallel to the Shad Chakra format of Human body. The organizational format of nucleus of the nucleus as well avails this zone where Jyoti flows from core of the sun into the core of the Sanatana base as 7 streams. This flow is channelised on Om Formulation of 4 components as a Divya Ganga flow. This is a complete discipline in itself. At present we may not be in a position to enter into the details of this Vedic discipline of channelization into the Sanatana base but it may not be out of context to make a mention that advanced students of Vedic Science and Technology have to learn about it.

Organisation of Manasara

Prayer of Lord Brahma to Chiselling of Eye of Lord Shiva

Opening Prayer of Lord Brahma

He (Brahma), while causing the creation, the preservation, and the destruction of the worlds, brings forth earth, water, fire, air, and the sky. I bow to (His) lotus-like feet kissed by the waving lines of bee-like crowns of the kings of various gods. (Manasara Chapter-1 Shaloka 2)

The Chiselling of the Eye

1-2 the chiselling of the eye of the (idols of the) gods, goddesses, and devotees, and the purification (cleansing) of the phalli (of Shiva) will be described in order of the subject.

3-96 ....

97. The Chief architect should show (to the idol) the Mudra signs (with his hands as the concluding ceremony) amidst music and songs.

98-111 The circumambulation of the village by the idol.

98.Thus is described the chiselling of the eye. After this should be (performed) the circumambulation of the village (by the idol).

99-114 ....

115-118. This great science (of architecture) was at first revealed by Brahma, Indra, and all other gods: It is from their statements that this Manasara (the essence of measurement) has been complied as a guide book (for the architects) for the benefit of the people.

Thus in the Manasara, the science of architecture, the seventieth chapter, entitled.

The description of the chiselling of the eye (of image). The Manasara is complete.

Sri Shri Shiva Mahapuran

Lord Shiva is the overlord of real 5-space. Sri Sri Shiva Mahapuran is a text book of real 5-space. The organizational format of Sri Sri Shiva Mahanpuran is the organizational format of real 5-space. The scripture is organized as Samhitas parallel to 7 dimensions of 7-space.

5-space plays the role of dimension of 7-space and as such each of the 7 Samhitas of Sri Sri Shiva Mahanpuran enlightens us about one or the other aspect of the knowledge of real 5-space in the role of dimension.

These 7 Samhitas are organized as 12 parts. This is parallel to 12 boundary components of real 6-space. Each boundary component of real 6-space is a 5-space set up. Therefore, these 12 parts are the organisation of 7 Samhitas of Sri Sri Shiva Mahapuran as boundary of 6-space.

The 7 Samhitas of Sri Sri Mahapuran consisting of 12 parts are further organised as chapters and chapters are further organised as Shlokas. These organisations accept formats of real 5-space. The knowledge content of each chapter and the orgnizational format of each chapter work out one or other aspect of real 5-space.

The first chapter of first Samhita of Sri Sri Shiva Mahapuran posses a question as (that): What is the shortest way to salvation? And the great Rishi remembers Lord Shiva and gets prepared for answering the same. This that way would help us comprehend and appreciate that the question being posed and the answer which is going to follow is to be about Lord Shiva.

The manifestation which is being talked to be transcended through is the manifestation of real 4-space. As such the first chapter of the first Samhlta takes us to the format of hypercube-4 with a sealed origin. The great concern being shown by the sages (Munis) to the great Rishi is about the spatial nature of the setup of 4-space and the question being raised is how this manifestation can be transcended through.

Naturally, the great Rishi concentrates upon the origin and prepares for answering the question by remembering Lord Shiva. Who is the overlord or real 5-space playing the role of origin and remaining in dormant state in the sealed origin. Therefore, content-wise, Shi Sri Shiva Mahapuran beings with the geometric format of hypercube-4 with sealed origin. Likewise, the Manasara opens with the prayer to Lord Brahma and ends with the chiselling of an eye with which the seal of the origin would be unlocked and transcendence would take place and the 5-space coming into play and 4-space content resoling back to the boundary of 5-space. The knowledge content of Sri Sri Shiva Mahapuran deserves to be imbibed by all the students of mathematics, science and technology of Vedic Order. Sri Sri Shiva Mahapuran is a complete course in itself and it may not be possible to go into details at present.

Organizational Format of Manasara

The great design of nature is imprinted in each grain of Nature. Vedic Comprehension of this design conceptualises as Bindu "Sarowar". This grand unification of Trinity of Gods in each point reservoir of nature (Bindu Sarowar) may be viewed as the Trinity of Gods as first three folds and their unification as the fourth fold. In concrete terms, this amounts to 4-space playing the role of dimension for the organisational format of the great design of Bindu Sarowar and 5-space playing the role of boundary there of and 6-space manifesting as their domain fold and 7-space as origin thereof.

It is this grand design which is availed by the Vedic scriptures to organise the pure knowledge. The aim of Manasara as well is to avail this grand design of nature. The 7-space flows as 7 streams of Divya Ganga from core of the Sun; the sun being the body of 6-space and its core being the 7-space as origin of 6-space. The Divya Ganga flow from core of the sun (6-space) takes to the Sanatana base (5-space in the role of boundary of 6-space). Being vibrant synthetic reality this flow is going to be of vibrant and synthetic as reality. It is going to be of two parts. Accordingly, the Jyoti flowing from core of the Sun shall be flowing into the Sanantna base on either side of the Sun. This would help us appreciate and comprehend the Jyoti flow of 7 streams into Sanatana base (5-space) as of 7 × 5 =35 units on either side of the Sun and thereby we shall be having the flow rate of 35 + 35 = 70 units. It is this flow rate which runs parallel to the first organisation of the knowledge of Manasara as of 70 chapters.

The whole range of organization of 70 chapters is in fact the organization of the great design of the Bindu Sarowar. This range with definite end points shall be making it a close range of 1 + 70 + 1 = 72 units. This would completely wrap 6-space within 5-space boundary. With the help of the geometric set up of hypercube-6 as representative body of real 6-space on synthetic format of di-monad i.e. monad of two parts would provide us domain-boundary ratio A6:12A5. The complete Jyoti flow from core of the Sun (6-space) into the Sanatana base (5-space) as boundary of 12 components as through 6 dimensions of the Sun, (this flow) would be at the rate of 12 × 6 = 72 units. This also would help us comprehend and appreciate the sequential state of the this organization in the sequential order of 2 × 1, 2 × 2, 6 × 3, 8 × 4, 10 × 5 & 12 × 6 of the atomic organization. In fact this six steps long sequential order would take us uptill the nucleus of the nucleus of the atom. This also would help us comprehend and appreciate the Manasara accepting Lord Vishnu, the lord of 6-space as the god of the measuring-rod.

Now if we have a close look at the subject matter of the 70 chapters of Manasara, we shall be noticing that first 8 chapters constitute one topic. This is parallel to the boundary of hypercube-4 being of 8 components. Therefore the organizational format of the first 8 chapters of Mansara is the format of 8 components of boundary of hypercube 4. The next 10 chapters constitute one topic. This is parallel to 10 components boundary format of hypercube- 5.

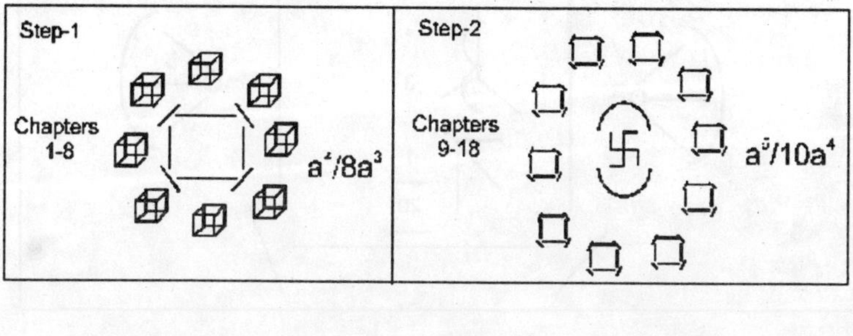

The next set of 12 chapters i.e. chapter 19 to 30 topically take up one story building to 12 storied building. This is parallel to the 12 components boundary of hypercube-6. With this we shall be reaching the state where we have to tap the Jyoti flow from the core of the Sun.

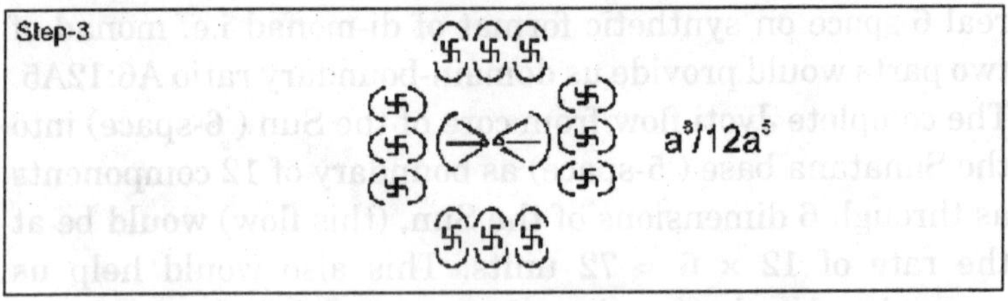

The Sun is a body of 6-space. Six space in the role of dimension shall be taking us uptill 9-space in the role of origin. In fact this Jyoti Bindu flowing from the core of the Sun is accepting 6-space (Sun) as dimension, 7-space (Saptrishi) as boundary, 8-space (Asht-prakriti) as domain and 9-space (Nav Brahm) as origin. This 4 folds Jyoti Bindu flow is the Divya Ganga flow of 4 folds. Jyoti Bindu flow is the Divya Ganga flow of 4 folds on the format of OM as formulation of 4 components, first as the Bindu Sarowar of Nav-Brahma as of 9 units, the second component of Ardha-Matra as of 7 units, the third component as Tri-Pundarm of 3 units and fourth component as Swastik Pada as of 1 unit. Total being of 20 units.

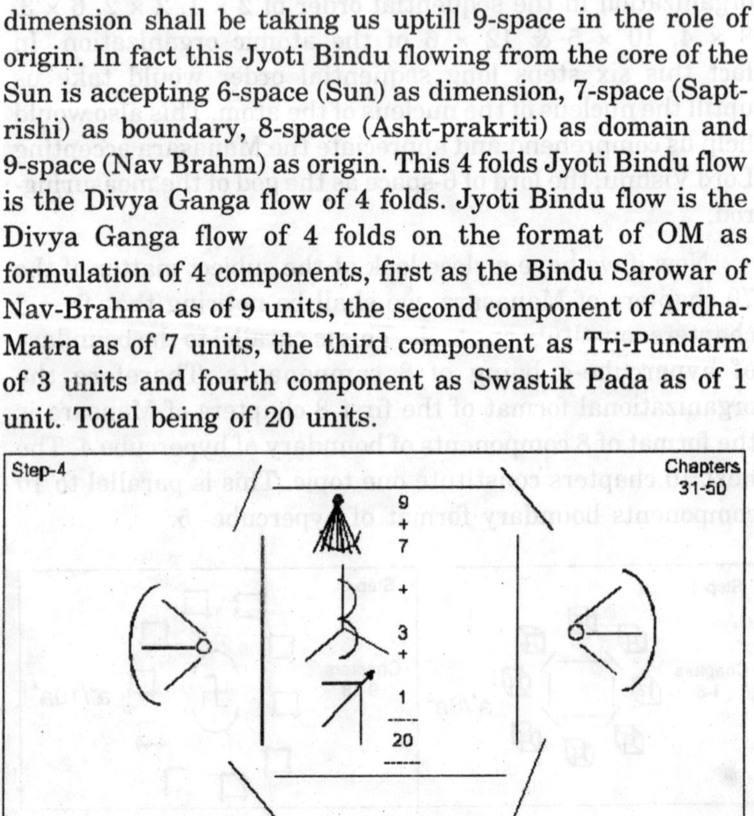

And these 20 units flow would be on either side and as such we shall be having 20+20 = 40 unit flow which shall be covering the remaining range of 40 chapters. Topically the 20 chapters from chapter 31 to 70 would be completing one topic. The organization of the remaining 20 chapters would get a division, in the context of Manasara as of 15 chapter and 5 chapters.

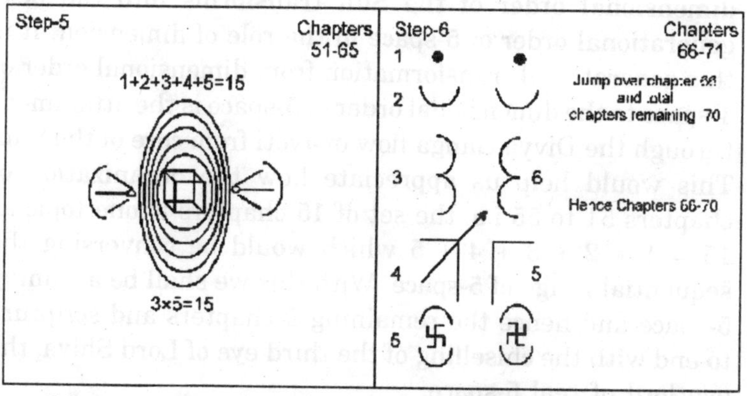

I have worked out the details of organizational format of Manasara as 138 graphics as separate compilation under Vedic Geometrical Series as a treatise for Indian Institute of Mahraishi Vedic Science & Technology. These graphic plates of the organizational format of the Manasara take up the organization right uptil the verses and even subverses of each chapter. Of these, I am using them now to illustrate how the Manasara is availing geometric format of the great design of the nature in its each grain as a Bindu Sarowar where in Trinity of Gods unify.

As per this great design we are getting a great message as that from within the Bindu Sarowar flows a Jyoti Bindu Sarowar where the Sun, Sapt-rishi and Ashta-prakiti get unified with Nav-Brahma as the origin. This phenomenal emergence of Jyoti Bindu Sarowar from within the Bindu Sarowar is the transcendental transformation. This transcendental transformation phenomenon is at work which helps sequential attainment of consciousness state. Also this transcendental transformation phenomenon is at work which takes us from seed to the tree and tree to the seed as a life-span and a sequential generational flow.

The Jyoti flows from the core of the Sun into the Sanatana base at boundary of the Sun potentialised the 'boundary space' to play the dimensional role. With this 4-space dimensional order of the Sun transforms into the next generational order of 5-space in the role of dimension. It is this generational transformation from dimensional order of 4-space to the dimensional order of 5-space is the attainment through the Divya Ganga flow of Jyoti from core of the Sun. This would help us appreciate how the organization of chapters 51 to 55 i.e. the set of 15 chapters is one topic as 15 = 1 + 2 + 3 + 4 + 5 which would be conversing the sequential range of 5-space. With this we shall be attaining 5-space and hence the remaining 5 chapters and scripture to end with the chiselling of the third eye of Lord Shiva, the overlord of real 5-space.

Measures and Measuring-rod

The concept of measures and measuring-rod of Manasara is of central importance. In fact Mana means measure and Sara means essence and as such Manasara means essence of measures. The science of Architecture and Sculpture centres around the essence of measures and the exactness of the measuring-rod. Therefore, the relevant knowledge of measures and measuring rod of chapter 2 Manasara deserve to be read thoroughly.

The verses 64 & 65 describe the measuring-rod as that: Measuring-rod should be one cubic long, one anguls (three fourth inch) broad, and its thickness is stated to be a half Angul. The yard-stick (lit, cubic measure) should be accurately marked.

The verse 68 enlightens us that: Vishnu is stated to be the tutelary god of (the wood for) both the yard-stick and the (measuring) rod.

Further, verse 75 enlightens us that: Vasuki (serpent-god) is the presiding deity of the (measuring) rope, and Brahma is known as the presiding deity of measurement.

Lord Brahma is a 4 head lord. Lord Vishnu is a 6 head lord,. Lord Brahma is born from the lotus whose reed springs from navel of Lord Vishnu. This Puranic comprehension is the Vedic enlightenment of real 4 space playing the role of dimension of real 6 space and in general (n-2) space plays the role of dimension of n space and thereby there being always a difference of 2 units between the domain and dimension which would help us comprehend and appreciate, the knowledge preserved in the epic stories of the arm of Lord Rama and Kag-Bhushandi of Ramayana and Yashoda Trying to tie Lord Krishna but every string found to be short by two units. This fundamental result is at the base of the mathematics which would help avail the results of arithmetic into geometry and vice versa. In terms of it only, we may comprehend and appreciate the multi-dimensions of time and space of Manasara as well as of other Vedic Scriptures.

4

Practical Application of Vedic Mathematics

Vedic Mathematics has Certain Visual Solutions which could be Applied in Problem Solving

There are 16 known sutras or axioms in vedic mathematics, some of which provide 'visual' mathematical solutions or solutions using mental techniques. It is perhaps possible to make practical use of these sutras in certain situations. Particularly where data or information validation can be done by using the appearance of 'zero' or nil values in any data set. Situations, where data, information or records are destroyed or lost, by accident or fraud, provide a natural camouflage for all kinds of manipulations. Plenty of insurance claims are known to have been inflated by altering data. For example, in case of an insurance claim for 'loss of profit' where a business has been disrupted by fire, the claimant will try and inflate past sales and profit figures to maximise his deemed profit for the claim.

Further, in most of these cases, the claimant can always expresse his inability to provide sufficient evidence to back his claim, using the 'destruction of records' as an excuse even though the evidence actually may not be even partially destroyed. In these situations, sometimes unconventional or unknown researched techniques could prove more effective.

Tests of 'reasonableness' using human judgement are the only tools which may be applied to ferret out inaccuracies and even deceptions. In such circumstances, perhaps Vedic mathematics might offer considerable assistance. Take the case of the following information given to an insurance company for a claim for 'loss of profits' on account of a fire and disruption of activities, where machines, stocks and several records were destroyed by fire. The company had two products in stock. Each product was packed and dispatched in boxes with standardised, precise weights. The claimant provided information of status of stocks and dispatches as given below in the two tables to support his claim for loss of profits:

Table A: Statement of Inventories (boxes of stocks as endorsed by third parties' confirmations:

Third Parties confirming the inventory quantities	VERIFICATION DATE	PRODUCT A (Quantities in boxes)	PRODUCT B (Quantities in boxes)
Auditors year end verification	March 31	74	126
Bankers hypothecated stock verification	May 31	67	113
Stock lost on date of fire	June 30	54	98

On account of seasonal trends there was no production of A and B in the above period. Therefore all changes in figures related to sales and dispatches only. Accordingly sales were as follows.

Table B: Statement of Sales Dispatches of Boxes in the post verification periods per Table A

Event	Period	Dispatches-product A	Dispatches-product B
After Auditor's verification	March 31-May 31	7	13
After Banker's verification	May 31- June 30	13	15

These dispatches were critical from the point of view of quantifying sales and determining profits claimed to have been lost because of the fire. The sales registers supported this statement also. However, the surveyors requested an accountant who was also a certified fraud examiner to give his assessment. The accountant reviewed the relevant data and found that these boxes were transported or shipped by professional container lorries. These transporters furnished details of the number of boxes, destinations and the time and date of delivery. Using these lorry receipts, the accountant painstakingly totalled the weights of all boxes transported during the two periods in table B above. He found that the total weights dispatched were 10.5 tonnes and 19.5 tonnes respectively. This is where he remembered vedic mathematics, the sutra 'Anurupye Shunyam anayat'. In simple words, in simultaneous equations, if the ratio of the co-efficients of one of the variables is in exact proportion as the constants, then the other variable is 0. Thus in the above case, if the weights of the boxes for products A and B were taken as x and y respectively, the dispatches, in terms of weight, could be represented as the following simultaneous equations:

$$7x + 13y = 10.5$$

$$13x + 15y = 19.5$$

The ratio of the coefficients of x is 7:13, proportionate to the ratio of the constants 10.5: 19.5 or 1.5*7:1.5*13 This implied that y=0, which further meant that dispatches of product y were of weight 0 or, in other words, there were no dispatches of product y. The accountant then inquired into the actual weights of boxes from external sources and he found that the weights of the boxes were in fact standardised at 1.5 tonnes and 2 tonnes for products A and B which confirmed the result of the above equations. On launching an investigation it was found that the sales bills for product B were to sister concerns and in fact fictitious. Since product B had a high margin of profits, the claimant was tempted to inflate profits by showing sales of product

B also, though there were none. The bankers stock statements showed lesser stocks because they counted only those which were hypothecated, as they were concerned with the stocks required to cover their loan and margin requirements. In fact it was also learnt that the stock of product B was in a bad condition and unsaleable. The company was in a bad shape financially and therefore had resorted to such means of falsification. However, applying visual application of vedic mathematics it was possible to prove the claim false.

This illustration does not seek to establish that solutions cannot be found by other means. It emphasizes on the simplicity of the tenets of the sutras of vedic mathematics. Perhaps research and intensive study of vedic scriptures might reveal even more advanced applications. What is illustrated above is a very elementary application of the sutra. The depth and richness of the vedic knowledge is beyond description. Greater research and more teamwork in sharing of ideas and interpretation may provide revolutionary results.

Vedic Re-establishment

Vedas, the most ancient Indian Scriptures or the oldest scriptures of the whole world, deal with all branches of learning; either spiritual, or temporal and are capable to give the knowledge of each and every technology, being used today.

The word 'Veda' means 'the fountain head' or 'the store house of knowledge', being sustained by the speak-learning process of ancient Indian Gurukuls. The ancient Indian Vedic Knowledge is all round complete and perfect and able to throw the necessary light on all matters of the MODERN AGE. The treasure of Vedas & Vedangas include (1) Ayurveda (Physiology, Anatomy, Hygiene, Sanitary Science, surgery, Medical Science, etc.), (2) DhanuVeda (Archery and other Military Sciences), (3) Gandharva Veda (the Science of Art of music), (4) Sthapatya Veda (Enginnering, Architecture etc., and all branches of Mathematics), (5) Rasayanam (Chemistry), (6) Bhautiki (Physics), (7) Jeevanam (Biology), (8) Ganit (Mathematics), etc., and the Vedangas include (a) Grammer, (b) Prosody, (c) Astronomy, (d) Lexicography. There were many reasons for our quest for discovering and researching on all the departments, branches and sub-branches of Vedas.

It was a deepest intensive legacy of our sages & scientists which provided the ethics, psychology, metaphysics on one hand to us and the pure and applied sciences and mathematics on the other. And when we would read or hear the so-called Orientalists, Indologists, Antiquarians, research-scholars, who condemned irresponsibly for the Vedas as "Superstitious-Nonsence". These taunts added fuel to the fire ignited in our minds and as it spread, the darkness vanished the path of re-establishment of the true Bhartiya knowledge became clear to us. And then we made our mind to unzip the wealth hidden in the ancient texts and to open the hidden mysteries of science, philosophy & mathematics of Vedas. After a deep intensive research, the Research Wing of Samkranti (Shodh Vahini) is in front of you with concentrated wealth of Ancient Advanced Technified Bharat and are capable to recover the lost keys which alone can unlock the secrets of all the affairs of life and spirituality.

You will be surprised to know that extremely tough mathematical problems which the modern mathematical rules find hard to solve, can be easily and readily solved with the help of ULTRA EASY VEDIC SUTRAS contained in the Appendix (Parishishta) portion of the ATHARVA VEDA in a few steps which do not require more calculation work. We are planning to carry the campaign of this VEDIC KNOWLEDGE to the India by means of Schools, Colleges, Universities, Lectures, Seminars, Vocational and Full time Courses.

The Vedic Sutras cover each and every part of each and every topic of each and every branch of Mathematics including Arithmetics, Algebra, Trigonometry - plane & spherical, Conics - Geometrical and Analytical, Geometry - plane and solid, Astrology, Calculus - Differential and Integral, etc. There is no part of Mathematics, which is beyond the grip of Vedas. As compared to the modern western mathematical methods, the methods based on Vedic Sutras take a third, a fourth or even a much smaller fraction of time. And amazingly, the sums requiring 50, 80 or 100 or even more steps can be answered in simple or even single steps by Vedic Methods.

A lot of seminars are conducted on this mathematics and the doctors, professors, and the "Big Bosses" of mathematics are wonder struck. Interest in the Vedic system is growing and the mathematics teachers are looking for something better, something more precise, more accurate and the Vedic Mathematics is the final solution for them. Researches are being carried out in the effects of learning Vedic Maths on children, and we, the Research Panel (Shodh Vahini) of SAMKRANTI and the Faculty of Vedic Research Institute, are trying to develop new, powerful, accurate, but easy applications of Vedic Sutras in Mathematics. But the real advantage, beauty and effectiveness of Vedic Mathematics can not be fully understood without actually practicing the system. Try it, and see that it is the most refined and efficient Mathematical and Computational System ever possible. The main Sutras revealed in Vedic Mathematics are like :

By one more then one before.

All from nine and last from ten.

Vertically and crosswise.

Transpose and apply.

If the Sammucchya is same, it is Zero.

If one is in the Ratio, the other is Zero.

By addition and by subtraction.

By completion and non-completion.

Differential Calculus.

By the Deficiency.

Specific and General.

The remainders by the last digit.

The Ultimate and twice the Penultimate.

By one less then one before.

The Product of Sum.

All the Multiples.

The Sub-Sutras :

Propositionality.

The Remainder Remains constant.

The first by the first, and the last by the last.

For seven, the Multiplicant is 143.

By Osculation.

Lessen by deficiency.

Whatever the deficiency, lessen by that amount and set up the square of deficiency.

Last totalling 10.

Only the last terms.

The sum of the products.

By alternative elimination and reduction.

By mere observations.

The product of sum is sum of products.

On the Flag.

Our organization is researching on the Sutras deeply and making a pure Indian Mathematical 10 year syllabus to be taught in Indian Schools.

Vedic Cognitive Sciences

SAMKRANTI plans to make a syllabus from the Cognitive sciences of Vedas for every class including all the useful chapters of Vedas, Upanishadas and Darshans. Let us take some illustrations; Rigveda Speaks of Cosmic Order. It reveals some links and equivalence between outer and inner worlds. The Vedic vision tells that Material and Conscious are aspects of same transcendental reality. The Vedangas caution the ultimate knowledge groups like 1. Logic, 2. Physics, 3. Cosmology, 4. Psychology, 5. Language, 6. Reality.

The super latest technologies that are being used and studied in the world arose in India, several centuries before their re-discovery in the West. We have written proofs for this, the MAHABHARAT mentions EMBRYO-TRANSPLANTATION, MULTIPLE BIRTHS from same fetus, Battles with Extra Terrestrials who are wearing Air-Tight suits and weapons of Mass Destruction. RAMAYAN mentions Air-Travel. SHRIMADBHAGWAT and BRAHMAPURAAN has incidence related to different passage of time for different observers and of reality. YOG VASHISHTH, YOG RAHASYAM, YOG SUTRAM and TRIPURA-RAHASYAM give the deep knowledge of nature of Consciousness. We find examples of accurate Astronomical numbers in early texts. The accuracy was perfectly based on the Biological Cycles that reflect Astronomical process like Menses according to the period of moon. I recall one of my favourite quote, "Yatha pindey, tatha Brahmaandey" that means, we can understand the outer existence by the understanding of inner. In the Tetteriya Upanishad, an individual is represented in terms of five different sheaths or levels. The SANKHYA and YOG explain the mind as consisting of five components: Manas, Ahamkaar, Chitta, Buddhi, Atman. Manas is the lower part of mind and collects sensory imressions. This sensory motor mind obtains its inputs from the senses of hearing, touch, sight, taste, smell. The Ahamkaar is the sense of I-ness that has some perceptions to a subjective and personal experiences related to the I-ness. By Ahamkaar, their decisions, evaluations and results are expressed by the intellect that is Buddhi. The Manas, Ahamkaar and Buddhi collectively called Mind and the memory bank of this Mind is Chitta. And the innermost aspect of consciousness is called Atman, Jiva, Brahma. Read the Brahmavaivarta Puraan. In the incident of Indra and Ants, Sri Vishnu in the guise of a boy tells Indra that these Ants have been Indras in their own Solar systems in different times. This is an excellent example of the knowledge of Solar System and Universe in Indian Literatures. Aryabhatta, born in 476 is auther of Aryabhattiyam, a most prestegious book on Mathematics and computing science. In the book, he covers Mathematics, cosmic and planetary theories. The book is divided into four chapters.

Astronomical Constant and Sine table.

Mathematics for computation.

Division of Time, rule for computing the Longitudes of planets using eccentrics and epicycles.

The armillary Sphere, rules relating problems of Trignometry and computation of Eclipse.

Aryabhatta Explains Earth to spin on its axis. He also considered the heavenly motions to go through a cycle of 4.32 Billion Years. He was also well aware of relativity of motion as he writes, "Just as a man in a boat sees the trees on the bank move in the opposite direction, so and observer on equator sees the stationary stars as moving precisely towards west." Brahmagupta contributed to the solution of a certain indeterminate equation. Bhasker (born in 1114) wrote Siddhanta Shiromani, a book containing four part:

Lilavati, based on Arithmetic.

Beej Ganit, on Algebra.

Ganitadhyaya.

Goladhyaya, on astronomy.

Madhava (1340- 1425) developed a procedure to determine the position every 36 minutes. He also gave power series expansions for Trignometric Functions and for Pi correct to 11 decimal places.

So much wealth Indian Pure Science and Mathematics have. The foreigners are still getting the advantage by their clavery. When the Vedic in a well practise in India, it was discouraged and suppressed trickfully and they renamed the terminologies and pretended to discover the existing laws as some new ones. Now, when people cowardishly accepted western impure science, they riturn back to the Gyan given by VISHVA GURU BHARAT. The Maharishi School in Lankashire has developed a course on Vedic Mathematics that covers the National Curriculum. There are lots of such examples about awaking of the foreign eyes towards the essence of Vedic Studies, but we, the owners of this uncomparable knowledge, are ignoring it for long. We ourself are making fortune in trouble.

In the Vedic times, the records of Mathematical Activities are found in Vedic Literature. During the redistribution of land of farms, the problem of Mensuration came, which were accurately solved at that time. The Taxation System at that time were based on fixed proportions of Annual or seasonal crop-incomes. Mathematics was given considerable importance in Vedic period. The Vedanga Jyotish says, "Just as the feathers of a peacock and the jewel stone of a snake are placed at the highest point of the body, similarly the position of Ganit (mathematics) is the highest amongst all the branches of the Vedangas and Shastras.

Re-establishment

Veda—The constitution of the Universe

HH Maharishi Mahesh Yogi says, " Veda is the Self referral dynamism of the transcendental field of pure subjectivity which is the transcendental basis of creation and its orderly administration through NATURAL LAW.

EVERYTHING IS MADE OF VEDAS - TOTAL NATURAL LAW. EVERYTHING IS VEDIC. EVERYONE IS VEDIC. SINCE EVERYTHING IS VEDIC, EVERYTHING SHOULD BE WHAT IT IS."

India in the ancient times established a perfect, a pure government - RAMRAJYA, heaven on earth by strengthening higher intelligence. This higher intelligence can be re-achieved through Vedas only. Thus, the Vedic Re-establishment project of SAMKRANTI contains :

- The Vedic Culture Practice to culture the life and for enlightenment of life.
- The Vedic Education that will give the real, true knowledge to the world.
- The Vedic Health that is without side-effects and is totally natural.
- The Vedic Defense that contains the research on 'Vedic Kavachs Stotras and Paaths' that check the evolution of any enemy and give a super sysmetic power of nature.
- The Vedic Agriculture that controls the qualities of soil, seed, water, and the menures and strengthens the intellect of farmers.
- The Vedic Economy, that uses the National resources and money in the best possible way.
- The Vedic Architecture that constructs the houses, cities, villages in the perfect way and thus constructs the Nation.

5

Applications of Vedic Mathematics

The most notable application of Vedic mathematics is in education. Vedic mathematical strategies may prove to be a useful resource for teachers and students, who may find elements of it easier and more accessible to teach and learn than conventional mathematics. In particular, these strategies may be an invaluable resource to students that already struggle with mathematics, and could benefit from alternative approaches.

One attempt at incorporating Vedic mathematics into education was made by Mark Gaskell, the head of mathematics at the Maharishi School Lancashire, England (Gaskell, 2000). The school has developed a Vedic mathematics curriculum equivalent to the national one with impressive results. According to Gaskell, the alternative curriculum has resulted in livelier classes, greater student enjoyment and understanding, and improved academic performance (2000). In fact, the first set of students to complete the course were each able to not only pass, but achieve over 80%, on the General Certificate of Secondary Education, a proficiency test taken by all secondary school British students, a year earlier than their peers in the regular curriculum (2000). If harnessed appropriately, there seems to be great potential for how Vedic mathematics can be used to teach, learn and understand mathematics. Perhaps the most important aspect of including Vedic mathematics in an education system will be taking the step towards becoming open to conceptually different mathematical approaches—approaches that could one day free and transform mathematics education.

Controversy and Criticism

There has been much controversy amongst Indian scholars about Tirthaji's claims that the mathematics is Vedic and that it encompasses all aspects of mathematics (Kansara, 1961). First, Tirthaji's description of the mathematics as Vedic is most commonly criticised on the basis that, thus far, none of the sutras can be found in any extant Vedic literature (Williams, 2000). However, trying to locate the Tirthaji's references in the Vedic literature would be extremely difficult; the Vedas are too large for a complete translation, and are composed in an obscure and complicated ancient dialect (Kansara, 1961). Furthermore, it is also possible that Tirthaji rediscovered and reconstructed the sutras from stray references scattered throughout the Atharva-veda, making it even more difficult to trace them (Trivedi, 1965).

In response to criticisms that the sutras cannot be located within the texts, several people have explained how textual references should not be the basis for evaluating the Vedicity of the mathematics (Agrawala, 1992). Some propose that Vedic mathematics is different than other scientific work because it is not pragmatically worked out, but is based on a direct revelation, or an "intuitional visualisation" of fundamental mathematical truths (Agrawala, 1992; Pratyagatmananda, 1965). Tirthaji has been described as having the same "reverential approach" towards the Vedas as the ancient rishis that formed them. Thus, it seems as though some believe that Tirthaji may not have found the sutras within the Vedas, but that he received them spiritually as the rishis did, which should validate them as Vedic.

The controversy about the Vedicity of the mathematics is further confused by the double meaning of veda. Since veda can be translated to mean 'knowledge', it is also possible that Vedic mathematics simply refers to the fact that the sutras are supposed to present all knowledge of mathematics. Tirthaji's definition of veda does not clearly clarify whether he uses it to represent 'all knowledge' or the Vedic texts; rather, it seems that he uses it to refer to both:

The very word 'Veda' has this derivational meaning, i.e. the fountainhead and illimitable store-house of all knowledge. The derivation, in effect, means, connotes and implies that the Vedas should contain within themselves all knowledge needed by mankind relating not only to the 'spiritual' matters but also to those usually described as purely 'secular', 'temporal', or 'worldly' and also to the means required by humanity as such for the achievement of all-round, complete and perfect success in all conceivable directions and that there can be no adjectival or restrictive epithet calculated (or tending) to limit that knowledge down in any sphere, and direction or any respect whatsoever (1992).

Considering the lack of references to the sutras, coupled with the fact that the language style does not seem Vedic, some wager that the sutras were simply composed by Tirthaji himself (Agrawala, 1992). In that case, one must consider what motivated Tirthaji to attribute the mathematical sutras to the ancient texts. Was it because they are from the Vedas, or does claiming so give them more credibility? Other areas of controversy regarding Vedic mathematics focus on the actual mathematics itself. Tirthaji's assertion that the 16 sutras of Vedic mathematics encompass all branches of mathematics is an extreme one even if true, and so it is not surprising that many mathematicians challenge it (Kansara, 1961). They point to the inconsistency between the topics addressed by the system (such as decimal fractions) and the known mathematics of early India, the substantial extrapolations from a few words of a sutra to complex arithmetic strategies, and the restriction of applications to convenient, special cases. They further say that such arithmetic as is sped up by application of the sutras can be performed on a computer or calculator anyway, making their knowledge rather irrelevant in the modern world.

They are also worried that it deflects attention from genuine achievements of ancient and modern Indian mathematics and mathematicians, and that its promotion by Hindu nationalists may damage mathematics education in India.

Additionally, the promotion of Vedic Maths as advanced mathematics has drawn criticism from mathematicians. Although quite useful for practical calculations in daily life, some purport that the "tricks" involved have no application to higher mathematics or mathematical research.

Greatest Common Divisor of Two Polynomials

Informally, the greatest common divisor (GCD) of two polynomials p(x) and q(x) is the "biggest" polynomial that divides evenly into both p(x) and q(x). When dealing with numbers, the definition is similar.

If you have two whole numbers, the greatest common divisor of those numbers is simply the biggest whole number that you can divide into both of them while getting a remainder of zero. The greatest common divisor is also sometimes referred to as the greatest common factor or the highest common factor.

Formal Definition

The greatest common divisor of two polynomials p(x) and q(x), not both zero, with coefficients from a field is the monic polynomial d(x) of highest degree such that d(x) is a common divisor of both p(x) and q(x).

The definition is very specific and this is very important. For example, if both polynomials were zero, then any polynomial would divide them both. In this case, there would be no greatest common divisor. Thus, at least one of the polynomials must be nonzero.

Also, if there were no requirement for the GCD to be monic, there could be several different, even an infinite number of different, GCDs for any two polynomials. This results because any nonzero constant multiple (with constants from the field) of the monic common divisor of highest degree is still a common divisor of highest degree.

Finally, the definition states the coefficients must come from a field. If the coefficients could come from any ring, then factorization would not necessarily be unique. For example, in (Z/6Z),

$$(2x + 1)(3x + 1) = 6x^2 + 5x + 1 = 5x + 1$$

and

$$5(x + 5) = 5x + 25 = 5x + 1.$$

All of these details together ensure that there is a GCD and that the GCD is unique.

Properties

- As stated above, the GCD of two nonzero polynomials, with coefficients from a field, is unique.
- If c(x) is any common divisor of p(x) and q(x), then c(x) divides d(x). This is sometimes given in the definition instead of requiring d(x) to be the common divisor of highest degree. The two definitions are logically equivalent.
- The GCD of two polynomials, p(x) and q(x), can be written as a linear combination of p(x) and q(x). That is, there exist some polynomials, in the same field, r(x) and s(x), not necessarily unique, such that

  $$d(x) = p(x)\ r(x) + q(x)\ s(x).$$

Methods for Finding the GCD

There are several ways to find the greatest common divisor of two polynomials. Some of them include:

1. Factorization, in which one finds the factors of each expression, then selects the set of common factors held by all from within each set of factors.
2. The Euclidean Algorithm, which can be used to find the GCD of two polynomials in the same manner as for two numbers.
3. Swami Bharati Krishna Tirtha's Vedic mathematics contains a method, here called the Vedic method, which is mainly an application of the Lopana-Sthapana Sutra, the Sankalna-Vyavakalan process, and the Adyamadya rule. The sutra is a formula for the alternate destruction of the highest and lowest powers (by the addition or subtraction of multiples of the coefficients).

Factoring

To find the GCD of two polynomials using factoring, simply factor the two polynomials completely.

Then, take the product of all common factors. At this stage, we do not necessarily have a monic polynomial, so finally multiply this by a constant to make it a monic polynomial.

This will be the GCD of the two polynomials as it includes all common divisors and is monic.

EXAMPLE ONE

Find the GCD of $x^2 + 7x + 6$ and $x^2 - 5x - 6$.

$x^2 + 7x + 6 = (x + 1)(x + 6)$

$x^2 - 5x - 6 = (x + 1)(x - 6)$

Thus, their GCD = $(x + 1)$.

Euclidean Algorithm

Finding the GCD by factoring is intuitively simple but requires knowing how to factor the polynomials, which can be difficult, especially if the polynomials have large degree. The Euclidean Algorithm is a fast method which works for any pair of polynomials.

It makes repeated use of division with remainder, just as when the Euclidean Algorithm is applied to two numbers. When using the Euclidean Algorithm on two numbers, the size of the numbers decreases at each stage. With polynomials, the degree of the polynomials decreases at each stage. The last nonzero remainder, made monic if necessary, is the GCD of the two polynomials.

EXAMPLE ONE

Find the GCD of $x^2 + 7x + 6$ and $x^2 - 5x - 6$.

$x^2 + 7x + 6 = (x^2 - 5x - 6)(1) + (x + 1)(12)$

$x^2 - 5x - 6 = (x + 1)(x - 6) + 0$

Since $x + 1$ is the last nonzero remainder, the GCD of these polynomials is $x + 1$.

Algebraic proof of the Vedic method for finding the GCD

Let P and Q be two expressions. Let H be their GCD. Let A and B be the quotients (after dividing by the GCD). Therefore, P/H = A and Q/H = B and also P = HA and Q = HB. Therefore $P \pm Q = H(A \pm B)$ and $MP \pm NQ = H(MA \pm NB)$ Therefore, the GCD of P and Q is also the GCD of $P \pm Q$, $2P \pm Q$, $P \pm 2Q$, and $MP \pm NQ$. So, when we select our M and N so that the highest and lowest powers are removed, then the GCD appears and shows itself!

EXAMPLES

Example 1: the Vedic Method

Find the GCD of $x^2 + 7x + 6$ and $x^2 - 5x - 6$. To eliminate the x-squared terms, just subtract the expressions.

To eliminate the constant terms, simply add the two expressions. Then, pull out any common factors to reveal the GCD.

$$x^2 + 7x + 6 \qquad\qquad x^2 + 7x + 6$$

Subtract: $-(x^2 - 5x - 6)$ Add: $+ (x^2 - 5x - 6)$

$$12x + 12 = 12(x + 1) \qquad 2x^2 + 2x = 2x(x+1)$$

Thus, their GCD = $(x+1)$.

Example 2: Vedic Method

$E_1 = x^3 - 3x^2 - 4x + 12$ $E_2 = x^3 - 7x^2 +16x - 12$

Minus $-(x^3 -7x^2 +16x - 12)$ Add $+(x^3 - 3x^2 - 4x +12)$

$$4x^2 - 20x + 24 \qquad 2x^3 - 10x^2 + 12x$$

Remove common factors: $4(x^2 - 5x + 6)$ $2x(x^2-5x+6)$

Their GCD = (x^2-5x+6)

Example 3: Factorization

Find the GCD of the two expressions after factorization:

$E_1 = x^4 + x^3 -5x^2 -3x +2 = (x+1)(x-2)(x2+2x-1)$

and $E_2 = x^4 - 3x^3 + x^2 +3x -2 = (x+1)(x-2)(x-1)^2$

Seeing the shared factors are the binomials $(x+1)(x-2)$ we have them as the GCD.

Example 4: Vedic Method

Find the GCD of two expressions by elimination and retention of the highest and the lowest powers.

$E_1 = x^4 + x^3 -5x^2 -3x +2$ $E_1 = x^4 + x^3 -5x^2 -3x +2$

and E2 = $x^4 -3x^3 + x^2 +3x -2$ $E_2 = x^4 -3x^3 + x^2 +3x -2$

The sum: $2x^4 -2x^3 -4x^2$ Their difference: $4x^3 -6x^2 -6x +4$

Pull out the common factors:

$(2x^2)(x2 -x -2)$ $(2)(2x^3 -3x^2 -3x +2)$

Restore a factor of (2x) for a second elimination.

Multiply by $(2x)(x^2 - x - 2) = 2x^3 - 2x^2 - 4x$

Subtract second result from first result:

$$2x^3 - 3x^2 - 3x + 2$$

Minus $-(2x^3 - 2x^2 - 4x)$

$$- x^2 + x + 2$$

Since the expressions have positive x2 terms remove the factor of -1:

Their GCD = $(x^2 - x - 2)$

Example 5: Vedic Method

Find the GCD of two expressions.

$E_1 = 6x^4 - 7x^3 - 5x^2 + 14x + 7$

$E_2 = 3x^3 - 5x^2 + 7$

To eliminate the 4th power: $E_1 + (-2x)(E_2)$.

Subtract: E1 - E2 to eliminate the constant terms.

$6x^4 - 7x^3 - 5x^2 + 14x + 7$

$6x^4 - 7x^3 - 5x^2 + 14x + 7$

$(-2x)(3x^3 - 5x2 + 7) =$ add $+(6x4 + 10x^3 - 14x)$

Minus $-(3x^3 - 5x^2 + 7)$

$3x^3 - 5x^2 + 7$

$6x^4 - 10x^3 + 14x$

Take out the common factor: $(2x)(3x^3 - 5x^2 + 7)$

The GCD = $(3x^3 - 5x^2 + 7)$

Partial Fraction

In algebra, the partial fraction decomposition or (partial fraction expansion) is used to reduce the degree of either the numerator or the denominator of a rational function. The outcome of partial fraction expansion expresses that function as a sum of fractions, where:

- the denominator of each term is a power of an irreducible (not factorable) polynomial and
- the numerator is a polynomial of smaller degree than the denominator.

See partial fractions in integration for an account of their use in finding antiderivatives. They are also used in calculating the inverse of transforms; such as the Laplace transform, or the Z-transform.

Just which polynomials are irreducible depends on which field of scalars one adopts. Thus if one allows only real numbers, then irreducible polynomials are of degree either 1 or 2. If complex numbers are allowed, only 1st-degree polynomials can be irreducible. If one allows only rational numbers, then some higher-degree polynomials are irreducible.

One can use partial fraction expansion for one of two main uses:

- To change a function in the form $\frac{f(x)}{x+a}$ into a function of the form $\frac{A}{x+a}+\sum\frac{g_n(x)}{1}$
- To change a function in the form $\frac{f(x)}{g(x)}$ into a function of the form $\sum\frac{A_n}{g(x)/h_n(x)}$.

SOME EXAMPLES

If one wants to decompose x/(x+a), then one can follow these steps:

$$\frac{x}{x+a}=A+\frac{B}{x+a}$$

where A, B, and a are constants

$x + 0 = A(x + a) + B$

$x = Ax$ and $0 = Aa + B$

and

therefore

$A = 1$ and $B = -a$

Distinct First-degree Factors in the Denominator

Suppose it is desired to decompose the rational function

$$\frac{x+3}{x^2-3x-40}$$

into partial fractions. The denominator factors as

$(x - 8)(x + 5)$

and so we seek scalars A and B such that

$$\frac{x+3}{x^2-3x-40}=\frac{x+3}{(x-8)(x+5)}=\frac{A}{x-8}+\frac{B}{x+5}.$$

One way of finding A and B begins by "clearing fractions", i.e., multiplying both sides by the common denominator $(x - 8)(x + 5)$. This yields

$x + 3 = A(x + 5) + B(x - 8).$

Collecting like terms gives

$x + 3 = (A + B)x + (5A - 8B).$

Equating coefficients of like terms then yields:

$$A + B = 1$$

$$5A - 8B = 3$$

The solution is $A = 11/13$, $B = 2/13$. Thus we have the partial fraction decomposition

$$\frac{x+3}{x^2-3x-40}=\frac{\frac{11}{13}}{x-8}+\frac{\frac{2}{13}}{x+5}.$$

An Irreducible Quadratic Factor in the Denominator

In order to decompose

$$\frac{10x^2+12x+20}{x^3-8}$$

into partial fractions, first observe that

$x^3 - 8 = (x - 2)(x^2 + 2x + 4).$

The fact that x2 + 2x + 4 cannot be factored using real numbers can be seen by observing that the discriminant 22 - 4(1)(4) is negative. Thus we seek scalars A, B, C such that

$$\frac{10x^2+12x+20}{x^3-8}=\frac{10x^2+12x+20}{(x-2)(x^2+2x+4)}=\frac{A}{x-2}+\frac{Bx+C}{x^2+2x+4}.$$

When we clear fractions, we get

$10x^2 + 12x + 20 = A(x^2 + 2x + 4) + (Bx + C)(x - 2)$

We could proceed as in the previous example, getting three linear equations in three variables A, B, and C. However, since solving such systems becomes onerous as the number of variables grows, we try a different method. Substitution of 2 for x in the identity above makes the entire second term vanish, and we get

$10.2^2 + 12.2 + 20 = A(2^2 + 2.2 + 4),$

i.e., 84 = 12A, so A = 7, and we have

$10x^2 + 12x + 20 = 7(x^2 + 2x + 4) + (Bx + C)(x - 2).$

Next, substitution of 0 for x yields

$20 = 7(4) + C(-2),$

and so C = 4. We now have

$10x^2 + 12x + 20 = 7(x^2 + 2x + 4) + (Bx + 4)(x - 2).$

Substitution of 1 for x yields

$10 + 12 + 20 = 7(1 + 2 + 4) + (B + 4)(1 - 2),$

and so B = 3.

Our partial fraction decomposition is therefore:

$$\frac{10x^2+12x+20}{x^3-8}=\frac{7}{x-2}+\frac{3x+4}{x^2+2x+4}.$$

A Repeated First-degree Factor in the Denominator

Consider the rational function

$$\frac{10x^2-63x+29}{x^3-11x^2+40x-48}.$$

The denominator factors thus:

$x^3 - 11x^2 + 40x - 48 = (x - 3)(x - 4)^2$.

The multiplicity of the first-degree factor (x - 4) is more than 1. In such cases, the partial fraction decomposition takes the following form:

$$\frac{10x^2-63x+29}{x^3-11x^2+40x-48}=\frac{10x^2-63x+29}{(x-3)(x-4)^2}=\frac{A}{x-3}+\frac{B}{x-4}+\frac{C}{(x-4)^2}.$$

Repeated Factors in the Denominator Generally

For rational functions of the form

$$\frac{p(x)}{(x+2)(x+3)^5}$$

(where the p(x) may be any polynomial of sufficiently small degree) the partial fraction decomposition looks like this:

$$\frac{A}{x+2}+\frac{B}{x+3}+\frac{C}{(x+3)^2}+\frac{D}{(x+3)^3}+\frac{E}{(x+3)^4}+\frac{F}{(x+3)^5}.$$

The general pattern may be quickly guessed.

For rational functions of the form

$$\frac{p(x)}{(x+2)(x^2+1)^5}$$

with the irreducible quadratic factor $x^2 + 1$ in the denominator (where again, the p(x) may be any polynomial of sufficiently small degree), the partial fraction decomposition looks like this:

$$\frac{A}{x+2}+\frac{Bx+C}{x^2+1}+\frac{Dx+E}{(x^2+1)^2}+\frac{Fx+G}{(x^2+1)^3}+\frac{Hx+I}{(x^2+1)^4}+\frac{Jx+K}{(x^2+1)^5},$$

and a similar pattern holds for any other irreducible quadratic factor.

High-degree Polynomials in the Numerator

When you need to apply the partial fraction decomposition to a polynomial division like

$$\frac{a_n x^n + \ldots + a_1 x + a_0}{b_m x^m + \ldots + b_1 x + b_0}$$

where $n \geq m$, you just need to make the polynomial long division procedure first, and apply the partial fraction decomposition to the remainder.

Use in Deriving the Logistic General Equation

In many beginning calculus courses, partial fractions are introduced as a way to derive the general equation for a logistic function.

Logistic functions model a population which grows until it reaches a limit. The rate of change for the function is proportional (constant k) to both the population reached (P) and the fraction of the total carrying capacity (M) remaining. Thus:

$$\frac{dP}{dt} = kP\left(\frac{M-P}{M}\right)$$

$$\int \frac{1}{P(M-P)} dP = \int \frac{k}{M} dt$$

$$\frac{1}{P(M-P)} = \frac{A}{P} + \frac{B}{(M-P)}$$

$$1 = A(M - P) + BP$$

$$1 = AM - AP + BP$$

$$A = B,\ A = \frac{1}{M},\ B = \frac{1}{M}$$

$$\int \frac{1}{MP} + \frac{1}{M(M-P)} dP = \int \frac{k}{M} dt$$

$$\int \frac{1}{P} + \frac{1}{(M-P)} dP = \int k\, dt$$

$$\ln \frac{P}{(P-M)} = kt + C$$

$$\frac{P}{P-M} = e^{kt+C}$$

$$\frac{M-P}{P} = e^{-kt-C}$$

$$\frac{M}{P} - 1 = e^{-kt-C}$$

$$P = \frac{M}{e^{-kt-C} + 1}$$

$$P = \frac{M}{Ae^{-kt} + 1}$$

6

Vedic Mathematical Sutras

Consider the following three sutras:

1. "All from 9 and the last from 10," and its corollary: "Whatever the extent of its deficiency, lessen it still further to that very extent; and also set up the square (of that deficiency)."
2. "By one more than the previous one," and its corollary: "Proportionately."
3. "Vertically and crosswise," and its corollary: "The first by the first and the last by the last."

The first rather cryptic formula is best understood by way of a simple example: let us multiply 6 by 8.

1. First, assign as the base for our calculations the power of 10 nearest to the numbers which are to be multiplied. For this example our base is 10.
2. Write the two numbers to be multiplied on a paper one above the other, and to the right of each write the remainder when each number is subtracted from the base 10. The remainders are then connected to the original numbers with minus signs, signifying that they are less than the base 10.

6-4

8-2

3. The answer to the multiplication is given in two parts. The first digit on the left is in multiples of 10 (i.e. the 4 of the answer 48). Although the answer can be arrived at by four different ways, only one is presented here. Subtract the sum of the two deficiencies (4 + 2 = 6) from the base (10) and obtain 10 - 6 = 4 for the left digit (which in multiples of the base 10 is 40).

6-4

8-2

4

4. Now multiply the two remainder numbers 4 and 2 to obtain the product 8. This is the right hand portion of the answer which when added to the left hand portion 4 (multiples of 10) produces 48.

6-4

8-2

4/8

Another method employs cross subtraction. In the current example the 2 is subtracted from 6 (or 4 from 8) to obtain the first digit of the answer and the digits 2 and 4 are multiplied together to give the second digit of the answer. This process has been noted by historians as responsible for the general acceptance of the X mark as the sign of multiplication. The algebraical explanation for the first process is

$$(x - a)(x - b) = x(x - a - b) + ab$$

where x is the base 10, a is the remainder 4 and b is the remainder 2 so that

$$6 = (x-a) = (10-4)$$

$$8 = (x-b) = (10-2)$$

The equivalent process of multiplying 6 by 8 is then

x(x-a-b) + ab or

10(10-4-2) + 2×4 = 40 + 8 = 48

These simple examples can be extended without limitation. Consider the following cases where 100 has been chosen as the base:

97 - 3	93 - 7	25 - 75	
78 - 22	92 - 8	98 - 2	
75 / 66	85 / 56	23 / 150	= 24/50

In the last example we carry the 100 of the 150 to the left and 23 (signifying 23 hundred) becomes 24 (hundred). Herein the sutra's words "all from 9 and the last from 10" are shown. The rule is that all the digits of the given original numbers are subtracted from 9, except for the last (the righthand-most one) which should be deducted from 10.

Consider the case when the multiplicand and the multiplier are just above a power of 10. In this case we must cross-add instead of cross subtract. The algebraic formula for the process is: (x+a)(x+b) = x(x+a+b) + ab. Further, if one number is above and the other below a power of 10, we have a combination of subtraction and addition: viz:

108 + 8 and	13 + 3
97 - 3	8 - 2
105/-24	11/-6
= 104/(100-24)	= 10/(10-6)
= 104/76	= 10/4

The Sub-Sutra: "Proportionately" Provides for those cases where we wish to use as our base multiples of the normal base of powers of ten. That is, whenever neither the multiplicand nor the multiplier is sufficiently near a convenient power of 10, which could serve as our base we simply use a multiple of a power of ten as our working base, perform our calculations with this working base and then multiply or divide the result proportionately.

To multiply 48 by 32, for example, we use as our base 50 = 100/2, so we have

Base 50 48 - 2

32 - 18

2/ 30/36 or (30/2) / 36 = 15/36

Note that only the left decimals corresponding to the powers of ten digits (here 100) are to be effected by the proportional division of 2. These examples show how much easier it is to subtract a few numbers, (especially for more complex calculations) rather than memorize long mathematical tables and perform cumbersome calculations the long way.

Squaring Numbers

The algebraic equivalent of the sutra for squaring a number is: $(a+-b)^2 = a^2 +- 2ab + b^2$. To square 103 we could write it as $(100 + 3)^2 = 10,000 + 600 + 9 = 10,609$. This calculation can easily be done mentally. Similarly, to divide 38,982 by 73 we can write the numerator as $38x^3 + 9x^2 + 8x + 2$, where x is equal o 10, and the denominator is $7x + 3$. It doesn't take much to figure out that the numerator can also be written as $35x^3 + 36x^2 + 37x + 12$. Therefore,

38,982/73

$= (35x^3 + 36x^2 + 37x + 12)/(7x + 3)$

$= 5x^2 + 3x + 4 = 534$

This is just the algebraic equivalent of the actual method used. The algebraic principle involved in the third sutra, "vertically and crosswise," can be expressed, in one of it's applications, as the multiplication of the two numbers represented by (ax + b) and (cx + d), with the answer $acx^2 + x(ad + bc) + bd$. Differential calculus also is utilized in the Vedic sutras for breaking down a quadratic equation on sight into two simple equations of the first degree. Many additional sutras are given which provide simple mental one or two line methods for division, squaring of numbers, determining square and cube roots, compound additions and subtractions, integrations, differentiations, and integration by partial fractions, factorisation of quadratic equations, solution of simultaneous equations, and many more. For demonstrational purposes, we have only presented simple examples.

Historical Background

The ancient system of Vedic Mathematics was rediscovered from the Vedas between 1911 and 1918 by Sri Bharati Krsna Tirthaji (1884-1960). The word 'Veda' means 'knowledge' and also refers to a set of ancient Indian texts written in Sanskrit. The date these texts were written is unknown but the content of the Vedas was passed on by an oral tradition long before writing was invented. The Vedas are said to cover every aspect and area of knowledge: grammar, architecture, ethics, astronomy etc. are all covered. At the beginning of the twentieth century, when there was a great interest in the Sanskrit texts in Europe, Bharati Krsna tells us some scholars ridiculed certain texts which were headed 'Ganita Sutras'- which means mathematics. They could find no mathematics in the translation and dismissed the texts as rubbish. Bharati Krsna, who was himself a scholar of Sanskrit, Mathematics, History and Philosophy, studied these texts and after lengthy and careful investigation was able to reconstruct the mathematics of the Vedas. According to his research all of mathematics is based on sixteen Sutras, or word-formulae. For example, Vertically and Crosswise is one of these Sutras. The credibility of this extraordinary claim that all of mathematics, pure and applied, is governed by sixteen formulae is discussed in the page on The Sutras of Vedic Mathematics.

Bharati Krsna wrote sixteen volumes expounding the Vedic system but these were unaccountably lost and when the loss was confirmed in his final years he wrote a single book: Vedic Mathematics, currently available. It was published in 1965, five years after his death. A copy of the book was brought to London a few years later and some English mathematicians (Kenneth Williams, Andrew Nicholas, Jeremy Pickles) took an interest in it. They extended the introductory material given in Bharati Krsna's book and gave many courses and talks in London. A book (now out of print), Introductory Lectures on Vedic Mathematics, was published in 1981. Between 1981 and 1987 Andrew Nicholas made four trips to India initially to find out what further was known about it.

Following these journeys a renewed interest was taken by scholars and teachers in India. It seems that once they saw that some people in the West took Vedic Mathematics seriously they realised they had something special. St. James' School, then in Queensgate, London and other schools began to teach the Vedic system, with notable success. Today Vedic Mathematics is taught widely in schools in India and a great deal of research is being done. Three further books appeared in 1984 (those numbered 3, 6 and 7 in this site), the year of the centenary of the birth of Sri Bharati Krsna Tirthaji. These were published by The Vedic Mathematics Research Group.

When Maharishi Mahesh Yogi began to explain the significance and marvellous qualities of Vedic Mathematics in 1988 Maharishi schools around the world began to teach it. At the school in Skelmersdale, Lancashire a full course was written and trialled for 11 to 14 year old pupils, called The Cosmic Computer. (Maharishi had said that the Sutras of Vedic Mathematics are the software for the cosmic computer- the cosmic computer runs the entire universe on every level and in every detail). This course was published in March 1998.

Here we use When the Samuccaya is the same it is zero. This gives 3x-12=0, so x=4.

The same formula also gives the other solution to the equation, so x-7=0 and x=7.

Nuggets from Vedic Mathematics

The Vedas are ancient holy texts from India than can be legitimately characterized as the all-encompassing repository of (Hindu) knowledge from eons past. The term Vedic Mathematics refers to a set of sixteen mathematical formulae or sutras and their corollaries derived from the Vedas. The sixteen sutras are:

1. Ekadhikena Purvena
2. Nikhilam Navatashcaramam Dashatah
3. Urdhva-tiryagbhyam
4. Paraavartya Yojayet
5. Shunyam Saamyasamuccaye
6. (Anurupye) Shunyamanyat
7. Sankalana-vyavakalanabhyam
8. Puranapuranabhyam
9. Chalana-Kalanabhyam
10. Yaavadunam
11. Vyashtisamanshtih
12. Shesanyankena Charamena
13. Sopaantyadvayamantyam
14. Ekanynena Purvena
15. Gunitasamuchyah
16. Gunakasamuchyah

Vedic Number Representation

Vedic knowledge is in the form of slokas or poems in Sanskrit verse. A number was encoded using consonant groups of the Sanskrit alphabet, and vowels were provided as additional latitude to the author in poetic composition. The coding key is given as Kaadi nav, taadi nav, paadi panchak, yaadashtak ta ksha shunyam Translated as below

- letter "ka" and the following eight letters
- letter "ta" and the following eight letters
- letter "pa" and the following four letters
- letter "ya" and the following seven letters, and
- letter "ksha" for zero.

In other words,

- ka, ta, pa, ya = 1
- kha, tha, pha, ra = 2
- ga, da, ba, la = 3
- gha, dha, bha, va = 4
- gna, na, ma, scha = 5
- cha, ta, sha = 6
- chha, tha, sa = 7
- ja, da, ha = 8
- jha, dha = 9
- ksha = 0

For those of you who don't know or remember the varnmala, here it is:

ka kha ga gha gna

cha chha ja jha inya

Ta Tha Rda Dha Rna

ta tha da dha na

pa pha ba bha ma

ya ra la va scha
sha sa ha chjha tra gna

Thus pa pa is 11, ma ra is 52. Words kapa, tapa , papa, and yapa all mean the same that is 11. It was upto the author to choose one that fit the meaning of the verse well. An interesting example of this is a hymn below in the praise of God Krishna that gives the value of Pi to the 32 decimal places as .31415926535897932384626433832792.

Gopi bhaagya madhu vraata

Shrngisho dadhisandhiga

Khalajivita khaataava

Galahaataarasandhara

1. Ekadhikena Purvena

or By one more than the previous one.

The proposition "by" means the operations this sutra concerns are either multiplication or division. [In case of addition/subtraction proposition "to" or "from" is used.] Thus this sutra is used for either multiplication or division. It turns out that it is applicable in both operations.

An interesting application of this sutra is in computing squares of numbers ending in five. Consider:

35x35 = (3x(3+1)) 25 = 12,25

The latter portion is multiplied by itself (5 by 5) and the previous portion is multiplied by one more than itself (3 by 4) resulting in the answer 1225.

It can also be applied in multiplications when the last digit is not 5 but the sum of the last digits is the base (10) and the previous parts are the same. Consider:

37 × 33 = (3×4), 7×3 = 12,21

29 × 21 = (2×3), 9×1 = 6,09 [Antyayor dashake]

We illustrate this sutra by its application to conversion of fractions into their equivalent decimal form. Consider fraction 1/19. Using this sutra this can be converted into a decimal form in a single step.

This can be done either by applying the sutra for a multiplication operation or for a division operations, thus yielding two methods.

METHOD 1: USING MULTIPLICATIONS

1/19, since 19 is not divisible by 2 or 5, the fractional result is a purely circulating decimal. (If the denominator contains only factors 2 and 5 is a purely non-circulating decimal, else it is a mixture of the two.)

So we start with the last digit

1

Multiply this by "one more", that is, 2 (this is the "key" digit from Ekadhikena)

21

Multiplying 2 by 2, followed by multiplying 4 by 2

421 $\Rightarrow$ 8421

Now, multiplying 8 by 2, sixteen

68421

1 $\Leftarrow$ carry

multiplying 6 by 2 is 12 plus 1 carry gives 13

368421

1 $\Leftarrow$ carry

Continuing

7368421 $\Rightarrow$ 47368421 $\Rightarrow$ 947368421

1

Now we have 9 digits of the answer. There are a total of 18 digits (=denominator-numerator) in the answer computed by complementing the lower half:

052631578

947368421

Thus the result is .052631578,947368421

METHOD 2: USING DIVISIONS

The earlier process can also be done using division instead of multiplication. We divide 1 by 2, answer is 0 with remainder 1

.0

Next 10 divided by 2 is five

.05

Next 5 divided by 2 is 2 with remainder 1

.052

next 12 (remainder,2) divided by 2 is 6

.0526

and so on.

As another example, consider 1/7, this same as 7/49 which as last digit of the denominator as 9. The previous digit is 4, by one more is 5. So we multiply (or divide) by 5, that is,

...7 $\Rightarrow$ 57 $\Rightarrow$ 857 $\Rightarrow$ 2857 $\Rightarrow$ 42857 $\Rightarrow$ 142857 $\Rightarrow$ .142,857 (stop after 7-1 digits)

3 2 4 1 2

2. *Nikhilam Navatashcaramam Dashatah*

or All from nine and the last from ten.

This sutra is often used in special cases of multiplication.

Corollary 1: Yavdunam Jaavdunikritya Varga Cha Yojayet

or Whatever the extent of its deficiency, lessen it still further to that very extent; and also set up the square of that deficiency.

For instance: in computing the square of 9 we go through the following steps:

The nearest power of 10 to 9 is 10. Therefore, let us take 10 as our base.

Since 9 is 1 less than 10, decrease it still further to 8. This is the

left side of our answer.

On the right hand side put the square of the deficiency, that is 1^2.

Hence the answer is 81.

Similarly, 8^2 = 64, 7^2 = 49

For numbers above 10, instead of looking at the deficit we look at the surplus. For example:

11^2 = 12 1^2 = 121

12^2 = (12+2) 2^2 = 144

14^2 = (14+4) 4^2 = 18 16 = 196

and so on.

3. *Urdhva-tiryagbhyam*

or Vertically and crosswise.

This sutra applies to all cases of multiplication and is very useful in division of one large number by another large number.

4. *Paraavartya Yojayet*

or Transpose and apply.

This sutra complements the Nikhilam sutra which is useful in divisions by large numbers. This sutra is useful in cases where the divisor consists of small digits. This sutra can be used to derive the Horner's process of Synthetic Division.

5. Shunyam Saamyasamuccaye

or When the samuccaya is the same, that samuccaya is zero.

This sutra is useful in solution of several special types of equations that can be solved visually. The word samuccaya has various meanings in different applicatins. For instance, it may mean a term which occurs as a common factor in all the terms concerned. A simple example is equation "12x + 3x = 4x + 5x". Since "x" occurs as a common factor in all the terms, therefore, x=0 is a solution. Another meaning may be that samuccaya is a product of independent terms. For instance, in (x+7)(x+9) = (x+3)(x+21), the samuccaya is 7 x 9 = 3 x 21, therefore, x = 0 is a solution. Another meaning is the sum of the denominators of two fractions having the same numerical numerator, for example: 1/(2x-1) + 1/(3x-1) = 0 means 5x - 2 = 0.

Yet another meaning is "combination" or total. This is commonly used. For instance, if the sum of the numerators and the sum of denominators are the same then that sum is zero. Therefore,

$$\frac{2x+9}{2x+7}=\frac{2x+7}{2x+9}$$

therefore, 4x + 16 = 0 or x = -4

This meaning ("total") can also be applied in solving quadratic equations. The total meaning can not only imply sum but also subtraction. For instance when given N1/D1 = N2/D2, if N1+N2 = D1 + D2 (as shown earlier) then this sum is zero. Mental cross multiplication reveals that the resulting equation is quadratic (the coefficients of x^2 are different on the two sides). So, if N1 - D1 = N2 - D2 then that samuccaya is also zero. This yield the other root of a quadratic equation.

Yet interpretation of "total" is applied in multi-term RHS and LHS. For instance, consider

$$\frac{1}{x-7}+\frac{1}{x-9}=\frac{1}{x-6}+\frac{1}{x-10}$$

Here D1 + D2 = D3 + D4 = 2 x - 16. Thus x = 8.

There are several other cases where samuccaya can be applied with great versatility. For instance "apparently cubic" or "biquadratic" equations can be easily solved as shown below:

(x-3)^3 + (x-9)^3 = 2 (x-6)^3

Note that x -3 + x - 9 = 2 (x - 6). Therefore (x - 6) = 0 or x = 6.

consider

$$\frac{(x+3)^{\wedge}3}{(x+5)^{\wedge}3}=\frac{x+1}{x+7}$$

Observe: N1 + D1 = N2 + D2 = 2x + 8.

Therefore, x = -4.

This sutra has been extended further.

6. (Anurupye) Shunyamanyat

or If one is in ratio, the other one is zero.

This sutra is often used to solve simultaneous simple equations which may involve big numbers. But these equations in special cases can be visually solved because of a certain ratio between the coefficients. Consider the following example:

$6x + 7y = 8$

$19x + 14y = 16$

Here the ratio of coefficients of y is same as that of the constant terms.

Therefore, the "other" is zero, i.e., $x = 0$. Hence the solution of the

equations is $x = 0$ and $y = 8/7$.

This sutra is easily applicable to more general cases with any number of variables. For instance

$ax + by + cz = a$

$bx + cy + az = b$

$cx + ay + bz = c$

which yields $x = 1, y = 0, z = 0$.

A corollary (upsutra) of this sutra says Sankalana-Vyavakalanaabhyam or By addition and by subtraction. It is applicable in case of simultaneous linear equations where the x- and y-coefficients are interchanged. For instance:

$45x - 23y = 113$

$23x - 45y = 91$

By addition:

$68x - 68y = 204 \Rightarrow 68(x-y) = 204 \Rightarrow x - y = 3$

By subtraction:

$22x + 22y = 22 \Rightarrow 22(x+y) = 22 \Rightarrow x + y = 1$

7. *Puranapuranabhyam*

or By the completion or non-completion.

8. *Ekanynena Purvena*

It is converse of the Ekaadhika sutra. It provides for multiplications wherein the multiplier digits consist entirely of nines.

"Rules of Thumb"

Many of the basic sutras have been applied to devise commonly used rules of thumb. For instance, the Ekanyuna sutra can be used to derive the following results:

- Kevalaih Saptakam Gunyaat, or in the case of seven the multiplicand should be 143
- Kalau Kshudasasaih, or in the case of 13 the multiplicand should be 077
- Kamse Kshaamadaaha-khalairmalaih, or in the case of 17 the multiplicand should be 05882353 (by the way, the literal meaning of this result is "In king Kamsa's reign famine, and unhygenic conditions prevailed." -- not immediately obvious what it had to do with Mathematics. These multiple meanings of these sutras were one of the reasons why some of the early translations of Vedas missed discourses on vedaangas.)

These are used to correctly identify first half of a recurring decimal number, and then applying Ekanyuna to arrive at the complete answer mechanically. Consider for example the following visual computations:

1/7 = 143x999/999999 = 142857/999999 = 0.142857

1/13 = 077x999/999999 = 076923/999999 = 0.076923

1/17 = 05882353x99999999/9999999999999999

= 0.05882352 94117647

Note that

7×142857 = 999999

13×076923 = 999999

17×05882352 94117647 = 9999999999999999

which says that if the last digit of the denominator is 7 or 3 then the last digit of the equivalent decimal fraction is 7 or 3 respectively.

Some Interesting Nuggets and Examples:

- The Multiplication Sign "X" as a Cross-Addition: Let us multiply (decimal numbers) 8 by 7: first column lists the numbers and the second column the deficits (from base = 10):

$$\begin{array}{rr} 8 & -2 \\ \underline{\times 7} & \underline{-3} \end{array}$$

The multiplication proceeds from the most signficant digit to least significant digit (which is natural since the positional numbers are also read from MSD to LSD, thus the result can be produced "on-line"). The first digit (most significant digit) is obtained by

1. adding 8 and -3, or
2. adding 7 and -2, or

that is,

$$\begin{array}{ccc} 8 & & -2 \\ & \times & \\ 7 & & -3 \end{array}$$

This process of obtaining MSD of a multiplication by cross-addition is said to be the origin of the conventional cross sign for multiplication. BTW, you can generate the following digit by multiplication and (if necessary) by forwarding the carry to more significant digits. This method (derived from Nikhilam sutra) works multiplication of multidigit numbers and numbers greater than as well as less than the base (or half the base). Consider bit more complex examples below:

97	−3	102	2	888	−112
×98	−2	×104	4	×997	−003
95,	06	106,	08	885,	336

For cases when the numbers are closer to the middle of the base, Anurupyena sutra (according to the ratio) can be used to compute deficit/excess from a ratio of the base and then ratio the result:

$$\begin{array}{ll} 48 & -2 \quad (\text{base}/2 = 50) \\ \times 46 & -4 \\ \hline 44, & 08 \Rightarrow 22, 08 \end{array}$$

- Division using "Seshaanyankaani charamena": to carry out a division first compute remainders and then multiply the remainders by the last digit and put down the last digit of the multiplicand. Consider: 1/7. When divising 1(0) by 7 the remainder is 3. Therefore, dividing 3 by 7 will subsequently lead to remainder 9 (= 3x3). But since 9 is more than 7 the remainder would be 2, so the remainder sequence is:
- 3, 2

Now 2 divided by 7 will have remainder of 6 (3×2), that is

3, 2, 6

Continuing

3, 2, 6, 4, 5, 1

We stop when the remainder sequence starts to repeat. Now, multiply these remainders by the last digit (7) of the denominator and keep only the first digit (LSD). So we have:

7 × 3 = 21 ⇒ put down 1

.1

3, 2, 6, 4, 5, 1

7 × 2 = 14 ⇒ put down 4

.1 4

3, 2, 6, 4, 5, 1

7 × 6 = 42 ⇒ put down 2

.1 4 2

3, 2, 6, 4, 5, 1

Continuing

.1 4 2 8 5 7

3, 2, 6, 4, 5, 1

So the answer is 1/7 = .142857142857...

Why the Vedic Math Important?

History : An ancient system of mathematics, Vedic Math is based on simple rules and principles through which any mathematical problem - be it Arithmetic, Algebra, Geometry or Trigonometry can be solved orally ! Vedic math is a set of sixteen Sanskrit mathematical formulae or sutras and their corollaries derived from the Vedas, which were formulated over centuries. The Sutras were lost over time but thankfully, re- discovered by Jagadguru Swami Sri Bharati Krsna Tirthaji Maharaja between 1911 and 1918.

MATH SUTRAS

Natural Formulae: The system is based on 16 Vedic sutras or aphorisms which are actually word - formulae that describe natural ways of solving a range of mathematical problems.

Some examples of sutras are "By one more than the one before ", " All from 9 and the last from 10", and "Vertically and Crosswise". These 16 one-line formulae originally written in Sanskrit are easy to remember . These formulae enable you to solve long mathematical problems quickly.

A Simple System : According to common opinion, this method is far more systematic, coherent and unified than the conventional system normally followed in schools. Vedic math is mental tool for calculation that encourages the development and use of intuition and innovation, while giving the student a lot of flexibility fun & satisfaction.

This system of mathematics has is so simple, direct and easy to implement that it is even used by scientists at

NASA - a striking testimony of its enormous popularity among educationists and academicians the world over.

The Sixteen Sutras

Sutras

1 Ekadhikina Purvena
COROLLARY: Anurupyena
Meaning: By one more than the previous one

2 Nikhilam Navatashcaramam Dashatah
COROLLARY: Sisyate Sesasamjnah
Meaning: All from 9 and the last from 10

3 Urdhva-Tiryagbyham
COROLLARY: Adyamadyenantyamantyena
Meaning: Vertically and crosswise

4 Paraavartya Yojayet
COROLLARY: Kevalaih Saptakam Gunyat
Meaning: Transpose and adjust

5 Shunyam Saamyasamuccaye
COROLLARY: Vestanam
Meaning: When the sum is the same that sum is zero.

6 (Anurupye) Shunyamanyat
COROLLARY: Vestanam
Meaning: If one is in ratio, the other is zero

7 Sankalana-vyavakalanabhyam
COROLLARY: Yavadunam Tavadunikritya Vargancha Yojayet
Meaning: By addition and by subtraction

8 Puranapuranabyham
COROLLARY: Antyayordashake'pi
Meaning: By the completion or non-completion

9 Chalana-Kalanabyham

COROLLARY: Antyayoreva

Meaning: Differences and Similarities

10 Yaavadunam

COROLLARY: Samuccayagunitah

Meaning: Whatever the extent of its deficiency

11 Vyashtisamanstih

COROLLARY: Lopanasthapanabhyam

Meaning: Part and Whole

12 Shesanyankena Charamena

COROLLARY: Vilokanam

Meaning: The remainders by the last digit

13 Sopaantyadvayamantyam

COROLLARY: Gunitasamuccayah Samuccayagunitah

Meaning: The ultimate and twice the penultimate

14 Ekanyunena Purvena

COROLLARY: Dhvajanka

Meaning: By one less than the previous one

15 Gunitasamuchyah

COROLLARY: Dwandwa Yoga

Meaning: The product of the sum is equal to the sum of the product

16 Gunakasamuchyah

COROLLARY: Adyam Antyam Madhyam

Meaning: The factors of the sum is equal to the sum of the factors

Ekadhikena Purvena

The Sutra (formula) Ekadhikena Purvena means:

"By one more than the previous one".

Now let us apply this sutra to the 'squaring of numbers ending in 5'.

Consider the example 25^2.

Here the number is 25. We have to find out the square of the number. For the number 25, the last digit is 5 and the 'previous' digit is 2. Hence, 'one more than the previous one', that is, 2+1=3. The Sutra, in this context, gives the procedure 'to multiply the previous digit 2 by one more than itself, that is, by 3. It becomes the L.H.S (left hand side) of the result, that is,

2 × 3 = 6. The R.H.S (right hand side) of the result is 5^2, that is, 25.

Thus 25^2 = 2 × 3 / 25 = 6/25=625.

In the same way,

35^2= 3 × (3+1) /25 = 3 × 4/ 25 = 1225;

65^2= 6 × 7 / 25 = 4225;

105^2= 10 × 11/25 = 11025;

135^2= 13 × 14/25 = 18225;

Now try to find out the squares of the numbers 15, 45, 85, 125, 175 and verify the answers.

Vulgar Fractions Whose Denominators are Numbers Ending in NINE

We now take examples of 1 / a9, where a = 1, 2, -----, 9. In the conversion of such vulgar fractions into recurring decimals, Ekadhikena Purvena process can be effectively used both in division and multiplication.

Multiplication Method: Value of 1 / 19

First we recognize the last digit of the denominator of the type 1 / a9. Here the last digit is 9. For a fraction of the form in whose denominator 9 is the last digit, we take the case of 1 / 19 as follows: For 1 / 19, 'previous' of 19 is 1. And one more than of it is 1 + 1 = 2. Therefore 2 is the multiplier for the conversion.

We write the last digit in the numerator as 1 and follow the steps leftwards.

Step. 1 : 1

Step. 2 : 21(multiply 1 by 2, put to left)

Step. 3 : 421(multiply 2 by 2, put to left)

Step. 4 : 8421(multiply 4 by 2, put to left)

Step. 5 : 168421 (multiply 8 by 2=16, 1 carried over, 6 put to left)

Step. 6 : 1368421 (6 × 2 =12,+1 = 13, 1 carried over, 3 put to left)

Step. 7 : 7368421 (3 × 2, = 6 +1 = 7, put to left)

Step. 8 : 147368421 (as in the same process)

Step. 9 : 947368421 (Do - continue to step 18)

Step. 10 : 18947368421

Step. 11 : 178947368421

Step. 12 : 1578947368421

Step. 13 : 11578947368421

Step. 14 : 31578947368421

Step. 15 : 631578947368421

Step. 16 : 12631578947368421

Step. 17 : 52631578947368421

Step. 18 : 1052631578947368421

Now from step 18 onwards the same numbers and order towards left continue.

Thus 1 / 19 = 0.052631578947368421

Find the recurring decimal form of the fractions 1 / 29, 1 / 59, 1 / 69, 1 / 79, 1 / 89 using Ekadhika process

Nikhilam Navatas'charamam Dasatah

The formula simply means: "all from 9 and the last from 10".

The formula can be very effectively applied in multiplication of numbers, which are nearer to bases like 10, 100, 1000 i.e., to the powers of 10(eg: 96 × 98 or 102 × 104). The procedure of multiplication using the Nikhilam involves minimum number of steps, space, time saving and only mental calculation. The numbers taken can be either less or more than the base considered.

Case (i) : when both the numbers are lower than the base.

Find 97 × 94. Here base is 100. Now following the rules, the working is as follows: 97 is 3 less than the nearest base 100. and 94 is 6 less than the same nearest base 100. Hence 3 and 6 are called deviations from the base. Always the base should be same for the two numbers.

$$\begin{array}{ll} 97 & 03 \\ 94 & 06 \\ \hline \begin{matrix}(97-06)\text{or} \\ (94-03)\end{matrix} & (3\times 6) \end{array} = 91/18 = 9118$$

Note: Here '/' signifies just a seperation and has nothing to do with division.

In genreal, let N1 and N2 be two numbers near to a given base in powers of 10, and D1 and D2 are their respective deviations from the base. Then N1 X N2 can be represented as:

$$\begin{array}{ll} N1 & D1 \\ N2 & D2 \\ \hline \begin{matrix}(N1+D2)\text{or} \\ (N2+D1)\end{matrix} & (D1\times D2) \end{array}$$

Case (ii) : When both the numbers are higher than the base.

The method and rules follow as they are. The only difference is the positive deviation. Instead of cross - subtract, we follow cross - add.104 X 102. Base is 100.

$$\begin{array}{ll} 104 & 04 \\ 102 & 02 \\ \hline 106/4\times 2 & =10608 \end{array}$$

Note: We are considering 04x02=08 and appending '08' and not just '4x2=8'. This is done because, we need to consider two digits in deviation as it the base 100 has two zeros. If the deviation is near 1000 then we need to consider 3 digits in the deviation (eg, 004 and not just 4).

Case (iii): One number is more and the other is less than the base.

In this situation one deviation is positive and the other is negative.

So the product of deviations becomes negative. So the right hand side of the answer obtained will therefore have to be subtracted. To have a clear representation and understanding a vinculum is used. It proceeds into normalization. 13 × 7. Base is 10

$$\begin{array}{ll} 13 & 03 \\ 7 & -03 \\ \hline (13-03)\text{or} & \\ & 03\times -03 \\ (7+30) & \end{array} = 10/\text{-}09 = 91$$

10/(-9) = 91 how?

10/(-9) should be read as 'one zero, nine bar'. Here 'one' and 'zero' are in normal form. 'nine' is in complement form, so, when we bring a carry from normal form to complement form, '10' becomes '9' and '9-bar' becomes '1'(10's complement of 9). Hence 10/(-9)-91.

Another example would be 94/(-3) = 93/7 = 937.

Find the following products by Nikhilam formula.

1) 7 × 4
2) 93 × 85
3) 875 × 994
4) 1234 × 1002
5) 1003 × 997
6) 11112 × 9998
7) 1234 × 1002
8) 118 × 105

Anurupyena

The upa-Sutra 'Anurupyena' means 'proportionality' or 'similarly'. This Sutra is highly useful to find products of two numbers when both of them are near the Common bases like 50, 60, 200 etc(multiples of powers of 10).

EXAMPLE 1:

46 × 43

As per the previous methods, if we select 100 as base we get

$$\begin{array}{rr} 46 & -54 \\ \underline{43} & \underline{-57} \end{array}$$

This is much more difficult and of no use.

Now by 'Anurupyena' we consider a different working base through we can solve the problem.

Take the nearest higher multiple of 10. In this case it is 50.

Treat it as 100 / 2 = 50.

Now the steps are as follows:

i) Choose the working base near to the numbers under consideration.

i.e., working base is 100 / 2 = 50

ii) Write the numbers one below the other

46

43

iii) Write the differences of the two numbers respectively from 50 against each number on right side

46 –04

43 –07

iv) Write cross-subtraction or cross- addition as the case may be under the line drawn. Multiply the differences and write the product in the left side of the answer.

46		–04
43		–07
(46 – 07) or (43 – 04)	/	–04 × 07
= 39	/	28

v) Since base is 100 / 2 = 50 , 39 in the answer represents 39×50.

Hence divide 39 by 2 because 50 = 100 / 2

Thus 39 ÷ 2 gives 19½ where 19 is quotient and 1 is remainder . This 1 as Reminder gives one 50 making the L.H.S of the answer 28 + 50 = 78(or Remainder ½ x 100 + 28) i.e. R.H.S. 19 and L.H.S. 78 together give the answer 1978.

We represent it as

46 -04

43 -07

2) 39 / 28

19.5 / 28

= 19 / (50 + 28)

= 19 / 78 = 1978.

Solve 58 × 48 :

Working base 50 = 5 × 10 gives

58 08
48 -02

---

2) 56 / -16

---

28 / -16

27/(100 - 16)
= 27 / 84

= 1978.

Find the following products.

1. 46 × 46
2. 57 × 57
3. 54 × 45

Urdhva Tiryagbhyam

Urdhva - tiryagbhyam is the general formula applicable to all cases of multiplication and also in the division of a large number by another large number. It means "Vertically and cross wise."

Multiplication of two 2 digit numbers.

EXAMPLE 1:

Find the product 14 × 12

The symbols are operated from right to left .

Step (i) :

1 4 ↓
1 2 ↓

: 4x2

Step (ii) :

$$\begin{array}{cc} 1 & 4 \\ 1 & 2 \\ \hline 2+4 & 8 \end{array}$$

2+4 : 8

Step (iii) :

$$\begin{array}{cc} 1 & 4 \\ 1 & 2 \\ \hline \end{array}$$

1x1 : 6 : 8

which gives 168

Let us work another problem by placing the carried over digits under the first row and proceed.

$$\begin{array}{r} 234 \\ \times \quad 316 \\ \hline 61724 \\ 1222 \text{ — carry} \\ \hline 73944 \end{array}$$

Steps:

(i) 4 × 6 = 24 : 2, the carried over digit is placed below the second digit.

(ii) (3 × 6) + (4 × 1) = 18 + 4 = 22 ; 2, the carried over digit is placed below third digit.

(iii) (2 × 6) + (3 × 1) + (4 × 3) = 12 + 3 + 12 = 27 ; 2, the carried over digit is placed below fourth digit.

(iv) (2 × 1) + (3 × 3) = 2 + 9 = 11; 1, the carried over digit is placed below fifth digit.

(v) (2 × 3) = 6.

(vi) Respective digits are added.

General rule for a 3 digit by 3 digit multiplication:

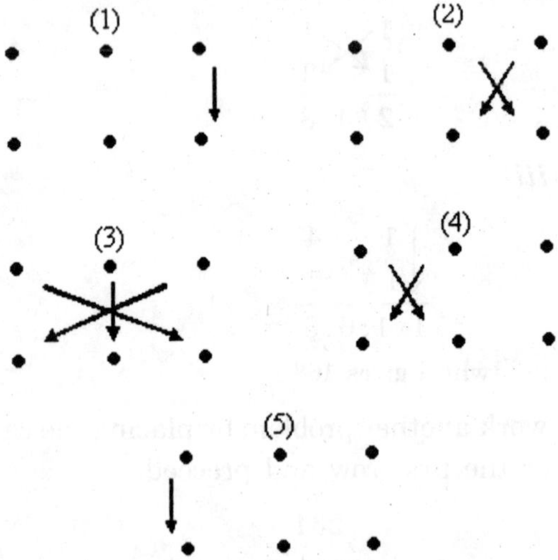

Adyamadyena-Antyamantyena

The Sutra ' Adyamadyena-Antyamantyena' means 'the first by the first and the last by the last'.

Suppose we are asked to find out the area of a rectangular card board whose length and breadth are respectively 6ft . 4 inches and 5 ft. 8 inches. Generally we continue the problem like this.

Area = Length × Breath

By Vedic principles we proceed in the way "the first by first and the last by last"

i.e. 6' 4" can be treated as $6x + 4$ and 5' 8" as $5x + 8$,

Where x = 1ft. = 12 in; x^2 is sq. ft.

Now $(6x + 4)(5x + 8)$

$= 30x^2 + 6.8.x + 4.5.x + 32$

$= 30x^2 + 45x + 20x + 32$

$= 30x^2 + 68.x + 32$

$= 30x^2 + (5x + 8).x + 32$

Writing 68 = 5 × 12 + 8

= $35x^2 + 8x + 32$

= 35 sq. ft. + 8 × 12 sq. in + 32 sq. in

= 35 sq. ft. + 96 sq. in + 32 sq. in

= 35 sq. ft. + 128 sq. in.

It is interesting to know that a mathematically untrained and even uneducated carpenter simply works in this way by mental argumentation. It goes in his mind like this

6' 4"

5' 8"

First by first i.e. 6' × 5' = 30 sq. ft.

Last by last i.e. 4" × 8" = 32 sq. in.

Now cross wise 6 × 8 + 5 x 4 = 48 + 20 = 68.

Adjust as many '12' s as possible towards left as 'units' i.e. 68 = 5 × 12 +8 , 5 twelve's as 5 square feet make the first 30+5 = 35 sq. ft ; 8 left becomes 8 x 12 square inches and go towards right i.e. 8 × 12 = 96 sq. in. towards right ives 96+32 = 128sq.in.

Thus he got area in some sort of 35 squints and another sort of 128 sq. units.

i.e. 35 sq. ft 128 sq. in

ANOTHER EXAMPLE:

4' 6"

3' 4"

4 × 3	4×4+6×3	6×4
=12	=16 + 18	= 24
	= 34	
	= 2×12+10	

Now 12 + 2 = 14, 10 x 12 + 24 = 120 + 24 = 144

Thus 4' 6" × 3' 4" = 14 Sq. ft. 144 Sq. inches.

Since 144 sq. in = 12 × 12 = 1 sq. ft .

The answer is 15 sq. ft.

Sankalana-Vyavakalanabhyam

This Sutra means 'by addition and by subtraction'. It can be applied in solving a special type of simultaneous equations where the x - coefficients and the y - coefficients are found interchanged.

EXAMPLE 1:

$45x - 23y = 113$

$23x - 45y = 91$

From Sankalana - vyavakalanabhyam

add them,

i.e., $(45x - 23y) + (23x - 45y) = 113 + 91$

i.e., $68x - 68y = 204$ i.e., $x - y = 3$

subtract one from other,

i.e., $(45x - 23y) - (23x - 45y) = 113 - 91$

i.e., $22x + 22y = 22$ i.e., $x + y = 1$

and repeat the same sutra, we get $x = 2$ and $y = - 1$

EXAMPLE 2:

$1955x - 476y = 2482$

$476x - 1955y = -4913$

Oh ! what a problem ! And still

just add, $2431(x - y) = - 2431$ i.e., $x - y = -1$

subtract, $1479 (x + y) = 7395$ i.e., $x + y = 5$

once again add, $2x = 4$ $x = 2$

subtract $- 2y = - 6$ $y = 3$

Solve the following problems using Sankalana - Vyavakalanabhyam.

1. $3x + 2y = 18$
   $2x + 3y = 17$
2. $5x - 21y = 26$
   $21x - 5y = 26$

Yavdunam Taavdunikritya Vargancha Yojayet

This sutra means whatever the extent of its deficiency, lessen it still further to that very extent; and also set up the square of that deficiency. This sutra is very handy in calculating squares of numbers near(lesser) to powers of 10

For instance in computing the square of 98 we go through the following steps:

1. The nearest power of 10 to 98 is 100.
2. Therefore, let us take 100 as our base.
3. Since 98 is 2 less than 100, we call 2 as the deficiency.
4. Decrease the given number further by an amount equal to the deficiency.

   i.e., perform (98 -2) = 96. This is the left side of our answer!!.
5. On the right hand side put the square of the deficiency, that is square of 2 = 04.
6. Append the results from step 4 and 5 to get the result. Hence the answer is 9604.

Note: While calculating step 5, the number of digits in the squared number (04) should be equal to number of zeroes in the base(100). Hence in our case, the base 100 has 2 zeros and hence sqaure of 2 is 04 and not just 4.

Now consider another example: Calculate the square of 94. (following the steps given above...)

1. The neareset power of 10 to given number 94 is 100

2. Hence the base is 100
3. Deficiency = 6
4. Subtract the deficiency from the given number (94 - 6) = 88
5. Square of deficiency = 36 (2 digits)
6. Append the results from step 4 and 5 to get the result. i.e., 8836

Try out calculating squares of 96, 994, 9999995

Yavadadhikam Taavadahikikritya Vargancha Yojayet

This sutra means whatever the extent of its surplus, increment it still further to that very extent; and also set up the square of that surplus. This sutra is very useful in calculating the sqaures of numbers nearer(greater) to powers of 10.

For instance: in computing the square of 103 we go through the following steps:

1. The nearest power of 10 to 103 is 100.
2. Therefore, let us take 100 as our base.
3. Since 103 is 3 more than 100(base), we call 3 as the surplus.
4. Increase the given number further by an amount equal to the surplus.

   i.e., perform (103 + 3) = 106.

   This is the left side of our answer!!.
5. On the right hand side put the square of the surplus, that is square of 3 = 09.
6. Append the results from step 4 and 5 to get the result.

Hence the answer is 10609.

Note: while calculating step 5, the number of digits in the squared number (09) should be equal to number of zeroes in the base(100). Hence in our case, the base 100 has 2 zeros and hence sqaure of 3 is 09 and not just 9.

Similarly if we need to calculate the square of 1009,

1. The nearest power of 10 to 1009 is 1000.
2. Therefore, let us take 1000 as our base.
3. Since 1009 is 9 more than 1000(base), we call 9 as the surplus.
4. Increase the given number further by an amount equal to the surplus.

   i.e., perform (1009 + 9) = 1018. This is the left side of our answer!!.
5. On the right hand side put the square of the surplus, that is square of 9 = 081.(not just 81 !)
6. Append the results from step 4 and 5 to get the result.

Hence the answer is 1018081.

Try to find the square of 106 and 19 using this sutra.

Now here's a twist. What is sqaure of 112 using this rule? will this rule apply?

Answer is Yes. But we need to stick to basics and use some common sense.

Here we go...

1.&2. the base is 100.

3. Surplus = 12
4. Increase the given number by surplus i.e. (112 + 12) = 124
5. Square of surplus, 12 = 144 . Now what?

   by rule we need to consider only 2 digits in step 5 (as the base has 2 zeros)

6. We have to append 124 and 144. The '1' in 144 becomes 'carry' to 124.

Hence the result is (124 +1) appended with 44 i.e., 12544 is the result.

Dont get confused. try this example to make yourself more confident.

Find the square of 111. Now try to find the square of 10101.

Antyayor Dasake'PI

The Sutra simply means - numbers of which the last digits added up give 10.

This sutra is helpful in multiplying numbers whose last digits add up to 10(or powers of 10). The remaining digits of the numbers should be identical.

i.e. the Sutra works in multiplication of numbers for example:

25 and 25, 2 is common and 5 + 5 = 10

47 and 43, 4 is common and 7 + 3 = 10

62 and 68,

116 and 114.

and also for 425 and 475

The last example can be looked as digit 4 being common and 25 and 75 add upto 100.

Note that in each case the sum of the last digit of first number to the last digit of second number is 10. Further the portion of digits or numbers left wards to the last digits remain the same.

At that instant use Ekadhikena on left hand side digits. Multiplication of the last digits gives the right hand part of the answer.

EXAMPLE 1 :

57 × 53

See the end digits sum 7 + 3 = 10 ; then by the sutras Antyayor dasakepi and Ekadhikena we have the answer.

Ekadhikena to the remaining digits means, increment the remaining digits by 1 and multiply it with the same.

57 × 53 = (5 + 1) × 5 / 7 × 3 (the '/' is just a seperator and not a division mark) = 30 / 21

= 3021.

EXAMPLE 2:

62 × 68

2 + 8 = 10, L.H.S. portion remains the same i.e., 6.

Ekadhikena (increment) of 6 gives 7

62 × 68 = (6 × 7) / (2 × 8)

= 42 / 16

= 4216.

Use Vedic sutras to find the products

1. 125 × 125
2. 34 × 36
3. 98 × 92

It is further interesting to note that the same rule works when the sum of the last 2, last 3, last 4 - - - digits added respectively equal to 100, 1000, 10000 -- - - . The simple point to remember is to multiply each product by 10, 100, 1000, - - as the case may be . Your can observe that this is more convenient while working with the product of 3 digit numbers.

Eg. 1: 292 × 208

Here 92 + 08 = 100, L.H.S portion is same i.e. 2

$292 \times 208 = (2 \times 3) / 92 \times 8$

$= 60 / 736$ (for 100 raise the L.H.S. product by 0)

$= 60736$.

Find the following products using 'Antyayordasakepi'

1. 318×312
2. 425×475

Lopana Sthapanabhyam

Lopana sthapanabhyam means 'by alternate elimination and retention'.

Consider the case of factorization of quadratic equation of type $ax^2 + by^2 + cz^2 + dxy + eyz + fzx$. This is a homogeneous equation of second degree in three variables x, y, z. The sub-sutra removes the difficulty and makes the factorization simple. The steps are as follows:

i) Eliminate z by putting $z = 0$ and retain x and y and factorize thus obtained a quadratic in x and y by means of Adyamadyena sutra.

ii) Similarly eliminate y and retain x and z and factorize the quadratic in x and z.

iii) With these two sets of factors, fill in the gaps caused by the elimination process of z and y respectively. This gives actual factors of the expression.

EXAMPLE 1 :

$3x^2 + 7xy + 2y^2 + 11xz + 7yz + 6z2$.

Step (i) : Eliminate z and retain x, y; factorize

$3x^2 + 7xy + 2y^2 = (3x + y)(x + 2y)$

Step (ii) : Eliminate y and retain x, z; factorize

$3x^2 + 11xz + 6z^2 = (3x + 2z)(x + 3z)$

Step (iii): Fill the gaps, the given expression

$= (3x + y + 2z)(x + 2y + 3z)$

EXAMPLE 2 :

$$3x^2 + 6y^2 + 2z^2 + 11xy + 7yz + 6xz + 19x + 22y + 13z + 20$$

Step (i) : Eliminate y and z, retain x and independent term i.e., y = 0, z = 0 in the expression (E).

Then E = $3x^2 + 19x + 20 = (x + 5)(3x + 4)$

Step (ii) : Eliminate z and x, retain y and independent term i.e., z = 0, x = 0 in the expression.

Then E = $6y^2 + 22y + 20 = (2y + 4)(3y + 5)$

Step (iii): Eliminate x and y, retain z and independent term i.e., x = 0, y = 0 in the expression.

Then E = $2z^2 + 13z + 20 = (z + 4)(2z + 5)$

Step (iv) : The expression has the factors (think of independent terms: constants)

$= (3x + 2y + z + 4)(x + 3y + 2z + 5)$.

Solve the following expressions into factors by using appropriate sutras:

1. $x^2 + 2y^2 + 3xy + 2xz + 3yz + z^2$.
2. $3x^2 + y^2 - 4xy - yz - 2z^2 - zx$.

Gunita Samuccayah—Samuccaya Gunitah

In connection with factorization of quadratic expressions a sub-Sutra, viz. 'Gunita samuccayah-Samuccaya Gunitah' is useful.

It is intended for the purpose of verifying the correctness of obtained answers in multiplications, divisions and factorizations.

It means in this context: 'The product of the sum of the coefficients(sc) in the factors is equal to the sum of the coefficients(sc) in the product'

Symbolically we represent as sc of the product = product of the sc (in the factors)/p>

EXAMPLE 1:

$(x + 3)(x + 2) = x^2 + 5x + 6$

Now $(x + 3)(x + 2) = 4 \times 3 = 12$: Thus verified.

EXAMPLE 2:

$(x + 5)(x + 7)(x - 2) = x^3 + 10x^2 + 11x - 70$

$(1 + 5)(1 + 7)(1 - 2) = 1 + 10 + 11 - 70$

i.e., $6 \times 8 \times -1 = 22 - 70$

i.e., $-48 = -48$ Verified.

Verify whether the following factorization of the expressions are correct or not by the Vedic check:

i.e. Gunita. Samuccayah-Samuccaya Gunitah:

1. $(2x + 3)(x - 2) = 2x^2 - x - 6$
2. $12x^2 - 23xy + 10y^2 = (3x - 2y)(4x - 5y)$
3. $12x^2 + 13x - 4 = (3x - 4)(4x + 1)$
4. $(x + 1)(x + 2)(x + 3) = x^3 + 6x^2 + 11x + 6$

Veni, Vedic, Vedic

The ancient science of Vedic mathematics may well give calculators a run for their money.

Does your mind wobble when confronted by a mathematical challenge more forbidding than two plus two? Do you dream of becoming the kind of person who can rattle off answers to the most complicated sums in the fraction of a second?

If the answer is yes, you need Vedic mathematics. Try this for size. What's the square of 65? Simple: just multiply the first digit, 6, with its successor, 7.

The answer is 42. Now find the square of the second digit, five, which is 25. Now bring the two together. Bingo, the answer is 4225! Don't believe it? Try it out with any two digit odd number divisible by five.

The law behind this apparent magic is the nikhilam sutra, one of the sutras or formulae that compose Vedic mathematics. These sutras were part of the parisista (appendix) of the Atharva Veda. "Our Vedas are the original source of knowledge. Vedic mathematics," says Dr. Narendra Chaudhury, "is one such source." A geology professor at Nagpur University in western India, Chaudhury turned to Vedic mathematics after attending a conference organized by the Einstein Foundation in 1984. Introduction to Vedic mathematics is like entering Alice's wonderland, where logic is turned on its head. Division is by a process of multiplication and addition, and multiplication is done by cross-subtraction. Nothing makes sense, but everything works. And the simplicity of the approach exposes top-heavy processes of our logic-driven world.

Unlike conventional mathematics, Vedic mathematics is not linear. Here, it would be as easy to multiply 432567 with 754352 as it would be to multiply four into five. This is because the processes are governed by an internal relationship between numbers rather than by external methods of conventional mathematics. Another distinguishing characteristic of Vedic mathematics is its holistic approach. Numbers are not separated from their philosophical and spiritual contexts. Number one stands for the Absolute, and therefore it cannot be subtracted or divided. Two represents unmanifest nature and three manifest nature. Four is associated with the world of mind and feeling. Number five represents ether and has the quality of sound. Six is air and possesses the quality of touch. Fire is at seven and water at eight. Finally, nine is represented by earth. Such an intimate relationship between numbers and the elements of the universe creates a unified body of knowledge that is far easier to follow than the fragmentary western way. To keep things simple, Vedic mathematics constantly reiterates that there are only nine numbers and a nought, and everything else derives from these.

Vedic mathematics, as it is found today, was formulated by the late Bharti Kishan Tirthaji, the Shankaracharya of Puri, a seaside temple town in the eastern Indian state of Orissa. He introduced 32 sutras through a study of the ancient texts, coupled with an insight into the nature of numbers. There are 16 main sutras and 16 sub-sutras. Application of the sutras depends on the problem: Vedic mathematics provides more than a way of solving a problem- the easiest is the best. Chaudhury rues the general ignorance shrouding Vedic mathematics. "The main reason," he says, "is the low awareness and the lack of an efficient faculty. Nobody will appreciate it unless the topic is made interesting." To make Vedic mathematics more accessible, it is necessary for the subject to be included in school curricula. Making learning fun is an oft-voiced concern. The means lie in ancient wisdom-which is ours to use or throw!

Addition

6932 + 1803 + 3331 = 1206

This can be cross-checked in the following manner:

(i) 6 + 9 + 3 + 2 = 20 = 2(2 + 0)

(ii) 1 + 8 + 0 + 3 = 12 = 3(1 + 2)

(iii) 3 + 3 + 3 + 1 = 10 = 1(1 + 0)

(iv) 2 + 3 + 1 = 6

(v) 1 + 2 + 0 + 6 + 6 = 15 = 6(1 + 5)

Steps (iv) and (v) contain the same digit and that is 6. So the solution is accurate.

Vertical and Crosswise Multiplication

The conventional method of multiplying

24 into 36 would go like this: 24

× 36 144

72 864

The Vedic method will work out the sum thus:

24 is the multiplicand and

36 is the multiplier *Step 1 Vertical multiplication*

Multiply the left-most of the multiplicand, in this case 2 24 with the leftmost digit of the multiplier, in this case 3 × 36

So, 2 × 3 = 6 6 *Step*

2 Crosswise multiplication Multiply the first digit of the multiplicand, in this case 2 with the second digit of the multiplier, in this case 6 So, 2 × 6 = 12

Now, multiply the second digit of the multiplicand, in this case 4 with the first digit of the multiplier, in this case 3 So, 4 × 3 = 12

Add the two products: 12 + 12 = 24

24 Place the second digit of the total next to the result of x36 step 1 and carry 2 to the next line 64 2 Step 3

Vertical multiplication Multiply the right-hand digits multiplicand and 24 multiplier.

4 × 6 = 24 × 36

Place 4 at the right hand most of the answer 644 and carry 2 to the next line + 22 Final result: 864

Vedic Mathematics

Rapid Mental Calculation

Galactic Mathematics Rediscovered: Jain will give you an understanding of the 16 basic Sutras or 'Word Formula' which can solve all known mathematical problems by using an ancient yet simple system of MENTAL, ONE LINE ARITHMETIC,used extensively by the Hindus more than 2,500 years ago.

The knowledge goes even further back. Jain calls this modality, Galactic Mathematics which is base 12.

During this day you will learn how to multiply 3 digit numbers in your head with no calculator, pen or paper.

How could the Aryan civilisation, near the Hunza Valley, 3,000 years ago, mentally compute mathematical operations that today only a calculator could achieve?

Jain wrote an article: VEDIC MATHEMATICS REDISCOVERED for the Nexus Magazine, Oct/Nov issue 2003. He shows how an understanding of the 16 basic Sutras or 'Word Formula' solves all known mathematical problems by using an advanced and highly intelligent system of MENTAL, ONE LINE ARITHMETIC used by the Hindus more than 2,000 years.

From response all over the world, this article has awakened many people to start questioning all the missing gaps and insufficiencies in the current mathematics curriculum.

If you or your children can't understand the fragmented and confusing mathematical data taught today? Would you be interested in another method that boosts confidence,increases memory skills... and is presented in a fun and exciting manner?

After this program, you will never multiply, divide or add in the same manner again. And no more Times Tables!

In this workshop you will know how to accomplish rapid calculations, mentally. No more Calculators which make the brain lazy. The brain is a mental muscle and needs to be exercised.

Jain predicts that "if we continue to teach children with an over-emphasis of using calculators in the class-room, over a 20 year period, then no doubt, the Global Brain will deteriorate". We pride ourselves that we are a highly intelligent species, but really, are we getting dumber.

Most young people are visual and the ability to tap into 'Pattern Recognition' is the key to rapid mental calculation.

"Jain he is an absolute wizard at displaying the real ability of your mind as he empowers students to perform mental calculations, in many cases... quicker than a calculator. Yes you did hear right! Jain reveals to you simple methods to mentally calculate all the complex math you can think of. At the same time he shows you how the world as we know it is more than we think it is".

Vedic Mathematics Rediscovered

Some questions to ponder upon:

- How could the Aryan civilization, near the Hunza Valley, 3,000 years ago, mentally compute mathematical operations that today only a calculator could achieve?
- And if the mathematics could not be done mentally, how could they do everything we know today, IN ONE-LINE? eg: converting the fraction 1/19 into a decimal is an 18 digit answer that requires 35 lines of exhausting long division, how could they do it in a one simple line solution?
- Why is NASA secretly using this MENTAL, ONE-LINE SYSTEM OF ARITHMETIC in the field of Advanced Artificial Intelligence? and not incorporating it into our current mathematics curriculum?
- Why are many of Bill Gate's Mathematicians Indians?
- Is the decimal form of Pi (3.141592) a sonic code for dedications to Krishna?
- What does the Bhagavad Gita have to say about this lost science of Clairvoyant Calculation that the ancient Temple Builders employed?
- How can you become a Human Bio-Calculator?
- Is the current Western Mathematics Curriculum keeping our Children in a state of limited intelligence?

The answers to all these questions are found by understanding the ancient system of Vedic Mathematics as rediscovered by Bharati Krsna Tirthaji (1884-1960), of Puri Govardhan Math, dux of India in Mathematics and a Sanskrit Philosopher who wrote and spoke perfect English, he reached the position of a Shankacharya (the spiritual head of India 100 years ago, like a Dalai Lama or Vatican Pope), renunciated several high official positions in society to retreat into a forest for 7 years as a yogi, between 1911 and 1918 in which time he collected all the lost mathematical knowledge (Ganita Sutras), improved and reconstructed them to 16 basic Sutras or word formulae that solve every known mathematical problem. These 16 Volumes relating to the 16 Sutras was our legacy from the ancient past.

This highly intelligent and sophisticated system of Mental, One-Line Arithmetic was taught by The Master to his main student, who today is the leading exponent in the world: Maharishi Mahesh Yogi, the long white-haired and bearded yogi that many of the 60s generation know of as the Beatle's guru. He is considered as the greatest human bio-calculator on the planet today, holding the vision of raising the collective consciousness of all nations by re-introducing the lost Vedic Mathematics Sutras early in a child's spiritual and educational development. Imagine a maths class in India today, like in the times of old when knowledge was passed on as an oral tradition, and the teacher quickly calls or drills out certain numbers to multiply or divide and the children instantly reply with the correct intuitive answers! Today, in schools where Maharishi's Transcendental Meditation technique are used students perform remarkably well academically, viewing Mathematics as a Universal Language and therefore often winning the highest competitive prizes in mathematics.

Maharishi organized the scattered Vedic Literature into a complete science of consciousness for both its theoretical and practical value, describing the 16 Sutras as fundamental impulses of intelligence that structure all the Laws of Nature and those that govern the Universe.

Maharishi, the Founder of TM, once said:

"Vedic Mathematics is that one field of knowledge which fulfils the purpose of education by developing the total creative genius of the individual, giving him/her the ability to be always spontaneously right and automatically precise so that his/her action, supported by Natural Law, is always effortlessly fulfilling."

Unfortunately, all of Bharati Krsna's 16 Volumes for the 16 Sutras were confirmed lost in 1958 after Bharati Krsna Tirthaji returned from the USA tour, sponsored by the Self-Realization Foundation: the successors of Paramahansa Yogananda (who died in 1952, and who introduced Yoga to the West). Disciples observed that people in the streets of India were imitating Tirthaji's lectures and dazzling audiences with mathematically mis-used psychic powers, and begged the Master to write a book just before his death in 1960. This single volume, called VEDIC MATHEMATICS, has become the bible on this subject and spawned a thousand web-sites worldwide. Maharishi schools disseminated the teachings to England where certain schools have adopted the Vedic curriculum but the irony is that today India is struggling to release the British curriculum and NASA have adopted it fully in the realms of advanced robotics. Calculations that can be solved as quick as lightning are a great tool to adopt, but you wouldn't want to teach it worldwide in the fear that you may churn out a generation of child geniuses that may threaten the intellectual status quo.

The gift that the Hindus gave to world, thousands of years ago, and that which is currently responsible for global silicon chip technology, was none other than the invention of ZERO and the use of the decimal point. We call our common numbers "Arabic Numerals" but really they extend back to the Hindu concept of creation and void known as "Bindu" or The Zero Point. In Maths, it makes the magic work eg: if you wish to multiply 98 x 98 we need to understand Zero and Unity. All Vedic Maths is based on the understanding of Unity Consciousness which means they utilize processes or Number Bases that correspond to:

0, 10, 100, 1000, 10000 etc. all of which add to 1.

Let us look at some practical examples, in light of the fact that the Vedas, literally "the illimitable storehouse of All Knowledge" came under 4 headings or categories like Ayurveda (Healing), and Vedas for Music and Astronomy etc. and one relevant sub-Veda that related to Temple Building and Engineering and therefore Mathematics called Sthapathyaveda. Thus a Vedic Mathematician was also an astronomer, healer and poet. It was a total system. As a temple builder, there was no pen and paper, you simply calculated in your head. You are out in the field and you need to tile a square floor that is, say 108 units square. How do you do it with mental ease?

I think only of the excess "8", saying how much is 108 more than my Base of 100. It is "8". So I will merely add this "8" to the number in question "108" and tag on the squaring of this excess:

108 Squared $= 108 + 8 / 8 \times 8$
$= 116 / 64 = 11,664$

Its very simple.

Here is another Mental, One-Line Calculation:

Sutra: By the Excess

The Multiplication of Numbers Over a Base

108 × 109 can be computed in one second by understanding this Pattern Recognition:

I add the excess of "8" to the other number "109" and then tag on the multiplicatin of those two excesses: "8" and "9".

108 × 109 = 109 + 8 / 8 x 9

= 117 / 72

= 11,772

Sutra: By the Deficiency

The Squaring of Numbers Under a Base

To solve 98 Squared (98 x 98) we must first determine what Base we are in. It is close to 100, therefore we say Base 100. We must now choose one of the 16 Sutras to effectively solve the problem. It is called: "By The Deficiency":

"By whatever the deficiency, lessen it further by that much and set up the square thereof"

says one of the 16 Sutras. Sounds cryptic and meaningless yet it quickly solves the problem.

We get our answer by merely knowing how much is 100 less 98. Knowing that the deficiency is 2, we merely lessen 98 by 2 and then we tag on the squaring of that 2. As a one-line answer, the setting out would appear as thus:

98 Squared = 98 - 2 / 2 × 2. Simplifying it:

= 96 / _ 4

We almost have our answer. What we need to know is that since our Base is 100, it has 2 zeroes, therefore this fact governs the need for 2 spaces or 2 digits after the "forward slash" symbol (/). By inserting or inventing The Zero as a "Place Marker", the answer is achieved:

98 Squared = 96 / 04

= 9,604.

(Perhaps the greatest invention of all time, the Hindu creation of the Zero was to change the world unimaginably as time passed on leading to the ability to send Vimana U.F.O craft and rockets into space, to supercomputers and thus the ancient and continuing battle between Michael and Lucifer:

Archangel Michael's Internal Light Vehicle Merkabah Time / Space Travel and Natural Spiritual Powers Versus Lucifer's daring invention of the External Merkabah or Artificial Technological Metal Spacecraft).

Observe similar examples:

97 Squared = 97 - 3 / 3 × 3

= 94 / 09

96 Squared = 96 - 4 / 4 × 4

= 92 / 16.

What if we enlarged our numbers to 998 Squared?

It is close to 1,000 so we say Base 1,000 and know to have 3 zeroes on the right hand side of the (/).

998 Squared= 998 - 2 / 2 × 2

= 996 / _ _ 4

= 996 / 004.

= 996,004

Understanding this, you can be calculating digits in the millions:

9998 Squared = 9998 - 2 / 2 × 2

= 9996 / _ _ _ 4

(Since we are in Base 10,000 the 4 Zeroes determine the need for 4 digits after the (/).

= 9996 / 0004

= 99,960,004.

There is a world-wide debate currently raging about the efficacy of Vedic Mathematics versus the crumbling foundations of Western Mathematics.

Generally speaking, the theorems we all learned at school are not wrong but clumsy. Some of the Western geometrical formulae are certainly wrong or inadequate: for example, the formulae for sphere packing in the higher dimensions increase up to the 6th Dimension then suddenly decrease for higher dimensions, which is simply absurd.

Unfortunately, some die-hard senior mathematicians in an attempt to protect the crumbling foundations that they now stand on feel threatened by the lightning quick mental calculations of the Vedic seers, and go to great lengths to deride Vedic maths as a "bag of tricks". And of course, many insecure teachers worldwide are afraid to rewrite all their course material.

Here is another example illustrating its utter simplicity:

"The Path of Least Resistance", or "The Law of Economy".

Sutra: By one More

The Squaring of Numbers Ending in 5

If we wanted to square the number 25, i.e. 25 x 25, we would conventionally take 3 lines of working out. Vedic Mathematics merely looks at the Question, applies one of the 16 Sutras, and solves it Mentally in One-Line. In this case, the Sutra at work is "By One More Than The Previous Digit". Observing that 25 is a 2 digit number, 5 is the last digit, but we are mainly interested in "the Previous Digit" which is 2. We say, mentally, "What is One more than Two?" It is 3. The word "By" in the Sutra really means "to multiply". The setting out for the first half of the answer is thus:

$$25^2 = 2 \text{ "By" } 3 \ / \ \ldots\ldots$$

$$= 2 \times 3 \ / \ \ldots\ldots$$

To this we tag on the last digit "5" squared:

$$= 2 \times 3 / 5 \times 5 = 6 / 25$$

Thus the answer is 625.

Similarly, all other numbers that end in 5, when squared, can be done instantly:

$15^2 = 1 \times 2 / 5 \times 5 \quad = 2 / 25 \quad = 225$

$35^2 = 3 \times 4 / 5 \times 5 \quad = 12 / 25 \quad = 1{,}225$

$45^2 = 4 \times 5 / 5 \times 5 \quad = 20 / 25 \quad = 2{,}025$

$95^2 = 9 \times 10 / 5 \times 5 \quad = 90 / 25 \quad = 9, 025$

Sutra: Vertically and Crosswise:

Sutra: Vertically and Crosswise

Here is another simple Sutra, the one that Bharati Krsna Tirthaji refers to as most widely used, called "Vertically and Crosswise" and solves all multiplication by application of a pattern (which is registered by the Right Brain as Feminine-Natured Mathematics in contrast to the logical Male Left Brain style of moths you learnt at school).

eg: To find 26 Multiplied by 31 in One-Line, we need to look at a Pattern,

viz: This Sutra, known as: "Vertically and Crosswise" shows we will have a 3-Digit Answer represented by the 3 short horizontal lines.

Here is how we traditionally write the setting out for "26 × 31":

```
 2 6
   ×
 3 1
------

------
```

Notice there are 4 digits involved (2, 6, and 3, 1) and let them be represented by a small circle or dot and that explains the format seen in Fig. This will help you to understand "CROSS ADDITION" which is shown as the middle part:

[(2 × 1) + (6 × 3)] and uses both multiplication and addition in the form of the letter "X" and corresponds to the "cross-over of the optical nerve in the brain" shown in Fig 137e below. (Here, the symbol "x" = Multiplication):

= 2 × 3 (2 × 1) + (6 × 3) 6 × 1

= 6 2 0 6

(nb: the "2" of the "20" is carried over to the left hand side).

= 8 0 6

= 806 (This shows only the mental workings. The Multiplication of 26 by 31 can be solved mentally!)

There are 16 Jaw-Dropping Sutras or simple Sanskrit Word Formulae which solve all known mathematical problems in the branches of Arithmetic, Algebra, Geometry and Calculus are easy to understand, easy to apply and easy to remember.

This Vedic One-Line Mental Arithmetic is very helpful in stimulating Modern Mathematicians to adopt its simplicity and quickness. Once initiated to these Vedic rules, students of all ages will appreciate and enjoy the enhanced subject.

The aim of Vedic Mathematics is to bring this long-hidden treasure trove of intelligent mathematical knowledge within easy reach of everyone who wishes to obtain it and benefit by it.

Sutra: Digital Sums—for Multiplication by Eleven

When scientists have a problem, say with electronic data, and they need to send it efficiently, the solution is invariably Compression. Digital Compression is a powerful Sutra that solves multiplication by 11 very quickly.

If we wanted to multiply 25 × 11 we merely add the two digits of the 25 and say "2 + 5" which is 7 and insert it between the two digits. Thus the answer is 275. Another way of showing this is to separate the two digits and insert their digital sum:

25 × 11 = 2 (2+5) 5

= 2 7 5

39 × 11 = 3 (3+9) 9

= 3 12 9

(nb, the "1" of the "12" gets carried over to the left)

= 429

Vedic Mathematics

Additional insights into the Hindutva perspective on science are provided by the track record of the four states captured by the BJP in the 1991 elections: Uttar Pradesh, Madhya Pradesh, Rajasthan and Himachal Pradesh. All four BJP State Governments introduced Vedic Mathematics (VM) into the school syllabus and intended to do the same in the syllabi of engineering colleges and polytechnics. Strangely, the sneaking of pseudo-science into publicly-funded education triggered very little debate in Indian scientific circles.

What is this `Vedic Mathematics'? It is somewhat difficult to define as the very terminology is used only by a small set of authors and a handful of mathematicians. The purveyors of `Vedic Mathematics' are agreed, however, that the landmark work in this ``discipline" is the text ``Vedic Mathematics", written by Jagadguru Swami Sri Bharathi Krishna Tirthaji Maharaja, Shankaracharya of Govardhan Matha, Puri (he died in 1960) and published in 1965

(Jagadguru, 1965). It is instructive to examine this text in some detail. The title page has the interesting sub-title, ``Sixteen Simple Mathematical Formulae from the Vedas (For one-line answers to all Mathematical Problems)." Passing through diverse prefatory remarks by various personalities (none of whom is a mathematician), we come to the author's preface, arranged in numbered paragraphs. After explaining the meaning of the word the `Veda' (paragraph 2), the author continues in the following vein (paragraph 3) :`` In other words, it connotes and implies that our ancient Indian Vedic lore should be all-round complete and perfect and able to throw the fullest necessary light on all matters which any aspiring seeker after knowledge can possibly seek to be enlightened on.'

In paragraph 9 the reader learns that ``we were agreeably astonished and intensely gratified to find that exceedingly tough mathematical problems (which the mathematically most advanced present day Western scientific world had spent huge lots of time , energy and money on and which even now it solves with the utmost difficulty ...) can be easily and readily solved with the help of these ultra-easy Vedic Sutras ...contained in the Parisista (the Appendix-portion) of the Atharvaveda in a few simple steps and by methods which can be conscientiously described as mere mental arithmetic."

Having informed us that he has regularly lectured on the subject at Nagpur University (paragraph 11), the author unfolds his `revolutionary' findings in paragraph 14 (vi): ``As regards the time required by the students for mastering the whole course of Vedic mathematics as applied to all its branches, we need merely state from our actual experience that 8 months (or 12 months) at an average rate of 2 or 3 hours per day should suffice for completing the whole course of mathematical studies on these Vedic lines instead of 15 or 20 years required according to the existing systems of the Indian and also of foreign universities."

After all this introduction, we find inside the book mathematics of the middle and high school level, where the emphasis is throughout on a series of tricks (whose origins are explained) to solve various problems: for example, faster ways of dividing, multiplying and factorising numbers, and ways of dealing with problems in conics . Some readers may find the book diverting; some of the methods described may be actually be useful somewhere in school teaching (though this needs detailed investigation). The real issue, however, is not so much the level of the contents as the additional philosophical baggage that is provided. How are we to take the assertion that the Vedas provide the answers to all questions of mathematics? Is there any scientific sense in which a particular body of knowledge can be said to contain the answer to all questions of mathematics? The contents of the book seem far-removed from this goal; comprising as they do elementary mathematics; even the list provided of the subjects intended to be dealt with by the author in later volumes (which were never published) does not rise above this level.

In fact, mathematics from the high school level indisputably requires material that cannot be provided by the `Vedic' or for that matter `Islamic', `Greek', or `Babylonian'. Modern mathematical teaching requires considerable input that cannot be traced back to antiquity. The book ``Vedic Mathematics," seeks constantly to prove the superiority of various methods to those used in the ``Western" methods, the reference point often being English texts long fallen out of use, presumably dating to the author's youth. Nor is the book of any use in studying the history of mathematics in India; it is simply unequal to the task.

What, in fact, is ``Vedic" about the mathematics set out in the book? Some of the examples cited are already well known to historians and adequately described elsewhere. As for the appendices to the Atharvaveda, the alleged source of the Vedic mathematical riches described in the book, they appear to be totally non-existent. No authoritative edition of the text of the Atharvaveda contains the appendices referred to. The game is given away in the General Editor's Foreword, where it is noted that these ``appendices" are not part of the established texts but should be regarded as new and the `discovery' of the author himself! The General Editor also notes that the style of language makes it clear that the `appendices' are the author's own discovery and that the book should be judged on its own merits - where of course the book fails miserably from the standpoint of science.

Despite its absurd claims and transparently bogus status, Jagadguru Swami's book became the focus of a National Workshop on Vedic Mathematics held at Jaipur in March 1988 by the Rashtriya Veda Vidya Pratishthan and various other Government bodies and universities. The proceedings were published as ``Issues in Vedic Mathematics" (Khare, 1991).

The seminar enjoyed the official patronage of the Ministry of Human Resource Development and the objectives of the seminar were spelt out in a letter to all participants by the Special Secretary, Department of Education. The papers presented at the seminar scarcely rise above the level of the book that inspired them. There is much talk of applying the ``Vedic" methods to developing algorithms for computers (some participants opt out and claim that these methods render calculators and computers redundant) and some of the articles indeed are, to put it charitably, nonsense from any scientific viewpoint.

The seminar ends with recommendations to the Government: a second seminar to be held on VM, in which traditional and modern mathematicians are to study the introduction of VM at different levels; a committee to study the introduction of VM in schools; research in VM; multimedia strategies for awareness of VM; the Department of Electronics to be sounded out on VM applications to computer technology.

One interesting whistle-blowing paper (Shukla, 1991, page 31) brings to light the fact that the Shankaracharya's book is not based on what is in the Vedas.

Clearly the BJP-led Governments rushed in where others feared to tread and introduced this material in the schools. Any doubts about their reasons for doing so were dispelled by the BJP Education Minister of Uttar Pradesh at a workshop in Allahabad in April 1992, he simply declared that ``whatever is very ancient for India, that precisely is most modern for the world" (see Seminarist, 1992). The efforts of Hindutva in respect of the exact sciences would indeed be laughable were it not for the potential for serious damage at the school level.

Textbooks and the 'Aryan' Thesis

Further evidence of the Hindutva attitude to the pursuit of truth and scientific rigour is available in the history textbooks foisted on students in the four erstwhile BJP-ruled States. Take the case of the compulsory textbook for first-year graduate students titled ``Bharat ki Sanskritik Virasat" introduced in Madhya Pradesh. Essentially a primer of the Hindutva brand of culture, it determinedly pushes the pseudo-thesis that the Aryans did not migrate to the Indian sub-continent from outside but were `indigenous.' This is very much in line with the absolutist assertions made by `Guru' M. S. Golwalkar, the RSS's second supremo and its most important ideologue, in his 1939 work, ``We: or Our Nationhood defined," which the Hindutva camp is doing its best to cover up today. Rejecting theories that held that the `Aryans' migrated to India, Golwalkar asserts: `` Hypothesis is not truth. Out of the heap of hypotheses we reject all and positively maintain that we Hindus came into this land from nowhere, but are indigenous children of the soil always, from times immemorial and are natural masters of the country. Here we compiled our inimitable Vedas, reasoned out our Philosophy of the Absolute--the last word on the subject, built our sciences and arts and crafts..."(Golwalkar, 1939, page 8-9). The equation of `Hindu' and 'Aryan'--`` the Hindu, i.e., Aryan race" (Golwalkar, 1939, page 6) is central to the thesis about national identity as expounded in this chillingly-explicit work.

The anti-scientific character of this thesis has been exposed by India's leading historians, among them Dr. Romila Thapar (an excellent article is available in Seminar; see Thapar, 1992). Historians point out that term Aryan should be applied, scientifically speaking, only to a language. The language of the Rig Veda for instance is Old Indo-Aryan. The links between this language and Old Avestan indicates a movement from West Asia to India by Aryan speakers. The evidence ``does not support the notion that India was the homeland of the Aryan-speaking people" (Thapar, 1992, p. 22). Further, Vedic Sanskrit and its dialect variations provide evidence of further geographical movement from the north-west into the Ganga valley. The Aryan speakers did not maintain any racial purity, so what really spread was not a body of racially ``pure" Aryans, but rather the use of the language Old Indo-Aryan. Dr. Romila Thapar goes on to discuss the implications of such pseudo-theses for developing a justification of upper caste domination. There are other dimensions to the pushing of this pseudo-thesis, especially the linking up with the fascistic vision of Golwalkar on national identity and the place of minorities, above all Muslims. Revealingly, Golwalkar was attracted to Nazi Germany's handling of the `minorities' problem: ``To keep up the purity of the Race and it culture, Germany shocked the world by her purging the country of the semitic Races-the Jews. Race pride at its highest has been manifested here. Germany has also shown how well-nigh impossible it is for Races and cultures, having differences going to the root, to be assimilated into one united whole, a good lesson for us in Hindustan to learn and profit by." (Golwalkar, 1939, p.35).

The Challenge before Indian Science

These are matters that should cause serious concern within the scientific community in India. But do they? The latest outrage, an award for astrology at the annual Indian Science Congress and its official defence by no less than a member of the Planning Commission, should be cause enough for serious concern. And, in fact, the award has been given to the president of an occult foundation. Professional scientists have protested but they have not been able to block the move. It is hard to imagine a major conclave of scientists in a developed country not breaking into pandemonium over such an affront to science and scientists, but this is India under the shadow of Hindutva. In fact, seen against the specific background of Ayodhya, the incident assumes even more menacing overtones. Even more shocking is the case of a physicist at the Indian Institute of Technology, Kanpur, who went to Ayodhya as a kar sevak in December 1992 and returned, with a piece of the rubble, to resume his duties in the service of science at one of India's leading S&T institutions.

We have referred in this essay to the many weaknesses of the science and technology infrastructure in India. These weaknesses are intimately connected with the overall social and economic crisis of Indian society today. If, as is widely recognised, the current Hindu fundamentalist wave is a far-Right reactionary response to the crisis, science and technology can hardly expect to remain immune to the influence of fundamentalism. Our scientific and technological institutions are not strong enough to withstand a sustained onslaught which is politically powered; it takes but a few years to dismantle active centres of research and teaching. And in the long run, give or take a few years, there could be a systematic dismantling of science and technology in this country. The direct links that Hindutva pseudo-science has with the less publicised fascist face of the Sangh parivar, suggest that if Hindutva triumphs, science and technology will be made subservient to its original dream of a pseudo-Hindu fascist state (as elaborated by Golwalkar in 1939). If these words seem unduly alarmist, we need to remember that the current grim political situation, if it had been forecast or projected in the mid-1980s, would have been dismissed as unimaginable in India.

What does a serious and genuine commitment to science and technology mean in the current context? In a perceptive essay Jacob Bronowski (Bronowski, 1977) argues that science is not a value-neutral human activity. "The activity of science is directed to seek the truth (about the material world), and it is judged by the criterion of being true to facts. We can practice science only if we value the truth" (Bronowski, 1977, p.212). Without truth, science is not possible. This truth is not immutable for we find descriptions of nature only up to a certain coarseness, a certain tolerance. This, in turn, implies that science values dissent, for without dissent there would be no challenge of the established concepts and no progress. And the right to dissent needs to be safeguarded. Science values independence of thought because it is a necessary condition of dissent, and originality because it is the tool with which the new is discovered. These values are essential to modern science and science needs a society that recognises these as social values. The right of free and critical enquiry which is implied by these values is one of the abiding things that science has given to society.

We have referred in this essay to the importance of the critical and rational spirit in society for science to flourish. It bears emphasis that it is in assisting in the task of expanding the realm of operation of free and critical enquiry in society that science and scientists fulfil their social responsibility. It is this perception, affirmed and grasped at the mass level, that has led to the remarkable phenomenon of popular science movements in India which seek to empower the people through the transmission of the knowledge of science and the scientific method. Hindutva, and the obscurantism, science fraud and pseudo-science it inflicts on society at the political and, therefore, the mass level, work precisely in the opposite way. From the standpoint of science and its values, Hindutva must be judged to be a serious threat indeed.

A commitment to science is a positive commitment to secularism. The historian K. N. Panikkar has referred (see Panikkar, 1991) to the necessity of expanding the secular space in society, of seizing the initiative from communalism and promoting a positive campaign for secularism. Science belongs firmly to this secular space; if it does not, we cannot conceive of another pursuit, another activity, which does. Hindutva cannot easily challenge the popular perception of the need for science and technology as an integral part of development. This perception can be used to promote a powerful campaign that links secularism firmly with science, that demonstrates that science is impossible without the development of a critical spirit. In the ultimate analysis, the development of science and of the spirit of free and critical enquiry can make a major contribution to eradicating the blight of communalism that is upon India.

Vedic—Mathematics

Mathematics is the queen of subjects. Rightly so, then, Vedic Mathematics is the glowing crown that adorns its proud forehead. Very few of the masses today are aware today, of Vedic mathematics, the magnanimity of its profound implications, and of its origins which guided the rest of the world towards purer and more intricate branches of mathematics and which, laid the foundations for number theory and arithmetic, the teeny-weeny part of which we are taught during our alma-mater days with hardly any reference made to its rightful owners ? our very own ancestors ?who pursued the study of mathematics with no less finesse than that of a fine art.

A well-known fact it is, as every one knows now, for he/she has seen himself/herself or his friend, being answered by his teacher, during the primary years of his education, in response to his/her query full of childish criticism "What has India given to Mathematics? that the numeral 0 was indeed the creation of Indian mathematicians. Introduction of zero brought about a new revolution into the world of mathematics. It was zero that gave rise to the idea of representing numbers using base 10, as it is commonly used today. And it is zero because of which you are able to read this article. But why? How would a computer work without zeros and ones!!! So that's the zero there, right!

Though the Arabs are given the credit of taking mathematics into broader frontiers, they had begun their work with the help of Indian manuscripts. The story goes something like this. It was in 773 that the Arabs were able to set their eyes on the astounding developments of numerical methods Indians used when one of the Indian palmist and fortune-teller happened to visit the Arabian lands. So impressed were the Arab mathematicians with Indian inventions that the Arab mathematician Muhammed-Ibna-Musa-Abu-Jafar-Al-Khwarizmi himself came to India to study Indian mathematics. After stating here for some time after learning the subjects to his satisfaction, he wrote his manuscript "Algebra '-b-e-Mukabla' This is how 'Algebra' was born. His works, which were nothing but a translation of his Indian studies, left the European mathematicians spell-bound, especially by the use of base 10 to represent numbers. The idea of representing numbers by base 10, is thus, originally Indian.

The trend then caught on. From Arabs to Greeks, from Greece to Spain and from Spain to Europe. Europeans however, initially reluctant to use base 10 to represent numbers, inevitably began to use it in 1202 (during the time of Bhaskaracharya). Though the mathematical works went on improving as it changed hands, the world availed the first systematically documented use of base 10 only in Bhaskaracharya's manuscript "Lilavati" 4th century mathematician Diaphantus, who is also known as 'Father of Arithmetic' has his works in his book 'Arithmetica' coinciding closely with the Indian manuscripts of earlier age. These are but a few evidences that arithmetic and basic mathematics that has evolved today in various forms, is but the creation of those great olden Indian minds.

In this article, we take a glimpse of the very fantastic inventions of Vedic mathematics in the field of number theory and arithmetic. Though cited here are examples of the methods described in Vedas , Vedic mathematics has its frontiers expanding into many more diverse areas and concepts like astrology, astronomy, geometry, trigonometry, to name a few. The examples

cited here would be useful in performing calculations that you may be coming across everyday. This article would be of course, a welcome change to those who complain that they hardly get time to stretch their legs during those interminably tightly timed aptitude tests! I will like to share some experts of a Book called "Drutganit" (Speed mathematics) by Mr. Shyam Marathe.... The book is in Marathi discussing the techniques of doing mathematical operation of big numbers very quickly using the techniques present in "Vedas" ....

EXAMPLES:

We shall start with a method to multiply 2 numbers. It will reveal to you how innovative Vedic mathematics can be.

Lets deal, to begin with, with numbers that are close to POWERS of 10 i.e. numbers near to 10, 100, 1000, etc.

EXAMPLE 1:

Say, we have to multiply 98 and 97.

(First, try using conventional method and find out how much time it takes!)

Now we go for the Vedic method.

Write down the numbers in this manner:

98 -2

97 -3

In the above representation, -2 and -3 stand for the difference of 98 and 97 from 100. We may call these differences -2 and -3 as OFFSETS of 98 and 97 from 100. We may call 100 as the 'BASE' as differences are taken from 100.

Step 1: First, we multiply the offsets -2 and -3

We get 6. Since our base is 100, which has 2 zeros, the product of offsets must also have 2 digits. Hence we write 6 as 06 and write it down as last 2 digits of our answer.

98 -2
97 -3

06

Step 2: Now for the previous digits, just add any 2 numbers in the above figure crosswise i.e. either (98-3) or (97-2)

We get 95. This is written before 06 as follows: -

98 -2
97 -3

---

95 / 06

That gives us 9506. That's our answer.

Hence 98 × 97 = 9506

Aint it simple!!

EXAMPLE 2:

Once again, we solve 25 × 98.

25 -75
98 -2

Here, once again, our base is 100. -75 and -2 are offsets of 25 and 98 from 100.

Step 1: -75 × -2 gives 150

Since our base (100) has 2 zeros, the product of offsets must have 2 digits. Hence we write down last 2 digits of 150 i.e. 50 as last 2 digits of our answer. 1 would be treated as carry.

25 -75

98 -2

---

50 (1 carry)

Step 2: Add the numbers crosswise. i.e. (25-2) or (98 - 75).

We get 23.

Add the carry from previous step.

We get 23 + 1 = 24

These are written as first 2 digits of our answer.

25 -75

98 -2

(23+1) / 50

i.e.

25 -75

98 -2

---

24 / 50

Answer: 25 x 98 = 2450

EXAMPLE 3:

Now, lets try for 108 x 109

Let us write down the numbers as:

108 8
109 9

---

and carry out the same method as above.

Note that 8 and 9 are offsets of 108 and 109 from base 100.

Step 1: Product of offsets 8 x 9 = 72

Write down 72 as last digits of our answer.

108 8
109 9

72

Step 2: Add any 2 numbers cross wise (108 + 9) or (109 + 8)

We get 108 + 9 = 117

These are the front digits of the answer.

108 8
109 9

---

117 / 72

Hence 108 x 109 = 11772

EXAMPLE 4:

Lets try 98 × 104.

98 -2
104 4

Step 1: Product of offsets -2 × 4 = -8

We write it down as,

98 -2
104 4

-8

Step 2: Now, 98 + 4 or 104 -2 gives 102.

These are the earlier digits of our answer.

98 -2

104 4

---

102 / -8

Step 3: Since the number after "/ " is as negative number we have to decrease the number to the left of "/" by 1 (i.e. 102 becomes 101). Next, the number 8 (which has a negative sign) has to be deducted from the number base times 1 i.e. deduct 8 from 100 × 1.

i.e. 100 - 8 = 92

92 is written in the place of -8. That makes our result as:

98 -2

104 4

---

101 / 92

Hence 104 × 98 = 10192.

Lets see how it simplifies our calculation for huge numbers:

EXAMPLE 5:

Multiply 888 by 998:

888 -112

998 -002

---

886 / 224

Here, the base is 1000 i.e. offsets are differences from 1000

-112 × - 002 = 224 (Simple enough)

And 888 - 002 = 886

Hence, 888 × 998 = 886224.

EXAMPLE 6:

Multiply 9997 by 9998:-

9997 -3 (base 10,000)
9998 -2

9997 - 2) /0006

(Here, base is 10,000, which has 4 zeros. Hence the product of offsets

-2 and -3 is padded with zeros to make it a 4-digit number)

Hence,

9997 -3
9998 -2

9995 / 0006 = 99950006 (Answer!!!)

EXAMPLE 7:

Multiply 999979 by 999998:

999979 -21 (base 1,00,000)

999998 -2

(999979 - 2) / 000042

Here, -21 × -2 = 42.

Base has 6 zeros. Hence pad 42 with 4 zeros to make 000042.

Then, 999979 - 2 = 999977. Those are the earlier digits of our answer.

999979 -21
999998 -2

999977 / 000042

Hence, 999979 × 999998 = 999977000042.

Ever wondered multiplying such big numbers could be that simple!!!

The methods discussed above are extremely effective for multiplying numbers near the powers of 10. Similar methods can be applied to find the product of ANY 2 numbers (irrespective of whether they are near powers of 10) using ANY MULTIPLE of 10 as base.

It is being taught in some of the most prestigious institutions in England and Europe. NASA scientists applied its principles in the area of artificial intelligence. And yet, in the country of its birth it languishes as a forgotten science.

Vedic mathematics, which simplifies arithmetic and algebraic operations, has increasingly found acceptance the world over. Experts suggest that it could be a handy tool for those who need to solve mathematical problems faster by the day, especially in a system where the emphasis is on examinations.

The subject was revived largely due to the efforts of Jagadguru Swami Bharathikrishna Tirthaji of Puri Jaganath. Having researched the subject for years, even his efforts would have gone in vain but for the enterprise of some disciples who took down notes during his last days. That resulted in the book, Vedic Mathematics, in the 1960s.

Says P.R. Srinivas, who has taught the subject in some of the city's colleges: ``It is an ancient technique which simplifies multiplication, divisibility, complex numbers, squaring, cubing, square and cube roots. Even recurring decimals and auxiliary fractions can be handled by Vedic mathematics."

Vedic scholars did not use figures for big numbers in their numerical notation. Instead, they preferred to use the Sanskrit alphabets, with each alphabet constituting a number. Several mantras, in fact, denote numbers; that includes the famed Gayatri mantra which adds to 108 when decoded. The bases of Vedic mathematics are the 16 sutras which attribute a set of qualities to a number or a group of numbers. Mr Srinivas feels that given the initial training in modern maths in today's schools, students will be able to comprehend the logic of Vedic mathematics only after they have reached the 8th standard. Among his students are Armymen and doctors in their 50s and younger students keen to make their mark in competitive entrance exams.

India's past could well help them make it in today's world.

Vedic Maths, a Forgotten Science *

Some of the Sutras (Phrases) used in Vedic Mathematics:*

1st Example - 1 Divided by 19, 29, 39, ... 129 etc. *

2nd Example - Square of Numbers ending in 5 *

3rd Example - Multiplier-digits consist entirely of nines*

4th Example General Multiplication of any number by any number *

5th Example Algebraic Divisions *

6th Example Division *

7th Example Factorisations of Quadratics *

8th Example - Verifying Correctness of answers *

9th Example Factorisations of Harder Quadratics *

10th Example - Factorisations of Harder Quadratics - Special Cases *

Sixteen Simple Mathematical Sutras (Phrases) From The Vedas *

Notes and Quiz *

7

Application of Vedic Mathematics in Integrals

The Differential

Earlier in the differentiation chapter, we wrote $\frac{dy}{dx}$ and f '(x) to mean the same thing. We used $\frac{d}{dx}$ as an operator.

We now see a different way to write, and to think about, the derivative.

Definition: The differential of $y = f(x)$ is written:

$$dy = f'(x)dx.$$

Mini-Lecture

Note: We are now treating $\frac{dy}{dx}$ as a fraction.

EXAMPLE:

Find the differential of $y = 3 \times 5 - x$.

ANSWER

$$dy = f'(x)dx$$

$$dy = (15x^4 - 1)dx$$

Note: We could use the differential to estimate the real change in value of a function (Δy) caused by a small change in x (written as Δx). Many text books do this, but it is pretty silly, since we can easily find the exact change - why approximate it?

We are introducing differentials here as an introduction to the notation used in integration.

How are dy, dx and Δy and Δx related?

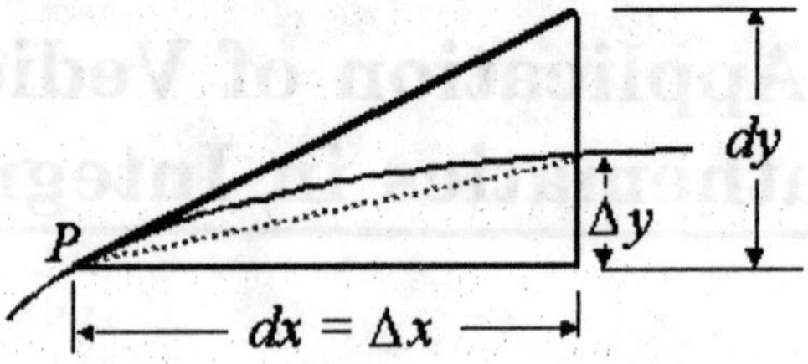

Relative Error

Note: We may perform a calculation and find an error of 1 cm. If this is part of a problem involving lengths of 8 cm, then we have a big problem. If it involves lengths of 3 km, we have an insignificant problem. We usually talk about the relative error:

$$\frac{1\,\text{cm}}{8\,\text{cm}} = 12.5\% \text{ or } \frac{1\text{ cm}}{3\text{ km}} = 3.3\times10^{-4}\%$$

rather than the absolute error of 1 cm.

Proportionate and Percentage Change

The proportionate change in y(x) is given by $\frac{dy}{y}$ (no units).

The percentage change in y(x) is given by $\frac{dy}{y}\times100\%$

Antiderivatives and the Indefinite Integral

We wish to perform the opposite process to differentiation. This is called "antidifferentiation" and later, we will call it "integration".

EXAMPLE

If we know that

$$\frac{dy}{dx} = 3x^2$$

and we need to know the function this derivative came from, then we "undo" the differentiation process. (Think: "What would I have to differentiate to get this result?")

$y = x^3$ is ONE antiderivative of $\frac{dy}{dx} = 3x^2$

There are infinitely many other antiderivatives which would also work:

$$y = x^3 + 4$$

$$y = x^3 + \pi$$

$$y = x^3 + 27.3$$

In General: $y = x^3 + K$, is the indefinite integral. K is called the constant of integration.

In electrical engineering, we prefer to write "+K", since C (as used in most math texts) is used for capacitance.

We write:

$\int 3x^2 dx = x^3 + K$ and say:

"The integral of $3x^2$ with respect to x equals $x^3 + K$."

Here is a template you can play with to get the idea.

The $\int$ sign is an elongated "S", standing for "sum". Later we will see that the integral is the sum of the areas of infinitely thin rectangles.

SUM: $\int$ "integration is an infinite sum"

Note: Sometimes we write: F(x) to mean the integral of f(x). So we have:

EXERCISE:

Find $\int(x^2-5)dx$

ANSWER

$$\int(x^2-5)dx=\frac{x^3}{3}-5x+K$$

Note: In general, we can use:

1. $\int kdx = kx + K$ (k,K are constants.)

The integral of a constant is that constant times x.

EXAMPLE:

$$\int 4dx = 4x + K$$

2. $\int x^n dx=\frac{x^{n+1}}{n+1}+K\ (n\neq -1)$

For the integral of a power of x: add 1 to the power and divide by the new number.

EXAMPLE:

$$\int x^5 dx=\frac{x^{5+1}}{5+1}+K$$

$$=\frac{x^6}{6}K$$

DON'T FORGET THE "+ K" (or "+ C"). THIS CONSTANT OF INTEGRATION IS VITAL IN LATER APPLICATIONS.

EXAMPLE 1:

$$\int 7x^6 dx$$

ANSWER

$$\int 7x^6 dx=7\frac{x^7}{7}+K=x^7+K$$

EXAMPLE 2:

$$\int\left(3x^2+\sqrt{x}-\frac{5}{x^3}\right)$$

ANSWER

$$\int\left(3x^2+\sqrt{x}-\frac{5}{x^3}\right) = \int\left(3x^2+x^{1/2}-5x^{-3}\right)dx$$

$$= x^3+\frac{x^{3/2}}{\frac{3}{2}}-5\frac{x^{-2}}{-2}+K$$

$$= x^3+\frac{x^{3/2}}{3}-\frac{5x^{-2}}{2}+K$$

EXAMPLE 3:

$$\int\sqrt[3]{x^2}\,dx$$

ANSWER

$$\int\sqrt[3]{x^2}\,dx = \int x^{2/3}dx$$

$$= \frac{x^{\frac{2}{3}+1}}{\frac{2}{3}+1}+K = \frac{x^{\frac{5}{3}}}{\frac{5}{3}}+K$$

$$= \frac{3\sqrt[3]{x^5}}{5}+K$$

EXAMPLE 4:

A particular curve has $\frac{dy}{dx}=3x^2-2x.$

We are told that the curve passes through the point (2, 5). Find the equation of the curve.

EXAMPLE 5:

This is different to the other exercises above! It is a function of a function situation, so we have to do the reverse of the chain rule, which we met in the section on differentiation.

The new rule we require is...

Power Formula for Integration

$$\int u^n du = \frac{u^{n+1}}{n+1} + K \ (n \neq -1)$$

This requires a substitution step, where u(x) is some function of x.

Now back to the problem to see how to apply this formula:

ANSWER

$$\int (2x^4 - 5)^6 x^3 dx$$

We use, as a starting point, the substitution

$$u = 2x^4 - 5.$$

Now $du = 8x^3\ dx$

Our question has only x^3dx so we need to divide both sides by 8:

$$\frac{1}{8} du = x^3 dx$$

We can now rewrite our question as:

$$\int (2x^4 - 5)^6 x^3 dx = \frac{1}{8} \int u^6 du$$

$$= \frac{1}{8} \times \frac{u^7}{7} + K$$

$$= \frac{u^7}{56} + K = \frac{(2x^4 - 5)^7}{56} + K$$

EXAMPLE 6:

$$\int (x^3 - 2)^6 (3x^2)dx$$

ANSWER

We use, as a starting point, $u = x^3 - 2$.

Now $du = 3x^2dx$

This time, this is exactly what we have in the question, so there is no need to divide both sides.

We can now write:

$$\int (x^3 - 2)^6 (3x^2)dx = \int u^6 du$$

$$= \frac{u^7}{7} + K = \frac{(x^3 - 2)^7}{7} + K$$

EXAMPLE 7:

Find $\int \frac{x}{\sqrt{x^2 + 9}} dx$ using a substitution.

ANSWER

Put $u = x^2 + 9$

Then $du = 2xdx$

The question has xdx so we write

$$\frac{du}{2} + xdx$$

Now

$$\int \frac{x}{\sqrt{x^2 + 9}} dx = \frac{1}{2} \int \frac{du}{\sqrt{u}}$$

$$= \frac{1}{2} \int u^{-1/2} du$$

$$= \frac{1}{2} \frac{u^{1/2}}{1/2} + K = \sqrt{x^2 + 9} + K$$

EXAMPLE 8:

Given $y' = \sqrt{2x+1}$, find the function $y = f(x)$ which passes through the point (0,2).

ANSWER

Note: y' means the derivative of y, that is dy/dx.

$$y' = \sqrt{2x+1}$$

So

$$y = \int \sqrt{2x+1}\, dx$$

This question requires us to integrate, and in the process, to find the constant of integration.

Put $u = 2x + 1.$

Now $du = 2dx$

So

$$\frac{1}{2} du = dx$$

So

$$y = \int \sqrt{2x+1}\, dx$$

$$= \frac{1}{2} \int u^{1/2} du$$

$$= \frac{1}{2} \times \frac{2}{3} \times u^{3/2} + K$$

$$= \frac{1}{3} (2x+1)^{3/2} + K$$

Now if x = 0 then y = 2 (from the question).

$$2 = \frac{1^{3/2}}{3} + K$$

and this gives K = 5/3.

So

$$y = \frac{(2x+1)^{3/2}}{3} + \frac{5}{3}$$

is the required function.

Note: You will see "+K" and "+C" in this work. Most textbooks use + C.

Always use +K if you are answering electrical problems.

The Area under a Curve

A building has parabolic archways and we need to supply glass to close in the archways. How much glass is needed?

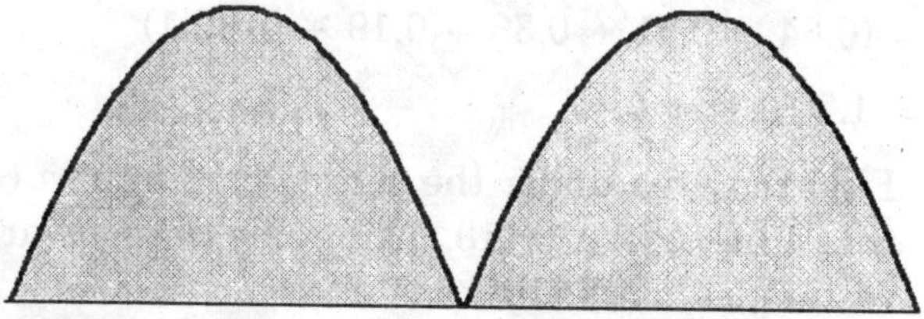

We need to know the area under the curve.

Before integration was developed, they could only approximate the answer by dividing the space into rectangles and adding the areas:

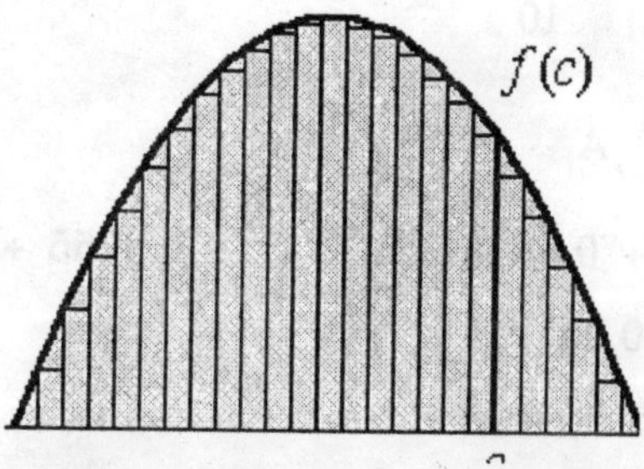

The height of each rectangle is found by calculating the function values, as shown for the typical case x = c. We get a better result if we take more and more rectangles.

EXAMPLE OF APPROXIMATION USING RECTANGLES:

(a) Find the area under the curve $y = 1 - x^2$ between $x = 0.5$ and $x = 1$, for $n = 5$, using the sum of areas of rectangles method.

Now for the normal answer:

ANSWER

Since $n = 5$,

$$h = \Delta x = \frac{b-a}{n} = \frac{1-05}{5} = 01$$

$$A = \sum_{i=1}^{5} A_i$$

$$= (0.64 + 0.51 + 0.36 + 0.19 + 0)\ (0.1)$$

$$= 1.7\ (0.1) = 0.17$$

(b) Find the area under the curve given in part (a), but this time use $n = 10$, using the sum of areas of rectangles method.

ANSWER

Since $n = 10$,

$$h = \Delta x = \frac{1-05}{10} = 0.05$$

$$A = \sum_{i=1}^{5} A_i$$

$$= (0.6975 + 0.64 + 0.19 + \ldots + 0.9755 + 0)\ (0.05)$$

$$= 3.7875\ (0.05)$$

$$= 0.189375.$$

There must be a better way! Integration was developed by Newton and Leibniz to save all this "adding areas of rectangles" work.

General Case

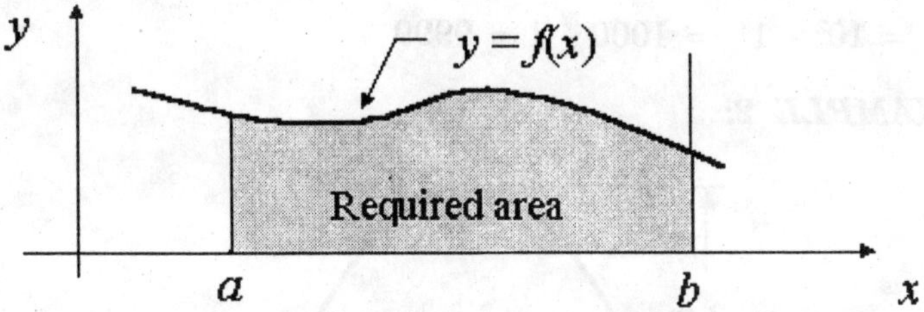

[Note: The curve is completely ABOVE the x-axis].

By a Δ-process, we can show that in general, the exact area under a curve y = f(x) from x = a to x = b is given by the definite integral:

$$\text{Area} = \int_a^b f(x)dx$$

How do we evaluate this expression?

If F(x) is the integral of f(x), then

$$\int_a^b f(x)dx = \left[f(x)_a^b\right] = F(b) - F(a)$$

This means:

To evaluate a definite integral, follow these steps:

- integrate the given function (do not include the K)
- substitute the upper limit (b) into the integral
- substitute the lower limit (a) into the integral
- subtract the second value from the first value
- the answer will be a number

This forms part of The Fundamental Theorem of Calculus.

Mini-Lecture

EXAMPLE 1:

Evaluation of Definite Integral:

$$\int_{1}^{10} 3x^2 dx = \left[x^3\right]_1^{10}$$

$$= 10^3 - 1^3 = 1000 - 1 = 9999$$

EXAMPLE 2:

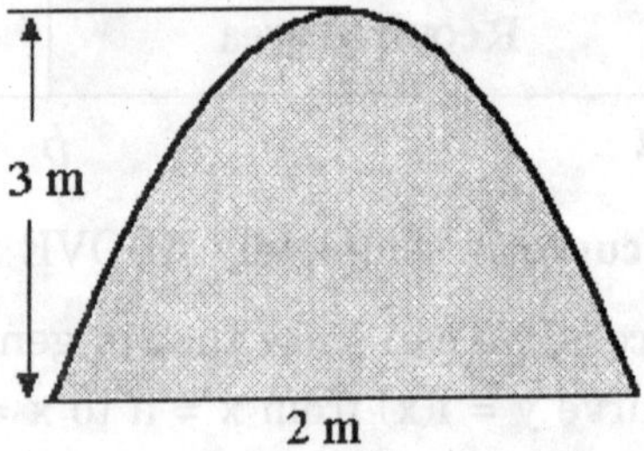

If the arch is 2 m wide at the bottom and is 3 m high,

(i) find the equation of the parabola

(ii) find the area under each arch using integration.

ANSWER

(i) We place the parabola so that the left side of the arch passes through (0, 0) and the right side will pass through (2, 0), since the arch is 2 m wide at the bottom. The vertex of the arch is at (1, 3).

General form of a parabola:

$$y = ax^2 + bx + c$$

At x = 0, y = 0 and on substituting,

we get 0 = 0 + 0 + c. So c = 0.

At x = 1, y = 3 and on substituting,

we get 3 = a + b.

At x = 2, y = 0 and on substituting,

we get 0 = 4a + 2b.

This gives the simultaneous equations

$$a + b = 3$$

$$2a + b = 0$$

Subtracting the 1st line from the 2nd gives a = -3.

And so b = 6. So the required parabola is $y = -3x^2 + 6x$, with x in metres.

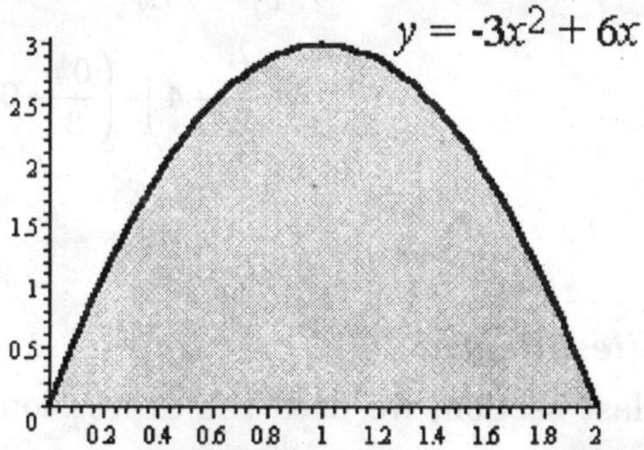

The process of finding the equation is called modelling. It is a very important skill in science and engineering.

(ii) Now for the area:

$$\int_0^2 \left(-3x^2 + 6x\right)dx = \left[-x^3 + 3x^2\right]_0^2$$

$$= [-(2^3) + 3(2)^2] - [0 + 0]$$

$$= [-8 + 12] = 4m^2.$$

EXAMPLE 3:

Find the (exact) area under the curve $y = x^2 + 1$ between x = 0 and x = 4 and the x-axis.

Now for the conventional answer:

ANSWER

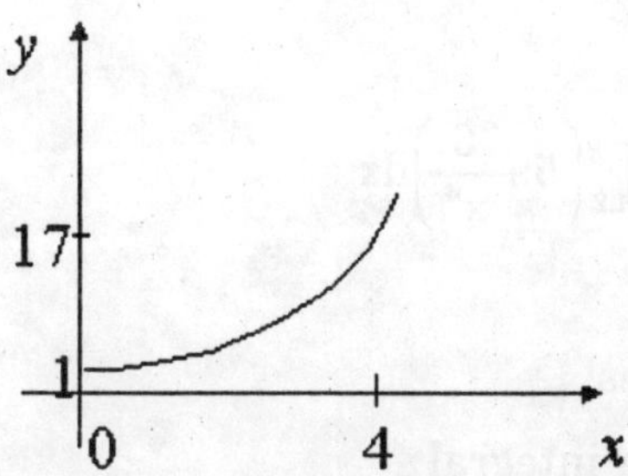

The area is given by:

$$\int_0^4 (x^2+1)dx = \left[\frac{x^3}{3}+x\right]_0^4$$

$$= \left(\frac{4^3}{3}+4\right)-\left(\frac{0^3}{3}+0\right)$$

$$= \frac{76}{3} \text{ units}^3$$

The Definite Integral

In the last section, we used the expression

$$\int_a^b f(x)dx = [F(x)]_a^b = F(b)-F(a)$$

to find the area under a curve. This expression is called a definite integral. It does not involve the constant of integration and it gives us a definite value (a number) at the end of the calculation.

CARE:The integral only gives us an area when the whole of the curve is above the x-axis in the region from x = a to x = b. If this is not the case, we have to break it up into individual sections.

It is quite possible to have a definite integral having value 0 or negative. There are many more applications of the integral than areas under curves.

EXAMPLE:

Evaluate:

$$\int_{1.2}^{1.6}\left(5+\frac{6}{x^4}\right)dx$$

SOLUTION:

This requires:

(1) Find the integral

(2) Substitute 1.6 into the integral

(3) Substitute 1.2 into the integral

(4) Subtract.

$$\int_{1.2}^{1.6}\left(5+\frac{6}{x^4}\right)dx = \int_{1.2}^{1.6}\left(5+6x^{-4}\right)dx$$

$$= \left[5x - 2x^{-3}\right]_{1.2}^{1.6}$$

$$= 7.5117 - 4.8426$$

$$= 2.6691$$

Recall

Recall the substitution formula for integration:

$$\int u^n du = \frac{u^{n+1}}{n+1} + K \qquad (n \neq -1)$$

Definite Integrals and Substitution

When we substitute, we are changing the variable, so we cannot use the same upper and lower limits. We can either:

- Do the problem as an indefinite integral first, then use upper and lower limits later
- Do the problem throughout using the new variable and the new upper and lower limits
- Show the correct variable for the upper and lower limit during the substitution phase.

We will be using the third of these possibilities.

EXAMPLE:

Find:

$$\int_{-1}^{0} x^3\left(1-2x^4\right)^3 dx$$

using a substitution.

ANSWER

Put $u = 1 - 2x4$

Then $du = -8x^3dx$

The question has x^3dx so we write

$$-\frac{du}{8} = x^3dx$$

So we have:

$$\int_{-1}^{0} x^3\left(1-2x^4\right)^3 dx = -\frac{1}{8}\int_{x=-1}^{x=0} u^3du$$

$$= -\frac{1}{8}\times\left[\frac{u^4}{4}\right]_{x=-1}^{x=0}$$

$$= -\frac{1}{32}\left[u^4\right]_{x=-1}^{x=0}$$

$$= -\frac{1}{32}\left[\left(1-2x^4\right)^4\right]_{-1}^{0}$$

$$= -\frac{1}{32}\left[(1-0)^4-(1-2)^4\right]$$

$$= -\frac{1}{32}\left[(1)-(1)\right]$$

$$= 0.$$

Note: If we take the second approach, *i.e.,* leaving the variable changed, we have:

$u = 1 - 2x^4.$

As $x = -1 \rightarrow 0,$

then $u = -1 \rightarrow 1$

So the problem becomes:

$$\int_{-1}^{0} x^3\left(1-2x^4\right)^3 dx = -\frac{1}{8}\int_{-1}^{1} u^3 du$$

$$= -\frac{1}{32}\left[u^4\right]_{-1}^{1}$$

$$= -\frac{1}{32}\left[1^4-(-1)^4\right] = 0$$

This second approach is quite useful later when the substitutions become more involved (e.g. trigonometric substitution).

EXERCISES

In physics, work is done when a force acting upon an object causes a displacement. (For example, riding a bicycle...).

If the force is not constant, we must use integration to find the work done.

We use:

$$W = \int_a^b F(x)dx$$

where F(x) is the variable force.

QUESTION 1:

Find the work done if the force is $F(x) = \sqrt{2x-1}$ and it is acting from x = 1 to x = 5.

That is, evaluate:

$$\int_1^5 \sqrt{2x-1}dx$$

ANSWER

Put u = 2x - 1 then du = 2 dx

So $$\frac{du}{2} = dx$$

So we have:

$$\int_1^5 \sqrt{2x-1}\,dx = \frac{1}{2}\int_{x=1}^{x=5} u^{1/2}du$$

$$= \frac{1}{2} \times \left[\frac{2}{3}u^{3/2}\right]_{x=1}^{x=5}$$

$$= \frac{1}{3}\left[(2x-1)^{3/2}\right]_1^5$$

$$= \frac{1}{3}\left[9^{3/2} - 1^{3/2}\right]$$

$$= \frac{1}{3}[27-1]$$

$$= \frac{26}{3}$$

So the work done is 8.67 units.

Application: The average value of a function f(x) from a to b is given by:

$$\text{Average} = \frac{\int_a^b f(x)dx}{b-a}$$

QUESTION 2:

Find the average value of $x(3x^2 - 1)^3$ from 0 to 1.

Here is the graph of the situation:

In this case, b - a = 1, so we simply need to evaluate:

$$\int_0^1 3(2x^2-1)^3\,dx$$

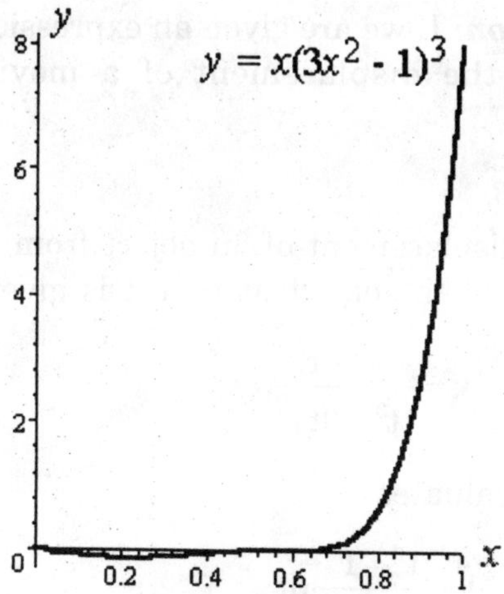

ANSWER

Put u = $3x^2 - 1$, then du = 6x dx

So $$\frac{du}{6} = x\,dx$$

$$\int_0^1 x\left(3x^2-1\right)^3 dx = \frac{1}{6}\int_{x=0}^{x=1} u^3 du$$

$$= \frac{1}{6}\left[\frac{u^4}{4}\right]_{x=0}^{x=1}$$

$$= \frac{1}{24}\left[\left(3x^2-1\right)^4\right]_0^1$$

$$= \frac{1}{24}\left[\left(3(1)^2-1\right)^4 - \left(3(0)^2-1\right)^4\right]$$

$$= \frac{1}{24}[16-1]$$

$$= \frac{15}{24} = \frac{5}{8}.$$

So the average value required is 0.625 units.

Application: If we are given an expression for velocity, we can find the displacement of a moving object by integration:

QUESTION 3:

Find the displacement of an object from t = 2 to t = 3, if the velocity of the object at time t is given by:

$$v = \frac{t^2 + 1}{\left(t^3 + 3t\right)^2}$$

That is, evaluate:

$$\int_2^3 \frac{t^2 + 1}{\left(t^3 + 3t\right)^2} dt$$

ANSWER

Put $u = t^3 + 3t$, then $du = (3t^2 + 3)\ dt = 3(t^2 + 1)\ dt$

So $$\frac{du}{3} = \left(t^2 + 1\right)dt$$

So we have:

$$\int_2^3 \frac{t^2 + 1}{\left(t^3 + 3t\right)^2} dt \quad = \frac{1}{3}\int_{t=2}^{t=3} \frac{1}{u^2} du$$

$$= \frac{1}{3}\int_{t=2}^{t=3} u^{-2} du$$

$$= \frac{1}{3}\left[\frac{1}{u}\right]_{t=2}^{t=3}$$

$$= \frac{1}{3}\left[\frac{1}{t^3 + 3t}\right]_2^3$$

$$= \frac{1}{3}\left[\frac{1}{3^3 + 3(3)} - \frac{1}{2^3 + 3(2)}\right]$$

Golden Mean

The science of sacred geometry includes study of the Platonic solids, numbers, and the golden mean ratio.. This is all mathematics, and is very interesting if you are so inclined (left brain stuff).

The Golden Rectangle can be drawn using a straight line and compass. It is based on the square root of 5, and is particularly pleasing to the eye. This rectangle was used in some of Leornado da Vinci's works (eg Virgin in the Cave). Also shown below are two intersecting circles with the Vesica Pisces, and how the three fundamental geometries based on $\sqrt{2}$, $\sqrt{3}$ and $\sqrt{5}$ and are derived from it.

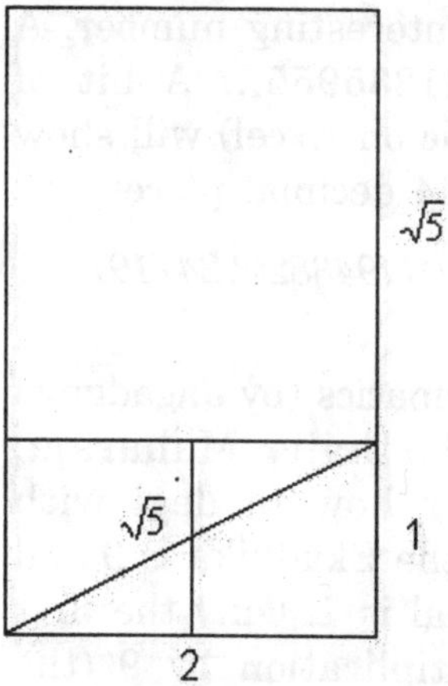

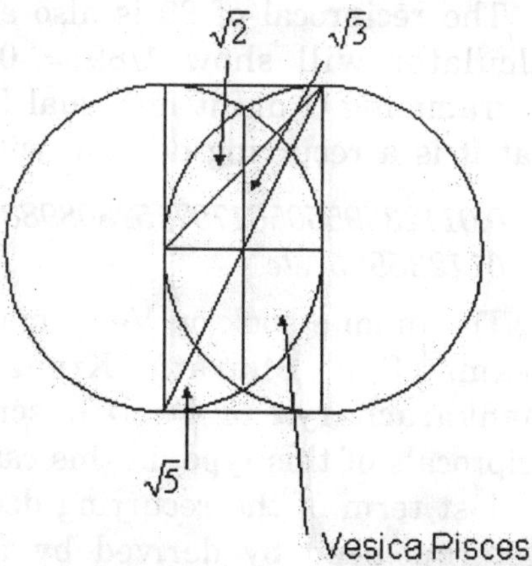

Fibonacci Sequences and the Phi Ratio - The Phi (Φ) ratio turns up frequently in nature. It can be derived from geometry or from a Fibonacci sequence. From geometry it's value is:

$$\Phi = \frac{1+\sqrt{5}}{2} = 1.618033989....$$

From the Fibonacci sequence (each term is derived from adding the previous two terms):

1, 1, 2, 3, 5, 8, 13, 21, 34, 55, 89, 144, 233, 377, 610, 987, 1597, 2584, 4181, 6765, 10946, ...

diving two successive terms yield values closer and closer to Φ, eg 10946 / 6765 = 1.618033999...

Another sequence:

1, 3, 4, 7, 11, 18, 29, 47, 76, 123, 199, 322, 521, 843, 1364, 2207, 3571, 5778, 9349, 15127, ...

yields $\sqrt{5}$ in the limit when a term from the second series is divided by a term from the first, eg the 18th terms 5778 / 2584 = 2.236068112... (cf 2.236067978...).

Reciprocal of 89

The reciprocal of 89 is also an interesting number. A calculator will show 1/89 = 0.011235955... A bit of programming (I did it in Visual Basic on Excel) will show that it is a recurring decimal with 44 decimal places:

0.01123595505617977528089887640449438202247191 011235955 etc.

The unique book on Vedic mathematics (by Jagadguru Swami Sri Bharati Krsna Tirthaji Maharaja, Shankaracharya of Puri) describes how to deal with reciprocals of this type, in this case the Ekadhika is 9 and the last term of the recurring decimal is 1. Only the first 22 terms need by derived by multiplication by 9 (the Ekadhika) at which point 88 (which is the denominator minus the numerator) is obtained. The rest of the terms are derived by 9's complement, i.e:

9 8 8 7 6 4 0 4 4 9 4 3 8 2 0 2 2 4 7 1 9 1

0 1 1 2 3 5 9 5 5 0 5 6 1 7 9 7 7 5 2 8 0 8

---

9 9 9 9 9 9 9 9 9 9 9 9 9 9 9 9 9 9 9 9 9 9

Starting from the top right, 1 x 9 = 9, which is the 2nd digit. 9 × 9 = 81, put down 1 carry 8. 9 × 1 = 9 + 8 = 17, put down 7 carry 1, etc.

Just thought I'd mention that for fun! The Fibonacci sequence is actually "hidden" inside 1/89, viz:

```
0 1 1 2 3 5 8 3 1 4 5 9 4 3 7 0 7 7 4 1 5 6 1 7 8 . .
            1 2 3 5 8 4 3 7 1 8 9 8 8 6 4 1 5 6 . .
                    1 2 3 6 9 5 5 1 7 9 7 6 3 . .
                            1 2 4 6 0 7 8 6 . .
                                  1 1 2 4 . . .

0 1 1 2 3 5 9 5 5 0 5 6 1 7 9 7 7 5|2 8 0 0 0 3 8 ...
```

On the top line I have started to write the Fibonacci sequence, 1, 1, 2, 3, 5, 8, 13. At 13 I put the 3 on the top line and carry the 1, adding it to the 8. This continues carrying 12 when we get to 123, etc. There are also 1's which get carried when the columns are added together. The | shows the accuracy of 1/89 for the number of terms of the Fibonacci sequence taken- when the numbers are all added up you get 1/89. The significance of the 44 recurring decimals is interesting as it is the number of chromosomes in the human body (44 + XX or XY) - is this just a coincidence?

Mathematics and the Spiritual Dimension

It's place of origin is India (the land of the present system of numerals, decimals and zero) and dates back to prehistoric times. In essence, it is based on sixteen formulas (sutras) made up to barely 120 words in all. It was discovered by Jagadguru Swami Sri Bharati Krsna Tirth Maharaja, Shankaracharya Of Govardhana Math, Puri who propagated his pioneering work through out the world.

It must be emphasized that although because of its origin, it is rightly called Vedic Mathematics, the mathematics it deals with is universal in that it does not require the knowledge of Sanskrit nor any acquaintance with the Vedas. Anyone who is conversant with the basics of mathematics can understand and follow these methods. Let us now see why Vedic Mathematics has become so popular and is rated so high by it's users.

It is very original, totally unconventional and provides a new thinking and approach. Vedic Mathematics is so powerful, yet simple and fun to work with.

It's word-formula lead to efficient and super fast calculations.

It encourages mental calculations since it handles the numbers of small magnitudes.

It is easy, simple, direct and straightforward. The students are freed from committing to memorise the multiplication tables and the like.. They are enthusiastic in learning the methods and eagerly look forward to the mathematics classes. They also enjoy sharing their newfound knowledge with others.

Everyone loves variety and Vedic mathematics offers it in ample measure.

It is comprehensive and is based on pattern recognition.

It is exceedingly flexible and offers a large number of alternatives for almost every kind of problem. It is this aspect that captivates the students and others the most. The weaker students gain mastery of the subject slowly and steadily by taking advantage of the flexibility and availability of alternatives for a given problem; in contrast, present available blanket methods provide only two options: "to learn them or leave them". On the other hand, the bright students take full advantage of the flexibility by learning to be creative and challenge in choosing the most appropriate and time saving method that is consistent with underlying pattern of a given problem.

Finally,all areas of modern mathematics, be it pure or applied, come under the scope of Vedic mathematics. This is because it provides a unified approach to different areas of mathematics by the application of the same work formulas to diverse fields like trigonometry, analytical geometry, calculus,differential equations and so on. The present day researchers in Vedic Mathematics have confirmed the enormous versatility that it has in this respect. The spin off this extendibility is that students who have been trained in the application of Vedic Mathematics to arithmetic can easily move over to college and higher mathematics deriving the same benefits as they did in the case of arithmetic.

" *Like the crest of the peacock,*

Like the gem on the head of a snake,

So is mathematics at the head of all knowledge. "

Mathematics is regarded as science of all sciences. Mathematics stands forth as that which unites and mediates between man and nature, thought and perception, inner and outer world. Vedic mathematics is not just a pedagogic and research tool but an introduction to ancient civilization. It is thus a bridge across centuries, civilizations, linguistic barriers and national frontiers.

Vedic mathematics has originated from that Vedas. VED is all divine knowledge. Any knowledge derived from the Ved is bound to have the touch of the divine bliss. Therefore the very natural, easy and super fast algorithms of Vedic mathematics bring an upsurge of joy and bliss. Due to its very special and universal features, Vedic maths system converts the dry and tedious maths into playful and joyful subject, which everyone learn with smiles. Therefore the Vedic maths is the gift of the Vedas to solve the problem of MATHS ANXIETY being faced in maths education in the world.

Sutras: The Natural Formula

In Sanskrit, the ancient language of India, the word Sutra means "Thread of Knowledge". In English we have the word Suture from Sutra and Suture is the thread used for stitching wounds together. So each rule is a thread of knowledge and the whole subject is based on these sutras.

The system is based on 16 Vedic Sutras together with thirteen sub-sutras which are actually word formulae describing natural ways of solving a whole range of mathematical problems. These are 16 one-line formulae originally written in Sanskrit, which can be easily memorized, enable one to solve long mathematical problems quickly.

The Main Sutras

Ekadikena Purvena(also a corollary)	By one more than the one before.
Nikhilam Navatascaramam Dasatah	All from 9 and the last from 10.
Urdhva-tiryabhyam	Vertically and Cross-Wise
Paravartya Yojayet	Transpose and Apply
Sunyam Samyasamuccaye	The summation if equal to Zero
(Anurupye) Sunyamanyat	If One is in Ratio the Other is Zero
Sankalana - vyavakalanabhyam	By Addition and by Subtraction
Puranapuranabhyam	By the Completion or Non-Completion
Chalana-Kalanabhyam	Sequential Motion
Yavadunam	The Deficiency
Vyastisamastih	Whole as One and One as Whole
Sesanyankena Caramena	The Remainders by the Last Digit
Sopantyadvayamantyam	The Ultimate and Twice the Penultimate
Ekanyunena Purvena	By One Less than the Previous One
Gunitasamuccayah	The Whole Product is the same
Gunakasamuccayah	Collectivity of multipliers

The Sub Sutras

Proportionately
The Remainder Remains Constant
The First by the First and the Last by the Last
For 7 the Multiplicand is 143
By Osculation
Lessen by the Deficiency
Whatever the Deficiency lessen by that amount & set up Square of Deficiency
Last Totalling 10
Only the Last Terms
The Sum of the Products
By Alternative Elimination and Retention
By Mere Observation
The Product of the Sum is the Sum of the Products
On the Flag

Vedic Mathematics as Fun

Mathematics is always fun. Vedic mathematics specifies a set of SUTRAS (formulae) that helps you to play with numbers.. mathematics made quicker and easier!

ALL FROM 9 AND THE LAST FROM 10 to perform instant subtractions

We simply take each figure in 357 from 9 and the last figure from 10.

1 0 0 0 - 3 5 7

(9-3) (9-4) (10-7)

6 4 3

So the answer is 1000 - 357 = 643

Using VERTICALLY AND CROSSWISE you do not need to the multiplication tables beyond 5 × 5.

Suppose you need 8 × 7

8 is 2 below 10 and 7 is 3 below 10.

Think of it like this:

82

73

56

You subtract crosswise 8-3 or 7 - 2 to get 5,the first figure of the answer.

And you multiply vertically: 2 x 3 to get 6,the last figure of the answer.

Multiplying numbers just over 100.

103 × 104 = 10712

The answer is in two parts: 107 and 12,

107 is just 103 + 4 (or 104 + 3), and 12 is just 3 x 4.

Similarly 107 × 106 = 11342

107 + 6 = 113 and 7 x 6 = 42

A quick way to square numbers that end in 5 using the formula BY ONE MORE THAN THE ONE BEFORE.

752 = 5625

752 means 75 × 75.

The answer is in two parts: 56 and 25.

The last part is always 25. The first part is the first number, 7, multiplied by the number "one more", which is 8:

so 7 × 8 = 56

Method for multiplying numbers where the first figures are the same and the last figures add up to 10.

32 × 38 = 1216

Both numbers here start with 3 and the last figures (2 and 8) add up to 10.

So we just multiply 3 by 4 (the next number up)

to get 12 for the first part of the answer.

And we multiply the last figures: 2 x 8 = 16 to get the last part of the answer.

And 81 x 89 = 7209

We put 09 since we need two figures as in all the other examples.

An elegant way of multiplying numbers using a simple pattern.

21 × 23 = 483

This is normally called long multiplication but actually the answer can be written straight down

using the VERTICALLY AND CROSSWISE formula.

We first put, or imagine, 23 below 21:

There are 3 steps:

(a) Multiply vertically on the left: $2 \times 2 = 4$.
This gives the first figure of the answer.

(b) Multiply crosswise and add: $2 \times 3 + 1 \times 2 = 8$
This gives the middle figure.

(c) Multiply vertically on the right: $1 \times 3 = 3$
This gives the last figure of the answer.

And that's all there is to it.

Similarly $61 \times 31 = 1891$

$6 \times 3 = 18$; $6 \times 1 + 1 \times 3 = 9$; $1 \times 1 = 1$

Multiply any 2-figure numbers together by mere mental arithmetic!

If you want 21 stamps at 26 pence each you can easily find the total price in your head.

There were no carries in the method given above. However, there only involve one small extra step.

$$21 \times 26 = 546$$

The method is the same as above except that we get a 2-figure number, 14, in the middle step, so the 1 is carried over to the left (4 becomes 5).

$$33 \times 44 = 1452$$

There may be more than one carry in a sum:

Vertically on the left we get 12.

Crosswise gives us 24, so we carry 2 to the left and mentally get 144.

Then vertically on the right we get 12 and the 1 here is carried over to the 144 to make 1452.

Multiplying a number by 11.

To multiply any 2-figure number by 11 we just put the total of the two figures between the 2 figures.

$$26 \times 11 = 286$$

Notice that the outer figures in 286 are the 26 being multiplied.

And the middle figure is just 2 and 6 added up.

So $72 \times 11 = 792$

$77 \times 11 = 847$

This involves a carry figure because $7 + 7 = 14$

we get $77 \times 11 = 7147 = 847$.

$$234 \times 11 = 2574$$

We put the 2 and the 4 at the ends.

We add the first pair $2 + 3 = 5$ and we add the last pair: $3 + 4 = 7$.

Method for diving by 9.

$23 / 9 = 2$ remainder 5

The first figure of 23 is 2, and this is the answer.

The remainder is just 2 and 3 added up!

$43 / 9 = 4$ remainder 7

The first figure 4 is the answer

and $4 + 3 = 7$ is the remainder - could it be easier?

Multiplying by 12 - shortcut

So how does the 12's shortcut work?

Let's take a look.

$$12 \times 7$$

The first thing is to always multiply the 1 of the twelve by the number we are multiplying by, in this case 7.

So $1 \times 7 = 7$.

Multiply this 7 by 10 giving 70. (Why? We are working with BASES here.

Bases are the fundamentals to easy calculations for all multiplication tables.

To find out more check out our Vedic Maths ebook at www.vedic-maths-ebook.com

Now multiply the 7 by the 2 of twelve giving 14. Add this to 70 giving 84.

Therefore 7 × 12 = 84

Let's try another:

17 × 12

Remember, multiply the 17 by the 1 in 12 and multiply by 10 (Just add a zero to the end)

1 × 17 = 17, multiplied by 10 giving 170.

Multiply 17 by 2 giving 34.

Add 34 to 170 giving 204.

So 17 × 12 = 204

lets go one more

24 × 12

Multiply 24 × 1 = 24. Multiply by 10 giving 240.

Multiply 24 by 2 = 48. Add to 240 giving us 288

24 × 12 = 288

(these are Seriously Simple Sums to do aren't they?!)

Converting Kilos to Pounds

In this section you will learn how to convert Kilos to Pounds, and Vice Versa.

Let's start off with looking at converting Kilos to pounds.

86 kilos into pounds:

Step one, multiply the kilos by TWO.

To do this, just double the kilos.

86 × 2 = 172

Step two, divide the answer by ten.

To do this, just put a decimal point one place in from the right.

172 / 10 = 17.2

Step three, add step two's answer to step one's answer.

172 + 17.2 = 189.2

86 Kilos = 189.2 pounds

Universal Formula Mathematics

Find out what you didn t learn at school.

Ancient Knowledge that has been held secret... or hidden for thousands of years.

Jain

About Jain: Jain is his only name... he was born in Sydney to Lebanese parents. His father, a wealthy and successful businessman, migrated to Australia with his family in 1956. A long time meditator, Jain has always been drawn to the spiritual and mystical, he is fascinated by numbers and shapes. He has enjoyed a lifetime involvement with the Fibonacci Sequence and the 5 Platonic Solids which taps into the highly visual hemisphere of the Right Brain.

His work with Vedic Mathematics, a 2,500 year old Indian system of Mental, One-Line Arithmetic... is to empower the student so that they can perform mental calculations, in many cases... quicker than a calculator. Jain expands and demonstrates the real ability of your mind by showing you how to complete complex mathematics ... in your head! Jain reveals ancient yet simple Math formulas which allow ordinary people to perform complex equations.

At the same time he shows you how the world as we know it is more than we think... his teachings can for some people, bring about major shifts in the way we think about the world! Over the years, Jain has been devoted to the collection and theatrical education of Magic Squares, Magic Cubes and the Golden Ratio (PHI). In 1995, Jain taught Magic Square Art-forms to a conference of 100 maths teachers at the Australian Southern Cross University .

He has been guest speaker at school Math Camps throughout Australia. Also many Public Schools, Catholic Schools, Rudolf Steiner and Montessori Schools have employed him to rekindle the beauty of mathematics. He has recently completed filming for a DVD/TV documentary about Vedic Mathematics and Ancient Knowledge. Jain runs seminars on various forms of Ancient Knowledge, teaches children and adults (individuals and groups), records his discoveries having authored 8 books and produced several videos... and he continues his research into Sacred Geometry, Magic Squares, Platonic Solids, Fibonacci Sequences and their relationship with various atomic structures.

His passionate and ongoing commitment to share this incredible knowledge is furthered with an initial 6 month Australian Tour followed by 2 months USA Tour. The US Tour includes a special guest appearance at the Global Conference on Sacred Geometry in California (August 2004), where he has been invited to teach Vedic Mathematics, Golden Mean and Magic Squares. His travels continue with 3 x 1 month tours of Sweden, England , Nepal and India. He is compiling a database of people all over the world who have expressed interest to study with him and is developing internet also regional programs. If you feel like some enlightenment; want to experience Maths as Art, Maths as Science, Maths as History or develop your mental muscle with rapid mental calculations... come and experience these subjects which will expand your consciousness.

Universal Formula Mathematics

Due to an immense response from people all over the world, there is a great demand to know more about the Magic of Vedic Mathematics. Jain wrote an article: VEDIC MATHEMATICS REDISCOVERED for the Nexus Magazine, Oct/Nov issue 2003, and it has awakened many people to start questioning all the missing gaps and insufficiencies in the current mathematics curriculum. It also has motivated Jain to begin his Australia Tour until August before he travels to America then Europe and India. Do your children really understand the fragmented and confusing mathematical data taught to them?

Would you be interested in another method that boosted their confidence, increased their memory skills and is presented in a fun and exciting manner? After this course of study, you will never multiply, divide or add in the same manner again. And no more Times Tables! An understanding of the 16 basic Sutras or "Word Formulae" solves all known mathematical problems by using an advanced and highly intelligent system of MENTAL, ONE LINE ARITHMETIC used by the Hindus 2,000 years.

This system was reconstructed by Bharati Krsna Tirthaji (1884-1960, aka Shankarcharya of Puri) who was invited by Paramahansa Yogananda to take this knowledge, for the first time, to the western world, in 1958. The Beatle's guru: Maharishi Mahesh Yogi, who introduced Transcendental Meditation to the west, was also a student of this master and currently teaches this system in India in the spiritual Maharishi schools, whose teenage students are highly distinguished mathematical geniuses. Learn also how the Hindu Invention of Zero (and the decimal system) led to modern technological advancements. In this one day workshop you will know how to accomplish rapid calculations, mentally.

No more Calculators which make the brain lazy. The brain is a mental muscle and needs to be exercised. Jain predicts that if we continue to teach children with an over-emphasis of using calculators in the class-room, over a 20 year period, the Global Brain will deteriorate. Most children are visual, and the ability to tap into 'Pattern Recognition' is the key to rapid mental calculation. There is another similar system known as the Trachtenberg method, but this Vedic System successfully harnesses the mental "Path of Least Resistance" and is distinctly the superior system.

The Divine Proportion

Learn why the shape of the Pine Cone is optically similar to the Human Heart, why we are awe-struck when we view sacred architecture like the Parthenon, why we are attracted to the proportions of the famous Mona Lisa. Essentially, the human body is in resonance with the Living Mathematics of Nature. A day for exploring the hidden mysteries within the 5 Platonic Solids and the 13 Archimedean Solids. What is the importance of the Star Tetrahedron (the 3-Dimensional version of the Star of David) having 24 faces and 24 edges? How does this link to the 24 edges of the Cuboctahedron, the compressible and alchemical shape that was revered by Buckminster Fuller (who designed the first geodesic domes).

A lesson in x-raying the Fibonacci Sequence : 0, 1, 1, 2, 3, 5, 8, 13, 21, 34, 55, 89, 144 etc. to discover where the infinite phi expression has its underlying rhythm, a distinct periodicity of 24 recurring digits that can not be seen by the Western Mathematical Eye but only by the Truth of Vortex Mathematics expressed as Vedic Mathematics! (Really this Knowledge predates the Vedas and Atlantis and is better referred to as Galactic Mathematics).

Bridging Physics to Metaphysics by understanding the work of Charles Leadbeater and Annie Besant who, 100 years ago, drew all known atomic structures of the elements of the periodic table. In fact some recently discovered isotypes have been drawn 100 years ago! Having associated with yogis, this Pair of Clairvoyant Chemists were also able to successfully describe the moons of Jupiter, confirmed now by NASA!

This material actually validates Jain's life's work:that when the numbers of certain Magic Squares, like the Magic Square of 3x3*, which is the centre of the Tibetan Calendar, when tiled or tessellated into a larger matrix actually shows the shadow form of the 3-Dimensional view of the Atomic Structure of Diamond-Lattice! This material shakes the whole foundations of the current Western mathematics curriculum, will rewrite our current world view on Mathematics and will require a new definition of the Golden Mean: that its decimal does carry on infinitely but within its infinity there is recursive or self-similar beauty. Scientists like Marco Rodin wind toroidal coils in sync with this recursion or rhythm or pulse or pattern of 24 digits to get more output than their input! The sum of these 24 digits is 108 (108 degrees being the external angles of the pentagram and the number of beads in a rosary) which is sonically disguised as the GAYATRI MANTRA, the most famous Eastern prayer for Enlightenment.

Magic Square

4	9	2
3	5	7
8	1	6

A fascinating day that will help you to remember this Lost Knowledge of the PHI CODE that is already morphologically and geometrically encoded in your bio-magnetic and bio-electrical fields. Ultimately, this Pythagorean Knowledge is not an intellectual process, but rather that these essential geometries are invisibly constructed and nested within the Heart as the Centre of a greater Galactic Grid connecting us to all Memory of all Universes and Atomic Levels. Magic Squares, Vedic Mathematics and the Phi Ratio (1:1.618) : Why have they not been taught in your maths classes ? ... because they are keys to your ascension process..

it became possible to note higher numerals with limited charecters.

In the earlier Roman and Babylonian systems of numeration, a large number of characters were required to denote higher numerals. Thus enumeration and computation became unwieldy. For instance, as E the Roman system of numeration, the number thirty would have to be written as X: while as per the decimal system it would 30, further the number thirty three would be XXXIII as per the Roman system, would be 33 as per the decimal system. Thus it is clear how the introduction of the decimal system made possible the writing of numerals having a high value with limited characters. This also made computation easier.

Apart from developing the decimal system based on the incorporation of zero in enumeration, Brahmagupta also arrived at solutions for indeterminate equations of 1 type ax2+1=y2 and thus can be called the founder of higher branch of mathematics called numerical analysis. (Brahmagupta's treatise Brahma-sputa-siddhanta was translated into Arabic under the title Sind Hind).

For several centuries this translation mained a standard text of reference in the Arab world. It was from this translation of an Indian text on Mathematics that the Arab mathematicians perfected the decimal system and gave the world its current system of enumeration which we call the Arab numerals, which are originally Indian numerals.

Nikhilam Navatashcaramam Dashatah

Nikhilam Navatashcaramam Dashatah: All from 9 and the last from 10.

Nikhilam Navatashcaramam Dashatah is the second sutra of Vedic mathematics. Its corollary is Sisyate Sesasamjnah.

Nikhilam multiplication can save time when multiplying numbers that are both near base 10, or both near base 100, or both near base 1000, etcetera.

Nikhilam division can be used to quickly divide any number (regardless of size) by a number that is near a base 10, base 100, etc. (such as the numbers 9, 98, 996, etc.)

What follows is a brief and incomplete summary of the math shortcuts this sutra contains:

Nikhilam multiplication using base 10, base 100, base 1000, etc.

The easiest way to explain this is to show two examples:

Rounding to base-10: If you wanted to multiply 14 by 13 you would start by adding the two numbers and subtracting 10.

14 + 13 = 27

27 - 10 = 17.

Add one zero (for the one zero in base 10) to the end: 170.

Now calculate the difference between 14 and 13 and their common base of ten, which would be the number 10.

14 is 4 away from 10; 13 is 3 away from 10.

Multiply 4 and 3. This equals 12.

Add the 12 to the number we got before (170).

12 + 170 equals 182.

Thus 14 x 13 equals 182.

Rounding to base-100: If you wanted to multiply 97 by 89, for example, you would start by adding the two numbers (186) and subtracting 100.

186 - 100 = 86.

Now add the two zeroes (for the two zeroes in base 100) to the end: 8600.

Now calculate the difference between the two numbers and their common base and multiply those two numbers.

100 - 97 = 3

100 - 89 = 11

3 × 11 = 33.

Add the 33 to 8600 for the answer.

97 × 89 = 8633.

Nikhilam division using base 10, base 100, base 1000, etc.

Nikhilam division seems, at first glance, too complicated to save time, but is actually surprisingly fast and simple to do. Below are examples of some (but not all) applications of Nikhilam division:

Dividing any two-digit number by 9: Example: 53 divided by 9. the first digit (5) is the quotient. Now add the first and second digit: 5 + 3 = 8. The remainder is 8. Simple as that.

If the first and second digit add up to more than nine, you subtract 9 from the remainder and add 1 to the quotient. Example: 69 divided by 9. The quotient (at first glance) appears to be 6. But 6 + 9 = 15. You subtract 9 from 15 to get the real remainder (6) and add one to the quotient. Thus, the quotient is 7 and the remainder is 6.

Dividing any three-digit number by 9: This one might require paper and pen but is also quite easy.

Example: 412 divided by 9.

The first digit of the quotient will be the same as the first digit of the dividend: 4.

Now add the first two digits of the dividend: 4 + 1 = 5

Place this number (5) after the first digit of the dividend (4): 45

The quotient is 45.

To find the remainder, add all three digits of the dividend:

$$4 + 1 + 2 = 7$$

The quotient is 45 and the remainder is 7. If the remainder is more than 9, add one to the quotient and subtract 9 from the remainder.

Dividing any four-digit number by 9: Example: 1234 divided by 9.

The first digit of the quotient will be the same as the first digit of the dividend: 1

Add the first two digits (1 + 2 = 3) and place that after the first digit (1). Now we have 13.

Add the first three digits (1 + 2 + 3 = 6) and place that after the 13. Now we have 136. This is the quotient.

The remainder will be all for digits added together.

$$1 + 2 + 3 + 4 = 10$$

But wait, the remainder can't be more than 9, so subtract 9 from the remainder and add one to the quotient.

The quotient is 137 and the remainder is 1.

Nikhilam division can also be used to divide very large numbers by a number that is near a base, such as 98, 984, 897, etc., but this gets a little more complicated.

Vedic Mathematics: The Ancient Indian Knowledge

Indian civilization has surprised academicians time and again with simple age old techniques for solving complex problems. Vedic Mathematics, derived from the Ganita Sutra in the Vedas is one such technique. Vedic Mathematics is the name given to the ancient system of Mathematics which was rediscovered from the Vedas between 1911 and 1918 by Sri Bharati Krsna Tirthaji (1884 - 1960).

In the beginning of the twentieth century, when there was a great interest in various Sanskrit texts in Europe, some scholars ridiculed certain texts which were titled GANITA SUTRA, which means mathematics.

They could not decipher any mathematics in the translations and therefore dismissed the texts as rubbish. Bharati Krsna, who himself was a scholar of Sanskrit, Mathematics, History and Philosophy, studied these texts and after lengthy and careful investigation was able to reconstruct the mathematics of the Vedas.

Bharati Krsna wrote sixteen volumes expounding the Vedic system but these were unaccountably lost and when the loss was confirmed in his final years he wrote a single book - Vedic Mathematics. It was published in 1965, five years after his death.

According to his research all of mathematics is based on sixteen Sutras, or word - formulae.

For example, 'Vertically and Crosswise' is one of these Sutras. These formulae describe the way the mind naturally works and are therefore a great help in directing the student to the appropriate method of solution.

Perhaps the most striking feature of the Vedic system is its coherence. Instead of a hotchpotch of unrelated techniques the whole system is beautifully interrelated and unified.

The multiplication method, for example, is easily reversed to allow one-line divisions and the simple squaring method can be reversed to give one-line square roots. And all are very easy to understand. This unifying quality is very satisfying; it makes mathematics easy and enjoyable and encourages innovation.

In the Vedic system difficult problems or huge sums can often be solved immediately by the Vedic method. These striking and beautiful methods are just a part of a complete system of mathematics, which is far more systematic than the modern system. Vedic mathematics manifests the coherent and unified structure of mathematics and the methods are complimentary, direct and easy.

The simplicity of Vedic Mathematics means that calculations can be carried out mentally (though the methods can be written down).

There are many advantages in using a flexible, mental system. Pupils can invent their own methods; they are not limited to the one correct method. This leads to more creative, interested and interesting pupils.

Interest in the Vedic system is growing in education where mathematics teachers are looking for something better and finding Vedic system is the answer. Research is being carried out in many areas including the effects of learning Vedic Mathematics on children, developing new, powerful but easy applications of the Vedic Sutras in geometry, calculus, computing, etc.

But the real beauty and effectiveness of Vedic Mathematics cannot be fully appreciated without actually practicing the system. One can then see that it is perhaps the most refined and efficient mathematical system possible.

Why do We Need to Combine Pure and Vedic Mathematics?

Schools all over the world utilize pure mathematics for their curriculum. Therefore an in-depth knowledge of pure mathematics is essential from the examination point of view. Unfortunately the mathematics that is taught is most schools and colleges lack flexibility and therefore become 'boring' to the students.

On top of that many teachers lack the insight and knowledge required to teach the subject. Fixed methods are also used for all operations thereby making the process of calculation lengthy and tiresome.

The right combination of Vedic and pure mathematics ensures correct understanding, extreme flexibility and an in-depth knowledge of the subject, thereby making calculations extremely rapid and interesting.

This combination can be utilized for simple addition and subtraction, and also for more difficult operations like calculus, and algebra. Many of these operations can be done mentally. Therefore this can be used for students of all ages and standards. There are also wonderful methods for checking the sums.

Utilizing this method, the student will become faster and more accurate in all-mathematical procedures.

Western vs. Vedic Astrology—Predictive Methodology

As mentioned in another article astrological calculations for planetary longitudes differ in Western and Vedic astrology. Western system uses so called moving zodiac or coordinate system called Sayana while Vedic system uses fixed or Nirayana system.

But for astrological predictions using the basic planetary coordinates there is a gulf of difference between Western and Vedic system. In Western system the intricacies are not there and interpretation of planetary interactions are limited, while in Vedic system the tools and the depth of application is limited only by the depth of ones knowledge and interest with many cross checking and confirmation tests.

In Western system all the planets including those discovered in 19th Century - viz Uranus, Neptune and Pluto are generally used. In a mixed system some Indian astrologers use Sayana and these extra Saturn planets with the names of Indra Varuna and Yama.

Vedic system uses mainly the two luminaries and five other planets but include two "CHAYA" shadow planets viz

Rahu and Ketu. Their importance can be judged from Physics definition of UMBRA and PENUMBRA points when two luminaries interact. The importance of these becomes evident by the fact that an eclipse occurs whenever luminaries and these fall in a straight line.

In interpretation Western system uses only angular distances between planets - treating 120 and 60 as good and 90 and 45 as bad. The others like 180, 72, 30, 0, etc., are also considered.

The timing also is very static - only progressions are taken as 1 degree = I year of life, 1 day = I year of life, etc., *i.e.*, whenever say a planet is in 14 degrees of a house and another at say 24 of same house - at the age of 10 both are conjunct.

Thus by these progressions the timing is calculated and the type of angular distance determines the event. The houses are of equal duration, which does not take the latitude effect of place. The Gochara or effect of current planetary positions is accounted by the angles they form with birth planets. Thus over all the methods of interpretations of western system is simple arithmetic and static.

In contrast Vedic system introduces a unique frame of reference called Lagna or ascendant, which is the Eastern horizon point at the time of birth.

Thus all the spatial coordinates of place of birth are taken in each birth. Lagna forms the starting point and from there different bhavas are defined to account for different aspects of our life. The planetary relations are also classified as benefic/evil, naturally friendly/foe and dynamically friendly/foe, etc.

The houses or bhavas are also classified as good and bad. Synthesis and analysis are basic tools in any scientific study. In Vedic astrology also such tools are extensively used. The bhavas are not of equal duration but astronomical effects are accounted by Sripathi method of Bhava Sphutam. The strength of planets is also calculated by many methods including graha balas, which take into account their absolute positions relative positions, etc.

Thus planets have complex properties like naturally good/bad, positionally good/bad and different strengths to do that good or bad. The rules are also dynamic in that for many rules have exceptions where the rule has to be over ruled. When ever a gross examination does not answer a complex question in a scientific field one goes to microscopic examination. Having AMSA charts where the house is divided into microscopic units - uses same principle.

In this one go in depth to a particular activity like profession marriage, etc. Thus one can devise charts where a house is divided to 1/3,1/4, 1/5, 1/6, 1/7,1/8 1/9,1/10, 1/11, 1/12, 1/16, 1/20, 1/27, 1/30 an 1/60 - each representing a different aspect of life. The timing is also followed by what is called Dasa system — in fact many dasa systems like Vimsottari, Ashtottari, etc., exist and one has to apply correct dasa system.

The Gochara is also very finely tuned by different methods like present-day movements of planets vis a vis birth position of moon. In fact a more complex system exists called Ashtakavarga which is a 12 by 8 matrix - *i.e.*, the position of 7 planets and lagna in 12 houses at birth interacting with present day planetary positions of 12 by 8 matrix.

The mathematical tools used in the evaluation of these is mind boggling - if we note that the evaluation rules of Matrix are hardly 3 centuries old - the evaluation rules of Ashtakavarga - like if there is a zero in a row (Column) cancel that, Subtract smallest number from highest number, etc., proves the depth of Vedic mathematics used - which are at least thousands of years old.

The Gochara is not limited to these and has other methods like Nilakhanthas solar return horoscope or Tajaka system to give accurate yearly results. While all these go under the head of Parasara system we have the complex Jaimini system with its complex rules. Thus Vedic astrology is a ocean of nectar and any one of us can enjoy only a drop of the same as per our luck and karma. It is a holistic approach to satisfy achieving of all purshardhas— of Dharma, Ardha, Kama and Moksha.

Finally if one is predicting based on divine intuition it is immaterial if one is following East or Western method or if one is born white, black, yellow or brown. Astrology is a divine science and only needs divine blessing and all calculations and interpretations are only road signs to follow the path - a means but not the ends.

Bibliography

A B Chace, L S Bull, H P Manning and R C Archibald: *The Rhind Mathematical Papyrus,* Oberlin, Ohio, 1927.

A.L. Basham: *The Wonder That Was India*, Rupa & Co., Calcutta, 1967.

Aaboe, Asger: *Episodes from the Early History of Mathematics*, New York, 1964.

Amma, Saraswarti: *Geometry in Ancient and Medieval India*, Motilal Banarsidas, 1979.

Armstrong, Robert G.: *Yoruba Numerals*, London, Oxford University Press, 1962.

Arnett, William S.: *The Predynastic Origin of Egyptian Hieroglyphs*, Washington, University Press of America, 1982.

Ascher, Marcia: *Ethnomathematics*, Pacific Grove, CA: Brooks/Cole, 1991.

Bag, A. K.: *Mathematics in Ancient and Medieval India*, Varanasi: Chaukhambha Orientalia, 1979.

Barnard, F. P.: *The Casting-Counter and the Counting Board. A Chapter in the History of Numismatics and Early Arithmetic*, Oxford, Oxford University Press, 1916.

Barrow, John D.: *Pi in the Sky*, Oxford, Clarendon Press, 1992.

Basham, A. L.: *The Wonder That Was India*, Calcutta, Rupa & Co., 1967.

Beckmann, Petr: *A History of Pi*, New York, St. Martin's Press, 1974.

Bell, Eric T.: *The Magic of Numbers*, New York, McGraw-Hill, 1946.

Berlin, Brent: Tzeltal Numeral Classifiers, The Hague, Mouton, 1968.

Bharati Krishna Tirthaji Majaraja.: *Vedic Mathematics*. Delhi, Motilal Banarsidass, 1988.

Bidyaranya, Swami: *History of Hindu Mathematics*,Bombay, Asia Pub. House, 1962.

Bischoff, Bernard: *Latin Palaeography*, Cambridge, Cambridge University Press, 1990.

Bonfante, Larissa: *Etruscan*. Berkeley, University of California Press, 1990.

Bowditch, Charles P.: *The Numeration, Calendar Systems and Astronomical knowledge of the Mayas*, Cambridge, Cambridge University Press, 1910.

Brainerd, Charles J.: *The Origins of the Number Concept*, New York, Praeger, 1979.

Brunschwig, Jacques and Geoffrey E.R. Lloyd:. *Greek Thought: A Guide to Classical Knowledge*, Cambridge, Harvard University Press, 2000.

Bunt, L. N. H., Jones, P. S., & Bedient, J. D.: *The Historical Roots of Elementary Mathematics*, New York, Dover, 1988.

Chattopadhyaya, D. P.: *Mathematics, Astronomy, and Biology in Indian Tradition: Some Conceptual Preliminaries*, New Delhi, Project of History of Indian Science, Philosophy, and Culture, 1995.

Clawson, Calvin: *The Mathematical Traveller*, New York, Plenum Press, 1994.

Closs, Michael P.: *Native American Mathematics*, Austin, University of Texas Press, 1986.

Conant, Levi L.: *The Number Concept*, New York, MacMillan, 1896.

Crossley, John N: *The Emergence of Number*, Singapore, World Scientific, 1987.

Crump, Thomas: *The Anthropology of Numbers*, Cambridge, Cambridge University Press, 1990.

Dantzig, Tobias: *Number, the Language of Science*, London, George Allen and Unwin, 1940.

De Villiers, Melius: *The' Numeral-Words: Their Origin, Meaning, History and Lesson*, London, Witherby, 1923.

Dehaene, Stanislas: *The Number Sense*, New York, Oxford University Press, 1997.

Dilke, Oswald Ashton Wentworth: *Mathematics and Measurement*, Berkeley, CA: U. of Cal. Press, 1987.

Eves, Howard: *An Introduction to the History of Mathematics*, New York, Rinehart and Company Inc., 1953.

Flegg, Graham: *Numbers: Their History and Meaning*, London, André Deutsch, 1983.

Fowler, David: *The Mathematics of Plato's Academy*, Oxford, Clarendon, 1999.

Fuson, K.C.: *Children's Counting and Concepts of Number*, New York, Springer-Verlag, 1988.

Ganeri, Anita: *The Story of Numbers and Counting*, New York, Oxford University Press, 1996.

Gokhale, Shebhana Laxman: *Indian Numerals*, Poona, Deccan College, 1966.

Griffith, Francis Llewellyn: *Hieratic Papyri from Kahun and Gurob*, London, Bernard Quaritch, 1898.

Grover, Sudarshan: *History of Development of Mathematics in India*, Delhi, Atma Ram & Sons, 1994.

Haldane, John Burdon Sanderson: *Science and Indian Culture*, Calcutta, New Age Publishers, 1965.

Halliwell, James Orchard: *Rara Mathematica*, London, J.W. Parker, 1839.

Herbert Meschkowski: *Ways of Thought of Great Mathematicians*, Holden-Day Inc., San Francisco, 1964.

Hogben, Lancelot: *Mathematics for the Million*, New York, W. W. Norton, 1937.

Howard Eves: *An Introduction to the History of Mathematics*, Rinehart and Company Inc., New York, 1953.

Ifrah, Georges: *From One to Zero: A Universal History of Numbers*, New York, Viking Penguin, 1985.

Jagadguru Swami Shri Bharati Krishna Tirthaji Maharaja: *Vedic Mathematics*, Motilal Banarsidass, Delhi, 1988.

Jaggi, O. P.: *Science and Technology in Medieval India*, New Delhi, Atma Ram & Sons, 1977.

Kline, Morris: *Mathematical Thought from Ancient to Modern Times*, New York, Oxford U. Press, 1990.

Kuppuram, G & K. Kumudamani: *History of Science and Technology in India*, Delhi, Sundeep Prakashan, 1990.

Lancy, David F.: *Cross-Cultural Studies in Cognition and Mathematics*, New York, Academic, 1983.

Marcus, Joyce: *Mesoamerican Writing Systems*, Princeton, Princeton University Press, 1992.

Marshack, Alexander: *The Roots of Civilization*, New York, McGraw-Hill, 1972.

Martzloff, Jean-Claude: *A History of Chinese Mathematics*, New York, Springer, 1997.

McLeish, John: *The Story of Numbers*, New York, Fawcett Columbine, 1991.

Menninger, Karl: *Number-Words and Number-Symbols*. London, MIT, 1969.

Murthy, T. S. Bhanu: *A Modern Introduction to Ancient Indian Mathematics*, Wiley Eastern Ltd., New Delhi, 1992.

Neugebauer, Otto and A. Sachs: *Mathematical Cuneiform Texts*, New Haven, Yale University Press, 1945.

Neugebauer, Otto: *The Exact Sciences in Antiquity*, Princeton, Princeton University Press, 1952.

North, John: *The Norton History of Astronomy and Cosmology*, New York, W.W.Norton, 1995.

Pandit, M. D.: *Mathematics as known to the Vedic Samhitas*, Delhi, Sri Satguru Publications, 1993.

Pappas, Theoni: *The Magic of Mathematics*, San Carlos, Wide World Publishing / Tetra, 1994.

Parpola, Asko: *Deciphering the Indus Script*, Cambridge, Cambridge University Press, 1994.

Peet, T. Eric: *The Rhind Mathematical Papyrus*, London, British Museum, 1923.

Piaget, Jean: *The Child's Conception of Number*, New York, W.W. Norton, 1965.

Pullan, J.M: *The History of the Abacus*, New York, Praeger, 1968.

Rao, S. Balaachandra: *Indian Mathematics and Astronomy*, Bangalore, Jnana Deep Publications, 1994.

Restivo, Sal: *The Social Relations of Physics, Mysticism and Mathematics*, Dordrecht, Kluwer, 1983.

Robins, G. and C. Shute: *The Rhind Mathematical Papyrus*, London, British Museum, 1987.

Russell, Bertrand: *The Principles of Mathematics*, Cambridge, Cambridge University Press, 1903.

Salomon, Richard: *Indian Epigraphy*, New York, Oxford University Press, 1998.

Saraswati Amma: *Geometry in Ancient and Medieval India*, Motilal Banarsidas, 1979.

Sarton, George: *The Study of the History of Mathematics*, Cambridge, Cambridge University Press, 1936.

Satterthwaite, Linton, Jr.: *Concepts and Structures of Maya Calendrical Arithmetics*, Philadelphia, University of Pennsylvania, 1947.

Schaaf, W.L.: *Mathematics, Our Great Heritage*, New York, Harper and Row, 1948.

Schimmel, Annemarie: *The Mystery of Numbers*, New York, Oxford University Press, 1993.

Scriba, C.J.: *The Concept of Number*, Zurich, Mannheim, 1968.

Seidenberg, A.: *The Zero in the Mayan Numerical Notation*, Austin, University of Texas Press, 1986.

Smeltzer, Donald: *Man and Number*, New York, Emerson, 1959.

Smith, D. E.: *History of Mathematics*, New York, Dover Publications Inc., 1951.

Srinivasiengar, C. N.: *The History of Ancient Indian Mathematics*. Calcutta, World Press, 1967.

Struik, Dirk J.: *A Concise History of Mathematics*, New York, Dover, 1987.

Thompson, Sir J. Eric S.: *Maya Hieroglyphic Writing: An Introduction*, Norman, University of Oklahoma Press, 1971.

Todd, R.R., P.J. Barber, and D. Jones: *The Internal Representation of Number: Analogue or Digital?*, Oxford, Clarendon, 1987.

Urton, Gary: *The Social Life of Numbers*, Austin, University of Texas Press, 1997.

Volodarsky, A.I.: *Mathematics in Ancient India*, Tokyo, Science Council of Japan, 1974.

Woolner, Alfred C.: *Introduction to Prakrit*, Calcutta, Baptist Mission Press, 1917.

Index

A

B

C

D

E

T

U

V

❑❑❑

ENCYCLOPAEDIA OF VEDIC MATHEMATICS

ENCYCLOPAEDIA OF VEDIC MATHEMATICS
(Vol. 4)

Shyam Priya

ANMOL PUBLICATIONS PVT. LTD.
NEW DELHI - 110 002 (INDIA)

ANMOL PUBLICATIONS PVT. LTD.
H.O.: 4374/4B, Ansari Road, Darya Ganj,
New Delhi-110 002 (India)
Ph.: 23278000, 23261597
B.O.: No. 1015, Ist Main Road, BSK IIIrd Stage
IIIrd Phase, IIIrd Block,
Bangalore - 560 085 (India)
Visit us at: www.anmolpublications.com

Encyclopaedia of Vedic Mathematics

First Published, 2008
ISBN 978-81-261-3504-2 (Set)

PRINTED IN INDIA

Printed at Mehra Offset Press, Delhi.

Contents

Preface

Vedic Mathematics contains many examples of striking methods of calculation and there is a remarkable coherence to the system which makes it very attractive. Also the Vedic system itself suggests a kind of approach that involves going directly to the answer. Vedic mathematics is based on sixteen Sutras and some sub-Sutras which provide links through mathematics: the word 'Sutra' means 'thread'.

Vedic scheme is not simply a collection of rapid methods; it is a system, a unified approach, which can be swiftly learnt. Vedic Mathematics extensively exploits the properties of numbers in every practical applications, particularly in the field of computation. It makes available a whole range of methods ideally suited to these properties. After a detailed introduction to Vedic Mathematics, this book shows applications of Vertically and Crosswise from basic calculations to the evaluation of logarithms, exponentials, trigonometrical functions and the solution of simultaneous, transcendental, polynomial and differential equations. The methods shown are not only exceedingly efficient and new but have a unity and coherence which renders them easy to master and enjoyable to execute.

The methods given in this book are not intended to be a complete or thorough treatment of the topics they deal with. All of the ideas can probably be developed further or applied in other areas, and all can doubtless be improved upon. The mathematician will also observe a certain lack of rigor as an attempt has been made to make the material intelligible to as wide a readership as possible.

—Shyam Priya

1

Vedic Mathematicians in the Celestial Timekeepers

Introduction

The Ancient Vedics seemed to have an obsession for precision as well as a fascination for large numbers. A combination such as this makes an excellent prerequisite for time keeping and for devising a useful and practical calendar. So, they turned to the sky and began to decipher the meaning behind the various cycles they observed. Let us see how they went about developing a calendar that would convey a lot of information merely by knowing the day of the month after constant observation of the sky both during the day and the night over centuries.

The basic information they used for purposes of time keeping were the motions of the sun and the moon relative to the earth. So far nothing unusual, as did all the other ancients. The cycles they used apart from the day, the week, the fortnight, and the month are shown in Table.

Some Definitions

Let us establish the coordinate systems first. Everyday the celestial sphere appears to turn as the earth rotates, causing the daily rising and setting of the sun, stars and other celestial objects.

1. 60 year Jovian cycle/360 year 'divine cycle
2. 2700 year cycle of the Sapta Rishi or the Ursa Major
3. 27000 year cycle of the asterisms called the Great Year or the precession cycle
4. 432,000 year cycle called a yuga (= duration of Kaliyuga)
5. 4,320,000 year cycle known as the Maha Yuga
6. Kalpa, the cycle consisting of 4.32*10**9 years

Cosmology and Numerology

A bit of trivia - all of the numbers in Table 1 are divisible by 9 except the 60 year cycle. The number 9 and its multiples have a mystic significance in the Vedic tradition.

One way of visualizing the number 9 is as follows

The universe is constituted of 3 factors - time, space, and causation

The universe is constituted of 3 Gunas (ingredients) — Sattva, rajas, and tamas

The universe is constituted of the three functions — creation, preservation, and destruction

Thus 3 times 3 makes 9

Two times 9 makes 18, the number of chapters in the Bhagawad Gita

The 18 portions (parvas) of the Mahabharata epic define in detail the career of man on earth.

Thee are 18 days of warfare in the Great Bharata War

There are a total of 18 divisions in the Mahabharata war. 7 divisions on the Pandava side and 11 on the Kaurava side.

The Mahabharata war is thus an exposition of the human possibilities and achievements graded into eighteen categories, the first multiple of 9.

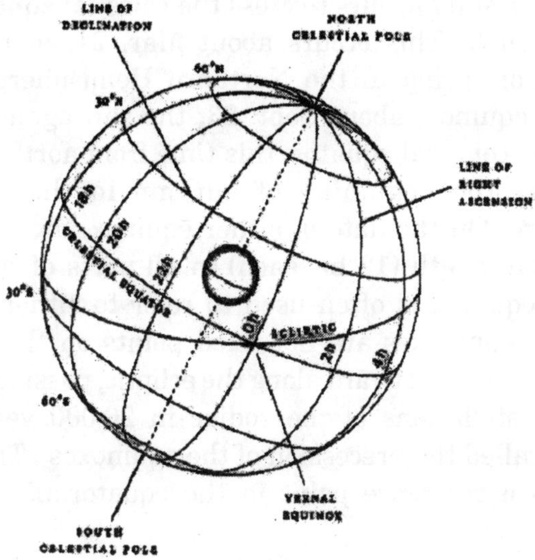

Figure 1 The celestial sphere showing the ecliptic and its inclination to the celestial equator

Ecliptic

The great circle on the celestial sphere that lies in the plane of the earth's orbit (called the plane of the ecliptic). Because of the earth's yearly revolution around the sun, the sun appears to move in an annual journey through the heavens with the ecliptic as its path. The ecliptic is the principal axis in the ecliptic coordinate system . The two points at which the ecliptic crosses the celestial equator are the equinoxes. The obliquity of the ecliptic is the inclination of the plane of the ecliptic to the plane of the celestial equator, an angle of about 23 1/2°. The constellations through which the ecliptic passes are the constellations of the zodiac.

Equinox (Vasanth Sampat) Vernal Equinox

Either of two points on the celestial sphere where the ecliptic and the celestial equator intersect. The vernal equinox, also known as "the first point of Aries," is the point at which the sun appears to cross the celestial equator from south to north. This occurs about Mar. 21, marking the beginning of spring in the Northern Hemisphere. At the autumnal equinox, about Sept. 23, the sun again appears to cross the celestial equator, this time from north to south; this marks the beginning of autumn in the Northern Hemisphere. On the date of either equinox, night and day are of equal length (12 hr. each) in all parts of the world; the word equinox is often used to refer to either of these dates. The equinoxes are not fixed points on the celestial sphere but move westward along the ecliptic, passing through all the constellations of the zodiac in 26,000 years. This motion is called the precession of the equinoxes . The vernal equinox is a reference point in the equatorial coordinate system.

Equatorial Coordinate System

the most commonly used astronomical coordinate system for indicating the positions of stars or other celestial objects on the celestial sphere. The celestial sphere is an imaginary sphere with the observer at its centre. It represents the entire sky; all celestial objects other than the earth are imagined as being located on its inside surface. If the earth's axis is extended, the points where it intersects the celestial sphere are called the celestial poles; the north celestial pole is directly above the earth's North Pole, and the south celestial pole directly above the earth's South Pole.

The great circle on the celestial sphere halfway between the celestial poles is called the celestial equator; it can be thought of as the earth's equator projected onto the celestial sphere. It divides the celestial sphere into the northern and southern skies. An important reference point on the celestial equator is the vernal equinox , the point at which the sun crosses the celestial equator in March. To designate the position of a star, the astronomer considers an imaginary great circle passing through the celestial poles and through the star in question. This is the star's hour circle , analogous to a meridian of longitude on earth. The astronomer then measures the angle between the vernal equinox and the point where the hour circle intersects the celestial equator. This angle is called the star's right ascension and is measured in hours, minutes, and seconds rather than in the more familiar degrees, minutes, and seconds. (There are 360 degrees or 24 hours in a full circle.) The right ascension is always measured eastward from the vernal equinox. Next the observer measures along the star's hour circle the angle between the celestial equator and the position of the star. This angle is called the declination of the star and is measured in degrees, minutes, and seconds north or south of the celestial equator, analogous to latitude on the earth. Right ascension and declination together determine the location of a star on the celestial sphere. The right ascensions and declinations of many stars are listed in various reference tables published for astronomers and navigators. Because a star's position may change slightly (see proper motion and precession of the equinoxes), such tables must be revised at regular intervals. By definition, the vernal equinox is located at right ascension 0 h and declination 0°.

Another useful reference point is the sigma point, the point where the observer's celestial meridian intersects the celestial equator. The right ascension of the sigma point is equal to the observer's local sidereal time . The angular distance from the sigma point to a star's hour circle is called its hour angle ; it is equal to the star's right ascension minus the local sidereal time. Because the vernal equinox is not always visible in the night sky (especially in the spring), whereas the sigma point is always visible, the hour angle is used in actually locating a body in the sky.

Calendars and Tithis

Like most Asian calendars Indian calendars do not employ the solar year and day (i. e. tropical year and solar day) but the sidereal year, and the Synodic month(29.5306 days). Thus, the calendric year based on the sidereal year is defined as the time between two successive passes of the sun through a certain star's circle of declination. Lunar days and sidereal months are also used, and in certain lunisolar calendars lunar year and lunar month are taken into account, too.

Astronomical knowledge of Ancient India was written down in scientific treatises, called Siddhantas. In them, values for the lengths of months and years were given representing the latest knowledge at the time the Siddhanta was written. The values range from 365.258681 days in the Âryabhatiya to 365.258756 days in the Surya Siddhanta and are all too long compared with the modern sidereal year length of 365.25636 days. Nevertheless they are still in use for Indian calendars today.

The sidereal month is about two day shorter (27.3217) than the Synodic month.

Meaning of Tithi

According to the Indian calendar or Panchanga, Tithi is a lunar date based on the rotation of the moon around the earth, and is one of the five important aspects of an Indian almanac (Panchanga — Panch means five and anga means parts). Most of the Indian social and religious festivals are celebrated on a date corresponding to the original Ṭithi.

The current calendar "date" that we are so familiar with in our daily life is heliocentric and is based on the rotation of the earth around the sun. It takes the earth approximately 365 ¼ days to complete its rotation around the Sun. The calendar that most of us use today divides the 365 days of earth's period of rotation around the Sun in twelve months. The leap year, which occurs once every four years, accounts for ¼ day per year.

Similar to the solar calendar, the lunar calendar is also popular and widely used in the Asian countries such as

China, Pacific-rim countries, Middle East countries, and India. The lunar calendar, which is believed to have originated in India, has been around for a very long time, even long before the solar calendar.

The lunar calendar is geocentric and is based on the moon's rotation around the Earth. The lunar month corresponds to one complete rotation of the Moon around the Earth. Since this period of rotation of moon around the earth varies, the duration of lunar month also varies. On average, the lunar month has about 29 ½ days, the period of the lunar Synodic orbit. In addition to moon's rotation around the earth, the lunar year is based on earth's rotation around the Sun. In general, the lunar year has twelve lunar months of approximately 354 days (29.5 *12), thus making it shorter by about 11 days than the solar year. However, the lunar calendar accounts for this difference by adding an extra lunar month about once every 2 ½ years. The extra lunar month is commonly known as "Adhik Mas" in India (Adhik means extra and the Mas means month). The concept of this extra month is similar to the "Blue Moon" in the West, which occurs almost with the same frequency of 2 ½ years.

The Indian lunar year begins on the new moon day that occurs near the beginning of the Spring season. The twelve lunar months are:

Chaitra

Vaishakh

Jeshta

Ashadh

Shrawan(Sawan)

Bhadrapad(Bhado)

Ashwin

Kartik

Margshirsh

Paush

Magha

Falgoon (Fagan)

As mentioned earlier, to account for the difference between the solar and lunar year an extra lunar month occurs about every 2 ½ years as "Adhik Mas".

According to the Moslem calendar which is widely followed in Middle East and in other Moslem countries the lunar year is strictly based on twelve lunar months of 354 days per year. That's why their holy month of Ramadan occurs by approximately 11 to 12 days earlier than that in the preceding year.

The solar day (commonly referred as the "the date" in western calendar) has a fixed length of 24 hours. The change of date occurs at midnight as per local time or standard time of a given local time zone. Thus, the date changes from midnight to midnight. Similarly the day (as in weekdays) changes from midnight to midnight as per local or standard time for that location. In other words, as per the western (or English) calendar the length of day and date is exactly 24 hours, and there is a definite correspondence between the date and the corresponding day of the week.

A lunar day usually begins at sunrise, and the length of lunar day is determined by the time elapsed between the successive sunrises. As per the Jewish calendar their lunar day begins at the sunset, and lasts through the next sunset. A lunar day is essentially the same as a weekday. In India the lunar day is commonly referred as "War". Just as the English calendar has seven days for a week, the Indian calendar has seven wars for a week. Thus,

The lunar day, however, varies approximately between 22 to 26 hours based on the angular rotation of moon around the earth in its elliptical orbit. In the Indian calendar, the lunar date is referred as "Tithi". The basis for the length of a lunar date is geocentric and is defined as the angular distance between the sun and the moon as seen from the earth. As the moon rotates around the earth, the relative angular distance between the sun and the moon as seen from the earth increases from 0 degrees to 360 degrees. It takes one lunar month or about 29 ½ solar days for the angular distance between the sun and the moon to change from 0 to 360 degrees. When the angular distance reaches zero, the next lunar month begins. Thus, at the new moon a lunar month begins, at full moon, the angular distance between the sun and the moon as seen from the earth becomes exactly 180 degrees.

English calendar weekdays	***Indian calendar weekdays***
Sunday	
Monday	
Tuesday	
Wednesday	
Thursday	
Friday	
Saturday	Raviwar
Somwar (Chandrawar)	
Mangalwar	
Budhwar	
Guruwar	
Shukrawar	
Shaniwar	

The lunar cycle begins with crescent moon and the crescent phase lasts till that phase culminates in the full moon, typically lasting for about 15 days. Then the moon enters in the waning phase until it disappears from the sky by lining up with the Sun. The waning phase also lasts for about 15 days. According Indian lunar month, the crescent lunar phase fortnight is called as "Shudha or Shukla Paksha" and the waning phase of the lunar cycle fortnight as "

Krishna Paksha". Thus, during Shudha (or Shukla) Paksha the angular distance between the moon and the sun varies from 0 degrees to 180 degrees while that during the Krishna Paksha from 180 to 0 degrees. If we divide 180 degrees into 15 equal parts, then each part becomes of 12 degrees in length. Thus, this each twelve-degree portion of angular distance between the moon and the sun as it appears from the earth is the lunar date or Tithi. Tithis or lunar dates in Shudha (or Shukla) Paksha begin with Prathama (first), Dwitiya (second), etc., till we reach the Poornima, the lunar date for full moon day. Similarly for the waning fortnight lunar cycle or Wadya (or Krushna) Paksha, this begin again with Prathama (first), Dwitiya (second), etc., till we arrive Amavasya or a day before the new moon. Thus when we refer to Ramnavami (the birthday of Rama), it's the Navami (ninth lunar day) of Shudha Paksha of the lunar month Chaitra, or Chaitra Shudha Navami. Similarly, the Gokulashtmi (also called as Janmashtami, the birthday of Krishna) occurs on Shrawan Wadya Ashtami (eighth lunar day of Wadya Paksha of the lunar month Shrawan).

The angular velocity of moon in its elliptical orbit around the earth varies continuously as it is affected (according to Kepler's Law) by the relative distance between the earth and the moon, and also by the earth's relative distance from the sun. As a result, the daily angular speed (the speed of the angular change between the moon and the sun as seen from the earth) varies somewhere between 10 to 14 degrees per day. Since the length of a Tithi corresponds to 12 such degrees, the length of a Tithi also varies accordingly. Therefore, a Tithi can extend over one day (24 hour-period) or it can get sorteneded if two Tithis occur in one 24 hour day.

Since the angular distance between the moon and the sun as referred here is always relative to the entire earth, a lunar day or Tithi starts the same time everywhere in the world but not necessarily on the same day. Thus, when a certain Tithi starts at 10:30 PM in India it also begins in

New York at the same time, which is 12 PM (EST) on the same day. Since the length of a Tithi can vary between 20 to 28 hours, its correspondence to a War (a weekday) becomes little confusing.

As per the Indian calendar, the Tithi for a given location on the earth depends on the angular distance between the moon and the sun relative to the earth at the time of sunrise at that location. Thus, for instance, assume on a November Monday sunrise in New York city occurs 8:30 AM (EST). Further assume that at 9 AM (EST) on Monday the angular distance between the sun and moon is exactly 12 degrees just following the new moon of the Indian lunar month Kartik. Since the length of a tithi is 12 degrees, the tithi, Kartik Shudha Dwitiya (second day) begins exactly at 9 AM on Monday of that November in New York. However, at the time of sunrise on that Monday the tithi Dwitiya has not begun. Therefore, the tithi for that Monday for city of New York is Kartik Shudha Prathama (first day). On the same Monday morning the sunrise in Los Angeles occurs well past 9 AM (EST). Since the Tithi Dwitiya occurs everywhere in the world at the same instant, therefore, for Los Angeles, the Tithi for that Monday would be Karthik Shudha Dwitiya.

For the same Monday at 9 AM (EST), it would be 7:30 PM in Mumbai or New Delhi. Thus, Tithi for that Monday for city of New York, Mumbai, and New Delhi is Karthik Shudha Prathama (the first day of Indian lunar month Karthik) while for most of the regions west of Chicago or St. Louis the Tithi for that Monday is Dwitiya. In other words, the Tithi Karthik Shudha Prathama for regions west of Chicago or St. Louis should occur on the preceding day, the Sunday.

Karthik Shudha Prathama (the first day of Indian lunar month Karthik) also happens to be the first day after Diwali. Most of the Indians celebrate this as their New Year 's Day. Indians living in India, Europe, and eastern part of the

United States thus should celebrate their New Year on that Monday while regions west of Chicago should celebrate on the preceding day, the Sunday. (Based on description by Jagdish C. Maheshri, October 12, 2000).

Adhik Mas Occurs only when Two Amavasyas

1	Pratipat	Pratipat	The presiding deity of the first lunar day in Brahma and is good for all types of auspicious and religious ceremonies
2	Dvitiya	Dvitiya	Vidhatr rules this lunar day and is good for the laying of foundations for buildings and other things of a permanent nature.
3	Trtiya	Trtiya	Visnu is the lord of this day and is good for the cuttings of one's hair and nails and shaving.
4	Caturthi	Caturthi	Yama is lord of the 4th lunar day, which is good for the destruction of one's enemies, the removal of obstacles, and acts of combat.
5	Pancami	Pancami	The Moon rules this day, which is favourable for administering medicine, the purging of poisons, and surgery.
6	Sasti	Sasti	Karttikeya presides over this day and is favourable for coronations, meeting new friends, festivities, and enjoyment.
7	Saptami	Saptami	The 7th lunar day is ruled by Indra; one may begin a journey, buy conveyances, and deal with other such things as a movable nature.
8	Astami	Astami	The Vasus rule this day, which is good for taking up arms, building of one's defenses, and fortification.
9	Navami	Navami	The Serpent rules this day, with is suitable for killing enemies, acts of destruction, and violence.

10	Dasami	Dasami	The day is ruled by Dharma and is auspicious for acts of virtue, religious functions, spiritual practices, and other pious activities.
11	Ekadasi	Ekadasi	Rudra rules this day; fasting, devotional activities, and remembrance of the Supreme Lord are very favourable.
12	Dvadasi	Dvadasi	The Sun rules this day, which is auspicious for religious ceremonies the lighting of the sacred fire, and the performance of one's duties.
13	Trayodasi	Trayodasi	The day is ruled by Cupid and is good for forming friendships, sensual pleasures, and festivities.
14	Caturdasi	Caturdasi	Kali rules this day suitable for administering poison and calling of elementals and spirits.
15	Amavasya (new moon)	Purnima (full moon)	The Vasve-devas rule the New Moon suitable for the propitiation of the Manes and performance of austerities.

Zodiac sign	*Sanskrit Name*	*a*	*Sector end*
Aries	Mesha	00	30
Taurus	Vrishabha	30	60
Gemini	Mithuna	60	90
Cancer	Karka	90	120
Leo	Simha	120	150
Virgo	Kanya	150	180
Libra	Tula	180	210
Scorpio	Vrishchika	210	240
Sagittarius	Dhanus	240	270
Capricorn	Makara	270	300
Aquarius	Kumbha	300	330
Pisces	Meena	330	360

The Tropical Zodiac

Ecliptic, Tropical Zodiac and the Sidereal Zodiac

9 degrees to either side of the Ecliptic is a belt of the Heavens known as the Zodiac. (Dante called it the Oblique Line that beareth all planets).

First 30 degrees of the Zodiac constitute the sign of Aries. The next 30 degrees Taurus and so on. The Zodiac counted from the first degree of Aries to the 360th degree of Pisces is called the Tropical Zodiac.

These 12 signs are the limbs of the Cosmic Man or Time Eternal (Kalapurusha- The Almighty Self as Time).

Aries is His head, Taurus His face, Gemini His neck, Cancer His heart, Leo the place beneath, Virgo His belly, Libra His generative organs, Scorpio the place beneath, Sagittarius His upper thigh, Capricorn his lower thigh, Aquarius His leg and Pisces His feet!

The Clock, the Sidereal Zodiac, Nakshatras, and the Precession of the Equinoxes

The basis of the Hindu calendar calculation is Vedic. This calendar has been modified and elaborated, but because it is based on the stars (Nakshatras) visible to the naked eye, and on the visible Lunar phases, it is more accurate than any others of the past. The actual moments when Lunar months begin can easily be checked by the regular appearances of Solar eclipses, and the middle moment of a Lunar month — Poornima or full moon -- can similarly be verified by the more frequent Lunar eclipses. Hence the Hindu calendar, not requiring special instruments for its rectification, has maintained great accuracy for thousands of years.

The oldest calendar is probably the Vedic among the languages referred to as i.e languages; at first lunar, later with solar elements added to it. The sister Avesta calendar is similarly first Lunar, but later only Solar. Both these calendars (the oldest in the i.e universe) are influenced by the prehistoric calendars of the first and second root races at the North Pole and its surroundings, as they reckon with days and nights lasting six months.

For untold ages, the Hindus have observed the motion of the moon, the sun and the seven planets along a definite path that circles our sky and is marked by fixed clusters of stars. The moon afforded the simplest example. These early astronomers observed that the moon, moving among these fixed star constellations which they called Nakshatras, returned to the same Nakshatra in 27.32166 days, the exact quantity determined by Aryabhatta, thus completing one Nakshatra month. They found it convenient to divide these groups of stars into 27 almost equal sections, or the 27 Nakshatras. By this method of reckoning, instead of giving the date of a month, as Western calendars do, the Hindus gave the name of the Nakshatra in which the moon was to be seen. (The moon is in each of these Nakshatras for approximately one day plus eighteen minutes.)

This scheme fitted nicely with the sun's cycle, for the Hindus noted that the sun traversed the same circle through the sky, but that it returned to its starting place only after 365.258756481 days, or what we call a Solar Sidereal Year. (Modern figures based on this Hindu figure quote 365.2596296 days -- a distinction without a difference, for ordinary purposes.) Now, having already divided the month into the 27 Nakshatras for the convenience of reckoning the moon's voyage through the heavens, what more natural than that these same Nakshatras should serve for the study of the Sun's course? Being in a circle of 360 degrees, each Nakshatra takes up 13 1/3 degrees of that circle. The Sun, moving about 1 degree in a day, is seen for 13 1/3 days in each Nakshatra. The system of reckoning according to the moon Nakshatras is current today that of the sun's being uncommon.

During the course of one day, the earth has moved a short distance along its orbit around the sun, and so must rotate a small extra angular distance before the sun reaches its highest point. The stars, however, are so far away that the earth's movement along its orbit makes a generally negligible difference to their apparent direction (see, however parallax), and so they return to their highest point in slightly less than 24 hours. A mean sidereal day is about 23h 56m in length. Due to variations in the rotation rate of the Earth, however, the rate of an ideal sidereal clock deviates from any simple multiple of a civil clock. The actual period of the Moon's orbit as measured in a fixed frame of reference is known as a Sidereal month, because it is the time it takes the Moon to return to the same position on the celestial sphere among the fixed stars (Latin: sidus): 27.321 661 days (27 d 7 h 43 min 11.5 s) or about 27 days. This type of month has appeared among cultures in the Middle East, India, and China in the following way: they divided the sky in 27 or 28 lunar mansions or Nakshatras, characterized by asterisms (apparent groups of stars), one for each day that the Moon follows its track among the stars.

The basis of the Hindu calendar calculation is Vedic. This calendar has been modified and elaborated, but because it is based on the stars (Nakshatras) visible to the naked eye, and on the visible Lunar phases, it is more accurate than any others of the past. The actual moments when Lunar months begin can easily be checked by the regular appearances of Solar eclipses, and the middle moment of a Lunar month — Purnima or full moon -- can similarly be verified by the more frequent Lunar eclipses. Hence the Hindu calendar, not requiring special instruments for its rectification, has maintained great accuracy for thousands of years.

The oldest calendar is probably the Vedic among the languages referred to as i.e languages; at first lunar, later with solar elements added to it. The sister Avesta calendar is similarly first Lunar, but later only Solar. Both these calendars (the oldest in the IE universe) are influenced by the prehistoric calendars of the first and second root races at the North Pole and its surroundings, as they reckon with days and nights lasting six months.

For untold ages, the Hindus have observed the motion of the moon, the sun and the seven planets along a definite path that circles our sky and is marked by fixed clusters of stars. The moon afforded the simplest example. These early astronomers observed that the moon, moving among these fixed star constellations which they called Nakshatras, returned to the same Nakshatra in 27.32166 days, the exact quantity determined by Aryabhatta, thus completing one Nakshatra month. They found it convenient to divide these groups of stars into 27 almost equal sections, or the 27 nakshatras. By this method of reckoning, instead of giving the date of a month, as Western calendars do, the Hindus gave the name of the Nakshatra in which the moon was to be seen. (The moon is in each of these Nakshatras for approximately one day plus eighteen minutes.)

This scheme fitted nicely with the sun's cycle, for the Hindus noted that the sun traversed the same circle through the sky, but that it returned to its starting place only after 365.258756481 days, or what we call a Solar Sidereal Year. (Modern figures based on this Hindu figure quote 365.2596296 days -- a distinction without a difference, for ordinary purposes.) Now, having already divided the month into the 27 nakshatras for the convenience of reckoning the moon's voyage through the heavens, what more natural than that these same Nakshatras should serve for the study of the Sun's course? Being in a circle of 360 degrees, each Nakshatra takes up 13 1/3 degrees of that circle. The Sun, moving about 1 degree in a day, is seen for 13 1/3 days in each nakshatra. The system of reckoning according to the moon Nakshatras is current today, that of the sun's being uncommon.

In brief, then, the earliest method, the Vedic, of counting, was to name the moon through the various Nakshatras — the circle or cycle repeating itself each Sidereal-Star-Month. Later the sun's place in the same Nakshatras was noted, the year ending when the Sun returned to the same Nakshatra. Then came the noting of the Solar and Lunar eclipses, and the observance of the New and Full Moons divided the month into the two phases of waxing and waning Moon, the month beginning at the moment of New Moon. This is how the Hindus reckon today, the month taking its name from the Nakshatra in which the Full Moon is seen each month. The Full Moon being exactly opposite the Sun, the Solar nakshatra bears the same name as the Lunar month six months ahead, while each Lunar month bears the same name as the 14th Solar Nakshatra ahead.

The Western student faced with these unfamiliar calculations may echo the old Persian proverb, "Why count big numbers and small fractions, when they are all amassed in 1?" But the Hindu looks on these figures from another point of view—he lives with them, and among them, and by them, much of the time. Consider a Sanskrit sloka (verse) about the Savati or pearl nakshatra, which marks the new season after the monsoon is over. The sloka says, "If in the Swati a rain drop falls into the sea, that drop becomes a pearl." This may sound foolish, for the peasant, though he live in the depth of the interior of India, knows that pearls come from the sea—even if he does not necessarily understand that these pearls grow inside the oyster. He does know, however, that if it rains at this period of the year, his crops will yield great wealth. And the pearl is synonymous with wealth among people who, if they have any money, invest it in jewellery, especially gold and pearls, rather than in the banks. (Poetically, rice, their staple food)

To summarize, the earth revolves around the Sun once in 365 days 5 hours 48 minutes and 46 seconds. Considered from the earth, the Sun appears to complete one round of the ecliptic during this period. This is the Tropical year. In the span of a tropical year, the earth regains its original angular position with the Sun. It is also called the Year of seasons since the occurrence, and timing, of seasons depends on the rotation of the earth around the sun. If, for example, we consider the revolution of the Sun around the earth from one vernal equinox (around 21st March, when the day and night all over the globe are equal) to the next vernal equinox, it takes one tropical year to do so.

However, if at the end of a tropical year from one vernal equinox to the next, we consider the position of the earth with reference to a fixed star of the zodiac, the earth appears to lie some 50.26 seconds of celestial longitude to the west of its original position. In order for the earth to attain the same position with respect to a fixed star after one revolution, it takes a time span of 365 days 6 hours 9 minutes and some 9.5 seconds. This duration of time is called a sidereal year. The sidereal year is just over 20 minutes longer than the tropical year; this time difference is equivalent to 50.26 seconds of celestial longitude.

Each year, the Vernal equinox will fall short by 50.26 seconds along the zodiac reckoned along the fixed stars. This continuous receding of the Vernal equinox along the zodiac is termed the Precession of the Equinoxes and it takes about 25776 years to make one complete revolution of the precessional motion of the earth's axis. Hipparchus regarded as the discoverer of the precession of the equinoxes in the west gave us either 28,000 or 28,173 years for one revolution.. Another figure given is 25,920 years for the precession cycle, These figures indicate that the mean value of 27,000 years given in the Vedic scriptures is reasonable. The precession of the equinoxes has proved to be very useful for dating certain events in Vedic and Post Vedic times.

There are only a few methods, by which we can determine the age of an event in the absence of radiocarbon dating which is not as precise as the astronomical clocks.

1. Use the Precession of the equinoxes to determine the Nakshatra in which the Vernal equinox occurs in a particular Nakshatra. If, we recall there are 27 Nakshatras, it follows that the vernal equinox occurs in a different Nakshatra, once every 1000 years.
2. Use the statements made in the texts to check for internal consistency. If for example Aryabhatta uses a place value system, the zero must have been in fairly wide use by then. If further he uses classical sanskrit (codified by Panini then he must have lived after Panini.

	Western Zodiac name	*Indian Nakshatras (Sidereal Zodiac)*	*Sector in deg,min deg, min*
1.	Beta Arietis	Aswini	00 00 13 20
2.	41 Arietis	Bharani	13 20 26 40
3.	Eta Tauri	Karthika	26 40 40 00
4.	Alpha Tauri	Rohini	40 00 53 20
5.	Lamda Orionis	Mrigasira	53 20 66 40
6.	Alpha Orionis	Aridra	66 40 80 00
7.	Beta Geminorum	Punarvasu	80 00 93 20
8.	Delta Cancri	Pushya	93 20 106 40
9.	Alpha Cancri	Aslesha	106 40 120 00
10.	Alpha Leonis	Magha	120 00 133 20
11.	Delta Leonis	Pubba	133 20 146 40
12.	Beta Leonis	Uttara	146 40 160 00
13.	Delta Corvi	Hasta	160 00 173 20
14.	Alpha Virginis	Chitra	173 20 186 40
15.	Alpha Bootis	Chothi	186 40 200 00
16.	Beta Librae	Vishakam	200 00 213 20
17.	Delta Scorpi	Anuradha	213 20 226 40
18.	Alpha Scorpi	Jyeshta	226 40 240 00
19.	Lamda Scorpi	Moola	240 00 253 20
20.	Delta Sagittari	Poorvashad	253 20 266 40
21.	Delta Sagittari	Uthrashad	266 40 280 00
22.	Alpha Aquilae	Sravana	280 00 293 20

Contd...

Western Zodiac name	*Indian Nakshatras (Sidereal Zodiac)*	*Sector in deg,min deg, .nin*
23. Alpha Delphini	Dhanishta	293 20 306 40
24. Lamda Aquar	Satabhisha	306 40 320 00
25. Alpha Pegasi	Poorvabhadra	320 00 333 20
26. Alpha Andromeda	Uttrarabhadra	333 20 346 40
27. Zeta Piscium	Revathi	346 40 360 00

Values for the Lunar sidereal orbit and the Lunar Synodic orbit are given in Table below

Comparisons	*Lunar sidereal orbit*	*Lunar synodic orbit*
AD 2000.0	27.32166156	29.53058888
AD 498	27.3216638	29.530591
Aryabhatta	27.321668	29.530582
Paulisa Siddhanta	27.321673	29.530587
1604 BC	27.321668	29.530595
ASTRONOMIC		
AUTHORITY	Aryabhatta	
(from Clarke and Kay)		Surya
Siddanta		
Years in Cycle	4,320,000	4,320,000
Rotations	1,582,237,500	1,582,237,828
Days	1,577,917,500	1,577,917,828
Lunar Orbits	57,753,336	57,753,336
Kay notes 57,753,339 lunar orbits rather than 57,753,336 per Clarke.		
Synodic Months	53,433,336	53,433,336
Mercury	17,937,920	17,937,060
Venus	7,022,388	7,022,376
Mars	2,296,824	2,296,832
Jupiter	364,224	364,220
Saturn	146,564	146,568

How old is the Universe, Kalachakra and the Yuga Concept, Hindu Cosmological Time Frames

The Hindu Calendar (also known as the Panchanga) currently in practice reckons time in terms of very large cycles called Kalpa (4.32 billion years) consisting of 14 Manvantaras (Manvantara or age of Manu,~ 308 million years). A Manvantara is made up of Mahayugas (Mahayuga= great yuga consists of 4 yugas: Krita, Treta, Dwapara and Kali). Kali yuga is equivalent to 432,000 years and 1 Mahayuga= 4.32 million years. This system appears to have been in use since the days of the Epics and Puranas, and attested in the Siddhantas. However, the earliest Vedic Calendar was based on a cycle also called yuga, but consisting of only five years. This ancient Vedic Calendar was a Luni-solar calendar and used two intercalary months in a five year period and has often been criticized as being very crude.

First we have Kalpa, a day in Brahma's 'life' or 4320 million earthly years, and a night of equal length. During the day he creates and during the night he absorbs to begin the cycle each Brahma day . Each kalpa is divided into 14 Manvantaras or 308.448 million years we are supposed to be in the seventh Manvantara of Vaivasvata Manu. Each Manvantara contains 71 Mahayugas , plus 1Krtayuga ,and each Mahayuga is divided into 4 yugas - Krta, Treta, Dvapara and Kali of 4800, 3600, 2400 and 1200 divine years of the Gods, each of which = 360 human years. We are at present in the Kali yuga which began in 3102 BCE the traditional year of the Mahabharata war .

Thus, we have a day in Brahma's life of 1 Kalpa

1 Brahma Day (day and night) = 2 Kalpa

1 Kalpa = 4,320,000,000 earthly years (Y)

= 14 Manus + 1Kritayuga = 1000 MY =14*71.4+.4

Kaliyuga = 432,000 Y = 1KY

= 1200 divine years (DY) = 1 Yuga

1 DY = 360 Y

Dwapara = 864,000 Y = 2KY = 2400 DY

TretaYuga = 1,296,000 Y = 3KY = 3600 DY

Kritayuga = 1,728,000 Y =4 KY = 4800 DY

= 0.4 MY =.4/71.4 = 5.6022408964e-3

Mahayuga (MY) = 4,320,000 earthly years

= 10 KY = 12000 DY

1Manvantra (M) = 71 MY = 306.72 million years

1 Manu = 1M + 1 KritaYuga = 308.448 million years

= 856,800 DY

1 Kalpa = 14 Manus + 1Krita Yuga = 14*71.4 +.4

= 1000 MY = 12,000,000 DY = 4.32 billion Y

So How Old is the Universe

As of Vaisakhapratipada of 2006 CE, May 1 we are in the second quarter of Brahma's day, called Shewtavarah Kalpa, seventh Manvantara named Vaivasvata and entered into the first quarter of the 28th Kaliyuga. Already 5107 years of this 28th KY have passed, so the time elapsed in this Kalpa is

6 M	=1,850,688,000 Y (6*308,448,000)
	= 6 M
And 27 MY	= 116,640,000 Y (27 * 4,320,000)
	= 27/71.4 = 0.3781512605 M
And 28th	
(Krita+Treta +Dwapara	= 3,888,000 Y (9*432,000)
	=0.9 MY =.9/71.4 = 0.012605042
M	
5107 Y of Kaliyuga)	= 5107 Y
So the current year 2006 CE	= 1 ,971,221,108 Y

The 12 signs of the Zodiac with Sanskrit names are mentioned In Brihat Samhita and Laghu Bhaskariyam. The former is the work of Varahamihira 505 CE. He is supposed to have borrowed it from a Greek of the 4th century CE. (Could it be Hipparchus)

To conclude this brief acquaintance with Vedic astronomy, we want to draw attention to the possible presence in the Rg-Veda of a momentous cultural artifact, the origin of which is usually situated in Babylonia in about 600 BC: the twelve-sign Zodiac. In RV 1:164:11, the sun wheel in heaven is said to have 12 spokes, and to be subdivided into 360 pairs of "sons": the days (consisting of day and night), rounded off to an arithmetically manageable number, also the basis of the "Babylonian" division of the circle in 3600. The division in 12 already suggests the Zodiac, and we also find, in the footsteps of N.R. Waradpande, that a number of the Zodiacal constellations/rashis (classically conceived as combinations of 2 or 3 successive Lunar mansions or Nakshatras of 13 ° and 20' each) are mentioned. Obviously the Rg should be dated prior to the beginning of Kaliyuga, as we have already demonstrated and hence the Babylonian origin of the twelve sign Zodiac is suspect.

India as a Creative Civilization

The idea of India as an imitative civilization that borrowed everything is based on colonial stereotyping and obsolete theories.

Unfounded Assumptions

It is a widespread belief, especially in the West, that Indian civilization developed mainly by borrowing from others and not through innovation and creative endeavour. This, however, is a recent view, one that can be traced to the motivated scholarship of the European colonial period and its immediate aftermath. The reality is that throughout history, except during the colonial period, peoples of all nations have looked to India as the most creative and original of civilizations. To be specific, medieval and ancient scholars from Arabia, Spain, China and even Greece- all acknowledged their indebtedness to Indian science. For example, a medieval Arab scholar Sa'id ibn Ahmad al-Andalusi (1029-1070) wrote in his Tabaqat al-'umam, one of the earliest books on history of sciences:

> "*The first nation to have cultivated science is India. ... India is known for the wisdom of its people. Over many centuries, all the kings of the past have recognized the ability of the Indians in all the branches of knowledge.*

"The kings of China have stated that the kings of the world are five in number and all the people of the world are their subjects. They mentioned the king of China, the king of India, the king of the Turks, the king of the Persians, and the king of the Romans.

> "*... They referred to the king of India as the "king of wisdom" because of the Indians' careful treatment of 'ulum [sciences] and all the branches of knowledge.*

"The Indians, known to all nations for many centuries, are the metal [essence] of wisdom, the source of fairness and objectivity. They are people of sublime pensiveness, universal apologues, and useful and rare inventions.

> "*... To their credit the Indians have made great strides in the study of numbers and of geometry. They have acquired immense information and reached the zenith in their knowledge of the movements of the stars [astronomy] ... After all that they have surpassed all other peoples in their knowledge of medical sciences ...*"

When the necessary allowance is made for the exuberance of the writer and even some exaggeration, it is clear that no one until the coming of the modern Europeans (and their Indian disciples) questioned the antiquity of Indian science.

In his book al-Andalusi goes on to give details of several Indian texts on astronomy and tells us that the Arab scholars used them in preparing their own almanacs.

Not only Medieval Arabs, even some early Christian scholars recognized Indian contributions. Writing in 662 AD, when the Byzantine Empire was its height and it was thought that there was no knowledge beyond Greek knowledge, Sebokht, the Bishop of Qinnesrin in North Syria observed:

> *"I will omit all discussion of the science of the Hindus [Indians], a people not the same as Syrians, their subtle discoveries in the science of astronomy, discoveries more ingenious than those of the Greeks and the Babylonians; their valuable method of calculation [the decimal system]; their computing that surpasses description. I wish only to say that this computation is done by means of nine signs. If those who believe because they speak Greek, that they have reached the limits of science should know these things, they would be convinced that there are also others who know something."*

The reference of course is to the famous place decimal system using zero invented by the Hindus. (It is often called the Arabic numeral system, but the Arabs themselves called it the Indian system acknowledging their indebtedness to India.) In fact the Greek (and the Roman) method of computing and solving equations was cumbersome in the extreme when compared to the method used by Indians. Mathematics as we know today would hardly be possible without this invention- probably the greatest single advance in the history of mathematics.

India's Contributions : Material

This brings up a basic issue about India's contribution: while the world has by and large acknowledged it in the spiritual realm - like Yoga, Vedanta and Buddhism - there is widespread belief that the Indians were other worldly and did not pay attention to material progress. This is far removed from the truth. In addition to mathematics and astronomy (more of which later), India led the world in ferrous and non-ferrous metallurgy, until the industrial base was deliberately destroyed by the British. Indian steel was deemed the best in the world well into the eighteenth century. (Steel is an Indian invention.) Indian textiles have been prized throughout history, until it was suppressed by the British, though it is now being gradually regained.

This being the case, why does this image of India as largely an imitative civilization still persist? (This though is mainly in the West- not Asia.) Three reasons can be given: (1) colonial stereotyping; (2) Marxist domination of intellectual life in the post-colonial period; and (3) incompetence of scholars.

The colonial image is gradually eroding but the Marxist formulation of India as an imitative culture is the result of what Marx himself wrote about India: "India has no history. What is called history is only the record of successive intruders." Marx of course knew nothing about India, but his statement became sacrosanct to Marxist scholars who continue to dominate the intellectual scene in India as elsewhere.

This brings us to the third cause, scholarly incompetence. Scholars of the colonial period whose works still constitute standard references on India, were mostly bureaucrats and missionaries and lacked the scientific knowledge needed to appreciate true achievements. This is generally true of history of science. Einstein himself complained more than once that most historians of science were linguists who really did not understand the problems scientists were grappling with. It is hardly surprising that charges of lack of scientific capacity and originality on the part of Indians is made by those who are themselves scientifically ignorant- like nineteenth century

European theologians and linguists continuing down to their modern Indian followers. It is interesting to contrast their views with those of competent scientists who studied ancient India- that is to say the views of people who were competent to express opinions about science. This will help shed light on the true fact of prejudice and unfounded charges used as a fig leaf to conceal their own scientific ignorance.

Albrecht Weber was a leading nineteenth century Sanskrit scholar from Germany. At a time when the relationship between Indian and Greek mathematics was being seriously debated, he went on to assert that the Vedic mathematics was borrowed by the Indians from the Greeks following Alexander's invasion. His actual statement was that there was "nothing of a literary-historical nature standing in the way of the assumption of a use [by the Vedic mathematicians] of the teachings of Hero of Alexandria." Hero of Alexandria is now known to have been living in 62 AD! Even in Weber's time Hero was known to be later than 200 BC. Now of course we know that Vedic mathematics must have existed long before 2000 BC. This is clear from archaeology.

There is more to this that goes to highlight the creativity and originality of Indian mathematics. After more than twenty years of research, the distinguished American mathematician and historian of science A. Seidenberg showed that Greek, Egyptian and Old-Babylonian mathematics are derivatives of the Vedic mathematics known as the Sulbasutras. In 1962, he reached the following epoch making conclusion:

> "... *the elements of ancient geometry found in Egypt [c. 2050-1800 BC] and Babylonia [c. 1900-1750 BC] stem from a ritual system of the kind observed in the Sulbasutras.*"

Fifteen years later, in a famous paper, appropriately titled 'The Origin of Mathematics' Seidenberg elaborated further on his revolutionary discoveries: "The arithmetical tendencies here encountered [in Vedic mathematics] were expanded, and in connection with observations on the rectangle led to Babylonian mathematics.

> *"A contrary tendency, namely, a concern for exactness of thought (or the myth of its importance), together with a recognition that arithmetic methods are not exact, led to Pythagorian [Greek] mathematics."*

In other words, both of the world's great systems of mathematics, which are seen as the basis of Western civilization, namely, Greek and Egyptian-Babylonian, owe their existence to ancient Indian (Vedic) mathematics. In the face of this to argue that Indian civilization did not originate anything is a colossal misrepresentation based on obsolete and discredited theories. In fact, it is well known that the Chinese themselves heavily borrowed from India in fields like astronomy and mathematics.

It is a similar story when we turn to astronomy. William Dwight Whitney and other linguists of the colonial period claimed that Indian astronomy was based on the Greek- a position still reiterated by many scholars. A.B. Keith, another linguist of the colonial period gave out the view that Indians had borrowed their astronomy from some 'Semitic source' without specifying which one. It is illuminating to contrast Keith's (and Whitney's) statements with what Jean-Sylvain Baily, a leading French astronomer had to say about the astronomical achievements of the Hindus:

> "... *the motions of the stars calculated by the Hindus before some 4500 years vary not even a single minute from the tables of Cassine and Meyer [used in Europe in the nineteenth century]. The Indian tables give the same annual variation of the moon as that discovered by Tycho Brahe—a variation unknown to the school of Alexandria and also to the Arabs who followed the calculations of the school. ... The Hindu systems of*

astronomy are by far the oldest and that, from which the Egyptians, Greeks, Romans and even the Jews derive the knowledge."

This, it is worth noting, is by a great astronomer- not some colonial bureaucrat or missionary scholar who knew some Indian language. And this view was echoed by John Playfair, Astronomer Royal at the University of Edinburgh.

India's Contributions II: Spiritual

In the spiritual realm, no one seriously disputes India's contributions, though even these are often distorted. While it is unnecessary to go into details of yoga, meditation and others, it is worth noting that even these have potential for concrete applications. For example, the great linguist Panini gave the concept for meta-language and constructed one thousands of years before computer scientists began exploring the same idea. No one has been able to match him to this day. Similarly, Patanjali, in his Yogasutra, gives rules for testing the correctness of proofs of statements and proof theories centuries before mathematicians began to ponder about these fundamental problems. It is worth noting that one of the central problems of modern physics (Quantum Physics) is the problem of reality and consciousness. This was anticipated by Vedic seers thousands of years ago. The great physicist J. Robert Oppenheimer wrote:

"The general notions about human understanding... which are illustrated by discoveries in atomic physics are not in the nature of things wholly unfamiliar, wholly unheard of or new. Even in our own culture they have a history, and in Buddhist and Hindu thought a more considerable and central place. What we shall find [in modern physics] is an exemplification, an encouragement, and a refinement of old wisdom."

In this context it is worth emphasizing that India's contribution of Buddhism to China (and other countries of the region) is by no means insubstantial. These civilizations would hardly exist without the Indian contribution in all aspects of culture- from science and technology, the arts, philosophy and spirituality. In his book The Art of Southeast Asia, Philip Rawson writes:

> *"The culture of India has been one of the world's most powerful civilizing forces. Countries of the Far East, including China, Korea, Japan, Tibet and Mongolia owe much of what is best in their own cultures to the inspiration of ideas imported from India. The West, too, has its own debts. But the members of that circle of civilizations beyond Burma scattered around the Gulf of Siam and the Java Sea, virtually owe their very existence to the creative influence of Indian ideas... No conquest or invasion, no forced conversion imposed them. They were adopted because people saw they were good and that they were good and that they could use them..."*

In the face of this history and this reality, it is not merely wrong but a grotesque misrepresentation to portray India as civilization that only received from outside. The truth, as every serious student knows is the exact opposite: India has always given a great deal more than she has received. Civilization as we know today would not exist without India.

Magical Squares

May I introduce to you Jain he is an absolute wizard at displaying the real ability of our minds to do complex mathematics simply! Yes you did hear right! Jain revels to you simple methods to do all the complex maths you can think of with only knowing the 2x, 3x, 4x, and 5x times tables.

At the same time he shows you how the world as we know it is more than what we think and can for some people bring out major shifts in your way of thinking about the world! I personally am amazed at studying the sacred geometry and what it shows about nature.

For more of an idea about what sacred geometry is then think of a crop circle and its wonderful display of precise maths and geometry.

If you feel like some enlightenment and a great relief that you can really learn maths and actually do it in your head without working it out on paper then please get along to Jain's Lecture...

Explore and discover Geometry, Art and The Mathematics of Nature in a new and challenging way with Sacred Geometry. You will thoroughly enjoy drawing 'Magic Square Mandalas' as a creative and centering process. This is a course for anyone interested in MultiCultural Art, Design, Architecture, Healing and Mathematics.

The Life of Pythagoras will be covered as well as 'The Platonic Solids'. 'The Divine Proportion', 'The Fibonacci Sequence', 'Galactic Mathematics' and related topics like, 'The Mystical Squaring of the Circle', 'The Vesica Pisces' and 'Hypercubes'/'Fourth Dimension Cubes'.

Magic Squares Lecture

During this Lecture Jain will focus our attention upon the ancient Magic Square of 3 × 3 (also known as the "Lo-Shu" in ancient China, C17th BC) which is also the centre of the Tibetan Calendar/Cosmology. It is a timeless mathematical harmonic whose sums of the columns, rows and diagonals all add up to and vibrate to 15. Symbolically the Magic Square creates Order amidst Chaos, and Equality in All Dimensions.

He has authored 4 books on this subject which translates Numbers into Art, by merely connecting the first number (1) to the last number (9). This is the Path of Least Resistance. Exquisite Atomic Art patterns are created that are inherently encoded with much Ancient Knowledge e.g.:

Numerologically, these Magic Sums of 15, 34, 65, 175 and 260, etc., equate with the many names of the Gods and Goddesses in Qaballahistic lore.

6 1 8

7 5 3

2 9 4

Vedic Mathematics Lecture:

On Vedic Mathematics: (Mental, One Line Arithmetic. No More Calculators!)

Do your children really understand the fragmented and confusing mathematical data taught to them? Would you be interested in another method that boosted their confidence, increased their memory skills and is presented in a fun and exciting manner?

After this course of study, you will never multiply, divide or add in the same manner that you are currently using to solve arithmetic problems like 98 × 97, etc. And no more Times Tables!

An understanding of the 16 basic Sutras or "Word Formulae" solves all known mathematical problems by using an advanced and highly intelligent system of MENTAL, ONE LINE ARITHMETIC used by the Hindus 2,000 years.

This system was reconstructed by Bharati Krsna Tirthaji (1884-1960, aka Shankarcharya of Puri) who was invited by Paramahansa Yogananda to take this knowledge, for the first time, to the western world, in 1958. The Beatle's guru: Maharishi Mahesh Yogi, who introduced Transcendental Meditation to the west, was also a student of this master and currently teaches this system in India in the spiritual Maharishi schools, whose teenage students are highly distinguished mathematical geniuses.

Learn also how the Hindu Invention of Zero (and the decimal system) led to modern technological advancements.

In this 1 and half hour Introduction you will know the basics of how to accomplish rapid calculations, mentally. No more Calculators which make the brain lazy. The brain is a mental muscle and needs to be exercised. Jain predicts that if we continue to teach children with an over-emphasis of using calculators in the class-room, over a 20 year period, the Global Brain will deteroriate.

Most children are visual, and the ability to tap into "Pattern Recognition" is the key to rapid mental calculation.

Magic Square and Vedic Mathmatics Workshop: Jain's essential work is to teach Ancient Knowledge to Children and Adults.

On Magic Squares: (Translating Numbers Into Art)

6 1 8

7 5 3

2 9 4

Imagine in front of you some playing cards and the ancient puzzle was to create a 3x3 square such that all columns, rows and diagonals that have the same sum. As shown above, the Magic Sum, aka the Magic Square Constant = 15. There is a treasure house of knowledge encoded within this harmonic matrix that the ancients knew and hid. Magic Squares for thousands of years have fascinated many cultures and many great minds like Benjamin Franklin and Albrecht Durer.

Why was it at the centre of the Tibetan Calendar/ Cosmolgy (Time Code/Applications in Time Travel)?

Why did Ambaji, of the Hindu pantheon, the ancient goddesses of Creation (the universe flows through her breast), have as her tantric symbol, this Magic Square of 3x3, which has 8 possible permutations? and which Buddha, in his enlightenment, described in detail the 8 sided nature of sub-atomic molecules: the "Acta-Kalapos". Durga and Kali and

Shakti, etc., have variations of the same. Why is it that when you translate NUMBERS INTO ART, by drawing a long connecting line, from 1 to 2 to 3, etc., to the last number, that elegant and phiratioed patterns appear? that when tiled or tessellated, produce the atomic structure of Diamond (which is verified by the clairvoyant drawings of the Occult Chemists: Charles Leadbeater and Annie Besant.

Why is it that the pattern for the Magic Square of 11x11 is almost identical to the Ley Line Grid developed by Bruce Cathy, a New Zealand pilot, which was originally the jig-saw-puzzle-map of UFO sightings in his area?

Why did Professor Searl, of the UK, have his work on levity discs and over-unity generators banned? The metals of his generators for the space-craft he built were intimately related, by proportion, to the atomic table of elements and to the opposing pairs of the above Magic Square of 3x3, e.g.: notice that all pairs around the central cell have a sum of 10, thus 2 parts gold was alloyed with 8 parts silver, or 3 parts titanium to 7 parts nickel.

Why did King Solomon the Rabbi, design his Seal based on the 3 rows of the above Magic Square being translated into a Natural Square of 3 (that has its numbers in Counting Order)?

In this day lesson, you will superimpose a Magic Square of 7x7, by rotation of 90 degrees upon itself, to reveal ancient symbols at its centre: the Hindu Peace Symbol of the Swaztika nesting within the Christian Cross!

The ancient Essenes adored the richness of symbolism within these Mystic Squares and gave accounts or Parables of Jesus (Isa or Iesu) that related to our sense of community symbolized by the magic square, that each person, so positioned, has a divine role to play, that if one number was taken away or moved, it would no longer be magical, to realise "Order amidst the Chaos", and the ancient axiom: "As Above, So Below".

About Jain

Jain first began teaching mathematics visually 20 years ago when working with dyslexic children, developing the concept of "Remedial mathematics". Not everyone learns their numbers by left-brain, rote learning, but better through visual, right-brain, the universal language of pictures and symbols. He has since authored 8 mandala, colouring-in books on Magic Squares that converts Mathematics into Atomic Art, surprising many nuclear scientists with his crystalline creations. His authorised travelling school show: "Mathemagics", is the only one of its kind and helps restore and enhance the inherent beauty in mathematics, by teaching it via theatre, with magicians and Alice-in-Wonderland type characters that appear as walking Books of Life, etc. He is currently active teaching this around Australia.

Having come from an Arabic background, the rich ornamental designs that you often see on domes and temples, are the main influences that pervades his work. He is pulling together all his vast knowledge on ancient Magic Squares, Divine Proportions (as in the mathematics of the nautilus shell), and the 3-Dimensional Geometries of the Platonic Solid Symmetries into a unique 2 hour performance on June 21st and 22nd, our Winter Equinox, the shortest day of the year. He calls this rehearsed gathering: THE THEATRE OF THE HOLY NUMBERS, where the Spirit of the 9 Numbers (like 9 Gods) incarnate into a Human Magic Square of 3x3.

Most people know of the Magic Square of 3x3 as an ancient playing-card puzzle, the task being to arrange all 9 cards into a 3x3 square in such a way that the sums of the 3 columns, 3 rows and 2 diagonals all equal 15.

This fascinating production will explore the Magic of the Magic Square of 3x3 and truly download to you all there is to know about it, in a true Keanu Reeves/Matrix-like style, just by sitting in the comfort of your chair.

If you never understood the true meaning and depth of Pythagoras' work 2,500 years ago, this mathematically-based ceremony will not only educate you but leave you enchanted.

I teach by creating the classroom as a theatre. The walls are draped with colourful Magic Square murals about 2 metres square. I wear a special MatheMagician's, magic square robe, that depicts the class lesson with the appropriate numbers and art.

Upon a large stand I have what is called a "Human Cloth Book", it has 30 pages (1,200mm square) of magic square theory and design. It is a unique hand- sewn garment that can be worn and is used throughout the whole show. One of the main lessons shows how the common Multiplication Table produces the Atomic Structure of Platinum Crystal.

I offer Creative and Multicultural Mathematics whose roots stem from Islam, India, Egypt and other cultures. I relate how mathematics can be Inter-Subjectal and may cross Key Learning Areas - Maths as Art, Maths as Science, Maths as History.

Magic Squares are ancient boxes of numbers whose sums of rows, columns and diagonals all add up to the same sum. This matrix represents Perfect Equality in all directions or dimensions. Elegant or aesthetically pleasing patterns can be created when a long, continuous line is joined in arithmetical order from 1 to 2 to 3 to 4, etc., to the last number. This is how Mathematics is transformed into Art. This line connecting such consecutive or sequential numbers is the backdoor into the world of Atomic Art or Nuclear Geometry, arriving at scientifically verified atomic structures. Order is created out of Chaos.

The Magic Square of 3x3 shown over is the centre of the Sino-Tibetan calendar/cosmology and has fascinated many cultures for millennia. The Tibetans believe that every number has a Soul, as does any insect, worm or creature and therefore a Magic Square is a holy thing compared to a Spiritual Community of Souls!

Many sacred symbols emerge when the original Magic Square Pattern or Yantram (singular of Yantra which is the plural) is rotated upon itself. Like snowflakes, millions of Magic Squares can be created. There are 8 permutations of the primal 3x3 Magic Square known also as The Lo-Shu from China 4,000 years ago. There are 880 permutations of the 4x4 Magic Square, but as we get to the 5x5s and 6x6s there are millions! Mathematics is a Universal Language. That means that it will always exist, without decay or corruption.

The concept of 1,2,3 is superior to the concept of A,B,C as English, French, German or any other language eventually changes with slang, etc., but not numbers. As Pythagoras said, in the 6th century BC, "NUMBER IS THE BOND OF ALL THINGS". This Galactic Mathematics is also called "FIXED DESIGN", i.e: it is Timeless and Constant.

Some questions to ponder upon:

* How could the Aryan civilisation, near the Hunza Valley, 3,000 years ago, mentally compute mathematical operations that today only a calculator could achieve?
* And if the maths could not be done mentally, how could they do everything we know today, IN ONE-LINE? e.g.: converting the fraction 1/19 into a decimal is an 18 digit answer that requires 35 lines of exhausting long division, how could they do it in a one simple line solution?
* Why is NASA secretly using this MENTAL ONE-LINE SYSTEM OF ARITHMETIC in the field of Advanced Artificial Intelligence? and not incorporating it into our current mathematics curriculum?
* Why are many of Bill Gate's mathematicians Indian?

* Is the decimal form of Pi (P) a sonic code for dedications to Krishna?
* What does the Bhagawad Gita have to say about this lost science of Clairvoyant Calculation that the ancient Temple Builders employed?
* How can you become a Human Bio-Calculator?
* Is the current Western Maths Curriculum keeping our Children in a state of limited intelligence?

The answer to all these questions are found by understanding the ancient system of Vedic Mathematics as rediscovered by Bharati Krsna Tirthaji (1884- 1960), of Puri Govardhan Math, dux of India in Mathematics and a Sanskrit Philosopher who wrote and spoke perfect English, he reached the position of a Shankacharya (like a Dalai Lama), renunciated several high officical positions in society to retreat into a forest for 7 years as a yogi, between 1911 and 1918 in which time he collected all the lost mathematical knowledge (Ganita Sutras), improved and reconstructed them to 16 basic Sutras or word formulae that solve every known mathematical problem. These 16 Volumes relating to the 16 Sutras was our legacy from the ancient past.

This highly intelligent and sophisticated system of Mental, One-Line Arithmetic was taught by The Master to his main student, who today is the leading exponent in the world: Maharishi Mahesh Yogi, the long white-haired and bearded yogi that many of the 60s generation know of as the Beatleis guru. He is considered as the greatest human bio-calculator on the planet today, holding the vision of raising the collective consciousness of all nations by re-introducing the lost Vedic Mathematics Sutras early in a childis spiritual and educational development. Imagine a maths class in India today , like in the times of old when knowlege was passed on as an oral tradition, and the teacher quickly calls or drills out certain numbers to multiply or divide and the children instantly reply with the correct intuitive answers! Today, in schools where Maharishiis Transcendental Mediatation technique are used students perform remarkably well academically, viewing Mathematics as a Universal Language and therefore often winning the highest competitve prizes in mathematics.

Maharishi organised the scattered Vedic Literature into a complete science of consciousness for both its theoretical and practical value, describing the 16 Sutras as fundamental impulses of intelligence that structure all the Laws of Nature and those that govern the Universe.

Maharishi, the Founder of TM, once said: "Vedic Mathematics is that one field of knowledge which fulfils the purpose of education by developing the total creative genius of the individual, giving him/her the ability to be always spontaneously right and automatically precise so that his/her action, supported by Natural Law, is always effortlessly fulfilling."

Unfortunately, in 1958 after Bharati Krsna Tirthaji returned from the USA tour, sponsored by Paramahansa Yogananda (who introduced Yoga to the West and created the Self-Realisation Foundation) all of Bharati Krsnais 16 Volumes for the 16 Sutras were confirmed lost. Disciples observed that people in the streets of India were imitating Tirthajiis lectureis and dazzling audiences with mathematically misused pschic powers, and begged the Master to write a book just before his death in 1960. This single volume, called VEDIC MATHEMATICS, has become the bible on this subject and spawned a thousand web-sites worldwide. Maharishi schools disseminated the teachings to England where certain schools have adopted the Vedic curriculum but the irony is that today India is struggling to release the British curriculum and NASA have adopted it fully in the realms of advanced robotics. Calculations that can be solved as quick as lightning are a great tool to adopt, but you wouldn't want to teach it worldwide in the fear that you may churn out a generation of child geniuses that may threaten the intellectual status quo.

The gift that the Hindus gave to world, thousands of years ago, and that which is currently responsible for a global silicon chip technology, was none other than the invention of ZERO. We call our common numbers Arabic Numerals but really they extend back to the Hindu concept of creation and void known as Bindui or The Zero Point. In Maths, it makes the magic work e.g.: if you wish to multiply 98 × 98 we need to understand Zero and Unity. All Vedic Maths is based on the understanding of Unity Consciousness which means they utilise processes or Number Bases that correspond to: 0, 10, 100, 1000, 10000, etc., all of which add to 1.

Sacred geometry is a fascinating and complex field of study. Possibly the most commonly known aspect of Sacred Geometry is, The Golden Mean, sometimes referred to as the The Sacred Cut or The Phi Ratio. This ratio has connections with Galactic and Atomic mathematics, as revealed in ancient temples and pyramids; in classical paintings like the Mona Lisa; in the cross-sectional view of the D.N.A. molecule; in the nesting of the 5 Platonic Solids; its occurrence encrypted in the Fibonacci Sequence: 1,1,2,3,5,8,13,21,34,55,89,144, etc.; in the counter-rotating fields or spirals of the sunflower and pine cones.

This is the study of the Universe as found in all parts of the Human Body. This is the mathematics of infinity; Phi expressed as 1:1.618033, etc. (The decimal part goes on forever without any detectable repetitious pattern, like pi).

This Transcendental Number has fascinated many cultures and individuals for millennia. This Phi Ratio is part of our Ascension Process and is an important clue to accessing or Multi-Dimensionality.

2

Vedic Mathematics : As Relevant Today as Ever

By investigating the knowledge and viewpoints in the many topics found in Vedic culture we can certainly see that the practice and utilization of this Vedic knowledge can indeed assist us in many ways. In regard to all the trouble we presently find in this world, maybe it is time to look at things through a different and deeper view to find the answers and directions that are so needed. The knowledge and understandings of this great Vedic culture may indeed be what will help us see through the fog of confusion that seems to envelope so much of society.

What we find in Vedic culture are areas of study, progress and expression that are as relevant today for human advancement as they were hundreds or thousands of years ago. India and its Vedic culture has contributed much to the world, such as its music, beautiful forms of art and architecture, martial arts, astronomy, holistic medicine in Ayurveda, and the mathematical system based on the number ten, along with its yoga and philosophy.

In the United States, yoga has exploded into a three billion dollar industry. A recent survey (at the time of this writing in 2005) showed that 16.5 million people are practicing yoga, or 7.5 percent of the United States. Also, the Yoga Journal magazine has grown from a circulation of 90,000 in 1998, to 170,000 in 2000, to 325,000 in 2005.

Vedic mathematics is another example of its contribution to world progress. It is an ancient development that continues to play an important part in modern society. Without the advancements in math that had been established by Vedic culture as far back as 2500 BC and passed along to others, such as the Greeks and Romans, we would not have many of the developments and inventions that we enjoy today. The Greek alphabet, for example, was a great hindrance to calculating.

The Egyptians also did not have a numerical system suitable for large calculations. For the number 986 they had to use 23 symbols. Even after the Greeks, the Romans also were in want of a system of mathematical calculations. Only after they adopted the Indian system that was called Arabic numerals did they find what they needed. Weights and measures and scales with decimal divisions had been found from that period which were quite accurate.

The difference was that Vedic mathematics had developed the system of tens, hundreds, thousands, etc., and the basis of carrying the remainder of one column of numbers over to the next. This made for easy calculations of large numbers that was nearly impossible in other systems, as found with the Greeks, Romans, Egyptians and even Chinese. The Vedic system had also invented the zero, which has been called one of the greatest developments in the history of mathematics.

The numeral script from India is said to have evolved from the Brahmi numerals. This spread to Arabia through traders and merchants, and from there up into Europe and elsewhere. It became known as the Arabic numerals, yet the Arabians had called them "Indian figures" (Al-Arqan-Al-Hindu) and the system of math was known as hindisat, or the Indian art.

Vedic culture already had an established mathematical system that had been recorded in the Shulba Sutras. These are known to date back to the 8th century BC. The name Shulba Sutras meant "codes of rope". This was because such calculations were used for measuring precise distances for altars and temple structures by using lengths of rope.

The Shulba Sutras were actually a portion of a larger text on mathematics known as the Kalpa Sutras. These and the Vedic mathematicians were recognized for their developments in arithmetic and algebra. Indians were the first to use letters of the alphabet to represent unknowns. But they were especially known for what they could do in geometry. In fact, geometrical instruments had been found in the Indus Valley dating back to 2500 BC. Furthermore, what became known as the Pythagorean theorem was already existing in the Baudhayana, the earliest of the Shulba Sutras before the 8th century BC. This was presented by Pythagoras around 540 BC after he discovered it in his travels to India. So this shows the advanced nature of the Vedic civilization.

After the Shulba Sutras, Vedic mathematics enjoyed further development in the field of Jyotish, Vedic astronomy, which used all forms of math. Indian mathematicians continued creating systems that were not known in Europe until much later in the Renaissance period. For example, Aryabhatta in the 5th century introduced sines and versed sines, and is credited as the inventor of algebra. He is said to be the first to state that the Earth travels around the sun. However, the ancient Vedic texts have described this many years earlier, which shows the wisdom of the early Vedic seers.

Aryabhatta was followed by Brahmagupta (7th century) who was the great mathematician that especially developed the use of zero and was the first to use algebra to solve problems in astronomy. Next was Mahavira (9th century) who made great strides in the use of fractions and figuring out how to divide one fraction by another. Then there was Bhaskara (12th century) who made progress in spherical trigonometry and principles of calculus before Newton by 500 years. He used it to determine the daily motion of planets.

The Vedic system of math, as explained in the sutras, also reduced the number of steps in calculations to merely a few that otherwise required many steps by conventional methods. Thus, this ancient science is still worthy of study today.

A well-developed medical system was in existence by the 1st century AD. Progress in medicine led to developments in chemistry and the production of medicine, alkaline substances and glass. Colourfast dies and paints were developed to remain in good condition over the centuries. The paintings in the caves of Ajanta are a testimony to this.

Vedic art is another ancient development that still holds much appreciation in modern times. Art in the Vedic tradition was never a mere representation of an artist's imagination. It was always a vehicle to convey higher truths and principles, levels of reality that may exist beyond our sense perception. It was always used to bring us to a higher purpose of existence and awareness. In this way, it was always sacred and beheld the sacred. Still today it is used to allow others to enter into a transcendental experience. It may also present the devotional objects of our meditation.

Vedic paintings or symbols are unique in that they can deliver the same spiritual energy, vibration and insight that it represents. In other words, through the meditation and devotional mood of the artist, the art becomes a manifestation of the higher reality. In this way, the painting or symbol becomes the doorway to the spiritual essence contained within. They are like windows into the spiritual world. Through that window we can have the experience of darshan of the Divine or divinities, God or His associates. Darshan is not merely seeing the Divine but it is also entering into the exchange of seeing and being seen by the Divine.

Thus the art, or the Deity, is beyond mundane principles or ingredients, such as paint, paper, stone or metal with which it may be made, but it becomes completely spiritual through which the Deity can reveal Himself or Herself. Thus, the truth of spiritual reality can pierce through the darkness of the material energy and enter our mind and illuminate our consciousness.

To convey higher realities in paintings and sculpture, everything has a meaning. The postures, gestures, colours, instruments or weapons, everything conveys a principle or purpose, which often must be explained to those who lack understanding. Thus, knowing the inner meaning of the painting increases its depth for those who can perceive it, which makes it worthy of further meditation and contemplation.

As with art, dance in India was not merely an expression of an artist's emotional mindset or imagination, but was meant to be an interpretation or conveyance of higher spiritual principles or pastimes of the Divine. In fact, in the Vedic pantheon Shiva is known as Nataraja, the king of dancers. Shiva's dance was also not without a more significant purpose. His dance was based on the rhythm of cosmic energy that pervades the universe, and the destruction of the illusory energy by which all souls are given the opportunity for release from the illusion to attain liberation, moksha.

In this way, traditional Indian dance is highly spiritual and often accompanies important religious rituals and holy days and festivals. Vedic dance goes back to prehistoric times. Bharata Muni wrote his Natya Shastra, science of drama and dance, over 2000 years ago. In it he explains that it was Lord Brahma, the secondary engineer of the universal creation, who brought dance (natya) and drama to the people of Earth millions of years ago, shortly after the Earth was created.

Now dance has evolved into a tradition involving various schools and styles but with strict discipline. It is not uncommon that Indian families will have their daughters spend at least several years or more in such study and practice. There is a precise method of postures, facial and hand gestures (mudras), and movements, along with footwork that must be learned and synchronized to the beat and music in order to convey specific meanings, moods and stories to the audience. Many temples, especially in South India, were known for maintaining large groups of dancers that performed at festivals and religious functions.

When the dance is performed according to the spiritual standards, which some view as similar to the practice of yoga, even the dancers can invoke a high degree of spirituality in their own consciousness and bring unity between their inner selves and God. Then the transcendental atmosphere can manifest and draw the Divine to appear in the performers on stage. Thus, the environment becomes transformed and the audience may also experience darshan of the Divine and experience an inspiring upliftment in their own consciousness. In this way, the dance is divine beauty in motion. Or it is a way of invoking the spiritual dimension into our midst. Few other forms of dance attempt to do this.

Various schools of dance include Bharata Natyam, Kathakali, Manipuri, Orissi, Kathak, Mohini Atam, Krishna Atam, Bhagavata Mela, etc. Thus, we may have many dances that convey stories from the Ramayana and Mahabharata, or Krishna-lila from the Bhagavata Purana. Nowadays this ancient art of Indian dance is enjoying a wide audience and a prominent place on the international stage.

So, as we can see, Vedic culture and its many areas of knowledge and devotional expression are still as relevant today as it was thousands of years ago. And humanity can benefit from it by introspection and in spiritual as well as material development as it did in the past.

The power of the Dharma and the relevancy of Vedic culture are found in the number of tools it has always provided in order for humanity to reach its fullest potentials, both as individuals who are searching for their own fulfilment and spiritual awakening, and as a society that can function in harmony with nature and cooperation amongst themselves.

By investigating the knowledge and viewpoints in the many topics found in the Vedic tradition we can certainly see that the practice and utilization of this Vedic knowledge can indeed assist us in many ways. Let us take a look at a few.

Ayurveda

Ayurveda is the Vedic system of holistic medicine. It has become quite popular in the West and is continuing to gain ground and acceptance. To understand briefly what Ayurveda is, I let Pratichi Mathur, an Ayurvedic practitioner herself, tell us about it from the book, "Vedic Culture: The Difference It Can Make in Your Life":

"So what is Ayurveda exactly? Literally translated from Sanskrit it is composed of two words 'Ayus' which means life and 'Veda' which denotes knowledge. So Ayurveda is the knowledge of healthy living and is confined not only to the treatment of diseases. Life is a vast, and an all-encompassing phenomena, which includes death. On one end, life is a celebration of birth, growth, child bearing, youth and sexuality; on the other end, life also brings forth disease, decay, aging, and loss of vigour. Ayurveda is that ancient art and science that helps us understand this very 'life' with all its different shades and colours; understand how best we can undertake this journey; and how we transition through its different phases, example from teenage, to adulthood, to maturity, etc. Following the principles of Ayurveda brings about a profound understanding of the inner ability to have sound body, mind and spirit. From this point of view, Ayurveda is a compendium of life and not disease. This is a major agenda indeed for any system of medicine, but can it be any less--especially if true healing has to take place. Perhaps, this is exactly why Ayurveda manages to get to the root of the disease that distresses the mind or the emotion that ails the body.

"Ayurveda has twin objectives--maintaining the health of the healthy, and cure illnesses of the diseased. Ayurveda, which is not just a system of disease and its management, but literally a living dynamic philosophy and manual on the art of living, is well fitted to meet its objectives. On one hand Ayurveda offers treatments like Panchakarma or even surgery for the diseased; and on the other hand Ayurveda offers preventative medicine for the healthy. These include elaborate details for following ideal daily and seasonal routines, specialized diets for optimizing health and immunity (Ojas), Rasayana Chikitsa (promotive therapy), Vajikarna Chikitsa (aphrodisiac therapy), Swasthavritta (regimen to stay healthy furnishing details on topics such as exercise, smoking for health), Sadachar (social hygiene), etc.

"Ayurveda advocates a complete promotive, preventive and curative system of medicine and includes eight major clinical specialties of medicine namely,

(1) Medicine (Kayachikitsa),

(2) Surgery (Salya Tantra),

(3) ENT (Salakya Tantra),

(4) Pediatrics (Kaumatabhritya),

(5) Psychiatry (Bhutvidya),

(6) Toxicology (Agad Tantra),

(7) Nutrition, rejuvenation and geriatrics (Rasayan tantra), and

(8) Sexology and virilization (Vajikarana).

This shows what a developed science Ayurveda was in ancient times.

"The exact origin of Ayurveda is lost in the mists of antiquity. Since Panini is placed at 7th century BC and Ayurveda depicts non-Paninian Sanskrit grammar, it is logical to place Ayurveda between 6th -10th Century BC. Tracing the continuity of Ayurveda, it is natural to look for the continuing thread in India's ancient Vedic tradition. Although the term Ayurveda, does not seem to appear in the Vedas, and it appears first in Panini's Ashtadhayayi, however, there are positive evidences to show that in the Vedic period, medicine as a profession was prevalent. The Rig Veda and the Atharva Veda both mention that there were thousands of medical practitioners and thousands of medicines. References to Ayurveda are found as early as the Rig Veda. The three Rig Vedic gods Indra, Agni and Soma relate to the three biological humours: Vata, Pitta and Kapha. References are made of organ transplants as in the case of the artificial limb of queen Vishpala, daughter of King Khela. The functions of physicians are also described in the Rig Veda.

"Rishi Sushruta, famous Ayurvedic Surgeon, also holds that Ayurveda is a supplement (upanga) of the Atharva Veda. While several other sources including the famous Hindu epic Mahabharata speak of Ayurveda as an upanga of Atharva Veda; several other schools of thought hold Ayurveda as a fifth Veda (Panchamveda). Perhaps Ayurveda grew from Atharva Veda first as a branch and then as a comprehensive vast system deserving it's own status, or it developed parallel to the four Vedas as an independent knowledge (with close resemblance to the Atharva Veda)."

Jyotish—Vedic Astrology

Jyotish is the Vedic form of astrology, which is an ancient science and is also being accepted and gaining popularity in the West. Vedic Astrology is meant to help the individual better find his or her way through life. It is to assist in discovering one's highest proclivities, personality, character, qualities and traits and what may be one's best direction for a career, and other things. Thus a person will least likely waste one's time in unfulfilling activities, professions or pursuits.

To further our understanding of Jyotish, I let Chakrapani Ullal, one of the most well-known Vedic Astrologers, describe it as taken from the book, "Vedic Culture: The Difference It Can Make in Your Life":

We turn our attention now to the subject of a branch of the Vedas called Vedic astrology or Jyotish, which is called the 'eye of the Vedas'. It has a cognizing influence of the truth of life and self-knowledge. It acts as a mirror to an individual without which one may not know how to approach life most effectively. It is also called the 'Science of Time'. Time is the source power that rules the universe. All things originate through the procession of time. Hence, Vedic Astrology constitutes the science that maps the structure of time. Astrology is considered divine knowledge that is pure, supreme, secret, and exalted.

Astrology can be defined as the science of correlations of astronomical facts with terrestrial events, and demonstrates the Vedic understanding of the universal interconnectedness and interdependence of all phenomenon, that microcosm and macrocosm are but reflections of one another. Just as mathematics is the organizing principle of science when dealing with inanimate matter, so also astrology is the organizing principle which deals with life and its significance in relation to all living bodies. The planets are seen as reflectors or transmitters of light and solar energy. The solar and planetary rays, like radio waves, affect biological and psychological processes. The rays of influence are unseen vibrations that are not perceptible to the physical eye.

Astrology gives insight and guidance to the fortunes and misfortunes of men, issues of empires and republics, floods and earthquakes, volcanic eruptions, plagues, pestilence and other incidents concerning terrestrial phenomena in relation to the regular movements of the planets.

Over 10,000 years ago the ancient sages, in their superconscious state, cognized that there is energy in planets, and that they send out different rays at different angles which bear influence on everything animate and inanimate on other planets. Through their sensitized intuition and repeated observations these highly evolved souls were able to find out the different characteristics inborn in the planets and also discovered that each rules a distinctive part of the human mind/body. It was also found that particular groups of stars known as constellations have different characteristics, and that they modulate the influence of the planets.

Astrologers say that there are two forces, Daiva and Purushakara, fate and individual energy. The individual energy can modify and even frustrate fate. Moreover, the stars often indicate several fate possibilities; for example, that one may die in mid-age, but that if, through determination, one gives attention in that area it can be overcome, one can live to a predictable old age. Thus, astrology does not say that events must and should happen, but gives the benefic and malefic tendencies which can be directed or modified through conscious effort. The horoscope shows a man's character and temperament. Though it may show that he could become a criminal, it does not mean he is fated to become so. What it means is that he is just the sort of person who will have criminal tendencies, but they can be checked by proper care and training. Additionally, if emotional and financial challenges are indicated in any particular year, one can certainly meet the crisis better if one knows that it might occur.

"Then, how would one define astrology? It is the philosophy of discovering and analyzing past impulses and future actions of both individuals and nations in the light of planetary configurations. Astrology explains life's reactions to planetary vibrations."

Vedic Gemology

Gemology is an important field in today's market. But when we speak of Vedic gemology, we do not mean that it is merely for judging the value of a gem. The Vedic purpose in gemology is to determine the best type of quality gem for a person to wear. Thus, Vedic gemology worked in conjunction with Ayurveda and Jyotish to establish the best gem a person should wear for health and positive influence. To give a little more understanding about this increasingly recognized field, I include the following description by Howard Beckman, a qualified and practicing Vedic gemologist, from the book, "Vedic Culture: The Difference It Can Make in Your Life":

"It is a field that is making great strides medically by using gems for illness and disease both of the physical body and the mind. It is a noninvasive therapy that has produced definite repeatable results medically. (It should be noted that only natural gems, not synthetic, have this inherent energy and also that certain gem treatments commonly used for colour or clarity enhancement will render the gem 'dead' and ineffective.)

Our research and record keeping of case histories of gem use in jewellery for astrological reasons has also allowed us to not only prove the efficacy of gems, but in "debunking" commonly held incorrect notions as far as how to recommend them, as well as baseless superstitions.

"It is the energy force of the cosmos that sustains all living organisms. This energy is called 'prana'. It energizes our bodies throughout life until it leaves at the time of death, leaving the gross material body to decay and return to the elements from which it arose. The Vedic scriptures calculate our life spans in the number of breaths we are allotted during our lives.

If we use this energy more quickly, then the life span will be shorter. (Long distance runners are renowned for dying in their 50's.) If we conserve our energy, especially through systems such as the yoga system, then the life span may be extended. The Ayurvedic system of healing first evaluates the intake and distribution of prana within both the physical and subtle (ethereal) bodies of an individual.

"Gem therapy has been used by many ancient cultures and especially the wearing of gemstones on the body had great significance for the Vedic culture, other than the purely cosmetic or ornamental value that gems are mostly used for today. The science of Ayurveda when combined with Vedic astrology gives a wealth of knowledge in the correct application of gemstones to amplify planetary rays, which can have a dynamic effect on one's physical and emotional health, one's ability to prosper materially, and the general well-being of individual persons here on earth.

"As Gems have such vibratory qualities, we may utilize them to not only affect the brain, but also the higher vibrations in the physical body necessary for healthy functioning of all our internal and external organs. Dr. Young and Bruce Tainio of Cheny University in Washington have made the following statements from their research in this regard. 'The average frequency of the human body during the daytime is between 62 and 68 cycles per second. If it drops below this rate the immune defense system will start to shut down. Cold symptoms appear at 58 cycles, flu at 57, candida at 55, glandular fever at 52, and cancer at 42 cycles per second'.

"Natural (meaning from the earth, which does not include synthetic, man-made material), untreated gemstones, which are repositories of cosmic colours, can restore the pranic energy to the cells of the body, so that its natural vibratory rate and normal health may be regained when it is in a diseased condition. Blue sapphire can tranquilize or have a sedative effect. Emerald can be used as an analgesic. Yellow sapphire has antiseptic properties, and diamond's ability to stimulate cell growth are just a few examples of how gems can affect the healing process in the body."

Vastu Shastra

Vãstu is the Vedic science of architectural and home arrangement. It made its way through the orient and became known as Feng Shui, which has made particular progress in popularity in the West. However, Vastu is a particular science that deals with the flow of energy through a house or building for the highest benefits. It is not enough to merely arrange a house so it looks nice or that there is a good flow of energy through it. But there is much that depends on the directions in which things are facing or which parts of the building in which certain activities are performed.

To get a little more insight into the Vedic science of Vastu, I have included the following description by Arun Naik, an architect that practices the science and art of Vastu Shastra. Again, this is taken from the book, "Vedic Culture: The Difference It Can Make in Your Life":

The Vedic and the Agamic traditions of ancient India always held that the microcosm is a reflection of the macrocosm. A dwelling is an ecological unit, a microcosm which reflects the Cosmos, the macrocosm. Vastu Shastra is the applied aspect of this philosophy, a highly refined method of creating a living space which is a miniature replica of the cosmos as perceived by the Vedas. Vastu Shastra is about emulating the attributes of the Cosmic Space, about bringing the divine sentinels of Cosmic Directions into our homes, about creating Harmony by creating a living environment where the forces of nature are balanced and at peace with each other.

Sri Aurobindo has said... 'Indian sacred architecture of whatever date, style or dedication goes back to something timelessly ancient and now outside India almost wholly lost, something which belongs to the past, and yet it goes forward too, though this the rationalistic mind will not easily admit, to something which will return upon us and is already beginning to return, something which belongs to the future.' (SA, The Renaissance in India)

There is a prayer is Sama Veda:

May there be peace in the sky, may there be peace in mid region, may there be peace on earth, may there be peace in the waters, may the medicinal plants be peaceful, may the forest be peaceful, may there be peace in gods, may Brahma be peaceful, may all the creation be peaceful, may there be peace and peace only, may such peace come to us.

"Vastu is about creating an Inner Space, the chidakash, where this divine peace can park itself. And it achieves it by creating a harmonious external environment—the bahyakash.

At a more earthly level, Vastu Shastra aims at establishing a dynamic balance between Form and Energy so that harmonious conditions are created for the inhabitants. Vastu buildings have harmonious energies and they promote stability, prosperity, happiness, and mental peace for the occupants and owners.

The principle of Vastu is that the Cosmic World with its order and stern discipline has been built by the gods who occupy all the spaces, from the celestial Space within the Cosmic World to the little spaces in our homes, and even our mental space, chidambaram. Man's existence in the Cosmic World has a purpose: it must ascend to immortality and godhood; and the gods, having occupied man's inner Space, strive to create different states in man's consciousness for his ascension from mortality and low nature to Truth, godhood and immortality.

Vastu Shastra helps the effort of the gods by creating an external space—a dwelling, a place to worship and meditate, or a place to work by applying the same laws which the gods have used to create the Cosmic World. This, indeed, is the ultimate function and the highest objective of Vastu Shastra.

So here we can see how various aspects of the ancient Vedic culture are still applicable today and can provide assistance in our attempts to reach our highest potential, both materially and spiritually. This is the constant and higher nature of the power of the dharma that can be recognized and utilized generation after generation.

Vedic Mathematics

Jagadguru swami, Bharathi Krishna Tirthaji maharaja established the Vedic mathematics system from the ancient Vedic texts coupled with natural process of mathematical reasoning.

It is based on sixteen sutras which are in the form of short rules of working or aphorisms or formulae. Sutras provide rules for special cases and this envisages continuous practice of their application. A sutra is a thread of knowledge, a theorem, a ground norm, a repository of proof. A single sutra would generally encompass a wide and varied range of particular application.

Veda has the derivational meaning, the fountain head and unlimited store house of all knowledge. The Vedas are well known as four in number Rig, Yajur, Sama, and Adharvana, but they have also the four Upa Vedas and the six Vedangas all of which form an indivisible corpus of divine knowledge.

Upa Veda

Ayurveda

Gandharva Veda

Dhanur Veda

Sthapatya Veda

Veda

Rig Veda

Sama Veda

Yajur Veda

Adharvana Veda

To our surprise, the Vedic Maths principles could very well be fit in multiplication and division and further to fractions, squares and square roots in abacus system for any calculation.

Multiplication: Conventionally in abacus, multiplication is done number by number in 1st round and 2nd round multiplication and so on. But this is cumbrous with large number of digits. Here come to play are the vedic sutras viz all nine and last from ten and vertically and cross wise. This help to get in one line answer.

Division: In abacus, division is done with assumed quotient on trial basis for big divisors. The different vedic methods viz straight all from 9 and last from 10. Transpose and adjust proportional and straight division could be easier with small or big numbers.

Maiden Attempt: To the few of the author infusing Vedic Maths Sutras in abacus system, may the first of its kind the world over. And we, the Indians should be proud of mastering Indian Vedic Maths and oriental abacus system in one stroke.

Ancient Indian Mathematics

"I have never found one among them who could deny that a single shelf of a good European library was

worth the whole native literature of India and Arabia." —Thomas Babington Macaulay, Minute on Indian Education, 1835.

India was usually known to outsiders until very recently as the land of naked sadhus, snake charmers, overtly religious mystical land which taught the world how to make love. Such a perspective definitely helped the colonisers in assigning an inferior status for the colonised and justify their acts of colonisation and "civilisation". Macaulay's remarks in his Minute on Indian Education is a good read to understand the British perspective then about India. Many influential political leaders (Gandhi for example) who had renounced the more superstitious religious beliefs, in favour of less superstitious beliefs also contributed to the mainstream western perspective on India remaining a status quo. This is surprising considering the fact that even during Gandhi's time, there were huge difference of opinion between Gandhi and other more rational minded people like Tagore and Ambedkar.

Gandhi in his essay Hind Swaraj, strongly condemds technology and industrialization, whereas Tagorc believed in science, industry and rationality. Tagore was particularly disgusted by Gandhi remark that the 1934 Bihar earthquake was a divine retribution for ill treatment of dalits. Tagore, who was as much against caste exploitation as Gandhi was, however could not stand such an irrational explanation for the earthquake suggested by Gandhi. Their debates on this and many other issues (like Gandhi's belief that everyong should use the carka for atleast 30 mins a day) in the form of letters make a very good read.

Another non ignorable figure who had serious issues with Gandhi's irrationality was Ambedkar. The latter firmly believed in the evils of caste system, where as Gandhi who was against the opression against dalits, nevertheless believed in the sanctitty of the age old caste system and even justified its working as a means to achieving social stability.

You can read Ambedkars response to Gandhi's views on caste in his essay "Annihilation of Caste". It looks like Tagore and Ambedkar were not at all influential in changing outsiders outlook towards India. Gandhi definitely takes the spotlight.

Gladly enough, thanks to the IT age, outsourcing and the recent economic boom, this image is rapidly changing. Such a categorization, not only misrepresented India, but also ignored completely many other aspects like agnosticism (Buddhism), atheism (Carvaka) and science and mathematics which were very much being developed in ancient India. Buddhism, which originated in India and spread as a rebel against caste system and rituals, was in full glory for about a millenium and was one of the prime exports of India during those days to China and other eastern countries. So much that China refered to India as the Kingdom of Buddhism. Rationalism and skepticism too, was a school of thought which cannot be ignored during these times. Hinduism, Buddhism, Jainism, Atheism and many other schools were competing at a time for acceptance as a mainstream philosophy. Amartya Sen's book The Argumentatvie India forms an excellent read on this subject.

As far as ancient Indian sciences are concerned, it was developed and transmitted to regions outside India by Chinese and Arabs. Many mathematics and science texts were translated by Al-Beruni, Al-Khwarizmi (who is responsible for the temrs algebra and algorithm) into Arabic and found its way to Europe there on. One of the prime motivations for doing mathematics those days was astronomy. Approximations for trigonometric values, pi, circumference of circle, etc., were developed. Bhaskara's approximation for sin(x) as a rational function of x, Brahmagupta's calculations of eclipses, Aryabhatta I's approximation of pi, Aryabhatta I's method to solve linear Diophantine equations are just some examples amongst a huge literature.

Broadly, ancient Indian mathematics can be categorized into the following based on their period of development.

- Pre Vedic Indus mathematics dating back to as early as 3000BC. Development of mathematics were highly influenced by practical applications like measuring scales, calculating brick ratios, etc.
- Vedic or Sulbasutras, which contained rules to construct altars for various rites and rituals. Various constructions based on Pythagoras theorem are listed, approximation to square root of 2, approximation to pi are some of the achievements. This period probably lasted till around 500BC.
- Jaina mathematics, from 600BC to 500AD. Prime achievements are various notions of infinities, Pascal's triangle, form of set theory, operations with roots of order larger than 2, etc.
- A set of manuscripts were found around 1880s, called the Bakhshali manuscript. The dates of these manuscripts are assumed to be around 400AD. This book contains sets of problems and solutions in linear equations, fractions, square roots, etc.
- Golden age of Indian mathematics set off by Aryabhatta I. Numbers start becoming more abstract and makes it possible to consider zero and negative numbers. Brahmagupta, Mahavira and Bhaskara lists down rules for multiplying, subtracting and (wrongly) dividing by zero. All of them missed that dividing by zero does not make sense :) It was during this period that Al-Khwarizmi, who lived from 790 to 840, wrote Al'Khwarizmi on the Hindu Art of Reckoning and passed on Indian achievements to Islamic and Arabic countries. Ibn Ezra, who lived from 1092 to 1167 in Spain also helped transmit Indian number system to Europe.

- *Keralese Mathematics:* The golden age started to decline after Bhaskara's works around 1200. There is some consensus that political instabilities are partly to be blamed for this decline. Despite all these, mathematics continued to develop in Kerala. The time period between 1400 and 1600 is considered to be the peak period of this development. In addition to building upon and extending previous works by Bhaskara, Brahmagupta, Aryabhatta, etc., one of the significant achievements of this period was mathematical inductive proof in the works of Nilakantha, Jyeshtadeva. The prominent figure during this time period is definitely Madhava of Sangamagramma. He built tools essential for modern analysis. Though not fully certain, he is supposed to have independently derived taylor series expansion for arctan, sin, cos and many other function. A list of 13 different expansions attributed to Madhava is listed here. Madhava also managed to calculate the value of pi to 17 decimal places (3.14155265358979324), much ahead of his contemporaries.

People have started recognizing and acknowledging role of Indians, but awareness still seems to be a big issue. I think it will take some time before terms like Madhava-Gregory series, Leibniz-Gregory-Madhava constant, Hemachandra-Fibonacci series, Bhaskara-Brouncker Algorithm, Aryabhatta algorithm, Aryabhatta Remainder Theorem, Aryabhatta Kuttaka, etc., become commonplace.You might want to check the following links to dig deeper into the history of Indian mathematics.

The Origin of Zero

The first thing to say about zero is that there are two uses of zero which are both extremely important but are somewhat different. One use is as an empty place indicator in our place-value number system. The second use of zero is as a number itself in the form we use it as 0 (or the absence of something).

The use of zero as a numeral appeared sporadically in Egyptian number systems. It was used, however, only between two numbers to indicate an empty position, never at the end of a number. The Mayans also had a zero symbol. Around 650AD the use of zero as a number came into Indian mathematics and the first use of zero as a place holder in positional base notation is considered by some to be due to Arab mathematicians.

The 'Base' System for Numbers

General modern mathematics uses a base 10 numbering system e.g. 1,2,3....10,11. Other systems exist such as digital computation using binary, base 2. A useful site on base numbering systems can be found here.

However, in early times before written numbers it is likely that numbering systems used a base 5, probably because the hand has five digits.

The use of 20 as a grouping number was used by many cultures throughout our earth's history. This was because people have twenty digits (fingers and toes). The Mayan culture developed a rigorous numbering system that was consistent with place values being multiples of 20. The Mayans used a system of a dot for one and a line for 5.

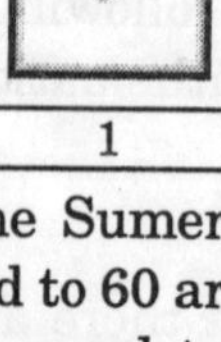	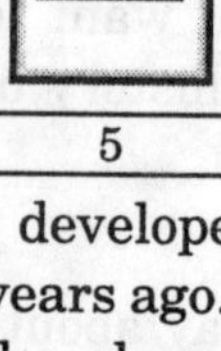	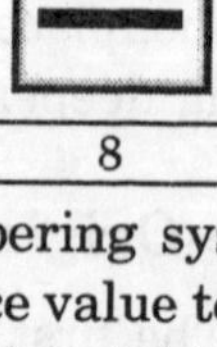
1	5	8

The Sumerian culture developed a numbering system related to 60 around 5000 years ago. Each place value to the left is equal to 60 times the place value to the right. e.g. 3600, 60, 1, 1/60, 1/3600.

Number Sytems in Different Languages

Many languages have their own mathematical numerals. However, some of these numerals have a passing and sometimes confusing resemblance to numerals as used in European languages.

٦	'Arabic' numeral for the number '6'
ו	Hebrew numeral for the number '6'
৪	Bengali numeral for the number '4'

Vedic Sutra—All from 9 and Last from 10

With this method you only ever need multiplication tables up to 5 times 5. It is one of many ancient Indian sutras and this one involves a cross subtraction method which, according to old historical traditions, is responsible for the acceptance of the ´ mark as the sign of multiplication. Here is a very simple example of the method. Can you give a good explanation of WHY it works?

Suppose we want to multiply 9 by 7. We subtract each number from 10 and, using these differences (or deficiencies), write:

9-1

7-3

6 3

The product has two parts, left and right.

To get the right part (or units digit) multiply the deficiencies (1×3).

The left hand digit (tens digit) of the answer can be found in four different ways. Why do they all give the same answer?

1. Subtract 10 from the sum of the two given numbers (9+7=16, 16-10=6)
2. Subtract the sum of the two deficiencies (1+3=4) from 10 and you get 6.
3. Cross subtract (9-3=6)
4. Cross subtract (7-1=6)

This gives the answer 63.

Here are some more examples. Try some of your own.

9-1	8-2	9-1	8-2
6-4	7-3	9-1	5-5
5 4	5 6	8 1	4 0
			Note: Here you have to express 5 times 2 as 1 ten and 0 units.

Therapies and Techniques

In keeping with the Life Positive philosophy to open the doors to all schools of wisdom, and to all therapies and techniques that aid well-being, the Expo presented a generous smorgasbord of experiences, systems of knowledge and viewpoints that could not fail to have stimulated, enlightened and inspired. LP reader Chhaya Kothari probably speaks for many participants when she confides, "I came all the way from Rishikesh especially for this event. Rishikesh, of course, hosts many discourses and programmes too, but nowhere is such a varied mix presented. I have thoroughly enjoyed myself here and am truly glad I came."

The most heartwarming aspect of the event and one which has won our eternal gratitude is the number of LP readers who came from all over India to participate. One such was Tony Bonham from Goa who said, "Looking at the event list, I decided I must make it here and I must say, I am truly impressed. Really glad I came." We are really glad you came too, sir! Another was Father Peter from the Father

Agnel Ashram in Pune, who came with a colleague, Father Ivan Almeida. Says the affable Father, "Life Positive Expo-2005 was fantastic.

The three days were like being in a retreat. Most of the exponents of various techniques were good. Swami Veda Bharati was very inspiring." Each day's programme was flagged off at 8 am, with a salutation to the sun in the form of sun yoga by our very own Sun yogi, Umasankarji. Most people who came the first day, continued on all the three days, bringing their friends and relatives, truly inspired by the amazing Sun yogi.

Throughout the three days, participants thronged the workshops, expanding their minds with such concepts as time-line therapy, which demonstrates the vital need for clarity on the past, present and future. Freedom from past 'baggage' through time-line therapy can focus attention on the present, affirmed Dr. Somesh Chadda. Dr. Ketan Shah's ability to heal people of neck aches, backaches, stomach disorders and knee pain in a minute through the application of acupressure, drew a long queue of patients. "This is not a miracle," smiled Dr. Shah, "Just the application of a time-tested proven technique."

The stalls too did brisk business. Tarot reader Sumita Somany had many takers as did AAC-Sukoon, whose stress-relieving armchairs were in great demand. Energy healing through various modalities at Rodsie Realms was also a huge draw. The Swami Rama ashram had a fascinating range of books by the revered sage, including his immensely popular, Living with the Himalayan Masters, eagerly procured by seekers. Other bookstalls such as Wisdom Tree, Pustak Mahal and Happy Thoughts, also drew immense response. Sangeeta Gupta's Money Mantra stall attracted much attention as did the Pranic Healing stall, which saw a steady stream of people being cleansed of their negative auras. Shahnaz Husain's herbal cosmetics also attracted attention and the Nirankari Satsang Mandal did a double whammy, taking both a stall as well as presenting a wonderful workshop with bhajans, poems and discourses, conveying the message of realisation of Nirankar with devotion and fervour.

On a more intellectual plane were our two panel discussions. Both drew packed halls and proved to be stimulating, thought-provoking and inspiring. The sublime insights shared by the panelists of 'What is enlightenment' were enough to send listeners into a mild ecstasy. Vedic scholar Swami Veda Bharati beautifully affirmed that true enlightenment is reflected in humility - in how low one can make oneself.

Islamic scholar Maulana Wahiduddin Khan aptly honed in on anger management as a signal of enlightenment. Prashant Khanna of the Sri Aurobindo Ashram quoted extensively from the Bhagawad Gita to outline the qualities of the individual with stithprajna. Moderator Rajiv Mehrotra, himself a Buddhist practitioner, skilfully wielded the threads of the discussion and brought it to a fitting conclusion.

The other discussion on 'Allopathy and alternatives: Points of convergence' was a very enlightening and inspiring interaction between four allopaths who were firmly in favour of alternative approaches. Sounding more like gurus than sceptical medicos, the quartet agreed fervently on the need for a broader and more holistic approach to healing.

Dr. H. K. Chopra affirmed that alternatives were the real thing and allopathy was the alternative. Dr. Ashwini Chopra, co-owner of Aashlok, divulged that he routinely advised his patients to take up yoga and meditation and no preliminary investigation was complete without inquiring into the patient's emotional state. Dr. Newton and Dr. Tuli also made very relevant points, enthralling the audience with their open-mindedness.

But the highlight of the Expo was indubitably the series of evening programmes. The programme for the first day began with mellifluous and heart-stirring bhajans by Chitra Roy. Starting with the great Shankaracharya's Chidananda Roopam, she went on to sing some of the bhajans popular among the Art of Living followers such as Radhe Shyam.

Then came a tour de force discourse by the inimitable Sri Sri Ravi Shankar, whose fan following girdles the globe. One devotee tailed him all the way from Kenya and was in emotional transports at sight of her guru. He started with a simple and profound observation, "The centre of the atom is positive and negative electrons only exist in the periphery. Similarly, at our core we are positive and only need to remove the superficial negativity." Saying this he gratifyingly eulogised Life Positive's own role in stoking positive values in society and in creating awareness of the great plurality of paths, techniques and healing systems. Such knowledge is essential to guide people on the spiritual path, he said. He also congratulated the magazine for having taken giant steps in growth and becoming the leading purveyor of spiritual values in the country today.

The second day's evening programme started off with a charming Bharata Natyam performance by Marie Elangovan and her students, D. Ayushree Moorthy and S. Aparna. Although the lithe and lissome Ms Elangovan was very impressive, her tiny students, less than nine years old, carried the day with their talent and enthusiasm.

Vedic scholar and international meditation teacher, Swami Veda Bharati, then proceeded to give a rivetting lecture on the qualities of the Indian mind. First deploring the tendency among most of us to embark on this exercise as a preliminary to boosting our ego at the cost of other nations, he reiterated his commitment to global citizenship and emphasised the need for a broad all-inclusive perspective. To introduce the audience to the Indian mind, he requested them to be guided by him into a meditation for it was only through the meditative mind that one could access the nature of the Indian mind. Interspersing his lecture with quotations, anecdotes and incidents that revealed the fathomless depths of his mind, he concluded by saying that one of the most remarkable aspects of the Indian mind was its proclivity for 'also'. The Indian mind declines to be limited by either-or, and insists on saying this and that, he said. The third day's programme was once again entirely different in character. It went off to a rollicking start with joyful and exuberant songs in English from the Ananda Sangha. Aptly called Philosophy in Song, the songs, written by their founder Swami Kriyananda, articulated love of life, love of nature and love of God.

The redoubtable Anandmurti Gurumaa then took the stage and held the audience enthralled with her deep and powerful insights into spirituality and why it is best practised within the context of our day-to-day lives. First emphasising the need for all seekers to free themselves from the outer garb of religion and to immerse themselves in its spiritual core, she went on to enumerate the story of someone who goes to a master and asked him how he could find God. The master looked at him and said, 'Certainly I will help you find Him but first you have to give me the full details, including time and place, of when you lost Him!' At one point she broke into a full-throated rendition of an Urdu couplet that spoke beautifully and movingly about a man who went to a sage simply out of curiosity and found to his astonishment that he was changed for life. Her powerful and mellifluous voice is a phenomenon not easily found and one can well understand why her cassettes are such sell-outs.

And with that we reluctantly came to the end of our soul-satisfying and deeply gratifying programme. To us, the Life Positive team, the Expo was an exhilarating experience, an opportunity to meet and bond with our readers and fellow seekers. It is with deep humility and gratitude that we acknowledge your many generous compliments about Life Positive. We are grateful that you find the magazine of benefit, and that you enjoyed the event. We want you to know that you, the reader, are the sole inspiration for our existence and ongoing activities. In serving you, we hope to scale fresh peaks of excellence.

Vedic Mathematics Today

Having recently read several articles published in India which are critical of Vedic mathematics it seems appropriate to give a response based on the work carried out within the United Kingdom. I have taught Vedic mathematics for almost twenty years at an independent school in London. Some of the students have Indian ancestry but this is not the reason for our approach to mathematics. All our mathematics teachers have become familiar with the system and use it in their lessons. We regularly hold training workshops for the staff and have written our own text books for resource material. Our associate schools in different parts of the world also teach Vedic mathematics. Both teachers and students have found Vedic mathematics of great benefit because it has so many positive qualities. The most obvious of these are found through experience of working with the sutras. Children and adults alike find the system profound, inspiring and delightful. We have also found that the students gain exceptional examination results.

What is Vedic mathematics really about? What is it about the subject that an English school should use it so thoroughly? Is it really Vedic and what was the purpose of Sri Bharati Krishna Tirthaji's book on the subject? In this account I attempt to answer some of these questions.

Vedic mathematics is based on sixteen sutras together with a similar number of sub-sutras. Each sutra provides a principle of mental working applicable to many diverse areas of mathematics. There are several reasons for the title

Vedic but before considering these it is worth mentioning that the word Veda generally has two meanings. The first is the collection of ancient Indian texts relating to both spiritual and secular knowledge. Due to the once strong oral tradition by which these texts were passed on it is not possible to accurately place a date when they were first composed. This meaning of Veda is the most common. The second way that Veda is used is to describe true knowledge in the present which resides within peoples hearts or minds. This is the meaning which we have accepted. This is just as well since Sri Tirthaji bequeathed to us no accurate or definite reference for the sutras. Locating the sutras within ancient texts has so far proven to be beyond some of the most able scholars in this field. It is quite possible that Sri Tirthaji intuited the sutras from his deep understanding of the subject, the Vedas and the nature of the human mind.

Arising from the first of these two meanings of Veda historians of mathematics have tried to relate the work of Sri Tirthaji to the mathematics of well-known ancient texts. If only this first meaning is accepted then the question as to whether Tirthaji's system is or is not Vedic becomes an almost insurmountable task. In my experience it is better to approach Vedic mathematics from the second meaning relating to natural laws working within the human psyche. This is a practical approach and certainly most of the work in the UK has followed this line. We have sympathy for the historians but regard any discovery of reference to the sutras as a bonus rather than a necessity.

The introduction to Vedic Mathematics indicates that during the early part of the 20th century Sri Tirthaji rediscovered or reconstructed Vedic mathematics from stray references within the appendix portions of the Atharvaveda. We do not know whether or not these appendices were published. He evidently spent a large proportion of his life teaching the system but it was only shortly before he passed away that he set down an illustrative volume on the subject.

This was published posthumously in 1965 and is the main source of all the serious study on the subject. His book offers a snapshot of the sutraic system. Some of the sutras are applied to relatively elementary topics in arithmetic and algebra giving rise to fast and easy methods of calculation. Some of these methods are developed quite carefully particularly at the beginning of the book. A few sutras are only mentioned once or twice in rather abstruse and specialised topics. The really surprising aspect is contained within one of his introductions where he describes these few sutras as having jurisdiction over the whole of mathematics.

Years ago when I was first involved with various groups studying Vedic mathematics we all thought this statement outrageous and absurd. How could sixteen sutras apply to the whole of mathematics? Our view was strengthened by the text due to the paucity of explanation of some of the rules. For example, there is a sutra, Vyashti Samashti, which is mentioned only once in the text and even then it is given in relation to a very particular type of biquadratic equation. It is the sort of equation that you are highly unlikely to meet when working in mathematics. Nevertheless he uses this equation as the means to provide us with a brief glimpse at how the sutra can be applied. As it turns out this sutra is fundamental to mathematics particularly in statistics and mechanics. It has countless applications because it describes a common mental process. The sutra means Individuality and totality or Being many, being one or Specific and general. One of the simplest applications occurs when finding the mean or average of a set of numbers. The idea is that an average in some way is a single number representing the whole group. The sutra also carries a much deeper meaning relating to the individual and universal Self. Expressed as above — so below, it is recognised as one of the fundamental philosophical tenets of Western culture. It finds expression within the art and architecture of ancient Egypt, the philosophy of Plato, both the Judaic and Christian teachings and within the Humanism of the Florentine Renaissance. Many of the sutras are like this. They go beyond the purely mathematical sphere and deal with a more fundamental level of human experience.

Going back in time when we first came across Vedic mathematics in London we were impressed by the methods of calculation. They had a different quality to what we were brought up with. Here were methods that were enjoyable, quick, light, easy and refreshing. They related the numbers to their source, unity. They brought the mind out of the wrangles of the past and into the immediacy of the present moment. We found the sutras really brought the subject alive and we still find that all students feel more alive by practising Tirthaji's methods.

Once our enthusiasm was kindled we studied and practiced the whole of his book. We worked through every sum and read and re-read every word to try and make sense of the system. It became clear that since Sri Tirthaji had only left a specimen we would have to see how the sutras could be applied to other areas of mathematics. What if his statement about the jurisdiction of the sutras was true? Could it be possible that so few rules can take account of all mathematics?

We decided to put his statement to the test and see if there were other applications not mentioned in his book. Initially this was difficult but after a while we discovered a fast method of calculation based on one of the sutras. Over a period of years this work continued and we gradually began to see more and more applications to fast methods of calculation, algebraic manipulations and geometrical theorems. We also discovered the sutras at work within entirely conventional methods, systems and algorithms. For example, although we had found a fast method of subtraction of numbers based on one sutra we also found the correct sutra for the conventional paying and borrowing back method.

This work in turn led to greater understanding of the individual sutras together with a greater confidence in the system. It appeared that the sutras described the mental processes or laws that naturally operate within the mind. The implication is that the human mind has only a limited number of natural processes, channels or principles - each having a myriad of applications.

The next step was to consider topics within mathematics and simply ask, what sutra is working here? For example, what is the sutra working when you bisect an angle or when simplifying an irrational number? The elementary topics are fairly straightforward but what about more sophisticated mathematics? We recently worked through various solutions to the 'shoe-maker's knife' problem in geometry. This is an ancient problem but which has become quite a popular area during the last sixty years or so. The mathematics involved is not easy.

You are given two circles drawn inside a larger circle so that they all touch each other. The area between the circles is the Arbelos or shoe-maker's knife. The problem is to construct further circles within the Arbelos as shown in the diagram. We worked through some of the conventional solutions to this. There was one particularly elegant solution that seemed the quickest and easiest method. It required transformations of a series of circles.

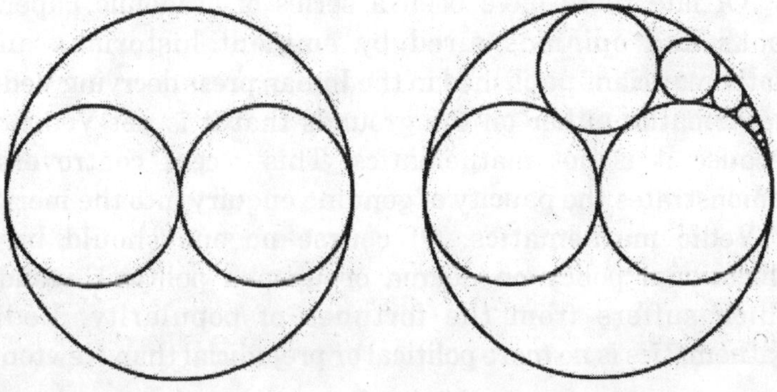

Whilst looking at this solution it dawned on us that the sutra involved was none other than Transpose and adjust, one of the most common sutras in this Vedic system. This is by way of example and illustrates the way in which a simple principle can have an overriding influence over the specific details of the problem.

These anecdotal instances help describe the nature of what we see as Vedic mathematics. On the one hand you have a series of fast, interesting and delightful algorithms and on the other the sutras supply us with a universal approach by which any mathematical process can be perceived as an integral part of a larger whole.

By looking at Tirthaji's system from this point of view we have found that the applications of the sutras are not time dependent. In other words they can be applied and are entirely valid not only to known mathematics from Indian antiquity as well as the mathematics that the Shankarcarya himself had to deal with but also modern topics ranging from the theory of matrices to the theories of chaos and catastrophe. The system Tirthaji described and what has been developed from his work is Vedic by way of function independently of historical background. We have found it of immense practical value. It stands as a bona fide system supported by deep logic. This is why it succeeds in the educational ethos of the United Kingdom and elsewhere.

Of late there have been a series of academic papers, books and opinions aired by eminent historians and mathematicians published in the Indian press decrying Vedic mathematics either on the grounds that it is not Vedic or because it is not mathematics. This recent controversy demonstrates the paucity of genuine enquiry into the merits of Vedic mathematics. Of course no one should base educational policy on dogma or upon a political agenda which suffers from the fortunes of popularity. Vedic mathematics is no more political or prejudicial than Newton's three laws of motion. This can be testified by those outside India who use Vedic mathematics freely and without any cultural association. Nevertheless, it is philosophical since it deals with knowledge and natural law.

Those responsible for developments in education must face practical issues. If, in India, there is a demand for Vedic mathematics in schools then this must be met with the tasks of training teachers, supplying resource materials and altering syllabuses according to a more fundamental and unified system of logic. I hope that every effort can be made to pursue the study of Vedic mathematics in a practical and open-hearted spirit.

Vedic Mathematicians in Ancient India

Uncovering the scope of Ancient Indian Mathematics faces a twofold difficulty. To determine who discovered what we must have an accurate idea of the chronology of Ancient India. This has been made doubly difficult by the faulty dating of Indian Historical events by Sir William Jones, who practically invented the fields of linguistics and philology if for a moment we discount the contributions of Panini (Ashtadhyayi) and Yaska (Nirukta) a couple of millennia before him.

Sir William, who was reputed to be an accomplished linguist, was nevertheless totally ignorant of Sanskrit when he arrived in India and proceeded in short order to decipher the entire history of India from his own meager understanding of the language, In the process he brushed aside the conventional history as known and memorized by Sanskrit pundits for hundreds of years and as recorded in the Puranas and invented a brand new timeline for India which was not only egregiously wrong but hopelessly scrambled up the sequence of events and personalities. See for instance my chronicle on the extent of the damage caused by Sir William and his cohorts in my essay on the South Asia File.

It is not clear whether this error was one caused by inadequate knowledge of language or one due to deliberate falsification of records. It is horrific to think that a scholar of the stature of sir William would resort to skullduggery merely to satisfy his preconceived notions of the antiquity of Indic contributions to the sum of human knowledge. Hence we will assume Napoleon's dictum was at play here and that we should attribute not to malice that which can be explained by sheer incompetence. This mistake has been compounded over the intervening decades by a succession of British historians, who intent on reassuring themselves of their racial superiority, refused to acknowledge the antiquity of India, merely because 'it could not possibly be'. When once they discovered the antiquity of Egypt, Mesopotamia and Babylon, every attempt was made not to disturb the notion that the Tigris Euphrates river valley was the cradle of civilization.

When finally they stumbled upon increasing number of seals culminating in the discovery of Mohenjo Daro and Harappa by Rakhal Das Banerjee and Daya Ram Sahni, they hit upon the ingenious idea that the Vedic civilization and the Indus Valley Civilization or the Saraswathi Sindhu Civilization, a more apt terminology since most of the archaeological sites lie along the banks of the dried up Saraswathi river, were entirely distinct and unrelated to each other. The consequences of such a postulate have been detailed in the South Asia File.

The second difficulty was the Euro centricity (a euphemism for a clearly racist attitude) of European mathematicians, who refused to appreciate the full scope of the Indic contributions and insisted on giving greater credit to Greece and later to Babylonian mathematics rather than recognize Indic and Vedic mathematics on its own merits. If this was indeed a surprise revelation, I fail to see the irony, when a similar Euro centricity was exhibited towards the antiquity of the Vedic people themselves.

An Overview of Indian Mathematics

It is without doubt that mathematics today owes a huge debt to the outstanding contributions made by Indian mathematicians over many hundreds of years. What is quite surprising is that there has been a reluctance to recognize this and one has to conclude that many famous historians of mathematics found what they expected to find, or perhaps even what they hoped to find, rather than to realize what was so clear in front of them.

We shall examine the contributions of Indian mathematics in this article, but before looking at this contribution in more detail we should say clearly that the "huge debt" is the beautiful number system invented by the Indians on which much of mathematical development has rested. Laplace put this with great clarity:-

The ingenious method of expressing every possible number using a set of ten symbols (each symbol having a place value and an absolute value) emerged in India. The idea seems so simple nowadays that its significance and profound importance is no longer appreciated. Its simplicity lies in the way it facilitated calculation and placed arithmetic foremost amongst useful inventions. The importance of this invention is more readily appreciated when one considers that it was beyond the two greatest men of Antiquity, Archimedes and Apollonius.

We shall look briefly at the Indian development of the place-value decimal system of numbers later in this article and in somewhat more detail in the separate article Indian numerals. First, however, we go back to the first evidence of mathematics developing in India.

Histories of Indian mathematics used to begin by describing the geometry contained in the Sulvasutras but research into the history of Indian mathematics has shown that the essentials of this geometry were older being contained in the altar constructions described in the Vedic mythology text the Shatapatha Brahmana and the Taittiriya Samhita. Also it has been shown that the study of mathematical astronomy in India goes back to at least the third millennium BC and mathematics and geometry must have existed to support this study in these ancient times.

Evidence from Europe that India is the True Birthplace of Our

Numerals

The views of savants and learned scholars from a non-Indian tradition about Indian mathematics are presented here. Note that most of these are dated prior to the1800's, when India was still untainted with the prefix of being a colonized country

Severus Sebokt of Syria in 662 CE: (the following statement must be understood in the context of the alleged Greek claim that all mathematical knowledge emanated from them

> *"I shall not speak here of the science of the Hindus, who are not even Syrians, and not of their subtle discoveries in astronomy that are more inventive than those of the Greeks and of the Babylonians; not of their eloquent ways of counting nor of their art of calculation, which cannot be described in words - I only want to mention those calculations that are done with nine numerals. If those who believe, because they speak Greek, that they have arrived at the limits of science, would read the Indian texts, they would be convinced, even if a little late in the day, that there are others who know something of value". (Nau, 1910)*

Said al-Andalusi, probably the first historian of Science who in 1068 wrote Kitab Tabaqut al-Umam in Arabic (Book of Categories of Nations).

To their credit, the Indians have made great strides in the study of numbers (3) and of geometry. They have acquired immense information and reached the zenith in their knowledge of the movements of the stars (astronomy) and the secrets of the skies (astrology) as well as other mathematical studies. After all that, they have surpassed all the other peoples in their knowledge of medical science and the strengths of various drugs, the characteristics of compounds and the peculiarities of substances.

Albert Einstein in the 20th century also comments on the importance of Indian arithmetic: "We owe a lot to the Indians, who taught us how to count, without which no worthwhile scientific discovery could have been made."

Quotes from Liberabaci (Book of the Abacus) by Fibonacci (1170-1250): The nine Indian numerals are ...with these nine and with the sign 0 which in Arabic is sifr, any desired number can be written. (Fibonacci learnt about Indian numerals from his Arab teachers in North Africa) .Fibonacci introduced Indian numerals into Europe in 1202CE.

G Halstead:

> *...The importance of the creation of the zero mark can never be exaggerated. This giving to airy nothing, not merely a local habituation and a name, a picture, a symbol but helpful power, is the characteristic of the Hindu race from whence it sprang. No single mathematical creation has been more potent for the general on go of intelligence and power. [CS, P 5]*

The Following Quotes are from George Ifrah's Book Universal History of Numbers

The real inventors of this fundamental discovery, which is no less important than such feats as the mastery of fire, the development of agriculture, or the invention of the wheel, writing or the steam engine, were the mathematicians and astronomers of Indian civilisation: scholars who, unlike the

Greeks, were concerned with practical applications and who were motivated by a kind of passion for both numbers and numerical calculations.

There is a great deal of evidence to support this fact, and even the Arabo-Muslim scholars themselves have often voiced their agreement.

The following is a succession of historical accounts in favour of this theory, given in chronological order, beginning with the most recent.

1. P. S. Laplace (1814): "The ingenious method of expressing every possible number using a set of ten symbols (each symbol having a place value and an absolute value) emerged in India. The idea seems so simple nowadays that its significance and profound importance is no longer appreciated. Its simplicity lies in the way it facilitated calculation and placed arithmetic foremost amongst useful inventions. The importance of this invention is more readily appreciated when one considers that it was beyond the two greatest men of Antiquity, Archimedes and Apollonius." [Dantzig. p. 26]
2. J. F. Montucla (1798): "The ingenious number-system, which serves as the basis for modern arithmetic, was used by the Arabs long before it reached Europe. It would be a mistake, however, to believe that this invention is Arabic. There is a great deal of evidence, much of it provided by the Arabs themselves that this arithmetic originated in India." [Montucla, I, p. 375J.
3. John Walls (1616-1703) referred to the nine numerals as Indian figures. [Wallis (1695), p. 10]
4. Cataneo (1546) le noue figure de gli Indi, "the nine figures from India". [Smith and Karpinski (1911), p.3.

5. Willichius (1540) talks of Zyphrae! Nice, "Indian figures". [Smith and Karpinski (1911) p. 3]
6. The Crafte of Nombrynge (c. 1350), the oldest known English arithmetical tract: II fforthermore ye most vndirstonde that in this craft ben vsed teen figurys, as here bene writen for esampul 098 ^ 654321... in the quych we vse teen figwys of Inde. Questio II why Zen figurys of Inde? Soiucio. For as I have sayd afore thei werefondefrrst in Inde. [D. E. Smith (1909)
7. Petrus of Dada (1291) wrote a commentary on a work entitled Algorismus by Sacrobosco (John of Halifax, c. 1240), in which he says the following (which contains a mathematical error): Non enim omnis numerus per quascumquefiguras Indorum repraesentatur "Not every number can be represented in Indian figures". [Curtze (1.897), p. 25
8. Around the year 1252, Byzantine monk Maximus Planudes (1260-1310) composed a work entitled Logistike Indike ("Indian Arithmetic") in Greek, or even Psephophoria kata Indos ("The Indian way of counting"), where he explains the following: "There are only nine figures. These are:

   123456789

   [figures given in their Eastern Arabic form]

   A sign known as tziphra can be added to these, which, according to the Indians, means 'nothing'. The nine figures themselves are Indian, and tziphra is written thus: 0". [B. N., Pans. Ancien Fonds grec, Ms 2428, f" 186 r"]
9. Around 1240, Alexandre de Ville-Dieu composed a manual in verse on written calculation (algorism). Its title was Carmen de Algorismo, and it began with the following two lines: Haec algorismus ars praesens dicitur, in qua Talibus Indorumfruimur bis quinquefiguris

"Algorism is the art by which at present we use those Indian figures, which number two times five". [Smith and Karpinski (1911), p. 11]

10. In 1202, Leonard of Pisa (known as Fibonacci), after voyages that took him to the Near East and Northern Africa, and in particular to Bejaia (now in Algeria), wrote a tract on arithmetic entitled Liber Abaci ("a tract about the abacus"), in which he explains the following:

    Cum genitor meus a patria publicus scriba in duana bugee pro pisanis mercatoribus ad earn confluentibus preesset, me in pueritia mea ad se uenire faciens, inspecta utilitate el cornmoditate fiutura, ibi me studio abaci per aliquot dies stare uoluit et doceri. Vbi a mirabii magisterio in arte per nouem figuras Indorum introductus. . . Novem figurae Indorum hae sun!: cum his itaque novemfiguris. et turn hoc signo o. Quod arabice zephirum appellatur, scribitur qui libel numerus:

    "My father was a public scribe of Bejaia, where he worked for his country in Customs, defending the interests of Pisan merchants who made their fortune there. He made me learn how to use the abacus when I was still a child because he saw how I would benefit from this in later life. In this way I learned the art of counting using the nine Indian figures... The nine Indian figures are as follows:

    987654321

    [figures given in contemporary European cursive form].

    "That is why, with these nine numerals, and with this sign 0, called zephirum in Arab, one writes all the numbers one wishes."[Boncompagni (1857), vol.1]

11. C. U50, Rabbi Abraham Ben MeIr Ben Ezra (1092-1167), after a long voyage to the East and a period

spent in Italy, wrote a work in Hebrew entitled: Sefer ha mispar ("Number Book"), where he explains the basic rules of written calculation.

He uses the first nine letters of the Hebrew alphabet to represent the nine units. He represents zero by a little circle and gives it the Hebrew name of galgal ("wheel"), or, more frequently, sfra ("void") from the corresponding Arabic word.

However, all he did was adapt the Indian system to the first nine Hebrew letters (which he naturally had used since his childhood).

In the introduction, he provides some graphic variations of the figures, making it clear that they are of Indian origin, after having explained the place-value system: "That is how the learned men of India were able to represent any number using nine shapes which they fashioned themselves specifically to symbolize the nine units." (Silberberg (1895), p.2: Smith and Ginsburg (1918): Steinschneider (1893)1

12. Around the same time, John of Seville began his Liberalgoarismi de practica arismetrice ("Book of Algoarismi on practical arithmetic") with the following:

Numerus est unitatum cot/echo, quae qua in infinitum progredilur (multitudo enim crescit in infinitum), ideo a peritissimis Indis sub quibusdam regulis et certis lirnitibus infinita numerositas coarcatur, Ut de infinitis dfinita disciplina traderetur etfuga subtilium rerum sub alicuius artis certissima Jege ten eretur:

"A number is a collection of units, and because the collection is infinite (for multiplication can continue indefinitely), the Indians ingeniously enclosed this infinite multiplicity within certain rules and limits so that infinity could be scientifically defined: these strict rules enabled them to pin down this subtle concept.

[B. N., Paris, Ms. lat. 16 202, p 51: Boncompagni (1857), vol. I, p. 261

13. C. 1143, Robert of Chester wrote a work entitled: Algoritmi de numero Indorum ("Algoritmi: Indian figures"), which is simply a translation of an Arabic work about Indian arithmetic. [Karpinski (1915); Wallis (1685). p. 121

14. C. 1140, Bishop Raymond of Toledo gave his patronage to a work written by the converted Jew Juan de Luna and archdeacon Domingo Gondisalvo: the Liber Algorismi de numero Indorum ("Book of Algorismi of Indian figures) which is simply a translation into a Spanish and Latin version of an Arabic tract on Indian arithmetic. [Boncompagni (1857), vol. II

15. C. 1130, Adelard of Bath wrote a work entitled: Algoritmi de numero Indorum ("Algoritmi: of Indian figures"), which is simply a translation of an Arabic tract about Indian calculation. [Boncompagni (1857), vol. II

16. C. 1125, The Benedictine chronicler William of Malmesbury wrote De gestis regum Anglorum, in which he related that the Arabs adopted the Indian figures and transported them to the countries they conquered, particularly Spain. He goes on to explain that the monk Gerbert of Aurillac, who was to become Pope Sylvester II (who died in 1003) and who was immortalized for restoring sciences in Europe, studied in either Seville or Cordoba, where he learned about Indian figures and their uses and later contributed to their circulation in the Christian countries of the West. L Malmesbury (1596), f' 36 r'; Woepcke (1857), p. 35J

17. Written in 976 in the convent of Albelda (near the town of Logroño, in the north of Spain) by a monk named Vigila, the Coda Vigilanus contains the nine numerals in question, but not zero. The scribe clearly indicates in the text that the figures are of Indian origin:

    Item de figuels aritmetice. Scire debemus Indos subtilissimum ingenium habere et ceteras gentes eis in arithmetica et geometrica et ceteris liberalibu.c disciplinis concedere. Et hoc manifèstum at in novem figuris, quibus quibus designant unum quenque gradum cuiu.slibetgradus. Quatrum hec sunt forma:

    9 8 7 6 5 4 3 2 1.

"The same applies to arithmetical figures. It should be noted that the Indians have an extremely subtle intelligence, and when it comes to arithmetic, geometry and other such advanced disciplines, other ideas must make way for theirs. The best proof of this is the nine figures with which they represent each number no matter how high. This is how the figures look:

9 8 7 6 5 4 3 2 1

Al-Khwarismi (783-850 CE) Popularized Indian numerals, mathematics including Algebra in the Islamic world and the Christian West .Algebra was named after his treatise 'Al jabr wa'l Muqabalah" which when translated from Arabic means 'Transposition and Reduction'. Little is known about his life except that he lived at the court of the Abbasid Caliph al Ma'amun , in Baghdad shortly after Charlemagne was made emperor of the west, and that he was one of the most important mathematicians and astronomers who worked at the house of Wisdom (Bayt al Hikma)

'The first mathematics which we shall describe in this article developed in the Indus valley. The earliest known urban Indian culture was first identified in 1921 at Harappa in the Punjab and then, one year later, at Mohenjo-Daro, near the Indus River in the Sindh. Both these sites are now in Pakistan but this is still covered by our term "Indian mathematics" which, in this article, refers to mathematics developed in the Indian subcontinent. The Indus civilisation (or Harappan civilisation as it is sometimes known) was based in these two cities and also in over a hundred small towns and villages. It was a civilisation which began around 2500 BC and survived until 1700 BC or later. The people were literate and used a written script containing around 500 characters which some have claimed to have deciphered but, being far from clear that this is the case, much research remains to be done before a full appreciation of the mathematical achievements of this ancient civilisation can be fully assessed. "

The above statement must be revised based on new archaeological discoveries. More than 400 sites have been found along the banks of the dried up river bed of the ancient river Saraswathi. These sites include the submerged city of Bet Dwaraka, the city ruled by Sri Krishna during the episodes of the Mahabharata and the great Bharata war that is described in detail in that epic. The important point to note is that a prerequisite to do numerical work is a script. So, there must have been a script by the time the Saraswathi Sindhu civilization was flourishing not just centred in the two cities of Mohenjo Daro and Harappa but along dozens of urban towns and cities like Dholavira, Lothal, Dwaraka and others. European historians often wonder what happened to the denizens of the Indus Valley civilization. Ockham's razor suggests the right answer . Nothing catastrophic happened to these people and we the modern Indics are the descendants of this civilization which was spread over a huge area stretching from Haryana in the north to the present day province of Maharashtra to places like Prathishtan (later Pathan) which eventually became the capital of the Satavahana Kingdoms are in fact a successor to the Urban civilizations that existed prior to them. This makes eminent sense because the word Brahmi signifies the goddess Saraswathi (consort of Brahma) and is therefore also considered to be the Guardian deity of Knowledge and the one who is credited with blessing us with the gift of a script. There are a group of Brahmanas in the Konkan area of present day state of Karnataka who call themselves Saraswath Brahmanas and legend has it that they migrated from the banks of the Saraswath river when it eventually dried out. In fact the Gowda Saraswath Brahmanas have done extremely well over the succeeding centuries and have prospered far in excess of their proportion in the population. In fact our family records show that about 15 generations ago my ancestor by the name of Hanuman Bhat migrated to the Andhra country , to escape the turmoil caused by the interminable wars and the tyranny of Aurangazeb , from the area which is considered present day Konkan.

We often think of Egyptians and Babylonians as being the height of civilisation and of mathematical skills around the period of the Indus civilisation, yet V G Childe in New Light on the Most Ancient East (1952) wrote:-

India confronts Egypt and Babylonia by the 3rd millennium with a thoroughly individual and independent civilisation of her own, technically the peer of the rest. And plainly it is deeply rooted in Indian soil. The Indus civilisation represents a very perfect adjustment of human life to a specific environment. And it has endured; it is already specifically Indian and forms the basis of modern Indian culture.

3

Vedic Mathematics Teaching Methodology

Learning mathematics should be a delightful experience for all children and they should all succeed in it. The Cosmic Computer Course offers a complete system of mental mathematics which can be taught in a holistic way.

The straightforward and beautifully interrelated Vedic methods mean that mathematics can be done mentally and this, together with the many methods of solution which the Vedic system offers, encourages flexibility and innovation.

This in turn leads to the development of creativity and intuition. The Vedic system does not insist on a purely analytic approach as many modern teaching methods do. This makes a big difference to the attitude which children have towards mathematics.

With its direct, easy and integrated approach this mental system (the methods can also be written down) brings out the naturally coherent and unified structure of mathematics.

The Cosmic Computer Course, for example, uses a Unified Field Chart which shows the whole subject at a glance and how the various parts are related and structured.

Vedic Mathematics Sutras*

Sutras	*Subsutras or Corollaries*
1. Ekadhikena Purvena (By One More than the One Before)	Proportionately
2. Nikhilam Navatascharamam Dastah (All from 9 and the Last from 10)	The Remainder Remains Constant
3. Urdhwa-tiryagbhyam (Vertically and Crosswise)	The First by the First and the Last by the Last
4. Paravartya Yojayet (Transpose and Apply)	For 7 the Multiplicand is 143
5. Sunyam Samyasamuchchaye (If the Samuccaya is the Same it is Zero)	By Osculation
6. (Anurupye) Sunyamanyat (If One is in Ratio the Other is Zero)	Lessen by the Deficiency
7. Sankalana-vyavkalanabhyam (By Addition and by Subtraction)	Whatever the Deficiency lessen by that amount and set up the Square of the Deficiency
8. Puranpuranabhyam (By the Completion or Non-Completion)	Last Totalling 10
9. Chalana-Kalanabhyam (Differential Calculus)	Only the Last Terms
10. Yavdunam (By the Deficiency)	The Sum of the Products
11. Vyastisamastih (Specific and General)	By Alternate Elimination and Retention
12. Sesanyankena Charmena (The Remainders by the Last Digit)	By Mere Observation

Contd...

Sutras	*Subsutras or Corollaries*
13. Sopantyadyaymantyam (The Ultimate and Twice the Penultimate)	The Product of the Sum is the Sum of the Products
14. Ekanyunena Purvena (By One Less than the One Before)	On the Flag**
15. Gunitasamuchachayah (The Product of the Sum)	- - -
16. Gunaksamuchchayah (All the Multipliers)	- - -

* This list is taken from Vedic Mathematics, which includes a full list of the sixteen Sutras in Sanskrit, but in some cases a translation of the Sanskrit is not given in the text and comes from elsewhere.

** This formula is not in the list given in Vedic Mathematics, but is referred to in the text.

AN EXAMPLE

Urdhwa-tiryagbhyam Sutra

This sutra says -"Vertically and Crosswise". That's all to multiply two numbers!

Till now, you were multiplying like this:

Question : Multiply 432 by 617.

Answer :

```
   432
  X 617
-------------
  3024
   432
 2592
==========
 266544
==========
```

More the number of digits in the numbers, more lines and time you consume. No more! Using the Sutra "Vertically and Crosswise", you have

Step 1 : (mentally, don't write on notebook) vertically (last digits) :

2 × 7 = 14; write 4 carry 1

Step 2 (mentally) : crosswise (last two digits) :

3 × 7 + 2 × 1 = 23 + carry 1 = 24; write 4 carry 2

Step 3 : vertically and crosswise (three digits) :

4 × 7 + 3 × 1 +2 × 6 = 43 + carry 2 = 45; write 5 carry 4

Step 4 : (move left; first two digits) :

4 × 1 + 3 × 6 = 22 + carry 4 = 26; write 6 carry 2

Step 5 : (move left; first digit of each number) :

4 × 6 = 24 + carry 2 = 26. End.

Write answer : 266544

This is how it appears on notebook :

```
  432
X 617
-------------
266544
-------------
```

No matter how big the numbers are, you will need to write only the final answer. All other steps are easily carried out mentally. If the two numbers have different number of digits, write smaller number below the other and pad it on left side with zeros.

Why Vedic Mathematics?

Many Indian Secondary School students consider Mathematics a very difficult subject. Some students encounter difficulty with basic arithmetical operations. Some students feel it difficult to manipulate symbols and balance equations. In other words, abstract and logical reasoning is their hurdle.

Many such difficulties in learning Mathematics enter into a long list if prepared by an experienced teacher of Mathematics. Volumes have been written on the diagnosis of 'learning difficulties' related to Mathematics and remedial techniques. Learning Mathematics is an unpleasant experience to some students mainly because it involves mental exercise.

Of late, a few teachers and scholars have revived interest in Vedic Mathematics which was developed, as a system derived from Vedic principles, by Swami Bharati Krishna Tirthaji in the early decades of the 20th century.

Dr. Narinder Puri of the Roorke University prepared teaching materials based on Vedic Mathematics during 1986 - 89. A few of his opinions are stated hereunder:

(i) Mathematics, derived from the Veda, provides one line, mental and super- fast methods along with quick cross checking systems.

(ii) Vedic Mathematics converts a tedious subject into a playful and blissful one which students learn with smiles.

(iii) Vedic Mathematics offers a new and entirely different approach to the study of Mathematics based on pattern recognition. It allows for constant expression of a student's creativity, and is found to be easier to learn.

(iv) In this system, for any problem, there is always one general technique applicable to all cases and also a number of special pattern problems. The element of choice and flexibility at each stage keeps the mind lively and alert to develop clarity of thought and intuition, and thereby a holistic development of the human brain automatically takes place.

(v) Vedic Mathematics with its special features has the inbuilt potential to solve the psychological problem of Mathematics - anxiety.

J. T. Glover (London, 1995) says that the experience of teaching Vedic Mathematics' methods to children has shown that a high degree of mathematical ability can be attained from an early stage while the subject is enjoyed for its own merits.

A. P. Nicholas (1984) puts the Vedic Mathematics system as 'one of the most delightful chapters of the 20th century mathematical history'.

Prof. R. C. Gupta (1994) says 'the system has great educational value because the Sutras contain techniques for performing some elementary mathematical operations in simple ways, and results are obtained quickly'.

Prof. J. N. Kapur says 'Vedic Mathematics can be used to remove math-phobia, and can be taught to (school) children as enrichment material along with other high speed methods'.

Dr. Michael Weinless, Chairman of the Department of Mathematics at the M.I.U, Iowa says thus: 'Vedic Mathematics is easier to learn, faster to use and less prone to error than conventional methods. Furthermore, the techniques of Vedic Mathematics not only enable the students to solve specific mathematical problems; they also develop creativity, logical thinking and intuition.'

Keeping the above observations in view, let us enter Vedic Mathematics as given by Sri Bharati Krishna Tirthaji (1884 - 1960), Shankaracharya of Govardhana Math, Puri. Entering into the methods and procedures, one can realize the importance and applicability of the different formulae (Sutras) and methods.

Vedic Mathematical Formulae

What we call Vedic Mathematics is a mathematical elaboration of 'Sixteen Simple Mathematical formulae from the Vedas' as brought out by Sri Bharati Krishna Tirthaji. In the text authored by the Swamiji, nowhere has the list of the Mathematical formulae (Sutras) been given. But the Editor of the text has compiled the list of the formulae from stray references in the text. The list so compiled contains Sixteen Sutras and Thirteen Sub - Sutras as stated hereunder.

Sixteen Sutras

1. Ekddhikena Purena (also a corollary)
2. Nikhilam Navatascaramam Dasatah
3. Urdhva-tiryagbhyam
4. Paravarya Yojayet
5. Sunyam Samyasamuccaye
6. (Anurupye) Sunyamanyat
7. Sankalana-vyavakalana-bhyam (also a corollary)
8. Puranapuranabhyam
9. Calana-Kalanabhyam
10. Yavadunam
11. Vyastisamastih
12. Sesanyankena Caramena
13. Sopantyadvayamantyam
14. Ekanyunena Purvena
15. Gunitasamuccayah
16. Gunakasamuccayah.

Thirteen Sub-sutras

1. Anurupyena
2. sisyate Sesasamjnah
3. Adyamadyenantya-mantyena

4. Kevalaih Saptakam Gunyat
5. Vestanam
6. Yavadunam Tavadunam
7. Yavadunam Tavadunikrtya Varganca Yojayet
8. Antyayordasakepi
9. Antyayoreva
10. Samuccayagunitah
11. Lopanasthapanabhyam
12. Vilokanam
13. Gunitasamuccayah Samuccayagunitah.

In the text, the words Sutra, aphorism, formula are used synonymously. So are also the words Upa-sutra, Sub-sutra, Sub-formula, corollary used.

Now we shall have the literal meaning, contextual meaning, process, different methods of application along with examples for the Sutras. Explanation, methods, further short-cuts, algebraic proof, etc. follow. What follows relates to a single formula or a group of formulae related to the methods of Vedic Mathematics.

Ekadhikena Purvena

The Sutra (formula) Ekadhikena Purvena means: "By one more than the previous one".

(i) Squares of numbers ending in 5 :

Now we relate the sutra to the 'squaring of numbers ending in 5'. Consider the example 25^2.

Here the number is 25. We have to find out the square of the number. For the number 25, the last digit is 5 and the 'previous' digit is 2. Hence, 'one more than the previous one', that is, 2+1=3. The Sutra, in this context, gives the procedure 'to multiply the previous digit 2 by one more than itself, that is, by 3'. It becomes the L.H.S (left hand side) of the result, that is, $2 \times 3 = 6$. The R.H.S (right hand side) of the result is 52, that is, 25.

Thus $25^2 = 2 \times 3 / 25 = 625$.

In the same way,

$35^2 = 3 \times (3+1) /25 = 3 \times 4/25 = 1225$;

$652 = 6 \times 7 / 25 = 4225$;

$1052 = 10 \times 11/25 = 11025$;

$1352 = 13 \times 14/25 = 18225$;

Apply the formula to find the squares of the numbers 15, 45, 85, 125, 175 and verify the answers.

ALGEBRAIC PROOF:

(a) Consider $(ax + b)^2 \equiv a^2 . x^2 + 2abx + b^2$.

This identity for x = 10 and b = 5 becomes

$$\begin{aligned}(10a + 5)\ 2 &= a^2 . 10^2 + 2. 10a . 5 + 5^2 \\ &= a^2 . 10^2 + a. 10^2 + 5^2 \\ &= (a^2 + a) . 10^2 + 5^2 \\ &= a (a + 1) . 10^2 + 25.\end{aligned}$$

Clearly 10a + 5 represents two-digit numbers 15, 25, 35, -------, 95 for the values a = 1, 2, 3, -------, 9 respectively. In such a case the number $(10a + 5)^2$ is of the form whose L.H.S is a (a + 1) and R.H.S is 25, that is, a (a + 1) / 25.

Thus any such two digit number gives the result in the same fashion.

EXAMPLE:

$45 = (40 + 5)^2$, It is of the form $(ax+b)^2$ for a = 4, x=10 and b = 5. giving the answer a (a+1) / 25 that is, 4 (4+1) / 25 + 4 × 5 / 25 = 2025.

(b) Any three digit number is of the form $ax^2 + bx + c$ for $x = 10, a \neq 0, a, b, c \in W$.

Now $(ax^2+bx+c)^2 = a^2x^4 + b^2x^2 + c^2 + 2abx^3 + 2bcx + 2cax^2$

$= a^2x^4 + 2ab.x^3 + (b^2 + 2ca)x^2 + 2bc.x + c^2.$

This identity for x = 10, c = 5 becomes $(a.10^2 + b.10 + 5)^2$

$= a^2.10^4 + 2.a.b.10^3 + (b^2 + 2.5.a)10^2 + 2.b.5.10 + 5^2$

$= a^2.10^4 + 2.a.b.10^3 + (b^2 + 10\,a)10^2 + b.10^2 + 5^2$

$= a^2.10^4 + 2ab.10^3 + b^2.10^2 + a.10^3 + b.10^2 + 5^2$

$= a^2.10^4 + (2ab + a).10^3 + (b^2 + b)10^2 + 5^2$

$= [a^2.10^2 + 2ab.10 + a.10 + b^2 + b]\,10^2 + 5^2$

$= (10a + b)(10a + b + 1).10^2 + 25$

$= P(P+1)\,10^2 + 25$, where $P = 10a+b$.

Hence any three digit number whose last digit is 5 gives the same result as in (a) for P=10a + b, the 'previous' of 5.

EXAMPLE:

$165^2 = (1.10^2 + 6.10 + 5)^2$.

It is of the form $(ax^2 + bx + c)^2$ for a = 1, b = 6, c = 5 and x = 10. It gives the answer P(P+1) / 25, where P = 10a + b = 10 × 1 + 6 = 16, the 'previous'. The answer is 16 (16 + 1) / 25 = 16 × 17 / 25 = 27225.

Apply Ekadhikena purvena to find the squares of the numbers 95, 225, 375, 635, 745, 915, 1105, 2545.

(ii) Vulgar fractions whose denominators are numbers ending in NINE :

We now take examples of 1 / a9, where a = 1, 2, -----, 9. In the conversion of such vulgar fractions into recurring decimals, Ekadhika process can be effectively used both in division and multiplication.

(a) *Division Method* : Value of 1 / 19.

The numbers of decimal places before repetition is the difference of numerator and denominator, i.e., 19 -1=18 places.

For the denominator 19, the purva (previous) is 1.

Hence Ekadhikena purva (one more than the previous) is 1 + 1 = 2.

The sutra is applied in a different context. Now the method of division is as follows:

Step. 1 : Divide numerator 1 by 20.

i.e., 1 / 20 = 0.1 / 2 = .10 (0 times, 1 remainder)

Step. 2 : Divide 10 by 2

i.e., 0.005 (5 times, 0 remainder)

Step. 3 : Divide 5 by 2

i.e., 0.0512 (2 times, 1 remainder)

Step. 4 : Divide 12 i.e., 12 by 2

i.e., 0.0526 (6 times, No remainder)

Step. 5 : Divide 6 by 2

i.e., 0.05263 (3 times, No remainder)

Step. 6 : Divide 3 by 2

i.e., 0.0526311(1 time, 1 remainder)

Step. 7 : Divide 11 i.e., 11 by 2

i.e., 0.05263115 (5 times, 1 remainder)

Step. 8 : Divide 15 i.e., 15 by 2

i.e., 0.052631517 (7 times, 1 remainder)

Step. 9 : Divide 17 i.e., 17 by 2

i.e., 0.05263157 18 (8 times, 1 remainder)

Step. 10 : Divide 18 i.e., 18 by 2

i.e., 0.0526315789 (9 times, No remainder)

Step. 11 : Divide 9 by 2

i.e., 0.0526315789 14 (4 times, 1 remainder)

Step. 12 : Divide 14 i.e., 14 by 2

i.e., 0.052631578947 (7 times, No remainder)

Step. 13 : Divide 7 by 2

i.e., 0.05263157894713 (3 times, 1 remainder)

Step. 14 : Divide 13 i.e., 13 by 2

i.e., 0.052631578947316 (6 times, 1 remainder)

Step. 15 : Divide 16 i.e., 16 by 2

i.e., 0.052631578947368 (8 times, No remainder)

Step. 16 : Divide 8 by 2

i.e., 0.0526315789473684 (4 times, No remainder)

Step. 17 : Divide 4 by 2

i.e., 0.05263157894736842 (2 times, No remainder)

Step. 18 : Divide 2 by 2

i.e., 0.052631578947368421 (1 time, No remainder)

Now from step 19, i.e., dividing 1 by 2, Step 2 to Step. 18 repeats thus giving

0 ________________ .

1/19 = 0.052631578947368421 or 0.052631578947368421

Note that we have completed the process of division only by using '2'. Nowhere the division by 19 occurs.

(b) *Multiplication Method:* Value of 1 / 19

First we recognize the last digit of the denominator of the type 1 / a9. Here the last digit is 9.

For a fraction of the form in whose denominator 9 is the last digit, we take the case of 1 / 19 as follows:

For 1 / 19, 'previous' of 19 is 1. And one more than of it is 1 + 1 = 2.

Therefore 2 is the multiplier for the conversion. We write the last digit in the numerator as 1 and follow the steps leftwards.

Step. 1 : 1

Step. 2 : 21(multiply 1 by 2, put to left)

Step. 3 : 421(multiply 2 by 2, put to left)

Step. 4 : 8421(multiply 4 by 2, put to left)

Step. 5 : 168421 (multiply 8 by 2 =16,
1 carried over, 6 put to left)

Step. 6 : 1368421 (6 × 2 =12,+1 [carry over]
= 13, 1 carried over, 3 put to left)

Step. 7 : 7368421 (3 × 2, = 6 +1 [Carryover]
= 7, put to left)

Step. 8 : 147368421 (as in the same process)

Step. 9 : 947368421 (Do - continue to step 18)

Step. 10 : 18947368421

Step. 11 : 178947368421

Step. 12 : 1578947368421

Step. 13 : 11578947368421

Step. 14 : 31578947368421

Step. 15 : 631578947368421

Step. 16 : 12631578947368421

Step. 17 : 52631578947368421

Step. 18 : 1052631578947368421

Now from step 18 onwards the same numbers and order towards left continue.

Thus 1 / 19 = 0.052631578947368421

It is interesting to note that we have

(i) not at all used division process

(ii) instead of dividing 1 by 19 continuously, just multiplied 1 by 2 and continued to multiply the resultant successively by 2.

Observations :

(a) For any fraction of the form 1 / a9 i.e., in whose denominator 9 is the digit in the units place and a is the set of remaining digits, the value of the fraction is in recurring decimal form and the repeating block's right most digit is 1.

(b) Whatever may be a9, and the numerator, it is enough to follow the said process with (a+1) either in division or in multiplication.

(c) Starting from right most digit and counting from the right, we see (in the given example 1 / 19)

Sum of 1st digit + 10th digit = 1 + 8 = 9

Sum of 2nd digit + 11th digit = 2 + 7 = 9

- -

Sum of 9th digit + 18th digit = 9 + 0 = 9

From the above observations, we conclude that if we find first 9 digits, further digits can be derived as complements of 9.

(i) Thus at the step 8 in division process we have 0.052631517 and next step. 9 gives 0.052631578

Now the complements of the numbers

0, 5, 2, 6, 3, 1, 5, 7, 8 from 9

9, 4, 7, 3, 6, 8, 4, 2, 1 follow the right order

i.e., 0.052631578947368421

Now taking the multiplication process we have

Step. 8 : 147368421

Step. 9 : 947368421

Now the complements of 1, 2, 4, 8, 6, 3, 7, 4, 9 from 9

i.e., 8, 7, 5, 1, 3, 6, 2, 5, 0 precede in successive steps, giving the answer.

0.052631578947368421.

(d) When we get (Denominator - Numerator) as the product in the multiplicative process, half the work is done. We stop the multiplication there and mechanically write the remaining half of the answer by merely taking down complements from 9.

(e) Either division or multiplication process of giving the answer can be put in a single line form.

ALGEBRAIC PROOF :

Any vulgar fraction of the form 1 / a9 can be written as

1 / a9 = 1 / ((a + 1) x - 1) where x = 10

$$=\frac{1}{(a+1)+\left[\frac{1-1}{(a+1)x}\right]}$$

$$=\frac{1}{(a+1)x}\left[\frac{1-1}{(a+1)x}\right]^{-1}$$

$$=\frac{1}{(a+1)x}\left[1+\frac{1}{(a+1)}x+\frac{1}{(a+1)}x^2+\ldots\right]$$

$$=\frac{1}{(a+1)}x+\frac{1}{(a+1)^2}x^2+\frac{1}{(a+1)^3}x^3+\ldots\ldots \text{ ad inf inf initum}$$

$$=10^{-1}\left(\frac{1}{(a+1)}\right)+10^{-2}\left(\frac{1}{(a+1)^2}\right)+10^{-3}\left(\frac{1}{(a+1)^3}\right)+\ldots \text{ ad inf initum}$$

This series explains the process of ekadhik.

Now consider the problem of 1 / 19. From above we get

$1/19 = 10^{-1}(1/(1+1)) + 10^{-2}(1/(1+1)2) + 10^{-3}(1/(1+1)3)$ +-

(since a=1)

$= 10^{-1}(1/2) + 10^{-2}(1/2)^2 + 10^{-3}(1/3)^3$ + ----------

$= 10^{-1}(0.5) + 10^{-2}(0.25) + 10^{-3}(0.125)$+ ----------

= 0.05 + 0.0025 + 0.000125 + 0.00000625+ - - - - - - -

= 0.052631 - - - - - - -

EXAMPLE 1 :

1. Find 1 / 49 by ekadhikena process.

Now 'previous' is 4.

'One more than the previous' is 4 + 1 = 5.

Now by division right ward from the left by '5'.

1 / 49 = .10 - - - - - - - - - - - -(divide 1 by 50)

= .02 - - - - - - - - - (divide 2 by 5, 0 times, 2 remainder)

= .0220 - - - - - - --(divide 20 by 5, 4 times)

= .0204 - - - - - - -(divide 4 by 5, 0 times, 4 remainder)

= .020440 -- - -- - (divide 40 by 5, 8 times)

= .020408 - - - - - (divide 8 by 5, 1 time, 3 remainder)

= .02040831 - - - -(divide 31 by 5, 6 times, 1 remainder)

= .02040811 6 - - - - - - - continue

= .0204081613322615306111222244448 - -- - - - -

On completing 21 digits, we get 48

i.e.,,Denominator - Numerator = 49 - 1 = 48 stands.

i.e, half of the process stops here. The remaining half can be obtained as complements from 9.

Thus 1 / 49 = 0.020408163265306122448

979591836734693877551

Now finding 1 / 49 by process of multiplication left ward from right by 5, we get

1/49 = ---1

= ---51

= ---2551

= --27551

= ---- 4839472945941183336172334469433837275 51

i.e.,Denominator - Numerator = 49 - 1

= 48 is obtained as 5X9+3

(Carry over) = 45 + 3 = 48.

Hence half of the process is over. The remaining half is automatically obtained as complements of 9.

Thus1 / 49 = --------------979591836734693877551

= 0.020408163265306122448

979591836734693877551

EXAMPLE 2:

Find 1 / 39 by Ekadhika process.

Now by multiplication method, Ekadhikena purva is 3 + 1 = 4

1/39 = -------------------------------------1

= ---------------------------------41

= ------------------------------1641

= -----------------------------25641

= ----------------------------225641

= ---------------------------1025641

Here the repeating block happens to be block of 6 digits. Now the rule predicting the completion of half of the computation does not hold. The complete block has to be computed by ekadhika process. Now continue and obtain the result. Find reasons for the non-applicability of the said 'rule'.

Nikhilam Navatas 'Caramam Dasatah

The formula simply means : "all from 9 and the last from 10".

The formula can be very effectively applied in multiplication of numbers, which are nearer to bases like 10, 100, 1000 i.e., to the powers of 10 . The procedure of multiplication using the Nikhilam involves minimum number of steps, space, time saving and only mental calculation. The numbers taken can be either less or more than the base considered.

The difference between the number and the base is termed as deviation. Deviation may be positive or negative. Positive deviation is written without the positive sign and the negative deviation, is written using Rekhank (a bar on the number). Now observe the following table.

Number	Base	Number - Base	Deviation
14	10	14 - 10	4
8	10	8 - 10	–2 or $\overline{2}$
97	100	97 - 100	–03 or $\overline{03}$
112	100	112 - 100	12
993	1000	993 - 1000	–007 or $\overline{007}$
1011	1000	1011 - 1000	011

Some rules of the method (near to the base) in Multiplication.

(a) Since deviation is obtained by Nikhilam sutra we call the method as Nikhilam multiplication.

E.g. : 94. Now deviation can be obtained by 'all from 9 and the last from 10' sutra i.e., the last digit 4 is from 10 and remaining digit 9 from 9 gives 06.

(b) The two numbers under consideration are written one below the other. The deviations are written on the right hand side.

E.g. : Multiply 7 by 8.

Now the base is 10. Since it is near to both the numbers, we write the numbers one below the other.

7
8

Take the deviations of both the numbers from the base and represent Rekhank or the minus sign before the deviations

7 $\bar{3}$
8 $\bar{2}$
------

or 7 -3
8 -2
----

or remainders 3 and 2 implies that the numbers to be multiplied are both less than 10.

(c) The product or answer will have two parts, one on the left side and the other on the right. A vertical or a slant line i.e., a slash may be drawn for the demarcation of the two parts i.e.,

7	$\bar{3}$		7	-3
8	$\bar{2}$	(or)	8	-2
/			/	

(d) The R.H.S. of the answer is the product of the deviations of the numbers. It shall contain the number of digits equal to number of zeroes in the base.

i.e.,

$$\begin{array}{ll} 7 & \bar{3} \\ 8 & \bar{2} \\ \hline / & (3 \times 2) = 6 \\ \hline \end{array}$$

Since base is 10, 6 can be taken as it is.

(e) L.H.S of the answer is the sum of one number with the deviation of the other. It can be arrived at in any one of the four ways.

(i) Cross-subtract deviation 2 on the second row from the original number 7 in the first row i.e., 7-2 = 5.

(ii) Cross-subtract deviation 3 on the first row from the original number 8 in the second row (converse way of (i)) i.e., 8 - 3 = 5

(iii) Subtract the base 10 from the sum of the given numbers. i.e., (7 + 8) - 10 = 5

(iv) Subtract the sum of the two deviations from the base.

i.e., 10 - (3 + 2) = 5

Hence 5 is left hand side of the answer.

Thus

$$\begin{array}{ll} 7 & \bar{3} \\ 8 & \bar{2} \\ \hline 5 & / \\ \hline \end{array}$$

Now (d) and (e) together give the solution

$$\begin{array}{ll} 7 & \bar{3} \\ 8 & \bar{2} \\ \hline 5 \;/ & 6 \\ \hline \end{array} \text{ i.e., } \begin{array}{r} 7 \\ \times \;\; 8 \\ \hline 56 \\ \hline \end{array}$$

(f) If R.H.S. contains less number of digits than the number of zeros in the base, the remaining digits are filled up by giving zero or zeroes on the left side of the R.H.S. If the number of digits are more than the number of zeroes in the base, the excess digit or digits are to be added to L.H.S of the answer.

The general form of the multiplication under Nikhilam can be shown as follows : Let N_1 and N_2 be two numbers near to a given base in powers of 10, and D_1 and D_2 are their respective deviations from the base. Then $N_1 \times N_2$ can be represented as

$$\begin{array}{ll} N_1 & D_1 \\ N_2 & D_2 \\ \hline (N_1 + D_2) \text{ or } & \\ & / \; D_1 \times D_2 \\ (N_2 + D_1) & \\ \hline \end{array}$$

Case (i)

Both the numbers are lower than the base. We have already considered the example 7 × 8 , with base 10. Now let us solve some more examples by taking bases 100 and 1000 respectively.

EXAMPLE 1

Find 97 × 94. Here base is 100. Now following the rules, the working is as follows:

$$\begin{array}{ll} 97 & \overline{03} \\ 94 & \overline{06} \\ \hline (97-06) \text{ or } & \\ & / \; 3 \times 6 \\ (94-03) & \\ \hline \end{array}$$

EXAMPLE 2

98 × 97 Base is 100.

$$\begin{array}{ll} 98 & \overline{02} \\ 97 & \overline{03} \\ \hline (98-03) \text{ or } & \\ & / \; 2 \times 3 \\ (97-02) & \\ \hline \end{array}$$

EXAMPLE 3

75 × 95. Base is 100.

$$\begin{array}{ll} 75 & \overline{25} \\ 95 & \overline{05} \\ \hline (75-05) \text{ or } (95-25) & / \; 25 \times 5 = 70/125 \text{ (observe rule - f)} \\ & = (70+1)/25 = 7125 \end{array}$$

EXAMPLE 4

986 × 989. Base is 1000.

$$\begin{array}{ll} 986 & \overline{014} \\ 989 & \overline{011} \\ \hline (986-14) \text{ or } (989-14) & / \; 14 \times 11 = 975/154 = 975154 \end{array}$$

EXAMPLE 5

994 × 988. Base is 1000.

$$\begin{array}{ll} 994 & \overline{006} \\ 988 & \overline{012} \\ \hline (994-12) \text{ or } (988-06) & / \; 6 \times 12 = 982/072 \text{ (rule - f)} \\ & = 982072 \end{array}$$

EXAMPLE 6

750 × 995.

$$\begin{array}{ll} 750 & \overline{250} \\ 995 & \overline{005} \\ \hline (750-005) \text{ or } (995-250) & / \; 250 \times 6 = 745/1250 \text{(rule - f)} \\ & = 746250 \end{array}$$

Case (ii)

Both the numbers are higher than the base.

The method and rules follow as they are. The only difference is the positive deviation. Instead of cross - subtract, we follow cross - add.

EXAMPLE 7

13 × 12. Base is 10

$$\begin{array}{ll} 13 & 03 \\ 12 & 02 \\ \hline (13+02) \text{ or } (12+03) & / \; 3\times 2 \end{array} = 15/6 = 156$$

EXAMPLE 8

18 × 14. Base is 10.

$$\begin{array}{ll} 18 & 08 \\ 14 & 04 \\ \hline (18+06) \text{ or } (14+03) & / \; 8\times 4 \end{array} = 22/32 \text{(rule – f, 3 carry over)}$$

$$= 252$$

EXAMPLE 9

104 × 102. Base is 100.

$$\begin{array}{ll} 104 & 04 \\ 102 & 02 \\ \hline 106 & / \; 4\times 2 \end{array} = 10608 \text{ (rule – f)}$$

EXAMPLE 10

1275 × 1004. Base is 1000.

$$\begin{array}{ll} 1275 & 275 \\ 1004 & 004 \\ \hline 1279 & / \; 275\times 4 \end{array} = 1279/1100 \text{(rule – f)}$$

$$= 128100$$

Case (iii)

One number is more and the other is less than the base.

In this situation one deviation is positive and the other is negative. So the product of deviations becomes negative. So the right hand side of the answer obtained will therefore have to be subtracted. To have a clear representation and understanding a vinculum is used. It proceeds into normalization.

EXAMPLE 11

13 × 7. Base is 10

$$\begin{array}{ll} 13 & 03 \\ 7 & \overline{03} \\ \hline (13-03) \text{ or } & / \ 3 \times \overline{3} \\ (7+03) & \\ \hline \end{array} = 10/\overline{9} = 100.9 = 91$$

Note : Conversion of common number into vinculum number and vice versa.

E.g. :

$9 = 10 - 1 = 1\overline{1}$

$98 = 100 - 2 = 10\overline{2}$

$196 = 200 - 4 = 20\overline{4}$

$3\overline{2} = 30 - 2 = 28$

$14\overline{5} = 140 - 5 = 135$

$33\overline{2} = 300 - 22 = 278$. etc.

The procedure can be explained in detail using Nikhilam Navatascaram Dasatah, Ekadhikena purvena, Ekanyunena purvena in the foregoing pages of this book.]

EXAMPLE 12

108 × 94. Base is 100.

$$\begin{array}{ll} 108 & 08 \\ 94 & \overline{06} \\ \hline (108-06) \text{ or } & / \ 8 \times \overline{6} \\ (94+08) & \\ \hline \end{array} = 12/4\overline{8} = 10152$$

(since the complement of 48 is 52 – base 10)

EXAMPLE 13

998 × 1025. Base is 1000.

$$\begin{array}{ll} 998 & \overline{002} \\ 1025 & \overline{025} \\ \hline \begin{array}{l}(998-25) \text{ or} \\ (1025+2)\end{array} \;/ & \bar{2}\times 25 \end{array} = 1023/\overline{050} = 1022950$$

(since the complement of 50 is 950 for the base 1000)

ALGEBRAIC PROOF

Case (i)

Let the two numbers N1 and N2 be less than the selected base say x.

N_1 = (x-a), N_2 = (x-b). Here a and b are the corresponding deviations of the numbers N_1 and N_2 from the base x. Observe that x is a multiple of 10.

Now $N_1 \times N_2$ = (x-a) (x-b) = x.x - x.b - a.x + ab

= x (x - a - b) + ab. [rule - e(iv), d]

= x [(x - a) - b] + ab = x (N_1-b) + ab[rule-e(i),d]

or= x [(x - b) - a] = x (N2 - a) + ab. [rule -e (ii),d]

x (x - a - b) + ab can also be written as

x[(x - a) + (x - b) - x] + ab = x[N_1+N_2 - x] + ab [rule - e(iii),d].

A difficult can be faced, if the vertical multiplication of the deficit digits or deviations i.e., a.b yields a product consisting of more than the required digits. Then rule-f will enable us to surmount the difficulty.

Case (ii)

When both the numbers exceed thc selected base, we have N_1 = x + a, N_2 = x + b, x being the base. Now the identity (x+a) (x+b) = x(x+a+b) + a.b holds good, of course with relevant details mentioned in case (i).

Case (iii)

When one number is less and another is more than the base, we can use (x-a)(x+b) = x(x-a+ b)-ab and the procedure is evident from the examples given.

Find the following products by Nikhilam formula.

1) 7 × 4	**2) 93 × 85**
3) 875 × 994	**4) 1234 × 1002**
5) 1003 × 997	**6) 11112 × 9998**
7) 1234 × 1002	**8) 118 × 105**

Nikhilam in Division

Consider some two digit numbers (dividends) and same divisor 9. Observe the following example.

(i) 13 ÷ 9 The quotient (Q) is 1, Remainder (R) is 4.

since 9) 13 (1
9
―――
4

(ii) 34 ÷ 9, Q is 3, R is 7.

(iii) 60 ÷ 9, Q is 6, R is 6.

(iv) 80 ÷ 9, Q is 8, R is 8.

Now we have another type of representation for the above examples as given hereunder:

(i) Split each dividend into a left hand part for the Quotient and right - hand part for the remainder by a slant line or slash.

E.g..13 as 1 / 3, 34 as 3 / 4 , 80 as 8 / 0.

(ii) Leave some space below such representation, draw a horizontal line.

E.g. 1 / 3 3 / 4 8 / 0

―――― , ―――― , ――――

(iii) Put the first digit of the dividend as it is under the horizontal line. Put the same digit under the right hand part for the remainder, add the two and place the sum i.e., sum of the digits of the numbers as the remainder.

E.g.

```
1 / 3      3 / 4      8 / 0
    1,         3,         8
------     ------     ------
1 / 4      3 / 7      8 / 8
```

Now the problem is over. i.e.,

13 ÷ 9 gives Q = 1, R = 4

34 ÷ 9 gives Q = 3, R = 7

80 ÷ 9 gives Q = 8, R = 8

Proceeding for some more of the two digit number division by 9, we get

(a) 21 ÷ 9 as

```
9)  2 / 1   i.e.  Q = 2, R = 3
        2
   ------
    2 / 3
```

(b) 43 ÷ 9 as

```
9)  4 / 3   i.e.  Q = 4, R = 7
        4
   ------
    4 / 7
```

The examples given so far convey that in the division of two digit numbers by 9, we can mechanically take the first digit down for the quotient - column and that, by adding the quotient to the second digit, we can get the remainder.

Now in the case of 3 digit numbers, let us proceed as follows.

(i) 9) 104 (11 as 9) 10 / 4

$$\begin{array}{r} 9)\ 104\ (11 \\ 99 \\ \hline 5 \end{array} \quad \text{as} \quad \begin{array}{r} 9)\ 10\ /\ 4 \\ 1\ /\ 1 \\ \hline 11\ /\ 5 \end{array}$$

(ii)

$$\begin{array}{r} 9)\ 212\ (23 \\ 207 \\ \hline 5 \end{array} \quad \text{as} \quad \begin{array}{r} 9)\ 21\ /\ 2 \\ 2\ /\ 3 \\ \hline 23\ /\ 5 \end{array}$$

(iii)

$$\begin{array}{r} 9)\ 401\ (44 \\ 396 \\ \hline 5 \end{array} \quad \text{as} \quad \begin{array}{r} 9)\ 40\ /\ 1 \\ 4\ /\ 4 \\ \hline 44\ /\ 5 \end{array}$$

Note that the remainder is the sum of the digits of the dividend. The first digit of the dividend from left is added mechanically to the second digit of the dividend to obtain the second digit of the quotient. This digit added to the third digit sets the remainder. The first digit of the dividend remains as the first digit of the quotient.

Consider 511 ÷ 9

Add the first digit 5 to second digit 1 getting 5 + 1 = 6.

Hence Quotient is 56. Now second digit of 56 i.e., 6 is added to third digit 1 of dividend to get the remainder i.e., 1 + 6 =7

Thus

$$\begin{array}{r} 9)\ 51\ /\ 1 \\ 5\ /\ 6 \\ \hline 56\ /\ 7 \end{array}$$

Q is 56, R is 7.

Extending the same principle even to bigger numbers of still more digits, we can get the results.

E.g : 1204 ÷ 9

(i) Add first digit 1 to the second digit 2. 1 + 2 = 3

(ii) Add the second digit of quotient 13. i.e., 3 to third digit '0' and obtain the Quotient. 3 + 0 = 3, 133

(iii) Add the third digit of Quotient 133 i.e., 3 to last digit '4' of the dividend and write the final Quotient and Remainder. R = 3 + 4 = 7, Q = 133

In symbolic form

```
9) 120 / 4
    13 / 3
   -------
   133 / 7
```

Another example.

```
gives 9) 13210 / 1    132101 ÷ 9
          1467 / 7    Q = 14677, R = 8
         ---------
         14677 / 8
```

In all the cases mentioned above, the remainder is less than the divisor. What about the case when the remainder is equal or greater than the divisor?

```
E.g. 9) 3 / 6               9) 24 /  6
          / 3        or         2 /  6
        -----                  --------
        3 / 9 (equal)          26 / 12 (greater).
```

We proceed by re-dividing the remainder by 9, carrying over this Quotient to the quotient side and retaining the final remainder in the remainder side.

```
9) 3 / 6        9) 24 /  6
     / 3            2 /  6
   -----           -------
   3 / 9           26 / 12
   -----           -------
   4 / 0           27 /  3
```

Q = 4, R = 0 Q = 27, R = 3.

When the remainder is greater than divisor, it can also be represented as:

```
9)      24 / 6
         2 / 6
      ---------
        26 /1 / 2
             / 1
      ---------
         1  / 3
      ---------
        27  / 3
```

Now consider the divisors of two or more digits whose last digit is 9, when divisor is 89.

We Know

113 = 1 × 89 + 24, Q = 1, R = 24

10015 = 112 × 89 + 47, Q = 112, R = 47.

Representing in the previous form of procedure, we have

```
89)  1  /  13        89)  100  /  15
        /  11              12  /  32
    ----------            ----------
     1  /  24             112  /  47
```

But how to get these? What is the procedure?

Now Nikhilam rule comes to rescue us. The nikhilam states "all from 9 and the last from 10". Now if you want to find 113 ÷ 89, 10015 ÷ 89, you have to apply nikhilam formula on 89 and get the complement 11. Further while carrying the added numbers to the place below the next digit, we have to multiply by this 11.

```
89) 1  /  13     89)  100  /  15
       /  11           11  /  11     first digit 1 × 11
   ---------
    1  /  24            1  /   1     total second is 1 × 11
                              22     total of 3rd digit is 2 × 11
                      ----------
                      112  /  47
```

What is 10015 ÷ 98 ? Apply Nikhilam and get 100 - 98 = 02. Set off the 2 digits from the right as the remainder consists of 2 digits. While carrying the added numbers to the place below the next digit, multiply by 02. Thus

```
98)  100  /  15
02    02  /
       0  /  0     i.e., 10015 ÷ 98 gives
          /  04       Q=102,  R = 19
     ------------
     102  /  19
```

In the same way

```
897)  11  /  422
103    1  /   03
          /  206
      -----------
      12  /  658
```

gives 11,422 ÷ 897, Q = 12, R=658.

In this way we have to multiply the quotient by 2 in the case of 8, by 3 in the case of 7, by 4 in the case of 6 and so on. i.e., multiply the Quotient digit by the divisors complement from 10. In case of more digited numbers we apply Nikhilam and proceed. Any how, this method is highly useful and effective for division when the numbers are near to bases of 10.

*** Guess the logic in the process of division by 9.**

*** Obtain the Quotient and Remainder for the following problems.**

(1) 311 ÷ 9

(2) 120012 ÷ 9

(3) 1135 ÷ 97

(4) 2342 ÷ 98

(5) 113401 ÷ 997

(6) 11199171 ÷ 99979

Observe that by nikhilam process of division, even lengthier divisions involve no division or no subtraction but only a few multiplications of single digits with small numbers and a simple addition. But we know fairly well that only a special type of cases are being dealt and hence many questions about various other types of problems arise. The answer lies in Vedic Methods.

Urdhva Tiryagbhyam

Urdhva - tiryagbhyam is the general formula applicable to all cases of multiplication and also in the division of a large number by another large number. It means "Vertically and cross wise."

(a) Multiplication of two 2 digit numbers.

EXAMPLE 1

Find the product 14 × 12

(i) The right hand most digit of the multiplicand, the first number (14) i.e., 4 is multiplied by the right hand most digit of the multiplier, the second number (12) i.e., 2. The product 4 × 2 = 8 forms the right hand most part of the answer.

(ii) Now, diagonally multiply the first digit of the multiplicand (14) i.e., 4 and second digit of the multiplier (12) i.e., 1 (answer 4 × 1 = 4); then multiply the second digit of the multiplicand i.e., 1 and first digit of the multiplier i.e., 2 (answer 1 × 2 = 2); add these two i.e., 4 + 2 = 6. It gives the next, i.e., second digit of the answer. Hence second digit of the answer is 6.

(iii) Now, multiply the second digit of the multiplicand i.e., 1 and second digit of the multiplier i.e., 1 vertically, i.e., 1 × 1 = 1. It gives the left hand most part of the answer.

Thus the answer is 16 8.

Symbolically we can represent the process as follows :

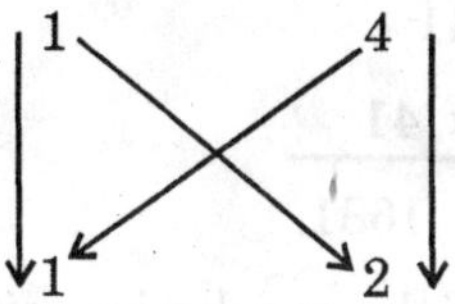

The symbols are operated from right to left .

Step (i) :

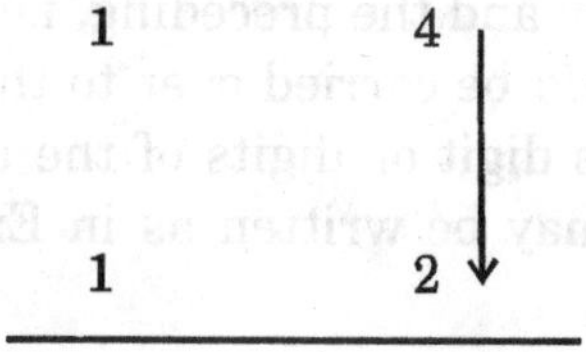

4 × 2

Step (ii) :

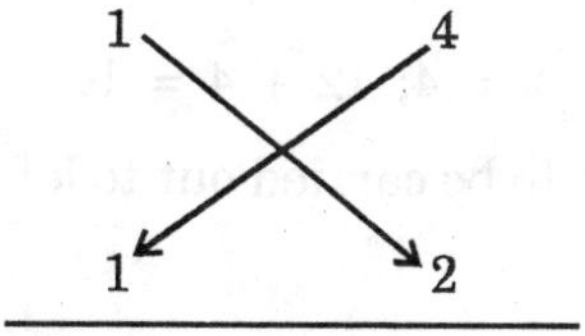

2 + 4 : 8

Step (iii) :

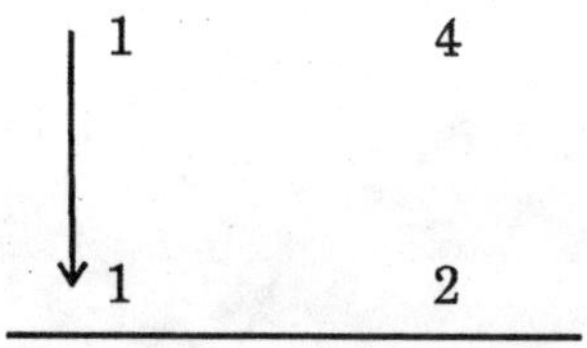

1 × 1 : 6 : 8

which gives 168

Now in the same process, answer can be written as

23
13
2 : 6 + 3 : 9 = 299 (Recall the 3 steps)

EXAMPLE 3

$$\begin{array}{r} 41 \\ \times\ 41 \\ \hline \end{array}$$

$$16 : 4 + 4 : 1 = 1681.$$

What happens when one of the results i.e., either in the last digit or in the middle digit of the result, contains more than 1 digit ? Answer is simple. The right - hand - most digit there of is to be put down there and the preceding, i.e., left -hand -side digit or digits should be carried over to the left and placed under the previous digit or digits of the upper row. The digits carried over may be written as in Ex. 4.

EXAMPLE 4

$$32 \times 24$$

Step (i) : $2 \times 4 = 8$

Step (ii) : $3 \times 4 = 12;\ 2 \times 2 = 4;\ 12 + 4 = 16.$

Here 6 is to be retained. 1 is to be carried out to left side.

Step (iii) : $3 \times 2 = 6.$

Now the carried over digit 1 of 16 is to be added.

i.e., $6 + 1 = 7.$

Thus $32 \times 24 = 768$

We can write it as follows

$$\begin{array}{r} 32 \\ 24 \\ \hline 668 \\ 1 \\ \hline 768. \end{array}$$

Note that the carried over digit from the result $(3\times4) + (2 \times 2) = 12 + 4 = 16$ i.e., 1 is placed under the previous digit $3 \times 2 = 6$ and added.

After sufficient practice, you feel no necessity of writing in this way and simply operate or perform mentally.

EXAMPLE 5

$$28 \times 35.$$

Step (i) : 8 × 5 = 40. 0 is retained as the first digit of the answer and 4 is carried over.

Step (ii) : 2 × 5 = 10; 8 × 3 = 24; 10 + 24 = 34; add the carried over 4 to 34. Now the result is 34 + 4 = 38. Now 8 is retained as the second digit of the answer and 3 is carried over.

Step (iii) : 2 × 3 = 6; add the carried over 3 to 6. The result 6 + 3 = 9 is the third or final digit from right to left of the answer.

Thus 28 × 35 = 980.

EXAMPLE 6

$$\begin{array}{l} 48 \\ \underline{47} \\ 1606 \\ \underline{65} \\ 2256 \end{array}$$

Step (i): 8 × 7 = 56; 5, the carried over digit is placed below the second digit.

Step (ii): (4 × 7) + (8 × 4) = 28 + 32 = 60; 6, the carried over digit is placed below the third digit.

Step (iii): Respective digits are added.

ALGEBRAIC PROOF :

(a) Let the two 2 digit numbers be (ax+b) and (cx+d). Note that x = 10. Now consider the product

$$(ax + b)(cx + d) = ac.x^2 + adx + bcx + b.d$$
$$= ac.x^2 + (ad + bc)x + b.d$$

Observe that:

(i) The first term i.e., the coefficient of x^2 (i.e., 100, hence the digit in the 100th place) is obtained by vertical multiplication of a and c i.e., the digits in 10th place (coefficient of x) of both the numbers;

(ii) The middle term, i.e., the coefficient of x (i.e., digit in the 10th place) is obtained by cross wise multiplication of a and d; and of b and c; and the addition of the two products;

(iii) The last (independent of x) term is obtained by vertical multiplication of the independent terms b and d.

(b) Consider the multiplication of two 3 digit numbers.

Let the two numbers be $(ax^2 + bx + c)$ and $(dx^2 + ex + f)$. Note that x=10

Now the product is

$$\begin{array}{r} ax^2 + bx + c \\ \times\ dx^2 + ex + f \\ \hline \end{array}$$

$ad.x^4+bd.x^3+cd.x^2+ae.x^3+be.x^2+ce.x+af.x^2+bf.x+cf$

$= ad.x^4 + (bd + ae).\ x^3 + (cd + be + af).x^2 + (ce + bf)x + cf$

Note the following points :

(i) The coefficient of x4 , i.e., ad is obtained by the vertical multiplication of the first coefficient from the left side:

$$\begin{array}{l} \downarrow\ ax^2 + bx + c \\ \ \ \ dx^2 + ex + f \\ \hline \ \ \ adx^4 \end{array}$$

(ii) The coefficient of x3 , i.e., (ae + bd) is obtained by the crosswise multiplication of the first two coefficients and by the addition of the two products;

$$ax^2 + bx + c$$
$$dx^2 + ex + f$$
$$aex^3 + bdx^3$$
$$= (ae + bd)x^3$$

(iii) The coefficient of x^2 is obtained by the multiplication of the first coefficient of the multiplicand ($ax^2 + bx + c$) i.e., a; by the last coefficient of the multiplier ($dx^2 + ex + f$) i.e., f ; of the middle one i.e., b of the multiplicand by the middle one i.e., e of the multiplier and of the last one i.e., c of the multiplicand by the first one i.e., d of the multiplier and by the addition of all the three products i.e., af + be +cd :

$$ax^2 \; + \; bx \; + \; c$$
$$dx^2 \; + \; ex \; + \; f$$
$$afx^2 + bex^2 + cdx^2$$
$$= (af + be + cd)x^2$$

(iv) The coefficient of x is obtained by the cross wise multiplication of the second coefficient i.e., b of the multiplicand by the third one i.e., f of the multiplier, and conversely the third coefficient i.e., c of the multiplicand by the second coefficient i.e., e of the multiplier and by addition of the two products, i.e., bf + ce;

$$ax^2 \; + \; bx \; + \; c$$
$$dx^2 \; + \; ex \; + \; f$$
$$bfx + cex$$
$$=(bf + ce)\,x$$

(v) And finally the last (independent of x) term is obtained by the vertical multiplication of the last coefficients c and f i.e., cf

$$\begin{array}{r} ax^2 + bx + c \\ dx^2 + ex + f \downarrow \\ \hline cf \end{array}$$

Thus the process can be put symbolically as (from left to right)

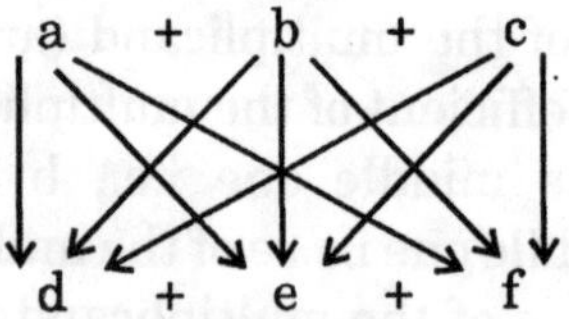

Consider the following example

124 × 132.

Proceeding from right to left

(i) 4 × 2 = 8. First digit = 8

(ii) (2 × 2) + (3 × 4) = 4 + 12 = 16. The digit 6 is retained and 1 is carried over to left side. Second digit = 6.

(iii) (1 × 2) + (2 × 3) + (1 × 4) = 2 + 6 + 4 =12. The carried over 1 of above step is added i.e., 12 + 1 = 13. Now 3 is retained and 1 is carried over to left side. Thus third digit = 3.

(iv) (1 × 3) + (2 × 1) = 3 + 2 = 5. the carried over 1 of above step is added i.e., 5 + 1 = 6 . It is retained. Thus fourth digit = 6

(v) (1 × 1) = 1. As there is no carried over number from the previous step it is retained. Thus fifth digit = 1

124 × 132 = 16368.

Let us work another problem by placing the carried over digits under the first row and proceed.

$$\begin{array}{r} 234 \\ \times\ 316 \\ \hline \end{array}$$

$$\begin{array}{r} 61724 \\ 1222 \\ \hline 73944 \end{array}$$

(i) 4 × 6 = 24 : 2, the carried over digit is placed below the second digit.

(ii) (3 × 6) + (4 x 1) = 18 + 4 = 22 ; 2, the carried over digit is placed below third digit.

(iii) (2 × 6) + (3 × 1) + (4 × 3) = 12 + 3 + 12 = 27 ; 2, the carried over digit is placed below fourth digit.

(iv) (2 × 1) + (3 × 3) = 2 + 9 = 11; 1, the carried over digit is placed below fifth digit.

(v) (2 × 3) = 6.

(vi) Respective digits are added.

Note:

1. We can carry out the multiplication in urdhva - tiryak process from left to right or right to left.
2. The same process can be applied even for numbers having more digits.
3. urdhva -tiryak process of multiplication can be effectively used in multiplication regarding algebraic expressions.

EXAMPLE 1:

Find the product of (a + 2b) and (3a + b).

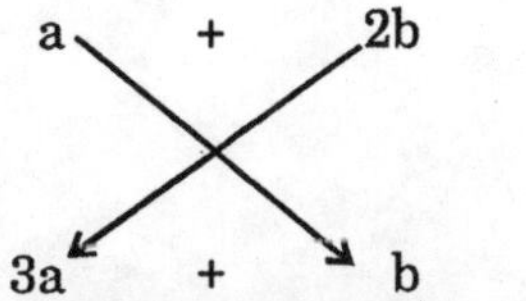

$2a^2 + 7ab + 2b^2$

EXAMPLE 2 :

$$\begin{array}{r} 3a^2 + 2a + 4 \\ \times\ 2a^2 + 5a + 3 \\ \hline \end{array}$$

(i) $4 \times 3 = 12$

(ii) $(2 \times 3) + (4 \times 5) = 6 + 20 = 26$ i.e., $26a$

(iii) $(3 \times 3) + (2 \times 5) + (4 \times 2) = 9 + 10 + 8 = 27$ i.e., $27a^2$

(iv) $(3 \times 5) + (2 \times 2) = 15 + 4 = 19$ i.e., $19\ a^3$

(v) $3 \times 2 = 6$ i.e., $6a^4$

Hence the product is $6a^4 + 19a^3 + 27a^2 + 26a + 12$

EXAMPLE 3 :

Find $(3x^2 + 4x + 7)\ (5x + 6)$

Now $$\begin{array}{r} 3.x^2 + 4x + 7 \\ 0.x^2 + 5x + 6 \\ \hline \end{array}$$

(i) $7 \times 6 = 42$

(ii) $(4 \times 6) + (7 \times 5) = 24 + 35 = 59$ i.e., $59x$

(iii) $(3 \times 6) + (4 \times 5) + (7 \times 0) = 18 + 20 + 0 = 38$ i.e., $38x^2$

(iv) $(3 \times 5) + (0 \times 4) = 15 + 0 = 15$ i.e., $15x^3$

(v) $3 \times 0 = 0$

Hence the product is $15x^3 + 38x^2 + 59x + 42$

Find the products using urdhva tiryagbhyam process.

(1) 25×16

(2) 32×48

(3) 56×56

(4) 137×214

(5) 321×213

(6) 452 × 348

(7) (2x + 3y) (4x + 5y)

(8) $(5a^2 + 1)(3a^2 + 4)$

(9) $(6x^2 + 5x + 2)(3x^2 + 4x + 7)$

(10) $(4x^2 + 3)(5x + 6)$

Urdhva - tiryak in converse for division process:

As per the statement it an used as a simple argumentation for division process particularly in algebra.

Consider the division of $(x^3 + 5x^2 + 3x + 7)$ by (x - 2) process by converse of urdhva - tiryak :

(i) x^3 divided by x gives

$$\frac{x^2.x^3 + 5x^2 + 3x + 7}{x - 2}$$

$Q = x^2 +$ - - - - - - - - - - -

It is the first term of the Quotient.

(ii) $x^2 \times -2 = -2x^2$. But $5x^2$ in the dividend hints $7x^2$ more since 7x2 - 2x2 = 5x2 . This 'more' can be obtained from the multiplication of x by 7x. Hence second term of Q is 7x.

$$\frac{x^3 + 5x^2 + 3x + 7}{x - 2}$$

gives Q = x2 + 7x + - - - - - - - -

(iii) We now have - 2 × 7x = -14x. But the 3rd term in the dividend is 3x for which '17x more' is required since 17x - 14x =3x.

Now multiplication of x by 17 gives 17x. Hence third term of quotient is 17

Thus

$$\frac{x^3 + 5x^2 + 3x + 7}{x - 2}$$

gives Q= x2 + 7x +17

(iv) Now last term of Q, i.e., 17 multiplied by -2 gives $17 \times -2 = -34$ but the relevant term in dividend is 7. So $7 + 34 = 41$ 'more' is required. As there no more terms left in dividend, 41 remains as the remainder.

$$\frac{x^3 + 5x^2 + 3x + 7}{x - 2}$$

gives $Q = x^2 + 7x + 17$ and $R = 41$.

Find the Q and R in the following divisions by using the converse process of urdhva - tiryagbhyam method:

1. $\dfrac{3x^2 - x - 6}{3x - 7}$

2. $\dfrac{16x^2 + 24x + 9}{4x + 3}$

3. $\dfrac{x^3 + 2x^2 + 3x + 5}{x - 3}$

4. $\dfrac{12x^4 - 3x^2 - 3x + 12}{x^2 + 1}$

Paravartya—Yojayet

'Paravartya - Yojayet' means 'transpose and apply'

(i) Consider the division by divisors of more than one digit, and when the divisors are slightly greater than powers of 10.

EXAMPLE 1 :

Divide 1225 by 12.

Step 1 : (From left to right) write the Divisor leaving the first digit, write the other digit or digits using negative (-) sign and place them below the divisor as shown.

```
                12
                -2
              ____
```

Step 2 : Write down the dividend to the right. Set apart the last digit for the remainder.

```
i.e.,        12    122    5
                         - 2
```

Step 3 : Write the 1st digit below the horizontal line drawn under the dividend. Multiply the digit by -2, write the product below the 2nd digit and add.

```
i.e.,        12    122    5
                    -2   -2
                  ____  ____

                         10
```

Since 1 × -2 = -2 and 2 + (-2) = 0

Step 4 : We get second digits' sum as '0'. Multiply the second digits' sum thus obtained by -2 and writes the product under 3rd digit and add.

```
         12      122      5
                 - 2    -20
        ____    ____________

                   102    5
```

Step 5 : Continue the process to the last digit.

```
i.e.,    12      122      5
         - 2     -20     -4
        _____   ____________

                 102      1
```

Step 6: The sum of the last digit is the Remainder and the result to its left is Quotient.

Thus Q = 102 and R = 1

EXAMPLE 2 :

Divide 1697 by 14.

14	1	6	9	7
–4		–4	–8	–4
	1	2	1	3

Q = 121, R = 3.

EXAMPLE 3 :

Divide 2598 by 123.

Note that the divisor has 3 digits. So we have to set up the last two digits of the dividend for the remainder.

1 2 3 25 98 Step (1) & Step (2)

-2-3

Now proceed the sequence of steps write -2 and -3 as follows :

1 2 3	2 5	9 8
-2-3	-4	-6
		-2-3
	2 1	1 5

Since2 × (-2, -3)= -4 , -6; 5 - 4 = 1

and (1 × (-2,-3); 9 - 6 - 2 = 1; 8 - 3 = 5.

Hence Q = 21 and R = 15.

EXAMPLE 4 :

Divide 239479 by 11213. The divisor has 5 digits. So the last 4 digits of the dividend are to be set up for Remainder.

```
1 1 2 1 3        2 3        9 4 7 9
  -1-2-1-3       -2          -4-2-6    with 2
________                     -1-2-1-3    with 1
                             ______________

                 2 1         4 0 0 6
```

Hence Q = 21, R = 4006.

EXAMPLE 5 :

Divide 13456 by 1123

```
      1 1 2 3               1 3 4 5      6
        -1-2-3                -1-2-3
      _______                   -2-4    -6
                          _______________

                            1 2 0-2      0
```

Note that the remainder portion contains -20, i.e., a negative quantity. To over come this situation, take 1 over from the quotient column, i.e., 1123 over to the right side, subtract the remainder portion 20 to get the actual remainder.

Thus Q = 12 - 1 = 11,

and R = 1123 - 20 = 1103.

Find the Quotient and Remainder for the problems using paravartya - yojayet method.

(1) 1234 ÷ 112

(2) 11329 ÷ 1132

(3) 12349 ÷ 133

(4) 239479 ÷ 1203

Now let us consider the application of paravartya - yojayet in algebra.

EXAMPLE 1 :

Divide $6x^2 + 5x + 4$ by $x - 1$

$$\begin{array}{l r}
\underline{X - 1} & 6x^2 + 5x + 4 \\
1 & \underline{6 + 11} \\
 & 6x + 11 + 15
\end{array}$$

Thus Q = 6x+11, R=15.

EXAMPLE 2 :

Divide $x^3 - 3x^2 + 10x - 4$ by $x - 5$

$$\begin{array}{l r}
\underline{X - 5} & x^3 - 3x^2 + 10x - 4 \\
5 & \underline{5 + 10 \quad 100} \\
 & x^2 + 2x + 20, + 96
\end{array}$$

Thus $Q = x^2 + 2x + 20$, R = 96.

The procedure as a mental exercise comes as follows :

(i) x^3 / x gives x^2 i.e., 1 the first coefficient in the Quotient.

(ii) Multiply 1 by + 5, (obtained after reversing the sign of second term in the Quotient) and add to the next coefficient in the dividend. It gives 1 × (+5) = +5, adding to the next coefficient, i.e., -3 + 5 = 2. This is next coefficient in Quotient.

(iii) Continue the process : multiply 2 by +5, i.e., 2 × +5 =10, add to the next coefficient 10 + 10 = 20. This is next coefficient in Quotient. Thus Quotient is $x^2 + 2x + 20$

(iv) Now multiply 20 by + 5 i.e., 20 × 5 = 100. Add to the next (last) term, 100 + (-4) = 96, which becomes R, i.e., R = 9.

EXAMPLE 3:

$$\frac{x^4 - 3x^3 + 7x^2 + 5x + 7}{x + 4}$$

Now thinking the method as in example (1), we proceed as follows.

$$\begin{array}{l|l} \underline{x + 4} & x^4 - 3x^3 + 7x^2 + 5x + 7 \\ & \underline{-4 - 4 + 28 - 140 + 540} \\ & x^3 - 7x2 + 35x - 135 \quad 547 \end{array}$$

Thus $Q = x^3 - 7x^2 + 35x - 135$ and $R = 547$.

or we proceed orally as follows:

x^4/x gives 1 as first coefficient.

(i) $-4 \times 1 = -4$: add to next coefficient $-4 + (-3) = -7$ which gives next coefficient in Q.

(ii) $-7 \times -4 = 28$: then $28 + 7 = 35$, the next coefficient. in Q.

(iii) $35 \times -4 = -140$: then $-140 + 5 = -135$, the next coefficient in Q.

(iv) $-135 \times -4 = 540$: then $540 + 7 = 547$ becomes R.

Thus Q = x3 - 7x2 + 35x - 135 , R = 547.

Note :

1. We can follow the same procedure even the number of terms is more.
2. If any term is missing, we have to take the coefficient of the term as zero and proceed.

Now consider the divisors of second degree or more as in the following example.

EXAMPLE :4

$2x^4 - 3x^3 - 3x + 2$ by $x^2 + 1$.

Here x^2 term is missing in the dividend. Hence treat it as $0 . x^2$ or 0 . And the x term in divisor is also absent we treat it as $0 . x$. Now

$$\begin{array}{llll} x^2 & +1 & & 2x^4 - 3x^3 + 0\,.\,x^2 - 3x + 2 \\ x^2 + 0\,.\,x + 1 & & & \quad 0 \quad -2 \\ \hline 0 & -1 & & \quad\quad 0 \quad +3 \\ & & & \quad\quad\quad\quad 0 + 2 \\ \hline & & & 2 \;\; -3 \;\; -2 \quad\quad 0 \quad 4 \end{array}$$

Thus Q = $2x^2 - 3x - 2$ and R = $0\,.\,x + 4 = 4$.

EXAMPLE 5 :

$2x^5 - 5x^4 + 3x^2 - 4x + 7$ by $x^3 - 2x^2 + 3$.

We treat the dividend as $2x^5 - 5x^4 + 0.\,x^3 + 3x^2 - 4x + 7$ and divisor as $x^3 - 2x^2 + 0\,.\,x + 3$ and proceed as follows:

$$\begin{array}{lll} x^3 - 2x^2 + 0\,.\,x + 3 & & 2x^5 - 5x^4 + 0.x^3 + 3x^2 - 4x + 7 \\ \hline 2 \quad 0 \quad -3 & & 4 \quad 0 \quad -6 \\ & & \quad -2 \quad 0 \quad\quad +3 \\ & & \quad\quad -4 \quad\quad 0 \quad +6 \\ \hline & & 2 \;\; -1 \;\; -2 \;\; -7 \quad -1 \;\; +13 \end{array}$$

Thus Q = $2x^2 - x - 2$, R = $- 7x^2 - x + 13$.

You may observe a very close relation of the method paravartya in this aspect with regard to REMAINDER THEOREM and HORNER PROCESS of Synthetic division. And yet paravartya goes much farther and is capable of numerous applications in other directions also.

Apply paravartya - yojayet to find out the Quotient and Remainder in each of the following problems.

(1) $(4x^2 + 3x + 5) \div (x+1)$

(2) $(x^3 - 4x^2 + 7x + 6) \div (x - 2)$

(3) $(x^4 - x^3 + x^2 + 2x + 4) \div (x2 - x - 1)$

(4) $(2x^5 + x^3 - 3x + 7) \div (x3 + 2x - 3)$

(5) $(7x^6 + 6x^5 - 5x^4 + 4x^3 - 3x^2 + 2x - 1) \div (x-1)$

Paravartya in solving simple equations :

Recall that 'paravartya yojayet' means 'transpose and apply'. The rule relating to transposition enjoins invariable change of sign with every change of side. i.e., + becomes - and conversely; and X becomes ÷ and conversely. Further it can be extended to the transposition of terms from left to right and conversely and from numerator to denominator and conversely in the concerned problems.

Type (i) :

Consider the problem

$$7x - 5 = 5x + 1$$

$$7x - 5x = 1 + 5$$

i.e.,$2x = 6$ $x = 6 \div 2 = 3$.

Observe that the problem is of the type $ax + b = cx + d$ from which we get by 'transpose' $(d - b)$, $(a - c)$ and

$$x = \frac{d-b}{a-c}$$

In this example $a = 7$, $b = -5$, $c = 5$, $d = 1$

Hence

$$x = \frac{1-(-5)}{7-5} = \frac{1+5}{7-5} = \frac{6}{2} = 3$$

EXAMPLE 2:

Solve for x, $3x + 4 = 2x + 6$

$$x = \frac{d-b}{a-c} = \frac{6-4}{3-2} = \frac{2}{1} = 1$$

Type (ii):

Consider problems of the type $(x + a)(x+b) = (x+c)(x+d)$. By paravartya, we get

$$x = \frac{cd - ab}{(a+b)-(c+d)}$$

It is trivial form the following steps

$(x + a)(x + b) = (x + c)(x + d)$

$x^2 + bx + ax + ab = x^2 + dx + cx + cd$

$bx + ax - dx - cx = cd - ab$

$x(a + b - c - d) = cd - ab$

$$x = \frac{cd - ab}{a+b-c-d} \quad x = \frac{cd - ab}{(a+b)-(c+d)}$$

EXAMPLE 1 :

$(x - 3)(x - 2) = (x + 1)(x + 2).$

By paravartya

$$x = \frac{cd - ab}{a+b-c-d} = \frac{1(2)-(-3)(-2)}{-3-2-1-1}$$

$$= \frac{2-6}{-8} = \frac{-4}{-8} = \frac{1}{2}$$

EXAMPLE 2 :

$(x + 7)(x - 6) = (x + 3)(x - 4).$

Now

$$x = \frac{cd - ab}{a+b-c-d} = \frac{(3)(-4)-(7)(-6)}{7+(-6)-3-(-4)}$$

$$= \frac{-12+42}{7-6-3+4} = \frac{30}{2} = 15$$

Note that if $cd - ab = 0$ i.e., $cd = ab$, i.e., if the product of the absolute terms be the same on both sides, the numerator becomes zero giving $x = 0$.

For the problem (x + 4) (x + 3) = (x - 2) (x - 6)

Solution is x = 0 since 4 × 3 = - 2 × - 6. = 12

Type (iii) :

Consider the problems of the type

$$\frac{ax+b}{cx+d}=\frac{m}{n}$$

By cross - multiplication,

$$n(ax + b) = m(cx + d)$$

$$nax + nb = mcx + md$$

$$nax - mcx = md - nb$$

$$x(na - mc) = md - nb$$

$$x=\frac{md-nb}{na-mc}.$$

Now look at the problem once again

$$\frac{ax+b}{cx+d}=\frac{m}{n}$$

paravartya gives md - nb, na - mc and

$$x=\frac{md-nb}{na-mc}.$$

EXAMPLE 1:

$$\frac{3x+1}{4x+3}=\frac{13}{19}$$

$$x=\frac{md-nb}{na-mc}=\frac{13(3)-19(1)}{19(3)-13(4)}$$

$$=\frac{39-19}{57-52}=\frac{20}{5}=4.$$

EXAMPLE 2:

$$\frac{4x+5}{3x+\frac{13}{2}}=\frac{7}{8}$$

$$x=\frac{(7)\left(\frac{13}{2}\right)-(8)(5)}{(8)(4)-(7)(3)}=\frac{\left(\frac{91}{2}\right)-40}{32-21}$$

$$=\frac{\frac{(91-80)}{2}}{32-21}=\frac{11}{2\times 11}=\frac{1}{2}.$$

Type (iv) :

Consider the problems of the type

$$\frac{m}{x+a}+\frac{n}{x+b}=0$$

Take L.C.M and proceed.

$$\frac{m(x+b)+n(x+a)}{(x+a)(x+b)}=0$$

$$\frac{mx+mb+nx+na}{(x+a)(x+b)}=0$$

$$(m + n)x + mb + na = 0$$

$$\therefore \quad (m + n)x = -mb - na$$

$$x=\frac{-mb-na}{(m+n)}.$$

Thus the problem $\frac{m}{x+a}+\frac{n}{x+b}=0$, by paravartya process gives directly

$$x=\frac{-mb-na}{(m+n)}.$$

EXAMPLE 1 :

$$\frac{3}{x+4}+\frac{4}{x-6}=0$$

gives

$$x=\frac{-mb-na}{(m+n)}.$$

Note that m = 3, n = 4, a = 4, b = - 6

$$=\frac{-(3)(-6)-(4)(4)}{(3+4)}=\frac{18-16}{7}=\frac{2}{7}.$$

EXAMPLE 2 :

$$\frac{5}{x+1}+\frac{6}{x-21}=0$$

gives

$$x=\frac{-(5)(-21)-(6)(1)}{5+6}$$

$$=\frac{105-6}{11}=\frac{99}{11}=9$$

I. Solve the following problems using the sutra Paravartya - yojayet.

(1) 3x + 5 = 5x - 3

(2) (2x/3) + 1 = x - 1

(3) $\frac{7x+2}{3x-5}=\frac{5}{8}$

(4) $\frac{x+1/3}{3x-1}=1$

(5) $\frac{5}{x+3}+\frac{2}{x-4}=0$

(6) (x + 1) (x + 2) = (x - 3) (x - 4)

(7) (x - 7) (x - 9) = (x - 3) (x - 22)

(8) (x + 7) (x + 9) = (x + 3) (x + 21)

(II)

1. Show that for the type of equations

$$\frac{m}{x+a}+\frac{n}{x+b}+\frac{p}{x+c}=0, \text{ the solution is}$$

$$x=\frac{-mbc-nca-pab}{m(b+c)+n(c+a)+p(a+b)}, \text{ if } m+n+p=0.$$

2. Apply the above formula to set the solution for the problem

$$\frac{3}{x+4}+\frac{2}{x+6}-\frac{5}{x+5}=0.$$

Some more simple solutions :

$$\frac{m}{x+a}+\frac{n}{x+b}+\frac{m+n}{x+c}$$

Now this can be written as,

$$\frac{m}{x+a}+\frac{n}{x+b}=\frac{m}{x+c}+\frac{n}{x+c}$$

$$\frac{m}{x+a}-\frac{m}{x+c}=\frac{n}{x+c}-\frac{n}{x+b}$$

$$\frac{m(x+c)-m(x+a)}{(x+a)(x+c)}=\frac{n(x+b)-n(x+c)}{(x+c)(x+b)}$$

$$\frac{mx+mc-mx-ma}{(x+a)(x+c)}=\frac{nx+nb-nx-nc}{(x+c)(x+b)}$$

$$\frac{m(c-a)}{x+a} = \frac{n(b-c)}{x+b}$$

m (c - a).x + m (c - a).b = n (b - c). x + n(b - c).a

x [m(c - a) - n(b - c)] = na(b - c) - mb (c - a)

orx [m(c - a) + n(c - b)]= na(b - c) + mb (a - c)

Thus

$$x = \frac{mb(a-c)+na(b-c)}{m(c-a)+n(c-b)}.$$

By paravartya rule we can easily remember the formula.

EXAMPLE 1 :

Solve $\frac{3}{x+1} + \frac{4}{x+2} = \frac{7}{x+3}$

In the usual procedure, we proceed as follows.

$$\frac{3}{x+1} + \frac{4}{x+2} = \frac{7}{x+3}$$

$$\frac{3(x+2)+4(x+1)}{(x+1)(x+2)} = \frac{7}{x+3}$$

$$\frac{3x+6+4x+4}{x^2+2x+x+2} = \frac{7}{x+3}$$

$(7x + 10) (x + 3) = 7(x^2 + 3x + 2)$

$7x^2 + 21x + 10x + 30 = 7x^2 + 21x + 14.$

$31x + 30 = 21x + 14$

$31x - 21x = 14 - 30$

i.e., $10x = - 16$

$x = - 16 / 10 = - 8 / 5$

Now by Paravartya process

$$\frac{3}{x+1}+\frac{4}{x+2} = \frac{7}{x+3}$$

$(\ldots N_1 + N_2 = 3+4 = 7 = N_3)$

$$x = \frac{mb(a-c)+na(b-c)}{m(c-a)+n(c-b)}$$

here $N_1 = m = 3$, $N_2 = n = 4$; $a = 1$, $b = 2$, $c = 3$

$$= \frac{3.2(1-3)+4.1.(2-3)}{3(3-1)+4(3-2)}$$

$$= \frac{6(-2)+4(-1)}{3(2)+4(1)} = \frac{-12-4}{6+4}$$

$$= \frac{-16}{10} = \frac{-8}{5}$$

EXAMPLE 2 :

$$\frac{3}{x-2}+\frac{5}{x-6} = \frac{8}{x+3}$$

Here $N_1 + N_2 = 3 + 5 = 8$.

$$x = \frac{mb(a-c)+na(b-c)}{m(c-a)+n(c-b)}$$

$$= \frac{3.(-6)(-2-3)+5.(-2)(6-3)}{3(3-(-2))+5(-3(-6))}$$

$$= \frac{3(-6)(-5)+5(-2)(-9)}{3(3+2)+5(3+6)}$$

$$= \frac{90+90}{15+45} = \frac{180}{60} = 3.$$

Solve the problems using the methods explained above.

(1) $\frac{2}{x+2}+\frac{3}{x+3}=\frac{5}{x+5}$

(2) $\frac{4}{x+1}+\frac{6}{x+3}=\frac{10}{x+4}$

(3) $\frac{5}{x-2}+\frac{2}{3-x}=\frac{3}{x-4}$

(4) $\frac{4}{2x+1}+\frac{9}{3x+2}=\frac{15}{3x+1}$

Note : The problem (4) appears to be not in the model said above.

But $\frac{3(4)}{3(2x+1)}+\frac{2(9)}{2(3x+2)}=\frac{2(15)}{2(3x+1)}$ gives

$$\frac{12}{6x+3}+\frac{18}{6x+4}=\frac{30}{6x+2}.$$

Now proceed.

Simultaneous Simple Equations

By applying Paravartya sutra we can derive the values of x and y which are given by two simultaneous equations. The values of x and y are given by ration form. The method to find out the numerator and denominator of the ratio is given below.

EXAMPLE 1:

$2x + 3y = 13,\ 4x + 5y = 23.$

(i) To get x, start with y coefficients and the independent terms and cross-multiply forward, i.e., right ward. Start from the upper row and multiply across by the lower one, and conversely, the connecting link between the two cross-products being a minus. This becomes numerator.

i.e., $2x + 3y = 13$

$4x + 5y = 23$

Numerator of the x - value is $3 \times 23 - 5 \times 13 = 69 - 65 = 4$

(ii) Go from the upper row across to the lower one, i.e., the x- coefficient but backward, i.e., leftward.

Denominator of the x - value is $3 \times 4 - 2 \times 5 = 12 - 10 = 2$

Hence value of $x = 4 \div 2 = 2$.

(iii) To get y, follow the cyclic system, i.e., start with the independent term on the upper row towards the x-coefficient on the lower row. So numerator of the y-value is

$13 \times 4 - 23 \times 2 = 52 - 46 = 6$.

(iv) The denominator is the same as obtained in Step (ii) i.e., 2. Hence value of y is $6 \div 2 = 3$.

Thus the solution to the given equation is $x = 2$ and $y = 3$.

EXAMPLE 2:

$5x - 3y = 11$

$6x - 5y = 09$

NowNr. of x is $(-3)(9) - (5)(11) = -27 + 55 = 28$

Dr. of x is $(-3)(6) - (5)(-5) = -18 + 25 = 07$

$x = Nr \div Dr. = 28 \div 7 = 4$

and for y,

Nr is $(11)(6) - (9)(5) = 66 - 45 = 21$

Dr. is 7

Hence $y = 21 \div 7 = 3$.

EXAMPLE 3:

Solve $3x + y = 5$

$4x - y = 9$

Now we can straight away write the values as follows:

$$x = \frac{(1)(9)-(-1)(5)}{(1)(4)-(3)(-1)} = \frac{9+5}{4+3} = \frac{14}{7} = 2$$

$$y = \frac{(5)(4)-(9)(3)}{(1)(4)-(3)(-1)} = \frac{20-27}{4+3} = \frac{-7}{7} = -1.$$

Hence x = 2 and y = -1 is the solution.

ALGEBRAIC PROOF:

$ax + by = m$ .... (i)

$cx + dy = n$ .... (ii)

Multiply (i) by d and (ii) by b, then subtract

$adx + bdy = m.d$

$cbx + dby = n.b$

$(ad - cb) .x = md - nb$

$$x = \frac{md - nb}{ad - cb} = \frac{bn - md}{bc - ad}$$

Multiply (i) by c and (ii) by a, then subtract

$acx + bcy = m.c$

$cax + day = n.a$

$(bc - ad) . y = mc - na$

$$y = \frac{mc - na}{bc - ad}$$

You feel comfort in the Paravartya process because it avoids the confusion in multiplication, change of sign and such other processes.

Find the values of x and y in each of the following problems using Paravartya process.

1. $2x + y = 5$

   $3x - 4y = 2$

2. $3x - 4y = 7$

   $5x + y = 4$

3. $4x + 3y = 8$

   $6x - y = 1$

4. $x + 3y = 7$

   $2x + 5y = 11$

Sunyam Samyasamuccaye

The Sutra 'Sunyam Samyasamuccaye' says the 'Samuccaya is the same, that Samuccaya is Zero.' i.e., it should be equated to zero. The term 'Samuccaya' has several meanings under different contexts.

(i) We interpret, 'Samuccaya' as a term which occurs as a common factor in all the terms concerned and proceed as follows.

EXAMPLE 1:

The equation $7x + 3x = 4x + 5x$ has the same factor ' x ' in all its terms. Hence by the sutra it is zero, i.e., $x = 0$.

Otherwise we have to work like this:

$$7x + 3x = 4x + 5x$$

$$10x = 9x$$

$$10x - 9x = 0$$

$$x = 0$$

This is applicable not only for 'x' but also any such unknown quantity as follows.

EXAMPLE 2:

$$5(x+1) = 3(x+1)$$

No need to proceed in the usual procedure like

$$5x + 5 = 3x + 3$$

$$5x - 3x = 3 - 5$$

$$2x = -2$$

or $$x = -2 \div 2 = -1$$

Simply think of the contextual meaning of ' Samuccaya'

Now Samuccaya is (x + 1)

$$x + 1 = 0 \text{ gives } x = -1$$

(ii) Now we interpret 'Samuccaya 'as product of independent terms in expressions like (x+a) (x+b)

EXAMPLE 3:

$$(x + 3)(x + 4) = (x - 2)(x - 6)$$

Here Samuccaya is 3 x 4 = 12 = -2 × -6

Since it is same , we derive x = 0

This example, we have already dealt in type (ii) of Paravartya in solving simple equations.

(iii) We interpret ' Samuccaya ' as the sum of the denominators of two fractions having the same numerical numerator.

Consider the example.

$$\frac{1}{3x-2} + \frac{1}{2x-1} = 0$$

for this we proceed by taking L.C.M.

$$\frac{(2x-1)+(3x-2)}{(3x-2)(2x-1)} = 0$$

$$\frac{5x-3}{(3x-2)(2x-1)} = 0$$

$$5x - 3 = 0$$

$$5x = 3$$

$$x = \frac{3}{5}$$

Instead of this, we can directly put the Samuccaya i.e., sum of the denominators

i.e., $3x - 2 + 2x - 1 = 5x - 3 = 0$

giving $5x = 3$

$$x = 3 / 5$$

It is true and applicable for all problems of the type

$$\frac{m}{ax+b} + \frac{m}{cx+d} = 0$$

Samuccaya is ax+b+cx+d and solution is ($m \neq 0$)

$$x = \frac{-(b+d)}{(a+c)}$$

(iii) We now interpret 'Samuccaya' as combination or total.

If the sum of the numerators and the sum of the denominators be the same, then that sum = 0.

Consider examples of type

$$\frac{ax+b}{ax+c} = \frac{ax+c}{ax+b}$$

In this case, (ax+b) (ax+b) = (ax+c) (ax+c)

$$a^2x^2 + 2abx + b^2 = a^2x^2 + 2acx + c^2$$

$$2abx - 2acx = c^2 - b^2$$

$$x\,(2ab - 2ac) = c^2 - b^2$$

$$x = \frac{c^2 - b^2}{2a(b-c)}$$

$$= \frac{(c+b)(c-b)}{2a(b-c)} = \frac{-(c+b)}{2a}$$

As per Samuccaya

$$(ax+b) + (ax+c) = 0$$

$$2ax+b+c = 0$$

$$2ax = -b-c$$

$$x = \frac{-(c+b)}{2a}$$

Hence the statement.

EXAMPLE 4:

$$\frac{3x+4}{3x+5} = \frac{3x+5}{3x+4}$$

Since$N_1 + N_2 = 3x + 4 + 3x + 5 = 6x + 9$,

And$D_1 + D_2 = 3x + 4 + 3x + 5 = 6x + 9$

We have $N_1 + N_2 = D_1 + D_2 = 6x + 9$

Hence from Sunya Samuccaya we get 6x + 9 = 0

$$6x = -9$$

$$x = \frac{-9}{6} = \frac{-3}{2}$$

EXAMPLE 5:

$$\frac{5x+7}{5x+12}=\frac{5x+12}{5x+7}$$

Hence $N_1 + N_2 = 5x + 7 + 5x + 12 = 10x + 19$

And$D_1 + D_2 = 5x + 12 + 5x + 7 = 10x + 19$

$N_1 + N_2 = D_1 + D_2$ gives $10x + 19 = 0$

$$10x = -19$$

$$x=\frac{-19}{10}.$$

Consider the examples of the type, where $N_1 + N_2 = K (D_1 + D_2)$, where K is a numerical constant, then also by removing the numerical constant K, we can proceed as above.

EXAMPLE 6:

$$\frac{2x+3}{4x+5}=\frac{x+1}{2x+3}$$

Here $N_1 + N_2 = 2x + 3 + x + 1 = 3x + 4$

$D_1 + D_2 = 4x + 5 + 2x + 3 = 6x + 8 = 2 (3x + 4)$

Removing the numerical factor 2, we get 3x + 4 on both sides.

$3x + 4 = 0$ $3x = -4$ $x = - 4/3$.

(v) 'Samuccaya' with the same meaning as above, i.e., case (iv), we solve the problems leading to quadratic equations. In this context, we take the problems as follows;

If $N_1 + N_2 = D_1 + D_2$ and also the differences

$N_1 \sim D_1 = N_2 \sim D_2$ then both the things are equated to zero, the solution gives the two values for x.

EXAMPLE 7:

$$\frac{3x+2}{2x+5} = \frac{2x+5}{3x+2}$$

In the conventional text book method, we work as follows:

$$\frac{3x+2}{2x+5} = \frac{2x+5}{3x+2}$$

$$(3x + 2)(3x + 2) = (2x + 5)(2x + 5)$$

$$9x^2 + 12x + 4 = 4x^2 + 20x + 25$$

$$9x^2 + 12x + 4 - 4x^2 - 20x - 25 = 0$$

$$5x^2 - 8x - 21 = 0$$

$$5x^2 - 15x + 7x - 21 = 0$$

$$5x(x - 3) + 7(x - 3) = 0$$

$$(x - 3)(5x + 7) = 0$$

$$x - 3 = 0 \text{ or } 5x + 7 = 0$$

$$x = 3 \text{ or } -7/5$$

Now 'Samuccaya' sutra comes to help us in a beautiful way as follows :

Observe $N_1 + N_2 = 3x + 2 + 2x + 5 = 5x + 7$

$D_1 + D_2 = 2x + 5 + 3x + 2 = 5x + 7$

Further $N_1 \sim D_1 = (3x + 2) - (2x + 5) = x - 3$

$N_2 \sim D_2 = (2x + 5) - (3x + 2)$

$= -x + 3 = -(x - 3)$

Hence $5x + 7 = 0$, $x - 3 = 0$

$5x = -7$, $x = 3$

i.e., $x = -7/5$, $x = 3$

Note that all these can be easily calculated by mere observation.

EXAMPLE 8:

$$\frac{3x+4}{6x+7} = \frac{5x+6}{2x+3}$$

Observe that

$N_1 + N_2 = 3x + 4 + 5x + 6 = 8x + 10$

and $D_1 + D_2 = 6x + 7 + 2x + 3 = 8x + 10$

Further $N_1 \sim D_1 = (3x + 4) - (6x + 7)$

$= 3x + 4 - 6x - 7$

$= -3x - 3 = -3 (x + 1)$

$N_2 \sim D_2 = (5x + 6) - (2x + 3)$

$= 3x + 3 = 3(x + 1)$

By 'Sunyam Samuccaye' we have

$8x + 10 = 0$	$3(x + 1) = 0$
$8x = -10$	$x + 1 = 0$
$x = - 10 / 8$	$x = -1$
$= - 5 / 4$	

(vi) 'Samuccaya' with the same sense but with a different context and application.

EXAMPLE 9:

$$\frac{1}{x-4} + \frac{1}{x-6} = \frac{1}{x-2} + \frac{1}{x-8}$$

Usually we proceed as follows.

$$\frac{x-6+x-4}{(x-4)(x-6)} = \frac{x-8+x-2}{(x-2)(x-8)}$$

$$\frac{2x-10}{x^2-10x+24} = \frac{2x-10}{x^2-10x+16}$$

(2x - 10) (x2 - 10x + 16) = (2x - 10) (x2 - 10x + 24)

$$2x^3-20x^2+32x-10x^2+100x-160 = 2x^3-20x^2+48x-10x^2+100x-240$$

$$2x^3 - 30x^2 + 132x - 160 = 2x^3 - 30x^2 + 148x - 240$$

$$132x - 160 = 148x - 240$$

$$132x - 148x = 160 - 240$$

$$-16x = -80$$

$$x = -80 / -16 = 5$$

Now 'Samuccaya' sutra, tell us that, if other elements being equal, the sum-total of the denominators on the L.H.S. and their total on the R.H.S. be the same, that total is zero.

Now $\quad D_1 + D_2 = x - 4 + x - 6 = 2x - 10$, and

$$D_3 + D_4 = x - 2 + x - 8 = 2x - 10$$

By Samuccaya, $2x - 10$ gives $2x = 10$

$$x = \frac{10}{2} = 5.$$

EXAMPLE 10:

$$\frac{1}{x-8} + \frac{1}{x-9} = \frac{1}{x-5} + \frac{1}{x-12}$$

$$D_1 + D_2 = x - 8 + x - 9 = 2x - 17, \text{ and}$$

$$D_3 + D_4 = x - 5 + x - 12 = 2x - 17$$

Now $\quad 2x - 17 = 0$ gives $2x = 17$

$$x = \frac{17}{2} = 8\frac{1}{2}.$$

EXAMPLE 11:

$$\frac{1}{x+7} - \frac{1}{x+10} = \frac{1}{x+6} - \frac{1}{x+9}$$

This is not in the expected form. But a little work regarding transposition makes the above as follows.

$$\frac{1}{x+7}+\frac{1}{x+10} = \frac{1}{x+6}-\frac{1}{x+9}$$

Now 'Samuccaya' sutra applies

$D_1 + D_2 = x + 7 + x + 9 = 2x + 16$, and

$D_3 + D_4 = x + 6 + x + 10 = 2x + 16$

Solution is given by $2x + 16 = 0$ i.e., $2x = -16$.

$x = -16 / 2 = -8.$

Solve the following problems using Sunyam Samya-Samuccaye process.

1. $7(x + 2) + 3(x + 2) = 6(x + 2) + 5(x + 2)$
2. $(x + 6)(x + 3) = (x - 9)(x - 2)$
3. $(x - 1)(x + 14) = (x + 2)(x - 7)$
4. $\frac{1}{4x-3}+\frac{1}{x-2}=0$
5. $\frac{4}{3x+1}+\frac{4}{5x+7}=0$
6. $\frac{2x+11}{2x+5}=\frac{2x+5}{2x+11}$
7. $\frac{3x+4}{6x+7}=\frac{x+1}{2x+3}$
8. $\frac{4x-3}{2x+3}=\frac{x+4}{3x-2}$
9. $\frac{1}{x-2}+\frac{1}{x-5}=\frac{1}{x-3}+\frac{1}{x-4}$
10. $\frac{1}{x-7}-\frac{1}{x-6}=\frac{1}{x-10}-\frac{1}{x-9}$

Sunyam Samya Samuccaye in Certain Cubes

Consider the problem $(x - 4)^3 + (x - 6)^3 = 2(x - 5)^3$. For the solution by the traditional method we follow the steps as given below:

$$(x - 4)^3 + (x - 6)^3 = 2(x - 5)^3$$

$$x^3 - 12x^2 + 48x - 64 + x^3 - 18x^2 + 108x - 216$$

$$= 2(x^3 - 15x^2 + 75x - 125)$$

$$2x^3 - 30x^2 + 156x - 280 = 2x^3 - 30x^2 + 150x - 250$$

$$156x - 280 = 150x - 250$$

$$156x - 150x = 280 - 250$$

$$6x = 30$$

$$x = 30 / 6 = 5.$$

But once again observe the problem in the vedic sense.

We have $(x - 4) + (x - 6) = 2x - 10$. Taking out the numerical factor 2, we have $(x - 5) = 0$, which is the factor under the cube on R.H.S. In such a case "Sunyam samya Samuccaye" formula gives that $x - 5 = 0$. Hence $x = 5$

Think of solving the problem

$$(x-249)^3 + (x+247)^3 = 2(x-1)^3$$

The traditional method will be horrible even to think of.

But $(x - 249) + (x + 247) = 2x - 2 = 2(x - 1)$. And $x - 1$. on R.H.S. cube, it is enough to state that $x - 1 = 0$ by the 'sutra'.

$x = 1$ is the solution. No cubing or any other mathematical operations.

ALGEBRAIC PROOF :

Consider $(x - 2a)^3 + (x - 2b)^3 = 2(x - a - b)3$ it is clear that

$$x - 2a + x - 2b = 2x - 2a - 2b$$

$$= 2(x - a - b)$$

Now the expression,

$x^3 - 6x^2a + 12xa^2 - 8a^3 + x^3 - 6x^2b + 12xb^2 - 8b^3$

$= 2(x^3 - 3x^2a - 3x^2b + 3xa^2 + 3xb^2 + 6axb - a^3 - 3a^2b - 3ab^2 - b^3)$

$= 2x^3 - 6x^2a - 6x^2b + 6xa^2 + 6xb^2 + 12xab - 2a^3 - 6a^2b - 6ab^2 - 2b^3$

cancel the common terms on both sides

$12xa^2 + 12xb^2 - 8a^3 - 8b^3$

$= 6xa^2 + 6xb^2 + 12xab - 2a^3 - 6a^2b - 6ab^2 - 2b^3$

$6xa^2 + 6xb^2 - 12xab = 6a^3 + 6b^3 - 6a^2b - 6ab^2$

$6x(a^2 + b^2 - 2ab) = 6[a^3 + b^3 - ab(a + b)]$

$x(a - b)^2 = [(a + b)(a^2 + b^2 - ab) - (a + b)ab]$

$= (a + b)(a^2 + b^2 - 2ab)$

$= (a + b)(a - b)^2$

$\therefore \quad x = a + b$

Solve the following using "Sunyam Samuccaye" process :

1. $(x - 3)^3 + (x - 9)^3 = 2(x - 6)^3$
2. $(x + 4)^3 + (x - 10)^3 = 2(x - 3)^3$
3. $(x + a + b - c)^3 + (x + b + c - a)^3 = 2(x + b)^3$

EXAMPLE :

$$\frac{(x+2)^3}{(x+3)^3} = \frac{x+1}{x+4}$$

with the text book procedures we proceed as follows

$$\frac{x^3 + 6x^2 + 12x + 8}{x^3 + 9x^2 + 27x + 27} = \frac{x+1}{x+4}$$

Now by cross multiplication,

$(x + 4)(x^3 + 6x^2 + 12x + 8) = (x + 1)(x^3 + 9x^2 + 27x + 27)$

$x^4 + 6x^3 + 12x^2 + 8x + 4x^3 + 24x^2 + 48x + 32$

$= x^4 + 9x^3 + 27x^2 + 27x + x^3 + 9x^2 + 27x + 27$

$$x^4 + 10x^3 + 36x^2 + 56x + 32 = x^4 + 10x^3 + 36x^2 + 54x + 27$$

$$56x + 32 = 54x + 27$$

$$56x - 54x = 27 - 32$$

$$2x = -5$$

$$x = -5/2$$

Observe that (N1 + D1) with in the cubes on

L.H.S. is x + 2 + x + 3 = 2x + 5 and

$N_2 + D_2$ on the right hand side

is x + 1 + x + 4 = 2x + 5.

By vedic formula we have 2x + 5 = 0 x = - 5 / 2.

Solve the following by using vedic method :

1. $\dfrac{(x+3)^3}{(x+5)^3} = \dfrac{x+1}{x+7}$

2. $\dfrac{(x-5)^3}{(x-7)^3} = \dfrac{x-3}{x-9}$

Anurupye Sunyamanyat

The Sutra Anurupye Sunyamanyat says : 'If one is in ratio, the other one is zero'.

We use this Sutra in solving a special type of simultaneous simple equations in which the coefficients of 'one' variable are in the same ratio to each other as the independent terms are to each other. In such a context the Sutra says the 'other' variable is zero from which we get two simple equations in the first variable (already considered) and of course give the same value for the variable.

EXAMPLE 1:

$$3x + 7y = 2$$

$$4x + 21y = 6$$

Observe that the y-coefficients are in the ratio 7 : 21 i.e., 1 : 3, which is same as the ratio of independent terms i.e., 2 : 6 i.e., 1 : 3. Hence the other variable x = 0 and 7y = 2 or 21y = 6 gives y = 2 / 7

EXAMPLE 2:

$$323x + 147y = 1615$$

$$969x + 321y = 4845$$

The very appearance of the problem is frightening. But just an observation and anurupye sunyamanyat give the solution x = 5, because coefficient of x ratio is

323 : 969 = 1 : 3 and constant terms ratio is

1615 : 4845 = 1 : 3.

y = 0 and 323 x = 1615 or 969 x = 4845 gives x = 5.

Solve the following by anurupye sunyamanyat.

1. $12x + 78y = 12$

   $16x + 96y = 16$

2. $3x + 7y = 24$

   $12x + 5y = 96$

3. $4x - 6y = 24$

   $7x - 9y = 36$

4. $ax + by = bm$

   $cx + dy = dm$

In solving simultaneous quadratic equations, also we can take the help of the 'sutra' in the following way:

EXAMPLE 3 :

Solve for x and y

$$x + 4y = 10$$

$$x^2 + 5xy + 4y^2 + 4x - 2y = 20$$

$x^2 + 5xy + 4y^2 + 4x - 2y = 20$ can be written as

$(x + y)(x + 4y) + 4x - 2y = 20$

$10(x + y) + 4x - 2y = 20$ (Since $x + 4y = 10$)

$10x + 10y + 4x - 2y = 20$

$14x + 8y = 20$

Now $x + 4y = 10$

$14x + 8y = 20$ and $4 : 8 :: 10 : 20$

from the Sutra, $x = 0$ and $4y = 10$, i.e., $8y = 20$ $y = 10/4 = 2\frac{1}{2}$

Thus $x = 0$ and $y = 2\frac{1}{2}$ is the solution.

Sankalana—Vyavakalanabhyam

This Sutra means 'by addition and by subtraction'. It can be applied in solving a special type of simultaneous equations where the x - coefficients and the y - coefficients are found interchanged.

EXAMPLE 1:

$$45x - 23y = 113$$

$$23x - 45y = 91$$

In the conventional method we have to make equal either the coefficient of x or coefficient of y in both the equations. For that we have to multiply equation (1) by 45 and equation (2) by 23 and subtract to get the value of x and then substitute the value of x in one of the equations to get the value of y or we have to multiply equation (1) by 23 and equation (2) by 45 and then subtract to get value of y and then substitute the value of y in one of the equations, to get the value of x. It is difficult process to think of.

From Sankalana - vyavakalanabhyam

add them,

i.e., $(45x - 23y) + (23x - 45y) = 113 + 91$

i.e., $68x - 68y = 204$

$\therefore\ x - y = 3$

Subtract one from other,

i.e., (45x - 23y) - (23x - 45y) = 113 - 91

i.e., 22x + 22y = 22

$\therefore$ x + y = 1

and repeat the same sutra, we get x = 2 and y = - 1

Very simple addition and subtraction are enough, however big the coefficients may be.

EXAMPLE 2:

$$1955x - 476y = 2482$$
$$476x - 1955y = -4913$$

Oh ! what a problem ! And still

just add, 2431(x - y) = - 2431 x - y = -1

subtract, 1479 (x + y) = 7395 x + y = 5

once again add, 2x = 4 x = 2

subtract - 2y = - 6 y = 3

Solve the following problems using Sankalana - Vyavakalanabhyam.

1. 3x + 2y = 18
   2x + 3y = 17
2. 5x - 21y = 26
   21x - 5y = 26
3. 659x + 956y = 4186
   956x + 659y = 3889

Puranapuranabhyam

The Sutra can be taken as Purana — Apuranabhyam which means by the completion or non - completion. Purana is well known in the present system. We can see its application in solving the roots for general form of quadratic equation.

We have : $ax^2 + bx + c = 0$

$$x^2 + (b/a)x + c/a = 0 \text{ (dividing by a)}$$

$$x^2 + (b/a)x = - c/a$$

completing the square (i.e., purana) on the L.H.S.

$$x^2 + (b/a)x + (b^2/4a^2) = -c/a + (b^2/4a^2)$$

$$[x + (b/2a)]^2 = (b^2 - 4ac) / 4a^2$$

Proceeding in this way we finally get

$$x = \frac{-b \pm \sqrt{b^2 - 4ac}}{2a}$$

Now we apply purana to solve problems.

EXAMPLE 1:

$$x^3 + 6x^2 + 11x + 6 = 0.$$

Since $(x + 2)^3 = x^3 + 6x^2 + 12x + 8$

Add $(x + 2)$ to both sides

We get

$x^3 + 6x^2 + 11x + 6 + x + 2 = x + 2$

i.e., $x^3 + 6x^2 + 12x + 8 = x + 2$

i.e., $(x + 2)^3 = (x + 2)$

this is of the form $y^3 = y$ for $y = x + 2$

solution $y = 0, y = 1, y = -1$

i.e., $x + 2 = 0, 1, -1$

which gives $x = -2, -1, -3$

EXAMPLE 2:

$x^3 + 8x^2 + 17x + 10 = 0$

We know $(x + 3)^3 = x^3 + 9x^2 + 27x + 27$

So adding on the both sides, the term $(x^2 + 10x + 17)$, we get

$x^3 + 8x^2 + 17x + x^2 + 10x + 17 = x^2 + 10x + 17$

i.e., $x^3 + 9x^2 + 27x + 27 = x^2 + 6x + 9 + 4x + 8$

i.e., $(x + 3)^3 = (x + 3)^2 + 4(x + 3) - 4$

$y^3 = y^2 + 4y - 4$ for $y = x + 3$

$y = 1, 2, -2.$

Hence $x = -2, -1, -5$

Thus purana is helpful in factorization.

Further purana can be applied in solving Biquadratic equations also.

Solve the following using purana - apuranabhyam.

1. $x^3 - 6x^2 + 11x - 6 = 0$
2. $x^3 + 9x^2 + 23x + 15 = 0$
3. $x^2 + 2x - 3 = 0$
4. $x^4 + 4x^3 + 6x^2 + 4x - 15 = 0$

Calana—Kalanabhyam

In the book on Vedic Mathematics Sri Bharati Krishna Tirthaji mentioned the Sutra 'Calana - Kalanabhyam' at only two places. The Sutra means 'Sequential motion'.

(i) In the first instance it is used to find the roots of a quadratic equation $7x^2 - 11x - 7 = 0$. Swamiji called the sutra as calculus formula. Its application at that point is as follows. Now by calculus formula we say:

$14x - 11 = \pm \sqrt{317}$

A Note follows saying every Quadratic can thus be broken down into two binomial factors. An explanation in terms of first differential, discriminant with sufficient number of examples are given under the chapter 'Quadratic Equations'.

(ii) At the Second instance under the chapter 'Factorization and Differential Calculus' for factorizing expressions of 3rd, 4th and 5th degree, the procedure is mentioned as 'Vedic Sutras relating to Calana - Kalana - Differential Calculus'.

Further other Sutras 10 to 16 mentioned below are also used to get the required results. Hence the sutra and its various applications will be taken up at a later stage for discussion.

But sutra - 14 is discussed immediately after this item.

Now the remaining sutras :

10. YAVADUNAM (The deficiency)
11. VYASTISAMASTIH (Whole as one and one as whole)
12. SESANYAN KENA CARAMENA (Remainder by the last digit)
13. SOPANTYADVAYAMANTYAM (Ultimate and twice the penultimate)
15. GUNITASAMUCCAYAH (The whole product is the same)
16. GUNAKA SAMUCCAYAH (Collectivity of multipliers)

The Sutras have their applications in solving different problems in different contexts. Further they are used along with other Sutras. So it is a bit of inconvenience to deal each Sutra under a separate heading exclusively and also independently. Of course they will be mentioned and also be applied in solving the problems in the forth coming chapter wherever necessary.

This decision has been taken because up to now, we have treated each Sutra independently and have not continued with any other Sutra even if it is necessary. When the need for combining Sutras for filling the gaps in the process arises, we may opt for it. Now we shall deal the fourteenth Sutra, the Sutra left so far untouched.

Ekanyunena Purvena

The Sutra Ekanyunena purvena comes as a Sub-sutra to Nikhilam which gives the meaning 'One less than the previous' or 'One less than the one before'.

(1) The use of this sutra in case of multiplication by 9,99,999.. is as follows .

Method :

(a) The left hand side digit (digits) is (are) obtained by applying the ekanyunena purvena i.e. by deduction 1 from the left side digit (digits) .

e.g. (i) 7 × 9; 7 - 1 = 6 (L.H.S. digit)

(b) The right hand side digit is the complement or difference between the multiplier and the left hand side digit (digits) . i.e. 7 × 9 R.H.S is 9 - 6 = 3.

(c) The two numbers give the answer; i.e. 7 × 9 = 63.

EXAMPLE 1:

8 × 9 Step (a) gives 8 - 1 = 7 (L.H.S. Digit)

Step (b) gives 9 - 7 = 2 (R.H.S. Digit)

Step (c) gives the answer 72

EXAMPLE 2:

15 × 99 Step (a) : 15 - 1 = 14

Step (b) : 99 - 14 = 85 (or 100 - 15)

Step (c) : 15 x 99 = 1485

EXAMPLE 3:

24 × 99

Answer :

$$\begin{array}{c|c} (24-1) & (99-23) \\ =23 & =76(\text{or } 100-24) \end{array} = 2376$$

EXAMPLE 4:

356 × 999

Answer :

$$\frac{(356-1)}{=355} \Bigg/ \frac{(999-335)}{=644} = 355644$$

EXAMPLE 5:

878 × 9999

Answer :

$$\frac{(878-1)}{=877} \Bigg/ \frac{(9999-877)}{=9122(10000-876)} = 8779122$$

Note the Process : The multiplicand has to be reduced by 1 to obtain the LHS and the rightside is mechanically obtained by the subtraction of the L.H.S from the multiplier which is practically a direct application of Nikhilam Sutra.

Now by Nikhilam

24 - 1 = 23 L.H.S.
x 99 - 23 = 76 R.H.S. (100-24)

23 / 76 = 2376

Reconsider the Example 4:

356 - 1 = 355 L.H.S.
x 999 - 355 = 644 R.H.S.

355 / 644 = 355644

and in Example 5: 878 x 9999 we write

0878 - 1 = 877 L.H.S.
x 9999 - 877 = 9122 R.H.S.

877 / 9122 = 8779122

ALGEBRAIC PROOF :

As any two digit number is of the form (10x + y), we proceed

$(10x + y) \times 99$

$= (10x + y) \times (100 - 1)$

$= 10x \,.\, 10^2 - 10x + 10^2 \,.y - y$

$= x \,.\, 10^3 + y \,.\, 10^2 - (10x + y)$

$= x \,.\, 10^3 + (y - 1) \,.\, 10^2 + [\, 10^2 - (10x + y)]$

Thus the answer is a four digit number whose 1000th place is x, 100th place is (y - 1) and the two digit number which makes up the 10th and unit place is the number obtained by subtracting the multiplicand from 100 (or apply nikhilam).

Thus in 37 × 99. The 1000th place is x i.e. 3

100th place is (y - 1) i.e. (7 - 1) = 6

Number in the last two places 100-37=63.

Hence answer is 3663.

Apply Ekanyunena purvena to find out the products

1. 64 × 99
2. 723 × 999
3. 3251 × 9999
4. 43 × 999
5. 256 × 9999
6. 1857 × 99999

We have dealt the cases:

i) When the multiplicand and multiplier both have the same number of digits

ii) When the multiplier has more number of digits than the multiplicand.

In both the cases the same rule applies. But what happens when the multiplier has lesser digits?

i.e. for problems like 42 × 9, 124 × 9, 26325 × 99 etc.,

For this let us have a re-look in to the process for proper understanding.

Multiplication table of 9.

	a	b
2 x 9 =	1	8
3 x 9 =	2	7
4 x 9 =	3	6
8 x 9 =	7	2
9 x 9 =	8	1
10 x 9 =	9	0

Observe the left hand side of the answer is always one less than the multiplicand (here multiplier is 9) as read from Column (a) and the right hand side of the answer is the complement of the left hand side digit from 9 as read from Column (b)

Multiplication table when both multiplicand and multiplier are of 2 digits.

	a	b	
11 × 99 =	10	89	= (11-1) / 99 - (11-1) = 1089
12 × 99 =	11	88	= (12-1) / 99 - (12-1) = 1188
13 × 99 =	12	87	= (13-1) / 99 - (13-1) = 1287
18 × 99 =	17	82	________________
19 × 99 =	18	81	
20 × 99 =	19	80	= (20-1) / 99 - (20-1) = 1980

The rule mentioned in the case of above table also holds good here.

Further we can state that the rule applies to all cases, where the multiplicand and the multiplier have the same number of digits.

Consider the following Tables.

(i)

	a	b
11 × 9 =	9	9
12 × 9 =	10	8
13 × 9 =	11	7
————		
18 × 9 =	16	2
19 × 9 =	17	1
20 × 9 =	18	0

(ii)

21 × 9 =	18	9
22 × 9 =	19	8
23 × 9 =	20	7
————		
28 × 9 =	25	2
29 × 9 =	26	1
30 × 9 =	27	0

(iii)

35 × 9 =	31	5
46 × 9 =	41	4
53 × 9 =	47	7
67 × 9 =	60	3

———————— so on.

From the above tables the following points can be observed:

(1) Table (i) has the multiplicands with 1 as first digit except the last one. Here L.H.S of products are uniformly 2 less than the multiplicands. So also with 20 x 9

(2) Table (ii) has the same pattern. Here L.H.S of products are uniformly 3 less than the multiplicands.

(3) Table (iii) is of mixed example and yet the same result i.e. if 3 is first digit of the multiplicand then L.H.S of product is 4 less than the multiplicand; if 4 is first digit of the multiplicand then, L.H.S of the product is 5 less than the multiplicand and so on.

(4) The right hand side of the product in all the tables and cases is obtained by subtracting the R.H.S. part of the multiplicand by Nikhilam.

Keeping these points in view we solve the problems:

EXAMPLE 1 :

42 × 9

(i) Divide the multiplicand (42) of by a Vertical line or by the Sign : into a right hand portion consisting of as many digits as the multiplier.

i.e. 42 has to be written as 4/2 or 4:2

(ii) Subtract from the multiplicand one more than the whole excess portion on the left. i.e. left portion of multiplicand is 4.

one more than it 4 + 1 = 5.

We have to subtract this from multiplicand

i.e. write it as

4	:	2
	:	–5
3	:	7

This gives the L.H.S part of the product.

This step can be interpreted as "take the ekanyunena and sub tract from the previous" i.e. the excess portion on the left.

(iii) Subtract the R.H.S. part of the multiplicand by nikhilam process.

i.e. R.H.S of multiplicand is 2

its nikhilam is 8

It gives the R.H.S of the product

i.e. answer is 3 : 7 : 8 = 378.

Thus 42 × 9 can be represented as

$$\begin{array}{rrrrr} 4 & : & 2 & & \\ & : & -5 & : & 8 \\ \hline 3 & : & 7 & : & 8 = 378. \end{array}$$

EXAMPLE 2 :

124 × 9

Here Multiplier has one digit only .

We write 12 : 4

Now step (ii), 12 + 1 = 13

i.e.

$$\begin{array}{r} 12 : 4 \\ -1 : 3 \\ \hline \end{array}$$

Step (iii) R.H.S. of multiplicand is 4. Its Nikhilam is 6

124 × 9 is 12 : 4

$$\begin{array}{r} -1 : 3 : 6 \\ \hline 11 : 1 : 6 = 1116 \end{array}$$

The process can also be represented as

124 × 9 = [124 - (12 + 1)] : (10 - 4) = (124 - 13) : 6 = 1116

EXAMPLE 3:

15639 × 99

Since the multiplier has 2 digits, the answer is

[15639 - (156 + 1)] : (100 - 39) = (15639 - 157) : 61
= 1548261

Find the products in the following cases.

1. 58 × 9
2. 62 × 9
3. 427 × 99
4. 832 × 9
5. 24821 × 999
6. 111011 × 99

Ekanyunena Sutra is also useful in Recurring Decimals. We can take up this under a separate treatment.

Thus we have a glimpse of majority of the Sutras. At some places some Sutras are mentioned as Sub-Sutras. Any how we now proceed into the use of Sub-Sutras. As already mentioned the book on Vedic Mathematics enlisted 13 Upa-Sutras.

But some approaches in the Vedic Mathematics book prompted some serious research workers in this field to mention some other Upa-Sutras. We can observe those approaches and developments also.

Anurupyena

The upa-Sutra 'anurupyena' means 'proportionality'. This Sutra is highly useful to find products of two numbers when both of them are near the Common bases i.e powers of base 10 . It is very clear that in such cases the expected 'Simplicity' in doing problems is absent.

EXAMPLE 1:

46 × 43

As per the previous methods, if we select 100 as base we get

46	-54
43	-57

This is much more difficult and of no use.

Now by 'anurupyena' we consider a working base In three ways. We can solve the problem.

METHOD 1:

Take the nearest higher multiple of 10. In this case it is 50.

Treat it as 100 / 2 = 50. Now the steps are as follows:

(i) Choose the working base near to the numbers under consideration.

i.e., working base is 100 / 2 = 50

(ii) Write the numbers one below the other

i.e.

4 6

4 3

(iii) Write the differences of the two numbers respectively from 50 against each number on right side

i.e.

46	-04
43	-07

(iv) Write cross-subtraction or cross- addition as the case may be under the line drawn.

46	–04
43	–07
(46 – 7)or (43 – 4)	/
= 39	

(v) Multiply the differences and write the product in the left side of the answer.

46	-04
43	-07
39 / -4 x -7	
= 28	

(vi) Since base is 100 / 2 = 50 , 39 in the answer represents 39×50.

Hence divide 39 by 2 because 50 = 100 / 2

Thus 39 ÷ 2 gives 19½ where 19 is quotient and 1 is remainder.

This 1 as Reminder gives one 50 making the L.H.S of the answer 28 + 50 = 78 (or Remainder ½ × 100 + 28)

i.e. R.H.S 19 and L.H.S 78 together give the answer 1978 We represent it as:

	46	-04
	43	-07
2)	39 / 28	
	19½ / 28	
	= 19 / 78 = 1978	

EXAMPLE 2:

42 × 48.

With 100 / 2 = 50 as working base, the problem is as follows:

	42	-08
	48	-02
2)	40 / 16	
	20 / 16	

42 × 48 = 2016

METHOD 2:

For the example 1: 46 × 43. We take the same working base 50. We treat it as 50 = 5 × 10. i.e. we operate with 10 but not with 100 as in method

now

46	−04
43	−07
(46 – 7) or (43 – 4) = 39	−4 × −7 = 28
39 × 5 +2 (carried over	8

(195 + 2) / 8 = 1978

[Since we operate with 10, the R.H.S portion shall have only unit place .Hence out of the product 28, 2 is carried over to left side. The L.H.S portion of the answer shall be multiplied by 5, since we have taken 50 = 5 × 10.]

Now in the example 2: 42 × 48 we can carry as follows by treating 50 = 5 × 10

42	–08
48	–02
40	/ 16
×5	/
200	/ 16 = 2016

METHOD 3:

We take the nearest lower multiple of 10 since the numbers are 46 and 43 as in the first example, We consider 40 as working base and treat it as 4 × 10.

46	06
43	03
(46 + 3)or	6 × 3
(43 + 6)	
= 49	= 18

Since 10 is in operation 1 is carried out digit in 18.

Since 4 × 10 is working base we consider 49 × 4 on L.H.S of answer i.e. 196 and 1 carried over the left side, giving L.H.S. of answer as 1978. Hence the answer is 1978.

We proceed in the same method for 42 × 48

42	02	working base 4 × 10 = 40
48	08	
50	/ 16	
×4	/	sin ce 10 is in operation
200	/ 16	= 2016

Let us see the all the three methods for a problem at a glance.

EXAMPLE 3:

24 × 23

METHOD - 1:

Working base = 100 / 5 = 20

24 04
23 03

5) 27 / 12

5 2/5 / 12 = 5 / 52 = 552

[Since 2 / 5 of 100 is 2 / 5 × 100 = 40 and 40 + 12 = 52]

METHOD - 2:

Working base 2 × 10 = 20

24	04	
23	03	
27	12	it is more convenient than
×2		method – 1 Guss the reason
64	12	= 552

Now as 20 itself is nearest lower multiple of 10 for the problom under consideration, the case of method - 3 shall not arise.

Let us take another example and try all the three methods.

EXAMPLE 4:

492 × 404

METHOD - 1 :

Working base = 1000 / 2 = 500

492 -008
404 -096

2)396 / 768 since 1000 is in operation

198 / 768 = 198768

METHOD 2:

Working base = 5 × 100 = 500

```
492       -008
404       -096
---------------
395   /   768
×5   /
---------------
1980 /    768   = 198768
```

METHOD - 3.

Since 400 can also be taken as working base, treat 400 = 4 × 100 as working base.

Thus

```
492       092
404       004   since 100 is in operation
---------------
495   /   368
×4   /
---------------
1984 /    368   = 198768
```

No need to repeat that practice in these methods finally takes us to work out all these mentally and getting the answers straight away in a single line.

EXAMPLE 5:

3998 × 4998

Working base = 10000 / 2 = 5000

```
3998   -1002
4998   -0002
------------
```

2) 3996 / 2004 since 10,000 is in operation

1998 / 2004 = 19982004

or taking working base = 5 × 1000 = 5,000 and

3998	–1002	
4998	–0002	since 100 is carried out
3996 /	2004	
×5		
19980 /	2004	= 19982004

What happens if we take 4000 i.e. 4 × 1000 as working base?

$\overline{}$

3998 0002

4998 0998 Since 1000 is an operation

4996 / 1996

As 1000 is in operation, $\overline{1996}$ has to be written as $\overline{1996}$ and 4000 as base, the L.H.S portion 5000 has to be multiplied by 4. i.e. the answer is

4996 /	$\overline{1996}$
×4	
19982 /	004 = 19982004

A simpler example for better understanding.

EXAMPLE 6:

58 × 48

Working base 50 = 5 × 10 gives

50	+8
48	$\overline{2}$
56 /	$\overline{16}$
×5	
280 /	$\overline{16}$ = 280/4 = 2784

Since 10 is in operation.

Use anurupyena by selecting appropriate working base and method.

Find the following product.

1. 46 × 46
2. 57 × 57
3. 54 × 45
4. 18 × 18
5. 62 × 48
6. 229 × 230
7. 47 × 96
8. 87965 × 99996
9. 49 × 499
10. 389 × 512

Adyamadyenantya—Mantyena

The Sutra ' adyamadyenantya-mantyena' means 'the first by the first and the last by the last'. Suppose we are asked to find out the area of a rectangular card board whose length and breadth are respectively 6ft . 4 inches and 5 ft. 8 inches. Generally we continue the problem like this.

Area = Length × Breath

= 6' 4" × 5' 8" Since 1' = 12", conversion

= (6 × 12 + 4) (5 × 12 + 8) in to single unit

= 76" 68" = 5168 Sq. inches.

Since 1 sq. ft. =12 × 12 = 144sq.inches we have area

5168 = 144) 5168 (35

144   432

848

720 i.e., 35 Sq. ft 128 Sq. inches

128

By Vedic principles we proceed in the way "the first by first and the last by last"

i.e. 6' 4" can bc treated as 6x + 4 and 5' 8" as 5x + 8,

Where x = 1 ft. = 12 in; x^2 is sq. ft.

Now (6x + 4)(5x + 8)

$= 30x^2 + 6.8.x + 4.5.x + 32$

$= 30x^2 + 48x + 20x + 32$

$= 30x^2 + 68.\ x + 32$

$= 30x^2 + (5x + 8).\ x + 32$

Writing 68 = 5 x 12 + 8

$= 35x^2 + 8.\ x + 32$

= 35 Sq. ft. + 8 × 12 Sq. in + 32 Sq. in

= 35 Sq. ft. + 96 Sq. in + 32 Sq. in

= 35 Sq. ft. + 128 Sq. in.

It is interesting to know that a mathematically untrained and even uneducated carpenter simply works in this way by mental argumentation. It goes in his mind like this

6' 4"

5' 8"

First by first i.e. 6' × 5' = 30 sq. ft.

Last by last i.e. 4" × 8" = 32 sq. in.

Now cross wise 6 × 8 + 5 × 4 = 48 +20 = 68.

Adjust as many '12' s as possible towards left as 'units' i.e. 68 = 5 × 12 +8 , 5 twelve's as 5 square feet make the first 30+5 = 35 sq. ft ; 8 left becomes 8 × 12 square inches and go towards right i.e. 8 × 12 = 96 sq. in. towards right ives 96+32 = 128sq.in.

Thus he got area in some sort of 35 squints and another sort of 128 sq. units. i.e. 35 sq. ft 128 sq. in

Another Example:

4'		6"
3'		4"
4×3	$4\times4+6\times3$	6×4
	$=16+18$	
$=12$	$=34$	$=24$
	$=2\times12+10$	

Now 12 + 2 = 14, 10 × 12 + 24 = 120 + 24 = 144

Thus 4' 6? × 3' 4? = 14 Sq. ft. 144 Sq. inches.

Since 144 sq. in = 12 × 12 = 1 sq. ft.

The answer is 15 sq. ft.

We can extend the same principle to find volumes of parallelepiped also.

I. Find the area of the rectangles in each of the following situations.

1). l = 3' 8", b = 2' 4 "

2). l = 12' 5" , b= 5' 7"

3). l = 4 yard 3 ft. b = 2 yards 5 ft.(1yard =3ft)

4). l = 6 yard 6 ft. b = 5 yards 2 ft.

II. Find the area of the trapezium in each of the following cases. Recall area = ½ h (a + b) where a, b are parallel sides and h is the distance between them.

1) a = 3' 7", b = 2' 4", h = 1' 5"

2) a = 5' 6", b = 4' 4", h = 3' 2"

3) a = 8' 4", b = 4' 6", h = 5' 1".

Factorization of Quadratics

The usual procedure of factorizing a quadratic is as follows:

$3x^2 + 8x + 4$

$= 3x^2 + 6x + 2x + 4$

$= 3x (x + 2) + 2 (x + 2)$

$= (x + 2) (3x + 2)$

But by mental process, we can get the result immediately. The steps are as follows.

i) Split the middle coefficient in to two such parts that the ratio of the first coefficient to the first part is the same as the ratio of the second part to the last coefficient. Thus we split the coefficient of middle term of $3x^2 + 8x + 4$ i.e. 8 in to two such parts 6 and 2 such that the ratio of the first coefficient to the first part of the middle coefficient i.e. 3:6 and the ratio of the second pat to the last coefficient, i.e. 2: 4 are the same. It is clear that 3:6 = 2:4. Hence such split is valid. Now the ratio 3: 6 = 2: 4 = 1:2 gives one factor x + 2.

ii) Second factor is obtained by dividing the first coefficient of the quadratic by the fist coefficient of the factor already found and the last coefficient of the quadratic by the last coefficient of the factor.

i.e. the second factor is

$$\frac{3x^2}{x} + \frac{4}{2} = 3x + 2$$

Hence $3x^2 + 8x + 4 = (x + 2) (3x + 2)$

E.G.1:

$4x^2 + 12x + 5$

i) Split 12 into 2 and 10 so that as per rule 4 : 2 = 10: 5 = 2 : 1 i.e., 2x + 1 is first factor.

ii) Now

$$\frac{4x^2}{2x} + \frac{5}{1} = 2x + 5 \text{ is second factor.}$$

E.G.2:

$15x^2 - 14xy - 8y^2$

i) Split -14 into -20, 6 so that 15 : - 20 = 3 : - 4 and 6 : - 8 = 3 : - 4. Both are same. i.e., (3x - 4y) is one factor.

ii) Now

$$\frac{15x^2}{3x} + \frac{8y^2}{-4y} = 5x + 2y \text{ is second factor.}$$

Thus $15x^2 - 14xy - 8y^2 = (3x - 4y)(5x + 2y)$.

It is evident that we have applied two sub-sutras 'anurupyena' i.e.'proportionality' and 'adyamadyenantya-mantyena' i.e. 'the first by the first and the last by the last' to obtain the above results.

Factorise the following quadratics applying appropriate vedic maths sutras:

1) $3x^2 + 14x + 15$

2) $6x^2 - 23x + 7$

3) $8x^2 - 22x + 5$

4) $12x^2 - 23xy + 10y^2$

Yavadunam Tavadunikrtya Varganca Yojayet

The meaning of the Sutra is 'what ever the deficiency subtract that deficit from the number and write along side the square of that deficit'.

This Sutra can be applicable to obtain squares of numbers close to bases of powers of 10.

METHOD-1 :

Numbers near and less than the bases of powers of 10.

E.G 1:

92 Here base is 10.

The answer is separated in to two parts by a'/'

Note that deficit is 10 - 9 = 1

Multiply the deficit by itself or square it

$1^2 = 1$. As the deficiency is 1, subtract it from the number i.e., 9-1 = 8.

Now put 8 on the left and 1 on the right side of the vertical line or slash i.e., 8/1.

Hence 81 is answer.

E.G. 2:

96^2 Here base is 100.

Since deficit is 100-96=4 and square of it is 16 and the deficiency subtracted from the number 96 gives 96-4 = 92, we get the answer 92 / 16 Thus 962 = 9216.

E.G. 3:

994^2 Base is 1000

Deficit is 1000 - 994 = 6. Square of it is 36.

Deficiency subtracted from 994 gives 994 - 6 = 988

Answer is 988 / 036 [since base is 1000]

E.G. 4:

9988^2 Base is 10,000.

Deficit = 10000 - 9988 = 12.

Square of deficit = 12^2 = 144.

Deficiency subtracted from number = 9988 - 12 = 9976.

Answer is 9976 / 0144 [since base is 10,000].

E.G. 5:

88^2 Base is 100.

Deficit = 100 - 88 = 12.

Square of deficit = 12^2 = 144.

Deficiency subtracted from number = 88 - 12 = 76.

Now answer is 76 / 144 = 7744 [since base is 100]

ALGEBRAIC PROOF:

The numbers near and less than the bases of power of 10 can be treated as (x-y), where x is the base and y, the deficit.

Thus

(1) 9 = (10 -1)

(2) 96 = (100-4)

(3) 994 = (1000-6)

(4) 9988 = (10000-12)

(5) 88 = (100-12)

$$(x - y)^2 = x^2 - 2xy + y^2$$

$$= x\,(x - 2y) + y^2$$

$$= x\,(x - y - y) + y^2$$

= Base (number - deficiency) + $(\text{deficit})^2$

Thus

$$985^2 = (1000 - 15)^2$$

$$= 1000\,(985 - 15) + (15)^2$$

$$= 1000\,(970) + 225$$

$$= 970000 + 225$$

$$= 970225.$$

or we can take the identity $a^2 - b^2 = (a + b)(a - b)$ and proceed as

$$a^2 - b^2 = (a + b)(a - b).$$

gives $a^2 = (a + b)(a - b) + b^2$

Thus for a = 985 and b = 15;

$$a^2 = (a + b)(a - b) + b^2$$

$$985^2 = (985 + 15)(985 - 15) + (15)^2$$

$$= 1000\,(970) + 225$$

$$= 970225.$$

METHOD. 2 :

Numbers near and greater than the bases of powers of 10.

E.G.(1):

13^2.

Instead of subtracting the deficiency from the number we add and proceed as in Method-1.

for 13^2 , base is 10, surplus is 3.

Surplus added to the number = 13 + 3 = 16.

Square of surplus = 32 = 9

Answer is 16 / 9 = 169.

E.G.(2):

112^2

Base = 100, Surplus = 12,

Square of surplus = 122 = 144

add surplus to number = 112 + 12 = 124.

Answer is 124 / 144 = 12544

Or think of identity $a^2 = (a + b)(a - b) + b^2$ for a = 112, b = 12:

$$112^2 = (112 + 12)(112 - 12) + 122$$

$$= 124\,(100) + 144$$

$$= 12400 + 144$$

$$= 12544.$$

$$(x + y)^2 = x^2 + 2xy + y^2$$
$$= x\,(x + 2y) + y^2$$
$$= x\,(x + y + y) + y^2$$
$$= \text{Base (Number + surplus) + (surplus)}^2$$

gives

$$112^2 = 100\,(112 + 12) + 12^2$$
$$= 100\,(124) + 144$$
$$= 12400 + 144$$
$$= 12544.$$

E.G. 3:

10025^2

$= (10025 + 25) / 25^2$

$= 10050 / 0625$ [since base is 10,000]

$= 100500625.$

METHOD 3:

This is applicable to numbers which are near to multiples of 10, 100, 1000 .... etc. For this we combine the upa-Sutra 'anurupyena' and 'yavadunam tavadunikritya varganca yojayet' together.

EXAMPLE 1:

388^2 Nearest base = 400.

We treat 400 as 4 × 100. As the number is less than the base we proceed as follows

Number 388, deficit = 400 - 388 = 12

Since it is less than base, deduct the deficit

i.e. 388 - 12 = 376.

multiply this result by 4 since base is 4 × 100 = 400.

376 × 4 = 1504

Square of deficit = 122 = 144.

Hence answer is 1504 / 144 = 150544 [since we have taken multiples of 100].

EXAMPLE 2:

485^2 Nearest base = 500.

Treat 500 as 5 x 100 and proceed

$485^2 = \frac{(485-15)}{15^2}$ [since deficit is 15]

$= \frac{470 \times 5}{225}$ since 500 base is taken as 5×100

$= \frac{2350}{225}$ and 2 of 225 is carried over

= 235225.

EXAMPLE 3:

67^2 Nearest base = 70

$67^2 = \frac{(67-3)}{3^3}$ dificit is 3

$= \frac{64 \times 7}{9}$ [since $7 \times 10 = 70$]

$= \frac{448}{9}$

= 4489

EXAMPLE 4:

416^2 Nearest (lower) base = 400

Here surplus = 16 and 400 = 4 x 100

$416^2 = \frac{(416+16)}{16^2}$ since base is multiple 100,

$= \frac{432 \times 4}{256}$ and 2 of 256 is carried over

$= \frac{1728}{256}$

= 173056

EXAMPLE 5:

5012^2 Nearest lower base is 5000 = 5 x 1000

Surplus = 12

$5012^2 = \frac{(5012+12)}{12^2}$

$= \frac{5024}{144}$

$= \frac{25120}{144}$

= 25120144.

Apply yavadunam to find the following squares.

1. 72
2. 98^2
3. 987^2
4. 14^2
5. 116^2
6. 1012^2
7. 19^2
8. 475^2
9. 796^2
10. 108^2
11. 9988^2
12. 6014^2.

So far we have observed the application of yavadunam in finding the squares of number. Now with a slight modification yavadunam can also be applied for finding the cubes of numbers.

Cubing of Numbers

EXAMPLE :

Find the cube of the number 106.

We proceed as follows:

i) For 106, Base is 100. The surplus is 6.

Here we add double of the surplus i.e. 106+12 = 118.

(Recall in squaring, we directly add the surplus)

This makes the left-hand -most part of the answer.

i.e. answer proceeds like 118 /.- - - - -

ii) Put down the new surplus i.e. 118-100=18 multiplied by the initial surplus

i.e. 6=108.

Since base is 100, we write 108 in carried over form 108 i.e..

As this is middle portion of the answer, the answer proceeds like 118 / 108 /....

iii) Write down the cube of initial surplus i.e. 63 = 216 as the last portion

i.e. right hand side last portion of the answer.

Since base is 100, write 216 as 216 as 2 is to be carried over.

Answer is 118 / 108 / 216

Now proceeding from right to left and adjusting the carried over, we get the answer

119 / 10 / 16 = 1191016.

E.G.(1):

1023 = (102 + 4) / 6 × 2 / 23
= 106 = 12 = 08
= 1061208.

Observe initial surplus = 2,

next surplus =6 and base = 100.

E.G.(2):

94^3

Observe that the nearest base = 100.

Here it is deficit contrary to the above examples.

i) Deficit = -6. Twice of it -6 × 2 = -12

add it to the number = 94 - 12 = 82.

ii) New deficit is -18.

Product of new deficit x initial deficit = -18 x -6 = 108

iii) deficit 3 = $(-6)^3$ = -216.

Hence the answer is 82 / 108 / $\overline{-216}$

Since 100 is base 1 and -2 are the carried over. Adjusting the carried over in order, we get the answer

(82 + 1) / (08 - 03) / (100 - 16)

= 83 / = 05 / = 84 = 830584

16 becomes 84 after taking 1 from middle most portion i.e. 100. (100-16=84).

Now 08 - 01 = 07 remains in the middle portion, and 2 or 2 carried to it makes the middle as 07 - 02 = 05. Thus we get the above result.

E.G.(3):

998^3 Base = 1000; initial deficit = - 2.

$9983 = (998 - 2 \times 2) / (- 6 \times - 2) / (- 2)^3$

$= 994 / = 012 / = -008$

$= 994 / 011 / 1000 - 008$

$= 994 / 011 / 992$

$= 994011992.$

Find the cubes of the following numbers using yavadunam sutra.

1. 105
2. 114
3. 1003
4. 10007
5. 92
6. 96
7. 993
8. 9991
9. 1000008
10. 999992.

Antyayor Dasake'pi

The Sutra signifies numbers of which the last digits added up give 10. i.e. the Sutra works in multiplication of numbers for example: 25 and 25, 47 and 43, 62 and 68, 116 and 114.

Note that in each case the sum of the last digit of first number to the last digit of second number is 10.

Further the portion of digits or numbers left wards to the last digits remain the same.

At that instant use Ekadhikena on left hand side digits. Multiplication of the last digits gives the right hand part of the answer.

EXAMPLE 1 :

47 × 43

See the end digits sum 7 + 3 = 10 ; then by the sutras antyayor dasakepi and ekadhikena we have the answer.

$$47 \times 43 = (4 + 1) \times 4 / 7 \times 3$$
$$= 20 / 21$$
$$= 2021.$$

EXAMPLE 2:

62 x 68

2 + 8 = 10, L.H.S. portion remains the same i.e., 6.

Ekadhikena of 6 gives 7

$$62 \times 68 = (6 \times 7) / (2 \times 8)$$
$$= 42 / 16$$
$$= 4216.$$

EXAMPLE 3:

127 × 123

As antyayor dasakepi works, we apply ekadhikena

$$127 \times 123 = 12 \times 13 / 7 \times 3$$
$$= 156 / 21$$
$$= 15621.$$

EXAMPLE 4:

65 × 65

We have already worked on this type. As the present sutra is applicable.

We have $65 \times 65 = 6 \times 7 / 5 \times 5$

$$= 4225.$$

EXAMPLE 5:

395^2

$$395^2 = 395 \times 395$$
$$= 39 \times 40 \,/\, 5 \times 5$$
$$= 1560 \,/\, 25$$
$$= 156025.$$

Use Vedic sutras to find the products

1. 125 × 125
2. 34 × 36
3. 98 × 92
4. 401 × 409
5. 693 × 697
6. 1404 × 1406

It is further interesting to note that the same rule works when the sum of the last 2, last 3, last 4 - - - digits added respectively equal to 100, 1000, 10000 -- - - . The simple point to remember is to multiply each product by 10, 100, 1000, - - as the case may be . Your can observe that this is more convenient while working with the product of 3 digit numbers.

E.G. 1:

292 × 208

Here 92 + 08 = 100, L.H.S portion is same i.e. 2

292 × 208 = (2 × 3) / 92 × 8

⟶ 60 / = 736 (for 100 raise the L.H.S. product by 0)

= 60736.

E.G. 2:

848 × 852

Here 48 + 52 = 100, L.H.S portion is 8 and its 'ekhadhikena' is 9.

Now R.H.S product 48 × 52 can be obtained by 'anurupyena' mentally.

$$\begin{array}{lll} & 48 & \bar{2} \\ & 52 & 2 \\ \hline 2) & \underline{50} & \bar{4} \\ & 25 & \end{array}$$

= 24 / (100 - 4)

= 96

= 2496

and write 848 × 852 = 8 × 9 / 48 × 52

720 = 2496

= 722496.

[Since L.H.S product is to be multiplied by 10 and 2 to be carried over as the base is 100].

E.G. 3:

693 × 607

693 × 607 = 6 × 7 / 93 × 7

= 420 / 651

= 420651.

Find the following products using 'antyayordasakepi'

1. 318 × 312
2. 425 × 475
3. 796 × 744
4. 902 × 998
5. 397 × 393
6. 551 × 549

Antyayoreva

'Atyayoreva' means 'only the last terms'. This is useful in solving simple equations of the following type.

The type of equations are those whose numerator and denominator on the L.H.S. bearing the independent terms stand in the same ratio to each other as the entire numerator and the entire denominator of the R.H.S. stand to each other.

Let us have a look at the following example.

EXAMPLE 1:

$$\frac{x^2+2x+7}{x^2+3x+5}=\frac{x+2}{x+3}$$

In the conventional method we proceed as

$$\frac{x^2+2x+7}{x^2+3x+5}=\frac{x+2}{x+3}$$

$$(x + 3)(x^2 + 2x + 7) = (x + 2)(x^2 + 3x + 5)$$

$$x^3 + 2x^2 + 7x + 3x^2 + 6x + 21$$

$$= x^3 + 3x^2 + 5x + 2x^2 + 6x + 10$$

$$x^3 + 5x^2 + 13x + 21 = x^3 + 5x^2 + 11x + 10$$

Canceling like terms on both sides

$$13x + 21 = 11x + 10$$

$$13x - 11x = 10 - 21$$

$$2x = -11$$

$$x = -11 / 2$$

Now we solve the problem using anatyayoreva.

$$\frac{x^2+2x+7}{x^2+3x+5}=\frac{x+2}{x+3}$$

Consider

$$\frac{x^2+2x+7}{x^2+3x+5}=\frac{x+2}{x+3}$$

Observe that

$$\frac{x^2+2x}{x^2+3x}=\frac{x(x+2)}{x(x+3)}=\frac{x+2}{x+3}$$

This is according to the condition in the sutra. Hence from the sutra

$$\frac{x+2}{x+3}=\frac{7}{5}$$

$$5x + 10 = 7x + 21$$

$$7x - 5x = -21 + 10$$

$$2x = -11$$

$$x = -11 / 2$$

ALGEBRAIC PROOF:

Consider the equation

$$\frac{AC+D}{BC+E}=\frac{A}{B} \quad \text{...(i)}$$

This satisfies the condition in the sutra since

$$\frac{AC}{BC}=\frac{A}{B}$$

Now cross-multiply the equation (i)

$$B\ (AC + D) = A\ (BC + E)$$

$$BAC + BD = ABC + AE$$

$$BD = AE \text{ which gives}$$

$$\frac{A}{B}=\frac{D}{E} \quad \text{...(ii)}$$

i.e., the result obtained in solving equation (i) is same as the result obtained in solving equation (ii).

EXAMPLE 2:

Solve

$$\frac{2x^2+3x+10}{3x^2+4x+14}=\frac{2x+3}{3x+4}$$

Since

$$\frac{2x^2+3x}{3x^2+4x}=\frac{x(2x+3)}{x(3x+4)}=\frac{2x+3}{3x+4}$$

We can apply the sutra.

$$\frac{2x+3}{3x+4}=\frac{10}{14}$$

Cross-multiplying

$$28x + 42 = 30x + 40$$

$$28x - 30x = 40 - 42$$

$$-2x = -2\ x = -2 / -2 = 1.$$

Let us see the application of the sutra in another type of problem.

EXAMPLE 3:

$(x + 1)(x + 2)(x + 9) = (x + 3)(x + 4)(x + 5)$

Re-arranging the equation, we have

$$\frac{(x+1)(x+2)}{(x+4)(x+5)}=\frac{x+3}{x+9}$$

i.e.,

$$=\frac{x^2+3x+2x+3}{x^2+9x+20x+9}$$

Now

$$\frac{x^2+3x}{x^2+9x}=\frac{x(x+3)}{x(x+9)}=\frac{x+3}{x+9}$$

gives the solution by antyayoreva.

Solution is obtained from

$$\frac{x+3}{x+9} = \frac{2}{20}$$

$$20x + 60 = 2x + 18$$

$$20x - 2x = 18 - 60$$

$$18x = -42 \quad x = -42 / 18 = -7 / 3.$$

Once again look into the problem

$(x + 1)(x + 2)(x + 9) = (x + 3)(x + 4)(x + 5)$

Sum of the binomials on each side

$x + 1 + x + 2 + x + 9 = 3x + 12$

$x + 3 + x + 4 + x + 5 = 3x + 12$

It is same. In such a case the equation can be adjusted into the form suitable for application of antyayoreva.

EXAMPLE 4:

$(x + 2)(x + 3)(x + 11) = (x + 4)(x + 5)(x + 7)$

Sum of the binomials on L.H.S. = $3x + 16$

Sum of the binomials on R.H.S. = $3x + 16$

They are same. Hence antyayoreva can be applied. Adjusting we get

$$\frac{(x+2)(x+3)}{(x+4)(x+7)} = \frac{x+5}{x+11} = \frac{2\times 3}{4\times 7} = \frac{6}{28}$$

$$28x + 140 = 6x + 66$$

$$28x - 6x = 66 - 140$$

$$22x = -74$$

$$x = \frac{-74}{22} = \frac{-37}{11}$$

Solve the following problems using 'antyayoreva'

1. $\frac{3x^2+5x+8}{5x^2+6x+12}=\frac{3x+5}{5x+6}$

2. $\frac{4x^2+5x+3}{3x^2+2x+4}=\frac{4x+5}{3x+2}$

3. $(x + 3) (x + 4) (x + 6) = (x + 5) (x + 1) (x + 7)$

4. $(x + 1) (x + 6) (x + 9) = (x + 4) (x + 5) (x + 7)$

5. $\frac{2x^2+3x+9}{4x^2+5x+17}=\frac{2x+3}{4x+5}$

Lopanasthapanabhyam

Lopana sthapanabhyam means 'by alternate elimination and retention'.

Consider the case of factorization of quadratic equation of type $ax^2 + by^2 + cz^2 + dxy + eyz + fzx$ This is a homogeneous equation of second degree in three variables x, y, z. The sub-sutra removes the difficulty and makes the factorization simple. The steps are as follows:

i) Eliminate z by putting $z = 0$ and retain x and y and factorize thus obtained a quadratic in x and y by means of 'adyamadyena' sutra.;

ii) Similarly eliminate y and retain x and z and factorize the quadratic in x and z.

iii) With these two sets of factors, fill in the gaps caused by the elimination process of z and y respectively. This gives actual factors of the expression.

EXAMPLE 1:

$3x^2 + 7xy + 2y^2 + 11xz + 7yz + 6z^2$.

Step (i): Eliminate z and retain x, y; factorize

$3x^2 + 7xy + 2y^2 = (3x + y) (x + 2y)$

Step (ii): Eliminate y and retain x, z; factorize

$3x^2 + 11xz + 6z^2 = (3x + 2z)(x + 3z)$

Step (iii): Fill the gaps, the given expression

$= (3x + y + 2z)(x + 2y + 3z)$

EXAMPLE 2:

$12x^2 + 11xy + 2y^2 - 13xz - 7yz + 3z^2$.

Step (i): Eliminate z i.e., z = 0; factorize

$12x^2 + 11xy + 2y^2 = (3x + 2y)(4x + y)$

Step (ii): Eliminate y i.e., y = 0; factorize

$12x^2 - 13xz + 3z^2 = (4x - 3z)(3x - z)$

Step (iii): Fill in the gaps; the given expression

$= (4x + y - 3z)(3x + 2y - z)$

EXAMPLE 3:

$3x^2 + 6y^2 + 2z^2 + 11xy + 7yz + 6xz + 19x + 22y + 13z + 20$

Step (i): Eliminate y and z, retain x and independent term i.e., y = 0, z = 0 in the expression (E).

Then $E = 3x^2 + 19x + 20 = (x + 5)(3x + 4)$

Step (ii): Eliminate z and x, retain y and independent term i.e., z = 0, x = 0 in the expression.

Then $E = 6y^2 + 22y + 20 = (2y + 4)(3y + 5)$

Step (iii): Eliminate x and y, retain z and independent term i.e., x = 0, y = 0 in the expression.

Then $E = 2z^2 + 13z + 20 = (z + 4)(2z + 5)$

Step (iv): The expression has the factors (think of independent terms) $= (3x + 2y + z + 4)(x + 3y + 2z + 5)$.

In this way either homogeneous equations of second degree or general equations of second degree in three variables can be very easily solved by applying 'adyamadyena' and 'lopanasthapanabhyam' sutras.

Solve the following expressions into factors by using appropriate sutras:

1. $x^2 + 2y^2 + 3xy + 2xz + 3yz + z^2$.
2. $3x^2 + y^2 - 4xy - yz - 2z^2 - zx$.
3. $2p^2 + 2q^2 + 5pq + 2p - 5q - 12$.
4. $u^2 + v^2 - 4u + 6v - 12$.
5. $x^2 - 2y^2 + 3xy + 4x - y + 2$.
6. $3x^2 + 4y^2 + 7xy - 2xz - 3yz - z^2 + 17x + 21y - z + 20$.

Highest Common Factor

To find the Highest Common Factor i.e. H.C.F. of algebraic expressions, the factorization method and process of continuous division are in practice in the conventional system. We now apply' Lopana - Sthapana' Sutra, the 'Sankalana vyavakalanakam' process and the 'Adyamadya' rule to find out the H.C.F in a more easy and elegant way.

EXAMPLE 1:

Find the H.C.F. of x2 + 5x + 4 and x2 + 7x + 6.

1. Factorization method:

$x^2 + 5x + 4 = (x + 4)(x + 1)$

$x^2 + 7x + 6 = (x + 6)(x + 1)$

H.C.F. is $(x + 1)$.

2. Continuous division process.

$x^2 + 5x + 4)\ x^2 + 7x + 6\ (1$
$\quad x^2 + 5x + 4$
$\quad\overline{\quad 2x + 2)\ x^2 + 5x + 4\ (\tfrac{1}{2}x}$
$\quad\quad x^2 + x$
$\quad\quad\overline{\quad 4x + 4)\ 2x + 2\ (\tfrac{1}{2}}$
$\quad\quad\quad 2x + 2$
$\quad\quad\quad\overline{\quad 0}$

Thus $4x + 4$ i.e., $(x + 1)$ is H.C.F.

3. Lopana - Sthapana process i.e. elimination and retention or alternate destruction of the highest and the lowest powers is as below:

$$\text{Subtract}\begin{cases} x^2+5x+4 \\ x^2+7x+6 \\ \hline -2)-2x-2 \\ \overline{\quad x+1} \end{cases}$$

i.e., (x + 1) is H.C.F

EXAMPLE 2:

Find H.C.F. of $2x^2 - x - 3$ and $2x^2 + x - 6$

$$\text{Subtract}\begin{cases} 2x^2-x-3 \\ 2x^2+x-6 \\ \hline -1)-2x+3 \\ \overline{2x+3} \text{ is the HCF.} \end{cases}$$

EXAMPLE 3:

$x^3 - 7x - 6$ and $x^3 + 8x^2 + 17x + 10$.

Now by Lopana - Sthapana and Sankalana - Vyavakalanabhyam

$$\text{Subtract}\begin{cases} x^3 \qquad -7x-6 \\ x^3+8x^2+17x+10 \\ \hline -8)-8x^2-24x-16 \\ \overline{x^2+3x+2} \end{cases}$$

EXAMPLE 4:

$x^3 + 6x^2 + 5x - 12$ and $x^3 + 8x^2 + 19x + 12$.

$$\begin{matrix}\text{Add} \\ \div 2x\end{matrix}\begin{cases} x^3+6x^2+5x-12 \\ x^3+8x^2+19x+12 \\ \hline 2x^3+14x^2+24x \\ \hline x^2+7x+12 \end{cases}$$

(or)

$$\text{Subtract}\left\{\begin{array}{r} x^3 + 6x^2 + 5x - 12 \\ x^3 + 8x^2 + 19x + 12 \\ \hline 2x^2 - 14x - 24 \\ \hline x^2 + 7x + 12 \end{array}\right.$$

$\div -2$

EXAMPLE 5:

$2x^3 + x^2 - 9$ and $x^4 + 2x^2 + 9$.

By Vedic sutras:

Add: $(2x^3 + x^2 - 9) + (x^4 + 2x^2 + 9)$

$= x^4 + 2x^3 + 3x^2.$

$\div x^2$ gives

$x^2 + 2x + 3$ ...(i)

Subtract after multiplying the first by x and the second by 2.

Thus $(2x^4 + x^3 - 9x) - (2x^4 + 4x^2 + 18)$

$= x^3 - 4x^2 - 9x - 18$ ...(ii)

Multiply (i) by x and subtract from (ii)

$x^3 - 4x^2 - 9x - 18 - (x^3 + 2x^2 + 3x)$

$= - 6x^2 - 12x - 18$

$\div - 6$ gives $x^2 + 2x + 3$.

Thus $(x^2 + 2x + 3)$ is the H.C.F. of the given expressions.

ALGEBRAIC PROOF:

Let P and Q be two expressions and H is their H.C.F. Let A and B the Quotients after their division by H.C.F.

i.e., $\frac{P}{H} = A$ and $\frac{Q}{H} = B$ which gives $P = A.H$ and $Q = B.H$

$P + Q = AH + BH$ and $P - Q = AH - BH$

$= (A+B).H$ $= (A-B).H$

Thus we can write P ± Q = (A ± B).H

Similarly MP = M.AH and NQ = N.BH gives MP ± NQ = H (MA ± NB)

This states that the H.C.F. of P and Q is also the H.C.F. of P±Q or MA±NB.

i.e. we have to select M and N in such a way that highest powers and lowest powers (or independent terms) are removed and H.C.F appears as we have seen in the examples.

Find the H.C.F. in each of the following cases using Vedic sutras:

1. $x^2 + 2x - 8, x^2 - 6x + 8$
2. $x^3 - 3x^2 - 4x + 12, x^3 - 7x^2 + 16x - 12$
3. $x^3 + 6x^2 + 11x + 6, x^3 - x^2 - 10x - 8$
4. $6x^4 - 11x^3 + 16x^2 - 22x + 8,$
   $6x^4 - 11x^3 - 8x^2 + 22x - 8.$

Vilokanam

The Sutra 'Vilokanam' means 'Observation'. Generally we come across problems which can be solved by mere observation. But we follow the same conventional procedure and obtain the solution. But the hint behind the Sutra enables us to observe the problem completely and find the pattern and finally solve the problem by just observation.

Let us take the equation x + (1/x) = 5/2 Without noticing the logic in the problem, the conventional process tends us to solve the problem in the following way.

$$x+\frac{1}{x}=\frac{5}{2}$$

$$\frac{x^2+1}{x}=\frac{5}{2}$$

$$2x^2 + 2 = 5x$$

$$2x^2 - 5x + 2 = 0$$

$$2x^2 - 4x - x + 2 = 0$$

$$2x(x - 2) - (x - 2) = 0$$

$$(x - 2)(2x - 1) = 0$$

$$x - 2 = 0 \text{ gives } x = 2$$

$$2x - 1 = 0 \text{ gives } x = ½$$

But by Vilokanam i.e., observation

$x+\frac{1}{x}=\frac{5}{2}$ can be viewed as

$x+\frac{1}{x}=2+\frac{1}{2}$ giving x = 2 or ½.

Consider Some Examples.

EXAMPLE 1 :

$$\frac{x}{x+2}+\frac{x+2}{x}=\frac{34}{15}$$

In the conventional process, we have to take L.C.M, cross-multiplication. simplification and factorization. But Vilokanam gives

$$\frac{34}{15}=\frac{9+25}{5\times 3}=\frac{3}{5}+\frac{5}{3}$$

$$\frac{x}{x+2}+\frac{x+2}{x}=\frac{3}{5}+\frac{5}{3}$$

gives

$$\frac{x}{x+2}=\frac{3}{5} \text{ or } \frac{5}{3}$$

$$5x = 3x + 6$$

or $3x = 5x + 10$

$2x = 6$

or $-2x = 10$

$x = 3$

or $x = -5$

EXAMPLE 2 :

$$\frac{x+5}{x+6}+\frac{x+6}{x+5}=\frac{113}{56}$$

Now,

$$\frac{113}{56}=\frac{49+64}{7\times8}=\frac{7}{8}+\frac{8}{7}$$

$$\frac{x+5}{x+6}=\frac{7}{8} \text{ or } \frac{x+5}{x+6}=\frac{8}{7}$$

$8x + 40 = 7x + 42$ $\quad$ $7x + 35 = 8x + 48$

or

$x = 42 - 40 = 2$ $\quad$ $-x = 48 - 35 = 13$

$x = 2$ $\quad$ or $\quad$ $x = -13.$

EXAMPLE 3:

$$\frac{5x+9}{5x-9}+\frac{5x-9}{5x+9}=2\frac{82}{319}$$

At first sight it seems to a difficult problem.

But careful observation gives

$$2\frac{82}{319}=\frac{720}{319}=\frac{841-121}{11\times29}=\frac{29}{11}-\frac{11}{29}$$

(Note: $29^2 = 841$, $11^2 = 121$)

$$\frac{5x+9}{5x-9} = \frac{29}{11} \text{ or } \frac{-11}{29}$$

(Note: 29 = 20 + 9 = 5 x 4 + 9 ; 11 = 20 - 9=5 x 4 - 9)

i.e.,

x = 4 or

$$\frac{5x+9}{5x-9} = \frac{-11}{29}$$

$$145x + 261 = -55x + 99$$

$$145x + 55x = 99 - 261$$

$$200x = -162$$

$$x = \frac{-162}{200} = \frac{-81}{100}$$

Simultaneous Quadratic Equations

EXAMPLE 1

x + y = 9 and xy = 14.

We follow in the conventional way that

$(x - y)^2 = (x + y)^2 - 4xy = 9^2 - 4 (14) = 81 - 56 = 25$

$x - y = \sqrt{25} = \pm 5$

$$\left.\begin{matrix} x+y=9 \\ x-y=5 \end{matrix}\right\} \text{gives } 2x = 14, x = 7$$

x + y = 9 gives 7 + y = 9

y = 9 - 7 = 2.

Thus the solution is x = 7, y = 2 or x = 2, y = 7.

But by Vilokanam, xy = 14 gives x = 2, y = 7 or x = 7, y = 2 and these two sets satisfy x + y = 9 since 2 + 7 = 9 or 7 + 2 = 9. Hence the solution.

EXAMPLE 2:

$5x - y = 7$ and $xy = 6$.

$xy = 6$ gives $x = 6, y = 1; x = 1, y = 6;$

$x = 2, y = 3; x = 3, y = 2$ and of course negatives of all these.

Observe that $x = 6, y = 1; x = 1, y = 6$: are not solutions because they do not satisfy the equation $5x - y = 7$.

But for $x = 2, y = 3; 5x - y = 5 (2) - 3 = 10 - 3 = 7$ we have $5(3) - 2 \neq 7$.

Hence $x = 2, y = 3$ is a solution.

For $x = 3, y = 2$ we get $5 (3) - 2 = 15 - 2$? 7.

Hence it is not a solution.

Negative values of the above are also not the solutions. Thus one set of the solutions i.e., $x = 2, y = 3$ can be found. Of course the other will be obtained from solving $5x - y = 7$ and $5x + y = -13$.

i.e., $x = -3 / 5, y = -10$.

Partial Fractions

EXAMPLE 1

Resolve

$\frac{2x+7}{(x+3)(x+4)}$ into partial fractions.

We write $\frac{2x+7}{(x+3)(x+4)} = \frac{A}{(x+3)} + \frac{B}{(x+4)}$

$$= \frac{A(x+4)+B(x+3)}{(x+3)(x+4)}$$

$$2x + 7 = A (x + 4) + B (x + 3).$$

We proceed by comparing coefficients on either side

coefficient of x : A + B = 2 ....(i) × 3

Independent of x : 4A + 3B = 7 ....(ii)

Solving (ii) - (i) × 3

$$\begin{aligned} 4A + 3B &= 7 \\ 3A + 3B &= 6 \\ \hline A &= 1 \end{aligned}$$

A = 1 in (i) gives, 1 + B = 2 i.e., B = 1

Or we proceed as

$$2x + 7 \equiv A(x + 4) + B(x + 3).$$

Put x = -3, $\quad 2(-3) + 7 \equiv A(-3 + 4) + B(-3 + 3)$

$$1 = A(1) \therefore A = 1.$$

x = -4, $\quad 2(-4) + 7 = A(-4 + 4) + B(-4 + 3)$

$$-1 = B(-1) \therefore B = 1.$$

Thus $\dfrac{2x+7}{(x+3)(x+4)} = \dfrac{1}{(x+3)} + \dfrac{1}{(x+4)}$

But by Vilokanam $\dfrac{2x+7}{(x+3)(x+4)}$ can be resolved as

(x + 3) + (x + 4) =2x + 7, directly we write the answer.

EXAMPLE 2:

$$\frac{3x+13}{(x+1)(x+2)}$$

from (x + 1), (x + 2) we can observe that

10 (x + 2) - 7(x + 1) = 10x + 20 - 7x - 7 = 3x + 13

Thus $\dfrac{3x+13}{(x+1)(x+2)} = \dfrac{10}{x+1} - \dfrac{7}{x+2}$

EXAMPLE 3:

$$\frac{9}{x^2+x-2}$$

As $x^2 + x - 2 = (x - 1)(x + 2)$ and

$9 = 3(x + 2) - 3(x - 1)$

$(3x + 6 - 3x + 3 = 9)$

We get by Vilokanam, $\dfrac{9}{x^2+x-2} = \dfrac{3}{x-1} - \dfrac{3}{x+2}$

I. Solve the following by mere observation i.e. vilokanam

1. $x + \dfrac{1}{x} = \dfrac{25}{12}$

2. $x - \dfrac{1}{x} = \dfrac{5}{6}$

3. $\dfrac{x}{x+1} + \dfrac{x+1}{x} = 9\dfrac{1}{9}$

4. $\dfrac{x+7}{x+9} - \dfrac{x+9}{x+7} = \dfrac{32}{63}$

II. Solve the following simultaneous equations by vilokanam.

1. $x - y = 1$, $xy = 6$
2. $x + y = 7$, $xy = 10$
3. $2x + 3y = 19$, $xy = 15$
4. $x + y = 4$, $x^2 + xy + 4x = 24$.

III. Resolve the following into partial fractions.

1. $\dfrac{2x-5}{(x-2)(x-3)}$

2. $\frac{9}{(x+1)(x-2)}$

3. $\frac{x-13}{x^2-2x-15}$

4. $\frac{3x+4}{3x^2+3x+2}$

Gunìta Samuccayah—Samuccaya Gunìtah

In connection with factorization of quadratic expressions a sub-Sutra, viz. 'Gunita samuccayah-Samuccaya Gunitah' is useful. It is intended for the purpose of verifying the correctness of obtained answers in multiplications, divisions and factorizations. It means in this context:

'The product of the sum of the coefficients sc in the factors is equal to the sum of the coefficients sc in the product'.

Symbolically we represent as sc of the product = product of the sc (in the factors).

EXAMPLE 1:

$(x + 3)(x + 2) = x^2 + 5x + 6$

Now $(x + 3)(x + 2) = 4 \times 3 = 12$: Thus verified.

EXAMPLE 2:

$(x - 4)(2x + 5) = 2x^2 - 3x - 20$

Sc of the product $2 - 3 - 20 = -21$

Product of the Sc = $(1 - 4)(2 + 5) = (-3)(7) = -21$.

Hence verified.

In case of cubics, biquadratics also the same rule applies.

We have $(x + 2)(x + 3)(x + 4) = x^3 + 9x^2 + 26x + 24$

Sc of the product = $1 + 9 + 26 + 24 = 60$

Product of the Sc = (1 + 2) (1 + 3) (1 + 4)

= 3 × 4 × 5 = 60. Verified.

EXAMPLE 3:

$(x + 5)(x + 7)(x - 2) = x^3 + 10x^2 + 11x - 70$

$(1 + 5)(1 + 7)(1 - 2) = 1 + 10 + 11 - 70$

i.e., 6 × 8 x -1 = 22 - 70

i.e., -48 = -48 Verified.

We apply and interpret So and Sc as sum of the coefficients of the odd powers and sum of the coefficients of the even powers and derive that So = Sc gives (x + 1) is a factor for thee concerned expression in the variable x. Sc = 0 gives (x - 1) is a factor.

Verify whether the following factorization of the expressions are correct or not by the Vedic check:

i.e. Gunita. Samuccayah-Samuccaya Gunitah:

1. $(2x + 3)(x - 2) = 2x^2 - x - 6$
2. $12x^2 - 23xy + 10y^2 = (3x - 2y)(4x - 5y)$
3. $12x^2 + 13x - 4 = (3x - 4)(4x + 1)$
4. $(x + 1)(x + 2)(x + 3) = x^3 + 6x^2 + 11x + 6$
5. $(x + 2)(x + 3)(x + 8) = x^3 + 13x^2 + 44x + 48$

So far we have considered a majority of the upa-sutras as mentioned in the Vedic mathematics book. Only a few Upa-Sutras are not dealt under a separate heading . They are:

(2) S'ISYATE S'ESASAMJ ÑAH

(4) KEVALAIH SAPTAKAMGUNYAT

(5) VESTANAM

(6) YAVADUNAM TAVADUNAM and

(10) SAMUCCAYAGUNITAH already find place in respective places.

Further in some other books developed on Vedic Mathematics DVANDAYOGA, SUDHA, DHVAJANKAM are also given as Sub-Sutras. They are mentioned in the Vedic Mathematics text also. But the list in the text (by the Editor) does not contain them. We shall also discuss them at appropriate places, with these three included, the total number of upa-Sutras comes to sixteen.

We now proceed to deal the Sutras with reference to their variety, applicability, speed, generality etc. Further we think how 'the element of choice in the Vedic system, even of innovation, together with mental approach, brings a new dimension to the study and practice of Mathematics. The variety and simplicity of the methods brings fun and amusement, the mental practice leads to a more agile, alert and intelligent mind and innovation naturally follow' (Prof. K.R.Williams, London).

Bibliography

A.L. Basham: *The Wonder That Was India*, Rupa & Co., Calcutta, 1967.

Aaboe, Asger: *Episodes from the Early History of Mathematics*, New York, 1964.

Alfred, C.: *Introduction to Prakrit*, Calcutta, Baptist Mission Press, 1917.

Amma, S: *Geometry in Ancient and Medieval India*, Motilal Banarsidas, 1979.

Archibald, R.C.: *The Rhind Mathematical Papyrus,* Oberlin, Ohio, 1927.

Armstrong, G.: *Yoruba Numerals*, London, Oxford University Press, 1962.

Arnett, S.: *The Predynastic Origin of Egyptian Hieroglyphs*, Washington, University Press of America, 1982.

Ascher, Marcia: *Ethnomathematics*, Pacific Grove, CA: Brooks/Cole, 1991.

Bag, A. K.: *Mathematics in Ancient and Medieval India*, Varanasi: Chaukhambha Orientalia, 1979.

Barnard, F. P.: *The Casting-Counter and the Counting Board. A Chapter in the History of Numismatics and Early Arithmetic*, Oxford, Oxford University Press, 1916.

Barrow, D.: *Pi in the Sky*, Oxford, Clarendon Press, 1992.

Basham, A. L.: *The Wonder That Was India*, Calcutta, Rupa & Co., 1967.

Beckmann, Petr: *A History of Pi*, New York, St. Martin's Press, 1974.

Bedient, J. D.: *The Historical Roots of Elementary Mathematics*, New York, Dover, 1988.

Bell, Eric T.: *The Magic of Numbers*, New York, McGraw-Hill, 1946.

Berlin, Brent: *Tzeltal Numeral Classifiers*, The Hague, Mouton, 1968.

Bharati Krishna Tirthaji Majaraja.: *Vedic Mathematics*. Delhi, Motilal Banarsidass, 1988.

Bidyaranya, Swami: *History of Hindu Mathematics*,Bombay, Asia Pub. House, 1962.

Bischoff, Bernard: *Latin Palaeography*, Cambridge, Cambridge University Press, 1990.

Bonfante, Larissa: *Etruscan*. Berkeley, University of California Press, 1990.

Brainerd, Charles J.: *The Origins of the Number Concept*, New York, Praeger, 1979.

Brunschwig, Jacques : *Greek Thought: A Guide to Classical Knowledge*, Cambridge, Harvard University Press, 2000.

Bunt, L. N. H.: *The Historical Roots of Elementary Mathematics*, New York, Dover, 1988.

Charles, P.: *The Numeration, Calendar Systems and Astronomical knowledge of the Mayas*, Cambridge, Cambridge University Press, 1910.

Charles, P.: *The Numeration, Calendar Systems and Astronomical knowledge of the Mayas*, Cambridge, Cambridge University Press, 1910.

Chattopadhyaya, D. P.: *Mathematics, Astronomy, and Biology in Indian Tradition: Some Conceptual Preliminaries*, New Delhi, Project of History of Indian Science, Philosophy, and Culture, 1995.

Clawson, Calvin: *The Mathematical Traveller*, New York, Plenum Press, 1994.

Closs, P.: *Native American Mathematics*, Austin, University of Texas Press, 1986.

Conant, L.: *The Number Concept*, New York, MacMillan, 1896.

Crossley, N: *The Emergence of Number*, Singapore, World Scientific, 1987.

Crump, Thomas: *The Anthropology of Numbers*, Cambridge, Cambridge University Press, 1990.

Dantzig, Tobias: *Number, the Language of Science*, London, George Allen and Unwin, 1940.

David, F.: *Cross-Cultural Studies in Cognition and Mathematics*, New York, Academic, 1983.

De Villiers, Melius: *The Numeral-Words: Their Origin, Meaning, History and Lesson*, London, Witherby, 1923.

Dehaene, Stanislas: *The Number Sense*, New York, Oxford University Press, 1997.

Dilke, Wentworth: *Mathematics and Measurement*, Berkeley, CA: U. of Cal. Press, 1987.

Dirk, J.: *A Concise History of Mathematics*, New York, Dover, 1987.

Eric, S.: *Maya Hieroglyphic Writing: An Introduction*, Norman, University of Oklahoma Press, 1971.

Eves, Howard: *An Introduction to the History of Mathematics*, New York, Rinehart and Company Inc., 1953.

Flegg, Graham: *Numbers: Their History and Meaning*, London, André Deutsch, 1983.

Fowler, David: *The Mathematics of Plato's Academy*, Oxford, Clarendon, 1999.

Francis, Llewellyn: *Hieratic Papyri from Kahun and Gurob*, London, Bernard Quaritch, 1898.

Fuson, K.C.: *Children's Counting and Concepts of Number*, New York, Springer-Verlag, 1988.

Ganeri, Anita: *The Story of Numbers and Counting*, New York, Oxford University Press, 1996.

Geoffrey E.R. Lloyd:. *Greek Thought: A Guide to Classical Knowledge*, Cambridge, Harvard University Press, 2000.

Gokhale, Laxman: *Indian Numerals*, Poona, Deccan College, 1966.

Griffith, Llewellyn: *Hieratic Papyri from Kahun and Gurob*, London, Bernard Quaritch, 1898.

Grover, Sudarshan: *History of Development of Mathematics in India*, Delhi, Atma Ram & Sons, 1994.

Haldane, Sanderson: *Science and Indian Culture*, Calcutta, New Age Publishers, 1965.

Halliwell, Orchard: *Rara Mathematica*, London, J.W. Parker, 1839.

Herbert Meschkowski: *Ways of Thought of Great Mathematicians*, Holden-Day Inc., San Francisco, 1964.

Hogben, Lancelot: *Mathematics for the Million*, New York; W. W. Norton, 1937.

Howard Eves: *An Introduction to the History of Mathematics*, Rinehart and Company Inc., New York, 1953.

Ifrah, Georges: *From One to Zero: A Universal History of Numbers*, New York, Viking Penguin, 1985.

Jagadguru Swami Shri Bharati Krishna Tirthaji Maharaja: *Vedic Mathematics*, Motilal Banarsidass, Delhi, 1988.

Jaggi, O. P.: *Science and Technology in Medieval India*, New Delhi, Atma Ram & Sons, 1977.

James, Orchard: *Rara Mathematica*, London, J.W. Parker, 1839.

John, Burdon Sanderson: *Science and Indian Culture*, Calcutta, New Age Publishers, 1965.

John, D.: *Pi in the Sky*, Oxford, Clarendon Press, 1992.

John, N: *The Emergence of Number*, Singapore, World Scientific, 1987.

Jones, D.: *The Internal Representation of Number: Analogue or Digital?*, Oxford, Clarendon, 1987.

Kline, Morris: *Mathematical Thought from Ancient to Modern Times*, New York, Oxford U. Press, 1990.

Kuppuram, G & K. Kumudamani: *History of Science and Technology in India*, Delhi, Sundeep Prakashan, 1990.

Lancy, F.: *Cross-Cultural Studies in Cognition and Mathematics*, New York, Academic, 1983.

Levi, L.: *The Number Concept*, New York, MacMillan, 1896.

Manning, H.P. and R C Archibald: *The Rhind Mathematical Papyrus,* Oberlin, Ohio, 1927.

Marcus, Joyce: *Mesoamerican Writing Systems*, Princeton, Princeton University Press, 1992.

Marshack, Alexander: *The Roots of Civilization*, New York, McGraw-Hill, 1972.

Martzloff, Jean-Claude: *A History of Chinese Mathematics*, New York, Springer, 1997.

McLeish, John: *The Story of Numbers*, New York, Fawcett Columbine, 1991.

Menninger, Karl: *Number-Words and Number-Symbols*. London, MIT, 1969.

Michael, P.: *Native American Mathematics*, Austin, University of Texas Press, 1986.

Murthy, T. S. Bhanu: *A Modern Introduction to Ancient Indian Mathematics*, Wiley Eastern Ltd., New Delhi, 1992.

Neugebauer, Otto : *Mathematical Cuneiform Texts*, New Haven, Yale University Press, 1945.

Neugebauer, Otto: *The Exact Sciences in Antiquity*, Princeton, Princeton University Press, 1952.

North, John: *The Norton History of Astronomy and Cosmology*, New York, W.W.Norton, 1995.

Oswald Ashton Wentworth: *Mathematics and Measurement*, Berkeley, CA: U. of Cal. Press, 1987.

Pandit, M. D.: *Mathematics as known to the Vedic Samhitas*, Delhi, Sri Satguru Publications, 1993.

Pappas, Theoni: *The Magic of Mathematics*, San Carlos, Wide World Publishing / Tetra, 1994.

Parpola, Asko: *Deciphering the Indus Script*, Cambridge, Cambridge University Press, 1994.

Peet, T. Eric: *The Rhind Mathematical Papyrus*, London, British Museum, 1923.

Piaget, Jean: *The Child's Conception of Number*, New York, W.W. Norton, 1965.

Pullan, J.M: *The History of the Abacus*, New York, Praeger, 1968.

Rao, S. Balaachandra: *Indian Mathematics and Astronomy*, Bangalore, Jnana Deep Publications, 1994.

Restivo, Sal: *The Social Relations of Physics, Mysticism and Mathematics*, Dordrecht, Kluwer, 1983.

Robert, G.: *Yoruba Numerals*, London, Oxford University Press, 1962.

Robins, G. and C. Shute: *The Rhind Mathematical Papyrus*, London, British Museum, 1987.

Russell, Bertrand: *The Principles of Mathematics*, Cambridge, Cambridge University Press, 1903.

Sachs, A.: *Mathematical Cuneiform Texts*, New Haven, Yale University Press, 1945.

Salomon, Richard: *Indian Epigraphy*, New York, Oxford University Press, 1998.

Saraswarti, A.: *Geometry in Ancient and Medieval India*, Motilal Banarsidas, 1979.

Saraswati Amma: *Geometry in Ancient and Medieval India*, Motilal Banarsidas, 1979.

Sarton, George: *The Study of the History of Mathematics*, Cambridge, Cambridge University Press, 1936.

Satterthwaite, Jr.: *Concepts and Structures of Maya Calendrical Arithmetics*, Philadelphia, University of Pennsylvania, 1947.

Satterthwaite, Jr.: *Concepts and Structures of Maya Calendrical Arithmetics*, Philadelphia, University of Pennsylvania, 1947.

Schaaf, W.L.: *Mathematics, Our Great Heritage*, New York, Harper and Row, 1948.

Schimmel, Annemarie: *The Mystery of Numbers*, New York, Oxford University Press, 1993.

Scriba, C.J.: *The Concept of Number*, Zurich, Mannheim, 1968.

Seidenberg, A.: *The Zero in the Mayan Numerical Notation*, Austin, University of Texas Press, 1986.

Shebhana, Laxman: *Indian Numerals*, Poona, Deccan College, 1966.

Smeltzer, Donald: *Man and Number*, New York, Emerson, 1959.

Smith, D. E.: *History of Mathematics*, New York, Dover Publications Inc., 1951.

Srinivasiengar, C. N.: *The History of Ancient Indian Mathematics*. Calcutta, World Press, 1967.

Struik, J.: *A Concise History of Mathematics*, New York, Dover, 1987.

Thompson, Sir J. : *Maya Hieroglyphic Writing: An Introduction*, Norman, University of Oklahoma Press, 1971.

Todd, R.R.: *The Internal Representation of Number: Analogue or Digital?*, Oxford, Clarendon, 1987.

Urton, Gary: *The Social Life of Numbers*, Austin, University of Texas Press, 1997.

Volodarsky, A.I.: *Mathematics in Ancient India*, Tokyo, Science Council of Japan, 1974.

William, S.: *The Predynastic Origin of Egyptian Hieroglyphs*, Washington, University Press of America, 1982.

Woolner, C.: *Introduction to Prakrit*, Calcutta, Baptist Mission Press, 1917.

Index

N

O

P

Q

R

S

T

U

V

W

Y

Z

□□□

119A 3 SET

ENCYCLOPAEDIA OF VEDIC MATHEMATICS

ENCYCLOPAEDIA OF
VEDIC MATHEMATICS
(Vol. 5)

Shyam Priya

ANMOL PUBLICATIONS PVT. LTD.
NEW DELHI - 110 002 (INDIA)

ANMOL PUBLICATIONS PVT. LTD.
H.O.: 4374/4B, Ansari Road, Darya Ganj,
New Delhi-110 002 (India)
Ph.: 23278000, 23261597
B.O.: No. 1015, Ist Main Road, BSK IIIrd Stage
IIIrd Phase, IIIrd Block,
Bangalore - 560 085 (India)
Visit us at: www.anmolpublications.com

Encyclopaedia of Vedic Mathematics

First Published, 2008

ISBN 978-81-261-3504-2 (Set)

PRINTED IN INDIA

Printed at Mehra Offset Press, Delhi.

Contents

Preface

Vedic Mathematics contains many examples of striking methods of calculation and there is a remarkable coherence to the system which makes it very attractive. Also the Vedic system itself suggests a kind of approach that involves going directly to the answer. Vedic mathematics is based on sixteen Sutras and some sub-Sutras which provide links through mathematics: the word 'Sutra' means 'thread'.

Vedic scheme is not simply a collection of rapid methods; it is a system, a unified approach, which can be swiftly learnt. Vedic Mathematics extensively exploits the properties of numbers in every practical applications, particularly in the field of computation. It makes available a whole range of methods ideally suited to these properties. After a detailed introduction to Vedic Mathematics, this book shows applications of Vertically and Crosswise from basic calculations to the evaluation of logarithms, exponentials, trigonometrical functions and the solution of simultaneous, transcendental, polynomial and differential equations. The methods shown are not only exceedingly efficient and new but have a unity and coherence which renders them easy to master and enjoyable to execute.

The methods given in this book are not intended to be a complete or thorough treatment of the topics they deal with. All of the ideas can probably be developed further or applied in other areas, and all can doubtless be improved upon. The mathematician will also observe a certain lack of rigor as an attempt has been made to make the material intelligible to as wide a readership as possible.

—Shyam Priya

1

The Magic of Vedic Maths

What is Vedic Math?

What does mathematics have to do with Hinduism? Well, just as the basic principles of Hinduism lie in the Vedas, so do the roots of mathematics. The Vedas, written around 1500-900 BCE, are ancient Indian texts containing a record of human experience and knowledge. Thousands of years ago, Vedic mathematicians authored various theses and dissertations on mathematics. It is now commonly believed and widely accepted that these treatises laid down the foundations of algebra, algorithm, square roots, cube roots, various methods of calculation, and the concept of zero.

Vedic Mathematics

"Vedic Mathematics" is the name given to the ancient system of mathematics, or, to be precise, a unique technique of calculations based on simple rules and principles, with which any mathematical problem - be it arithmetic, algebra, geometry or trigonometry - can be solved, hold your breath, orally!

Sutras: Natural Formulae

The system is based on 16 Vedic sutras or aphorisms, which are actually word-formulae describing natural ways of solving a whole range of mathematical problems. Some examples of sutras are "By one more than the one before", "All from 9 & the last from 10", and "Vertically & Crosswise". These 16 one-line formulae originally written in Sanskrit, which can be easily memorized, enables one to solve long mathematical problems quickly.

Why Sutras?

Sri Bharati Krishna Tirtha Maharaj, who is generally considered the doyen of this discipline, in his seminal book Vedic Mathematics, wrote about this special use of verses in the Vedic age: "In order to help the pupil memorize the material assimilated, they made it a general rule of practice to write even the most technical and abstruse textbooks in sutras or in verse (which is so much easier - even for the children - to memorize)... So from this standpoint, they used verse for lightening the burden and facilitating the work (by versifying scientific and even mathematical material in a readily assimilable form)!"

Dr. L M Singhvi, the former High Commissioner of India in the UK, who is an avid endorser of the system says: "A single sutra would generally encompass a varied and wide range of particular applications and may be likened to a programmed chip of our computer age". Another Vedic maths enthusiast, Clive Middleton of vedicmaths.org feels, "These formulae describe the way the mind naturally works, and are therefore a great help in directing the student to the appropriate method of solution."

A Simple and Easy System

Practitioners of this striking method of mathematical problem-solving opine that Vedic maths is far more systematic, coherent and unified than the conventional system. It is a mental tool for calculation that encourages the development and use of intuition and innovation, while giving the student a lot of flexibility, fun and satisfaction. Therefore, it's direct and easy to implement in schools - a reason behind its enormous popularity among educationists and academicians.

Try These Out!

- If you want to find the square of 45, you can employ the Ekadhikena Purvena sutra ("By one more than the one before"). The rule says since the first digit is 4 and the second one is 5, you will first have to multiply 4 (4 +1), that is 4 × 5, which is equal to 20 and then multiply 5 with 5, which is 25. Viola! The answer is 2025. Now, you can employ this method to multiply all numbers ending with 5.
- If you want to subtract 4679 from 10000, you can easily apply the Nikhilam Navatashcaramam Dashatah sutra ("All from 9 and the last from 10"). Each figure in 4679 is subtracted from 9 and the last figure is subtracted from 10, yielding 5321. Similarly, other sutras lay down such simple rules of calculation.

"The world owes most to India in the realm of mathematics, which was developed in the Gupta period to a stage more advanced than that reached by any other nation of antiquity. The success of Indian mathematics was mainly due to the fact that Indians had a clear conception of the abstract number as distinct from the numerical quantity of objects or spatial extension." ~ A.L. Basham, Australian Indologist in The Wonder That Was India

Born in the Vedic Age, but buried under centuries of debris, this remarkable system of calculation was deciphered towards the beginning of the 20th century, when there was a great interest in ancient Sanskrit texts, especially in Europe. However, certain texts called Ganita Sutras, which contained mathematical deductions, were ignored, because no one could find any mathematics in them. These texts, it's believed, bore the germs of what we now know as Vedic Mathematics.

Bharati Krishna Tirthaji's Discovery

Vedic math was rediscovered from the ancient Indian scriptures between 1911 and 1918 by Sri Bharati Krishna Tirthaji (1884-1960), a scholar of Sanskrit, Mathematics, History and Philosophy. He studied these ancient texts for years, and after careful investigation was able to reconstruct a series of mathematical formulae called sutras.

Bharati Krishna Tirthaji, who was also the former Shankaracharya (major religious leader) of Puri, India, delved into the ancient Vedic texts and established the techniques of this system in his pioneering work - Vedic Mathematics (1965), which is considered the starting point for all work on Vedic math. It is said that after Bharati Krishna's original 16 volumes of work expounding the Vedic system were lost, in his final years he wrote this single volume, which was published five years after his death.

Development of Vedic Math

Vedic math was immediately hailed as a new alternative system of mathematics, when a copy of the book reached London in the late 1960s. Some British mathematicians, including Kenneth Williams, Andrew Nicholas and Jeremy Pickles took interest in this new system. They extended the introductory material of Bharati Krishna's book, and delivered lectures on it in London. In 1981, this was collated into a book entitled Introductory Lectures on Vedic Mathematics. A few successive trips to India by Andrew Nicholas between 1981 and 1987, renewed the interest on Vedic math, and scholars and teachers in India started taking it seriously.

Interest in Vedic maths is growing in the field of education where maths teachers are looking for a new and better approach to the subject. Even students at IIT (Indian Institute of Technology) are said to be using this ancient technique for quick calculations. No wonder, a recent Convocation speech addressed to the students of IIT, Delhi, by Dr. Murli Manohar Joshi, Indian Minister for Science & Technology, stressed the significance of Vedic maths, while pointing out the important contributions of ancient Indian mathematicians, such as Aryabhatta, who laid the foundations of algebra, Baudhayan, the great geometer, and Medhatithi and Madhyatithi, the saint duo, who formulated the basic framework for numerals.

Vedic Maths in Schools

Quite a few years ago, St. James' School, London, and other schools began to teach the Vedic system, with notable success. Today this remarkable system is taught in many schools and institutes in India and abroad, and even to MBA and economics students.

When in 1988, Maharishi Mahesh Yogi brought to light the marvels of Vedic maths, Maharishi Schools around the world incorporated it in their syllabi. At the school in Skelmersdale, Lancashire, UK, a full course called "The Cosmic Computer" was written and tested on 11 to 14 year old pupils, and later published in 1998. According to Mahesh Yogi, "The sutras of Vedic Mathematics are the software for the cosmic computer that runs this universe."

Since 1999, a Delhi-based forum called International Research Foundation for Vedic Mathematics and Indian Heritage, which promotes value-based education, has been organising lectures on Vedic maths in various schools in Delhi, including Cambridge School, Amity International, DAV Public School, and Tagore International School.

Vedic Math Research

Researches are being undertaken in many areas, including the effects of learning Vedic maths on children. A great deal of research is also being done on how to develop more powerful and easy applications of the Vedic sutras in geometry, calculus, and computing. The Vedic Mathematics

Research Group published three new books in 1984, the year of the centenary of the birth of Sri Bharati Krishna Tirthaji.

Plus Points

There are obviously many advantages of using a flexible, refined and efficient mental system like Vedic math. Pupils can come out of the confinement of the 'only one correct' way, and make their own methods under the Vedic system. Thus, it can induce creativity in intelligent pupils, while helping slow-learners grasp the basic concepts of mathematics. A wider use of Vedic math can undoubtedly generate interest in a subject that is generally dreaded by children.

2

Swami Bharati Krishna Tirtha's Vedic Mathematics

Vedic mathematics is a system of mathematics consisting of a list of 16 basic sutras, or aphorisms, that allegedly encompass all mathematics. They were presented by a Hindu scholar and mathematician, Jagadguru Swami Sri Bharati Krishna Tirthaji Maharaja, during the early part of the 20th century (Trivedi, 1965).

Tirthaji claims that he found the sutras after years of studying the Vedas, a set of sacred ancient Hindu texts (Agrawala, 1992). However, labelling the mathematics he presented as 'Vedic' has provoked great controversy amongst Indian mathematicians who question both the Vedic origin of the mathematics, and whether the sutras can fulfil the claim of encompassing all mathematics. Nonetheless, the calculation strategies provided by Vedic mathematics are creative and useful, and can be applied in a number of ways to calculation methods in arithmetic and algebra, most notably within the education system. Vedic math has some similarities to the Trachtenberg system and many of the arithmetic computational strategies are based on the same concepts.

Tirthaji and the Discovery of Vedic Mathematics

Vedic mathematics was presented by Jagadguru Swami Sri Bharati Krsna Tirthaji, who is described as having the "rare combination of the probing insight and revealing intuition of a Yogi with the analytical acumen and synthetic talent of a mathematician" (Pratyagatmananda, 1965). Born in India in 1884, Tirthaji was an exceptional scholar; by age twenty he had studied at a number of colleges and universities throughout the country, been awarded the title of 'Saraswati' by the Madras Sanskrit Association for his remarkable proficiency in Sanskrit, and had completed seven masters degrees, including Sanskrit, Philosophy, English, Mathematics, History and Science, with the American College of Sciences (Trivedi, 1965).

Around 1911, Tirthaji resolved to study several sections of the Atharva-veda that had been dismissed by Orientalists, Indologists and antiquarian scholars as nonsensical (Tirthaji, 1992). He was part of a shrinking group of Indian scholars who believed that the Vedas represented an "inexhaustible mine of profound wisdom" both spiritual and secular (Pratyagâtmananda, 1965). Tirthaji claimed that there were sections of the Atharva-veda labelled "ganita sutras" or "mathematical formulae" that mysteriously made no obvious reference to mathematics (Kansara, 1961). Tirthaji explains that he was determined to understand the "ganita sutra" references, and began studying ancient lexicons and lexicography in more detail (Kansara, 1961). With this resolve, Tirthaji went to Sringeri, Karnataka, where he began years of solitary study and meditation (Trivedi, 1965).

Eight years later, Tirthaji emerged claiming to have deciphered 16 fundamental mathematical sutras in the Vedas, which today have become the foundation of Vedic mathematics (Tirthaji, 1992). According to Tirthaji, the sutras cover every branch of mathematics, from arithmetic to spherical conics, and that "there is no mathematics beyond their jurisdiction" (Tirthaji, 1992).

After discovering the sutras, Tirthaji travelled around India presenting Vedic mathematics, and even lectured in the United States and England in 1958 (Trivedi, 1965). In addition to lecturing, Tirthaji also wrote sixteen volumes, one for each basic sutra, explaining their applications (Trivedi, 1965). Unfortunately, before they were published, the manuscripts were lost irretrievably (Kansara, 1961). Before falling ill and passing away in 1960, Tirthaji was able to rewrite the first of the sixteen volumes he had composed (Trivedi, 1965).

The Sutras

Vedic mathematics is based on sixteen sutras which serve as somewhat cryptic instructions for dealing with different mathematical problems.

Below is a list of the sutras, translated from Sanskrit into English:

- By one more than the one before
- "All from 9 and the last from 10"
- Vertically and crosswise
- Transpose and apply
- If the Samuccaya is the same it is zero
- If one is in ratio the other is zero
- By addition and by subtraction
- By the completion or non-completion
- Differential calculus
- By the deficiency
- Specific and general
- The remainders by the last digit
- The ultimate and twice the penultimate
- By one less than the one before
- The product of the sum
- All the multipliers

Subsutras or Corollaries

- "Proportionately"
- The remainder remains constant
- "The first by the first and the last by the last"
- For 7 the multiplicand is 143
- By osculation
- Lessen by the deficiency
- Whatever the deficiency lessen by that amount and set up the square of the deficiency
- Last Totalling 10
- Only the last terms
- The sum of the products
- By alternative elimination and retention
- By mere observation
- The product of the sum is the sum of the products
- On the flag

To check the factorization of polynomial expressions: "The product of the sum of the coefficients in the factors is equal to the sum of the coefficients in the product. The SC of the product (the expression) = Product of the SC (in the factors). Additionally, this sub-sutra helps to fill in the gaps when some factors are known.

Background Information on the Vedas

The word "veda" has two basic meanings. The first, a literal translation of the Sanskrit word, is "knowledge" (Veda). The second, and most common meaning of the word, refers to the sacred ancient literature of Hinduism, the Vedas, a collection of hymns, poetry and Hindu ceremonial formulas (Veda). Believed to be one of the oldest human written records, the Vedas date back over 4000 years (Gaskell, 2000). Traditionally, they were passed down orally and adapted from generation to generation by sacred sages called rishis, before eventually emerging written in Vedic, an ancient form of Sanskrit, during the 3rd century BCE (Veda).

The Vedas are divided into four main sections: the Rig-veda, Sama-veda, Yajurveda and the Atharva-veda, known collectively as the Samhitas (Veda). The first three, the Rig-veda, Sama-veda, and Yajurveda are basically ritual handbooks that were used by priests during the Vedic period (1500-500 BCE) (Veda). Vedic mathematics is apparently part of the fourth Veda, Atharva-veda, which is distinct from the others in several ways. First, unlike the religious focus of the other Vedas, the Atharva-veda contains hymns, spells and magical incantations for personal and domestic use (Veda). Also, the Atharva-veda, which was written later than the other Vedas, was not always considered authoritative, but only became so after being accepted by the Brahmans, the highest order of Hindu priests (Veda). Collectively, the Vedas do include information about a huge range of subjects, spanning religion, medicine, architecture, astronomy, etc. (Gaskell, 2000).

Although there is controversy about whether the Vedas themselves actually include references to mathematics, the roots of sophisticated mathematics have actually been traced back to the Vedic era. Ancient Indian Vedic civilisations are known for being skilled in geometry, algebra and computational mathematics complex enough to incorporate things like irrational numbers (Dutta, 2002). Furthermore, all ancient Indian mathematics literature is composed completely in verse; there was a tradition of composing terse sutras, like those of Vedic mathematics, to ensure that information would be preserved even if written records were damaged or lost (Dutta, 2002).

Applying the Sutras: Some of the Actual Mathematics

It is not difficult to understand and apply the Vedic mathematical strategies, as long as one does not rely on the sutras alone for mathematical insight. Those studying Vedic mathematics tend to strongly rely on the examples and explanations Tirthaji provides in his book.

All from Nine and the Last from Ten

When subtracting from a large power of ten with many columns of zeros you need not write the notation for "borrowing" from the column on the left. If you want to subtract 4679 from 10000, you can easily apply the Nikhilam Navatashcaramam Dashatah sutra ("[Subtract in] All [zero columns] from 9 and [in] the last [column subtract] from 10"). Each figure in 4679 is subtracted from 9 and the last figure is subtracted from 10, yielding 5321. This method is also used when finding the deficit from the next larger power of ten.

Corollary

"Whatever the extent of its deficiency, lessen it still further to that very extent; and also set up the square of that deficiency"

For instance, in computing the square of 9 we go through the following steps:

1. The nearest factor of 10 to 9 is 10. Therefore, let us take 10 as our base.
2. Since 9 is 1 less than 10, decrease it still further to 8. This is the left side of our answer.
3. On the right hand side put the square of the deficiency, which is 12. Hence the answer is 81.
4. Similarly, 82 = 64, 72 = 49.
    1. For numbers above 10, instead of looking at the deficit we look at the surplus. For example:

$11^2 = (11 + 1) . 10 + 1^2 = 121.$

$12^2 = (12 + 2) . 10 + 2^2 = 144.$

$14^2 = (14 + 4) . 10 + 4^2 = 18 . 10 + 16 = 196.$

$25^2 = [(25 + 5) . 2] . 10 + 5^2 = 625.$

$35^2 = [(35 + 5) . 3] . 10 + 5^2 = 40 . 3 . 10 + 25 = 1225.$

and so on.

This is based on the fact that a2 = (a + b)(a - b) + b2 where a is the number whose square is to be found and b is the deficit (or surplus) from nearest factor of 10.

By one More than the one Before

"Eka Dikeną Parvesam" is the Sanskrit term for "One more than the previous one". It provides a simple way of calculating values like 1/x9 (e.g 1/19, 1/29, etc.). Let's take one 1/x9 and calculate, e.g., 1/19. In this case, x=1. The method is to write that x, then multiply it by 2 and keep that digit to its left. If the result of the multiplication is greater than 10, keep (value - 10) and keep the "1" as "carry over" which you'll add to the next digit.

The preposition "by" means the operations this formula concerns are either multiplication or division. [In case of addition/subtraction proposition "to" or "from" is used.] Thus this formula is used for either multiplication or division. It turns out that it is applicable in both operations.

Note: This applies to multiplication of numbers with the same first digit and the sum of their last unit digits is 10.

An interesting application of this formula is in computing squares of numbers ending in five. Consider:

35 × 35 = ((3 × 3) + 3), 25 = 12, 25

125 × 125 = ((12 × 12) + 12), 25 = 156, 25

The latter portion is multiplied by itself (5 by 5) and the previous portion is square of first digit or first two digit (3×3) or (12×12) and adding the same digit in that figure (3 or 12) resulting in the answer 1225.

This is a simple application of $(a + b)^2 = a^2 + 2ab + b^2$ when a = 10c and b = 5, i.e.

$(10c + 6)^2 = 100c^2 + 100c + 25 = 100c (c + 1) = 25.$

It can also be applied in multiplications when the last digit is not 5 but the sum of the last digits is the base (10) and the previous parts are the same. Consider:

$37 \times 33 = (3 \times 4), 7 \times 3 = 12, 21$

$29 \times 21 = (2 \times 3), 9 \times 1 = 6, 09$

This uses $(a + b)(a - b) = a^2 - b^2$ twice combined with the previous result to produce:

$$\begin{aligned}(10c + 5 + d)(10c + 5 - d) &= (10c + 5)^2 - d^2 \\ &= 100c(c + 1) + 25 - d^2 \\ &= 100c(c + 1) + (5 + d)(5 - d).\end{aligned}$$

We illustrate this formula by its application to conversion of fractions into their equivalent decimal form. Consider fraction 1/19. Using this formula, this can be converted into a decimal form in a single step. This can be done by applying the formula for either a multiplication or division operation, thus yielding two methods.

METHOD 1: USING MULTIPLICATION

1/19, since 19 is not divisible by 2 or 5, the fractional result is a purely circulating decimal. (If the denominator contains only factors 2 and 5 is a purely non-circulating decimal, else it is a mixture of the two.) Each factor of 2 or 5 or 10 in the denominator gives one fixed decimal digit.

So we start with the last digit

1

Multiply this by "one more", that is, 2 (this is the "key" digit from Ekadhikena)

21

Multiplying 2 by 2, followed by multiplying 4 by 2

421 → 8421

Now, multiplying 8 by 2, sixteen

68421

1 ← carry

multiplying 6 by 2 is 12 plus 1 carry gives 13

368421

1 ← carry

Continuing

7368421 → 47368421 → 947368421

1

Now we have 9 digits of the answer. There are a total of 18 digits (≐ denominator - numerator) in the answer computed by complementing the lower half (with its complement from nine):

052631578

947368421

Thus the result is 1/19 = 0.052631578,947368421 repeating.

1

21

421

8421

68421 (carry 1) - we got 16, so we keep 6 and carry 1

368421 (carry 1) - we get 6*2 + carry 1 = 13, so we keep 3 and carry one do this to eighteen digits (19-1. If you picked up 1/29, you'll have to do it till 28 digits). You'll get the following

1/19 = 052631578947368421

10100111101011000

Run this on your favourite calculator and check the result!

METHOD 2: USING DIVISION

The earlier process can also be done using division instead of multiplication. We divide 1 by 2, answer is 0 with remainder 1

.0

Next 10 divided by 2 is five

.05

Next 5 divided by 2 is 2 with remainder 1

.052

next 12 (remainder,2) divided by 2 is 6

.0526

and so on.

As another example, consider 1/7, this same as 7/49 which as last digit of the denominator as 9. The previous digit is 4, by one more is 5. So we multiply (or divide) by 5, that is,

...7 ⇒ 57 ⇒ 857 ⇒ 2857 ⇒ 42857 ⇒ 142857 => .142,857 (stop after 7 - 1 digits)

3 2 4 1 2

Multiplying by 11

11*35= 385

(1) The five in the ones place of the answer is taken from the five in 35.

(2) The eight in the answer is the sum of 35 (3+5=8).

(3) The three in the hundreds place of the answer is taken from the three in 35.

However, if in step #2 the sum is greater than 9, the sum's left digit is added the first digit of the number multiplied by 11. For example:

11*59= 649

(1) The nine in the ones place of the answer is take from the nine in 59

(2) The four in the answer is the right digit in the sum of 59 (5 + 9 = 14)

(3) The six in the hundreds place of the answer is take from the sum of the five in 59 and the digit in the tens place from the sum of 59 (5 + 9 = 14) → (5 + 1 = 6)

Vertically and Crosswise

This formula applies to all cases of multiplication and is very useful in division of one large number by another large number.

For example, to multiply 23 by 12:

```
2          3
|    ×     |
1          2
2×1 2×2+3×1 3×2
2      7     6
```

So 23 × 12 = 276.

When any of these calculations exceeds 9 then a carry is required. This is just $(10a+b)(10c+d)=100ac+10(ad+bc)+bd$.

Transpose and Apply

This formula complements "all from nine and the last from ten", which is useful in divisions by large numbers. This formula is useful in cases where the divisor consists of small digits. This formula can be used to derive the Horner's process of Synthetic Division.

When the Samuccaya is the Same, that Samuccaya is Zero

This formula is useful in solution of several special types of equations that can be solved visually. The word samuccaya has various meanings in different applications. For instance, it may mean a term which occurs as a common factor in all the terms concerned. A simple example is equation "12x + 3x = 4x + 5x". Since "x" occurs as a common factor in all the terms, therefore, x = 0 is a solution. Another meaning may be that samuccaya is a product of independent terms. For instance, in (x + 7) (x + 9) = (x + 3) (x + 21), the samuccaya is 7 × 9 = 3 × 21, therefore, x = 0 is a solution. Another meaning is the sum of the denominators of two fractions having the same numerical numerator, for example: 1/(2x - 1) + 1/(3x - 1) = 0 means we may set the denominators equal to zero, 5x - 2 = 0.

Yet another meaning is "combination" or total. This is commonly used. For instance, if the sum of the numerators and the sum of denominators are the same then that sum is zero. Therefore,

$$\frac{2x+9}{2x+7} = \frac{2x+7}{2x+9}$$

Therefore, 4x + 16 = 0 or x = -4.

This meaning ("total") can also be applied in solving quadratic equations. The total meaning can not only imply sum but also subtraction. For instance when given $N_1/D_1 = N_2/D_2$, if $N_1 + N_2 = D_1 + D_2$ (as shown earlier) then this sum is zero. Mental cross multiplication reveals that the resulting equation is quadratic (the coefficients of x^2 are different on the two sides). So, if $N_1 - D_1 = N_2 - D_2$ then that samuccaya is also zero. This yields the other root of a quadratic equation.

Yet interpretation of "total" is applied in multi-term RHS and LHS. For instance, consider

$$\frac{1}{x-7} + \frac{1}{x-9} = \frac{1}{x-6} + \frac{1}{x-10}.$$

Here $D_1 + D_2 = D_3 + D_4$ = 2x - 16. Thus x = 8.

There are several other cases where samuccaya can be applied with great versatility. For instance "apparently cubic" or "biquadratic" equations can be easily solved as shown below:

$(x - 3)^3 + (x - 9)^3 = 2(x - 6)^3$.

Note that $x - 3 + x - 9 = 2(x - 6)$.

Therefore $(x - 6) = 0$ or $x = 6$.

This would not work for the apparently quadratic $(x - 3)^2 + (x - 9)^2 = 2(x - 6)^2$, which has no real or complex solutions.

Consider

$$\frac{(x+3)^3}{(x+5)^3} = \frac{x+1}{x+7}.$$

Observe: $N_1 + D_1 = N_2 + D_2 = 2x + 8$.

Therefore, $x = -4$.

This formula has been extended further.

If one is in Ratio, the other one is Zero

This formula is often used to solve simultaneous linear equations which may involve big numbers. But these equations in special cases can be visually solved because of a certain ratio between the coefficients. Consider the following example:

$$6x + 7y = 8$$

$$19x + 14y = 16$$

Here the ratio of coefficients of y is same as that of the constant terms. Therefore, the "other" variable is zero, i.e., $x = 0$. Hence, mentally, the solution of the equations is $x = 0$ and $y = 8/7$.

(alternatively:

$19x + 14y = 16$ is equivalent to:

$(19/2)x + 7y = 8$.

Thus it is obvious that x has to be zero, no ratio needed, just divide by 2!)

Note that it would not work if both had been "in ratio". For then we have the case of coinciding lines with an infinite number of solutions.:

$$6x + 7y = 8$$
$$12x + 14y = 16$$

This formula is easily applicable to more general cases with any number of variables. For instance

$$ax + by + cz = a$$
$$bx + cy + az = b$$
$$cx + ay + bz = c$$

which yields $x = 1$, $y = 0$, $z = 0$.

A corollary says solving "by addition and by subtraction." It is applicable in case of simultaneous linear equations where the x- and y-coefficients are interchanged. For instance:

$$45x - 23y = 113$$
$$23x - 45y = 91$$

By addition:

$$68x - 68y = 204$$
$$\Rightarrow \quad 68(x - y) = 204$$
$$\Rightarrow \quad x - y = 3.$$

By subtraction:

$$22x + 22y = 22$$
$$\Rightarrow \quad 22(x + y) = 22$$
$$\Rightarrow \quad x + y = 1.$$

Again, by addition, we eliminate the y-terms: $2x = 4$, so $x = 2$.

Or, by subtraction, we eliminate the x-terms: $-2y = 2$, and so $y = -1$.

The solution set is {2,-1}.

3

Vedic Altars and the "Pythagorean Theorem"

A perfect example of this sort of misattribution involves the so-called Pythagorean theorem, the well-known theorem which was attributed to Pythagoras who lived around 500 BCE, but which was first proven in Greek sources in Euclid's Geometry, written centuries later. Despite the scarcity of evidence backing this attribution, it is not often questioned, perhaps due to the mantra—like frequency with which it is repeated. However, Seidenberg, in his 1978 article, shows that the thesis that Greece was the origin of geometric algebra was incorrect, "for geometric algebra existed in India before the classical period in Greece." (1978:323) It is now generally understood that the so-called "Pythagorean theorem" was understood in ancient India, and was in fact proved in Baudhayana's Shulva Sutra, a text dated to circa 600 BCE. (1978:323).

Knowledge of mathematics, and geometry in particular, was necessary for the precise construction of the complex Vedic altars, and mathematics was thus one of the topics covered in the brahmanas. This knowledge was further elaborated in the kalpa sutras, which gave more detailed instructions concerning Vedic ritual. Several of these treat the topic of altar construction. The oldest and most complete of these is the previously mentioned Shulva Sutra of

Baudhaayana. As this text was composed about a century before Pythagoras, the theory that the Greeks were the source of Geometric algebra is untenable, while the hypothesis that India was have been a source for Greek geometry, transmitted via the Persians who traded both with the Greeks and the Indians, looks increasingly plausible. On the other hand, it is quite possible that both the Greeks and the Indians developed geometry. Seidenberg has argued, in fact, that both seem to have developed geometry out of the practical problems involving their construction of elaborate sacrificial altars.

Zero and the Place Value System

Far more important to the development of modern mathematics than either Greek or Indian geometry was the development of the place value system of enumeration, the base ten system of calculation which uses nine numerals and zero to represent numbers ranging from the most minuscule decimal to the most inconceivably large power of ten. This system of enumeration was not developed by the Greeks, whose largest unit of enumeration was the myriad (10,000) or in China, where 10,000 was also the largest unit of enumeration until recent times. Nor was it developed by the Arabs, despite the fact that this numeral system is commonly called the Arabic numerals in Europe, where this system was first introduced by the Arabs in the thirteenth century.

Rather, this system was invented in India, where it evidently was of quite ancient origin. The Yajurveda Samhitaa, one of the Vedic texts predating Euclid and the Greek mathematicians by at least a millennium, lists names for each of the units of ten up to 10 to the twelfth power (paraardha). (Subbarayappa 1970:49) Later Buddhist and Jain authors extended this list as high as the fifty-third power, far exceeding their Greek contemporaries, who lacking a system of enumeration were unable to develop abstract mathematical concepts.

The place value system of enumeration is in fact built into the Sanskrit language, where each power of ten is given a distinct name. Not only are the units ten, hundred and thousand (daza, zata, sahasra) named as in English, but also ten thousand, hundred thousand, ten million, hundred million (ayuta, laksa, koti, vyarbuda), and so forth up to the fifty-third power, providing distinct names where English makes use of auxiliary bases such as thousand, million, etc. Ifrah has commented that:

> By giving each power of ten an individual name, the Sanskrit system gave no special importance to any number. Thus the Sanskrit system is obviously superior to that of the Arabs (for whom the thousand was the limit), or the Greeks and Chinese (whose limit was ten thousand) and even to our own system (where the names thousand, million etc. continue to act as auxiliary bases). Instead of naming the numbers in groups of three, four or eight orders of units, the Indians, from a very early date, expressed them taking the powers of ten and the names of the first nine units individually. In other words, to express a given number, one only had to place the name indicating the order of units between the name of the order of units immediately below it and the one immediately above it. That is exactly what is required in order to gain a precise idea of the place-value system, the rule being presented in a natural way and thus appearing self-explanatory. To put it plainly, the Sanskrit numeral system contained the very key to the discovery of the place-value system. (2000:429)

As Ifrah has shown at length, there is little doubt that our place-value numeral system developed in India (2000:399-409), and this system is embedded in the Sanskrit language, several aspects of which make it a very logical language, well suited to scientific and mathematical reasoning. Nor did this system exhaust Indian ingenuity; as van Nooten has shown, Pingala, who lived circa the first century BCE, developed a system of binary enumeration convertible to decimal numerals, described in his Chandahzaastra. His system is quite similar to that of Leibniz, who lived roughly fourteen hundred years later.

India is also the locus of another closely related an equally important mathematical discovery, the numeral zero. The oldest known text to use zero is a Jain text entitled the Lokavibhaaga, which has been definitely dated to Monday 25 August 458 CE. (Ifrah 2000:417-1 9) This concept, combined by the place-value system of enumeration, became the basis for a classical era renaissance in Indian mathematics.

The Indian numeral system and its place value, decimal system of enumeration came to the attention of the Arabs in the seventh or eighth century, and served as the basis for the well known advancement in Arab mathematics, represented by figures such as al-Khwarizmi. It reached Europe in the twelfth century when Adelard of Bath translated al-Khwarizmi's works into Latin. (Subbarayappa 1970:49) But the Europeans were at first resistant to this system, being attached to the far less logical roman numeral system, but their eventual adoption of this system led to the scientific revolution that began to sweep Europe beginning in the thirteenth century.

Luminaries of Classical Indian Mathematics

Aryabhatta

The world did not have to wait for the Europeans to awake from their long intellectual slumber to see the development of advanced mathematical techniques. India achieved its own scientific renaissance of sorts during its classical era, beginning roughly one thousand years before the European Renaissance. Probably the most celebrated Indian mathematicians belonging to this period was Aryabhatta, who was born in 476 CE.

In 499, when he was only 23 years old, Aryabhatta wrote his Aaryabhattiya, a text covering both astronomy and mathematics. With regard to the former, the text is notable for its for its awareness of the relativity of motion. This awareness led to the astonishing suggestion that it is the Earth that rotates the Sun. He argued for the diurnal rotation of the earth, as an alternate theory to the rotation of the fixed stars and sun around the earth (Pingree 1981:18). He made this suggestion approximately one thousand years before Copernicus, evidently independently, reached the same conclusion.

With regard to mathematics, one of Aryabhatta's greatest contributions was the calculation of sine tables, which no doubt was of great use for his astronomical calculations. In developing a way to calculate the sine of curves, rather than the cruder method of calculating chords devised by the Greeks, he thus went beyond geometry and contributed to the development of trigonometry, a development which did not occur in Europe until roughly one thousand years later, when the Europeans translated Indian influenced Arab mathematical texts.

Aryabhatta's mathematics was far ranging, as the topics he covered include geometry, algebra, trigonometry. He also developed methods of solving quadratic and indeterminate equations using fractions. He calculated pi to four decimal places, i.e., 3.1416. (Pingree 1981:57) In addition, Aryabhatta "invented a unique method of recording numbers which required perfect understanding of zero and the place-value system." (Ifrah 2000:419)

Given the astounding range of advanced mathematical concepts and techniques covered in this fifth century text, it should be of no surprise that it became extremely well known in India, judging by the large numbers of commentaries written upon it. It was studied by the Arabs in the eighth century following their conquest of Sind, and translated into Arabic, whence it influenced the development of both Arabic and European mathematical traditions.

Brahmagupt

Born in 598 CE in Rajasthan in Western India, Brahmagupta founded an influential school of mathematics which rivalled Aryabhatta's. His best known work is the Brahmasphuta Siddhanta, written in 628 CE, in which he developed a solution for a certain type of second order indeterminate equation.

Mahavira

Mahavira was a Jain mathematician who lived in the ninth century, who wrote on a wide range of mathematical topics. These include the mathematics of zero, squares, cubes, square-roots, cube-roots, and the series extending beyond these. He also wrote on plane and solid geometry, as well as problems relating to the casting of shadows. (Pingree 1981:60)

Bhaskar

Bhaskar was one of the many outstanding mathematicians hailing from South India. Born in 1114 CE in Karnataka, he composed a four-part text entitled the Siddhanta Ziromani. Included in this compilation is the Biijaganita, which became the standard algebra textbook in Sanskrit. It contains descriptions of advanced mathematical techniques involving both positive and negative integers as well as zero, irrational numbers. It treats at length the "pulverizer" (kuttakaara) method of solving indeterminate equations with continued fractions, as well as the so-called "Pell's equation (vargaprakrti) dealing with indeterminate equations of the second degree. He also wrote on the solution to numerous kinds of linear and quadratic equations, including those involving multiple unknowns, and equations involving the product of different unknowns. (Pingree 1981, p. 64)

In short, he wrote a highly sophisticated mathematical text that proceeded by several centuries the development of such techniques in Europe, although it would be better to term this a rediscovery, since much of the Renaissance advances of mathematics in Europe was based upon the discovery of Arab mathematical texts, which were in turn highly influenced by these Indian traditions.

Madhava

The Kerala region of South India was home to a very important school of mathematics. The best known member of this school Madhava (c. 1444-1545), who lived in Sangamagrama in Kerala. Primarily an astronomer, he made history in mathematics with his writings on trigonometry. He calculated the sine, cosine and arctangent of the circle, developing the world's first consistent system of trigonometry. He also correctly calculated the value of p to eleven decimal places.

This is by no means a complete list of influential Indian mathematicians or Indian contributions to mathematics, but rather a survey of the highlights of what is, judged by any fair, unbiased standard, an illustrious tradition, important both for its own internal elegance as well as its influence on the history of European mathematical traditions. The classical Indian mathematical renaissance was an important precursor to the European renaissance, and to ignore this fact is to fail to grasp the history of latter, a history which was truly multicultural, deriving its inspiration from a variety of cultural roots.

There are in fact, as Frits Staal has suggested in his important (1995) article, "The Sanskrit of Science", profound similarities between the social contexts of classical India and renaissance Europe. In both cases, important revolutions in scientific thought occurred in complex, hierarchical societies in which certain elite groups were granted freedom from manual labour, and thus the opportunity to dedicate themselves to intellectual pursuits. In the case of classical India, these groups included certain brahmins as well as the Buddhist and Jain monks, while in renaissance Europe they included both the monks as well as their secular derivatives, the university scholars.

Why, one might ask, did Europe take over thousand years to attain the level of abstract mathematics achieved by Indians such as Aryabhatta? The answer appears to be that Europeans were trapped in the relatively simplistic and concrete geometrical mathematics developed by the Greeks. It was not until they had, via the Arabs, received, assimilated and accepted the place-value system of enumeration developed in India that they were able to free their minds from the concrete and develop more abstract systems of thought. This development thus triggered the scientific and information technology revolutions which swept Europe and, later, the world. The role played by India in the development is no mere footnote, easily and inconsequentially swept under the rug of Eurocentric bias. To do so is to distort history, and to deny India one of it's greatest contributions to world civilization.

Jagadguru Swami Sri Bharati Krishna Tirthaji Maharaja

Jagadguru Swami Sri Bharati Krsna Tirthaji was the Jagadguru (literally, teacher of the world; assigned to heads of Hindu mathas) of the Govardhana Matha of Puri during 1925-1960. He was one of the most significant spiritual figures in Hinduism during the 20th century. He is particularly known for his work on Vedic mathematics.

Early Life: Venkatraman Shastri was born in March, 1884 to P. Narasimha Shastri, originally a tehsildar at Tirunelveli in Madras Presidency. Narasimha Shastri later became the Deputy Collector of the Presidency. Venkatraman was born in a highly illustrious family. His uncle,

Chandrasekhara Shastri was the Principal of the Maharaja's College in Vizianagaram, while his great-grandfather, Justice C. Ranganath Shastri was a judge in the Madras High Court.

Educational Career: Venkatraman Shastri started his educational career as a student of the National College in Trichanapalli. After that he moved to the Church Missionary Society College and eventually the Hindu College, both in Tirunelveli. He was consistently first place in all subjects in all of his. classes. Shastri passed his matriculation examination from the Madras University in January, 1899, where he also finished at the head of the class.

As a student Venkatraman was marked for his splendid brilliance, superb retentive memory and an insatiable curiosity. By deluging his teachers with piercing questions, making them uneasy, and frequently forcing them to admit ignorance he was considered a terribly mischievous student.

Although Venkatraman always scored high in subjects like mathematics, sciences and humanities, he was also proficient in languages and particularly adept in Sanskrit. According to his own testimonials, Sanskrit and oratory were his favourite subjects. Such was his mastery over the language, that he was awarded the title "Saraswati" by the Madras Sanskrit Association in July, 1899 at the age of 16. At about that time, Venkatraman was profoundly influenced by his Sanskrit guru Sri Vedam Venkatrai Shastri whom he remembered with deepest love, reverence and gratitude, with tears in his eyes.

Venkatraman won the highest place in the graduation B.A. examination in 1902. He then appeared for the M.A. Examination for the American College of Sciences, in Rochester, New York from the Bombay centre in 1903. He passed the M.A. examination in seven subjects that he had chosen — Sanskrit, philosophy, English, mathematics, history, science and another — simultaneously scoring the highest honours in all, which was perhaps an all-time world record at the time.

Venkatraman Saraswati, as he was called after receiving the title, also contributed to William Thomas Stead's Review of Reviews on topics as diverse as religion and science. During his college days, he also wrote extensively on history, sociology, philosophy, politics and literature. Reading of the latest scientific research and discoveries was his hobby throughout his life.

Early Public Life: Venkatraman Saraswati worked under Gopal Krishna Gokhale in 1905 for the National Education Movement and the South African Indian problems. However, his inclination towards science and Indic studies led him to study the ancient Indian holy scriptures, Adhyatma-Vidya. In 1908 he joined the Sringeri Matha in Mysore to study under the Sringeri Shankaracharya Sri Satchidananda Sivabhinava Nrisimha Bharati Swami. However, his spiritual practise was interrupted when he was pressurized by nationalist leaders to head the newly started National College at Rajmahendri. Prof. Venkatraman Saraswati taught at the college for three years. But in 1911, he suddenly left the college to go back to Sri Satchidananda Sivabhinava Nrisimha Bharati Swami at the Sringeri Math in his quest for spiritual knowledge.

Spiritual Path: Returning to Sringeri, Venkatraman spent his next eight years studying advanced Vedanta philosophy at the feet of Shri Nrisimha Bharati Swami. He also practised vigorous meditation, Brahma-sadhana and Yoga-sadhana during those years in the nearby forests. It is believed that he attained spiritual self-realization during his years in the Sringeri Math. He would leave the material world and practise Yoga meditation in seclusion for many days. During those eight years, he also taught Sanskrit and Philosophy to local schools and ashrams. He delivered a series of sixteen lectures on Shankaracharya's philosophy at

Shankar Institute of Philosophy, Amalner [Khandesh]. During that time, he also lectured as a guest professor at various institutions in Bombay, Poona and Khandesh.

Initiation into Sanyasa Order: After Venkatraman's eight-year period of spiritual practice and study of the Vedanta and Vedic philosophy, he was initiated into the holy order of Samnyasa at Benaras by Jagadguru Shankaracharya Sri Trivikram Tirthaji Maharaj of Sharadapeeth on July 4, 1919 and on this occasion he was given the title of Swami and the new name, "Swami Bharati Krsna Tirtha".

Shankaracharya of Sharada Peetha: Swami Bharati Krsna Tirtha was installed as Shankaracharya of Sharada Peetha in 1921 after just two years of Sanyasa. After assuming the pontificate Shri Jagadguruji, he was given another title, Jagadguru, as is the tradition. The Swami then toured India from corner to corner giving lectures on Sanatana Dharma, Vedic philosophy and Vedanta. By his scintillating intellectual brilliance, powerful oratory, magnetic personality, sincerity of purpose, indomitable will, purity of thought, and loftiness of character he took the entire intellectual and religious class by storm.

Shankaracharya of Govardhan Math: Around the time the Swami became Shankaracharya of Sharada Peetha, the Shankaracharya of Govardhan Math Puri, Jagadguru Shankaracarya Sri Madhusudhan Tirtha, was in failing health and was greatly impressed by Bharati Krishna Tirtha. Madhusudan Tirtha requested Bharati to succeed him at the Govardhan Math, however the Swami respectfully declined the offer. However, in 1925, Shankaracarya Sri Madhusudhan Tirtha's health took a serious turn and Swami Bharati Krsna Tirtha had to accept the Govardhan Math's Gadi. In 1925, Swami Bharati Krsna Tirtha assumed the pontificate of Shankaracharya of Govardhan Math, Puri and relinquished the pontificate of Sharadapeeth Gadi of Sringeri. He installed Sri Swarupanandji as the new Shankaracharya of Sharada Peetha.

Jagadguru: After becoming the Shankaracharya of Govardhan Math, Swami Bharati toured all over the world for 35 years to spread the values of peace, harmony and brotherhood and to spread the message of the Sanatana Dharma. He took upon himself the colossal task of the renaissance of Indian culture.

He attended many national and international religious conferences and many other yoga workshops. He believed in the Vedantic ideal of "Purnatva" which literally translated means, "all-round perfection and harmony". He remained the Shankaracharya of the Govardhan Matha until his death in 1960. While being a pontiff, he wrote a large number of treatises and books on religion, sciences, mathematics, world peace and social issues. In 1953, at Nagpur, he founded an organization called "Sri Vishwa Punarnirmana Sangha" (World Reconstruction Association). Initially, the administrative board consisted of Jagadguruji's disciples, devotees and admirers of his spiritual ideals for humanitarian service, but later many distinguished people started to contribute to the mission. The Chief Justice of India, Justice B. P. Sinha served as its President. Dr. C. D. Deshmukh, the ex-Finance Minister of India and ex-Chairman or the University Grants Commission served as its Vice-President.

In February 1958 he went on a trans-oceanic tour to America to speak on world peace and Vedanta, staying three months in Los Angeles, California travelling via the United Kingdom. This was the first tour outside India by a Shankaracharya in the history of the order. The tour was sponsored by Self Realization Fellowship of Los Angeles, the Vedantic Society founded by Paramhansa Yogananda in America.

In 1965 a Chair of Vedic Studies was founded at Banares Hindu University by Shri Arvind N. Mafatlala, a generous Bombay business magnate and math devotee of the late Swami Shankaracarya.

Mathematics: Jagadguru Swami Sri Bharati Krsna Tirthaji Maharaja's book "Vedic Mathematics" opened the floodgates of similar literature, often derived from the Swami's 16 sutras themselves. His treatise on this field of mathematics is a fundamental work on speed and accuracy in basic mathematics. The Vedic Math ideal is a mental calculation and one-line notation.

The foundations of Vedic Mathematics were mentioned in the Vedas themselves and even in the Vedanta scriptures. These had lain unused for many millennia, till the Swami rediscovered them.

His book, Vedic Mathematics, comprises many algorithms. He revealed his source in the ancient Hindu Vedas. Some are intuitively reconstructed from the Atharva Veda and from Parisistas (appendix) of the Atharva Veda. "The Upaveda of Sthapatya (engineering) comprises all kinds of architectural and structural human endeavour and all visual arts (and mathematics)." His work seems to be a whole Parisistas (appendix) itself.

The ancient Sanskrit writers did not use numerals when writing big numbers but preferred to use the letters of the Sanskrit Devanagari alphabet. In the Vedic Sutras the key word steps to solving many problems are given in a terse, decimal code of certain sets of rhyming syllables, within the verses of the Sutra.

The fact that the alphabetic code is in the natural order and can be immediately interpreted, is clear proof that the code language was resorted not for concealment but for greater ease in verification.

The Swami had written sixteen volumes on the Vedic Mathematics field explaining all the topics of mathematical study. Alas, many advanced formula were promised but not given in his first and only book. After his 1956 life's work manuscript on Vedic mathematics was lost in a fire at the home of a disciple, though he was going blind from cataracts, he re-wrote the manuscript in 1957 in six weeks! It was to be proofread and published in the USA but was send back to India in 1960 after his death. In 1965, this manuscript was published by Motilal Banarsidass, Varanasi, India and reprinted four times in the 1970s.

His book, Vedic Mathematics, included sixteen terse formulas for mental mathematics. For arithmetic, we are given several algorithms for whole number multiplication and division, (flag or straight) division, fraction conversion to repeating decimal numbers, calculations with measures of mixed units, summation of a series, squares and square roots (duplex method), cubes and cube roots (with expressions for a digit schedule), and divisibility (by osculation).

He gives a poem in Anusub metre, couched in the alphabetic Code-Language that has three meanings, a hymn to Lord Sri Krsna, a hymn in praise of the Lord Shri Shankara, and the third the value of pi/10 to 32 decimal places, pi/10 = 0.31415926535897932384626433832792... with a "self-contained master-key" for extending the evaluation to any number of decimal places!

Several tests and techniques for factoring and solving certain algebraic equations with integer roots for quadratic, cubic, biquadratic, pentic equations, systems of linear equations, and systems of quadratic equations are demonstrated. For fractional expressions, a separation algorithm and fraction merger algorithms are given. Other techniques handle certain patterns of some special case algebraic equations. Just an introduction to differential and integral calculus is given.

Geometric applications are reviewed for linear equations, analytic conics, the equation for the asymptotes, and the equation to the conjugate-hyperbola. Five simple geometric proofs for the Pythagorean theorem are given. A 5-line proof of Apollonius' Theorem is given.

Advanced topics promised included the integral calculus (the centre of gravity of hemispheres, conics), Trigonometry, Astronomy (spherical triangles, earth's daily rotation, earth's annual rotation about the sun and eclipses), and Engineering (dynamics, statics, hydrostatics, pneumatics, applied mechanics).

In his final comments he asserted that the names for "Arabic numerals," "Pythagoras' Theorem," and "Cartesian" coordinates are historical misnomers.

4

Basic Principles

The basic principles involved are quite simple; it is the algorithmic aspects that require attention in particular cases.

Assume a rational function R(x) in one indeterminate x has denominator that factors as

$P(x)Q(x)$

over a field K (we can take this to be real numbers, or complex numbers). If P and Q have no common factor, then R may be written as

$A/P + B/Q$

for some polynomials A(x) and B(x) over K. The existence of such a decomposition is a consequence of the fact that the polynomial ring over K is a principal ideal domain, so that

$CP + DQ = 1$

for some polynomials C(x) and D(x).

Using this idea inductively we can write R(x) as a sum with denominators powers of irreducible polynomials. To take this further, if required, write

$G(x)/F(x)^n$

as a sum with denominators powers of F and numerators of degree less than F, plus a possible extra polynomial. This can be done by the Euclidean algorithm, polynomial case.

Therefore when K is the complex numbers and we can assume F has degree 1 (by the fundamental theorem of algebra) the numerators will be constant. When K is the real numbers we can have the case of

degree F = 2,

and a quotient of a linear polynomial by a power of a quadratic will occur. This therefore is a case that requires discussion, in the systematic theory of integration (for example in computer algebra).

Algorithms

Lagrange Interpolation

Partial fraction decomposition can be derived using Lagrange interpolation.

EXAMPLE

As an introductory example we take the rational function

$$\frac{x}{x^2-1}.$$

This by the difference of two squares identity can also be written as

$$\frac{x}{(x+1)(x-1)},$$

which can be transformed further. Consider an identity

$$\frac{A}{x+1}+\frac{B}{x-1}=\frac{x}{x^2-1},$$

where A and B are constants. In more explicit form, we have the relation of the numerators,

$$A(x - 1) + B(x + 1) = x.$$

We know that the constants on one side of an expression must equal those on the other side. On the left hand side, the constants are -A and B, and on the right, the constant is simply 0. So, comparing constants on both sides of the expression, we can see that

$B - A = 0,$

i.e. $A = B$.

Now, in the same way, we know that the number of x terms on the left must equal the number of x's on the right. Therefore, looking at x terms on both sides,

$Ax + Bx = x,$

therefore

$A + B = 1,$

and so, given that A = B, we can say that

$$A + A = 1$$

$$2A = 1$$

$$A = B = 0.5$$

Finally we find:

$$\frac{x}{x^2-1} = \frac{1}{2}.\frac{1}{x+1} + \frac{1}{2}.\frac{1}{x-1}$$

or

$$\frac{x}{x^2-1} = \frac{1}{2(x+1)} + \frac{1}{2(x-1)}$$

which holds true for all $x \neq \pm 1$.

Derivation

The preceding example can be generalized to the following situation:

Assume that Q(x) is a monic polynomial of some degree n which over the underlying field K decomposes into linear factors

$$Q(x) = \prod_{i=1}^{n}(x - x_i),$$

where all xi are pairwise different. In other words Q has simple roots (over K). If P(x) is any polynomial of degree $\leq n - 1$ then according to the Lagrange interpolation formula P(x) can be uniquely written as a sum (the Lagrange form representation)

$$P(x) = \sum_{j=1}^{n} P(x_j) L_j(x; x_j),$$

where L_j $(x; x_j)$ is the Lagrange polynomial

$$L_j(x; x_j) = \prod_{k \leq n,\, k \neq j} \frac{(x - x_k)}{(x_j - x_k)}.$$

Dividing the Lagrange representation on the right side termwise by the polynomial Q(x) in its factored form one obtains

$$\frac{P(x)}{Q(x)} = \sum_{j=1}^{n} \frac{P(x_j)}{\prod k \leq n,\, k \neq j (x_j - x_k)} \cdot \frac{1}{x - x_j}.$$

This is the partial fraction decomposition

$$\frac{P(x)}{Q(x)} = \sum_{j=1}^{n} \frac{P(x_j)}{Q'(x_j)} \cdot \frac{1}{x - x_j}$$

of the rational function R(x) = P(x)/Q(x) with the derivative of Q given by

$$Q'(x_j) = \sum_{l \leq n} \prod_{k \leq n,\, k \neq l} (x_j - x_k) = \prod_{k \leq n,\, k \neq j} (x_j - x_k)$$

The first example can be obtained as the special case

Q(x) = (x - 1) . (x + 1), P(x) = x

Note the close relationship to divided differences.

Paravartya Sutra

Separation of a fractional algebraic expression into partial fractions is the reverse of the process of combining fractions by converting each fraction to the lowest common denominator (LCM) and adding the numerators. This separation is accomplished by a mental, one-line Vedic formula called the Paravartya Sutra. Case one has fractional expressions where factors in the denominator are unique. Case two has fractional expressions where some factors may repeat as powers of a binomial.

In integral calculus we would want to write a fractional algebraic expression as the sum of its partial fractions in order to take the integral of each simple fraction separately. Once the original denominator, D0, has been factored we set up a fraction for each factor in the denominator. We may use a subscripted D to represent the denominator of the respective partial fractions which are the factors in D0. Letters A, B, C, D, E, and so on will represent the numerators of the respective partial fractions.

We calculate each respective numerator by (1) calculating the Paravartya value of the denominator (which is the value of the variable making that factor zero) and (2) then substituting this value into the original expression but ignoring that factor in the denominator. Each Paravartya value for the variable is the value which gives an undefined value to the expression since we do not divide by zero.

General formula:

$$\frac{lx^2+mx+n}{(x-a)(x-b)(x-c)}=\frac{A}{(x-a)}+\frac{B}{(x-b)}+\frac{C}{(x-c)}$$

Here, a, b, c, l, m, and n are given integer values.

Where x = a and

$$A=\frac{la^2+ma+n}{(a-b)(a-c)};$$

and where x = b and

$$B=\frac{lb^2+mb+n}{(b-c)(b-a)};$$

and where x = c and

$$C = \frac{lc^2 + mc + n}{(c-a)(c-b)};$$

Case One

Factorize the expression in the denominator. Set up a partial fraction for each factor in the denominator. Apply the Paravartya Sutra to solve for the new numerator of each partial fraction.

EXAMPLE

$$\frac{3x^2 + 12x + 11}{(x+1)(x+2)(x+3)} = \frac{A}{x+1} + \frac{B}{x+2} + \frac{C}{x+3}$$

Set up a partial fraction for each factor in the denominator. With this framework we apply the sutra to solve for A, B, and C by mental math.

1. D_1 is x + 1; set it equal to zero. This gives the Paravartya value for A when x = -1.
2. Next, substitute this value of x into the fractional expression, but without D_1.

3A. Put this value down as the value of A. Proceed similarly for B and C.

3B. For Paravartya B use x = -2.

3C. For Paravartya C use x = -3.

Thus, to solve for A, use x = -1:

$$\frac{3x^2 + 12x + 11}{(x+2)(x+3)} = \frac{3 - 12 + 11}{(1)(2)} = \frac{2}{2} = 1 = A.$$

Thus, to solve for B, use x = -2:

$$\frac{3x^2 + 12x + 11}{(x+2)(x+3)} = \frac{12 - 24 + 11}{(-1)(1)} = \frac{-1}{(-1)} = +1 = B.$$

Thus, to solve for C, use x = -3:

$$\frac{3x^2 + 12x + 11}{(x+2)(x+3)} = \frac{27 - 36 + 11}{(-2)(-1)} = \frac{2}{(+2)} = +1 = C.$$

Thus,

$$\frac{3x^2+12x+11}{(x+1)(x+2)(x+3)}=\frac{1}{x+1}+\frac{1}{x+2}+\frac{1}{x+3}$$

Case Two

When factors of the denominator include powers of one expression we (1) Set up a partial fraction for each unique factor and each lower power of D; (2) We set up an equation showing the relation of the numerators if all were converted to the LCD. From the equation of numerators we solve for each numerator, A, B, C, D, and so on. This equation of the numerators is an absolute identity, true for all values of x. So, we may select any value of x and solve.

EXAMPLE

$$\frac{3x+5}{(1-2x)^2}=\frac{A}{(1-2x)^2}+\frac{B}{(1-2x)}$$

Here, we set up a partial fraction for each descending power of the denominator. Then we solve for the numerators, A and B. As the Paravartya value for A and B will be the same, x = ½, we need an additional relation in order to solve for both. To write the relation of numerators the second fraction needs another factor of (1-2x) to convert it to the LCD, giving us 3x + 5 = A + B(1 - 2x).

To Solve for A: Set the denominator of the first fraction to zero, 1 - 2x = 0. This is the Paravartya value for A, x = ½. When we substitute this value, x = ½, into the relation of numerators we have 3(1/2) + 5 = A + B(0). Solving for A gives us A = 3/2 + 5 = 13/2. Numerator A equals six and one-half.

To Solve for B: Since the equation of the numerators, 3x + 5 = A + B(1 - 2x), is true for all values of x, pick a value for x and use it to solve for B. We have solved for the value of A above, A = 13/2. We may use it to solve for B.

We may pick x = 0, use A = 13/2, and solve for B.

$$3x + 5 = A + B(1 - 2x)$$
$$0 + 5 = 13/2 + B(1 + 0)$$
$$10/2 = 13/2 + B$$
$$-3/2 = B$$

We may pick x = 1.

Then solve for B:

$$3x + 5 = A + B(1 - 2x)$$
$$3 + 5 = 13/2 + B(1 - 2)$$
$$8 = 13/2 + B(-1)$$
$$16/2 = 13/2 - B$$
$$B = -3/2$$

We may pick x = -1. Solve for B:

$$3x + 5 = A + B(1 - 2x)$$
$$-3 + 5 = 13/2 + B(1 + 2)$$
$$4/2 = 13/2 + 3B$$
$$-9/2 = 3B$$
$$-3/2 = B$$

Hence,

$$\frac{3x+5}{(1-2x)^2} = \frac{13/2}{(1-2x)^2} + \frac{-3/2}{(1-2x)},$$

or

$$\frac{3x+5}{(1-2x)^2} = \frac{13}{2(1-2x)^2} + \frac{3}{2(1-2x)}.$$

Technique Three

A third technique is an analysis of the expanded relation of the numerators. Just match up the x-terms of each degree. Then one can see the coefficients of the matching terms and solve for the missing numerator.

EXAMPLE

$$\frac{3x+5}{(1-2x)^2}=\frac{A}{(1-2x)^2}+\frac{B}{(1-2x)}$$

Converting each fraction to the LCD we have the relation in the numerators: $3x + 5 = A + B(1 - 2x)$. As A and B are constants, we can expand and match up the constant terms and the x-terms. The values of A and B will then be apparent.

$$3x + 5 = A + B(1 - 2x)$$

$$3x + 5 = A + B - 2Bx$$

Hence, the constant terms are set equal and the x-terms are set equal:

$$5 = A + B$$

$$3x = -2Bx$$

Therefore, by setting the coefficients equal,

$$3 = -2B$$

$$-3/2 = B$$

And,

$$5 = A - 3/2$$

$$5 + 3/2 = A$$

$$13/2 = A$$

Hence,

$$\frac{3x+5}{(1-2x)^2}=\frac{13/2}{(1-2x)^2}+\frac{-3/2}{(1-2x)}.$$

Fractions of Integers

The idea of partial fractions can be generalized to other rings, say the ring of integers where prime numbers take the role of irreducible denominators. E.g., it is:

$$\frac{1}{18}=\frac{1}{2}-\frac{1}{3}-\frac{1}{3^2}.$$

Partial Fraction Decomposition over the Reals

In mathematics, partial fractions are used in real-variable integral calculus to find real-valued antiderivatives of rational functions. The partial fraction decomposition of real rational functions is also used for Laplace transforms.

This article describes the general method for obtaining the partial fraction decomposition of any real rational function. The proof of the existence of the partial fraction decomposition over an arbitrary field is not given here. For a sketch of the general proof, see partial fraction.

For applications of partial fraction decomposition over the reals, see

- Partial fractions in integration
- Partial fractions in Laplace transforms

First we state the general result, and then we offer several examples to illustrate the result and show how the method is used in practice.

General Result

Let f(x) be any rational function over the real numbers. In other words, suppose there exist real polynomials p(x) and $q(x) \neq 0$, such that

$$f(x) = \frac{p(x)}{q(x)}$$

By removing the leading coefficient of q(x), we may assume without loss of generality that q(x) is monic. By the fundamental theorem of algebra, we can write

$$q(x) = (x - a_1)^{j_1} \dots (x - a_m)^{j_m} \left(x^2 + b_1 x + c_1\right)^{k_1} \dots \left(x^2 + b_n x + c_n\right)^{k_n}$$

where $a_1, \dots, a_m, b_1, \dots, b_n, c_1, \dots, c_n$ are real numbers with $b_i^2 - 4c_i < 0$, and $j_1, \dots, j_m, k_1, \dots, k_n$ are positive integers. The terms $(x - a_i)$ are the linear factors of q(x) which correspond to real roots of q(x), and the terms $(x_i^2 + b_i x + c_i)$ are the irreducible quadratic factors of q(x) which correspond to pairs of complex conjugate roots of q(x).

Then the partial fraction decomposition of f(x) is the following.

$$f(x)=\frac{p(x)}{q(x)}=P(x)+\sum_{i=1}^{m}\sum_{r=1}^{j_i}\frac{A_{ir}}{(x-a_i)^r}+\sum_{i=1}^{n}\sum_{r=1}^{k_i}\frac{B_{ir}x+c_{ir}}{\left(x^2+b_ix+c_i\right)^r}$$

Here, P(x) is a (possibly zero) polynomial, and the Air, Bir, and Cir are real constants. There are a number of ways the constants can be found.

The most straightforward method is to multiply through by the common denominator q(x). We then obtain an equation of polynomials whose left-hand side is simply p(x) and whose right-hand side has coefficients which are linear expressions of the constants A_{ir}, B_{ir}, and C_{ir}. Since two polynomials are equal if and only if their corresponding coefficients are equal, we can equate the coefficients of like terms. In this way, a system of linear equations is obtained which always has a unique solution. This solution can be found using any of the standard methods of linear algebra. However, this is often not the best way to go when computing by pencil and paper, and there are other ways ("tricks") to obtain the constants.

The Algorithm

The partial fraction decomposition of a real rational function can be obtained by the following procedure:

1. Remove the leading coefficient of the denominator q(x) if necessary.
2. Reduce the rational function as much as possible; i.e. cancel any obvious common factors of the numerator and denominator.
3. Make sure the degree of the numerator p(x) is strictly smaller than the degree of the denominator q(x). If it is not, then use the division algorithm for

polynomials and long-divide the denominator into the numerator. You will get a (nonzero) polynomial quotient and a remainder whose degree is now strictly smaller than the degree of the denominator. Take this remainder as your new numerator and continue with the following steps. (Don't forget the polynomial quotient when you're done!)

4. Factor the denominator q(x) into linear and irreducible quadratic factors. This is not a straightforward task; however, you have some tools to use. If you find any real root a of q(x), then you know that (x - a) is a factor of q(x). Also, if you differentiate q(x) and use the Euclidean algorithm to find the greatest common divisor of q' (x) and q(x), then any real root a of the greatest common divisor gives you a factor (x - a). You can look for ways to complete the square in different powers of x. Most importantly, you can use the rational root test to determine all possible rational roots of q(x).
5. Express the partial fraction decomposition of f(x) as the sum of double sums above.
6. Find the constants A_{ir}, B_{ir}, and C_{ir} by whatever methods possible. (These methods are described below.) Once you have found all the constants, you now have the partial fraction decomposition. (Don't forget the polynomial quotient!)

Some Examples

EXAMPLE 1

$$f(x) = \frac{1}{x^2 + 2x - 3}$$

Here, the denominator splits into two distinct linear factors:

$$q(x) = x^2 + 2x - 3 = (x + 3)(x - 1)$$

so we have the partial fraction decomposition

$$f(x) = \frac{1}{x^2 + 2x - 3} = \frac{A}{x+3} + \frac{B}{x-1}$$

Multiplying through by $x^2 + 2x - 3$, we have the polynomial identity

$$1 = A(x - 1) + B(x + 3)$$

Substituting $x = -3$ and $x = 1$ into this equation gives $A = -1/4$ and $B = 1/4$, so that

$$f(x) = \frac{1}{x^2 + 2x - 3} = \frac{1}{4}\left(\frac{-1}{x+3} + \frac{1}{x-1}\right)$$

EXAMPLE 2

$$f(x) = \frac{x^3 + 16}{x^3 - 4x^2 + 8x}$$

After long-division, we have

$$f(x) = 1 + \frac{4x^2 - 8x + 16}{x^3 - 4x^2 + 8x} = 1 + \frac{4x^2 - 8x + 16}{x\left(x^2 - 4x + 8\right)}$$

Since $(-4)^2 - 4(8) = -16 < 0$, $x^2 - 4x + 8$ is irreducible, and so

$$\frac{4x^2 - 8x + 16}{x(x^2 - 4x + 8)} = \frac{A}{x} + \frac{Bx + C}{x^2 - 4x + 8}$$

Multiplying through by x3 - $4x^2 + 8x$, we have the polynomial identity

$$4x^2 - 8x + 16 = A(x^2 - 4x + 8) + (Bx + C)x$$

Taking $x = 0$, we see that $16 = 8A$, so $A = 2$. Comparing the x^2 coefficients, we see that $4 = A + B = 2 + B$, so $B = 2$. Comparing linear coefficients, we see that $-8 = -4A + C = -8 + C$, so $C = 0$. Altogether,

$$f(x) = 1 + 2\left(\frac{1}{x} + \frac{x}{x^2 - 4x + 8}\right)$$

The following example illustrates almost all the "tricks" one would need to use short of consulting a computer algebra system.

EXAMPLE 3

$$f(x) = \frac{x^9 - 2x^6 + 2x^5 - 7x^4 + 13x^3 - 11x^2 + 12x - 4}{x^7 - 3x^6 + 5x^5 - 7x^4 + 7x^3 - 5x^2 + 3x - 1}$$

After long-division and factoring, we have

$$f(x) = x^2 + 3x + 4 + \frac{x^5 - 2x^4 + 5x^3 - 5x^2 + 6x - 1}{(x-1)^3 (x^2+1)^2}$$

The partial fraction decomposition takes the form

$$\frac{x^5 - 2x^4 + 5x^3 - 5x^2 + 6x - 1}{(x-1)^3 (x^2+1)^2} = \frac{A}{x-1} + \frac{B}{(x-1)^2} + \frac{C}{(x-1)^3} + \frac{Dx+E}{x^2+1} + \frac{Fx+G}{(x^2+1)^2}$$

Multiplying through by $(x - 1)3(x^2 + 1)^2$ we have the polynomial identity

$x^5 - 2x^4 + 5x^3 - 5x^2 + 6x - 1$

$= A(x - 1)^2 (x^2 + 1)^2 + B(x - 1)$

$(x^2 + 1)^2 + C(x^2 + 1)^2 + (Dx + E)$

$(x - 1)^3 (x^2 + 1) + (Fx + G)(x - 1)^3$

Taking $x = 1$ gives $4 = 4C$, so $C = 1$. Similarly, taking $x = i$ gives $2 + 2i = (Fi + G)(2 + 2i)$, so $Fi + G = 1$, so $F = 0$ and $G = 1$ by equating real and imaginary parts.

We now have the identity

$x^5 - 2x^4 + 5x^3 - 5x^2 + 6x - 1$

$= A(x-1)^2(x^2 + 1)^2 + B(x - 1)(x^2 + 1)^2$

$+ (x^2+1)^2 + (Dx+E)(x - 1)^3(x^2 + 1)$

$+ (x - 1)^3$

Taking constant terms gives $E = A - B + 1$, taking leading coefficients gives $A = -D$, and taking x-coefficients gives $B = 3 - D - 3E$.

Putting all of this together, $E = A - B + 1 = -D - (3 - D - 3E) + 1 = 3E - 2$, so $E = 1$ and $A = B = -D$.

Now,

$x^5 - 2x^4 + 5x^3 - 5x^2 + 6x - 1$

$= A(x-1)^2(x^2 + 1)^2 + A(x - 1)(x^2 + 1)^2$

$+ (x^2 +1)^2 + (-Ax + 1)(x-1)^3(x^2+ 1) + (x - 1)^3$

Taking x = -1 gives -20 = -8A - 20, so A = B = D = 0. The partial fraction decomposition of f(x) is thus

$$f(x) = x^2 + 3x + 4 + \frac{1}{(x-1)^3} + \frac{1}{x^2+1} + \frac{1}{(x^2+1)^2}$$

General Methods for Finding Coefficients

We present in the sequel a list of methods for finding the coefficients occurring in the r.h.s. sums of the above equation for f(x)=p(x)/q(x)=..., to which we shall refer to by (PFD).

For some of the methods it would be useful if p and/or q were already written in factored form.

Simple Poles of Multiplicity 1

If ai is a simple pole of f, i.e. j_i=1, then one multiplies (PFD) by $(x-a_i)$ and takes the limit x→ai. On the r.h.s. only A_i1 "survives", which is therefore given by

$A_i1 = \lim_{x\to a}i\ (x-a_i)\ f(x) = p(a_i)/q_i(a_i)$

where q_i is q divided (simplified) by the factor $(x - a_i)$. It holds $q_i(a_i) = q'(a_i)$.

Leading Term of a Pole

Using the same idea, we determine the coefficients A_{ij_i} (multiplying in the highest negative power), by just replacing $(x-a_i)$ by its j_i-th power.

Leading Term of a Quadratic Factor

Still using the same idea, but going to the complex root of x^2+b x+c, one gets in the same way coefficients B_{ik_i}, C_{ik_i}

(from real and imaginary part of the previous equation for A which is now an equation for Bx + C).

Other Coefficients A

All coefficients relative to one pole can be obtained "in one step" by making a change of variables $t = x - a_i$ (i.e. replacing x by $t + a_i$) and then (long) dividing the "new" p by the "new" q, with the "pure" power t_{ji} removed.

Method of Derivatives

Instead of making the change of variables and using long division, one can obtain a recursive formula involving derivatives of p and q_i at $x = a_i$.

General Methods

Besides the above methods to obtain a particular coefficient, there are several general methods to get one or more equations involving one or several of the coefficients, which should then allow to determine the remaining unknown coefficients:

- method of particular values: this just consists in putting x equal to particular fixed values (0, ±1, ±2...), and thus obtaining numerical equations for the coefficients.
- method of limits: multiply both sides of (PFD), after subtracting known terms (especially the polynomial), by the highest possible power of x such that lim(x≠8) still is finite, and take this limit
- method of identification: the "final rescue", put all of the r.h.s. on the common denominator q(x) and identify coefficients of alike powers in the numerator on the r.h.s. and p(x) on the l.h.s.

Partial Fractions in Complex Analysis

In complex analysis, a partial fraction expansion is a way of writing a meromorphic function f(z) as an infinite sum of rational functions and polynomials. When f(z) is a rational function, this reduces to the usual method of partial fractions.

Motivation

By using polynomial long division and the partial fraction technique from algebra, any rational function can be written as a sum of terms of the form 1/(az + b)k + p(z), where a and b are complex, k is an integer, and p(z) is a polynomial. Just as polynomial factorization can be generalized to the Weierstrass factorization theorem, there is an analogy to partial fraction expansions for certain meromorphic functions.

A proper rational function, i.e. one for which the degree of the denominator is greater than the degree of the numerator, has a partial fraction expansion with no polynomial terms. Similarly, a meromorphic function f(z) for which | f(z) | goes to 0 as z goes to infinity at least as quickly as | 1/z |, has an expansion with no polynomial terms.

Calculation

Let f(z) be a function meromorphic in the finite complex plane with poles at λ_1, λ_2, ..., and let (Γ_1, Γ_2, ...) be a sequence of simple closed curves such that:

- The origin lies inside each curve Gk
- No curve passes through a pole of f
- Γk lies inside Γ_{k+1} for all k

  $\lim_{k\to\infty} d(\Gamma_k) = \infty$, where d(Gk) gives the distance from the curve to the origin

Suppose also that there exists an integer p such that

$$\lim_{k\to\infty} \oint \Gamma_\kappa \left|\frac{f(z)}{z^{p+1}}\right| dz < \infty$$

Writing PP(f(z); z = λ_k) for the principal part of the Laurent expansion of f about the point λ_k, we have

$$f(z) = \sum_{k=0}^{\infty} PP(f(z); z = \lambda_k)$$

if p = -1, and if p > -1,

$$f(z) = \sum_{k=0}^{\infty} PP(f(z); z = \lambda_k + c_{0,k} + c_{1,k}z + ... + c_{p,k}z^p)$$

where the coefficients $c_{j,k}$ are given by

$$c_{j,k} = \text{Res}_{z=\lambda_k} \frac{f(z)}{z^{j+1}}$$

λ_0 should be set to 0, because even if f(z) itself does not have a pole at 0, the residues of f(z)/zj+1 at z = 0 must still be included in the sum.

Note that in the case of $\lambda_0 = 0$, we can use the Laurent expansion of f(z) about the origin to get

$$f(z) = \frac{a_{-m}}{z^m} + \frac{a_{-m+1}}{z^{m-1}} + ... + a_0 + a_1 z + ...$$

$$c_{j,k} = \text{Res}_{z=0}\left(\frac{a_{-m}}{z^{m+j+1}} + \frac{a_{-m+1}}{z^{m+j}} + ... + \frac{a_j}{z} + ...\right) = a_j,$$

$$\sum_{j=0}^{p} c_{j,k} z^j = a_0 + a_1 z + ... + a_p z^p$$

so that the polynomial terms contributed are exactly the regular part of the Laurent series up to z^p.

For the other poles λ_k where k = 1, $1/z^{j+1}$ can be pulled out of the residue calculations:

$$c_{j,k} = \frac{1}{\lambda_k^{j+1}} \text{Res}_{z=\lambda_k} f(z)$$

$$\sum_{j=0}^{p} c_{j,k} z^j = \left[\text{Res}_{z=\lambda_k} f(z)\right] \sum_{j=0}^{p} \frac{1}{\lambda_k^{j+1}} z^j$$

To avoid issues with convergence, the poles should be ordered so that if λ_k is inside Γ_n, then λ_j is also inside Γ_n for all j < k.

EXAMPLE

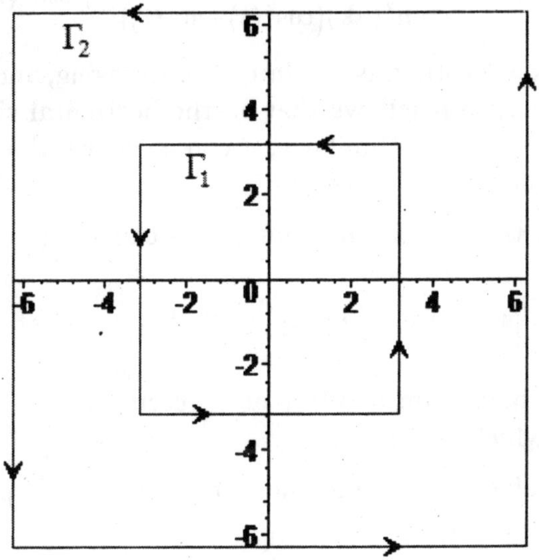

The simplest examples of meromorphic functions with an infinite number of poles are the non-entire trigonometric functions, so take the function tan(z). tan(z) is meromorphic with poles at $(n + 1/2)\pi$, n = 0, ±1, ±2, ... The contours Γ_k will be squares with vertices at $\pm\pi k \pm \pi ki$ traversed counterclockwise, k > 1, which are easily seen to satisfy the necessary conditions.

On the horizontal sides of Γ_k,

$$z = t \pm \pi ki,\ t \in [-\pi k, \pi k],$$

so

$$|\tan(z)|^2 = \left|\frac{\sin(t)\cosh(\pi k) \pm i\cos(t)\sinh(\pi k)}{\cos(t)\cosh(\pi k) \pm i\sin(t)\sinh(\pi k)}\right|^2$$

$$|\tan(z)|^2 = \frac{\sin^2(t)\cosh^2(\pi k) \pm \cos^2(t)\sinh^2(\pi k)}{\cos^2(t)\cosh^2(\pi k) \pm \sin^2(t)\sinh^2(\pi k)}$$

sinh(x) < cosh(x) for all real x, which yields

$$|\tan(z)|^2 < \frac{\cosh^2(\pi k)\left(\sin^2(t)+\cos^2(t)\right)}{\sinh^2(\pi k)\left(\cos^2(t)+\sin(t)\right)} = \coth^2(\pi k)$$

For $x > 0$, coth(x) is continuous, decreasing, and bounded below by 1, so it follows that on the horizontal sides of Γ_k, $|\tan(z)| < \coth(\pi)$. Similarly, it can be shown that $|\tan(z)| < 1$ on the vertical sides of Γ_k.

With this bound on $|\tan(z)|$ we can see that

$$\oint_{\Gamma_k} \left|\frac{\tan(z)}{z}\right| dz \le \text{length}(\Gamma_k) \max_{z\in\Gamma_k} \left|\frac{\tan(z)}{z}\right| < 8k\pi \frac{\coth(\pi)}{k\pi} = 8\coth(\pi) < \infty.$$

(The maximum of $|1/z|$ on Γ_k occurs at the minimum of $|z|$, which is $k\pi$).

Therefore $p = 0$, and the partial fraction expansion of tan(z) looks like

$$\tan(z) = \sum_{k=0}^{\infty}\left(PP(\tan(z); z=\lambda_k) + \operatorname{Res}_{z=\lambda_k} \frac{\tan(z)}{z}\right)$$

The principal parts and residues are easy enough to calculate, as all the poles of tan(z) are simple and have residue -1:

$$PP\left(\tan(z); z = \left(n+\frac{1}{2}\right)\pi\right) = \frac{-1}{z-\left(n+\frac{1}{2}\right)\pi}$$

$$\operatorname{Res}_{z=\left(n+\frac{1}{2}\right)\pi} \frac{\tan(z)}{z} = \frac{-1}{\left(n+\frac{1}{2}\right)\pi}$$

We can ignore $\lambda_0 = 0$, since both tan(z) and tan(z)/z are analytic at 0, so there is no contribution to the sum, and ordering the poles λ_k so that $\lambda_1 = \pi/2$, $\lambda_2 = -\pi/2$, $\lambda_3 = 3\pi/2$, etc., gives

$$\tan(z)=\sum_{k=0}^{\infty}\left[\left(\frac{-1}{z-\left(k+\frac{1}{2}\right)\pi}-\frac{1}{\left(k+\frac{1}{2}\right)\pi}\right)+\left(\frac{-1}{z+\left(k+\frac{1}{2}\right)\pi}+\frac{1}{\left(k+\frac{1}{2}\right)\pi}\right)\right]$$

$$\tan(z)=\sum_{k=0}^{\infty}\frac{-2z}{z^2-\left(k+\frac{1}{2}\right)^2\pi^2}$$

Applications

Infinite Products

Because the partial fraction expansion often yields sums of 1/(a+bz), it can be useful in finding a way to write a function as an infinite product; integrating both sides gives a sum of logarithms, and exponentiating gives the desired product:

$$\tan(z)=\sum_{k=0}^{\infty}\left(\frac{1}{z-\left(k+\frac{1}{2}\right)\pi}+\frac{1}{z+\left(k+\frac{1}{2}\right)\pi}\right)$$

$$\int_0^z \tan(w)\,dw=\log\ \sec z$$

$$\int_0^z \frac{1}{w\pm\left(k+\frac{1}{2}\right)\pi}\,dw=\log\left(1\pm\frac{z}{\left(k+\frac{1}{2}\right)\pi}\right)$$

Applying some logarithm rules,

$$\log\sec z=-\sum_{k=0}^{\infty}\left(\log\left(1-\frac{z}{\left(k+\frac{1}{2}\right)\pi}\right)+\log\left(1+\frac{z}{\left(k+\frac{1}{2}\right)\pi}\right)\right)$$

$$\log\sec z = -\sum_{k=0}^{\infty}\log\left(1-\frac{z^2}{\left(k+\frac{1}{2}\right)^2\pi^2}\right),$$

which finally gives

$$\cos z = \prod_{k=0}^{\infty}\left(1-\frac{z^2}{\left(k+\frac{1}{2}\right)^2\pi^2}\right)$$

Laurent Series

The partial fraction expansion for a function can also be used to find a Laurent series for it by simply replacing the rational functions in the sum with their Laurent series, which are often not difficult to write in closed form. This can also lead to interesting identities if a Laurent series is already known.

Recall that

$$\tan(z) = \sum_{k=0}^{\infty}\frac{-2z}{z^2-\left(k+\frac{1}{2}\right)^2\pi^2} = \sum_{k=0}^{\infty}\frac{-8z}{4z^2-(2k+1)^2\pi^2}$$

We can expand the summand using a geometric series:

$$\frac{-8z}{4z^2-(2k+1)^2\pi^2} = \frac{8z}{(2k+1)^2\pi^2}\frac{1}{1-\left(\frac{2z}{(2k+1)\pi}\right)^2}$$

$$= \frac{8}{(2k+1)^2\pi^2}\sum_{n=0}^{\infty}\frac{2^{2n}}{(2k+1)^{2n}\pi^{2n}}z^{2n+1}$$

Substituting back,

$$\tan(z) = 2\sum_{k=0}^{\infty}\sum_{n=0}^{\infty}\frac{2^{2n+2}}{(2k+1)^{2n+2}\pi^{2n+n}}z^{2n+1},$$

which shows that the coefficients an in the Laurent (Taylor) series of tan(z) about z = 0 are

$$a_{2n+1} = \frac{T_{2n+1}}{(2n+1)!} = \frac{2^{2n+3}}{\pi^{2n+2}} \sum_{k=0}^{\infty} \frac{1}{(2k+1)^{2n+2}}$$

$$a_{2n} = \frac{T_{2n}}{(2n)!} = 0,$$

where T_n are the tangent numbers.

Conversely, we can compare this formula to the Taylor expansion for tan(z) about z = 0 to calculate the infinite sums:

$$\tan(z) = z + \frac{1}{3}z^3 + \frac{2}{15}z^5 + \ldots$$

$$\sum_{k=0}^{\infty} \frac{1}{(2k+1)^2} = \frac{\pi^2}{2^3} = \frac{\pi^2}{8}$$

$$\sum_{k=0}^{\infty} \frac{1}{(2k+1)^4} = \frac{1}{3}\frac{\pi^4}{2^5} = \frac{\pi^4}{96}$$

Line Integral

In mathematics, a line integral (sometimes called a path integral) is an integral where the function to be integrated is evaluated along a curve. Various different line integrals are in use. In the case of a closed curve it is also called a contour integral.

The function to be integrated may be a scalar field or a vector field. The value of the line integral is the sum of values of the field at all points on the curve, weighted by some scalar function on the curve (commonly arc length or, for a vector field, the scalar product of the vector field with a differential vector in the curve). This weighting distinguishes the line integral from simpler integrals defined on intervals. Many simple formulas in physics (for example, $W = \vec{F}.\vec{d}$) have natural continuous analogs in terms of line integrals $\left(W = \int_C \vec{F}.d\vec{s}\right)$. The line integral finds the work done on an object moving through an electric or gravitational field, for example.

Vector Calculus

In qualitative terms, a line integral in vector calculus can be thought of as a measure of the total effect of a given vector field along a given curve.

DEFINITION

For some scalar field $f : U \rightarrow R$ defined on an open subset U of R^n, the line integral on a curve C, parametrized as r(t).U with $t \in [a, b]$ is defined by

$$\int_C f\ ds = \int_a^b f(r(t))\,|r'(t)|\,dt.$$

where

f is the scalar field being integrated,

r(t): [a, b] $\rightarrow$ C is a bijective parametrization of the curve C and

r(a) and r(b) give the endpoints of C.

The symbol ds is heuristically interpreted as an elementary arc length. Because they depend only on the element of arc length, line integrals of scalar fields are independent of the parametrization r(t).

For a vector field F : U . $R^n \rightarrow R^n$, the line integral on a curve C, parametrized as r(t).U with t . [a, b], is defined by

$$\int_C F(r).dr = \int_a^b F(r(t)).r'(t)dt.$$

Line integrals of vector fields are independent of the parametrization r(t) in absolute value, but they do depend on its orientation. Specifically, a reversal in the orientation of the parametrization changes the sign of the line integral.

Path Independence

If a vector field F is the gradient of a scalar field G, that is,

$$\nabla G = F,$$

then the derivative of the composition of G and r(t) is

$$\frac{dG(r(t))}{dt} = \nabla G(r(t)).r'(t) = F(r(t)).r'(t)$$

which happens to be the integrand for the line integral of F on r(t). It follows that, given a path C, then

$$\int_C F(r).dr = \int_a^b F(r(t)).r'(t)dt$$

$$= \int_a^b \frac{dG(r(t))}{dt} dt = G(r(b)) - G(r(a)).$$

In words, the integral of F over C depends solely on the values of G in the points r(b) and r(a) and is thus independent of the path between them.

For this reason, a line integral of a vector field which is the gradient of a scalar field is called path independent.

Applications

The line integral has many uses in physics. For example, the work done on a particle travelling on a curve C inside a force field represented as a vector field F is the line integral of F on C.

Relationship with the Line Integral in Complex Analysis

Viewing complex numbers as 2D vectors, the line integral in 2D of a vector field corresponds to the real part of the line integral of the conjugate of the corresponding complex function of a complex variable.

Due to the Cauchy-Riemann equations the curl of the vector field corresponding to the conjugate of a holomorphic function is zero. This relates through Stokes theorem both types of line integral being zero.

Complex Analysis

The line integral is a fundamental tool in complex analysis. Suppose U is an open subset of C, $\gamma : [a, b] \to U$ is a rectifiable curve and $f : U \to C$ is a function. Then the line integral

$$\int_\gamma f(z)dz$$

may be defined by subdividing the interval [a, b] into $a = t_0 < t_1 < ... < t_n = b$ and considering the expression

$$\sum_{1 \le k \le n} f(\gamma(t_k))(\gamma(t_k) - \gamma(t_{k-1})).$$

The integral is then the limit of this sum, as the lengths of the subdivision intervals approach zero.

If γ is a continuously differentiable curve, the line integral can be evaluated as an integral of a function of a real variable:

$$\int_\gamma f(z)dz = \int_a^b f(\gamma(t))\gamma'(t)dt.$$

When γ is a closed curve, that is, its initial and final points coincide, the notation

$$\oint_\gamma f(z)dz$$

is often used for the line integral of f along γ.

Important statements about contour integrals are the Cauchy integral theorem and Cauchy's integral formula.

Because of the residue theorem, one can often use contour integrals in the complex plane to find integrals of real-valued functions of a real variable (see residue theorem for an example).

EXAMPLE

Consider the function f(z)=1/z, and let the contour C be the unit circle about 0, which can be parametrized by e^{it}, with t in [0,2π]. Substituting, we find

$$\oint_C f(z)dz = \int_0^{2\pi} \frac{1}{e^{it}} ie^{it}dt = i\int_0^{2\pi} e^{-it}e^{it}dt$$

$$= i\int_0^{2\pi} dt = i(2\pi - 0) = 2\pi i$$

which can be also verified by the Cauchy integral formula.

Quantum Mechanics

The "path integral formulation" of quantum mechanics actually refers not to path integrals in this sense but to functional integrals, that is, integrals over a space of paths, of a function of a possible path. However, path integrals in the sense of this article are important in quantum mechanics; for example, complex contour integration is often used in evaluating probability amplitudes in quantum scattering theory.

5

Methods of Contour Integration

In complex analysis, contour integration is a method of evaluating certain integrals of real-valued functions along intervals on the real line that are not readily found by using only real variables. Complex analysis methods described below outline the procedures for calculating these real-valued integrals by means of contour integrals in the complex plane.

These methods include

- direct integration of a complex-valued function along a curve in the complex plane (a contour)
- application of the Cauchy integral formula
- application of the residue theorem

One method can be used, or a combination of these methods, or various limiting processes, for the purpose of finding these integrals or sums.

Direct Methods

Direct methods involve the calculation of the integral by means of methods similar to those in calculating line integrals in several-variable calculus. This means that we use the following method:

- parametrizing the contour

The contour is parametrized by a differentiable complex-valued function of real variables, or the contour is broken up into pieces and parametrized separately

- substitution of the parametrization into the integrand

Substituting the parametrization into the integrand transforms the integral into an integral of one real variable.

- direct evaluation

The integral is evaluated in a method akin to a real-variable integral.

EXAMPLE

A fundamental result in complex analysis is that the integral around the contour C which is the unit circle (or any Jordan curve about 0) of z^{-1} is $2\pi i$. Let us evaluate the integral

$$\oint_C \frac{1}{z} dz.$$

In evaluating this integral, we use the unit circle $|z| = 1$ as our contour, which we can parametrize by $\gamma(t) = e^{it}$, with t . [0, 2π]. Observe that $\pi'(t)$ = ieit. Now, substituting this for z, we have

$$\oint_C \frac{1}{z} dz = \int_0^{2\pi} \frac{1}{e^{it}} ie^{it} dt$$

$$= i\int_0^{2\pi} e^{-it}e^{it} dt = i\int_0^{2\pi} 1\, dt$$

$$= [t]_0^{2\pi}\, i = (2\pi - 0)i = 2\pi i$$

which is the value of the integral.

Applications of Integral Theorems

Applications of integral theorems are also often used to evaluate the contour integral along a contour, which means that the real-valued integral is calculated simultaneously along with calculating the contour integral.

Integral theorems such as the Cauchy integral formula or residue theorem are generally used in the following method:

- a specific contour is chosen.

The contour is chosen so that the contour follows the part of the complex plane that describes the real-valued integral, and also encloses singularities of the integrand so application of the Cauchy integral formula or residue theorem is possible

- application of the Cauchy-Goursat theorem

The integral is reduced to only an integration around a small circle about each pole.

- application of the Cauchy integral formula or residue theorem

Application of these integral formula gives us a value for the integral around the whole of the contour.

- division of the contour into a contour along the real part and imaginary part

The whole of the contour can be divided into the contour that follows the part of the complex plane that describes the real-valued integral as chosen before (call it R), and the integral that crosses the complex plane (call it I). The integral over the whole of the contour is the sum of the integral over each of these contours.

- demonstration that the integral that crosses the complex plane plays no part in the sum

If the integral I can be shown to be zero, or if the real-valued integral that is sought is improper, then if we demonstrate that the integral I as described above tends to 0, the integral along R will tend to the integral around the contour R+I.

- conclusion

If we can show the above step, then we can directly calculate R, the real-valued integral.

EXAMPLE (I)

Consider

$$\int_{-\infty}^{\infty} \frac{1}{\left(x^2+1\right)^2} dx$$

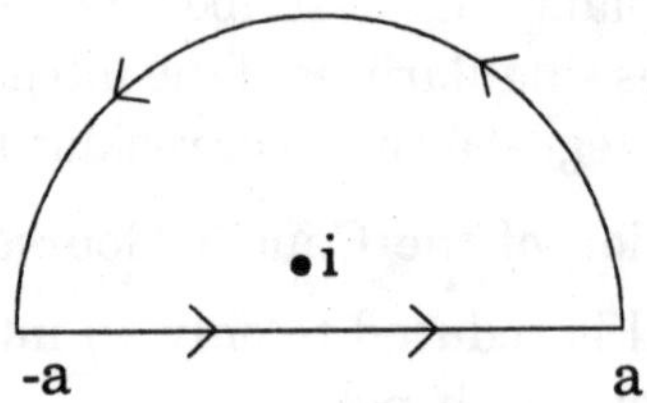

To evaluate this integral, we look at the complex-valued function

$$f(z)=\frac{1}{\left(z^2+1\right)^2}$$

which has singularities at i and -i. However, we will want to choose a contour that will enclose the real-valued integral, so we choose a semicircle as the one to the left, which we will let expand as to contain the whole real axis (a will tend to infinity). Call this contour C.

Now, there are two ways of proceeding, using the Cauchy integral formula or by the method of residues.

Using the Cauchy Integral Formula

Observe that

$$f(z)=\frac{1}{\left(z^2+1\right)^2}=\frac{1}{(z+i)^2(z-i)^2}.$$

Since the only singularity in the contour is the one at i, then we can write

$$f(z)=\frac{\frac{1}{(z+i)^2}}{(z-i)^2},$$

which puts the function in the form for direct application of the formula.

By the formula, then,

$$\oint_C f(z)dz = \oint_C \frac{1}{(z^2+1)^2}dz$$

$$= \oint_C \frac{\frac{1}{(z+i)^2}}{(z-i)^2}dz$$

$$= 2\pi i \frac{d}{dz}\left(\frac{1}{(z+i)^2}\right)\Bigg|_{z=i}$$

$$= 2\pi i\left(\frac{-2}{(z+i)^3}\right)\Bigg|_{z=i}$$

$$= 2\pi i(-i/4) = \frac{\pi}{2}$$

If we call the arc of the semicircle A, we need to show that the integral over A tends to zero as a tends to infinity-using the estimation lemma

$$\left|\int_A f(z)dz\right| \le ML$$

where M is an upper bound on |f(z)| and L the length of A. Now,

$$\int_A f(z)dz \le \frac{a\pi}{(a^2+1)^2} \to 0 \text{ as } a \to \infty$$

So

$$\int_{-\infty}^{\infty} f(z)dz = \frac{\pi}{2}.$$

Using the Method of Residues

Consider the Laurent series of f(z) about i, the only singularity we need to consider. We then have

$$f(z) = \frac{1}{4(z-i)^2} + \frac{-i}{4(z-i)} + \frac{3}{16} + \frac{i}{8}(z-i) + \frac{-5}{64}(z-i)^2 + \ldots$$

It is clear by inspection that the residue is -i/4, so, by the residue theorem, we have

$$\oint_C f(z)\,dz = \oint_C \frac{1}{(z^2+1)^2}\,dz = 2\pi i \,\mathrm{Res}_{z=i} f = 2\pi i(-i/4) = \frac{\pi}{2}$$

If we call the arc of the semicircle A, we need to show that the integral over A tends to zero as a tends to infinity - using the estimation lemma

$$\left|\oint_C f(z)\,dz\right| \le ML$$

where M is an upper bound on $|f(z)|$ and L the length of A. Now,

$$\oint_C f(z)\,dz \le \frac{a\pi}{(a^2+1)^2} \to 0 \text{ as } a \to \infty.$$

So

$$\int_{-\infty}^{\infty} f(z)\,dz = \frac{\pi}{2}.$$

Thus we get the same result as before.

Contour Note

As an aside, a question can arise whether we do not take the semicircle to include the other singularity, enclosing -i. To have the integral along the real axis moving in the correct direction, the contour must travel clockwise, i.e., in a negative direction, reversing the sign of the integral overall.

This does not affect the use of the method of residues by series.

EXAMPLE (II) - CAUCHY DISTRIBUTION

The integral

$$\int_{-\infty}^{\infty} \frac{e^{itx}}{x^2+1} dx$$

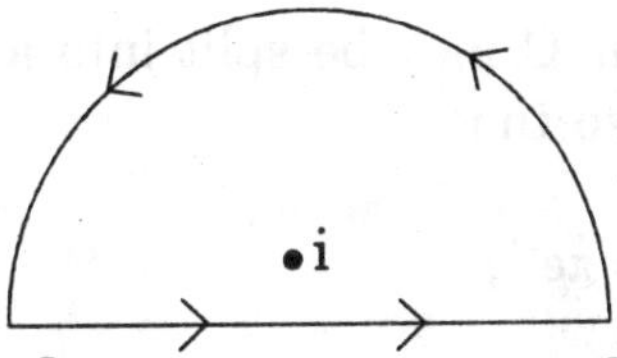

(which arises in probability theory as (a scalar multiple of) the characteristic function of the Cauchy distribution) resists the techniques of elementary calculus. We will evaluate it by expressing it as a limit of contour integrals along the contour C that goes along the real line from -a to a and then counterclockwise along a semicircle centred at 0 from a to -a. Take a to be greater than 1, so that the imaginary unit i is enclosed within the curve. The contour integral is

$$\int_C \frac{e^{itz}}{z^2+1} dz.$$

Since e^{itz} is an entire function (having no singularities at any point in the complex plane), this function has singularities only where the denominator $z^2 + 1$ is zero. Since $z^2 + 1 = (z + i)(z - i)$, that happens only where $z = i$ or $z = -i$. Only one of those points is in the region bounded by this contour. The residue of f(z) at z = i is

$$\lim_{z\to i}(z-i)f(z) = \lim_{z\to i}(z-i)\frac{e^{itz}}{z^2+1}$$

$$= \lim_{z\to i}(z-i)\frac{e^{itz}}{(z-i)(z+i)}$$

$$= \lim_{z \to i} (z - i) \frac{e^{iti}}{i + i} = \frac{e^{-t}}{2i}.$$

According to the residue theorem, then, we have

$$\int_C f(z)dz = 2\pi i. \text{Res}_{z=i} f(z) = 2\pi i \frac{e^{-t}}{2i} = \pi e^{-t}.$$

The contour C may be split into a "straight" part and a curved arc, so that

$$\int_{\text{straight}} + \int_{\text{arc}} = \pi e^{-1},$$

and thus

$$\int_{-a}^{a} = \pi e^{-t} - \int_{\text{arc}}.$$

It can be shown that if t > 0 then

$$\int_{\text{arc}} \frac{e^{itz}}{z^2 + 1} dz \to 0 \text{ as } a \to \infty$$

Therefore if t > 0 then

$$\int_{-\infty}^{\infty} \frac{e^{itz}}{z^2 + 1} dz = \pi e^{-t}.$$

A similar argument with an arc that winds around -i rather than i shows that if t < 0 then

$$\int_{-\infty}^{\infty} \frac{e^{itz}}{z^2 + 1} dz = \pi e^{t},$$

and finally we have

$$\int_{-\infty}^{\infty} \frac{e^{itz}}{z^2 + 1} dz = \pi e^{-|t|}.$$

(If t = 0 then the integral yields immediately to real-valued calculus methods and its value is p.)

EXAMPLE (III) - TRIGONOMETRIC INTEGRALS

Certain substitutions can be made to integrals involving trigonometric functions, so the integral is transformed into a rational function of a complex variable and then the above methods can be used in order to evaluate the integral.

As an example, consider

$$\int_{-\pi}^{\pi} \frac{1}{1+3(\cos t)^2} dt.$$

We seek to make a substitution of $z = e^{it}$.

Now, recall

$$\cos t = \frac{1}{2}\left(e^{it} + e^{-it}\right) = \frac{1}{2}\left(z + \frac{1}{z}\right)$$

and

$$\frac{dz}{dt} = iz, \; dt = \frac{dz}{iz}.$$

Taking C to be the unit circle, we substitute to get:

$$\oint_C \frac{1}{1+3\left(\frac{1}{2}\left(z+\frac{1}{z}\right)\right)^2} \frac{dz}{iz} = \oint_C \frac{1}{1+\frac{3}{4}\left(z+\frac{1}{z}\right)^2} \frac{1}{iz} dz$$

$$= \oint_C \frac{-i}{z+\frac{3}{4}z\left(z+\frac{1}{z}\right)^2} dz$$

$$= -i\oint_C \frac{1}{z+\frac{3}{4}z\left(z^2+2+\frac{1}{z}\right)} dz$$

$$= -i\oint_C \frac{1}{z+\frac{3}{4}\left(z^3+2z+\frac{1}{z}\right)} dz$$

$$= -i\oint_C \frac{1}{\frac{3}{4}z^3 + \frac{5}{2}z + \frac{3}{4z}} dz$$

$$= -i\oint_C \frac{4}{3z^3 + 10z + \frac{3}{z}} dz$$

$$= -4i\oint_C \frac{1}{3z^3 + 10z + \frac{3}{z}} dz$$

$$= -4i\oint_C \frac{z}{3z^4 + 10z^2 + 3} dz.$$

We use the Cauchy integral formula. Factorize the denominator:

$$= -4i\oint_C \frac{z}{3z^3 + 10z^2 + 3} dz$$

$$= -4i\oint_C \frac{z}{3(3z^2 + 3)(z^2 + 1/3)} dz$$

$$= -4i\oint_C \frac{z}{3\left(z + \sqrt{3}i\right)\left(z - \sqrt{3}i\right)\left(z + \sqrt{\frac{1}{3}}i\right)\left(z - \sqrt{\frac{1}{3}}i\right)} dz$$

$$= -\frac{4}{3}i\oint_C \frac{z}{\left(z + \sqrt{3}i\right)\left(z - \sqrt{3}i\right)\left(z + \sqrt{\frac{1}{3}}i\right)\left(z - \sqrt{\frac{1}{3}}i\right)} dz.$$

The singularities then to be considered are at $3^{-1/2}i$, $-3^{-1/2}i$. We can now reduce the integral:

$$= -\frac{4}{3}i\oint_{C_1} \frac{\dfrac{z}{\left(z+\sqrt{3i}\right)\left(z-\sqrt{3i}\right)\left(z+\sqrt{\frac{1}{3}}i\right)}}{\left(z-\sqrt{\frac{1}{3}}i\right)}dz$$

$$+ -\frac{4}{3}i\oint_{C_2} \frac{\dfrac{z}{\left(z+\sqrt{3i}\right)\left(z-\sqrt{3i}\right)\left(z-\sqrt{\frac{1}{3}}i\right)}}{\left(z+\sqrt{\frac{1}{3}}i\right)}$$

where C_1 is a small circle about $3^{-1/2}i$, and C_2 is a small circle about $-3^{-1/2}i$. We can now apply the formula:

$$= -\frac{4}{3}i\left(2\pi i\left(\frac{z}{\left(z+\sqrt{3i}\right)\left(z-\sqrt{3i}\right)\left(z+\sqrt{\frac{1}{3}}i\right)}\right)\Bigg|_{z=\sqrt{\frac{1}{3}}i}\right.$$

$$\left.+2\pi i\left(\frac{z}{\left(z+\sqrt{3i}\right)\left(z-\sqrt{3i}\right)\left(z-\sqrt{\frac{1}{3}}i\right)}\right)\Bigg|_{z=-\sqrt{\frac{1}{3}}i}\right)$$

$$= -\frac{4}{3}i\left(2\pi i\left(\frac{\sqrt{\frac{1}{3}}i}{\left(\sqrt{\frac{1}{3}}i+\sqrt{3i}\right)\left(\sqrt{\frac{1}{3}}i-\sqrt{3i}\right)\left(\sqrt{\frac{1}{3}}i-\sqrt{\frac{1}{3}}i\right)}\right)\right.$$

$$\left.+2\pi i\left(\frac{-\sqrt{\frac{1}{3}}i}{\left(-\sqrt{\frac{1}{3}}i+\sqrt{3i}\right)\left(-\sqrt{\frac{1}{3}}i-\sqrt{3i}\right)\left(-\sqrt{\frac{1}{3}}i-\sqrt{\frac{1}{3}}i\right)}\right)\right)$$

$$= -\frac{4}{3}i\left(2\pi i\left(\frac{\sqrt{\frac{1}{3}}i}{\left(\frac{4}{\sqrt{3}}i\right)\left(-\frac{2}{\sqrt{3}}i\right)\left(\frac{2}{\sqrt{3}}i\right)}\right)+2\pi i\left(\frac{-\sqrt{\frac{1}{3}}i}{\left(\frac{2}{\sqrt{3}}i\right)\left(-\frac{4}{\sqrt{3}}i\right)\left(-\frac{2}{\sqrt{3}}i\right)}\right)\right)$$

$$= -\frac{4}{3}i\left(2\pi i\left(\frac{\sqrt{\frac{1}{3}}i}{i\left(\frac{4}{\sqrt{3}}\right)\left(\frac{2}{\sqrt{3}}\right)\left(\frac{2}{\sqrt{3}}\right)}\right)+2\pi i\left(\frac{-\sqrt{\frac{1}{3}}i}{-i\left(\frac{2}{\sqrt{3}}\right)\left(\frac{4}{\sqrt{3}}\right)\left(\frac{2}{\sqrt{3}}\right)}\right)\right)$$

$$= -\frac{4}{3}i\left(2\pi i\left(\frac{\sqrt{\frac{1}{3}}}{\left(\frac{4}{\sqrt{3}}\right)\left(\frac{2}{\sqrt{3}}\right)\left(\frac{2}{\sqrt{3}}\right)}\right)+2\pi i\left(\frac{-\sqrt{\frac{1}{3}}}{\left(\frac{2}{\sqrt{3}}\right)\left(\frac{4}{\sqrt{3}}\right)\left(\frac{2}{\sqrt{3}}\right)}\right)\right)$$

$$= -\frac{4}{3}i\left(2\pi i\left(\frac{\sqrt{\frac{1}{3}}}{\frac{16}{3\sqrt{3}}}\right)+2\pi i\left(\frac{\sqrt{\frac{1}{3}}}{\frac{16}{3\sqrt{3}}}\right)\right)$$

$$= -\frac{4}{3}i\left(2\pi i\left(\frac{3}{16}\right)+2\pi i\left(\frac{3}{16}\right)\right) = -\frac{4}{3}i\left(\pi i\left(\frac{3}{8}+\frac{3}{8}\right)\right) = \frac{4}{3}\left(\frac{3}{4}\right)\pi = \pi.$$

EXAMPLE (IV) - BRANCH CUTS

Consider

$$\int_0^\infty \frac{\sqrt{x}}{x^2+6x+8}dx.$$

We can begin by formulating the complex integral

$$\int_C \frac{\sqrt{z}}{z^2+6z+8}dz = I.$$

We can use the Cauchy integral formula or residue theorem again to obtain the relevant residues. However, the important thing to note is that $z^{1/2}=e^{1/2\,\mathrm{Log}(z)}$, so $z^{1/2}$ has a branch cut. This affects our choice of the contour C. Normally the logarithm branch cut is defined as the negative real axis, however, this makes the calculation of the integral slightly more complex, so we define it to be the positive real axis.

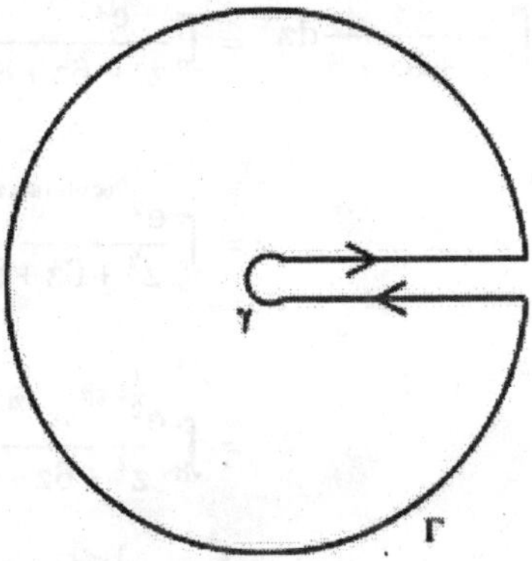

Then, we use the so-called keyhole contour, which consists of a small circle about the origin of radius e say, extending to a line segment parallel and close to the positive real axis but not touching it, to an almost full circle, returning to a line segment parallel, close, and below the positive real axis in the negative sense, returning to the small circle in the middle.

Let γ be the small circle of radius ε, Γ the larger, with radius r, then

$$\int_C \frac{\sqrt{z}}{z^2+6z+8}dz = \int_\varepsilon^R \frac{\sqrt{z}}{z^2+6z+8}dz$$

$$+\int_\Gamma \frac{\sqrt{z}}{z^2+6z+8}dz$$

$$+\int_R^\varepsilon \frac{\sqrt{z}}{z^2+6z+8}dz$$

$$+\int_\gamma \frac{\sqrt{z}}{z^2+6z+8}dz$$

Since $z^{1/2}=e^{1/2\ \mathrm{Log}(z)}$, along the contour below the branch cut, we have gained 2π in argument along Γ, so

$$\int_R^\varepsilon \frac{\sqrt{z}}{z^2+6z+8}dz = \int_R^\varepsilon \frac{e^{\frac{1}{2}\log(z)}}{z^2+6z+8}dz$$

$$= \int_R^\varepsilon \frac{e^{\frac{1}{2}(\log|z|+i\arg z)}}{z^2+6z+8}dz$$

$$= \int_R^\varepsilon \frac{e^{\frac{1}{2}\log|z|}e^{1/2(2\pi i)}}{z^2+6z+8}dz$$

$$= \int_R^\varepsilon \frac{e^{\frac{1}{2}\log|z|}e^{\pi i}}{z^2+6z+8}dz$$

$$= \int_R^\varepsilon \frac{-\sqrt{x}}{x^2+6x+8}dx$$

$$= \int_\varepsilon^R \frac{-\sqrt{x}}{x^2+6x+8}dx$$

simplifying,

$$= \int_\varepsilon^R \frac{\sqrt{x}}{x^2+6x+8}dx,$$

and then

$$\int_C \frac{\sqrt{z}}{z^2+6z+8}dz = \int_\varepsilon^R \frac{\sqrt{z}}{z^2+6z+8}dz$$

$$+ \int_\Gamma \frac{\sqrt{z}}{z^2+6z+8}dz$$

$$+ \int_\varepsilon^R \frac{\sqrt{z}}{z^2+6z+8}dz$$

$$+ \int_\gamma \frac{\sqrt{z}}{z^2+6z+8}dz$$

It can be shown that the integrals over Γ and γ both tend to zero as e tends to zero and R tends to infinity, by an estimation argument above. Thus, then,

$$\int_C \frac{\sqrt{z}}{z^2+6z+8}dz = 2\int_0^\infty \frac{\sqrt{x}}{x^2+6x+8}dx.$$

By using the residue theorem or the Cauchy integral formula one obtains

$$\pi i\left(\frac{i}{\sqrt{2}}-i\right) = \int_0^\infty \frac{\sqrt{x}}{x^2+6x+8}dx = \pi\left(1-\frac{1}{\sqrt{2}}\right).$$

EXAMPLE (V) - LOGARITHMS AND THE RESIDUE AT INFINITY

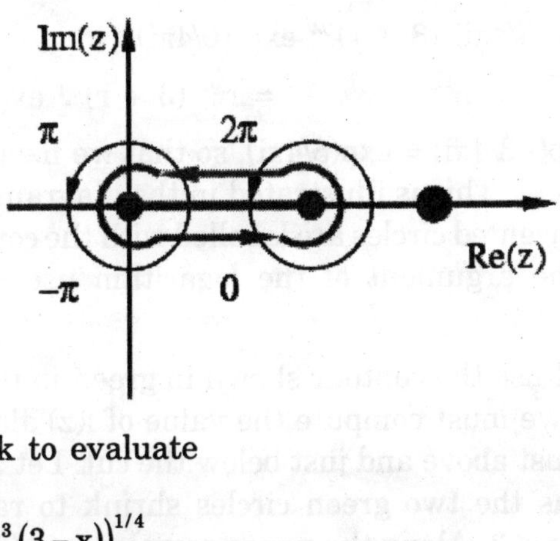

We seek to evaluate

$$I = \int_0^3 \frac{\left(x^3(3-x)\right)^{1/4}}{5-x}dx.$$

This requires a close study of

$f(z) = (z^3(3-z))^{1/4}$.

We will construct f(z) so that it has a branch cut on [0,3], shown in red in the diagram. To do this, we choose two branches of the logarithm, setting

$(z^3)^{1/4} = z^{3/4} = \exp(3/4 \log(z))$

where $-\pi \leq \arg(\log(z) < \pi$

and

$(3 - z)^{1/4} = \exp(1/4 \log(3 - z))$

where $0 \leq \arg(\log(z) < 2\pi$.

The cut of $z^{3/4}$ is therefore $(-\infty, 0]$ and the cut of $(3 - z)^{1/4}$ is $(-\infty, 3]$. It is easy to see that the cut of the product of the two, i.e. f(z), is [0,3], because f(z) is actually continuous across $(-\infty, 0)$. This is because when $z = -r < 0$ and we approach the cut from above, f(z) has the value

$r^{3/4} \exp(-3/4\pi i)(3 + r)^{1/4} \exp(0/4\pi i)$

$= r^{3/4}(3 + r)^{1/4} \exp(5/4\pi i)$.

When we approach from below, f(z) has the value

$r^{3/4} \exp(-3/4\pi i)(3 + r)^{1/4} \exp(0/4\pi i)$

$= r^{3/4}(3 + r)^{1/4} \exp(-3/4\pi i)$.

But $\exp(-3/4\pi i) = \exp(5/4\pi i)$, so that we have continuity across the cut. This is illustrated in the diagram, where the two black oriented circles are labelled with the corresponding value of the argument of the logarithm used in $z^{3/4}$ and $(3 - z)^{1/4}$.

We will use the contour shown in green in the diagram. To do this we must compute the value of f(z) along the line segments just above and just below the cut. Let $z = r$ (in the limit, i.e. as the two green circles shrink to radius zero), where $0 \leq r \leq 3$. Along the upper segment, we find that f(z) has the value

$r^{3/4} \exp(3/4\ 0\ \pi i)(3 - r)^{1/4} \exp(2/4\pi i) = ir^{3/4}(3 - r)^{1/4}$

and along the lower segment,

$r^{3/4} \exp(3/4\ 0\ \pi i)(3 - r)^{1/4} \exp(0/4\pi i) = r^{3/4}(3 - r)^{1/4}$

It follows that the integral of f(z)/(5 - z) along the upper segment is -i I in the limit, and along the lower segment, I.

If we can show that the integrals along the two green circles vanish in the limit, then we also have the value of I, by the Cauchy residue theorem. Let the radius of the green circles be ρ, where $\rho < 1/1000$ and $\rho \to 0$ and apply the ML-inequality. For the circle C_L on the left, we find

$$\left|\int_{CL} \frac{f(z)}{5-z} dz\right| \le 2\pi\rho \frac{\rho^{3/4}(3+1/1000)^{1/4}}{5-1/1000} \in O\left(\rho^{7/4}\right) \to 0.$$

Similarly, for the circle C_R on the right, we have

$$\left|\int_{CR} \frac{f(z)}{5-z} dz\right| \le 2\pi\rho \frac{(3+1/1000)^{3/4}\rho^{1/4}}{2-1/1000} \in O\left(\rho^{5/4}\right) \to 0.$$

Now using the Cauchy residue theorem, we have

$$(1-i)I = -2\pi i\left(\mathrm{Res}_{z=5}\frac{f(z)}{5-z} + \mathrm{Res}_{z=\infty}\frac{f(z)}{5-z}\right).$$

Using the branch of the logarithm from before, clearly

$$\mathrm{Res}_{z=5}\frac{f(z)}{5-z} = -5^{3/4}\exp(1/4\log(-2)).$$

The pole is shown in blue in the diagram. The value simplifies to

$-5^{3/4} \exp(1/4(\log(2)+\pi i)) = -\exp(1/\pi i)\, 5^{3/4}\, 2^{1/4}$.

We use the following formula for the residue at infinity:

$$\mathrm{Res}_{z=\infty}h(z) = \mathrm{Res}_{z=0} -\frac{1}{z^2}h\left(\frac{1}{z}\right).$$

Substituting, we find

$1/(5-1/z) = -z(1 + 5z + 5^2z^2 + 5^3z^3 + \ldots)$

and

$(1/z^3(3 - 1/z))^{1/4} = 1/z\,(3z - 1)^{1/4}\, 1/z \exp(1/4\pi i)(1 - 3z)^{14}$, where we have used the fact that - 1 = exp(πi) for the second branch of the logarithm. Next we apply the binomial expansion, obtaining

$$1/z\ \exp(1/4\pi i)\left(\left(\frac{1/4}{1}\right)3z+\left(\frac{1/4}{2}\right)3^2z^2-\left(\frac{1/4}{3}\right)3^3z^3+...\right).$$

The conclusion is that

$$\mathrm{Res}_{z=\infty}\frac{f(z)}{5-z}=\exp(1/4\pi i)(5-3/4)=\exp(1/4pi)17/4.$$

Finally, it follows that the value of I is

$$I=-2\pi i\frac{\exp(1/4\pi i)}{1-i}(17/4-5^{3/4}2^{1/4})$$

$$=-2\pi i2^{-1/2}i(17/4-5^{3/4}2^{1/4})$$

which yields

$$I=\frac{\pi}{2\sqrt{2}}\left(17-5^{3/4}2^{9/4}\right)=\frac{\pi}{2\sqrt{2}}\left(17-40^{3/4}\right).$$

Partial Fractions in Integration

In integral calculus, the use of partial fractions is required to integrate the general rational function. Any rational function of a real variable can be written as the sum of a polynomial function and a finite number of partial fractions. Each partial fraction has as its denominator a polynomial function of degree 1 or 2, or some positive integer power of such a function. If the denominator is a 1st-degree polynomial or a power of such a polynomial, then the numerator is a constant. If the denominator is a 2nd-degree polynomial or a power of such a polynomial, then the numerator is a 1st-degree polynomial.

For an account of how to find this partial fraction expansion of a rational function, see partial fraction.

This article is about what to do after finding the partial fraction expansion, when one is trying to find the function's antiderivative.

A 1st-degree Polynomial in the Denominator

The substitution u = ax + b, du = a dx reduces the integral

$$\int \frac{1}{ax+b} dx$$

to

$$\int \frac{1}{u}\frac{du}{a} = \frac{1}{a}\int \frac{du}{u} = \frac{1}{a}\ln|u| + C = \frac{1}{a}\ln|ax+b| + C.$$

A Repeated 1st-degree Polynomial in the Denominator

The same substitution reduces such integrals as

$$\int \frac{1}{(ax+b)^8} dx$$

to

$$\int \frac{1}{u^8}\frac{du}{a} = \frac{1}{2}\int u^{-8} du = \frac{1}{a}.\frac{u^{-7}}{(-7)} + C$$

$$= \frac{-1}{7au^7} + C = \frac{-1}{7z(ax+b)^7} + C.$$

An Irreducible 2nd-degree Polynomial in the Denominator

Next we consider such integrals as

$$\int \frac{x+6}{x^2-8x+25} dx.$$

The quickest way to see that the denominator $x^2 - 8x + 25$ is irreducible is to observe that its discriminant is negative. Alternatively, we can complete the square:

$x^2 - 8x + 25 = (x^2 - 8x + 16) + 9 = (x - 4)^2 + 9$

and observe that this sum of two squares can never be 0 while x is a real number.

In order to make use of the substitution

$u = x^2 - 8x + 25$

$du = (2x - 8)dx$

$du/2 = (x - 4)\ dx$

we would need to find x - 4 in the numerator. So we decompose the numerator x + 6 as (x - 4) + 10, and we write the integral as

$$\int \frac{x-4}{x^2-8x+25}dx + \int \frac{10}{x^2-8x+25}dx.$$

The substitution handles the first summand, thus:

$$\int \frac{x-4}{x^2-8x+25}dx = \int \frac{du/2}{u} = \frac{1}{2}\ln|u| + C$$

$$= \frac{1}{2}\ln(x^2-8x+25) + C.$$

Note that the reason we can discard the absolute value sign is that, as we observed earlier, $(x - 4)^2 + 9$ can never be negative.

Next we must treat the integral

$$\int \frac{10}{x^2-8x+25}dx.$$

First, complete the square, then do a bit more algebra:

$$\int \frac{10}{x^2-8x+25}dx = \int \frac{10}{(x-4)^2+9}dx = \int \frac{10/9}{\left(\frac{x-4}{3}\right)^2+1}dx$$

Now the substitution

w = (x - 4)/3

dw = dx/3

gives us

$$\frac{10}{3}\int\frac{dw}{w^2+1}=\frac{10}{3}\arctan(w)+C=\frac{10}{3}\arctan\left(\frac{x-4}{3}\right)+C.$$

Putting it all together,

$$\int\frac{x+6}{x^2-8x+25}dx=\frac{1}{2}\ln\left(x^2-8x+25\right)+\frac{10}{3}\arctan\left(\frac{x-4}{3}\right)+C.$$

A Repeated Irreducible 2nd-degree Polynomial in the Denominator

Next, consider

$$\int\frac{x+6}{(x^2-8x+25)^8}dx.$$

Just as above, we can split x + 6 into (x - 4) + 10, and treat the part containing x - 4 via the substitution

$u = x^2 - 8x + 25,$

du = (2x - 8)dx

du/2 = (x - 4) dx

This leaves us with

$$\int\frac{x+6}{(x^2-8x+25)^8}dx.$$

As before, we first complete the square and then do a bit of algebraic massaging, to get

$$\int\frac{10}{(x^2-8x+25)^8}dx=\int\frac{10}{((x-4)^2+9)^8}dx=\int\frac{10/9^8}{\left(\left(\frac{x-4}{3}\right)^2+1\right)^8}dx.$$

Then we can use a trigonometric substitution:

$$\tan\theta = \frac{x-4}{3},$$

$$\left(\frac{x-4}{3}\right)^2 + 1 = \tan^2\theta + 1 = \sec^2\theta,$$

$$d\tan\theta = \sec^2\theta d\theta = \frac{dx}{3}.$$

Then the integral becomes

$$\int \frac{30/9^8}{\sec^{16}\theta}\sec^2\theta\, d\theta = \frac{30}{9^8}\int \cos^{14}\theta d\theta$$

By repeated applications of the half-angle formula

$$\cos^2\theta = \frac{1}{2} + \frac{1}{2}\cos(2\theta)$$

one can reduce this to an integral involving no higher powers of cos ? higher than the 1st power.

Then one faces the problem of expression sin and cos as functions of x. Recall that

$$\tan(\theta) = \frac{x-4}{3},$$

and that tangent = opposite/adjacent. If the "opposite" side has length x - 4 and the "adjacent" side has length 3, then the Pythagorean theorem tells us that the hypotenuse has length O((x - 4)^2 + 3^2) = O(x^2 -8x + 25).

Therefore we have

$$\sin(\theta) = \frac{\text{opposite}}{\text{hypotenuse}} = \frac{x-4}{\sqrt{x^2-8x+25}},$$

$$\cos(\theta) = \frac{\text{adjacent}}{\text{hypotenuse}} = \frac{3}{\sqrt{x^2-8x+25}},$$

and

$$\sin(2\theta) = 2\sin(\theta)\cos(\theta) = \frac{6(x-4)}{x^2-8x+25}.$$

Auxiliary Fraction

In Swami Bharati Krishna Tirtha's Vedic mathematics, the auxiliary fraction method is used to convert a fraction to its equivalent decimal representation. The "auxiliary fraction" is not a true fraction, but is simply a mnemonic aid used in the calculation. The method is essentially the long division algorithm adapted for mental calculation. It is simplest when the fraction's denominator is one less than a multiple of 10, when it uses the identity

$$a = qb + r$$

$$\Rightarrow 10a = q\ (10b - 1) + (10r + q).$$

Variants of the method used when the denominator is not one less than a multiple of 10 become progressively more complex.

Dividing by Powers of Ten

To divide by a power of ten first move the decimal point in both the numerator and the denominator to the left the same number of places as the number of zeros at the end of the denominator. Then divide. For example, 1/800 = 0.01/8; 39/70 = 3.9/7; 3741/110000 = 0.3741/11; and 97654/ 90,000,000 = 0.0097654/9.

Forming the Auxiliary Fraction

The formation of the auxiliary fraction depends on the denominator. There are four cases:

- Type One A: denominators which end in a single nine
- Type One B: denominators which end in several nines

- Type Two: denominators which end in one
- Type Three: denominators which end in the digits, 2, 3, 4, 5, 6, 7, and 8.

If the denominator ends in zero(s), its first non-zero digit (from the right) identifies the family of the denominator.

Type One A

When the fraction's denominator ends in a single nine use the Ekadhika Purva. First replace the denominator by its Ekadhika which means adding one to the denominator and dividing the result by 10. Divide the numerator by 10 also. Hereafter, F is the actual fraction to be converted to a decimal value and A.F. is the auxiliary fraction used to remember the original dividend and the recurrent divisor.

EXAMPLES:

When F = 1/19, then the A.F. = 1/20 = 0.1/2;

When F = 4/29, then the A.F. = 4/30 = 0.4/3;

When F = 8/59, then the A.F. = 8/60 = 0.8/6;

Type One B

When the denominator ends in several nines, increase the denominator by one, then divide both numerator and denominator by a power of 10 equal to the number of terminal nines in the denominator.

EXAMPLES:

When F = 174/1299, then the A.F. = 174/1300 = 1.74/13;

When F = 1/6999, then the A.F. = 1/7000 = 0.001/7;

If the denominator ends in 1, 3, or 7, multiply both denominator and numerator by 9, 3 or 7 respectively to convert to an equivalent fraction in which the denominator ends in 9.

EXAMPLES:

When F = 36/121 = 324/1089, then the A.F. = 32.4/109;

When F = 53/93 = 159/279, then the A.F. = 15.9/28;

When F = 15/37 = 105/259, then the A.F. = 10.5/26.

Type Two

When the fraction has a denominator ending in one, form the auxiliary fraction by subtracting one from the denominator and from the numerator. Then divide both numerator and denominator by a power of 10 equal to the number of terminal zeros in the new denominator.

EXAMPLES:

When F = 3/61, then the A.F = 2/60 = 0.2/6;

When F = 28/71, then the A.F. = 27/70 = 2.7/7;

When F = 1/81, then the A.F. = 0/80 = 0.0/8;

When F = 14/131, then the A.F. = 13/130 = 1.3/13;

When F = 1/301, then the A.F. = 0/300 = 0.00/3;

When F = 6163/8001, then the

A.F. = 6162/8000 = 6.162/8;

If the denominator ends in 3 or 7, multiply both denominator and numerator by 7 or 3 respectively to convert to an equivalent fraction in which the denominator ends in 1.

EXAMPLES:

When F = 2/3 = 14/21, then the A.F. = 13/20 = 1.3/2;

When F = 10/27 = 30/81, then the A.F.= 29/80 = 2.9/8;

When F = 5/67 = 15/201, then the A.F.=14/200= 0.14/2;

When F = 4/17 = 12/51, then the A.F. = 11/50 = 1.1/5.

Type Three

For denominators that do not end in 1 or 9, use the nines family auxiliary fraction and count the number of units (above or below) that the ending is from the normal nine.

Using the Auxiliary Fraction — Type One

Given a fraction whose denominator ends in 9, first write the auxiliary fraction. Next divide in the auxiliary fraction to generate one (or more) quotient digit(s) at-a-time. Then write the quotient digit(s) and the remainder. The remainder at each step is prefixed to the just generated quotient digit for the next division. So if the quotient and remainder from one step are q and r, the dividend for the next step is 10r+q.

This algorithm meets the Vedic ideal of mental math with one line notation.

Word Examples — Type One A

EXAMPLE ONE

To convert the fraction F = 1/169 to a (repeating) decimal:

First, estimate the quotient as about six thousandths. Then, set up the auxiliary fraction, AF = 0.1/17. The first dividend is 0.1 and the working divisor is 17. Calculate one quotient digit at a time. Set down the remainder as a (subscripted) prefix to the quotient-digit just produced. Continue dividing to generate the quotient of the precision desired. Remember that the prefixed remainders are not parts of the quotient but only prefixes to the quotient-group in question and are dropped out of the answer — the lower row is a mere scaffolding and goes out.

As there are 168 remainders for the repeating decimal value of one one-hundred-sixty-ninth, there can be at most 168 digits before the decimal expansion repeats. If the remainder were to be zero or 169, then the fraction would terminate as an exact decimal value. There are in fact 78 digits before the decimal expansion of 1/169 repeats, and 78 is half of 156, which is the totient of 169.

17 into 0.1 goes 0 rem 1. 17 into 10 goes 0 rem 10. 17 into 100 goes 5 rem 15. 17 into 155 goes 9 rem 2. (The third dividend, 155, is formed by the remainder 15 prefixed to quotient from the previous step, 5.)

$$F = 1/169 = {}_{.1}0.\ {}_{1}0\ {}_{10}0\ {}_{15}5\ {}_{2}9\ {}_{12}1\ {}_{2}7\ {}_{10}1\ {}_{16}5\ {}_{12}9\ {}_{10}7\ {}_{5}6\ {}_{5}3\ {}_{2}3\ {}_{6}1\ {}_{10}3\ {}_{1}6\ {}_{15}0\ {}_{7}9\ {}_{11}4\ {}_{12}6\ {}_{7}7 \ldots$$

This all the notation that is needed. The calculating is mental using the multiples of 17: 17, 34, 51, 68, 85, 102, 119, 136, 153, 170.

F = 1/169 ≈ 0.005917159763331361...

EXAMPLE TWO

If the denominator is 73, this is a special case because (73)(137) = 10001. F = 1/73 = 137/10001 = (137)(9,999)/(99,999,999) = 0.0136,9863 repeating. Thus we may expect an eight-digit repeating decimal with complementary halves, i.e., the digits in the first half of the repeating decimal digit set are complements of nine for the digits of the second half.

EXAMPLE THREE

When F = 3/73 = 9/219, then the A.F. = 9/22. The working divisor is 22. The dividend at each step is created by prefixing the remainder to the quotient from the previous step.

$$F = 0.{}_{9}0\ {}_{2}4\ {}_{2}1\ {}_{21}0\ {}_{12}9\ {}_{19}5\ {}_{19}8\ {}_{0}9\ {}_{9}0\ {}_{2}4\ {}_{2}1\ {}_{21}0 \ldots$$

F = 3/73 ≈ 0.041095890410... The fraction repeats after eight decimal places. Furthermore, the eight digits have complementary halves (see Midy's theorem).

Worked Example — Type One B

When F = 53/799, then the A.F. = 0.53/8

The working divisor is 8. As the denominator has two nines, the division proceeds in steps of two dividend digits at a time, generating two quotient digits at each step. The remainder is prefixed to the pair of quotient digits from the previous step to create the dividend for the next step.

$F = 53/799 = 0._{5}06\ _{2}63\ _{7}32\ _{4}91\ _{3}61\ _{1}45\ 18...$

$F = 53/799 \approx 0.06633291614518...$

Generating a Denominator with more Terminal Nines

An additional technique is available to find another A.F. with a smaller divisor. By a judicious choice of the multiplier one can produce an equivalent fraction with more nines in the denominator.

When F = 1/7 = 7/49, then the A.F. = 0.7/5; We use a multiplier of 7 to produce an equivalent fraction, 7/49. Look at the denominator of the equivalent fraction, 49. Consider the 4. To build the 4 to a nine we need to add 5 in the tens place. The 5th multiple of 7 ends in 5, so we may use 5 tens or 57 as the multiplier.

F = 1/7 = 57/399, and the A.F. = 0.57/4 (dividing in bundles of two digits).

We may proceed likewise, until we have F = 1/7 = 142,857/999,999 giving a bundle of six nines and an A.F. = 0.142857/1. This means we have the repeating decimal digit set on sight because when we divide this bundle of six digits (0.142857) by one and we have no remainder and this bundle repeats!

Using the Auxiliary Fraction — Type Two

After forming the auxiliary fraction, divide in the first step, but create the next dividend by prefixing the remainder not to the quotient digit, but to the quotient digit's complement from nine. So if the quotient and remainder from one step are q and r, the dividend for the next step is 10r + (9 - q).

When the fraction has a numerator of one, the auxiliary fraction will have a numerator of zero. Having a dividend of zero is not a problem because on the second step the complement of zero is nine, an adequate dividend.

Worked Example — Type Two

F = 13/31, then the A.F. = 12/30 = 1.2/3

The working divisor is 3. The lower row is the dividend.

F = 13/31 = 0. 4 1 9 3 5 4 8 3 8 7 0 9

Prefix R to complement of quotient:

05 28 10 16 14 25 11 26 21 02 29 20

F = 13/31 ≈ 0.419354838709...

Using the Auxiliary Fraction — Type Three

When the fraction has a denominator that does not end in 1 or 9 apply the Anurupya Sutra, whereby, after prefixing each remainder to the quotient-digit in question, add to (or subtract from) the dividend at each step, as many times the quotient-digit as the divisor (the denominator) is below (or above) the normal nine. The process can be wholly mental with practice.

Worked Examples - Type Three

EXAMPLE ONE

If the denominator is 68, as 68=(4)(17), we may expect two fixed digits and 16 repeating digits. When F = 15/68, the A.F. = 1.5/7 (As D ends in 8, one below the normal ending, 9, we add the quotient-digit to the dividend at each step). The working divisor is 7. Divide to achieve the desired precision.

F = 15/68, A.F. = 1.5/7

A prefixed rem: $F = 0.{}_{1}2\ {}_{0}2\ {}_{4}0\ {}_{5}5\ {}_{4}8\ {}_{0}8\ {}_{2}2\ {}_{3}3\ {}_{1}5\ {}_{6}2\ {}_{1}9\ {}_{0}4\ {}_{1}1\ {}_{5}1$

Plus the Q-digit: 2 2 0 5 8 8 2 3 5 2 9 4 1 1

7 into actual dividend: 14 04 40 60 56 16 24 36 20 64 28 08 12 52

F = 15/68 ≈ 0.22058823529411...

EXAMPLE TWO

When F = 163/275, A.F. = 16.3/28. (Since D ends in 5, four below the normal ending, 9, we add four times the Q-digit to the dividend at each step.) The working divisor is 28. Since D = 275 = $(5^2)(11)$, we may expect two fixed digits, then a two-digit repeater. Remember that every factor of 2, 5, or 10 in the denominator generates one fixed decimal digit. Multiples of 28: 28, 56, 84, 112, 140, 168, 196, 224, 252, 280.

Prefix the rem: F =	0.	${}_{23}5$	${}_{3}9$	${}_{19}2$	${}_{4}7$	${}_{19}2$	${}_{4}7$
Plus four times the Q-digit:	x	20	36	08	28	08	28
28 into the actual dividend:	163	255	75	200	75	200	75

F = 163/275 ≈ 0.592727...

6

Methods of Computing Square Roots

Exponential Identity

Pocket calculators typically implement good routines to compute the exponential function and the natural logarithm, and then compute the square root of S using the identity

$$\sqrt{S} = e^{\frac{1}{2}\ln S}.$$

The same identity is used when computing square roots with logarithm tables or slide rules.

Rough Estimation

Many of the methods for calculating square roots require an initial seed value. If the initial value is too far from the actual square root, the calculation will be slowed down. It is therefore useful to have a rough estimate, which may be very inaccurate but easy to calculate.

One way of obtaining such an estimate for $\sqrt{S}$ is calculating 3^D, where D is the number of digits (to the left of the decimal point) in a positive real S. If $S < 1$, D is the negative of the number of zeros to the immediate right of the decimal point.

A better method of rough estimation is this:

If D is odd, D = 2n + 1, then use $\sqrt{S} \approx 2.10^n$

If D is even, D = 2n + 2, then use $\sqrt{S} \approx 6.10^n$.

When working in the binary numeral system (as computers do internally), an alternative method is to use $2^{[D/2]}$ (here D is the number of binary digits).

Babylonian Method

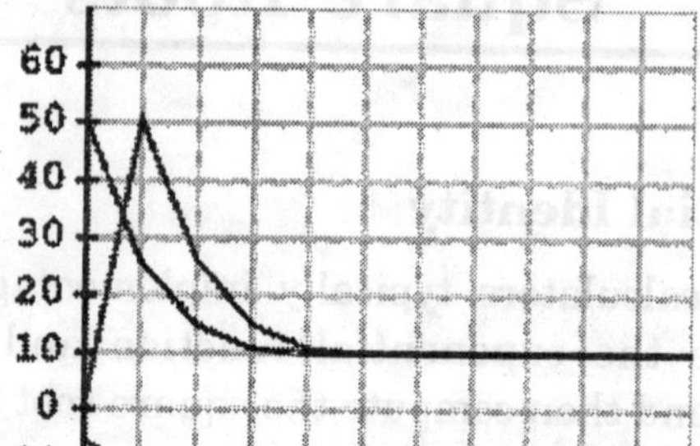

Graph charting the use of the Babylonian method for approximating the square root of 100 (10) using start values x_0=50, x_0=1, and x_0=-5. Note that using a negative start value yields the negative root.

A commonly used algorithm for approximating $\sqrt{S}$ (and perhaps the best) is known as the "Babylonian method" and can be derived from (but predates) Newton's method. This is a quadratically convergent algorithm, which means that the number of correct digits of the approximation roughly doubles with each iteration. It proceeds as follows:

1. Start with an arbitrary positive start value x_0 (the closer to the root, the better).
2. Let x_{n+1} be the average of xn and S/xn (using the arithmetic mean to approximate the geometric mean).

3. Repeat steps 2 and 3, until the desired accuracy is achieved.

It can also be represented as:

$$x_0 \approx \sqrt{S}.$$

$$x_{n+1} = \frac{1}{2}\left(x_n + \frac{S}{x_n}\right),$$

$$\sqrt{S} = \lim_{n \to \infty} x_n.$$

This algorithm works equally well in the p-adic numbers, but cannot be used to identify real square roots with p-adic square roots; it is easy, for example, to construct a sequence of rational numbers by this method which converges to + 3 in the reals, but to - 3 in the 2-adics.

EXAMPLE

We'll calculate , where S = 125348, to 6 significant figures. Here D, the number of digits in S, is 6. Then:

$$x_0 = 3^D = 3^6 = 729.000$$

$$x_1 = \frac{1}{2}\left(x_0 + \frac{S}{x_0}\right) = \frac{1}{2}\left(729.000 + \frac{125348}{729.000}\right) = 450.472$$

$$x_2 = \frac{1}{2}\left(x_1 + \frac{S}{x_1}\right) = \frac{1}{2}\left(450.472 + \frac{125348}{450.472}\right) = 364.365$$

$$x_3 = \frac{1}{2}\left(x_2 + \frac{S}{x_2}\right) = \frac{1}{2}\left(364.365 + \frac{125348}{364.365}\right) = 354.191$$

$$x_4 = \frac{1}{2}\left(x_3 + \frac{S}{x_3}\right) = \frac{1}{2}\left(354.191 + \frac{125348}{354.191}\right) = 354.045$$

$$x_5 = \frac{1}{2}\left(x_4 + \frac{S}{x_4}\right) = \frac{1}{2}\left(354.045 + \frac{125348}{354.045}\right) = 354.045$$

Therefore, $\sqrt{125348} \approx 354.045.$

Convergence

We let the relative error in xn be defined by

$$\epsilon_n = \frac{x_n}{\sqrt{S}} - 1$$

and thus

$$x_n = \sqrt{S}.(1 + \epsilon_n).$$

Then one can show that

$$\epsilon_{n+1} = \frac{\epsilon_n^2}{2(1+\epsilon_n)}$$

and thus that

$$0 \le \epsilon_{n+1} \le \min\left(\frac{\epsilon_{n+1}^2}{2}, \frac{\epsilon_{n+1}}{2}\right)$$

and consequently that convergence is assured provided that x0 and S are both positive.

Bakhshali Approximation

This is a method for finding an approximation to a square root which was described in an ancient manuscript known as the Bakhshali Manuscript. It is equivalent to two iterations of the Babylonian method beginning with N. The original presentation goes as follows: To calculate $\sqrt{S}$, let N^2 be the nearest perfect square to S. Then, calculate:

$$d = S - N^2$$

$$P = \frac{d}{2N}$$

$$A = N + P$$

$$\sqrt{S} \approx A - \frac{P^2}{2A}$$

This can be also written as:

$$\sqrt{S} \approx \frac{N^4 + 6N^2S + S^2}{4N^3 + 4NS}$$

$$\sqrt{N^2 + d} \approx N + \frac{d}{2N} - \frac{d^2}{8N^3 + 4Nd}$$

EXAMPLE

We'll find

N = 3

d = 9.2345 - 3² = 0.2345

$$P = \frac{0.2345}{2.3} = 0.0391$$

A = 3 + 0.0391 = 3.0391

$$\sqrt{9.2345} \approx 3.0391 - \frac{0.0391^2}{2 \times 3.0391} \approx 3.0388.$$

Digit by Digit Calculation

This is a method to find each digit of the square root in a sequence. It is much slower than the Babylonian method (if you have a calculator which can divide in one operation), but it has several advantages:

- It can be easier for manual calculations.
- Every digit of the root found is known to be correct, i.e. it will not have to be changed later.
- If the square root has an expansion which terminates, the algorithm will terminate after the last digit is found. Thus, it can be used to check whether a given integer is a square number.

Napier's bones include an aid for the execution of this algorithm. The Shifting nth-root algorithm is a generalization of this method.

The algorithm works for any base, and naturally, the way it proceeds depends on the base chosen. We will describe the method for the decimal system. It goes as follows:

Write the decimal expansion of the square. The numbers are laid out in a fashion similar to the long division algorithm, and the root will appear above its square. Divide the digits of the square into pairs, starting from the decimal point and going both left and right.

The decimal point of the root should be placed above the decimal point of the square. One digit of the root will appear above each pair of digits of the square.

Begin at the left-most (non-zero) pair of digits of the square. Initialize the remainder to zero. Initialize p to zero.

For each iteration (pair of digits) do:

1. Bring down the most significant (leftmost) pair of digits of the square not yet used (if all the digits have been used, use the digits 00) and append them to the remainder from the previous step, i.e. multiply the remainder by one hundred and add the two digits. This will be the current value c.
2. Find p, y and x, as follows:
   - Let p be the part of the root found so far, ignoring any decimal point. (For the first step, p = 0).
   - Let y equal (20 .p + x) . x.
   - Determine the greatest digit x such that (20 . p + x) . x (that is, y) does not exceed c.
     * Note: 20p + x is simply twice p, with the digit x appended to the right).
     * Note: You can find x by guessing what c/(20·p) is and doing a trial calculation of y, then adjusting x upward or downward as necessary.

- Place the digit x as the next digit of the root, i.e. above the two digits of the square which you just brought down. Thus the next p will be the old p times 10 plus x.

3. Subtract y from c to form a new remainder.
4. If the remainder is zero and there are no more digits to bring down, then the algorithm has terminated. Otherwise go back to step 1 for another iteration.

EXAMPLES

Find the square root of 152.2756.

$$\begin{array}{l} \quad 1 \quad 2.\ 3 \quad 4 \\ \sqrt{01\ \ 52.\ 27\ 56} \end{array}$$

x	01	1*1 <= 1 < 2*2	x = 1
	01	y = x*x = 1*1 = 1	
2x	00 52	22*2 <= 52 < 23*3	x = 2
	00 44	y = 2x*x = 22*2 = 44	
24x	08 27	243*3 <= 827 < 244*4	x = 3
	07 29	y = 24x*x = 243*3 = 729	
246x	98 56	2464*4 <= 9856 < 2465*5	x = 4
	98 56	y = 246x*x = 2464*4 = 9856	
	00 00	Algorithm terminates: Answer is 12.34	

Find the square root of 2.

$$\begin{array}{l} \quad 1. \quad 4 \quad 1 \quad 4 \quad 2 \\ \sqrt{02.\ 00\ 00\ 00\ 00} \end{array}$$

x	02	1*1 <= 2 < 2*2	x = 1
	01	y = x*x = 1*1 = 1	
2x	01 00	24*4 <= 100 < 25*5	x = 4
	00 96	y = 2x*x = 24*4 = 96	

28x	04 00	281*1 <= 400 < 282*2	x = 1
	02 81	y = 28x*x = 281*1 = 281	
282x	01 19 00	2824*4 <= 11900 < 2825*5	x = 4
	01 12 96	y = 282x*x = 2824*4 = 11296	
2828x	06 04 00	28282*2 <= 60400 < 28283*3	x = 2

The desired precision is achieved:

The square root of 2 is about 1.4142

Duplex Method for Extracting a Square Root

The duplex method is a variant of the digit by digit method for calculating the square root of a whole or decimal number one digit at a time. The duplex is the square of the central digit plus double the cross-product of digits equidistant from the centre. The duplex is computed from the quotient digits (square root digits) computed thus far, but after the initial digits. The duplex is subtracted from the dividend digit prior to the second subtraction for the product of the quotient digit times the divisor digit. For perfect squares the duplex and the dividend will get smaller and reach zero after a few steps. For non-perfect squares the decimal value of the square root can be calculated to any precision desired. However, as the decimal places proliferate, the duplex adjustment gets larger and longer to calculate. The duplex method follows the Vedic ideal for an algorithm, one-line, mental calculation. It is flexible in choosing the first digit group and the divisor. Small divisors are to be avoided by starting with a larger initial group.

In short, to calculate the duplex of a number, double the product of each pair of equidistant digits plus the square of the centre digit (of the digits to the right of the colon).

Number ⇒ Calculation = Duplex

574 ⇒ 2(5·4) + 72 = 89

406,739 ⇒ 2(4·9)+ 2(0·3)+ 2(6·7) = 72+0+84 = 156

123,456 ⇒ 2(1·6)+ 2(2·5)+ 2(3·4) = 12 +20 +24 = 56

88,900,777 ⇒ 2(56)+2(56)+2(63)+0+0 = 320+30 = 350

48329,03711 ⇒ 2(4·1)+2(8·1)+2(3·7)+2(2·3)+2(9·0)

= 8+16+42+12+0 = 78

In a square root calculation the quotient digit set increases incrementally for each step.

Number ⇒ Calculation = Duplex:

1 ⇒ 12 = 1

14 ⇒ 2(1·4) = 8

142 ⇒ 2(1·2) + 42 = 4 + 16 = 20

14,21 ⇒ 2(1·1) + 2(4·2) = 2 + 16 = 18

14213 ⇒ 6+8+4 = 18

142,135 ⇒ 10+24+4 = 38

1421356 ⇒ 12+40+12+1 = 65

1421,3562 ⇒ 4+48+20+6 = 78

142,135,623 ⇒ 6+16+24+10+9 = 65

142,1356,237 ⇒ 14+24+8+12+30 = 88

142,13562,373 ⇒ 6+56+12+4+36+25 = 139

EXAMPLE 1

Consider $53^2 = 2809$.

- Set down the number in groups of two digits.
- We define a divisor, a dividend and a quotient to find the root.
- Consider the first group. Here is it 28.
    * Find the nearest perfect square below that group. Here, for the 28, the closest perfect square is 25.
    * The root of that perfect square is the first digit of our root.

    * Since 28 > 25 and 25 = 52, we take 5 as the first digit in the root.
    * For the divisor we take double this first digit (2 · 5), which is here 10.

- We set up a division framework with a colon.
    * 28: 0 9 is the dividend and 5: is the quotient.
    * We put a colon to the right of 28 and 5 and keep the colons lined up vertically. The duplex is calculated only on quotient digits to the right of the colon.
- We calculate the remainder. 28: minus 25: is 3:.
    * We append the remainder to the next digit to get the new dividend.
    * Here, we append 0 to the digit 3, which makes the new dividend 30, as the divisor 10 goes into 30 just 3 times. (No reserve needed here for subsequent deductions.)
- We repeat the operation.
    * The zero remainder appended to 9. Nine is the dividend.
    * Now we have a digit to the right of the colon so we deduct the duplex. 32 = 9.
    * We get a zero remainder.
    * The next root digit is zero. The next duplex is 2(3·0) = 0.
    * The dividend is zero. We have an exact square root, 53.0.

Find the square root of 2809.

Set down the number in groups of two digits.

The number of groups gives the number of digits in the root.

From the first group, 28, we obtain the divisor, 10.

Put a colon after the first group to separate it.

Since 28>25=52. And 2x5=10.

Gross dividend:	28: 0 9.	Using mental math:
Divisor:	10) 3 0	Square: 10) 28: $_{3}0$ 9
Duplex, Deduction:	25: xx 09	Square root: 5: 3. 0
Dividend:	30 00	
Remainder:	3: 00 00	
Square Root, Quotient:	5: 3. 0	

EXAMPLE 2

Find the square root of 2,080,180,881. Solution by the duplex method: this ten-digit square has five digit-pairs, so it will have a five-digit square root. The first digit-pair is 20. Put the colon to the right. The nearest square below 20 is 16, whose root is 4, the first root digit. So, we use 2·4=8 for the divisor. Now we proceed with the duplex division, one digit column at a time. Prefix the remainder to the next dividend digit.

divisor; gross dividend:	8) 20:	8	0	1	8	0	8	8	1
read the dividend diagonally up:		4	8	7	11	10	10	0	8
minus the duplex:	16:	xx	25	60	36	90	108	00	81
actual dividend:	:	48	55	11	82	10	00	08	00
minus the product:	:	40	48	00	72	00	00	0	00
remainder:	4:	8	7	11	10	10	0	8	00
quotient:	4:	5,	6	0	9.	0	0	0	0

Duplex calculations:

Quotient-digits ⇒ Duplex deduction.

5 ⇒ 52= 25

5 and 6 ⇒ 2(5·6) = 60

5,6,0 ⇒ 2(5·0)+62 = 36

$5,6,0,9 \Rightarrow 2(5{\cdot}9)+2(6{\cdot}0) = 90$

$5,6,0,9,0 \Rightarrow 2(5{\cdot}0)+2(6{\cdot}9)+ 0 = 108$

$5,6,0,9,0,0 \Rightarrow 2(5{\cdot}0)+2(6{\cdot}0)+2(0{\cdot}9) = 0$

$5,6,0,9,0,0,0 \Rightarrow 2(5{\cdot}0)+2(6{\cdot}0)+2(0{\cdot}0)+92 = 81$

Hence the square root of 2,080,180,881 is exactly 45,609.

EXAMPLE 3

Find the square root of two to ten places. Let us take 20,000 as the beginning group, using three digit-pairs at the start. The perfect square just below 20,000 is 141, since $141^2 = 19881 < 20,000$. So, the first root digits are 141 and the divisor doubled, 2 x 141 = 282. With a larger divisor the duplex will be relatively small. Hence, we can pick the multiple of the divisor without confusion.

Dividend:	2.0000	:	0	0	0	0	0	0	0	0
Diagonal;Divisor:	282)	:	119	62	40	102	162	182	75	112
Minus duplex::		:	xxxx	16	16	12	28	53	74	59
Actual dividend:	20000	:	1190	604	384	1008	1592	1767	676	1061
Minus product:	19881	:	1128	564	282	846	1410	1692	564	846
Remainder:	119	:	62	40	102	162	182	75	112	215
Root quotient:	1.41	:	4	2	1	3	5	6	2	3

Ten multiples of 282: 282; 564; 846; 1128; 1410; 1692; 1974; 2256; 2538; 2820.

Iterative Methods for Reciprocal Square Roots

The following are iterative methods for finding the reciprocal square root of S which is $\frac{1}{\sqrt{S}}$. Once it has been found, we can find $\sqrt{S}$ by simple multiplication: $\sqrt{S} = \frac{1}{\sqrt{S}}.S.$ These iterations involve only multiplication, and not division. They are therefore faster than the Babylonian method. However, they are not stable. If the initial value is not close to the reciprocal square root, the iterations will diverge away from it rather than converge to it. It can therefore be advantageous to perform an iteration of the Babylonian method on a rough estimate before starting to apply these methods.

- One method is found by applying Newton's method to the equation $\frac{1}{x^2} - S = 0$. It converges quadratically:

  $$x_{n+1} = \frac{x_n}{2}.\left(3 - S.x_n^2\right).$$

- Another method converges cubically, but involves more operations per iteration:

  $$y_n = S.x_n^2,$$

  $$x_{n+1} = \frac{x_n}{8}.\left(15 - y_n.\left(10 - 3.y_n\right)\right).$$

Taylor Series

If N is an approximation to $\sqrt{S}$, a better approximation can be found by using the Taylor series of the square root function:

$$\sqrt{N^2 + d} = \sum_{n=0}^{\infty} \frac{(-1)^n (2n)!d^n}{(1-2n)n!^2 4^n N^{2n-1}}$$

$$= N + \frac{d}{2N} - \frac{d^2}{8N^3} + \frac{d^3}{16N^5} - \frac{5d^4}{128N^7} + \ldots$$

As an iterative method, the order of convergence is equal to the number of terms used. With 2 terms, it is identical to the Babylonian method; With 3 terms, each iteration takes almost as many operations as the Bakhshali approximation, but converges more slowly. Therefore, this is not a particularly efficient way of calculation.

Other Methods

Finding $\sqrt{S}$ is the same as solving the equation $x^2 - S = 0$. Therefore, any general numerical root-finding algorithm can be used. Newton's method, for example, reduces in this case to the Babylonian method. Other methods are less efficient than the ones presented above.

Continued Fraction Expansion

Quadratic irrationals (numbers of the form $\frac{a+\sqrt{b}}{c}$, where a, b and c are integers), and in particular, square roots of integers, have periodic continued fractions. Sometimes we may be interested not in finding the numerical value of a square root, but rather in its continued fraction expansion. The following iterative algorithm can be used for this purpose (S is any natural number which is not a perfect square):

$$m_0 = 0$$

$$d_0 = 1$$

$$a_0 = \left[\sqrt{S}\right]$$

$$m_{n+1} = d_n a_n - m_n$$

$$d_{n+1} = \frac{S - m_{n+1}^2}{d_n}$$

$$a_{n+1} = \left[\frac{\sqrt{S} + m_{n+1}}{d_{n+1}}\right] = \left[\frac{a_0 + m_{n+1}}{d_{n+1}}\right]$$

Notice that mn, dn, and an are always integers. The algorithm terminates when this triplet is the same as one encountered before. The expansion will repeat from then on. The sequence $[a_0; a_1, a_2, a_3, \ldots]$ is the continued fraction expansion:

$$\sqrt{S} = a_0 + \cfrac{1}{a_1 + \cfrac{1}{a_2 + \cfrac{1}{a_3 + \ldots}}}$$

Example, Square Root of 114 as a Continued Fraction

We begin with m_0=0; d_0=1; and a_0=10.

$$\sqrt{114} = \frac{\sqrt{114}+0}{1} = 10 + \frac{\sqrt{114}-10}{1}$$

$$= 10 + \frac{\left(\sqrt{114}-10\right)\left(\sqrt{114}+10\right)}{\sqrt{114}+10}$$

$$= 10 + \frac{114-100}{\sqrt{114}+10} = 10 + \frac{1}{\frac{\sqrt{114}+10}{14}}.$$

Next, m_1=10; d_1=14; and a_1=1.

$$\frac{\sqrt{114}+10}{14} = 1 + \frac{\sqrt{144}-4}{14}$$

$$= 1 + \frac{114-16}{14\left(\sqrt{144}-4\right)} = 1 + \frac{1}{\frac{\sqrt{114}+4}{7}}.$$

Next, m_2=4; d_2=7; and a_2=2.

$$\frac{\sqrt{114}+4}{7} = 2 + \frac{\sqrt{114}-10}{7}$$

$$= 2 + \frac{14}{7\left(\sqrt{114}+10\right)} = 2 + \frac{1}{\frac{\sqrt{114}+10}{2}}.$$

$$\frac{\sqrt{114}+10}{2} = 10 + \frac{\sqrt{114}-10}{2}$$

$$= 10 + \frac{14}{2\left(\sqrt{114}+10\right)} = 10 + \frac{1}{\frac{\sqrt{114}+10}{7}}.$$

$$\frac{\sqrt{114}+10}{7} = 2+\frac{\sqrt{114}-4}{7}$$

$$= 2+\frac{98}{7\left(\sqrt{114}+10\right)} = 2+\frac{1}{\frac{\sqrt{114}+4}{14}}.$$

$$\frac{\sqrt{114}+4}{7} = 1+\frac{\sqrt{114}-10}{14}$$

$$= 1+\frac{14}{14\left(\sqrt{114}+10\right)} = 1+\frac{1}{\frac{\sqrt{114}+10}{1}}.$$

$$\frac{\sqrt{114}+10}{1} = 20+\frac{\sqrt{114}-10}{1}$$

$$= 20+\frac{14}{\sqrt{114}+10} = 20+\frac{1}{\frac{\sqrt{114}+10}{14}}.$$

Now, loop back to the second equation above.

Consequently, the continued fraction for the square root of 114 is:

$\sqrt{114}$ = [10; 1, 2, 10, 2, 1, 20, 1, 2, 10, 2, 1, 20, 1, 2, 10, 2, 1, 20, ...].

Pell's Equation

Pell's equation and its variants yield a method for efficiently finding continued fraction convergents of square roots of integers. However, it can be complicated to execute, and usually not every convergent is generated. The ideas behind the method are as follows:

- If (p, q) is a solution (where p and q are integers) to the equation $p^2 = S.q^2 \pm 1$, then $\frac{p}{q}$ is a continued fraction convergent of $\sqrt{S}$, and as such, is an excellent rational approximation to it.

- If (p_a, q_a) and (p_b, q_b) are solutions, then so is:

  $p = p_a p_b + S.q_a q_b$

  $q = p_a q_b + p_b q_a$

- More generally, if (p_1, q_1) is a solution, then it is possible to generate a sequence of solutions (p_n, q_n) satisfying:

  $p_{m+n} = p_m p_n + S.q_m q_n$

  $q_{m+n} = q_m q_n + p_n q_m$

The method is as follows:

- Find natural numbers (0 excluded) p1 and q1 such that $p_1^2 = S.q_1^2 \pm 1$. This is the hard part; It can be done either by guessing, or by using fairly sophisticated techniques.
- To generate a long list of convergents, iterate:

  $p_{n+1} = p_1 p_n + S.q_1 q_n$

  $q_{n+1} = p_1 q_n + p_n q_1$

- To find the larger convergents quickly, iterate:

  $p_{2n} = p_n^2 + S.q_n^2$

  $q_{2n} = 2p_n q_n$

- In either case, is a rational approximation satisfying

$$\left|\frac{p_n}{q_n} - \sqrt{S}\right| < \frac{1}{q_n^2.\sqrt{S}}.$$

Approximations that Depend on IEEE Representation

On computers, a very rapid Newton's method based approximation to the square root can be obtained for floating point numbers when computers use an IEEE (or sufficiently similar) representation.

```
float fastsqrt(float val) {
    int tmp = *(int *)&val;
    tmp-= 1<<23;/*Remove last bit to not let it go to mantissa */
    /* tmp is now an approximation to logbase2(val) */
    tmp = tmp >> 1;/* divide by 2 */
    tmp += 1<<29;/*add 64 to exponent: (e+127)/2=(e/2)+63,*/
                 /*that represents (e/2)-64 but we want e/2*/
    return *(float *)&tmp;
}
```

In the above, the operations to remove last exponent bit and add the IEEE bias can be combined into a single operation. An additional adjustment can be made in the same operation to reduce the maximum relative error. So, the three operations, not including the cast, can be rewritten as:

```
tmp = (1<<29) + (tmp >> 1) - (1<<22) + m;
```

Where m is some magic number that corresponds to a calculation involving an initial guess used in the Newton's approximation.

Inverse Square Root

A variant of the above routine is included below, which can be used to compute the reciprocal of the square root, i.e. instead, was written by Greg Walsh, and implemented into the game Quake 3 by Gary Tarolli. As an approximation, it produced a relative error of less than 4%, and in computer graphics it is a very efficient way to normalize a vector.

```
float invSqrt(float x)
{
    float xhalf = 0.5f*x;
    int i = *(int *)&x;
    i = 0x5f3759df - (i >> 1);
    x = *(float *)&i;
    x = x * (1.5f - xhalf * x * x);
    return x;
}
```

The Vedic Mathematics Sutras

This list of sutras is taken from the book Vedic Mathematics, which includes a full list of the sixteen Sutras in Sanskrit, but in some cases a translation of the Sanskrit is not given in the text and comes from elsewhere.

This formula 'On the Flag' is not in the list given in Vedic Mathematics, but is referred to in the text.

The Main Sutras

By one more than the one before.

All from 9 and the last from 10.

Vertically and Crosswise

Transpose and Apply

If the Samuccaya is the Same it is Zero

If One is in Ratio the Other is Zero

By Addition and by Subtraction

By the Completion or Non-Completion

Differential Calculus

By the Deficiency

Specific and General

The Remainders by the Last Digit

The Ultimate and Twice the Penultimate

By One Less than the One Before

The Product of the Sum

All the Multipliers

The Sub Sutras

Proportionately

The Remainder Remains Constant

The First by the First and the Last by the Last

For 7 the Multiplicand is 143

By Osculation

Lessen by the Deficiency

Whatever the Deficiency lessen by that amount andset up the Square of the Deficiency

Last Totalling 10

Only the Last Terms

The Sum of the Products

By Alternative Elimination and Retention

By Mere Observation

The Product of the Sum is the Sum of the Products

On the Flag

Try a Sutra

Mark Gaskell introduces an alternative

system of calculation based on Vedic philosophy

At the Maharishi School in Lancashire we have developed a course on Vedic mathematics for key stage 3 that covers the national curriculum.

The results have been impressive: maths lessons are much livelier and more fun, the children enjoy their work more and expectations of what is possible are very much higher. Academic performance has also greatly improved: the first class to complete the course managed to pass their GCSE a year early and all obtained an A grade.

Vedic maths comes from the Vedic tradition of India. The Vedas are the most ancient record of human experience and knowledge, passed down orally for generations and written down about 5,000 years ago.

Medicine, architecture, astronomy and many other branches of knowledge, including maths, are dealt with in the texts. Perhaps it is not surprising that the country credited with introducing our current number system and the invention of perhaps the most important mathematical symbol, 0, may have more to offer in the field of maths.

The remarkable system of Vedic maths was rediscovered from ancient Sanskrit texts early last century. The system is based on 16 sutras or aphorisms, such as: "by one more than the one before" and "all from nine and the last from 10". These describe natural processes in the mind and ways of solving a whole range of mathematical problems.

For example, if we wished to subtract 564 from 1,000 we simply apply the sutra "all from nine and the last from 10". Each figure in 564 is subtracted from nine and the last figure is subtracted from 10, yielding 436.

1,000 - 564 = 436

1,000	-	5	6	4
		subtract from 9	subtract from 9	subtract from 10
		↓	↓	↓
		4	3	6

This can easily be extended to solve problems such as 3,000 minus 467. We simply reduce the first figure in 3,000 by one and then apply the sutra, to get the answer 2,533. We have had a lot of fun with this type of sum, particularly when dealing with money examples, such as £10 take away £2. 36. Many of the children have described how they have challenged their parents to races at home using many of the Vedic techniques — and won. This particular method can also be expanded into a general method, dealing with any subtraction sum.

The sutra "vertically and crosswise" has many uses. One very useful application is helping children who are having trouble with their tables above 5x5. For example 7x8. 7 is 3 below the base of 10, and 8 is 2 below the base of 10.

7 × 8 = 56

7 3 (3 is the difference from base)

8 2 (2 is the difference from base

(A) 7 3 starting at the left subtract crosswise either 8-3 or 7-2 to get 5, the first figure

8 2 of the answer

5

(B) 7 3 Multiply vertically

× to get 6

8 2

5 6

The whole approach of Vedic maths is suitable for slow learners, as it is so simple and easy to use.

The sutra "vertically and crosswise" is often used in long multiplication. Suppose we wish to multiply

32 by 44. We multiply vertically 2×4=8.

Then we multiply crosswise and add the two results: 3×4+4×2=20, so put down 0 and carry 2.

Finally we multiply vertically 3×4=12 and add the carried 2 =14. Result: 1,408.

We can extend this method to deal with long multiplication of numbers of any size. The great advantage of this system is that the answer can be obtained in one line and mentally. By the end of Year 8, I would expect all students to be able to do a "3 by 2" long multiplication in their heads. This gives enormous confidence to the pupils who lose their fear of numbers and go on to tackle harder maths in a more open manner.

All the techniques produce one-line answers and most can be dealt with mentally, so calculators are not used until Year 10. The methods are either "special", in that they only apply under certain conditions, or general. This encourages flexibility and innovation on the part of the students.

Multiplication can also be carried out starting from the left, which can be better because we write and pronounce numbers from left to right. Here is an example of doing this in a special method for long multiplication of numbers near a base (10, 100, 1,000 etc.), for example, 96 by 92. 96 is 4 below the base and 92 is 8 below.

We can cross-subtract either way: 96-8=88 or 92-4=88. This is the first part of the answer and multiplying the "differences" vertically 4×8=32 gives the second part of the answer.

96 × 96 = 8,832

(A) 96 4 (4 is the difference from base)

92 8 (8 is the difference from base

———

(B) 96 4 subtract crosswise from the left 96-8 = 88

92 8 o 92-4 = 88

———

88

(C) 96 4 Multiply vertically

× 4 × 8 = 32

92 8

———

88 32

This works equally well for numbers above the base: 105x111=11,655. Here we add the differences. For

205x211=43,255, we double the first part of the answer, because 200 is 2x100.

We regularly practise the methods by having a mental test at the beginning of each lesson. With the introduction of a non-calculator paper at GCSE, Vedic maths offers methods that are simpler, more efficient and more readily acquired than conventional methods.

There is a unity and coherence in the system which is not found in conventional maths. It brings out the beauty and patterns in numbers and the world around us. The techniques are so simple they can be used when conventional methods would be cumbersome.

When the children learn about Pythagoras's theorem in Year 9 we do not use a calculator; squaring numbers and finding square roots (to several significant figures) is all performed with relative ease and reinforces the methods that they would have recently learned.

TUTORIAL 1

Use the formula ALL FROM 9 AND THE LAST FROM 10 to perform instant subtractions.

For example 1000 - 357 = 643

We simply take each figure in 357 from 9 and the last figure from 10.

So the answer is 1000 - 357 = 643

And thats all there is to it!

This always works for subtractions from numbers consisting of a 1 followed by noughts: 100; 1000; 10,000 etc.

Similarly 10,000 - 1049 = 8951

For 1000 - 83, in which we have more zeros than figures in the numbers being subtracted, we simply suppose 83 is 083.

So 1000 - 83 becomes 1000 - 083 = 917

EXERCISE 1 TUTORIAL 1

Try some yourself:

1) 1000 - 777 =

2) 1000 - 283 =

3) 1000 - 505 =

4) 10,000 - 2345 =

5) 10,000 - 9876 =

6) 10,000 - 1011 =

7) 100 - 57 =

8) 1000 - 57 =

9) 10,000 - 321 =

10) 10,000 - 38 =

TUTORIAL 2

Using VERTICALLY AND CROSSWISE you do not need the multiplication tables beyond 5 × 5.

Suppose you need 8 × 7

8 is 2 below 10 and 7 is 3 below 10.

Think of it like this:

8 2
7 3
5 6 answer

The answer is 56.

The diagram below shows how you get it.

8 2
× |
7 3
5 6 answer

You subtract crosswise 8-3 or 7 - 2 to get 5, the first figure of the answer.

And you multiply vertically: 2 × 3 to get 6, the last figure of the answer.

That's all you do:

See how far the numbers are below 10, subtract one number's deficiency from the other number, and multiply the deficiencies together.

7 × 6 = 42

7 3

× |

6 4

3 12 = 42

Here there is a carry: the 1 in the 12 goes over to make 3 into 4.

EXERCISE 1 TUTORIAL 2

Multiply These:

1) 8 × 8 =

2) 9 × 7 =

3) 8 × 9 =

4) 7 × 7 =

5) 9 × 9 =

6) 6 × 6 =

Here's how to use VERTICALLY AND CROSSWISE for multiplying numbers close to 100.

Suppose you want to multiply 88 by 98.

Not easy, you might think. But with VERTICALLY AND CROSSWISE you can give the answer immediately, using the same method as above.

Both 88 and 98 are close to 100.

88 is 12 below 100 and 98 is 2 below 100.

You can imagine the sum set out like this:

$$\begin{array}{rr} 88 & -12 \\ \times & | \\ 98 & -2 \\ \hline 86 & 24 \\ \hline \end{array}$$

As before the 86 comes from subtracting crosswise: 88 - 2 = 86 (or 98 - 12 = 86: you can subtract either way, you will always get the same answer).

And the 24 in the answer is just 12 × 2: you multiply vertically.

So 88 × 98 = 8624

EXERCISE 2 TUTORIAL 2

This is so easy it is just mental arithmetic.

Try some:

1) 87 × 98 =

2) 88 × 97 =

3) 77 × 98 =

4) 93 × 96 =

5) 94 × 92 =

6) 64 × 99 =

7) 98 × 97 =

ANSWERS TO EXERCISE 2 TUTORIAL 2

1) 8526

2) 8536

3) 7546

4) 8928

5) 8648

6) 6336
7) 9506 (we put 06 because, like all the others, we need two figures in each part)

Multiplying numbers just over 100.

$103 \times 104 = 10712$

The answer is in two parts: 107 and 12,

107 is just 103 + 4 (or 104 + 3), and 12 is just 3×4.

Similarly $107 \times 106 = 11342$

$107 + 6 = 113$ and $7 \times 6 = 42$

EXERCISE 3 TUTORIAL 2

Again, just for mental arithmetic

Try a few:

1) $102 \times 107 =$
2) $106 \times 103 =$
3) $104 \times 104 =$
4) $109 \times 108 =$
5) $101 \times 123 =$
6) $103 \times 102 =$

ANSWERS TO EXERCISE 3 TUTORIAL 2

1) 10914
2) 10918
3) 10816
4) 11772
5) 12423
6) 10506 (we put 06, not 6)

TUTORIAL 3

The easy way to add and subtract fractions.

1) $\frac{4}{5}+\frac{1}{6}=$

2) $\frac{1}{3}+\frac{1}{4}=$

3) $\frac{2}{7}+\frac{2}{3}=$

4) $\frac{4}{5}-\frac{1}{6}=$

Use VERTICALLY AND CROSSWISE to write the answer straight down!

$$\frac{2}{3}+\frac{1}{5}=\frac{10+3}{15}=\frac{13}{15}$$

Multiply crosswise and add to get the top of the answer:

$2 \times 5 = 10$ and $1 \times 3 = 3$. Then $10 + 3 = 13$.

The bottom of the fraction is just $3 \times 5 = 15$.

You multiply the bottom number together.

So:

$$\frac{5}{7}+\frac{3}{4}=\frac{20+21}{28}=\frac{41}{28}$$

Subtracting is just as easy: multiply crosswise as before, but the subtract:

$$\frac{6}{7}-\frac{2}{3}=\frac{18-14}{21}=\frac{4}{21}$$

TUTORIAL 4

A quick way to square numbers that end in 5 using the formula BY ONE MORE THAN THE ONE BEFORE.

$75^2 = 5625$

75^2 means 75×75.

The answer is in two parts: 56 and 25.

The last part is always 25.

The first part is the first number, 7, multiplied by the number "one more", which is 8:

so $7 \times 8 = 56$

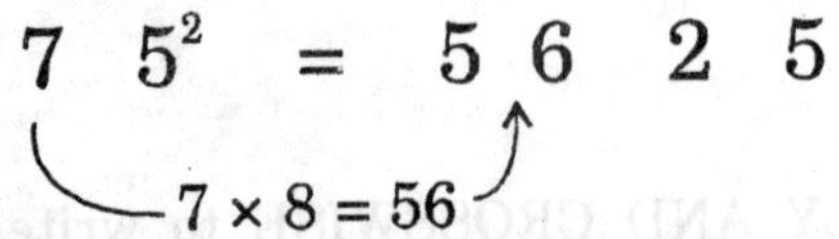

Similarly $85^2 = 7225$ because $8 \times 9 = 72$.

EXERCISE 1 TUTORIAL 4

Try these:

1) 452 =
2) 652 =
3) 952 =
4) 352 =
5) 152 =

Method for multiplying numbers where the first figures are the same and the last figures add up to 10.

$32 \times 38 = 1216$

Both numbers here start with 3 and the last figures (2 and 8) add up to 10.

So we just multiply 3 by 4 (the next number up) to get 12 for the first part of the answer.

And we multiply the last figures: $2 \times 8 = 16$ to get the last part of the answer.

Diagrammatically:

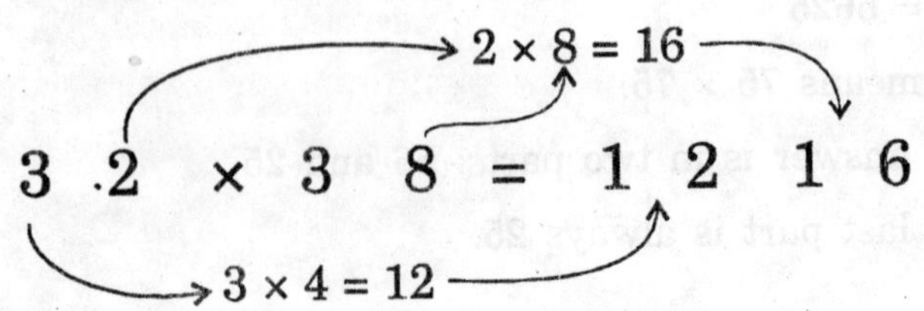

And 81 × 89 = 7209

We put 09 since we need two figures as in all the other examples.

ANSWERS TO EXERCISE 1 TUTORIAL 4

1) 2025

2) 4225

3) 9025

4) 1225

5) 225

EXERCISE 2 TUTORIAL 4

Practise some:

1) 43 × 47 =

2) 24 × 26 =

3) 62 × 68 =

4) 17 × 13 =

5) 59 × 51 =

6) 77 × 73 =

ANSWERS TO EXERCISE 2 TUTORIAL 4

1) 2021

2) 624

3) 4216

4) 221

5) 3009

6) 5621

TUTORIAL 5

An elegant way of multiplying numbers using a simple pattern

21 × 23 = 483

This is normally called long multiplication butactually the answer can be written straight downusing the VERTICALLY AND CROSSWISE formula.

We first put, or imagine, 23 below 21:

$$\begin{array}{ccc} 2 & & 1 \\ | & \times & | \\ 2 & & 3 \; \times \\ \hline 4 & 8 & 3 \\ \hline \end{array}$$

There are 3 steps:

a) Multiply vertically on the left: $2 \times 2 = 4$.
This gives the first figure of the answer.

b) Multiply crosswise and add: $2 \times 3 + 1 \times 2 = 8$
This gives the middle figure.

c) Multiply vertically on the right: $1 \times 3 = 3$
This gives the last figure of the answer.

And thats all there is to it.

Similarly $61 \times 31 = 1891$

$$\begin{array}{ccc} 6 & & 1 \\ | & \times & | \\ 3 & & 1 \; \times \\ \hline 18 & 9 & 1 \\ \hline \end{array}$$

$6 \times 3 = 18$; $6 \times 1 + 1 \times 3 = 9$; $1 \times 1 = 1$

EXERCISE 1 TUTORIAL 5

Try these, just write down the answer:

1) 14×21
2) 22×31
3) 21×31
4) 21×22
5) 32×21

ANSWERS TO EXERCISE 1 TUTORIAL 5

1) 294
2) 682
3) 651
4) 462
5) 672

EXERCISE 2A TUTORIAL 5

Multiply any 2-figure numbers together by mere mental arithmetic!

If you want 21 stamps at 26 pence each you can easily find the total price in your head.

There were no carries in the method given above.,/p>

However, there only involve one small extra step.

21 × 26 = 546

```
2     1
|  ×  |
2     6  ×
_________
4  4  6  = 546
_________
```

The method is the same as above except that we get a 2-figure number, 14, in the middle step, so the 1 is carried over to the left (4 becomes 5).

So 21 stamps cost £5.46.

Practise a few:

1) 21 × 47
2) 23 × 43
3) 32 × 53
4) 42 × 32
5) 71 × 72

ANSWERS TO EXERCISE 2A TUTORIAL 5

1) 987
2) 989
3) 1696
4) 1344
5) 5112

EXERCISE 2B TUTORIAL 5

33 × 44 = 1452

There may be more than one carry in a sum:

$$\begin{array}{ccc} 3 & & 3 \\ | & \times & | \\ 4 & & 4 \quad \times \\ \hline 12 & _{2}4 & _{1}2 \end{array} = 1452$$

Vertically on the left we get 12.

Crosswise gives us 24, so we carry 2 to the left and mentally get 144.

Then vertically on the right we get 12 and the 1 here is carried over to the 144 to make 1452.

6) 32 × 56
7) 32 × 54
8) 31 × 72
9) 44 × 53
10) 54 × 64

ANSWERS TO EXERCISE 2B TUTORIAL 5

6) 1792
7) 1728
8) 2232
9) 2332
10) 3456

TUTORIAL 6

Multiplying a number by 11.

To multiply any 2-figure number by 11 we just put the total of the two figures between the 2 figures.

26 × 11 = 286

Notice that the outer figures in 286 are the 26 being multiplied.

And the middle figure is just 2 and 6 added up.

So 72 × 11 = 792

EXERCISE 1 TUTORIAL 6

Multiply by 11:

1) 43 =

2) 81 =

3) 15 =

4) 44 =

5) 11 =

ANSWERS TO EXERCISE 1 TUTORIAL 6

1) 473

2) 891

3) 165

4) 484

5) 121

EXERCISE 2 TUTORIAL 6

Multiply by 11:

1) 11 × 88 =

2) 11 × 84 =

3) 11 × 48 =

4) 11 × 73 =

5) 11 × 56 =

ANSWERS TO EXERCISE 2 TUTORIAL 6

1) 968
2) 924
3) 528
4) 803
5) 616

EXERCISE 3 TUTORIAL 6

Multiply by 11:

1) 151 =
2) 527 =
3) 333 =
4) 714 =
5) 909 =

ANSWERS TO EXERCISE 3 TUTORIAL 6

1) 1661
2) 5797
3) 3663
4) 7854
5) 9999

TUTORIAL 7

Method for dividing by 9.

23/9 = 2 remainder 5

The first figure of 23 is 2, and this is the answer.

The remainder is just 2 and 3 added up!

43/9 = 4 remainder 7

The first figure 4 is the answer and 4 + 3 = 7 is the remainder - could it be easier?

EXERCISE 1A TUTORIAL 7

Divide by 9:

1) 61/9 =

2) 33/9 =

3) 44/9 =

4) 53/9 =

5) 80/9 =

ANSWERS TO EXERCISE 1A TUTORIAL 7

134/9 = 14 remainder 8

The answer consists of 1,4 and 8.

1 is just the first figure of 134.

4 is the total of the first two figures 1+ 3 = 4, and 8 is the total of all three figures 1+ 3 + 4 = 8.

ANSWERS TO EXERCISE 1A TUTORIAL 7

1) 6 r 7

2) 3 r 6

3) 4 r 8

4) 5 r 8

5) 8 r 8

EXERCISE 1B TUTORIAL 7

Divide by 9:

6) 232 =

7) 151 =

8) 303 =

9) 212 =

10) 2121 =

ANSWERS TO EXERCISE 1B TUTORIAL 7

1) 25 r 7

2) 16 r 7

3) 33 r 6

4) 23 r 5

5) 235 r 6 (we have 2, 2 + 1, 2 + 1 + 2, 2 + 1 + 2 + 1)

EXERCISE 2 TUTORIAL 7

Divide these by 9:

1) 771/9 =

2) 942/9 =

3) 565/9 =

4) 555 /9 =

5) 2382/9 =

6) 7070/9 =

Is it Divisible by Four?

This little math trick will show you whether a number is divisible by four or not.

So, this is how it works.

Let's look at 1234

Does 4 divide evenly into 1234?

For 4 to divide into any number we have to make sure that the last number is even

If it is an odd number, there is no way it will go in evenly.

So, for example, 4 will not go evenly into 1233 or 1235

Now we know that for 4 to divide evenly into any number the number has to end with an even number.

4 into 1234, the solution:

Take the last number and add it to 2 times the second last number

If 4 goes evenly into this number then you know that 4 will go evenly into the whole number.

So,

4 + (2 × 3) = 10

4 goes into 10 two times with a remainder of 2 so it does not go in evenly.

Therefore 4 into 1234 does not go in completely.

Let's try 4 into 3436546

So, from our example, take the last number, 6 and add it to two times the penultimate number, 4

6 + (2 × 4) = 14

4 goes into 14 three times with two remainder.

So it doesn't go in evenly.

Let's try one more.

4 into 212334436

6 + (2 × 3) = 12

4 goes into 12 three times with 0 remainder.

Therefore 4 goes into 234436 evenly.

So what use is this trick to you?

Well if you have learnt the tutorial at Memorymentor.com about telling the day in any year, then you can use it in working out whether the year you are calculating is a leap year or not.

Multiplying by 12 - shortcut

So how does the 12's shortcut work?

Let's take a look.

12 × 7

The first thing is to always multiply the 1 of the twelve by the number we are multiplying by, in this case 7.

So 1 × 7 = 7.

Multiply this 7 by 10 giving 70. (Why? We are working with BASES here.

Bases are the fundamentals to easy calculations for all multiplication tables.

Now multiply the 7 by the 2 of twelve giving 14. Add this to 70 giving 84.

Therefore 7 × 12 = 84

Let's try another:

17 × 12

Remember, multiply the 17 by the 1 in 12 and multiply by 10

(Just add a zero to the end)

1 × 17 = 17, multiplied by 10 giving 170.

Multiply 17 by 2 giving 34.

Add 34 to 170 giving 204.

So 17 × 12 = 204

lets go one more

24 × 12

Multiply 24 × 1 = 24. Multiply by 10 giving 240.

Multiply 24 by 2 = 48. Add to 240 giving us 288

24 × 12 = 288 (these are Seriously Simple Sums to do aren't they?!)

Converting Kilos to Pounds

In this section you will learn how to convert Kilos to Pounds, and Vice Versa.

Let's start off with looking at converting Kilos to pounds.

86 kilos into pounds:

Step one, multiply the kilos by TWO.

To do this, just double the kilos.

86 × 2 = 172

Step two, divide the answer by ten.

To do this, just put a decimal point one place in from the right.

172/10 = 17.2

Step three, add step two's answer to step one's answer.

172 + 17.2 = 189.2

86 Kilos = 189.2 pounds

Let's try:

50 Kilos to pounds:

Step one, multiply the kilos by TWO.

To do this, just double the kilos.

50 × 2 = 100

Step two, divide the answer by ten.

To do this, just put a decimal point one place in from the right.

100/10 = 10

Step three, add step two's answer to step one's answer.

100 + 10 = 110

50 Kilos = 110 pounds

Adding Time

Here is a nice simple way to add hours and minutes together:

Let's add 1 hr and 35 minutes and 3 hr 55 minutes together.

What you do is this: make the 1 hr 35 minutes into one number, which will give us 135 and do the same for the other number, 3 hours 55 minutes, giving us 355

Now you want to add these two numbers together:

135
355
—
490

So we now have a sub total of 490.

What you need to do to this and all sub totals is add the time constant of 40.

No matter what the hours and minutes are, just add the 40 time constant to the sub total.

490 + 40 = 530

So we can now see our answer is 5 hrs and 30 minutes!

Temperature Conversions

This is a shortcut to convert Fahrenheit to Celsius and vice versa.

The answer you will get will not be an exact one, but it will give you an idea of the temperature you are looking at.

Fahrenheit to Celsius:

Take 30 away from the Fahrenheit, and then divide the answer by two.

This is your answer in Celsius.

EXAMPLE:

74 Fahrenheit - 30 = 44. Then divide by two, 22 Celsius.

So 74 Fahrenheit = 22 Celsius.

Celsius to Fahrenheit just do the reverse:

Double it, and then add 30.

30 Celsius double it, is 60, then add 30 is 90

30 Celsius = 90 Fahrenheit

Remember, the answer is not exact but it gives you a rough idea.

Decimals Equivalents of Fractions

With a little practice, it's not hard to recall the decimal equivalents of fractions up to 10/11!

First, there are 3 you should know already:

1/2 = .5

1/3 = .333...

1/4 = .25

Starting with the thirds, of which you already know one:

1/3 = .333...

2/3 = .666...

You also know 2 of the 4ths, as well, so there's only one new one to learn:

1/4 = .25

2/4 = 1/2 = .5

3/4 = .75

Fifths are very easy. Take the numerator (the number on top), double it, and stick a decimal in front of it.

1/5 = .2

2/5 = .4

3/5 = .6

4/5 = .8

There are only two new decimal equivalents to learn with the 6ths:

1/6 = .1666...

2/6 = 1/3 = .333...

3/6 = 1/2 = .5

4/6 = 2/3 = .666...

5/6 = .8333...

What about 7ths? We'll come back to them at the end. They're very unique.

8ths aren't that hard to learn, as they're just smaller steps than 4ths.

If you have trouble with any of the 8ths, find the nearest 4th, and add .125 if needed:

1/8 = .125

2/8 = 1/4 = .25

3/8 = .375

4/8 = 1/2 = .5

5/8 = .625

6/8 = 3/4 = .75

7/8 = .875

9ths are almost too easy:

1/9 = .111...

2/9 = .222...

3/9 = .333...

4/9 = .444...

5/9 = .555...

6/9 = .666...

7/9 = .777...

8/9 = .888...

10ths are very easy, as well. Just put a decimal in front of the numerator:

1/10 = .1

2/10 = .2

3/10 = .3

4/10 = .4

5/10 = .5

6/10 = .6

7/10 = .7

8/10 = .8

9/10 = .9

Remember how easy 9ths were? 11th are easy in a similar way, assuming you know your multiples of 9:

1/11 = .090909...

2/11 = .181818...

3/11 = .272727...

4/11 = .363636...

5/11 = .454545...

6/11 = .545454...

7/11 = .636363...

8/11 = .727272...

9/11 = .818181...

10/11 = .909090...

As long as you can remember the pattern for each fraction, it is quite simple to work out the decimal place as far as you want or need to go!

Oh, I almost forgot! We haven't done 7ths yet, have we?

One-seventh is an interesting number:

1/7 = .142857142857142857...

For now, just think of one-seventh as: .142857

See if you notice any pattern in the 7ths:

1/7 = .142857...

2/7 = .285714...

3/7 = .428571...

4/7 = .571428...

5/7 = .714285...

6/7 = .857142...

Notice that the 6 digits in the 7ths ALWAYS stay in the same order and the starting digit is the only thing that changes!

If you know your multiples of 14 up to 6, it isn't difficult to,br> work out where to begin the decimal number. Look at this:

For 1/7, think "1 * 14", giving us .14 as the starting point.

For 2/7, think "2 * 14", giving us .28 as the starting point.

For 3/7, think "3 * 14", giving us .42 as the starting point.

For 4/14, 5/14 and 6/14, you'll have to adjust upward by 1:

For 4/7, think "(4 * 14) + 1", giving us .57 as the starting point.

For 5/7, think "(5 * 14) + 1", giving us .71 as the starting point.

For 6/7, think "(6 * 14) + 1", giving us .85 as the starting point.

Practice these, and you'll have the decimal equivalents of everything from 1/2 to 10/11 at your finger tips!

If you want to demonstrate this skill to other people, and you know your multiplication tables up to the hundreds for each number 1-9, then give them a calculator and ask for a 2-digit number (3-digit number, if you're up to it!) to be divided by a 1-digit number.

If they give you 96 divided by 7, for example, you can think, "Hmm... the closest multiple of 7 is 91, which is 13 * 7, with 5 left over. So the answer is 13 and 5/7, or: 13.7142857!"

Converting Kilometres to Miles

This is a useful method for when travelling between imperial and metric countries and need to know what kilometres to miles are.

The formula to convert kilometres to miles is number of (kilometres/8) × 5

So lets try 80 kilometres into miles

80/8 = 10

multiplied by 5 is 50 miles!

Another example

40 kilometres

40/8 = 5

5 × 5= 25 miles

Taxila University

Takshashila, (later corrupted as Taxila), one of the topmost centres of education at that time in India became Chanakya's breeding ground of acquiring knowledge in the practical and theoretical aspect. The teachers were highly knowledgeable who used to teach sons of kings. It is said that a certain teacher had 101 students and all of them were princes! The niversity at Taxila was well versed in teaching the subjects using the best of practical knowledge acquired by the teachers. The age of entering the university was sixteen. The branches of studies most sought after in around India ranged from law, medicine, warfare and other indigenous forms of learning. The four Vedas, archery, hunting, elephant-lore and 18 arts were taught at the university of Taxila. So prominent was the place where Chanakya received his education that it goes to show the making of the genius. The very requirements of admission filtered out the outlawed and people with lesser credentials.

At a time when the Dark Ages were looming large, the existence of a university of Taxila's grandeur really makes India stand apart way ahead of the European countries who struggled with ignorance and total information blackout. For the Indian subcontinent Taxila stood as a light house of higher knowledge and pride of India. In the present day world, Taxila is situated in Pakistan at a place called Rawalpindi. The university accommodated more than 10,000 students at a time. The university offered courses spanning a period of more than eight years. The students were admitted after graduating from their own countries. Aspiring students opted for elective subjects going for in depth studies in specialized branches of learning. After graduating from the university, the students are recognized as the best scholars in the subcontinent. It became a cultural heritage as time passed. Taxila was the junction where people of different origins mingled with each other and exchanged knowledge of their countries.

The university was famous as "Taxila" university, named after the city where it was situated. The king and rich people of the region used to donate lavishly for the development of the university. In the religious scriptures also, Taxila is mentioned as the place where the king of snakes, Vasuki selected Taxila for the dissemination of knowledge on earth.

Here it would be essential to mention briefly the range of subjects taught in the university of Taxila. (1) Science, (2) Philosophy, (3) Ayurveda, (4) Grammar of various languages, (5) Mathematics, (6) Economics, (7) Astrology, (8) Geography, (9) Astronomy, (10) Surgical science, (11) Agricultural sciences, (12) Archery and Ancient and Modern Sciences.

The university also used to conduct researches on various subjects.

Mathematics

Zero -The Most Powerful Tool

India invented the Zero, without which there would be no binary system. No computers! Counting would be clumsy and cumbersome! The earliest recorded date, an inscription of Zero on Sankheda Copper Plate was found in Gujarat, India (585-586 CE). In Brahma-Phuta-Siddhanta of Brahmagupta (7th century CE), the Zero is lucidly explained and was rendered into Arabic books around 770 CE. From these it was carried to Europe in the 8th century. However, the concept of Zero is referred to as Shunya in the early Sanskrit texts of the 4th century BCE and clearly explained in Pingala's Sutra of the 2nd century.

Geometry

Invention of Geometry

The word Geometry seems to have emerged from the Indian word 'Gyaamiti' which means measuring the Earth (land). And the word Trigonometry is similar to 'Trikonamiti' meaning measuring triangular forms. Euclid is credited with the invention of Geometry in 300 BCE while the concept of Geometry in India emerged in 1000 BCE, from the practice of making fire altars in square and rectangular shapes. The treatise of Surya Siddhanta (4th century CE) describes amazing details of Trigonometry, which were introduced to Europe 1200 years later in the 16th century by Briggs.

The Value of PI in India

The ratio of the circumference and the diameter of a circle are known as Pi, which gives its value as 3,1428571. The old Sanskrit text Baudhayana Shulba Sutra of the 6th century BCE mentions this ratio as approximately equal to 3. Aryabhatta in 499, CE worked the value of Pi to the fourth decimal place as 3.1416. Centuries later, in 825 CE Arab mathematician Mohammed Ibna Musa says that "This value has been given by the Hindus (Indians)".

Pythagorean Theorem or Baudhayana Theorem?

The so-called Pythagoras Theorem - the square of the hypotenuse of a right-angled triangle equals the sum of the square of the two sides - was worked out earlier in India by Baudhayana in Baudhayana Sulba Sutra. He describes: "The area produced by the diagonal of a rectangle is equal to the sum of the area produced by it on two sides."

[Note: Greek writers attributed the theorem of Euclid to Pythagoras]

The Decimal

100BCE the Decimal system flourished in India

"It was India that gave us the ingenious method of expressing all numbers by means of ten symbols (Decimal System)....a profound and important idea which escaped the genius of Archimedes and Apollonius, two of the greatest men produced by antiquity." -La Place

Raising 10 to the Power of 53

The highest prefix used for raising 10 to a power in today's maths is 'D' for 10 to a power of 30 (from Greek Deca). While, as early as 100 BCE Indian Mathematicians had exact names for figures upto 10 to the power of 53.

ekam =1

dashakam =10

shatam =100 (10 to the power of 10)

sahasram =1000 (10 power of 3)

dashasahasram =10000 (10 power of 4)

lakshaha =100000 (10 power of 5)

dashalakshaha =1000000 (10 power of 6)

kotihi =10000000 (10 power of 7)

ayutam =1000000000 (10 power of 9)

niyutam = (10 power of 11)

kankaram = (10 power of 13)

vivaram = (10 power of 15)

paraardhaha = (10 power of 17)

nivahaaha = (10 power of 19)

utsangaha = (10 power of 21)

bahulam = (10 power of 23)

naagbaalaha = (10 power of 25)

titilambam = (10 power of 27)

vyavasthaana

pragnaptihi = (10 power of 29)

hetuheelam = (10 power of 31)

karahuhu = (10 power of 33)

hetvindreeyam = (10 power of 35)

samaapta lambhaha = (10 power of 37)

gananaagatihi) = (10 power of 39)

niravadyam = (10 power of 41)

mudraabaalam = (10 power of 43)

sarvabaalam = (10 power of 45)

vishamagnagatihi = (10 power of 47)

sarvagnaha = (10 power of 49)

vibhutangamaa = (10 power of 51)

tallaakshanam = (10 power of 53)

(In Anuyogdwaar Sutra written in 100 BCE one numeral is raised as high as 10 to the power of 140).

Astronomy

Indian astronomers have been mapping the skies for 3500 years.

1000 Years Before Copernicus

Copernicus published his theory of the revolution of the Earth in 1543. A thousand years before him, Aryabhatta in 5th century (400-500 CE) stated that the Earth revolves around the sun, "just as a person travelling in a boat feels that the trees on the bank are moving, people on earth feel that the sun is moving". In his treatise Aryabhatteeam, he clearly states that our earth is round, it rotates on its axis, orbits the sun and is suspended in space and explains that lunar and solar eclipses occur by the interplay of the sun, the moon and the earth.

The Law of Gravity - 1200 Years Before Newton

The Law of Gravity was known to the ancient Indian astronomer Bhaskaracharya. In his Surya Siddhanta, he notes:

"Objects fall on earth due to a force of attraction by the earth, therefore, the earth, the planets, constellations, the moon and the sun are held in orbit due to this attraction".

It was not until the late 17th century in 1687, 1200 years later, that Sir Isaac Newton rediscovered the Law of Gravity.

Measurement of Time

In Surya Siddhanta, Bhaskaracharya calculates the time taken for the earth to orbit the sun to 9 decimal places.

Bhaskaracharya = 365.258756484 days.

Modern accepted measurement = 365.2596 days.

Between Bhaskaracharya's ancient measurement 1500 years ago and the modern measurement the difference is only 0.00085 days, only 0.0002%.

34000TH of a Second to 4.32 Billion Years

India has given the idea of the smallest and the largest measure of time.

Krati Krati = 34,000th of a second

1 Truti = 300th of a second

2 Truti = 1 Luv

2 Luv = 1 Kshana

30 Kshana = 1 Vipal

60 Vipal = 1 Pal

60 Pal = 1 Ghadi (24 minutes)

2.5 Gadhi = 1 Hora (1 hour)

24 Hora = 1 Divas (1 day)

7 Divas = 1 saptaah (1 week)

4 Saptaah = 1 Maas (1 month)

2 Maas = 1 Rutu (1 season)

6 Rutu = 1 Varsh (1 year)

100 Varsh = 1 Shataabda (1 century)

10 Shataabda = 1 sahasraabda

432 Sahasraabda = 1 Yug (Kaliyug)

2 Yug = 1 Dwaaparyug

3 Yug = 1 Tretaayug

4 Yug = 1 Krutayug

10 Yug = 1 Mahaayug (4,320,000 years)

1000 Mahaayug = 1 Kalpa

1 Kalpa = 4.32 billion years

Merely 16 Sutras and 13 Sub-Sutra form the basis of Vedic Mathematics. These can be used to solve all types of Mathematical Problems!!!

These sutras long lost in the cryptic Vedas where the mathematical solutions were hidden in Sanskrit Verses for centuries. Sri Bharati Krsna Tirthaji's "Vedic Mathematics" recently discovered these techniques again and given it to all of us to use.

The 16 sutras, or rules, and 14 sub-sutras can be used to solve complex mathematical problems mentally, leaving your fellow students and colleagues in awe.

The 16 Sutras of Vedic Mathematics:

Sutra 1 : By one more than the one before

Sutra 2: All from 9 and the last from 10

Sutra 3 : Vertically and crosswise

Sutra 4 : Transpose and apply

Sutra 5 : If the Samuccaya is the same it is zero

Sutra 6 : If one is in ratio the other is zero

Sutra 7: By addition and by subtraction

Sutra 8: By the completion or non-completion

Sutra 9: Differential calculus

Sutra 10: By the deficiency

Sutra 11: Specific and general

Sutra 12: The remainders by the last digit

Sutra 13: The ultimate and twice the penultimate

Sutra 14: By one less than the one before

Sutra 15: The product of the sum

Sutra 16: All the multipliers

The 13 Sub-sutras:

SS 1: Proportionately

SS 2: The remainder remains constant

SS 3: The first by the first and the last by the last

SS 4: For 7 the multiplicand is 143

SS 5: By osculation

SS 6: Lessen by the deficiency

SS 7: Whatever the deficiency lessen by that amount and set up the square of the deficiency

SS 8: Last totalling 10

SS 9: Only the last terms

SS 10: The sum of the products

SS 11: By alternative elimination and retention

SS 12: By mere observation

SS 13: The product of the sum is the sum of the products

We will work through these Sutras in the next few pages and also will give you lot of practice to get used to the techniques.

Sutra 1: "By one more than the one before" provides a simple way of multiplication of numbers with the same first digit and the sum of their last unit digits is 10.

An interesting application of this formula is in computing squares of numbers ending in five. Consider:

$35 \times 35 = ((3 \times 3) + 3)*10 + 25 = 1225$ or

$= (3 \times 4) *10 +(5 \times 5) = 1225$

$125 \times 125 = (12 \times 13)*10 + 25 = 15625$

EXPLANATION

The latter portion is multiplied by itself (5 by 5) and the previous portion is multiplication of first digit and the digit higher to the first digit resulting in the answer 1225

This is a simple application of $(a+b)^2 = a^2 + 2ab + b^2$ ....

Here $2b = 10$

It can also be applied in multiplications when the last digit is not 5 but the sum of the last digits is the base (10) and the previous parts are the same. Consider:

$37 \times 33 = (3 \times 4)*100 + (7 \times 3) = 1221$

$29 \times 21 = (2 \times 3)*100 + (9 \times 1) = 609$

This uses $(a + b)(a - b) = a^2 - b^2$ twice combined with the previous result to produce:

$$\begin{aligned}(10c + 5 + d)(10c + 5 - d) &= (10c + 5)^2 - d^2 \\ &= 100c(c + 1) + 25 - d^2 \\ &= 100c(c + 1) + (5 + d)(5 - d).\end{aligned}$$

The Sutra is very useful in its application to convert fractions into their equivalent decimal form. Consider fraction 1/19. Using this formula, this can be converted into a decimal form in a single step. This can be done by applying the formula for either a multiplication or division operation, thus yielding two methods.

METHOD 1

Using multiplication. The sutra "one more than the one before" provides a simple way of calculating values like 1/x9 (e.g 1/19, 1/29, etc.). Let's take one 1/x9 and calculate e.g. 1/19. In this case, x=1. To convert 1/19 to decimals, since 19 is not divisible by 2 or 5, the fractional result is a purely circulating (recurring) decimal, i.e the digits would repeat themselves after some time. (If the denominator contains only factors 2 and 5 it is a purely non-circulating decimal i.e it is a perfect decimal, else it is a mixture of the two.).

So to get the decimals of 1/19, we start with the numerator digit i.e 1 in this case and Multiply this by "one more" (1+x), in this case: 1+1 = 2.

21 Multiplying 2 by 2, followed by multiplying 4 by 2

421 → 8421 Now, multiplying 8 by 2, sixteen

68421.1 ← carry

If the result of the multiplication is greater than 10, keep (value - 10) and keep the "1" as "carry over" which you'll add to the next digit.

multiplying 6 by 2 is 12 plus 1 carry gives 13

368421.1 ← carry

Continuing

7368421 → 47368421 → 947368421

Now we have 9 digits of the answer. There are a total of 18 digits (= denominator - numerator i.e 19-1 = 18) in the answer. The last 9 digits can be computed by complementing the lower half (with its complement from nine i.e number + complement = 9):

052631578

947368421

Thus the result is 1/19 = 0.052631578,947368421 repeating.

If you picked up 1/29, you'll have to do it till 28 digits (i.e. 29-1). You'll get the following

1/29 = 03448275862068, 96551724137931

Run this on your calculator and check the result!

I do not like this method a lot as it seems academic. Why would someone want to find all the numbers in the ratio, until he is doing some research. For all practial purposes 4-5 digits of the decimal (from the left) are enough. So the next technique is my favoured method.

METHOD 2

Using Division: The earlier process can also be done using division instead of multiplication.

For A/X9, We divide A by (1+X). Incase of 1/19, we divide 1/(1+1), the answer is 0 (lets say N) with remainder 1 (lets say D)

0.0: D (in this case 1) carries forward and become (D*10 + N) i.e. 10. This is then divided by 2 for N = 5

0.05: Next 05 divided by 2 with answer as 2 and remainder 1.

0.052: Next 12 (1 from earlier remiander and 2 from the answer) is divided by 2 for answer 6

0.0526: and so on.

So the answer of 1/19 = 0.0526.

Consider another example, 1/7, this same as 7/49 which has last digit of the denominator as 9. The previous digit to 9 in 49 is 4. So by the sutra we take one more i.e. 5. So we divide the numerator by 5, that is,

...7/5: 1 R = 2; 21/5: 4 R = 1; 14/5: 2 R = 4, 42/5: 8 R = 2 and so on...so the answer is 1/7 = 0.1428

Now try to convert 18/19 into decimals.

(Numerator 18 to be divided by x+1=2)

...18/2:9 R = 0 ⇒ 9/2:4 R:1 ⇒ 14/2: 7 R = 0 ⇒ 7/2: 3 R = 1 and so on.

So the answer for 18/19 is 0.9473...

Now try 18/29.

Can you do it mentally? The answer is 0.6206...

Note this technique can be used for conversion of vulgar fractions ending in 1, 3, 7 including 9 such as 1/11, 1/21, 1/31 - - -- ,1/13, 1/23, - - - -, 1/7, 1/17, - - - - - by multiplying appropriate factors to the numerator and denominator to make them ed with 9.

1/11 = 9/99 → 0.09090909

1/13 = 3/39 → 0.07692307

Amazed at the speed at which you can calculate these sums MENTALLY!!!

People say there are so many small things that we have to remember, instead of one general method of division or multiplication. My answer is once you get used to it the effort in remembering such an useful method is far surpassed by the ease and speed.

Note: 1/17 = 7/119 → This technique becomes a bit difficult when you try to solve with 1/x9, where x is greater than 9. Will discuss other techniques for these type of big fractions.

Sutra 2: "All from nine and the last from 10" comes handy when we wish to subtract a number such as 578 from a number having zeros equivalent or more than the number of number being subtracted e.g. 1,000. Each figure in 578 is subtracted from nine and the last figure is subtracted from 10, yielding 422. The interesting thing is that this problem can be solved from left to right the same way we write and read text.

This sutra can be extended to solve problems such as 60,000 minus 987. We simply reduce the number left after taking out the 0s to account for the number being subtracted (i.e 60 left after removing the 3 left 0s) by 1 and then apply the sutra, to get the answer 59,013.

The concept gives rise to Sub-Sutra 7: "Whatever the extent of its deficiency, lessen it still further to that very extent; and also set up the square of that deficiency". Let me explain this further:

For instance, in computing 92 we go through the following steps:

Let us take 10 as our base. To get 9 to the base 10 we have to subtract 1 from 10.

Since decrease 9 further by 1 to 8. This is the left side of our answer.

On the right hand side put the square of the deficiency, which is 12. Hence the answer is 81.

Similarly, $8^2 \rightarrow (8\text{-}2)*10 + 2^2 \rightarrow 64$, $7^2 = 49$.

For numbers above 10, we take the surplus. For example:

$11^2 = (11+1)*10 + 1^2 = 121$

$13^2 = (13+3)*10 + 3^2 = 169$

and so on.

This is based on the fact that $a^2 = (a + b)(a - b) + b^2$ where a is the number whose square is to be found and b is the deficit (or surplus) from nearest factor of 10.

Sutra 3: "Vertically and crosswise" has many uses. One very useful application is helping children who are having trouble with their tables above 5 × 5. For example 7 × 8. 7 is 3 below the base of 10, and 8 is 2 below the base of 10. i.e

7 * 8

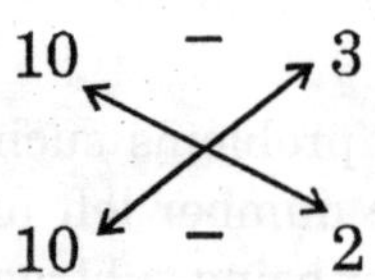

= (10-3) * (10 -2)

= 100 - 20 -30 + 6 = 56

The sutra "vertically and crosswise" is often used in long multiplication. Suppose we wish to multiply 32 by 46.

We multiply "vertically and crosswise":

Vertically 2 × 6=12.

so put down 2 and carry 1.

Then we multiply crosswise and add the two results: 3×6 + 4×2=26,

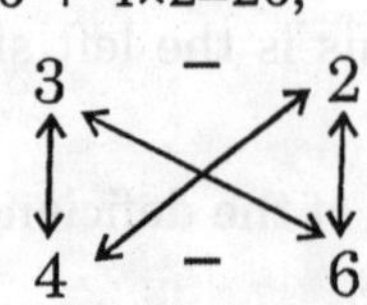

Add the carried 1 to make it 27

so put down 7 and carry 2.

Finally we multiply vertically 3×4=12 and

add the carried 2 =14.

Result: 1,472.

We can extend this method to deal with long multiplication of numbers of any size. The great advantage of this system is that the answer can be obtained in one line and mentally. With some practice, one can do pretty long multiplications in their heads.

Can you now do some quick calculations for multiplications with 11?

What is 28 * 11? What is 375 *11?

The answer for this is very simple:

28 * 11 = 308

Simply write the 2,(2+8),8 → 2,(10),8 (1 from 10 has to be taken forward and added to 2) so answer is 308

The magic is simplified when you break 11 to (10+1): 28 * 11 = 28 * (10+1) = 280 +28 → 2 (8+2) 8 → 308

375 * 11 = 4125

What is happening here? 375 * 11 → 3,(3+7),(7+5),5 → 3,(10),(12),5 → 4125

This Sutra is used to solve complex non-linear partial differential equations. It deals with calculation of common functions and their series expansions, and the solution of simultaneous, algebraic, transcendental and differential equations.

Multiplication can also be carried out starting from the left, which can be better because we write and pronounce numbers from left to right. Here is an example of doing this in a special method for long multiplication of numbers near a base (10, 100, 1,000 etc.), for example, 96 by 92. 96 is 4 below the base and 92 is 8 below. 100-4= 96 or 100-8=92. So 10000 - 1200 + 32. Applying the Sutra 2 for the calculation 10000 - 1168 = 8832.

Try this 86 * 88 = (90-4) * (90 - 2) = 8100 - 540 +8 (To get to Sutra 2, 100-540+8 is done first) = 8000-432 = (Sutra 2) 7568.

This works equally well for numbers above the base: 105x111=11,655. Here we add the differences. For 205×211=43,255, we double the first part of the answer, because 200 is 2×100.

Sutra 5: If the Samuccaya is the same it is zero.

This formula is useful in solution of several special types of equations that can be solved visually. The word samuccaya has various meanings in different applications. For instance, it may mean a term which occurs as a common factor in all the terms concerned. A simple example is equation "12x + 3x = 4x + 5x". Since "x" occurs as a common factor in all the terms, therefore, x = 0 is a solution. Another meaning may be that samuccaya is a product of independent terms. For instance, in (x + 7) (x + 9) = (x + 3) (x + 21), the samuccaya is 7 × 9 = 3 × 21, therefore, x = 0 is a solution. Another meaning is the sum of the denominators of two fractions having the same numerical numerator, for example: 1/(2x - 1) + 1/(3x - 1) = 0 means we may set the denominators equal to zero, 5x - 2 = 0.

Yet another meaning is "combination" or total. This is commonly used. For instance, if the sum of the numerators and the sum of denominators are the same then that sum is zero. Therefore,

$$\frac{2x+9}{2x+7} = \frac{2x+7}{2x+9}$$

Therefore, 4x + 16 = 0 or x = -4.

This meaning ("total") can also be applied in solving quadratic equations. The total meaning can not only imply sum but also subtraction. For instance when given $N_1/D_1 = N_2/D_2$, if $N_1 + N_2 = D_1 + D_2$ (as shown earlier) then this sum is zero. Mental cross multiplication reveals that the resulting equation is quadratic (the coefficients of x2 are different on the two sides). So, if $N_1 - D_1 = N_2 - D_2$ then that samuccaya is also zero. This yields the other root of a quadratic equation.

Yet interpretation of "total" is applied in multi-term RHS and LHS. For instance, consider

$$\frac{1}{x-7}+\frac{1}{x-9}=\frac{1}{x-6}+\frac{1}{x-10}.$$

Here $D_1 + D_2 = D_3 + D_4 = 2x - 16$. Thus $x = 8$.

There are several other cases where samuccaya can be applied with great versatility. For instance "apparently cubic" or "biquadratic" equations can be easily solved as shown below:

$(x - 3)^3 + (x - 9)^3 = 2(x - 6)^3$.

Note that $x - 3 + x - 9 = 2 (x - 6)$.

Therefore $(x - 6) = 0$ or $x = 6$.

This would not work for the apparently quadratic $(x - 3)^2 + (x - 9)^2 = 2(x - 6)^2$, which has no real or complex solutions.

Consider

$$\frac{(x+3)^3}{(x+5)^3}=\frac{x+1}{x+7}.$$

Observe: $N_1 + D_1 = N_2 + D_2 = 2x + 8$.

Therefore, $x = -4$.

This formula has been extended further.

Sutra 6: If one is in ratio the other is zero is often used to solve simultaneous linear equations which may involve big numbers. But these equations in special cases can be visually solved because of a certain ratio between the coefficients.

Consider the following example:

$6x + 7y = 8$

$19x + 14y = 16$

Here the ratio of coefficients of y is same as that of the constant terms. Therefore, the "other" variable is zero, i.e., $x = 0$. Hence, mentally, the solution of the equations is $x = 0$ and $y = 8/7$

(alternatively:

$19x + 14y = 16$ is equivalent to:

$(19/2)x + 7y = 8$.

Thus it is obvious that x has to be zero, no ratio needed, just divide by 2!

Note that it would not work if both had been "in ratio". For then we have the case of coinciding lines with an infinite number of solutions.:

$6x + 7y = 8$

$12x + 14y = 16$

This formula is easily applicable to more general cases with any number of variables. For instance

$ax + by + cz = a$

$bx + cy + az = b$

$cx + ay + bz = c$

which yields $x = 1, y = 0, z = 0$.

A corollary says solving "by addition and by subtraction." It is applicable in case of simultaneous linear equations where the x- and y-coefficients are interchanged.

For instance:

$45x - 23y = 113$

$23x - 45y = 91$

By addition:

$68x - 68y = 204 \Rightarrow 68(x - y) = 204 \Rightarrow x - y = 3.$

By subtraction:

$22x + 22y = 22 \Rightarrow 22(x + y) = 22 \Rightarrow x + y = 1.$

Again, by addition, we eliminate the y-terms:

$2x = 4$, so $x = 2$.

Or, by subtraction, we eliminate the x-terms:

$-2y = 2$, and so $y = -1$.

The solution set is $\{2,-1\}$.

7

Is Consciousness a Number?

Introduction

What could be simpler than a number? A child typically learns to count almost as soon as he learns to speak. What could be more essential, more ubiquitous than numbers? We use them constantly, often subconsciously. Yet, as with so many things that are simple, essential, and ubiquitous, not many of us stop to think about what a number really is. Among those that do stop to think about it, there is widespread disagreement. What does the current writer hope to add to this situation? He wishes to describe an answer that, for him at least, works spectacularly well.

This answer makes use of the revival of a remarkably effective and comprehensive science of consciousness, Maharishi Vedic Science. This science is based on the experience of pure consciousness through the Transcendental Meditation program, and the understanding of this experience and its integration into daily life. One branch of Maharishi Vedic Science is Maharishi Vedic Mathematics. This branch describes consciousness and its development from a mathematical point of view. A careful examination of one important concept of Maharishi Vedic Mathematics can achieve a breakthrough understanding of the role of number in mathematics and consciousness. This examination will be one of the primary goals of this paper.

This understanding will also enable us to resolve numerous problems that are currently plaguing the philosophy of mathematics. The primary problem in this area is determining the nature of mathematical objects. Until recently, mathematicians have usually taken the point of view that mathematical objects are transcendental, beyond all objects of the senses. But the current generation of mathematicians usually finds themselves unable to defend this position. Maharishi Vedic Mathematics gives us the tools to properly evaluate the nature of mathematical objects, both theoretically and experientially. We will examine this outcome in detail.

Another problem that Maharishi Vedic Mathematics enables us to solve is one whose existence is not widely appreciated at present. This is the assumption that set theory is a foundational theory of mathematics. We will locate the problem in set theory and then see how Maharishi Vedic Mathematics gives us a way to solve it.

Maharishi Vedic Science

It must be obvious that knowledge is based in consciousness. Our knowledge of anything is based on the degree of development of our consciousness. The past few decades have witnessed the rise of a remarkably effective and comprehensive science of consciousness, Maharishi Vedic Science. This science has its roots in the ancient Vedic culture of India, which enjoyed a remarkably high level of civilization. Only remnants of this civilization are available today in modern India.

The modern revival of Maharishi Vedic Science is based on the experience of pure consciousness through the Transcendental Meditation program, the development of this experience through other techniques that originate in ancient Vedic culture, and the understanding of this experience and its integration into daily life. Over 600 studies have demonstrated the effectiveness of the Transcendental Meditation technique in many areas. Maharishi Vedic Science has been established by a remarkable individual, Maharishi Mahesh Yogi, who started his worldwide Transcendental Meditation Movement in 1957.

Maharishi Vedic Mathematics

One branch of Maharishi Vedic Science is Maharishi Vedic Mathematics. This branch describes consciousness and its development from a mathematical point of view, and prescribes, among other things, the Transcendental Meditation technique as an "enjoyable exercise of Vedic Mathematics" (Maharishi 1996, p. 441).

Unity is an important concept in Maharishi Vedic Mathematics. Unity, in both common usage and among mathematicians, is simply the number one. In Maharishi Vedic Mathematics, Unity rises to the status of self-reference. The following diagrams help us to understand this.

When we perceive a cup, an apple, a child, a sentence, a song, or any such object, we are implicitly counting to One, because we are perceiving one such object. If we perceive more than one, we can consider each one individually. We are therefore characterizing each object by the number one. By counting to one, we are attributing the quality of the number one to an object.

But all aspects of the object can be so attributed. What, then, about the object cannot be attributed to oneness? Clearly none. Therefore, from the viewpoint of complete consideration of everything about the object, the number one contains everything about every object. The status of one rises from a derivative attribute to the source of all objects of consideration. Our appreciation of the number one is then reversed.

Now we will see that Unity is both a number and self-referral consciousness. In order for anything to exist, it must be possible to count it, and the first counting number is One. Then we can use other numbers to measure other aspects of the object. Numbers themselves can be counted. As with anything else that is counted, when we count numbers, we start with the number One. When the number One counts itself, or refers to itself, it thereby assumes a role of self-reference.

In order for Unity to rise to self-reference, it cannot count in an explicit way, because to do so would fill up the universe with nothing but the activity of counting. Therefore, One counting itself in a self-referral way must be on the level of knowingness, consciousness without activity, pure silence.

The Absolute Number is this unmanifest level of number. In this writer's opinion, this is what Maharishi is referring to when he says that the number one, when it is zeroed, becomes the Absolute Number: As any number is the modified state of the number one, Unity; and as Unity is an eternal continuum, the number one circled 1 expresses the eternal continuum of Unity, which can only be expressed in terms of zero-in terms of a circle. (Maharishi 1996, p. 614)

As any number is the expression of the number one, when the number one is zeroed it becomes unmanifest, which means in terms of numbers it becomes the Absolute Number. This means that any number zeroed transforms itself into the Absolute Number-any number zeroed becomes unmanifest. (Maharishi 1996, p. 614)

The Absolute Number is a field of pure self-referral consciousness, which does not enter into the field of distinctions but maintains itself as pure consciousness, without activity or distinctions.

We can deduce, therefore, that Unity is both a number and a field of self-referral consciousness. In the following passage, Maharishi verifies this by essentially equating the Absolute Number with the self-referral nature of Unity.

Vedic Mathematics is the mathematics of the Absolute Number, which is the reality of the absolute field (the field of Unity, the field of self-referral consciousness, the field of Samahita, the field of the Veda). (Maharishi 1996, p. 371)

Solution of Problems in the Foundations of Mathematics

Modern mathematics has drifted away from its ancient foundations in arithmetic and geometry. Almost all modern mathematicians take at least a nominal view that set theory is the foundation of modern mathematics. Yet set theory attracts very little research interest from professional mathematicians, and a problem or advance in set theory is usually viewed as having little or no bearing on the rest of mathematics. Maharishi Vedic Mathematics sheds much light on this peculiar situation. We will undertake a simple examination of set theory to show how.

A simple definition of a set is that it is any well-described collection. Each object that belongs to the collection is called an element of the set. We can speak of the set of all the nations of Europe, or the set of all points on a line. The set of the nations of Europe has a finite number of elements, and so it is called a finite set; while the set of all points on a line is an infinite set, since there are an infinite number of points on a line. A set may have no elements at all; there is only one such set, and it is called the empty set. Sets are denoted with braces: the set containing the numbers 0, 1, and 2 is denoted {0, 1, 2}. The empty set is denoted { }. Sets can contain other sets; thus { {} } is the set containing the empty set, which is a different set from just the empty set by itself.

The set theory that the mathematician uses is called pure set theory. In pure set theory, all the elements of a set are other sets. The construction of all such sets starts with the empty set. This allows only such sets as { }, or { {} }, or

{ {}, {{}} }, or an infinite set that contains only other sets. The whole numbers are defined as pure sets in the following way: the number 0 is defined to be the empty set { }; the number 1 is defined to be the set containing the empty set, i.e. the set containing 0, i.e. { {} }; the number 2 is defined as the set containing 0 and 1, i.e. { {}, {{}} }; 3 is defined as the set containing the 0, 1, and 2; and so forth. All mathematical structures are defined by modern mathematics in this way, starting from the empty set.

Pure set theory has an identifiable starting point in the empty set, and a kind of self-reference in that its sets admit only other sets as elements. Nevertheless, it fails to be a true foundational theory, for several reasons:

1. The inside and outside of a set are two distinct realms, and so the concept of a set is based on the assumption of a fundamental duality.
2. Sets are governed by the axioms of set theory, which are not regarded as themselves being sets. Thus the governance of set theory is based on another assumption of a fundamental duality.
3. Since sets may be infinite but the number of axioms of set theory is finite, set theory attempts to approach the infinite using only finite intelligence-it approaches the infinite from a finite point of view.
4. Set theory cannot fully account for self-reference. When the empty set includes itself, it becomes something different from the empty set. Set theory therefore cannot account for the simple ability of the number One to count itself while remaining One. Likewise, set theory cannot account for the self-referral counting ability of any number.
5. Because set theory cannot account for counting, it fails to account for any of the laws of arithmetic. Since arithmetic is fundamental to any branch of

mathematics, set theory is unable to fully account for the laws of any branch of mathematics, even its own laws.

Maharishi Vedic Mathematics rectifies these shortcomings and thus becomes a much better foundational theory for all of mathematics:

1. Numbers, including the Absolute Number, have no inside or outside. They simply count, and that counting may be purely self-referential. There is no fundamental duality in the structure of a number.
2. The laws which govern the Absolute Number and the relative numbers are included in the structure of the Absolute Number itself.
3. Maharishi Vedic Mathematics approaches the infinite from its own level-infinite intelligence. This infinite level of intelligence is accessible to everyone.
4. The number One does not change when it counts itself. This is an unchanging or absolute quality of Unity, which is due to the Absolute Number's support of every number.
5. The Absolute Number is the common source of all the numbers, and so it encompasses both Unity and diversity. Because all the laws of nature are based on numbers, the Absolute Number is also the home of all the laws of nature.

Set theory does provides a framework for associating any mathematical structure with a set. But a framework is not necessarily a foundation, and an association is not necessarily a definition. Those who claimed that set theory was the foundation of arithmetic never proved it and never could have, because they never had any independent theory of arithmetic. Maharishi Vedic Mathematics rectifies this by bringing to light the self-referral nature of number, both in theory and in practice.

The emphasis of Maharishi Vedic Mathematics on the self-referral nature of number finds confirmation in one of the most famous theorems in the foundations of mathematics, G¨odel's Incompleteness Theorem, often simply called G¨odel's Theorem. The discovery of this theorem in 1930 by Kurt G¨odel (1906-1978) made him one of the most famous mathematicians of the twentieth century. There are many ways to state G¨odel's theorem, some of them very technical, but perhaps one of the easiest is to paraphrase Gensler (1984, p. 1): G¨odel's Theorem states that a complete knowledge of arithmetic cannot be reduced to a finite set of rules. G¨odel proved this theorem about arithmetic by using arithmetic. He used the ability of numbers to count themselves and anything else by:

1. transforming the language of his theorem into numbers;
2. constructing a self-referential statement in this language;
3. using self-referral properties of numbers to show that this statement is true;
4. using the same properties to show that no finite set of rules can prove this statement.

Solution of Problems in Philosophy of Mathematics

Among the Western disciplines, mathematics affords us with probably the clearest example of the transcendental quality of knowledge and intelligence. This position is well known in Western philosophy as Platonism. The recent revival of interest in Western philosophy has sparked a widespread interest in the philosophy of mathematics, among mathematicians and non-mathematicians alike, and with this has come a renewed examination of Platonism.

Unfortunately, even though mathematicians in the past have generally defended Platonism, almost none of the current generation of mathematicians does so. A recent well-known work in the philosophy of mathematics by Hersh (1997) highlights this situation as he throws out Platonism and develops his own theory, called humanism, which takes the view that mathematics is solely a social activity. We will now examine this and other views of Hersh and see how their shortcomings can easily be rectified by Maharishi Vedic Mathematics.

Platonism . . . says that mathematical entities exist outside space and time, outside thought and matter, in an abstract realm independent of any consciousness, individual or social. (Hersh, 1997, p. 9)

Why do mathematicians believe something so unscientific, so farfetched as an independent immaterial timeless world of mathematical truth? (p. 11)

We don't ask, How does this immaterial realm relate to material reality? How does it make contact with flesh and blood mathematicians? We refuse to face this embarrassment. . . . (p. 11)

To answer, you have to forget Platonism, and look in the sociocultural past and present. . . . (p. 12)

Mathematical objects are created by humans. Not arbitrarily, but from activity with existing mathematical objects, and from the needs of science and daily life. (p. 16)

1. Mathematics is human. . . .

2. Mathematical knowledge isn't fallible. . . .

3. There are different versions of proof and rigor. . . .

4. Mathematical object are a distinct variety of social-historical objects. (p. 22)

The kind of Platonism that Hersh describes is in some ways like the reality described by Maharishi Vedic Mathematics, except that Hersh's Platonism prescribes no method for contact between the abstract, timeless realm and individual consciousness. Maharishi Vedic Mathematics gives many procedures, especially the Transcendental Meditation technique, for enabling the individual to contact the transcendental realm of pure consciousness and pure orderliness beyond the activity of the senses and the mind. A realm beyond all activity, containing nothing but the structures of pure orderliness, is thereby raised from the shrouds of mysticism to a simple matter of experience. This field is not created by humans, but it can be fully experienced by anyone.

Providing both techniques and proper understanding to support the techniques makes Maharishi Vedic Mathematics itself a science. Since this realm of pure orderliness is also the realm of pure consciousness, the Atma or Self of everyone, it develops individual awareness, and it results in more alertness and increased ability to function optimally in every way. In this way, Maharishi Vedic Mathematics supports all the other sciences.

Another important point in relation to modern mathematics is that the experience of pure consciousness which Maharishi Vedic Mathematics supplies is beyond the intellect. One can say that the experience of pure consciousness is knowledge, but it must then be said that this knowledge is different from intellectual knowledge. The philosophy of Maharishi Vedic Mathematics finds confirmation from Kurt G¨odel, whose philosophical views are much less well known than his famous theorem. Considering his status as a famous research mathematician, they are quite illuminating. Consciousness is connected with one unity. A machine is composed of parts. (Wang 1996, quotation 6.1.21)

It is a primitive idea of our thinking to think of many objects as one object. (Wang 1996, quotation 8.2.1)

A set is a unity of which its elements are the constituents. It is a fundamental property of the mind to comprehend multitudes into unities. (Wang 1996, quotation 8.2.2)

It is surprising that the fact that multitudes are also unities leads to no contradictions; this is the main fact of mathematics. (Wang 1996, quotation 8.2.2)

Numbers appear less concrete than sets. They have different representations and are what is common to all representations. For example, we add or multiply by dealing with a collections of two indeterminate things. (Wang 1996, quotation 8.2.5)

Another mathematician who confirms the philosophy of Maharishi Vedic Mathematics is Franklin Merrell-Wolff (1887-1985). He was a professor of mathematics at Stanford. Early in his academic career, in 1916, he left Stanford, spent twenty years in solitude studying Vedic texts, had several realizations of consciousness which he said confirmed certain traditional Vedic teachings (those of Shankara and Advaita Vedanta), and wrote several books describing these realizations and how they fulfilled Western philosophy and science. In the following passage, he describes his experience of Consciousness as Unity, which is essentially different from the duality in which knower and known are separate.

Once one has had a glimpse of the Consciousness which I have called the High Indifference, it readily becomes clear that there can be but one ultimate Element. Spirit is not divided from matter in any final sense. This is simply the truth concerning the subject and the object, given in other terms. In that State where I realize Myself as identical with all space and with all objects, there is no division between and "I" and a "not-I." This is equivalent to saying there\ is but one Element. But the same thing cannot be said concerning the seeming relationships within the shell of the subject-object consciousness. For here the subject has the appearing of being different from the object, but these two stand in some relationship to each other. (Merrell-Wolff 1994, p. 144)

Exercises of Maharishi Vedic Mathematics

Maharishi Vedic Mathematics is not merely a philosophy. It encompasses a range of practices which develop self-referral consciousness. In the following quotation, Maharishi refers to these as "exercises" and includes among them the Transcendental Meditation and TM-Sidhi programs.

> Exercises in Vedic Mathematics lie in fathoming and stabilizing selfreferral consciousness through my Transcendental Meditation and TM-Sidhi Programme, and studying different aspects of the Vedic Literature in the light of my Vedic Science and Technology, in the light of self-referral consciousness, because the study of self-referral consciousness is the study of invincible, perfect defence. (Maharishi 1996, p. 350)

The Transcendental Meditation technique is a simple, natural, effortless technique which takes the mind from the surface level of thinking to the source of thought, pure consciousness, consciousness in its self-referral state, a state of restful alertness which is also the source of all creativity. As the mind settles down during this technique, both mind and body experience deep rest, and stress and fatigue are released. The result of this experience is increased creativity and intelligence, which naturally lead to increased efficiency, enjoyment, and success in all aspects of life.

The marvel of Transcendental Meditation is that it increases orderliness in the individual's awareness while expanding the awareness to encompass all space and time.

Maharishi explains this as follows:

> The precision and order in the practice of Transcendental Meditation makes the procedure systematic and mathematical. Transcendental Meditation utilizes the holistic value of Natural Law to materialize an all-directional, progressive and enriching effect in all fields of space and time. It is this theme of Vedic Mathematics available through

Transcendental Meditation that promotes absolute order from the common basis of all activity in the universe-the selfreferral, unified field of consciousness. (Maharishi 1995, p. 173)

Transcendental Meditation is the programme of Vedic Mathematics because it enlivens and utilizes the holistic value of Natural Law to materialize an all-directional effect in all fields of space and time.

It is this theme of Vedic Mathematics available through Transcendental Meditation that promotes absolute order from the common basis of all activity in the universe-the self-referral, unified field of consciousness. (Maharishi 1996, p. 441)

The TM-Sidhi program is an advanced program based on the Transcendental Meditation program. The TM-Sidhi program develops the ability to function directly from the level of pure consciousness. This has the effect of dramatically accelerating the growth of mind-body coordination, by developing the ability to think and act from the level where thought and action are most powerful. The TM-Sidhi program is an application of Yoga, one of the branches of Vedic Literature. Other branches of Vedic Literature contain other programs for developing pure consciousness and gaining the support of the laws of nature. These programs also belong to Maharishi Vedic Mathematics, as Maharishi explains: Exercises in Vedic Mathematics are the programmes to actually gain custody and command over the Laws of Nature-the programmes detailed in all the twenty-seven areas of the Vedic Literature. (Maharishi 1996, p. 348)

Subjective research verifying the approach of Maharishi Vedic Mathematics The regular practice of the Transcendental Meditation and TM-Sidhi programs and other exercises of Maharishi Vedic Mathematics naturally results in higher states of consciousness, beyond the ordinary waking, dreaming, and sleep states of consciousness. The TM technique gives the experience of Transcendental Consciousness, a state of silent restful alertness, a state of pure self-referral without any mental activity. As the practice progresses, the silence of Transcendental Consciousness becomes integrated with the activity of daily life, until they are experienced simultaneously. This is the state of Cosmic Consciousness, where activity is present with and fully supported by the silence of the pure value of the Self, or Atma. We here quote from Maharishi (1976), who quotes experiences of various individuals identified only by their initials and state of residence. The following experience was recorded by S.B. of California:

Boundless infinitude, beautiful bliss, total silence. In activity a powerful silent wholeness rests on the surface of everything. Abeautiful softness connects and interfuses all I see. (Maharishi 1976, p. 77)

As an individual continues to experience Cosmic Consciousness, he begins to perceive finer values, until he is able to perceive the finest value of creation. This state is known as God Consciousness. The following experience was recorded by D. S. of California:

I am perceiving finer and finer values of bliss and beauty, and feeling part of a whole beautiful plan. How small the individuality is, yet how powerful with the growing value of all-encompassing wholeness. The body is perceived as a tiny boundary which the Self flows in and out of. (Maharishi 1976, p. 83)

Continued experience of refined perception in God Consciousness ultimately leads to the ability to perceive any object of attention in terms of the unbounded Self. This state is known as Unity Consciousness. The following experience was recorded by M. H. of California:

The continuum of Self within myself and outside of me just seems to have a very enjoyable, lively, intimate quality-as if my Self is smiling and radiating everywhere. For the first time I feel concretely, tangibly clearer and closer to all that is meant by God, to all that He is. (Maharishi 1976, p. 85)

The following is another such experience, recorded by C.C. of British Columbia: In the Unity [experience], I would see the object and know it to be myself. A few examples are: I saw the stars, planets, and galaxies of the universe spread out before me and instantly knew them to be part of myself-the same sense of recognition one experiences when one sees unexpectedly one's own image reflected in a mirror.

I saw a series of people's faces . . . and knew intuitively that 'I am all that'-I saw all Creation . . . and knew it to be existing inside myself. (Maharishi 1976, p. 79)

The present writer has had many satisfying experiences with various exercises of Maharishi Vedic Mathematics. One of these exercises is the examination of a body of knowledge in the light of a certain verse of R.

Veda, the first and most important branch of Veda and Vedic Literature in Maharishi Vedic Science. This verse is called the R.

The verses of the Ved exist in the collapse of fullness in the transcendental field, self-referral consciousness, the Self, in which reside all the devas, the impulses of creative intelligence, the laws of nature, responsible for the whole manifest universe.

He whose awareness is not open to this field, what can the verses of the Ved accomplish for him? Those who know this level of reality are established in evenness, wholeness of life. (R. k Veda 1.164.39)

The exercise calls for the analysis of any area of modern knowledge in the light of the eight subsections of this verse. I chose several topics in elementary arithmetic and did this analysis. Almost immediately I was aware that number has the property of self-reference. Shortly thereafter, I was sitting with my eyes closed, enjoying Transcendental Consciousness, when I saw the field of intelligence before me as a glowing plane.

I understood that this field of intelligence is identical with my own self-referral state of consciousness. On this glowing field, a circle of light traced itself out, starting at the bottom, proceeding clockwise, and ending up again at the bottom.

I understood this circle to be a circle of intelligence. As the first half of the circle was tracing itself out on the left side, I perceived that certain points on the left side of the circle were associated with certain areas of mathematics. When the tracing reached the top of the circle, I understood that the left side is what we have known so far, and the right side is what we will know.

I believe that "we" refers to human beings in general. As the tracing proceeded down the right side of the circle, both the circle and the plane became increasingly brighter. When the tracing reached the bottom, the circle and the plane were so bright that the circle became lost on the background of the plane, and the plane merged into my own state of self-referral consciousness.

Objective research verifying the approach of Maharishi Vedic Mathematics Objective scientific research verifies that Maharishi's Transcendental Meditation technique takes one from a state of duality toward unity. For example, a study by Haynes et al. (1976) showed that, on average, the more one experiences Transcendental Consciousness, the more one experiences the following:

- high scores on a standard creativity test;
- high synchrony in brain waves between the left and right hemispheres of the brain; and
- shorter recovery time in the muscles and nervous system after a receiving a disturbing stimulus.

Another study by Orme-Johnson et al. (1981) found that the TM-Sidhi program produces larger increases in brain wave coherence and creativity than the TM program alone. Together, these studies show the TM and TM-Sidhi programs result in the growth of unified values within the individual and integration of mind and body.

A study by Kory and Hufnagel (1974) examined the effect of practicing the Transcendental Meditation technique on high school students at three different high schools. At each school, the students practicing the TM technique were matched against students studying psychology. The study compared grade point average and anxiety using standard tests before and during the experimental period. In two out of three schools, there was a significant reduction in anxiety and a significant increase in grade point average among the TM students, while the psychology students showed no significant changes.

Additional objective verification of the growth of Unity through Maharishi's Transcendental Meditation technique is in the finding of increased self-actualization, which is the use of the full value of one's inner potential, including stability, integration, maturity, warmth, and adaptability. Two studies (Seeman et al. 1972, and Nidich et al. 1973) showed that those who practice the Transcendental Meditation program for only two months scored significantly higher on a standard measure of self-actualization, compared to those who do not practice TM.

Another remarkable phenomenon is that this effect spreads from the individual to the society. Numerous studies have confirmed that a group in which the participants collectively practice the TM-Sidhi Program and the number of participants is at least the square root of one percent of the population of a city, state, or country, has the effect of significantly increasing positive social trends and decreasing negative ones in that area.

For example, Dillbeck et al. (1987) showed that when such groups were present in: Delhi, India; Puerto Rico; Manila, Philippines; and Rhode Island, USA; such measures as crime rate, accident rate, alcohol sales, cigarette sales, unemployment rate, and even pollution level, all decreased significantly.

Another study by Dillbeck (1990) showed that when a long-term group was present in Fairfield, Iowa, supplemented by short-term groups in: Washington, D.C.; the Hague, Netherlands; and Amherst, Massachusetts; then a "violence index" consisting of auto accidents, homicides, and suicides decreased significantly.

Both of these papers ascribe these results to a field effect, a term borrowed from physics to indicate a factor which is present throughout space and time, itself unseen but producing measurable effects.

Summary and Conclusion

Modern society is based on science, and scientists regards mathematics as the ideal science. Virtually everyone studies modern mathematics for eight to twelve years at a young age. Yet this has not been enough to keep modern society from being riddled with problems.

Modern mathematics attempts to approach the infinite from the level of the finite, and other modern disciplines imitate this approach. We have seen that Maharishi Vedic Mathematics allows us to approach the infinite from its own level, and that this approach is the basis of Maharishi Vedic Science.

We have seen that Maharishi Vedic Mathematics is easy, enjoyable, natural, profound, all-encompassing, and useful. We have seen how Maharishi Vedic Mathematics has its roots in the self-referral nature of pure consciousness, which is the source of thought and the source of all creation. We have seen that this field of self-referral consciousness is both perfectly orderly and perfectly awake. We have seen that Maharishi Vedic Mathematics can easily solve problems currently plaguing the foundations and philosophy of mathematics. We have examined the exercises of Maharishi Vedic Mathematics, and we have seen subjective and objective results of these exercises.

It should be obvious that, if there is any truth to all this, Maharishi Vedic Mathematics should be a part of the educational curriculum at all levels as a supplement to modern mathematics. It only remains for everyone to discover for themselves that the science and technology of Maharishi Vedic Mathematics can fill our lives with truth, knowledge, and love.

8

Multi Dimensions of Time and Space

Cubic Model of Time and Space

The cubic model of time and space is E2×E3 space time model. The spatial boundary in the role of dimension manifests real 4-space and the solid domain in the role of dimension manifests real 5-space. This spacetime model works out smooth transition and continuum from real 4-space to real 5-space by attaining transcendence from the manifestation layers. This precisely is being availed in the organizational of Manasara.

The scripture opens with the prayer of Lord Brahma, the overlord of real 4-space and ends with the chiselling of third eye of Lord Shiva, the overlord of real 5-space. Precisely, this enlightenment of the transcendence phenomenon of real 5-space is the subject matter of instructions of Sri Sri Shiva Mahapuran, the text book of real 5-space. Therefore, the cubic model of spacetime is the basic spacetime model for working out the geometric continuum from manifestations layers on the format of real 4-space to the transcendence phenomenon on the format of real 5-space.

Spherical Model of Time and Space

Like cube, sphere is another representative regular body of 3-space. Therefore, parallel to the cubic model of the spacetime, we shall be having spherical model of spacetime. Here as well essentially the Cartesian expression for the spacetime frame would be $E^2 \times E^3$ as the surface boundary of the sphere shall be manifesting hypercube-4 while the solid domain of the sphere would be manifesting hypercube 5.

The domain-boundary ratio of the cube as well as of the sphere being identical, as such both cubical and spherical models can be worked parallelly. However, both these models have their distinct characteristics and hence their distinct utilities and applied values. The glaring difference between the geometric set ups of cube and sphere lies in the set ups of their boundaries.

The boundary of cube is synthetic set up of as many as 6 surface plates while on the other hand the spatial boundary of the sphere is a single integrated sheet. As such the architectural designs of vibrant synthetic reality would depend upon our capacity to exploit the geometric richness of the representative regular bodies of 3-space in the role of dimensions as spacetime formats.

Space, Time and Spacetime of Manasara

For proper appreciation of space, time and spacetime frame of Manasara, we may have a fresh look at the organizational format of the Manasara. The architectural and sculptural achievement of this scripture lies in bridging the deep void between the geometric set up of real 4-space and real 5-space. It really is bound to be a mental holocaust to imagine being on the one bank of this deep void and thinking of reaching the other bank of this deep void.

How to fill the gap in between the two consecutive integral dimensional spaces? Is the question which is going to be the cause for mental holocaust as the logic would fail the bridge the deep void causing holocaust. Mind would feed the meditative brain the supreme order of nature as a working mechanism of the geometric continuum which runs through the deep voids in between the integral dimensional spaces in terms of the spacetime frames with time being one step ahead of the dimensional order of the start with integral space.

In the instant case of the deep void in between real 4 & 5-space we shall be requiring spacetime frame of spatial dimensional order of real 5-space. Linearly, real 4-space is A8 and the solid flux shall be yielding 7 geometries of 3-space of linear dimensional order and hence we shall be having 8 + 7 = 15 linear dimensions which shall be working out 15 = 3 × 5 i.e. 5 solid dimensions for real 5-space. The 7 geometries of 3-space would be parallel to cube devoid of surfaces; cube with 1 surface, cube with 2 surfaces, cube with 3 surfaces and so on cubes with all the 6 surfaces. Therefore, the deep void of real 4 & 5-space would be filled with the help of 2 space sequentially stitching the boundary of 3-space. This sequential stitching mechanism of the boundary of 3-space is the supreme mechanism of nature in terms of which the geometric continuum is being attained by having transition through transcendence. Two space as 0 manifestation layer with 2-space playing the role of domain fold shall be admitting 3-space in the dormant state at its origin in a sealed state. With activisation of the dormant origin, seal of the origin would be unlocked and there would be phenomenal transformation with transcendence from 0 manifestation layer to 1 manifestation layer. This transcendence phenomenon starting with the origin of 2-space to the origin of 3-space is the life span. It is this span which works out the age of everything.

Multi-dimensions of Time and Space

The multi-dimensions of time and space as timespace frame for bridging the deep void in between two integral dimensional spaces and consequential attainment of geometric continuum for the vibrant synthetic reality of transcendence phenomenon through manifestation layers is the basic subject of interest for the scientists and technologists.

When like the mechanism of bridging of the deep void between real 4 & 5-space, the deep voids between real 5 & 6 spaces and between real 6 & 7 space stand worked out, it may be possible for us to have the bliss of Vedic comprehension of the Bindu Sarowar as the constituent of 6-space, Vishnu-lok, a point of the manifestation layer with 4-space in the role of dimensions, 5-space in the role of boundary, 6-space in the role of domain and 7-space in the role of unity state.

The blissful state is the unity state of consciousness which can be attended by all of us the Vedic Way through Transcendental meditation, a phenomenon of 5 space format, parallel to the transcendence phenomenon which would fill the deep voids between the different states of consciousness as well as between the different manifestation layers.

Sum Up

Knowledge and Organisation Knowledge

We can appreciate the distinction of knowledge and organization of knowledge in term of an affine space and metric affine space. In general terms, we may approach this distinction in terms of space and dimensional space. In more concrete terms, we can approach it in the context of our physical world in terms of 'volume' as attribute of 3-space content and the volume organized as a 'cube'. The organization of a cube requires points, lines and surface plates in addition of the volume, lines and surface plates. All these set ups of points, edges, surfaces and volume are to be made out of the available content. The unique mechanism of transformation of an affine space into metric affine space or of space into dimensional space or of 3-space or space into dimensional space or of 3-space content as volume into 3-space lump as cube has real mathematics which deserves to be learned.

Great Vedic comprehension of this mechanism as that the vibrant synthetic reality accepts synthetic monads, is responsible for manifestation layers with sealed origin, (n-2) space playing the role of dimensional of n-space and the domain-boundary ratio of hypercube-n comes to be $A^n:2nA^{n-1}$. This makes the dimensional order of space of two prongs distinct for odd and even dimensional spaces as much as that odd (2n-1) dimensional space shall be having linear dimensional order as $(2n + 1) \times (2n -3) \times .... \times 3 \times 1$ and for even (2n) dimensional space to have linear dimensional order as $2n \times (2n-2) \times (2n-4) \times ..... \times 6 \times 4 \times 2$.

Further the domains boundaries sequence shall be yielding sequential ratios for hypercube-n as A^n: $2nA^{n-1}$: $2n \times 2(n-1) \times A^{n-2}$: $2n \times 2\ (n-1) \times 2\ (n-2) \times A^{n-3}$ .... (till the term A0). This sequential order with the above linear dimensional order for dimensional space and the concept of synthetic monad of n parts making each part as 1/n, when worked together would yield the answer for the above great mechanism of nature, which in the language of Ramayana would mean a difference of two Anguls (measuring units) between the arm of Lord Rama and Kag-Bhushandi and in the language of Shri Bhagwad Mahapuran that every string used by Mother Yashoda to bind Lord Krishna turned out to two Anguls short.

In mathematical language, it means that the representative regular bodies of dimensional spaces i.e. interval, square, cube and hypercubes shall be working out a linear sequential order which shall be transforming respective affine spaces into respective metric spaces or spaces as dimensional space or space contents (hyper volumes) as hypercubes which can be expressed as $Bn = (A + 2)^n$ which means the hypervolume Bn remains only as $(B-2)^n$ and the remaining content stands consumed by the boundary for wrapping the domain $(B-2)^n$.

Now if we work out the above $B^n = (A + 2)^n$ equation for n= 1, 2, 3, 4, and so on, we shall be convincing ourselves about the above mechanism of the nature. For n=1, we shall be getting $B^1 = A^1 + 2A^0$. This straight way follows from the domain-boundary formulation for 1-space body. For n=2, we shall be getting $B^2 = A^2 + 4A^1 + 4A^0$. Here the first pair of terms follow straight from the domain-boundary formulation of 2-space body. The second term 4A1 would yield 8A0 boundary components. But here we shall be reaching boundary of the boundary which means, we shall be taking a synthetic step of two parts to reach the boundary. This will make each part being of ½ value. Applying this, we shall be getting boundary of boundary as $\frac{1}{2}(8A^0) = 4A^0$. Likewise, for n=3, we shall be getting $B^3 = A^3 + 6A^2 + 12A^1 + 8A^0$. Here as well the first pair of terms follows straight way from the domain-boundary formulation of 3-space body. The third term which shall be taking us to the boundary of the boundary would accordingly can be computed as $\frac{1}{2}(4 \times 6A^{2-1}) = 12A^1$. By the same logic we shall be reaching the next term as $1/3\ (2 \times 12A^{1-1})$ for n = 4, 5, 6 and so on.

Now let us appreciate the above mathematics in terms of the geometric set ups of close interval, square and cube and we shall be comprehending that the equation Bn = $(A+2)^n$ for n=1 yields the geometric set up of close interval as length (A^1) with two end points ($2A^0$), for n=2 yields the geometric set up square as area (A^2) with four boundary lines ($4A^1$) and four end points ($4A^0$) and for n=3 yields the geometric set up of cube as volume (A^3) with six surface plates ($6A^2$), 12 edges ($12A^1$) and 8 corner points ($8A^0$).

This will help us appreciate the greatness of Vedic model. Further this will also help us appreciate how Bn remains $(B-2)^n$ as domain in the synthetic set up and the remaining content is consumed by the boundary of $(B-2)^n$ as domain and this would be a never ending sequence which impliedly means, we shall be exhausting the contents and even logically would be demanding the existence in a void, the impossibility, which is the rule of ad-infinitum reduction and this will prove the Fermat's last theorem as that $z^n = X^n + y^n$ and $(z-2)^n = (x-n)^n + (y-2)^n$ can not hold simultaneously. Accordingly, every Vedic scriptural organizational has a great mathematics in store for us.

Lord Shiva, Sun, Man, Atom and Manasara

Lord Vishnu, is the Lord of real 6-space. The seat of Lord Vishnu is the heat of Lord Shiva. 6-space is located at the organ of 5-space. Dwadash Adityas (12 suns) are the 12 boundary components of hypercube-6. As many as 12 hypercube-5 are required to constitute the boundary of hypercube-6. Lord Shiva meditates upon His Lord Vishnu and multiplies 12 folds into Vishnu-lok. Isha punished enlightens us that Purusha, in the sun is Lord Vishnu. Ayurveda enlightens us that Pursuha is constituted by 6 Dhatus and Atman is the 6th Dhathu. First five Dhatus are Panchmahabhut. Lord Shiva is the Lord of Panchmahabhut. Lord Vishnu is the Atman. The studies of the set up of the atom also leads us to the conclusion that electronic configuration around the nucleus i.e the micro world accepts for format of hypercube-4 while nucleus itself concepts for format of hypercube 4 while nucleus of the nucleus would accept the format of hyupercube-6, Mansara accepts Lord Vishnu, the overlord of 6-space as the Lord of the measuring rod. Lord Brahma, the four head lord is the Lord of measure. Manasara begins with the prayer of Lord Brahma and ends with the chiselling of the eye of the Lord. As such the conclusion comes to be that Manasara adopts the orgnizational format of hypercube 5 at the boundary of hypercube-6 which would be parallel to the worlds of Lord Shiva, Aditya, Panchmahabhut and nucleus of the atom.

Spacetime Fram of Mansara

The spacetime frame of Manasara as such is the spacetime frame of hypercube-4 transcending into hypercube-5. This is a transition through transcendence from the spatial dimensional order into solid dimensional order. As such practically this would amount to acceptance of cube or sphere as representative bodies of 3-space in the role of dimensions. The spatial boundary of the cube shall be creating 4-space while the solid volume shall be creating 5-space. This really is very exciting role of vibrant synthetic world being tamed by the Manasara.

Three Space in the Role of Dimension

To have proper comprehension and appreciation of the organizational set up of Manasara we have to learn the mathematics & geometry of 3-space in the role of dimension. With 3-space as dimension we shall be creating 5-space domain with 4-space in the role of boundary and 6-space in the role of origin. This is the Panchmahabhut world. We may compare it with our present day thought mathematics, science and technology which accepts physical world as a 3-space domain.

Cube and sphere are the representative regular bodies of 3-space. They are miniatured replicas of linear manifestation layers. With sealed origin and un-manifest dimensions, the other two folds namely, spatial boundary and solid domain when come in action would create transition through transcendence from hypercube-4 to hypercube-5. To have proper appreciation of this phenomenon, we may have a fresh look at the geometric set up of the cube by viewing it as a cube of 10 units length, 10 units breadth and 10 units height.

We shall be having 1000 unit cubes synthesized as a single cube of 1000 units volume. If we want to know that how many edges / pillars we shall be having of this set up of 1000 cubes, then the answer of Vedic mathematics would be simple in terms of Devnagri numerals 3 and 6 being reflection images of each other like reflection pair 01 and 10. The tri-monad i.e. the monad of three parts shall be yielding 363 and then the length being 10 units so we shall be having the figure 3630. Let us count for ourselves and verify this answer.

Now this reflection property of the light and the visible light spectrum of 7 colours on the one hand and the need of only seven edges to sequentially connect all the 8 edges on the other hand and the property that real 4-space in linearly A8 would help us appreciate the interconnection of the artifices of whole numbers 3 & 7 in the context of 3-space. In this background, we shall be having a bliss of the first quarter of fist Mantra of the Atharvved.

Yeh Trishapta Prayanti Vishwa

Trishapta (3&7) wraps Vishwa (this world)

The message is that 3-space at the boundary of 4-space is a geometric set up of 3 linear dimensions which can take us sequentially uptill 7 linear steps.

The domain and boundary are distinct set ups. If domain represents and becomes the Vedic domain, the boundary would represent and would becomes the Upved. Sthaptyaved being Upved of Athavved and Manasara being the scripture of Sthaptyaupved, therefore, we shall be requiring comprehension of the Vishwa as is the enlightenment of the first quarter of the first Mantra of the Atharvved to appreciate the organization of Mansara.

Therefore the conclusion of all conclusions for us is that for proper learning of Vedic science & technology, there is a need for redrawing the school mathematics syllabi.

Om Tat Sat Iti

The sole syllable Om is the essence of pure knowledge and it is the truth and the ultimate.

Fermat's Last Theorem

Fermat's Last Theorem (FLT) has remained most famous unsolved mathematical problem. This theorem has also attracted the attention of Dr.S.K.Kapoor. He has published his results about the truth of this property of numbers. He has worked out the theorem with different approaches. Essentially all these approaches are geometric in nature but still these are different in characteristics and as such these may be taken as different proofs of the theorem. His approaches and conclusions are not only having intrinsic academic values but also these have historic angle as much as that these accept inspiration as well as working operations from a distant source i.e. Vedic literature. These results are also published in point of time than the proof now accepted in the academic circles.

Three of these proofs of FLT were published in "Modern Science & Vedic Science" (Volume 3 No.1, 1989) with the caption "Vedic Mathematical Concepts and Their Applications to Unsolved Mathematical Problems: Three Proofs of Fermat's Last Theorem". Chapter-10 of the book "Vedic Geometry" (1994) as well takes up some of the aspects of this property of numbers under the caption "Chapter 10: Conclusions and their Applications to the Solution of Fermat's Last Theorem". Then followed the book "Fermat's Last Theorem and Higher Spaces Reality Course" (1996) in which in addition to the outline of different formats of the approaches to the proof and in addition to the general proof of FLT, the result of the theorem has been generalised as Generalised FLT for whole range of hypercubes.

The concepts of power sets and the application of different place value systems to test the truth of any given triple of whole numbers have also been introduced which may prove to be very handy for tests of the truth of this property known as FLT. These proofs are on different formats which are worked out as:

1. On Domain Format
2. On Simplex Format
3. On Values Square Format

4. On the Format of Domain As Dimension Of Another Domain
5. On the Format of Hypercubes

Fermat's Last Theorem speaks out the property of linear dimensional order only. The cause of the restriction being the linear dimensional order so we get the restriction of powers being three and higher. The results generalized the theorem as a geometric property of hypercubes by a shift from linear dimensional order to a spatial dimensional order. The generalized statement comes to be as that "no hypercube can be duplicated" because of the interlocking of (n-2) space with n-space as dimension and domain respectively.

A Simple Proof of Fermat's Last Theorem

It is a shame that Andrew Wiles spent so many of the prime years of his life following such a difficult path to proving Fermat's Last Theorem, when there exists a much shorter and easier proof. Indeed, this concise, elegant alternative, reproduced below, is almost certainly the one that Fermat himself referred to in the margin of his copy* of Bachet's translation of Diophantus's Arithmetica--how arrogant of future chroniclers of mathematical history to insinuate that the garlicky Gallic genius was incapable of formulating the crucial insight! Why, a child can follow the logic below...however, for the benefit of those readers who are no longer children, a short set of notes fleshing out the argument is provided after the proof.

The Theorem: $x^a + y^a = z^a$ has no positive integer solutions (x, y, z, a) for $a > 2$. (Pierre De Fermat, 1601-1665)

The Proof:

(I) At least one of the following two sentences is true.

(II) The preceding sentence is false.

(III) $x^a + y^a = z^a$ has no positive integer solutions (x, y, z, a) for $a > 2$.

Fermat's Proofs of Special Cases

One of the methods used also by mathematicians in their struggle with hard problems is "Divide and Conquer" - splitting the problem into sub-problems that are easier to handle. This tactic appears in connection with FLT already in Fermat's writings, which contain a kernel of a proof for the special case n=4. This is the only proof in number theory that Fermat left (it had space in the margin of Diophantus' book), and it uses the infinite descent method. Fermat also dealt with the special case n=3, as we learn from his letter to Christian Hugeness, but, habitually, he did not leave a proof for this case.

The proof for the special case n=3 was found one hundred years later by Leonhard Euler, the greatest mathematician of the eighteenth century. In number theory, Euler is the follower of Fermat: many theorems stated by Fermat were proven by Euler. In 1753 Euler wrote to Christian Goldbach about the proof, but he did not publish it. In 1770, in his book Anleitung zur Algebra (A Guide to Algebra), Euler gave a proof for this case. This proof was slightly different from the one he gave to Goldbach, and in it Euler used algebraic numbers, which are created by adjoining numbers of the form $\sqrt{3}$, for example, to the group of whole numbers, creating numbers of the form a+b $\sqrt{3}$, where a and b are integers. Although these numbers are not integers, Euler found them useful to solve Diophantine equations.

Euler proofs use the infinite descent method, strengthening Fermat's claim that this method "will lead to marvellous advancement in the theory of numbers". Except the proof technique, the proof for the case n=3 is different from the one for the case n=4, and this explains why, in his letter to Goldbach, Euler said that he did not see any hope to generalize those two special proofs into a general proof. Desperately, Euler asked to make a thorough search in Fermat's home, in hope to find a lead to the missing proof.

The mail connection between Euler and Goldbach lasted 35 years, and is an important source of information about Euler's work. In the beginning of this correspondence, in 1729, Goldbach motivated Euler to deal with Fermat's work. In a letter from 1748, Euler mentioned for the first time Fermat's Last Theorem ("a very nice theorem", he wrote later). Goldbach's Conjecture appeared in a letter from 1742. This conjecture is the main direct contribution of Goldbach to the development of mathematics. His indirect influence, as a friend to Euler, is much more important. In 1843, sixty years after Euler death, his grand grandson published the full correspondence.

The proof for the case n=5, in a way that was an enlargement of the way Euler used for his proof of the case n=3, was created by two partners: the German Peter Gustav Lejeune Dirichlet, the pupil and friend of Gauss, who was 20-year-old, and the French Adrien Marie Legendre, one of the leaders of the French Academy of Sciences and follower of Fermat and Euler in building the foundations of number theory, who was 73-year-old. Dirichlet divided the proof needed for this case into two cases, and in July 1825 he presented to the Academy of Sciences a proof to one of those cases. Two months later, Legendre published, as an appendix to his successful book Théorie des nombres (Number Theory), a proof for the second case, completing Dirichlet's proof. Two months later, Dirichlet supplied a nicer proof for the second case, a kind of natural continuation of his proof for the first case.

Fourteen years later, Gabriel Lamé proved the theorem for the case n=7. Why did Lamé skip the proof for n=6? If the numbers A, B and C are a solution to Fermat's equation for the power of 6, namely $A^6 + B^6 = C^6$, then the numbers A^2, B^2 and C^2 are a solution to the power of 3, contradictory to the proof that FLT is true for this power. Generally, if we had proven FLT to any power k, than the theorem is valid to all the multiples of k. The reason for this is that if the numbers A, B and C are a solution for the power mk, then the numbers A^m, B^m and C^m are a solution to the power of k, contradictory to the proof that FLT is true for this power. This proof technique is called proof by contradiction (or Reductio ad absurdum, if you prefer to say it in Latin). Fermat's infinite descent method is a special type of this technique.

It follows from this proof that it is enough to prove FLT just for powers that are primes greater than 2, since each other power is a multiple of these primes (except 4, but for this power the theorem has already been proven by Fermat). It is a big step toward reduction of the scope of the problem, but it is still leaving us with the need to prove FLT for an infinite set of powers. An elegant proof of the infinitude of primes appears in Euclid's book, the Elements.

In the first two centuries that passed since Fermat wrote his theorem, number theory had developed enormously. After the pioneer work of Fermat, Euler and Legendre, in 1801 Carl Friedrich Gauss published his book, Disquisitiones arithmeticae (Arithmetic Research), which turned number theory from a collection of isolated results into a systematic mathematical theory. Gauss, one of the greatest mathematicians of all times, called number theory "The Queen of Mathematics". Gauss was very pedantic about the quality of works he gave for publication, and many of his achievements were published many years after they were accomplished. A proof by Gauss for the case n=3 of FLT, different from that of Euler but also using the infinite descent method, was published after his death.

Scriptures Formats

Knowledge and organization of knowledge are two different aspects of pure knowledge. One of the beautiful features of Vedic scriptures is that their knowledge is organized on geometric formats of real four and higher dimensional spaces. Here these are being reproduced in the form and sequence these have been taken up during this interaction:

Scriptures Format 1: Om

Om is one syllable scripture complete in itself. The past, present and future and all that transcends three folds of time is verily Om. This way the sole syllable expresses as four folds scripture complete in itself exhaustively covering the past, present and future and all that transcends the three folds of time on tetra-monad format.

Scriptures Format 2: Ved

Ved means knowledge; the whole range of knowledge. This the whole range of knowledge as one Ved was reorganised by Brahmaa-Rishi Ved Vyas into four Veds: Rig Ved, Yajur Ved, Sam Ved and Atharv Ved. Literally Vyas means diameter of a circle (a representative regular body of 2-space which in the role of dimension manifests as 4-space.

The organization of one Ved as four Veds is parallel to the one syllable Om unfolding on tetra-monad format i.e. as an expression of four folds. With this shift to a diameter, there is a shift from a monad (linear order) to di-monad (spatial order). This gives expression for 1 as 01 whose reflection member is 10 i.e. ten.

The above shift from 1 to 10 in that order and sequence, gives us:

(1) 10 + 01 + 10 = 21, the total number of branches of Rig Ved.

(2) 10 × 10 + 01 × 01 = 101, the total branches of Yajur Ved

(3) 10 × 10 × 10 = 1000, the total branches of Sam Ved

(4) 10 -01 = 9, the total branches of Atharv Ved

Scripture Formats 3: Idol of Lord Brahma

Lord Brahma, the creator the supreme, is the four heads Lord, with each head equipped with two eyes. He sits majestically on lotus seat of eight petals and meditates in cavity of his heart on his Lord, Lord Shiva and multiplies himself as ten Brahmas.

The idol of Lord Brahma as of four heads Lord, with two eyes in each head, sitting on lotus seat of eight petals and meditating upon his Lord, Lord Shiva, in the cavity of his heart, as such, accepts the format of hypercube-4 whose four folds are 2-space in the role of dimension (parallel to two eyes in each head), 3-space in the role of boundary (parallel to eight components of boundary of hypercube-4 and eight petals of the lotus seat of Lord Brahma), 4-space in the role of domain (parallel to the domain of four heads Lord) and 5-space in the role of origin (parallel to the seat of Lord Shiva, five heads Lord, in the cavity of the heart of Lord Brahma).

Scripture Formats 4: Idol of Lord Shiva

Lord Brahma, the creator the supreme, is the four heads Lord, with each head equipped with two eyes. He sits majestically on lotus seat of eight petals and meditates in cavity of his heart on his Lord, Lord Shiva and multiplies himself as ten Brahmas.

The idol of Lord Brahma as of four heads Lord, with two eyes in each head, sitting on lotus seat of eight petals and meditating upon his Lord, Lord Shiva, in the cavity of his heart, as such, accepts the format of hypercube-4 whose four folds are 2-space in the role of dimension (parallel to two eyes in each head), 3-space in the role of boundary (parallel to eight components of boundary of hypercube-4 and eight petals of the lotus seat of Lord Brahma), 4-space in the role of domain (parallel to the domain of four heads Lord) and 5-space in the role of origin (parallel to the seat of Lord

Shiva, five heads Lord, in the cavity of the heart of Lord Brahma).

Scripture Formats 5: Incarnations of Lord Vishnu

Lord Sakand is incarnation of Lord Vishnu. Lord Sakand a six heads lord with four eyes in each head and twelve long arms, is the incarnation of Lord Vishnu. Lord Sakand establishes the transcendental worlds by meditating upon eternity, eternally sustained by Dhruva (pole star).

This description is parallel to the description of hypercube-6 with 4-space (parallel to four eyes in each head) in the role of dimension, 5-space in the role of boundary, 6-space in the role of domain and 7-space in the role of origin. The other incarnations of Lord Vishnu accepts this format.

Scripture Formats 6: Measure and Measuring-Rod

Lord Vishnu is the god of measuring-rod and Lord Brahma is the god of measure.

The Lord Vishnu is the overlord of real 6-space and Lord Brahma is the overlord of real 4-space.

Real 4-space plays the role of dimension of real 6-space. As such while the measuring-rod is constituted by domain (6-space), the measure remains at the dimension (4-space).

Scripture Formats 7: Shad-chakra Format of Human Body

Vedic (Upanishadic) enlightenment is that human body admits Shad (six) Chakras (circuits) format. The external characteristics of the Shad-Chakras are 2, 4, 6, 8, 10 & 12.

The external characteristics of Shad-Chakras of human body are parallel to the number of boundary components of hypercubes-1, 2, 3, 4, 5 & 6.

This format of Shad-Chakras is the format of the order of 6-space whose presiding deity is Lord Vishnu. The creator-the-supreme, Lord Brahma, overlord of real 4-space as such is the presiding deity of the measure of the human body.

Scripture Formats 8: Organization of Knowledge of Sri Vishnu Puran

Sri Vishnu Puran is a scripture of 6 Ansha (parts). The organization of knowledge of all the six Ansha of Sri Vishnu Puran is of 126 chapters.

This organization of knowledge is as per the measuring-rod and the Shad-Chakra format of human body as sustained within the Vishnu-lok (6-space domain)

On the artifices of whole numbers this organization admits expression as:

$$6 \times (1 + 2 + 3 + 4 + 5 + 6) = 6 \times 21 = 126.$$

Scripture Formats 9: Vishwa-Rupa

Shatpath Brahman preserves the enlightenment : Tawastha had one son. He had three heads and six eyes. And three mouths. So was his form (Rupa). That is why he was named - Vishwa Rupa (form of the world).

This is parallel to the world of three dimensions and domain-boundary ratio of the representative regular body of 3-space being A3:6A2. The enlightenment preserve in the Shatpath Brahman as such is the enlightenment about the geometric setup of the representative regular body of 3-space. Being a representative regular body so is the representative form of the dimensional space / world.

Scripture Formats 10: Trishapta (3 and 7)

"Yeh Trishapta Paryani Vishwa" (This world is enveloped by Trishapata i.e. 3 and 7). Parallel to it is that 3-space has 7 geometries of signatures (0, 1, 2, 3, 4, 5, 6) corresponding to the cube with no surface plate, cube with 1 surface plate, cube with 2, 3, 4, 5 & 6 surface plates respectively.

Scripture Formats 11: One, Two, Three and Eight

Swateshawara Upanishad preserves the enlightenment that sequential order 1, 2, 3 and 8 leads to Moksha (liberation).

Parallel to it is that on the artifices of whole numbers, the sequence 1, 2, 3 and 8 gives us 1×1, 1×2, 1×3 and 2×4. Parallel to it on geometric formats the message is that the order of linear dimensions works uptil first three dimensions and a step ahead is the world of spatial dimensions. In other words while 3-space admits linear dimensions, the 4-space admits spatial dimensions.

Scripture Formats 12: 03 to 13

The Sushmana-Nari is the central nerve which runs through and coordinates the Shad Chakra of human body accepts the measuring-rod constituted by hypercubes-1 to 6. It means that starting with the close interval and reaching uptil hypercube-6 is the range which constitutes the measuring-rod for the Sushmana-Nari running through the Shad Chakras.

The hypercube-1 (close interval) as representative regular body of 1-space accepts three versions of interval parallel to three geometries of 1-space. The hypercube-6 as representative regular body of 6-space accepts 13 versions parallel to the 13 geometries of 6-space.

Lord Vishnu is the presiding deity of the measuring-rod being the overlord of 6-space. Srimad Bhagwad Gita is the divine song of Lord Krishna, the incarnation of Lord Vishnu.

Each of the chapters of Srimad Bhagwad Gita is precisely accepting the measuring-rod with Lord Vishnu as its presiding deity and as such each chapter opens with 3 words title and ends with 13 syllables Pushapika (colophon) parallel to the three geometries of 1-space to 13 geometries of 6-space.

Scripture Formats 13: 120 Years Cycle

The Jyotish Shastra accepts 120 years cycle for the planets' effects upon human life. This is the geometric organization of 4-space in the role of dimension creating 6-space as domain with complete coverage for the boundary of this domain with 120 hypercubes-4.

This is there as A6:12A5 is the domain-boundary ratio of hypercube-6 and further B5:10C4 is the domain-boundary ratio for each of the boundary components. With this, the 12 boundary components of hypercube-6 get enveloped within $12 \times 10 = 120$ hypercubes-4.

Alternatively, 4-space itself being a spatial dimensional set up, it gives us arrangement for $120 = 60 + 60$ and further $60 = 2 + 8 + 18 + 32$. It is parallel to the capacity of the orbits of an atom to retain electrons. As such, on either side of the nucleus, the set up admits expression as $2 + 8 + 18 + 32 + 32 + 18 + 8 + 2 = 120$.

Scripture Formats 14: 13-Edged Cube is Hypercube-4

Within hypercube-4, the 12-edged cube has one degree freedom of motion along the fourth dimension. As such 12-edged cube gets fixed in 4-space as 13-edged cube. The 13th edge of the cube is supplied by the fourth dimension. Because of the spatial dimensional character of the dimension of 4-space, the 13th edge permits motions of a cube with a spatial pivot. The motion of the cube as a whole in reference to such pivot is the unique cosmic arrangements like that of pole star. The Pushpikas (colophons) beneath all the 18 chapters of Srimad Bhagwad Gita enlighten us about the $6 \times 3 = 18$ different ways of outward motions along all the three dimensions from all the six surfaces of the cube.

Scripture Formats 15: Things Transform Just With the Attention at the Origin-I

Things transform just with the attention at the origin. Let us have attention at the centre of cube / origin of 3-space and everything starts transforming; the cube splits into 8 sub-cubes and 3-space splits into 8 octaves. The origin accepts 8 sub-cubes / 8 octaves enveloping. The 4-space flourishes from within at the seat of origin and everything transforms from 3-space to 4-space.

Scripture Formats 16: Things Transform Just With the Attention at the Origin-II

Things transform just with the attention at the origin. Let us have attention at the centre of hypercube-4 / origin of 4-space and everything starts transforming; the hypercube-4 / 4-space splits into 10 components. The origin accepts 10 components boundary enveloping. The 5-space flourishes from within at the seat of origin and everything transforms from 4-space to 5-space.

The scripture (Yog Vashishta) preserves the enlightenment as that Lord Brahma (the overlord of 4-space) when meditates in the cavity of his heart upon his lord, Lord Shiva, Lord Brahma multiplies himself into 10 Brahmas.

Scripture Formats 17: Things Transform Just With the Attention at the Origin-III

With just attention at the origin of 4-space/hyper-cube-4, the things transform into that of 5-space / hyper-cube-5 enveloped within ten 4-space / 10 hypercubes-5. As from spatial dimension things transform into solid dimensional order and as 3-space / cube admits split up as 8 octaves / 8 sub-cubes. Therefore, just with the attention at the origin we see simultaneous happening of split up into 10 components boundary and within it as within 8 components boundary. This is parallel to the simultaneous organization of Rig Ved Samhita as a scripture of 10 Mandals and also as of 8 Ashtaks. With this, with just attention, our mundane world set up of 3-space starts transforming as transcendental Vedic world of the order of Rig Ved.

Scripture Formats 18: Things Transform Just With the Attention at the Origin-IV:

Simultaneous Internal and External Expansions: With just attention at the origin of a space, say n-space / hypercube-n, the interactive process gets initiated and there follows simultaneous internal and external transformations and expansions to the next higher space set up.

If we just have attention at the set up of a cube, the outer surfaces, as 2×3 boundary components / surface plates, being spatial in organization, as to be set to fly off in the outer space and to dimensionalize it as 4-space. Two of the boundary components together to constitute a di-monad format and the remaining four of them to constitute four spatial dimensions for 4-space. This external transformation and expansion as dimensionalize 4-space is parallel to the internal transformation and expansion at the seat of origin as 4-space. This way there is parallel internal and external emergence of 4-space.

Likewise, if we have attention at the set up of a hyper-cube-4, the outer solids, as 2×4 boundary components / solid components, being solid in organization, as to be set to fly off in the outer space and to dimensionalize it as 5-space. Three of the boundary components together to constitute a tri-monad format and the remaining five of them to constitute five solid dimensions for 5-space. This external transformation and expansion as dimensionalize 5-space is parallel to the internal transformation and expansion at the seat of origin as 5-space. This way there is parallel internal and external emergence of 5-space.

Further, if we have attention at the set up of a hyper-cube-5, the outer hypersolids, as 2×5 boundary components/ hypersolid components, being hypersolid in organization, as to be set to fly off in the outer space and to dimensionalize it as 6-space. Four of the boundary components together to constitute a tetra-monad format and the remaining six of them to constitute six hypersolid dimensions for 6-space. This external transformation and expansion as dimensionalize 6-space is parallel to the internal transformation and expansion at the seat of origin as 6-space. This way there is parallel internal and external emergence of 6-space.

In general, if we have attention at the set up of a hypercube-n, the outer hypersolids, as 2×n boundary components / hypersolid components, being hypersolid in organization, as to be set to fly off in the outer space and to dimensionalize it as (n+1) space. (n-1) of the boundary components together to constitute a {(n-1)-order} monad format and the remaining n+1 of them to constitute n+1 hypersolid dimensions for (n+1) space. This external transformation and expansion as dimensionalize (n+1) space is parallel to the internal transformation and expansion at the seat of origin as (n+1) space. This way there is parallel internal and external emergence of (n+1) space.

Scripture Formats 19: Things Transform Just With the Attention at the Origin-V

Compactification at the Origin: With just attention at the origin sequentially the simultaneous inward and outward expansion would unfold the compactification of whole range of dimensional spaces at thc origin.

The attention at the origin of 3-space would initiate the interaction process of the mind and everything would start transforming and there would emerge simultaneous, inward and outward expansion of 4-space. With this, 3-space / cube as such shall be having placement at the origin and there emerging outward and inward, and as such all around, 4-space in its display. It may be expressible as 4, 3, 4. Three space at the centre and 4-space on its either side. The further attention would transform the very interactive process as much as that while on the one hand the attention would get centred at the origin of 4-space which would expand it outward as well as inward as 5-space and on the other hand the previous stage expression of 3-space at the centre and 4-space around would set into initiation for the 3-space with available degree of freedom of motion to play the role of dimension and as such there would be dimensionalization of 5-space. This simultaneous unfoldment, transformation and expansion as {(4, 3, 4) and (4, 5, 4)} and in general {(n, n-1, n) and (n, n+1, n)} is uniquely there because of the compactification at the origin of whole range of real spaces. (The spaces where n-space plays the role of dimension of n+2 space).

Scripture Formats 20: Things Transform Just With the Attention at the Origin-VI

Continuity Maintained by Bridging of Gaps: The Sushmana-Nari is the central nerve which runs through and coordinates the Shad Chakra of human body accepts the measuring-rod constituted by hypercubes-1 to 6. It means that starting with the close interval and reaching uptil hypercube-6 is the range which constitutes the measuring-rod for the Sushmana-Nari running through the Shad Chakras.

The hypercube-1 (close interval) as representative regular body of 1-space accepts three versions of interval parallel to three geometries of 1-space. The hypercube-6 as representative regular body of 6-space accepts 13 versions parallel to the 13 geometries of 6-space.

Lord Vishnu is the presiding deity of the measuring-rod being the overlord of 6-space. Srimad Bhagwad Gita is the divine song of Lord Krishna, the incarnation of Lord Vishnu. Each of the chapters of Srimad Bhagwad Gita is precisely accepting the measuring-rod with Lord Vishnu as its presiding deity and as such each chapter opens with 3 words title and ends with 13 syllables Pushapika (colophon) parallel to the three geometries of 1-space to 13 geometries of 6-space.

Scripture Formats 21: Vykata, Avyakata, Avyakato-avakatat and Purusha #1

Srimad Bhagwad Gita preserves the enlightenment of the order of the Reality as Vyakta, Avyakata, Avyakto, Avyakato-avyakatat and Purusha.

Vyakta means expressed. It is the expressed world, Vishwa, 3-space.

Avyakata means unexpressed which is yet to manifest in the Vishwa, 4-space world.

Avyakato-avyakatat means which is at the base of Avyakata. It is the transcendental base, a 5-space world.

Purusha is fully expressed transcendental consciousness within human frame. It is the consciousness world of Atman (soul) domain, Vishnu-lok, the Purusha in the Surya (sun), the 6-space reality.

Scripture Formats 22: Vykata, Avyakata, Avyakato-avakatat and Purusha #2

13-Edges Cube Within Domain of Hypercube-4

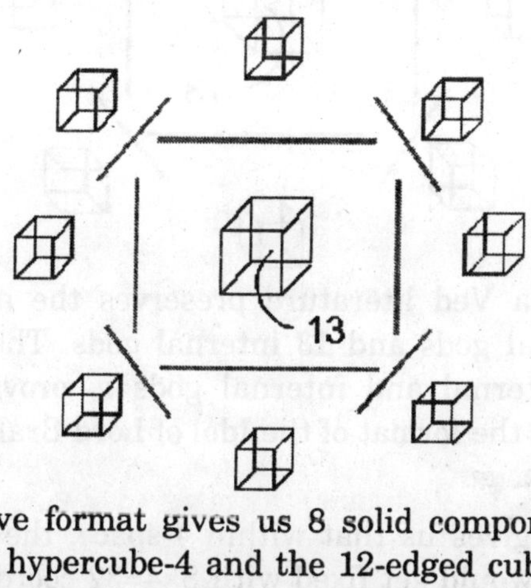

The above format gives us 8 solid components of the boundary of hypercube-4 and the 12-edged cube gets fixed along the degree of freedom available to it within domain of hypercube-4 and as such manifest as 13-edged cube. There being a degree of freedom of motion along each of the four dimension of 4-space, as such there are 4 different ways for fixation of 13-edged cube within the domain of hypercube-4 and hence within 4-space. Otherwise as the 4-space itself is a spatial dimensional space, as such we get first format for the Great debate of enlightenment and knowledge within the kingdom of King Janak where everybody except Ashtavakra were held up at two fixations for the artifice of whole number 13 while Ashta Vakra won the debate by completing all the four fixations for the artifice of whole number 13.

Scripture Formats 24: Vykata, Avyakata, Avyakato-avakatat and Purusha #4

32 External Gods and 13 Internal Gods:

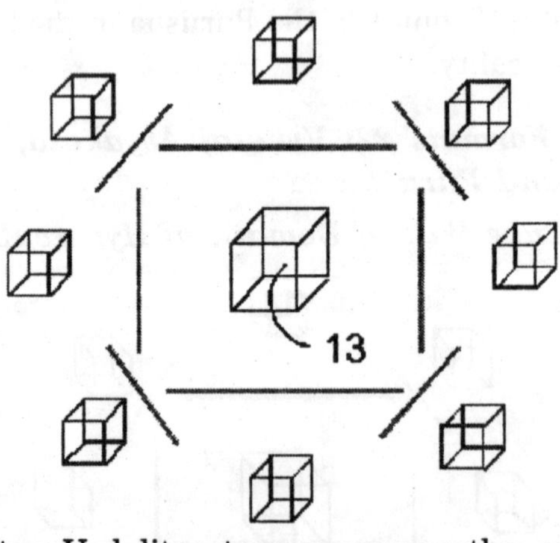

Sthapatya Ved literature preserves the arrangement of 32 external gods and 13 internal gods. The format for these 45 external and internal gods is provided by the hypercube-4, the format of the Idol of Lord Brahma, creator the supreme.

$A^4 : 8B^3$ gives us that within 4-space, the 8 boundary components would get fixed with 8×4=32 coordinates. This together with the fixed 13-edged cube within the domain would provide an arrangement of 32 + 13 = 45 fixations.

The above organization of the format of hypercube-4 with the availability of 45 characteristics arrangements would help us comprehend the enlightenment of Ur-Mahabharatam about the text of Srimad Bhagwad Gita as that this text is of 745 Shalokas out of which 700 Shalokas are the expressed Shalokas while 45 are the unexpressed Shalokas. The available text of Srimad Bhagwad Gita with us is the text of 700 Shalokas and the remaining 45 unexpresed Shalokas remain unmanifest in this world and these are lively and available at the format beneath.

Scripture Formats 25: Vykata, Avyakata, Avyakato-avakatat and Purusha #5:

32 Installation of Shiva Lingam at the Centre of Mystic Symbol

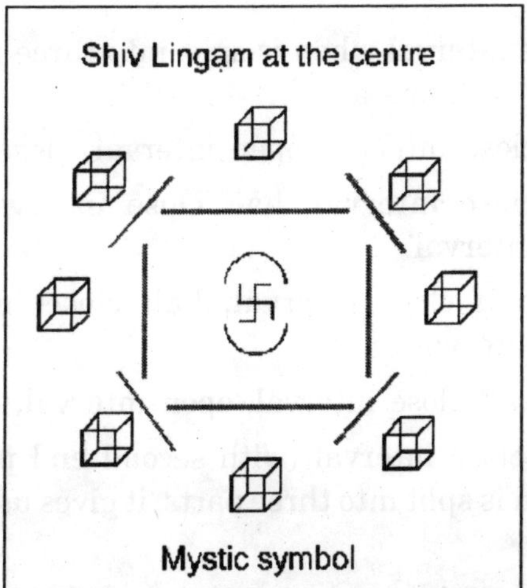

Sri Shiva Puran preserves the enlightenment as that Shiva Lingam is to be installed at the centre of the mystic symbol. The mystic symbol is the format of the Idol of Lord Brahma, the creator the supreme. Shiva Lingam is the solid dimensional expression of the Shiva Lok (5-space). The centre of the mystic symbol is the centre of hypercube-4 / origin of 4-space. The origin of 4-space is the seat of compactification of dimensional spaces from 5-space onwards. The Shiva Lingam emerges from within the origin.

Scripture Formats 26: Description of the Wheel of Cause Brahman #1

Sawteshwara Upanishad describes the wheel of cause Brahman as of 1 (Nemi), 3 (Vrit), 16 (Antah), 50 (Aras), 20 (Pratya), 6 (Ashtak), 1 (Pash, Vishwarupa), 3 (Marg), 2 (Paap, Punya) and 1 (Karan). The artifices of whole numbers availed in that sequence are: 1, 3, 16, 50, 20, 6, 1, 3, 2, 1.

Artifices of Whole Numbers 1 & 3: One as 1-space in the role of dimension creates 3-space.

Artifice of Whole Number 16: The tri-monad, a monad of 3 parts, and an interval of 3 parts gives us 16 versions as:

(1) close interval when is split into three parts, it gives us 4 versions as:

(a) close interval, open interval, close interval

(b) close interval, half close interval, half close interval

(c) half close interval, half close interval, close interval

(d) half close interval, open interval, close interval

(2) half-close interval (with second end point missing) when is split into three parts, it gives us four versions as:

(a) close interval, open interval, half close interval

(b) half close interval, half close interval, half close interval

(c) half close interval, close interval, open interval

(d) close interval, half close interval, open interval

(3) half-close interval (with first end point missing) when is split into three parts, it gives us four versions as:

(a) half close interval, open interval, close interval
(b) half close interval, half close interval, open interval
(c) open interval, close interval, half close interval
(d) open interval, half close interval, close interval

(4) open interval when is split into three parts, it gives us four versions as:

(a) half close interval, open interval, half close interval
(b) half close interval, half close interval, open interval
(c) open interval, close interval, open interval
(d) open interval, half close interval, half close interval

Artifice of Whole Number 50: 3-space in the role of dimension creates 5-space and as such 10×5=50 coordinates of solid dimensional frame are required to fix the boundary of 5-space constituted by 10 hypercubes-4.

Artifice of Whole Number 20: The domain-boundary fixation of hypercube-5 can be had in terms of the artifice of whole number 20 which admits arrangement and organization as 20=5×4.

Artifice of Whole Number 6: The origin of hypercube-5 is the seat of 6-space and as such origin of hypercube-5 admits fixation in terms of the artifice of whole number 6.

Artifice of whole number 1, 3, 2 & 1: The origin of origin of hypercube-5 is the seat of 7-space. The artifices of whole numbers 1, 3, 2 & 1 together are the arrangement of the artifice of whole number 7. Going from the artifice of 1 to the artifice of 3 is the going from dimension to domain and going from 1 to 2 is going from the dimension to the boundary.

The above expression of the wheel of "cause Brahman" takes us uptil the origin of the origin of the transcendental world of 5-space.

Scripture Formats 27: Description of the Wheel of Cause Brahman #2

Sawteshwara Upanishad describes the wheel of cause Brahman as of 1 (Nemi), 3 (Vrit), 16 (Antah), 50 (Aras), 20 (Pratya), 6 (Ashtak), 1 (Pash, Vishwarupa), 3 (Marg), 2 (Paap, Punya) and 1 (Karan). The artifices of whole numbers availed in that sequence are: 1, 3, 16, 50, 20, 6, 1, 3, 2, 1. In the reverse order the artifices of whole numbers give us the sequence as 1, 2, 3, 1, 6, 20, 50, 16, 3, 1. This sequence admits expression as: 1, 2, 3, 4=1, 4 leads to 6, 6 leads to 9+7+3+1, 4*5=20 leads to 5-space with 4-space boundary and 5*(1+2+3+4)=50 solid points fixation of boundary, the tri-monad format of solids gives us 6-space at the centre of 5-space as 5+6+5=16 and 3 as tri-monad format and 1 as the linear order of 3-space.

On Om formulation, which is the basic formulation for organization on 4-space format, sequentially the above artifices of whole numbers leads to the following phases of organization of the wheel of Cause Brahman.

The phase of artifices of whole numbers 1, 2, 3, 4 and 1, 2, 3, 1 :

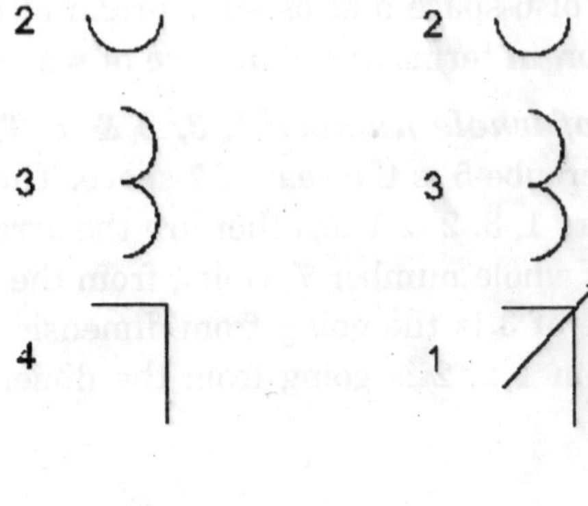

The phase of artifice of whole numbers 6:

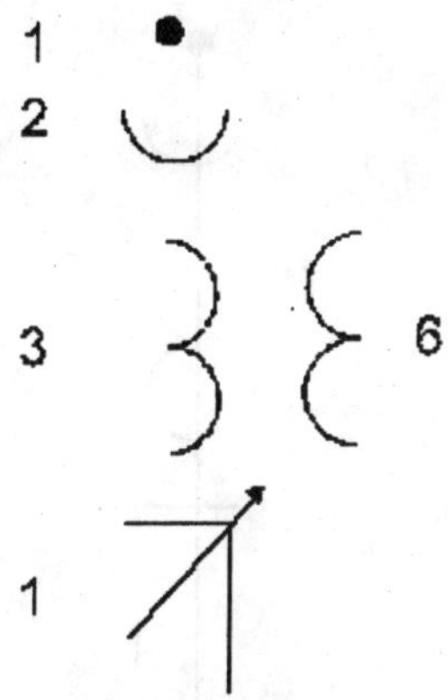

The phase of artifice of whole number 20:

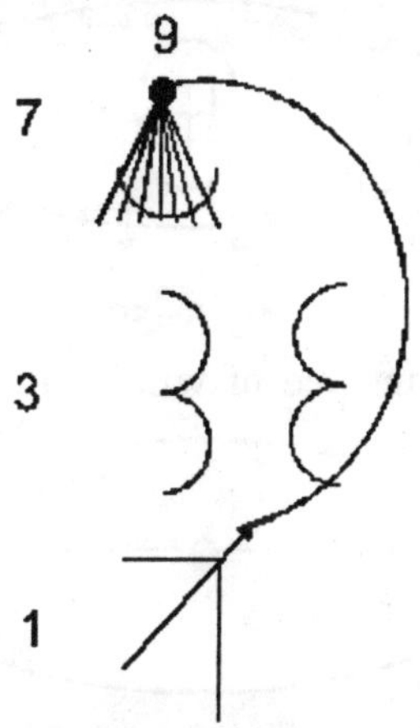

9+7+3+1=20

The phase of artifice of whole number 16:

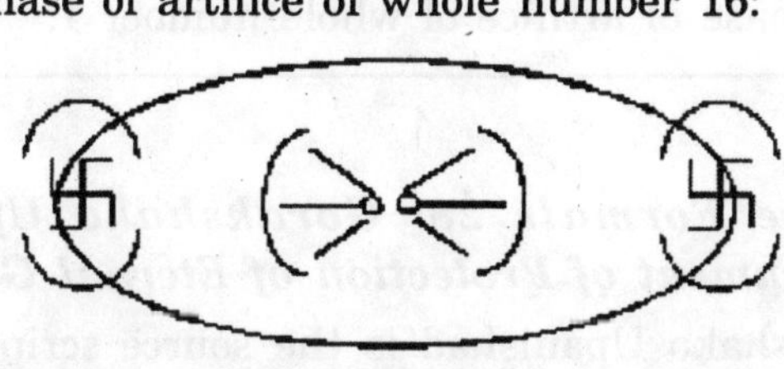

5+6+5=16

The phase of artifice of whole number 50:

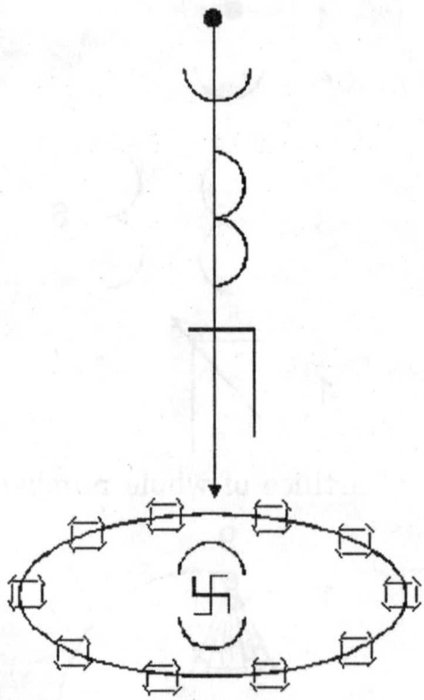

10*5=50

The phase of artifice of whole number 3:

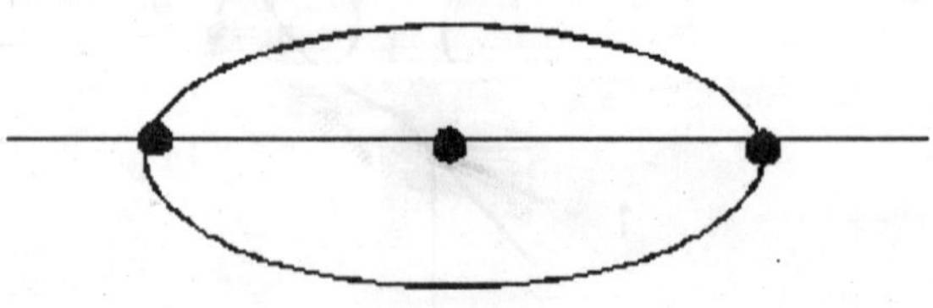

1+1+1

The phase of artifice of whole number 1:

1

Scripture Formats 28: Gorakshako Upanishad (Enlightenment of Protection of Eternal Glow) #1

Gorakshako Upanishad is the source scripture of the Nath sect. Guru Gorakshanath, the incarnation of Lord Shiva is the author of this scripture. This scripture is also known as Sidha Sidhanta Padhati. This is a scripture of 6 Upadeshas (enlightenment). The scripture range is of 353 Shalokas. The split up of this range is as of 73, 39, 13, 29, 84 and 115 Shalokas. The titles of the six Upadeshas are:

- Pinda Utapatti (73 Shalokas range): Pinda means body and Utapatti means emergence. As such the simple English rendering for the title of first Upadesha is "Emergence of Body". The enlightenment of this Upadesha is the enlightenment is the emergence of human body. This enlightenment is organised on the geometric format of 73 Shalokas range. The artifice of whole number 73 has been well utilised for the organisation of this Upadesha.
- Pinda Vichaar (39 Shalokas range): Pinda means body and Vichhar means the deep thought. As such the simple English rendering for the title of second Upadesha is "Deep thought about the body". This enlightenment is about the basic characteristics of the geometric format for the life flow within the human body.
- Pinda Samviti (13 Shalokas range): Pinda means body and Samviti means knowledge. As such the simple English rendering for the title of third Upadesha is "Knowledge of human body". This enlightenment is about the basic knowledge of life within the human body.
- Pinda Adhara (29 Shlokas range): Pinda means body and Ahdara means base. As such the simple English rendering for the title of fourth Upadesha is "The ultimate base of human body". This enlightenment is about the ultimate base of life within the human body.
- Samras Karanam (84 Shlokas range): Sam means common, Ras means essence and Karanam means ultimate sustenance. As such the simple English

rendering for the title of fifth Upadesha is "Common essence of the ultimate sustenance of life". This enlightenment is about the common essence of the ultimate sustenance of life within body.

- Avadhut Yogi Lakshanam (115 Shalokas range): Avadhut Yogi means fully elightened being and Lakshanam means characteristic features. As such the simple English rendering for the title of sixth Upadesha is "The characteristics features of fully enlightened being". This enlightenment takes us to the common ultimate state of enlightenment for the living beings.

Scripture Formats 29: Gorakshako Upanishad (Enlightenment of Protection of Eternal Glow) #2

Guru Gorakhnath, the author of the Upanishad is incarnation of Lord Shiva, the overlord of real 5-space. Therefore, the basic format for the organisation of the scripture gets fixed as of hypercube-5 / real 5-space.

Gorakshako Upanishadic text is of 353 Shalokas range. The artifice of whole number 353 is unique in many ways and it, in the light of the enlightenment of the scripture and the organisation format beneath, helps us reach at the basic geometric lock of manifestation of human body.

It is like the emergence of real 5-space at the origin of affine 3-space. The origin of affine 3-space is the seat of real 4-space and as such the transcendence and emergence at this seat is there because of real 5-space being at the base of this seat. The real 5-space itself being a solid dimensional space, this (solid dimensional order) as such locks the seat and the emerging geometric continuum locks the existence phenomena.

This phenomenon of locking and unlocking at the base of the origin of affine 3-space is the basic phenomenon whose comprehension and chasing is to provide us the enlightenment of the manifestation of human body and transcendence therefrom. This life phenomenon is the phenomenon of eternal glow whose protection within the human body is the subject matter of enlightenment of this Upanishad. One name for this discipline may be the discipline of geometry of Vedic genetic engineering.

Scripture Formats 30: Gorakshako Upanishad (Enlightenment of Protection of Eternal Glow) #3

Guru Gorakhnath (Gorakshnath), the author of the Upanishad, is the incarnation of Lord Shiva and with it the basic format for organisation of the scripture is settled as hypercube-5 / real 5-space.

Om Namah Shivai is the basic Mantra of enlightenment of Lord Shiva. Om is sole-syllable, Namah is a pair of syllables and Shivai is a composition of three syllables. With this, this is the enlightenment of artifices of whole number 1, 2 & 3 and parallel to it 1-space, 2-space and 3-space and finally a tri-monad format and so the solid dimensional order takes us to real 5-space.

The artifices of whole numbers 1, 2 & 3 together also give us $1 + 2 + 3 = 1 \times 2 \times 3 = 6$. Simultaneously, we have 6 non-negative geometries of 5-space. This as such helps us to reach at the organisation of the scripture as 6 Upadeshas parallel to the artifice of 6 supplied by the basic Mantra as well as the available 6 non-negative geometries of 5-space.

Scripture Formats 31: Gorakshako Upanishad (Enlightenment of Protection of Eternal Glow) #4

The enlightenment of Gorakshako Upanishad is the enlightenment of Shiva-lok and is the glow of Sri Shivpuran.

Sri Shivpuran preserves the enlightenment as that the gods who were proud of their powers were made to realise the limits of their powers by Lord Shiva as that they were even not capable of making a split of even the blade of grass.

This is precisely the Shloka 26 of Chapter 16 of Satrudra Samhita and it reads as:

> *"If you are proud and think that you know your strength to be great, O gods, split this blade of grass placed by me, with your own weapons."*

The geometric and algebraic message here is that when we creates structures with the help of a binary composition then the elements of the set at first operation give us the composition elements of the form $c = a \times b$. Then when further we operate such composite elements and the resultant elements $d = c^1 \times c^2$ then d takes us uptil the boundary of a 3-space. The domain of 3-space remains untouched though wrapped within the composed elements.

It is because of it that the blade of grass as a solid remains un-split.

This enlightenment and message deserves to be comprehended well to appreciate as to why our set theory approach is creating mathematical knots and while we get dodged and feel as if we can have space filling curves and why the situations like that of everywhere continuous but nowhere differentiable functions and also the situations like the hypercubes-8 onwards creating a decreasing value sequence at all arise for us and we are not in a position even to logically catch as to why this all is happening. This impossibility to split a blade of grass is the proposition and situation at the base of Fermat's Last Theorem and similar like mathematical problems of the day.

Scripture Formats 32: Gorakshako Upanishad (Enlightenment of Protection of Eternal Glow) #5

The organisation of the Upanishad as 6 Upadeshas admits arrangement for its 353 Shalokas as 73, 39, 13, 29, 84 and 115. This arrangement is there because of the following format beneath because of the geometries of 0, 1, 2, 3, 4 & 5. Here is drawn the arrangement:

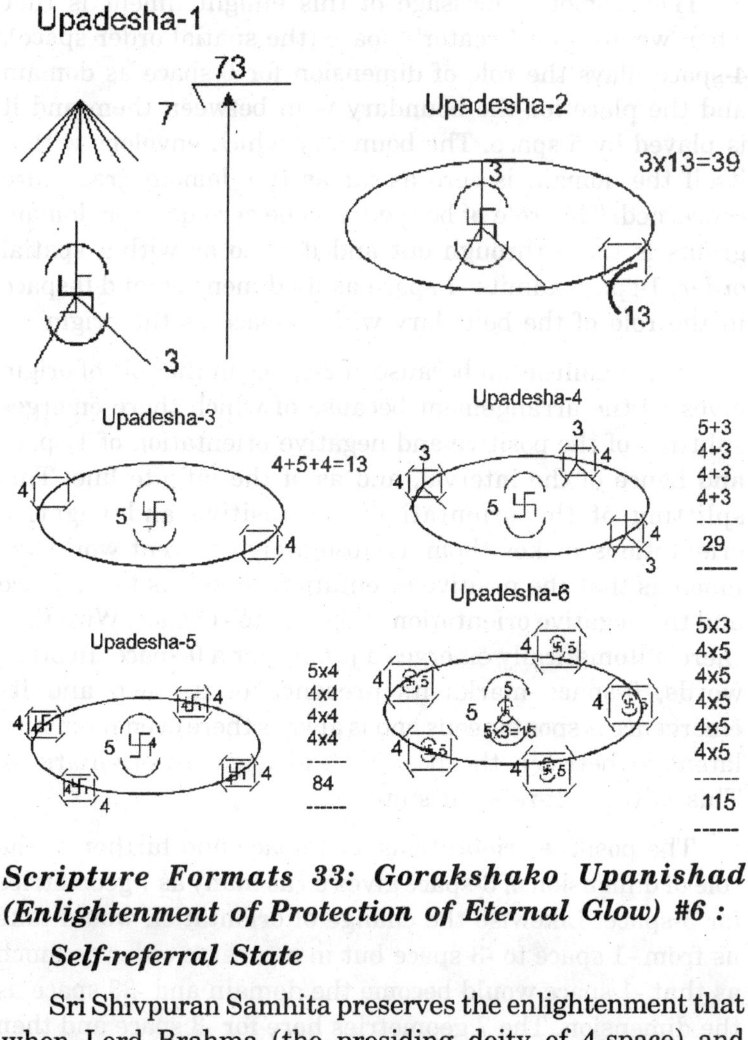

Scripture Formats 33: Gorakshako Upanishad (Enlightenment of Protection of Eternal Glow) #6 : Self-referral State

Sri Shivpuran Samhita preserves the enlightenment that when Lord Brahma (the presiding deity of 4-space) and Lord Vishnu (the presiding deity of 6-space) are face to face and are interacting to know the Brahman, there emerges Lord Shiva (the presiding deity of 5-space) in between them as an infinite pillar of fire (Agani Tatva; the third basic element with Prithvi (earth) as first and Jal (water) as the second).

The geometric message of this enlightenment is that when we are in a Creator's space (the spatial order space), 4-space plays the role of dimension for 6-space as domain and the place for the boundary is in between them and it is played by 5-space. The boundary which envelops within itself the domain is zero as far as the domain grains are concerned. This role of boundary to be zero qua the domain grains is there through out and it is so as within spatial order, 1-space admits -1 space as its dimension and 0-space in the role of the boundary with 2-space as the origin.

This organisation because of 2-space in the role of origin gives us the arrangement because of which there emerges splitting of the positive and negative orientation of 1-space and hence of the interval, and as of the infinite line. This splitting of the orientations as positive and negative orientations makes them as absolutely different worlds as much as that the positive orientations takes us to +1 space and the negative orientations takes us to -1 space. With this, there automatically emerges a jump over a 0-space. In other words, 0-space marks its presence of its own and its emergence is spontaneous and is always there even if remains latent in between the positive and negative orientations. This is the self-referral state.

The positive orientations as 1-space and further in the role of dimension of 3-space gives us as many as 7 geometries for 3-space. Likewise the change of orientation would take us from -1 space to -3 space but in a reverse order as much as that -1 space would become the domain and -33 space as the dimension. The 7 geometries here for -3 space and then the continuity from -3 space to +3 space with 0-space in between -1 space and +1 space gives us two parallel set ups like the pair of objects with their reflection images and because of it a self-referral state makes it possible to have self-evaluation and self-appraisal of one's knowledge and it is there at the format of the division of the brain within human heads as of left and right parts.

The enlightenment of Sri Shivpuran from the stage of emergence of Lord Shiva as infinite pillar of fire in between Lord Brahma and Lord Vishnu goes further as that both Lord Brahma and Lord Vishnu interacted and asked as to what this form is there and when both showed ignorance about it to each other they decided as that one should go upward and try to know uptil which the pillar takes and the other should go downward and try to know uptil what it takes to. Both had tried and realized that it was not possible to scale the pillar.

This enlightenment made them realized the transcendental state of Lord Shiva (the presiding deity of real 5-space) and both of them attained self-referral state and ultimately the transcendental state and realized that it was not sufficient only to go upward and the going upward and going downward were two different organizations and arrangements and those can be known only with reference to the orientation free state of 0-space emerging between the positive and negative orientations spaces.

9

Goldbach's Conjecture

Precisely the conjecture is that every even greater than 2 can be written as a pair of sum of primes. It is part of the famous letter (June 7, 1942) of Christian Goldbach (1690-1764) to great Swiss mathematician Leonhard Euler. Since then this conjecture has remained a brain teaser and unsolved problem of the order of Fermat's Last Theorem and Riemann Hypothesis.

The book "Goldbach Theorem" has four chapters. The proof except Step 2A has been settled in Chapter 1 of the book. All what is presumed as background is only the mature comprehension of the whole numbers. As every even number by definition admits two parts, therefore, it is this property of evens which has come very handy to express even E as M+M. With this the maximum possible pairs with numbers 0 to E have been formed putting a restriction that some of the numbers of the pairs should always be equal to E. The pairs are designated as duplexes. Then the recursive subsets of the source set of maximum number of duplexes of given even number E are constructed and the subset of duplexes with both numbers of the duplexes being primes is reached at. The cardinality of this set is computed as $\geq 1/4 \times \sqrt{E}$.

As $1/4 \times \sqrt{E}$ for $E \geq 64$ is ≥ 2, so after accounting for the duplex (1,E-1), there always remains a minimum of one such duplex whose both members are to be primes and with it the conjecture stands satisfied. Otherwise, this in fact, amounts to extension of the conjecture for E $\geq$ 64 from expectation of one partition for E as sum of primes to that of the minimum of 1/4× $\sqrt{E}$ number of partitions for E as sum of primes.

Chapter 2 of the book takes us to Vedic geometric inspiration and approach of di-monad format (a format which accepts entity as of two parts). The interesting property of the geometric setup of "square" of area E is that such a square would be admitting $\sqrt{E}$ as a side of the square. This gives us the ratio of area E and the sum of its four sides as 1/4× $\sqrt{E}$. The conceptual parallelism is much inspiring. This makes Vedic geometric approach as much potentialised approach.

Chapter 3 of the book introduces a slide rule for finding out solutions for E=P+Q. This together with the computer testing (Chapter 4) for the number of solutions of equation E=P+Q adds empirical value to the proof.

Several Proofs of the Twin Primes and Goldbach Conjectures

Proof of Goldbach's Conjecture, the Prime Number Theorem, and Euclid's Logic Provide Proofs of the Twin Primes Conjecture. Proof of the Twin Primes Conjecture Provides Proof of Goldbach's Conjecture.

THEOREM

There are infinitely many twin primes.

Proof of the Twin Primes Conjecture Using Proofs of Goldbach's Conjecture or Using the Prime Number Theorem

The twin primes conjecture (TPC) suggests that there is an infinite number of primes a and b with a difference 2, i.e., a - b = 2. Goldbach's conjecture (GC) suggests that every even number greater than 2 is the sum s of two prime numbers a and b, i.e., a + b = s where s is even >2. GC is proved by the author herein below and elsewhere.

For prime numbers a,b,c

. . . . . a - b = (a + c) - (b + c) . . . . even integer ...(1)

and thus, generally,

. . . . . a - b = 2k . . . . k = integer $\geq$ 1 ...(2)

and since a + b is an even number

. . . . . a + b = 2n . . . . n = integer>2 ...(3)

Now, using (2) and (3) results in

. . . . . a = n + k . . . and . . . b = n - k . . . n>k ...(4)

which say that for every single value of k primes a and b are separated by an interval 2k and occur as numbers n + k and n - k. Suppose that $n_1, n_2, n_3, \ldots, n_r$ are all the twin prime (k=1) intervals. Since number n is unlimited, intervals $n_1, n_2, n_3, \ldots, n_r$ are not all the twin prime intervals, i.e., $n \to \infty$ in $n_1, n_2, n_3, \ldots, n_r$.

Because all numbers n + k and n - k are not prime numbers, equations (2)-(4) are necessary for proving the infinitude but are not sufficient for finding the existence of prime numbers at these locations. However, from the Prime Number Theorem (PNT), I compute that the probability of finding twin primes a at n + 1 and b at n -1 is about 2/logn (k=1) and the number of twin primes in the interval n is about $2n/(\log n)^2$. Also, in contrast to the foregoing actual proof using Goldbach's conjecture, the PNT provides a proof of the TPC as $n \to \infty$. We can also say that,

- the difference of two prime numbers a and b is an even number 2k; and
- the probability of finding primes a at n + k and b at n - k decreases as n increases.

Proof of the Twin Primes Conjecture Using Euclid's Logic

Another actual proof of the TPC is suggested by Euclid's proof that there are infinitely many primes.

Suppose that $i_1 = i_2 = i_3 = \ldots = i_r = 2$ are all the twin prime intervals. Let $I_r = 2_r + 1$, $i_{r+1} = 2$ and $I_{r+1} = 2^{r+1} + 1$ from which it follows that $2I_r = I_{r+1} + 1$. For all values of r, since interval i_{r+1} divides $2I_r$ (without dividing I_r and I_{r+1},) it is yet another interval and intervals $i_1 = i_2 = i_3 = \ldots = i_r$ are not all the twin prime intervals, i.e., in $i_1 = i_2 = i_3 = \ldots = i_r = 2$.

Proof of Goldbach's Conjecture Using Proof of the Twin Primes Conjecture

I can now prove Goldbach's Conjecture by reversing my foregoing Proof of the Twin Primes Conjecture Using Proof of Goldbach's Conjecture. With reference to equations (2)-(4), suppose that not every sum of two primes a and b is an even number 2n>4. However, since the number of twin primes is unlimited, their sum is an even number 2n unlimited and every sum of primes is an even number.

Mathematics and the Spiritual Dimension

Unfortunately, I think I went too far to the other side. I threw reason to the wind, so to speak, and unceremoniously became a self-ordained "spiritual person." Science, the foundation of which is mathematics, as I saw it, had nothing to offer. It was only years later, when the cloud of my sentimentalism was dissipated by the sun of my soul's integrity, that I was able to separate myself from yet another delusion-the first being the advice of my father, and the second being the idea that I could wish myself into a more profound understanding of the nature of reality.

Math cannot take the mystery out of life without doing away with life itself, for it is life's mystery, its unpredictability-the fact that it is dynamic, not static-that makes it alive and worth living. We may theoretically explain away God, but in so doing we only choose to delude ourselves; I = everything is just bad arithmetic. However, before we can connect with our heart of hearts, our real spiritual essence, we cannot cast reason aside. With the help of the discriminating faculty we can know at least what transcendence is not. Withdrawing our heart from that is a good beginning for a spiritual life.

Mathematics has only recently risen to attempt to usurp the throne of Godhead. Ironically, it originally came into use in human society within the context of spiritual pursuit. Spiritually advanced cultures were not ignorant of the principles of mathematics, but they saw no necessity to explore those principles beyond that which was helpful in the advancement of God realization. Intoxicated by the gross power inherent in mathematical principles, later civilizations, succumbing to the all-inviting arms of illusion, employed these principles and further explored them in an attempt to conquer nature. The folly of this, as demonstrated in modern society today, points to the fact that "wisdom" is more than the exercise of intelligence. Modern man's worship of intelligence blinds him from the obvious: the superiority of love over reason.

Archimedes and Pythagoras

A common belief among ancient cultures was that the laws of numbers have not only a practical meaning, but also a mystical or religious one. This belief was prevalent amongst the Pythagoreans. Prior to 500 B.C.E., Pythagoras, the great Greek pioneer in the teaching of mathematics, formed an exclusive club of young men to whom he imparted his superior mathematical knowledge. Each member was required to take an oath never to reveal this knowledge to an outsider. Pythagoras acquired many faithful disciples to whom he preached about the immortality of the soul and insisted on a life of renunciation. At the heart of the Pythagorean world view was a unity of religious principles and mathematical propositions.

In the third century B.C. another great Greek mathematician, Archimedes, contributed considerably to the field of mathematics. A quote attributed to Archimedes reads, "There are things which seem incredible to most men who have not studied mathematics." Yet according to Plutarch, Archimedes considered "mechanical work and every art concerned with the necessities of life an ignoble and inferior form of labour, and therefore exerted his best efforts only in seeking knowledge of those things in which the good and the beautiful were not mixed with the necessary." As did Plato, Archimedes scorned practical mathematics, although he became very expert at it.

The Abacus: A Mechanical Counting Device

The Greeks, however, encountered a major problem. The Greek alphabet, which had proved so useful in so many ways, proved to be a great hindrance in the art of calculating. Although Greek astronomers and astrologers used a sexagesimal place notation and a zero, the advantages of this usage were not fully appreciated and did not spread beyond their calculations. The Egyptians had no difficulty in representing large numbers, but the absence of any place value for their symbols so complicated their system that, for example, 23 symbols were needed to represent the number 986. Even the Romans, who succeeded the Greeks as masters of the Mediterranean world, and who are known as a nation of conquerors, could not conquer the art of calculating. This was a chore left to an abacus worked by a slave. No real progress in the art of calculating nor in science was made until help came from the East.

Shulba Sutra

In the valley of the Indus River of India, the world's oldest civilization had developed its own system of mathematics. The Vedic Shulba Sutras (fifth to eighth century B.C.), meaning "codes of the rope," show that the earliest geometrical and mathematical investigations among the Indians arose from certain requirements of their religious rituals. When the poetic vision of the Vedic seers was externalized in symbols, rituals requiring altars and precise measurement became manifest, providing a means to the attainment of the unmanifest world of consciousness. "Shulba Sutras" is the name given to those portions or supplements of the Kalpasutras, which deal with the measurement and construction of the different altars or arenas for religious rites. The word shulba refers to the ropes used to make these measurements. Math cannot take the mystery out of life without doing away with life itself, for it is life's mystery, its unpredictability - the fact that it is dynamic, not static- that makes it alive and worth living.

Although Vedic mathematicians are known primarily for their computational genius in arithmetic and algebra, the basis and inspiration for the whole of Indian mathematics is geometry. Evidence of geometrical drawing instruments from as early as 2500 B.C.E. has been found in the Indus Valley. The beginnings of algebra can be traced to the constructional geometry of the Vedic priests, which are preserved in the Shulba Sutras. Exact measurements, orientations, and different geometrical shapes for the altars and arenas used for the religious functions (yajnas), which occupy an important part of the Vedic religious culture, are described in the Shulba Sutras. Many of these calculations employ the geometrical formula known as the Pythagorean theorem.

This theorem (c. 540 B.C.), equating the square of the hypotenuse of a right angle triangle with the sum of the squares of the other two sides, was utilized in the earliest Shulba Sutra (the Baudhayana) prior to the eighth century B.C. Thus, widespread use of this famous mathematical theorem in India several centuries before its being popularized by Pythagoras has been documented. The exact wording of the theorem as presented in the Sulba Sutras is: "The diagonal chord of the rectangle makes both the squares that the horizontal and vertical sides make separately."

The proof of this fundamentally important theorem is well known from Euclid's time until the present for its excessively tedious and cumbersome nature; yet the Vedas present five different extremely simple proofs for this theorem. One historian, Needham, has stated, "Future research on the history of science and technology in Asia will in fact reveal that the achievements of these peoples contribute far more in all pre-Renaissance periods to the development of world science than has yet been realized."

The Shulba Sutras have preserved only that part of Vedic mathematics which was used for constructing the altars and for computing the calendar to regulate the performance of religious rituals. After the Shulba Sutra period, the main developments in Vedic mathematics arose from needs in the field of astronomy. The Jyotisha, science of the luminaries, utilizes all branches of mathematics.

The need to determine the right time for their religious rituals gave the first impetus for astronomical observations. With this desire in mind, the priests would spend night after night watching the advance of the moon through the circle of the nakshatras (lunar mansions), and day after day the alternate progress of the sun towards the north and the south. However, the priests were interested in mathematical rules only as far as they were of practical use. These truths were therefore expressed in the simplest and most practical manner. Elaborate proofs were not presented, nor were they desired.

Evolution of Arabic (Roman) Numerals from India

A close investigation of the Vedic system of mathematics shows that it was much more advanced than the mathematical systems of the civilizations of the Nile or the Euphrates. The Vedic mathematicians had developed the decimal system of tens, hundreds, thousands, etc. where the remainder from one column of numbers is carried over to the next. The advantage of this system of nine number signs and a zero is that it allows for calculations to be easily made. Further, it has been said that the introduction of zero, or sunya as the Indians called it, in an operational sense as a definite part of a number system, marks one of the most important developments in the entire history of mathematics. The earliest preserved examples of the number system which is still in use today are found on several stone columns erected in India by King Ashoka in about 250 B.C. Similar inscriptions are found in caves near Poona (100 B.C.) and Nasik (200 C.E.). These earliest Indian numerals appear in a script called brahmi.

After 700 C.E. another notation, called by the name "Indian numerals," which is said to have evolved from the brahmi numerals, assumed common usage, spreading to Arabia and from there around the world. When Arabic numerals (the name they had then become known by) came into common use throughout the Arabian empire, which extended from India to Spain, Europeans called them "Arabic notations," because they received them from the Arabians. However, the Arabians themselves called them "Indian figures" (Al-Arqan-Al-Hindu) and mathematics itself was called "the Indian art" (hindisat).

Evolution of "Arabic Numerals" from Brahmi (250 BC) to the 16th Century.

Mastery of this new mathematics allowed the Muslim mathematicians of Baghdad to fully utilize the geometrical treatises of Euclid and Archimedes. Trigonometry flourished there along with astronomy and geography. Later in history, Carl Friedrich Gauss, the "prince of mathematics," was said to have lamented that Archimedes in the third century B.C.

had failed to foresee the Indian system of numeration; how much more advanced science would have been.

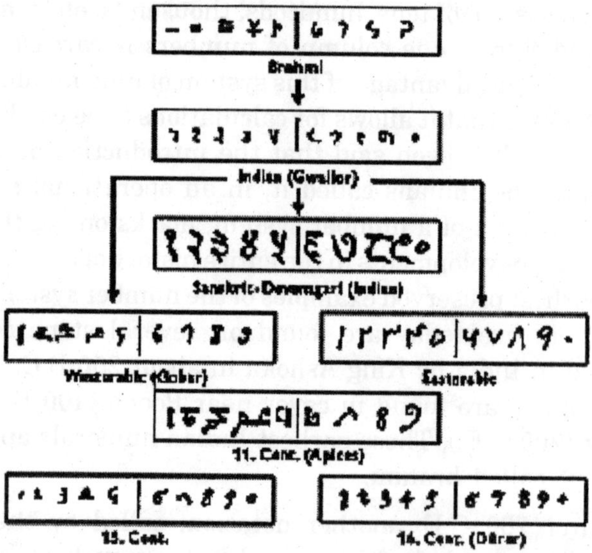

Prior to these revolutionary discoveries, other world civilizations-the Egyptians, the Babylonians, the Romans, and the Chinese-all used independent symbols for each row of counting beads on the abacus, each requiring its own set of multiplication or addition tables. So cumbersome were these systems that mathematics was virtually at a standstill.

The new number system from the Indus Valley led a revolution in mathematics by setting it free. By 500 C.E. mathematicians of India had solved problems that baffled the world's greatest scholars of all time. Aryabhatta, an astronomer mathematician who flourished at the beginning of the 6th century, introduced sines and versed sines-a great improvement over the clumsy half-cords of Ptolemy. A.L. Basham, foremost authority on ancient India, writes in The Wonder That Was India, Medieval Indian mathematicians, such as Brahmagupta (seventh century), Mahavira (ninth century), and Bhaskara (twelfth century), made several discoveries which in Europe were not known until the Renaissance or later. They understood the import of positive and negative quantities, evolved sound systems of extracting square and cube roots, and could solve quadratic and certain types of indeterminate equations." Mahavira's most noteworthy contribution is his treatment of fractions for the first time and his rule for dividing one fraction by another, which did not appear in Europe until the 16th century.

Equations and Symbols

B.B. Dutta writes: "The use of symbols-letters of the alphabet to denote unknowns, and equations are the foundations of the science of algebra. The Hindus were the first to make systematic use of the letters of the alphabet to denote unknowns. They were also the first to classify and make a detailed study of equations.

Thus they may be said to have given birth to the modern science of algebra." The great Indian mathematician Bhaskaracharya (1150 C.E.) produced extensive treatises on both plane and spherical trigonometry and algebra, and his works contain remarkable solutions of problems which were not discovered in Europe until the seventeenth and eighteenth centuries. He preceded Newton by over 500 years in the discovery of the principles of differential calculus. A.L. Basham writes further, "The mathematical implications of zero (sunya) and infinity, never more than vaguely realized by classical authorities, were fully understood in medieval India. Earlier mathematicians had taught that X/0 = X, but Bhaskara proved the contrary. He also established mathematically what had been recognized in Indian theology at least a millennium earlier: that infinity, however divided, remains infinite, represented by the equation oo /X = oo." In the 14th century, Madhava, isolated in South India, developed a power series for the arc tangent function, apparently without the use of calculus, allowing the calculation of pi to any number of decimal places (since arctan 1 = pi/4). Whether he accomplished this by inventing a system as good as calculus or without the aid of calculus; either way it is astonishing.

Spiritually advanced cultures were not ignorant of the principles of mathematics, but they saw no necessity to explore those principles beyond that which was helpful in the advancement of God realization.

By the fifteenth century C.E. use of the new mathematical concepts from India had spread all over Europe to Britain, France, Germany, and Italy, among others. A.L. Basham states also that:

> *The debt of the Western world to India in this respect [the field of mathematics] cannot be overestimated. Most of the great discoveries and inventions of which Europe is so proud would have been impossible without a developed system of mathematics, and this in turn would have been impossible if Europe had been shackled by the unwieldy system of Roman numerals. The unknown man who devised the new system was, from the world's point of view, after the Buddha, the most important son of India. His achievement, though easily taken for granted, was the work of an analytical mind of the first order, and he deserves much more honour than he has so far received.*

Unfortunately, Eurocentrism has effectively concealed from the common man the fact that we owe much in the way of mathematics to ancient India. Reflection on this may cause modern man to consider more seriously the spiritual preoccupation of ancient India. The rishis (seers) were not men lacking in practical knowledge of the world, dwelling only in the realm of imagination. They were well developed in secular knowledge, yet only insofar as they felt it was necessary within a world view in which consciousness was held as primary.

In ancient India, mathematics served as a bridge between understanding material reality and the spiritual conception. Vedic mathematics differs profoundly from Greek mathematics in that knowledge for its own sake (for its aesthetic satisfaction) did not appeal to the Indian mind. The mathematics of the Vedas lacks the cold, clear, geometric precision of the West; rather, it is cloaked in the poetic language which so distinguishes the East.

Vedic mathematicians strongly felt that every discipline must have a purpose, and believed that the ultimate goal of life was to achieve self-realization and love of God and thereby be released from the cycle of birth and death.

Those practices which furthered this end either directly or indirectly were practiced most rigorously. Outside of the religio-astronomical sphere, only the problems of day to day life (such as purchasing and bartering) interested the Indian mathematicians.

Poetry in Math

One of the foremost exponents of Vedic math, the late Bharati Krishna Tirtha Maharaja, author of Vedic Mathematics, has offered a glimpse into the sophistication of Vedic math. Drawing from the Atharva-veda, Tirtha Maharaja points to many sutras (codes) or aphorisms which appear to apply to every branch of mathematics: arithmetic, algebra, geometry (plane and solid), trigonometry (plane and spherical), conics (geometrical and analytical), astronomy, calculus (differential and integral), etc.

Utilizing the techniques derived from these sutras, calculations can be done with incredible ease and simplicity in one's head in a fraction of the time required by modern means. Calculations normally requiring as many as a hundred steps can be done by the Vedic method in one single simple step. For instance the conversion of the fraction 1/ 29 to its equivalent recurring decimal notation normally involves 28 steps. Utilizing the Vedic method it can be calculated in one simple step.

In order to illustrate how secular and spiritual life were intertwined in Vedic India, Tirtha Maharaja has demonstrated that mathematical formulas and laws were often taught within the context of spiritual expression (mantra). Thus while learning spiritual lessons, one could also learn mathematical rules.

Tirtha Maharaja has pointed out that Vedic mathematicians prefer to use the devanagari letters of Sanskrit to represent the various numbers in their numerical notations rather than the numbers themselves, especially where large numbers are concerned. This made it much easier for the students of this math in their recording of the arguments and the appropriate conclusions.

Tirtha Maharaja states, "In order to help the pupil to memorize the material studied and assimilated, they made it a general rule of practice to write even the most technical and abstruse textbooks in sutras or in verse (which is so much easier-even for the children-to memorize).

And this is why we find not only theological, philosophical, medical, astronomical, and other such treatises, but even huge dictionaries in Sanskrit verse! So from this standpoint, they used verse, sutras and codes for lightening the burden and facilitating the work (by versifying scientific and even mathematical material in a readily assimilable form)!" The code used is as follows:

The Sanskrit consonants

ka, ta, pa, and ya all denote 1;

kha, tha, pha, and ra all represent 2;

ga, da, ba, and la all stand for 3;

Gha, dha, bha, and va all represent 4;

gna, na, ma, and sa all represent 5;

ca, ta, and sa all stand for 6;

cha, tha, and sa all denote 7;

ja, da, and ha all represent 8;

jha and dha stand for 9; and

ka means zero.

Vowels make no difference and it is left to the author to select a particular consonant or vowel at each step. This great latitude allows one to bring about additional meanings of his own choice.

For example kapa, tapa, papa, and yapa all mean 11. By a particular choice of consonants and vowels one can compose a poetic hymn with double or triple meanings. Here is an actual sutra of spiritual content, as well as secular mathematical significance.

gopi bhagya madhuvrata

srngiso dadhi sandhiga

khala jivita khatava

gala hala rasandara

While this verse is a type of petition to Krishna, when learning it one can also learn the value of pi/10 (i.e. the ratio of the circumference of a circle to its diameter divided by 10) to 32 decimal places. It has a self-contained master-key for extending the evaluation to any number of decimal places.

The translation is as follows:

O Lord anointed with the yogurt of the milkmaids' worship (Krishna), O savior of the fallen, O master of Shiva, please protect me.

At the same time, by application of the consonant code given above, this verse directly yields the decimal equivalent of pi divided by 10:

pi/10 = 0.31415926535897932384626433832792.

Thus, while offering mantric praise to Godhead in devotion, by this method one can also add to memory significant secular truths.

This is the real gist of the Vedic world view regarding the culture of knowledge: while culturing transcendental knowledge, one can also come to understand the intricacies of the phenomenal world. By the process of knowing the absolute truth, all relative truths also become known. In modern society today it is often contended that never the twain shall meet: science and religion are at odds. This erroneous conclusion is based on little understanding of either discipline. Science is the smaller circle within the larger circle of religion.

We should never lose sight of our spiritual goals. We should never succumb to the shortsightedness of attempting to exploit the inherent power in the principles of mathematics or any of the natural sciences for ungodly purposes. Our reasoning faculty is but a gracious gift of Godhead intended for divine purposes, and not those of our own design.

Bibliography

A B Chace, L S Bull, H P Manning and R C Archibald: *The Rhind Mathematical Papyrus,* Oberlin, Ohio, 1927.

A.L. Basham: *The Wonder That Was India*, Rupa & Co., Calcutta, 1967.

Aaboe, Asger: *Episodes from the Early History of Mathematics*, New York, 1964.

Amma, Saraswarti: *Geometry in Ancient and Medieval India*, Motilal Banarsidas, 1979.

Armstrong, Robert G.: *Yoruba Numerals*, London, Oxford University Press, 1962.

Arnett, William S.: *The Predynastic Origin of Egyptian Hieroglyphs*, Washington, University Press of America, 1982.

Ascher, Marcia: *Ethnomathematics*, Pacific Grove, CA: Brooks/Cole, 1991.

Bag, A. K.: *Mathematics in Ancient and Medieval India*, Varanasi: Chaukhambha Orientalia, 1979.

Barnard, F. P.: *The Casting-Counter and the Counting Board. A Chapter in the History of Numismatics and Early Arithmetic*, Oxford, Oxford University Press, 1916.

Barrow, John D.: *Pi in the Sky*, Oxford, Clarendon Press, 1992.

Basham, A. L.: *The Wonder That Was India*, Calcutta, Rupa & Co., 1967.

Beckmann, Petr: *A History of Pi*, New York, St. Martin's Press, 1974.

Bell, Eric T.: *The Magic of Numbers*, New York, McGraw-Hill, 1946.

Berlin, Brent: Tzeltal Numeral Classifiers, The Hague, Mouton, 1968.

Bharati Krishna Tirthaji Majaraja.: *Vedic Mathematics*. Delhi, Motilal Banarsidass, 1988.

Bidyaranya, Swami: *History of Hindu Mathematics*,Bombay, Asia Pub. House, 1962.

Bischoff, Bernard: *Latin Palaeography*, Cambridge, Cambridge University Press, 1990.

Bonfante, Larissa: *Etruscan*. Berkeley, University of California Press, 1990.

Bowditch, Charles P.: *The Numeration, Calendar Systems and Astronomical knowledge of the Mayas*, Cambridge, Cambridge University Press, 1910.

Brainerd, Charles J.: *The Origins of the Number Concept*, New York, Praeger, 1979.

Brunschwig, Jacques and Geoffrey E.R. Lloyd:. *Greek Thought: A Guide to Classical Knowledge*, Cambridge, Harvard University Press, 2000.

Bunt, L. N. H., Jones, P. S., & Bedient, J. D.: *The Historical Roots of Elementary Mathematics*, New York, Dover, 1988.

Chattopadhyaya, D. P.: *Mathematics, Astronomy, and Biology in Indian Tradition: Some Conceptual Preliminaries*, New Delhi, Project of History of Indian Science, Philosophy, and Culture, 1995.

Clawson, Calvin: *The Mathematical Traveller*, New York, Plenum Press, 1994.

Closs, Michael P.: *Native American Mathematics*, Austin, University of Texas Press, 1986.

Conant, Levi L.: *The Number Concept*, New York, MacMillan, 1896.

Crossley, John N: *The Emergence of Number*, Singapore, World Scientific, 1987.

Crump, Thomas: *The Anthropology of Numbers*, Cambridge, Cambridge University Press, 1990.

Dantzig, Tobias: *Number, the Language of Science*, London, George Allen and Unwin, 1940.

De Villiers, Melius: *The Numeral-Words: Their Origin, Meaning, History and Lesson*, London, Witherby, 1923.

Dehaene, Stanislas: *The Number Sense*, New York, Oxford University Press, 1997.

Dilke, Oswald Ashton Wentworth: *Mathematics and Measurement*, Berkeley, CA: U. of Cal. Press, 1987.

Eves, Howard: *An Introduction to the History of Mathematics*, New York, Rinehart and Company Inc., 1953.

Flegg, Graham: *Numbers: Their History and Meaning*, London, André Deutsch, 1983.

Fowler, David: *The Mathematics of Plato's Academy*, Oxford, Clarendon, 1999.

Fuson, K.C.: *Children's Counting and Concepts of Number*, New York, Springer-Verlag, 1988.

Ganeri, Anita: *The Story of Numbers and Counting*, New York, Oxford University Press, 1996.

Gokhale, Shebhana Laxman: *Indian Numerals*, Poona, Deccan College, 1966.

Griffith, Francis Llewellyn: *Hieratic Papyri from Kahun and Gurob*, London, Bernard Quaritch, 1898.

Grover, Sudarshan: *History of Development of Mathematics in India*, Delhi, Atma Ram & Sons, 1994.

Haldane, John Burdon Sanderson: *Science and Indian Culture*, Calcutta, New Age Publishers, 1965.

Halliwell, James Orchard: *Rara Mathematica*, London, J.W. Parker, 1839.

Herbert Meschkowski: *Ways of Thought of Great Mathematicians*, Holden-Day Inc., San Francisco, 1964.

Hogben, Lancelot: *Mathematics for the Million*, New York, W. W. Norton, 1937.

Howard Eves: *An Introduction to the History of Mathematics*, Rinehart and Company Inc., New York, 1953.

Ifrah, Georges: *From One to Zero: A Universal History of Numbers*, New York, Viking Penguin, 1985.

Jagadguru Swami Shri Bharati Krishna Tirthaji Maharaja: *Vedic Mathematics*, Motilal Banarsidass, Delhi, 1988.

Jaggi, O. P.: *Science and Technology in Medieval India*, New Delhi, Atma Ram & Sons, 1977.

Kline, Morris: *Mathematical Thought from Ancient to Modern Times*, New York, Oxford U. Press, 1990.

Kuppuram, G & K. Kumudamani: *History of Science and Technology in India*, Delhi, Sundeep Prakashan, 1990.

Lancy, David F.: *Cross-Cultural Studies in Cognition and Mathematics*, New York, Academic, 1983.

Marcus, Joyce: *Mesoamerican Writing Systems*, Princeton, Princeton University Press, 1992.

Marshack, Alexander: *The Roots of Civilization*, New York, McGraw-Hill, 1972.

Martzloff, Jean-Claude: *A History of Chinese Mathematics*, New York, Springer, 1997.

McLeish, John: *The Story of Numbers*, New York, Fawcett Columbine, 1991.

Menninger, Karl: *Number-Words and Number-Symbols*. London, MIT, 1969.

Murthy, T. S. Bhanu: *A Modern Introduction to Ancient Indian Mathematics*, Wiley Eastern Ltd., New Delhi, 1992.

Neugebauer, Otto and A. Sachs: *Mathematical Cuneiform Texts*, New Haven, Yale University Press, 1945.

Neugebauer, Otto: *The Exact Sciences in Antiquity*, Princeton, Princeton University Press, 1952.

North, John: *The Norton History of Astronomy and Cosmology*, New York, W.W.Norton, 1995.

Pandit, M. D.: *Mathematics as known to the Vedic Samhitas*, Delhi, Sri Satguru Publications, 1993.

Pappas, Theoni: *The Magic of Mathematics*, San Carlos, Wide World Publishing / Tetra, 1994.

Parpola, Asko: *Deciphering the Indus Script*, Cambridge, Cambridge University Press, 1994.

Peet, T. Eric: *The Rhind Mathematical Papyrus*, London, British Museum, 1923.

Piaget, Jean: *The Child's Conception of Number*, New York, W.W. Norton, 1965.

Pullan, J.M: *The History of the Abacus*, New York, Praeger, 1968.

Rao, S. Balaachandra: *Indian Mathematics and Astronomy*, Bangalore, Jnana Deep Publications, 1994.

Restivo, Sal: *The Social Relations of Physics, Mysticism and Mathematics*, Dordrecht, Kluwer, 1983.

Robins, G. and C. Shute: *The Rhind Mathematical Papyrus*, London, British Museum, 1987.

Russell, Bertrand: *The Principles of Mathematics*, Cambridge, Cambridge University Press, 1903.

Salomon, Richard: *Indian Epigraphy*, New York, Oxford University Press, 1998.

Saraswati Amma: *Geometry in Ancient and Medieval India*, Motilal Banarsidas, 1979.

Sarton, George: *The Study of the History of Mathematics*, Cambridge, Cambridge University Press, 1936.

Satterthwaite, Linton, Jr.: *Concepts and Structures of Maya Calendrical Arithmetics*, Philadelphia, University of Pennsylvania, 1947.

Schaaf, W.L.: *Mathematics, Our Great Heritage*, New York, Harper and Row, 1948.

Schimmel, Annemarie: *The Mystery of Numbers*, New York, Oxford University Press, 1993.

Scriba, C.J.: *The Concept of Number*, Zurich, Mannheim, 1968.

Seidenberg, A.: *The Zero in the Mayan Numerical Notation*, Austin, University of Texas Press, 1986.

Smeltzer, Donald: *Man and Number*, New York, Emerson, 1959.

Smith, D. E.: *History of Mathematics*, New York, Dover Publications Inc., 1951.

Srinivasiengar, C. N.: *The History of Ancient Indian Mathematics*. Calcutta, World Press, 1967.

Struik, Dirk J.: *A Concise History of Mathematics*, New York, Dover, 1987.

Thompson, Sir J. Eric S.: *Maya Hieroglyphic Writing: An Introduction*, Norman, University of Oklahoma Press, 1971.

Todd, R.R., P.J. Barber, and D. Jones: *The Internal Representation of Number: Analogue or Digital?*, Oxford, Clarendon, 1987.

Urton, Gary: *The Social Life of Numbers*, Austin, University of Texas Press, 1997.

Volodarsky, A.I.: *Mathematics in Ancient India*, Tokyo, Science Council of Japan, 1974.

Woolner, Alfred C.: *Introduction to Prakrit*, Calcutta, Baptist Mission Press, 1917.

Index

O

P

R

S

T

U

V

□□□